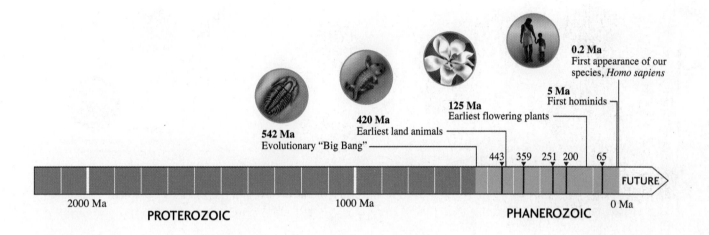

542 Ma
Evolutionary "Big Bang"

420 Ma
Earliest land animals

125 Ma
Earliest flowering plants

5 Ma
First hominids

0.2 Ma
First appearance of our species, *Homo sapiens*

443   359   251   200   65

2000 Ma

1000 Ma

FUTURE

0 Ma

PROTEROZOIC

PHANEROZOIC

The Essential Earth

The Essential Earth

SECOND EDITION

Thomas H. Jordan
University of Southern California

John Grotzinger
California Institute of Technology

W. H. FREEMAN AND COMPANY
New York

Senior Developmental Editor:	Randi Blatt Rossignol
Executive Editor:	Anthony Palmiotto
Senior Media and Supplements Editor:	Amy Thorne
Marketing Manager:	Alicia Brady
Editorial Assistants:	Nicholas Ciani, Heidi Bamatter
Marketing Assistant:	Joanie Rothschild
Photo Editor:	Ted Szczepanski
Cover Designer:	Diana Blume
Text Designer:	Cambraia Fernandes
Senior Project Editor:	Mary Louise Byrd
Illustrations:	Precision Graphics
Illustration Coordinator:	Bill Page
Production Coordinator:	Ellen Cash
Composition:	Sheridan Sellers, W. H. Freeman and Company, Electronic Publishing Center
Printing and Binding:	Quad Graphics

Library of Congress Control Number: 2011934107

ISBN-13: 9781429255240
ISBN-10: 1-4292-5524-2

First printing

W. H. Freeman and Company
41 Madison Avenue, New York, NY 10010
Houndmills, Basingstoke RG21 6XS, England

www.whfreeman.com

We dedicate this book to Frank Press and Ray Siever,
pioneering educators in the era of modern geology.
This book was possible only because they led the way.

TOM JORDAN is a geophysicist interested in the composition, dynamics, and evolution of the solid Earth. He has conducted research into the nature of deep subduction, the formation of thickened keels beneath the ancient continental cratons, and the question of mantle stratification. He has developed a number of seismological techniques for investigating Earth's interior that bear on geodynamic problems. He has also worked on modeling plate movements, measuring tectonic deformation, quantifying seafloor morphology, and characterizing large earthquakes. His current research focuses on predicting earthquakes and their effects. He received his Ph.D. in geophysics and applied mathematics at the California Institute of Technology (Caltech) in 1972 and taught at Princeton University and the Scripps Institution of Oceanography before joining the Massachusetts Institute of Technology (MIT) faculty as the Robert R. Schrock Professor of Earth and Planetary Sciences in 1984. He served as the head of MIT's Department of Earth, Atmospheric and Planetary Sciences for the decade 1988–1998. He moved from MIT to the University of Southern California (USC) in 2000, where he is University Professor, W. M. Keck Professor of Earth Sciences, and Director of the Southern California Earthquake Center.

Dr. Jordan received the Macelwane Medal of the American Geophysical Union in 1983, the Woollard Award of the Geological Society of America in 1998, and the Lehmann Medal of the American Geophysical Union in 2005. He is a member of the American Academy of Arts and Sciences, the U.S. National Academy of Sciences, and the American Philosophical Society.

JOHN GROTZINGER is a field geologist interested in the evolution of Earth's surface environments and biosphere. His research addresses the chemical development of the early oceans and atmosphere, the environmental context of early animal evolution, and the geologic factors that regulate sedimentary basins. He has contributed to developing the basic geologic framework of a number of sedimentary basins and orogenic belts in northwestern Canada, northern Siberia, southern Africa, and the western United States. He received a B.S. in geoscience from Hobart College in 1979, an M.S. in geology from the University of Montana in 1981, and a Ph.D. in geology from Virginia Polytechnic Institute and State University in 1985. He spent three years as a research scientist at the Lamont-Doherty Geological Observatory before joining the MIT faculty in 1988. From 1979 to 1990, he was engaged in regional mapping for the Geological Survey of Canada. He has worked as a geologist on the Mars Exploration Rover team, the first mission to conduct ground-based exploration of the bedrock geology of another planet, which has resulted in the discovery of sedimentary rocks formed in aqueous depositional environments. He currently is the Chief Scientist for the Mars Science Laboratory Mission, due to launch in November 2011 and land in August 2012.

In 1998, Dr. Grotzinger was named the Waldemar Lindgren Distinguished Scholar at MIT, and in 2000, he became the Robert R. Schrock Professor of Earth and Planetary Sciences. In 2005, he moved from MIT to Caltech, where he is the Fletcher Jones Professor of Geology. He received the Presidential Young Investigator Award of the National Science Foundation in 1990, the Donath Medal of the Geological Society of America in 1992, and the Henno Martin Medal of the Geological Society of Namibia in 2001. He is a member of the American Academy of Arts and Sciences and the U.S. National Academy of Sciences.

BRIEF CONTENTS

CONTENTS

CHAPTER 4 EARTH MATERIALS: MINERALS AND ROCKS

PREFACE

A NOTE FROM THE AUTHORS

THE FIRST EDITION OF *THE ESSENTIAL EARTH* was successful in offering an effective, reinvented approach to the physical geology course. Adopters embraced its emphasis on the process of science and the societal impact of the field, and appreciated its unique balance of depth and breadth. In the second edition, we continue to focus on our goal of providing instructors and students with a scientifically sound text that will produce meaningful and memorable learning experiences for non-scientists. Drawing on our teaching experiences as well as extensive feedback from adopters, reviewers, and others in the scientific community, we have maintained the core value of *The Essential Earth* and have enriched it with powerful and relevant topical coverage and a wealth of active learning opportunities.

The Essential Earth is now established as a significant resource in the teaching of physical geology. We hope that the changes we have made in this new edition, based on extensive feedback, have further improved this textbook.

FOCUS ON ESSENTIALS

In 14 chapters, *The Essential Earth* covers the fundamental concepts of physical geology and closely related subjects such as climate science. We introduce plate tectonics right up front in Chapter 3, so that we can use its basic ideas throughout our development of physical geology. We introduce the student to Earth's climate in Chapter 10, before we tackle the various subjects of surface processes in Chapters 11 and 12. To create a textbook that is manageable in depth and breadth, **we have integrated topics that are bound together by common geologic processes.** Thus, igneous rocks and volcanoes are discussed together in Chapter 5, "Igneous Processes and Volcanism."

Similarly, Chapter 7, "Deformation and Metamorphism," examines the basic processes that deform and metamorphose continental crust. The integration of these topics will help students see that deformation usually accompanies metamorphism and that both are intimately associated with plate tectonic processes.

Chapter 9, "History of Earth," weaves together several topics that provide context for an understanding of deep time. These topics include the origin of the solar system, the formation of Earth, and how Earth differs from other planets. The chapter also tells the remarkable story of how continents were formed and how they have changed over geologic time. It ends with a view of biological development across the span of Earth history, showing how organisms and their environment have interacted, sometimes in disastrous ways.

Chapter 10, "The Climate System and Glaciation," examines the geologic record of past climate changes, particularly the ice ages that sent vast glaciers across the northern continents, and how these changes have modified the landscape. Many spectacular features of the landscapes we see today show the student why glaciation is one of the most important processes of the climate system.

Chapter 12, "Shaping Earth's Surface: Streams, Coastlines, and Wind," discusses how water and air currents erode rock and transport sediments. The chapter addresses questions such as Why do deserts form? How do streams form, and why do they flow to the edges of continents? Why does the coastline have such a distinctive shape? This diverse list of topics has a common thread: moving currents of air and water.

THREE THEMES: FOCUS ON THE INTERACTIONS OF GEOLOGY AND SOCIETY

Societal impacts of geology are present everywhere around us. From the outcrops seen on a well-traveled road, to the energy used to power cars and homes, to a river spilling over its banks, students constantly encounter and interact with geologic processes. A guiding principle of *The Essential Earth* is to make those connections apparent for students, and provide opportunities for instructors to build on them in their classrooms. Toward that end, we have focused and enhanced our coverage of three major themes:

 Natural Resources — building an understanding of how the activities and knowledge of geologists provide our society with energy and crucial materials.

 Natural Hazards — illuminating the risks and impacts of dangerous geologic processes on our lives, economy, and future.

 Natural Environment — discussing in a balanced way the effects of human activity, both positive and negative, on our physical environment.

In each chapter, we focus on one or more of these areas within the context of the key geologic concepts, calling it out with a heading and graphical icon. Our goal is to highlight for students the importance of the geologic concepts for these very human-facing areas, showing them the relevance and scientific implications of the material in the chapter.

Chapter 1, "Why We Study Earth," launches our discussion of these themes. We explore geology's role in solving some of the great problems faced by civilization. A large part of Chapter 1 is devoted to society's need for sustainable **Natural Resources** — energy, minerals and other raw materials, and water. We focus on our consumption of both renewable and nonrenewable resources, as well as on our resource exploration and production capabilities.

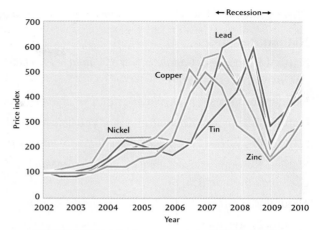

FIGURE 1.15 Metal prices rose substantially during the early 2000s, fueled by increasing global demand. Prices dropped considerably in 2007 because of the worldwide recession but are rising again. [*London Metal Exchange.*]

A section on **Natural Hazards** follows the section on Natural Resources. This section exposes students to extreme geologic events, including recent episodes such as the volcanic eruption in Iceland and the earthquake and tsunami that devastated parts of Japan.

FIGURE 1.33 Iceland's Eyjafjallajökull volcano erupting on April 17, 2010. The ash cloud from this eruption disrupted air traffic across northern Europe. [*Joanna Vestey/Corbis.*]

The final third of Chapter 1 discusses the **Natural Environment,** including Earth's biosphere, and human impacts, including global climate change.

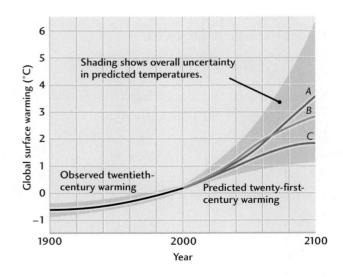

FIGURE 1.42 Global warming, past and future. [*IPCC, Climate Change 2007: The Physical Scientific Basis. Cambridge University Press, 2007.*]

Throughout the text the three themes are reinforced.

Natural Resources

Natural Resources: *Concentration of Valuable Elements in the Continental Crust (Ch. 2)*

Natural Resources: *Mineral Concentrations at Plate Boundaries (Ch. 3)*

Natural Resources: *Concentrations of Valuable Minerals (Ch. 4)*

Natural Resources: *Fractional Crystallization, Porphyries, and Veins (Ch. 5)*

Natural Resources: *Sandstone as a Source of Water, Oil and Natural Gas, and Uranium (Ch. 6)*

Natural Resources: *Coal, Oil, and Natural Gas (Ch. 6)*

Natural Resources: *Using Geologic Maps to Find Oil and Gas (Ch. 7)*

Natural Resources: *Ages of Petroleum Source Rocks (Ch. 8)*

Natural Resources: *Archean Greenstone Belts (Ch. 9)*

Natural Resources: *Who Should Get Water? (Ch. 11)*

Natural Resources: Using Geologic Maps to Find Oil and Gas

Crude oil, or *petroleum* (from the Latin words for "rock oil"), has been collected from natural seeps at Earth's surface since ancient times. The foul-smelling, tarry substance was used as boat caulking, wheel grease, and medicine, but not commonly as a fuel until the process of oil refining was developed in the 1850s. Demand skyrocketed at that time, primarily because oil from whale blubber, the best fuel then available for lamps, had became terribly expensive ($60 per gallon in today's dollars!) as overfishing decimated whale populations.

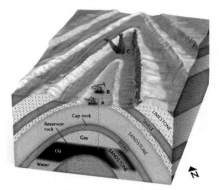

FIGURE 7.16 Oil is often found in large reservoirs along the fold of axes of anticlines. This figure illustrates a typical anticlinal trap.

Natural Hazards

Natural Hazards: *Extreme Events in the Geologic Record (Ch. 2)*

Natural Hazards: *The Pacific Ring of Fire (Ch. 3)*

Natural Hazards: *The Effects of Volcanoes (Ch. 5)*

Natural Hazards: *The Dangers of El Niño (Ch. 10)*

Natural Hazards: *Droughts (Ch. 11)*

Natural Hazards: *Floods (Ch. 12)*

Natural Hazards: *Hurricanes and Coastal Storm Surges (Ch. 12)*

Natural Hazards: *Earthquake Destructiveness (Ch. 13)*

Natural Hazards: *Can Earthquakes Be Predicted? (Ch. 13)*

Natural Hazards: Floods

There are benefits to living on or near a floodplain, but cities built on floodplains are also prone to destructive floods. Floods occur when a stream's **discharge**— the volume of water that passes a given point at a given time—increases such that more water flows into the stream than can flow out. When this happens the excess water spills over the banks.

Streams flood regularly, some at infrequent intervals, others almost every year. Some floods are large, with very high water levels lasting for days. At the other extreme are minor floods that barely break out of the channel before they recede. Small floods are more frequent, occurring every 2 or 3 years on average. Large floods are generally less frequent, usually occurring every 10, 20, or 30 years.

FIGURE 12.22 Floodwaters from the Red and Wild Rice rivers take over the Forst River area of Fargo, North Dakota, on April 18, 1997. [*Jim Mone/AP Photo.*]

Natural Environment

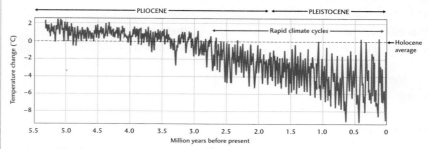

Natural Environment: Clocking the Climate System

The Pliocene and Pleistocene epochs were times of rapid and dramatic global climate change. We can chart these climate changes from the isotopes contained in shelly fossils buried in deep-sea sediments. Deep-sea drilling vessels such as the *JOIDES Resolution* (see Figure 3.13) have taken cores from sedimentary beds around the world's oceans. Geologists can use the carbon-14 dating method to estimate when the shells recovered from these sediment cores were formed, and they can measure the stable isotopes of oxygen to estimate temperature of the seawater in which the shell-producing organisms lived.

FIGURE 8.17 Changes in Earth's average surface temperature (jagged blue line) during the Pliocene and Pleistocene epochs, measured from temperature indicators in well-dated oceanic sediments. [*Courtesy of L. E. Lisiecki and M. E. Raymo.*]

CHAPTER OPENING STORIES

Illustrate how humans have affected geology and how geology has affected humans. From the De Beers diamond marketing blitz (Chapter 4) to the risks and rewards of deep-sea oil drilling (Chapter 6) to water management in California (Chapter 11), these stories touch on economic, social, and political aspects of geology. They also show ways in which geologists explore Earth: using technology to measure remote areas of Earth (Chapter 2); surveying the ocean floor in rough seas (Chapter 3); blasting into the sky on space missions (Chapter 9); and studying nature in the harshest environments (Chapter 10).

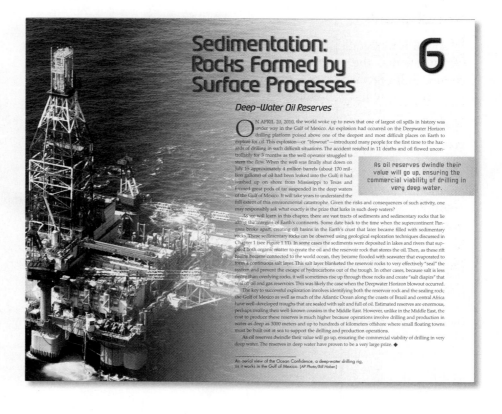

Sedimentation: Rocks Formed by Surface Processes

6

Deep-Water Oil Reserves

ON APRIL 20, 2010, the world woke up to news that one of largest oil spills in history was under way in the Gulf of Mexico. An explosion had occurred on the Deepwater Horizon drilling platform poised above one of the deepest and most difficult places on Earth to explore for oil. This explosion—or "blowout"—introduced many people for the first time to the hazards of drilling in such difficult situations. The accident resulted in 11 deaths and oil flowed uncontrollably for 3 months as the well operator struggled to stem the flow. When the well was finally shut down on July 15 approximately 4 million barrels (about 170 million gallons) of oil had been leaked into the Gulf. It had washed up on shore from Mississippi to Texas and formed great pods of tar suspended in the deep waters of the Gulf of Mexico. It will take years to understand the full extent of this environmental catastrophe. Given the risks and consequences of such activity, one may responsibly ask what exactly is the prize that lurks in such deep waters?

> As oil reserves dwindle their value will go up, ensuring the commercial viability of drilling in very deep water.

As we will learn in this chapter, there are vast tracts of sediments and sedimentary rocks that lie along the margins of Earth's continents. Some date back to the time when the supercontinent Pangaea broke apart, creating rift basins in the Earth's crust that later became filled with sedimentary rocks. These sedimentary rocks can be observed using geological exploration techniques discussed in Chapter 1 (see Figure 1.11). In some cases the sediments were deposited in lakes and rivers that supplied both organic matter to create the oil and the reservoir rock that stores the oil. Then, as these rift basins became connected to the world ocean, they became flooded with seawater that evaporated to form a continuous salt layer. This salt layer blanketed the reservoir rocks to very effectively "seal" the system and prevent the escape of hydrocarbons out of the trough. In other cases, because salt is less dense than overlying rocks, it will sometimes rise up through those rocks and create "salt diapirs" that seal off oil and gas reservoirs. This was likely the case when the Deepwater Horizon blowout occurred.

The key to successful exploration involves identifying both the reservoir rock and the sealing rock; the Gulf of Mexico as well as much of the Atlantic Ocean along the coasts of Brazil and central Africa have well-developed troughs that are sealed with salt and full of oil. Estimated reserves are enormous, perhaps rivaling their well-known cousins in the Middle East. However, unlike in the Middle East, the cost to produce these reserves is much higher because operations involve drilling and production in water as deep as 3000 meters and up to hundreds of kilometers offshore where small floating towns must be built out at sea to support the drilling and production operations.

As oil reserves dwindle their value will go up, ensuring the commercial viability of drilling in very deep water. The reserves in deep water have proven to be a very large prize. ◆

An aerial view of the Ocean Confidence, a deep-water drilling rig, as it works in the Gulf of Mexico. [*AP Photo/Bill Haber.*]

ACTIVE LEARNING: FOCUS ON ENGAGEMENT AND THINKING LIKE A GEOLOGIST

Our experience has shown that students working actively on interesting geological problems enjoy a more successful and rewarding learning experience. To foster this level of engagement, we have included more dynamic resources within the textbook.

Google Earth Projects

In order to create a more active, practical approach to learning the material, and to take advantage of a popular and useful multimedia resource, we have designed focused explorations of geologic locations. Using Google Earth—by far the most widely used virtual globe browser—students are first guided through observations rooted in geologic processes. They then work through a series of progressively intensive questions aimed at producing a unique and self-acquired knowledge. After navigating to the appropriate location and checking their position with the provided image, students may answer the questions in a free response format or within the text's accompanying learning system, GeologyPortal, which can automatically store and grade student responses.

Visual Literacy Tasks

In addition to significant general art enhancements, the Second Edition continues its focus on active learning by enriching core graphics in unique and effective Visual Literacy Tasks. Created in collaboration with geology education researchers Karen Kortz and Jessica Smay, these brief exercises explore a key illustration or photo through a series of conceptual questions aimed at building a comprehensive understanding. The complexity of the interaction increases as students work through it, and the result is a thorough, cumulative knowledge that at once focuses on a concept and enhances the students' overall ability to interpret and discuss visuals, an important skill that helps them do what geologists do.

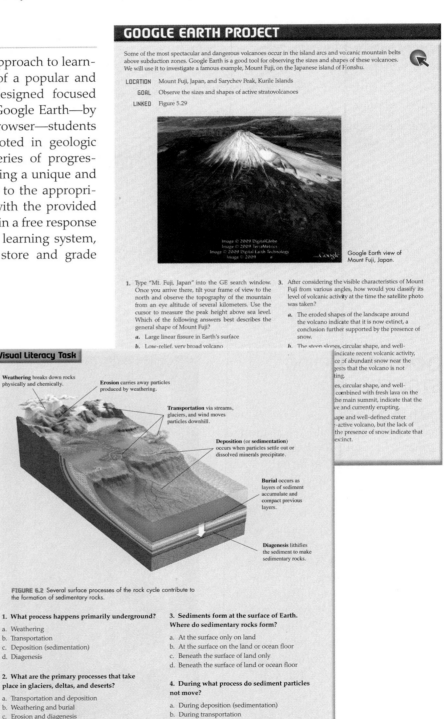

GOOGLE EARTH PROJECT

Some of the most spectacular and dangerous volcanoes occur in the island arcs and volcanic mountain belts above subduction zones. Google Earth is a good tool for observing the sizes and shapes of these volcanoes. We will use it to investigate a famous example, Mount Fuji, on the Japanese island of Honshu.

LOCATION   Mount Fuji, Japan, and Sarychev Peak, Kurile Islands

GOAL   Observe the sizes and shapes of active stratovolcanoes

LINKED   Figure 5.29

Image © 2009 DigitalGlobe
Image © 2009 TerraMetrics
Image © 2009 Digital Earth Technology
Image © 2009

Google Earth view of Mount Fuji, Japan.

1. Type "Mt. Fuji, Japan" into the GE search window. Once you arrive there, tilt your frame of view to the north and observe the topography of the mountain from an eye altitude of several kilometers. Use the cursor to measure the peak height above sea level. Which of the following answers best describes the general shape of Mount Fuji?

   a. Large linear fissure in Earth's surface
   b. Low-relief, very broad volcano

3. After considering the visible characteristics of Mount Fuji from various angles, how would you classify its level of volcanic activity at the time the satellite photo was taken?

   a. The eroded shapes of the landscape around the volcano indicate that it is now extinct, a conclusion further supported by the presence of snow.
   b. The steep slopes, circular shape, and well-
   indicate recent volcanic activity,
   ce of abundant snow near the
   gests that the volcano is not
   ting.
   es, circular shape, and well-
   combined with fresh lava on the
   he main summit, indicate that the
   ve and currently erupting.
   ape and well-defined crater
   -active volcano, but the lack of
   the presence of snow indicate that
   extinct.

Visual Literacy Task

Weathering breaks down rocks physically and chemically.

Erosion carries away particles produced by weathering.

Transportation via streams, glaciers, and wind moves particles downhill.

Deposition (or sedimentation) occurs when particles settle out or dissolved minerals precipitate.

Burial occurs as layers of sediment accumulate and compact previous layers.

Diagenesis lithifies the sediment to make sedimentary rocks.

FIGURE 6.2 Several surface processes of the rock cycle contribute to the formation of sedimentary rocks.

1. What process happens primarily underground?
   a. Weathering
   b. Transportation
   c. Deposition (sedimentation)
   d. Diagenesis

2. What are the primary processes that take place in glaciers, deltas, and deserts?
   a. Transportation and deposition
   b. Weathering and burial
   c. Erosion and diagenesis

3. Sediments form at the surface of Earth. Where do sedimentary rocks form?
   a. At the surface only on land
   b. At the surface on the land or ocean floor
   c. Beneath the surface of land only
   d. Beneath the surface of land or ocean floor

4. During what process do sediment particles not move?
   a. During deposition (sedimentation)
   b. During transportation
   c. During erosion
   d. During diagenesis

WHAT GEOLOGISTS DO AND HOW THEY DO IT

Chapter 2, "How We Study Earth," was written with the hope that all students will come away from this course with an appreciation for **how the scientific process works** and **what geologists do.** This chapter starts with the scientific method and shows how scientists gather and interpret information about our planet. It illustrates how the scientific method has been applied to discover some of Earth's most basic features: its shape and internal layering. Photos of geologists at work help to show students that geology is a "hands-on" science with its own particular style and outlook. They will learn how scientists combine field studies of Earth processes in the natural environments with controlled laboratory experiments and computer simulations to make major discoveries about the Earth system.

(a)

(b)

(c)

FIGURE 1.1 (a) Geologists examine rocks inside Spider Cave at Carlsbad Caverns, New Mexico. (b) A geochemist readies a rock sample for analysis by mass spectrometer. (c) Geophysicists deploy instruments to measure the underground activity of a volcano. [(a) AP Photo/Val Hildreth-Werker. (b) John McLean/Photo Researchers. (c) Hawaiian Volcano Observatory/USGS.]

FOCUS ON GEOLOGY AS A VISUAL SCIENCE

There is probably no science more visually compelling than geology. We hope to convey that to students by liberally illustrating the text with carefully developed figures and carefully selected photos of geologic structures, geologic processes, and geologists doing geology.

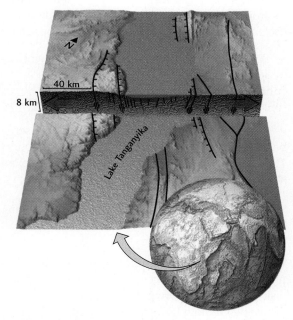

FIGURE 7.18 In East Africa, tensional forces are pulling the Somali subplate away from the African Plate creating rift valleys bounded by normal faults.

FIGURE 12.14 Potholes in river rock along McDonald Creek, Glacier National Park, Montana. The pebbles rotate inside the potholes, grinding deep holes in the bedrock. [Carr Clifton/Minden Pictures.]

MEDIA AND SUPPLEMENTS

The Essential Earth eBook

http://ebooks.bfwpub.com/essentialearth2e

This online version of *The Essential Earth* combines the text, student media resources, and additional study features, including highlighting, bookmarking, note taking, Google-style searching, in-text definitions of terms, and outlinks to Google. The eBook is available both as part of the GeologyPortal (see below) and as a stand-alone product. Visit http://ebooks.bfwpub.com/essentialearth2e to preview a sample chapter.

GeologyPortal for *The Essential Earth*

http://courses.bfwpub.com/essentialearth2e

GeologyPortal, W. H. Freeman's nationally hosted learning management solution, is available in a version created specifically to accompany *The Essential Earth*, Second Edition. It incorporates *The Essential Earth* eBook, all student media resources, and powerful learning management capabilities to form a rich, fully integrated online learning environment. The ready-to-use course template is fully customizable. The Portal contains the following features:

- Interactive eBook

- LearningCurve student quizzing system to help students master key topics from each chapter. A personalized study plan is provided at the end of each exercise to show students areas for improvement.

- Self-graded, editable quizzes and homework assignments

- Expeditions video clips—exclusive videos of Earth in action, filmed by Dr. Jerry Magloughlin of Colorado State University

- Video exercises based on the Expeditions series, which can be assigned to students and reported to the course gradebook. The exercise questions include multiple-choice, matching, and short-answer questions.

- Animations of key concepts, with full audio

- Key term flashcards

- Assignments and resources can easily be added via a simple, user-friendly process. For a preview of GeologyPortal, visit http://courses.bfwpub.com/essentialearth2e and click on "Preview GeologyPortal as a Student."

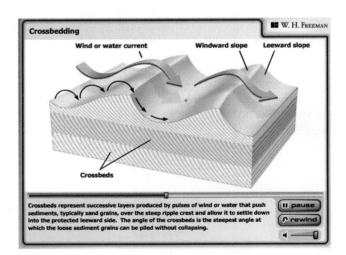

Companion Web Site

The Companion Web Site, at www.whfreeman.com/essentialearth2e, contains quizzes and key term flashcards to help students prepare for exams. For instructors, the Web site contains the same JPEGs and PowerPoint presentations available on the Instructor's Resource CD-ROM (see below).

Computerized Test Bank CD-ROM

The computerized test bank CD-ROM (with Windows and Mac versions on one disk) offers hundreds of multiple-choice questions. The CD format makes it easy to select, add, edit, and resequence questions, as well as to deliver and grade assignments and exams.

Instructor's Resource CD-ROM

To help instructors create their own Web sites and orchestrate dynamic lectures, this CD-ROM offers all art from the text in PowerPoint and JPEG formats, along with animations of key concepts.

Expeditions in Geology DVD, Volumes 1 and 2

Available to adopters, these two DVDs contain 25 brief videos that explore geological phenomena. These 3- to 7-minute videos are filmed and narrated by Dr. Jerry Magloughlin of Colorado State University.

Online Course Materials (WebCT, Blackboard)

As a service for adopters, we will provide content files in the appropriate online course format, including instructor and student resources for this text.

ACKNOWLEDGMENTS

We have been very fortunate to collaborate with some exceptionally talented people at W. H. Freeman. Since our first involvement with the third edition of *Understanding Earth*, which began in 2001, we have worked closely with Randi Rossignol as our superb developmental editor and Anthony Palmiotto as our thoughtful executive editor. We really appreciate their encouragement and sustained efforts in producing this new textbook. We also wish to thank our imaginative designer, Cambraia Fernandes; our valiant senior project editor, Mary Louise Byrd; our outstanding copy editor, Penny Hull; and our resourceful photo editor, Ted Szczepanski, as well as other capable members of Freeman's staff. In particular, with deadlines looming, Sheridan Sellers has shouldered heavy responsibilities for book composition with considerable grace and cheer, and Ellen Cash has been our production coordinator.

It is a challenge to both geology instructors and authors of geology textbooks to make decisions about what topics to teach, and at what level, and at the same time to inspire interest and enthusiasm in our students. To meet this challenge, we have called on the advice of many colleagues to guide us as we wrote and rewrote the chapters. We have relied on this consensus of views to shape the first and second editions of *The Essential Earth*. Our sincerest thanks go to our colleagues named here.

Reviewers of the Second Edition

R. Scott Babcock, *Western Washington University*
David L. Barbeau Jr., *University of Arizona*
Margaret Boettcher, *University of New Hampshire*
Saugata Datta, *Kansas State University*
Stewart S. Farrar, *Eastern Kentucky University*
Kurt Frankel, *Georgia Institute of Technology*

Lauren K. Heerschap, *Fort Lewis College*
Daniel I. Hembree, *Ohio University*
J. Bradford Hubeny, *University of Rhode Island*
Theodore C. Labotka, *University of Tennessee*
Steppen W. Murphy, *Central Piedmont Community College*
Nicole R Myers, *Sonoma State University*

Russell Perkins, *Plymouth State University*
Alycia Stigall, *Ohio University*
James H. Stout, *University of Minnesota*
Joyashish Thakurta, *Ohio University*
S. White, *University of South Carolina*
Michael Wysession, *Washington University in St. Louis*

Reviewers of the First Edition

Stephen Altaner, *University of Illinois at Urbana-Champaign*
Cathy Baker, *Arkansas Tech University*
Solweig Balzer, *University of Alberta*
E. Erik Bender, *Orange Coast College*
Larry Benninger, *University of North Carolina*
J Bret Bennington, *Hofstra University*
Barbara L. Brande, *University of Montevallo*
Susan H. Butts, *Southern Connecticut State University*
Wang-Ping Chen, *University of Illinois at Urbana-Champaign*
Kelly Dilliard, *University of South Dakota*
Grenville Draper, *Florida International University*

Aley El-Shazly, *Marshall University*
Stewart S. Farrar, *Eastern Kentucky University*
P. Jay Fleisher, *SUNY College at Oneonta*
Tracy Furutani, *North Seattle Community College*
Doug Haywick, *University of South Alabama*
Christi Hill, *Grossmont College*
Tami J. Jovanelly, *Berry College*
Nazrul I. Khandaker, *York College of CUNY*
Brian E. Lock, *University of Louisiana at Lafayette*
Ryan Mathur, *Juniata College*
Katherine Milla, *Florida A&M University*
Otto H. Muller, *Alfred University*
Alfred H. Pekarek, *St. Cloud State University*
Michael S. Petronis, *New Mexico Highlands University*
Emma C. Rainforth, *Ramapo College of New Jersey*
Brady P. Rhodes, *California State University, Fullerton*
Joseph M. Smoak, *University of South Florida*
Leif Tapanila, *Idaho State University*
Stephen Taylor, *Western Oregon University*
Paul Umhoefer, *Northern Arizona University*
Marilyn Velinsky Rands, *Lawrence Tech University*
Gary S. Zumwalt, *Louisiana Tech University*

The Essential Earth

Why We Study Earth

Greetings from Los Angeles!

OUR NAMES ARE TOM JORDAN AND JOHN GROTZINGER, and we are the authors of this textbook. We're geologists who have studied Earth from top to bottom and from the inside out. Throughout our professional lives—for seven decades between us—we have traveled around the world in the search for answers to fascinating scientific questions. We've enjoyed many adventures along the way, on lush islands of the South Pacific, in the wilds of Africa, and across the bleak tundra of the Canadian Arctic.

We now reside and teach in metropolitan Los Angeles, where civilization confronts the natural world on an astounding scale. In Southern California, we can see evidence of nature's bounty all around us, and we sometimes feel its wrath. LA was built on a fertile coastal plain underlain by rich reservoirs of life-sustaining water and energy-producing oil. The mighty geologic forces that formed those natural resources are also pushing rocks upward, creating a mountainous breeding ground for natural disasters. In the ranges that rise above us, we can see the scars of earthquakes, wildfires, landslides, and floods.

> **Our future as a civilization depends on how well we understand Earth and its challenges.**

We view this world from a geological perspective, continually amazed by what we observe: an active planet with a turbulent, glowing-hot interior; an ever-changing surface of oceans and continents; a thin layer between Earth and sky teeming with life; and a human society altering this environment on a global scale. We have written this textbook because we believe our future as a civilization depends on how well we understand Earth and its challenges.

We want to give you, the student, new eyes to look at the world around you. We also want to describe the scientific methods and tools that you, the next generation, will need to face the challenges. ◆

Geologic forces have pushed the San Gabriel Mountains to 3000 m above Los Angeles. Rocks have been brought down by earthquakes, landslides, and floods to form a fertile coastal plain that contains rich natural resources. [Richard Price/Getty Images.]

In this opening chapter, we will show you why Earth science is so essential to so many urgent human needs, from developing natural resources and dealing with natural hazards to preserving our environment and adapting to global change. We will also introduce you to the geological quest for fundamental knowledge about the natural world.

THE TERM **GEOLOGY** (from the Greek words for "earth" and "knowledge") was coined by scientific philosophers more than 200 years ago to describe the study of rock formations and the puzzling evidence from fossils. Through careful observations and reasoning, their successors developed the theories of biological evolution, continental drift, and plate tectonics—major topics of this textbook. Today, *geology* identifies the branch of Earth science that studies all aspects of the planet: its history, its composition and internal structure, and its surface features.

Geologists use many methods and tools to improve our scientific understanding of Earth (**Figure 1.1**). They study rock formations in remote regions with rucksacks on their backs, hammers and maps in their hands. They build precise laboratory instruments to investigate the physics and chemistry of Earth materials. They download huge datasets from satellites and networks of ground-

(a)

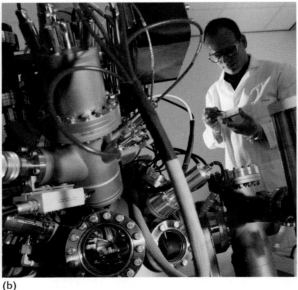
(b)

(c)

FIGURE 1.1 (a) Geologists examine rocks inside Spider Cave at Carlsbad Caverns, New Mexico. (b) A geochemist readies a rock sample for analysis by a mass spectrometer. (c) Geophysicists deploy instruments to measure the underground activity of a volcano. [(a) AP Photo/Val Hildreth-Werker. (b) John McLean/Photo Researchers. (c) Hawaiian Volcano Observatory/USGS.]

based sensors, process them on supercomputers, and work hand-in-hand with other Earth scientists to understand our planet as a whole.

Fields of Earth science closely related to geology include *meteorology,* the study of the atmosphere; *oceanography,* the study of the oceans; and *ecology,* which concerns the abundance and distribution of life. *Geophysics, geochemistry,* and *geobiology* are subfields of geology that apply the methods of physics, chemistry, and biology to geologic problems.

Geologists seek answers to many basic questions. Of what material is the planet composed? Why are there continents and oceans? How did the Himalaya, Alps, and Rocky Mountains rise to their great heights? Why are some regions subject to earthquakes and volcanic eruptions while others are not? How did Earth's surface environment and the life it contains evolve over billions of years? What changes are likely in the future?

We think you will find the answers to such questions fascinating. Welcome to the science of geology!

SPACESHIP EARTH

Spacecraft launched in the 1960s changed our view of Earth. For the first time, we could look from the outside at our small blue planet spinning through the solar system (**Figure 1.2**). We began to imagine Earth itself as a spaceship, stocked with limited provisions and furnished with

FIGURE 1.3 An astronaut checks out instrumentation for monitoring Earth's surface. [*StockTrek/SuperStock.*]

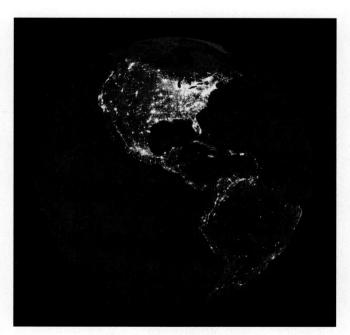

FIGURE 1.2 North and South America at night. [*Image and data processing by NOAA's National Geophysical Data Center, Earth Observation Group (http://www.ngdc.noaa.gov/dmsp).*]

delicate life-support systems. We became more aware that the fast-paced development of modern civilization might eventually deplete those resources and stress our life-support systems to the breaking point.

Half a century into the space age, we have gained a clearer view of our planet. We have been especially successful in understanding our **environment**—the uniquely habitable region near Earth's surface that humans share with all other organisms. In thinking about how we interact with our surroundings on Spaceship Earth, we sometimes distinguish the *built environment* of civilization—what humans have constructed from steel and concrete—from the *natural environment* of the world around us.

Both land-based and space-based instruments now track changes in the atmosphere, the oceans, and the land surface of Earth (**Figure 1.3**). We have become much better equipped to assess the effects of human activities on the natural environment and anticipate future change.

What changes can be expected? At this point in the voyage of Spaceship Earth, our environment is not in good health, and its condition appears to be worsening. Substantial increases in atmospheric concentrations of carbon dioxide (CO_2) and other "greenhouse" gases, primarily from the burning of fossil fuels, are warming the climate, especially near the poles. Some of this additional atmospheric CO_2 is dissolving into the oceans, making them more acidic and less hospitable to marine life. The loss of natural habitat, primarily to agriculture and urbanization, is resulting in the extinction of many plant and animal species. What midcourse corrections are available to steer us in a more favorable direction?

In 1963, the architect Buckminster Fuller wrote *An Operating Manual for Spaceship Earth,* in which he noted the need for more knowledge about how Earth works as a mechanical system: "You know that you're either going to have to keep the machine in good order or it's going to be in trouble and fail to function. We have not been seeing our Spaceship Earth as an integrally-designed machine [that] must be comprehended and serviced in total . . . no instruction book came with it."

Figuring out how to operate Spaceship Earth requires a scientific understanding of how the contraption works. We hope that some of the pages you read here will be added to your own personal manual for operating Spaceship Earth.

The space-age view of our planet as a big, complex machine—the *Earth system*—has stimulated a new way of doing Earth science, which we will explain in Chapter 2. It has also become a potent metaphor for people concerned about **sustainable development.** That concept was defined in a famous 1987 United Nations report as "development that meets the needs of the present without compromising the ability of future generations to meet their own needs." It offers an appealing if utopian vision: a promise that the current generation will carefully manage its interactions with Earth to ensure a hospitable environment for future generations.

Sustainability involves many economic and political issues about which many nations do not agree, so forging a global strategy that moves civilization toward this goal will not be easy. People must learn how to work together to accomplish a number of goals that will be emphasized throughout this text:

> *Sustainable development meets the needs of the present without compromising the ability of future generations to meet their own needs.*

1. **Natural resources:** Manage our natural resources in a way that can continue to provide society with adequate energy, water, and raw materials.

2. **Natural hazards:** Prepare society for inevitable natural disasters such as earthquakes, hurricanes, floods, and volcanic eruptions.

3. **Natural environment:** Repair the environmental damage we have already done and maintain a habitable environment on Spaceship Earth over the long haul.

These three goals are interrelated, and they involve political and economic issues that extend far beyond Earth science. But reaching them will depend on a scientific understanding of the natural world and on the new technology that science will help us create.

In this chapter, we sketch a broad picture of the practical issues addressed by Earth science. Later chapters will color in the details by describing the geologic processes that govern the natural environment, produce natural resources, and generate natural hazards.

NATURAL RESOURCES

Human populations and their effects are multiplying at phenomenal rates. Between 1927 and 1999—in the span of just one human lifetime—the global population rose threefold, from 2 billion to 6 billion, and is now increasing by 79 million people per year. The view of Earth from space shows a glowing lattice of civilization spreading rapidly across Earth's surface (see Figure 1.2). The number of people will reach 7 billion in 2011 and probably exceed 8 billion by 2030.

> *Human effects on our environment are growing with the human population, which now exceeds 6 billion.*

This global human society is consuming vast amounts of **natural resources.** This general term refers to the energy, water, and raw materials that are available from the natural environment. The demand for natural resources is skyrocketing as civilization expands and people around the world strive to improve the quality of their lives. Geology provides the scientific knowledge needed to find, exploit, and conserve natural resources.

Resources and Reserves

The supply of any material mined from the ground is finite. Its availability depends on its distribution in accessible deposits, as well as on how much we are willing to pay to get it out of the ground. Geologists use two measures of supply. **Reserves** are supplies that have already been discovered and can be exploited economically and legally at the present time. In contrast, *resources* constitute the entire amount of a given material, including the amount that may become available for use in the future. Resources include reserves plus known but currently unrecoverable supplies plus undiscovered supplies that geologists think may eventually be found (**Figure 1.4**).

In many cases, resources that are too poor in quality or quantity to be worth exploiting now or that are too difficult to retrieve become profitable when new technology is developed or prices rise. A recent example is the production of oil and gas from large reservoirs in the Gulf of Mexico at water depths greater than 300 m. Improved drilling techniques on offshore platforms were

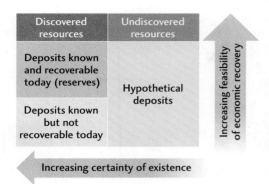

FIGURE 1.4 Resources include reserves plus known but currently unrecoverable deposits plus undiscovered deposits that geologists think may eventually be found.

Reserves are known supplies of natural resources that could be exploited economically under current conditions.

able to increase the annual deep-water oil production from 21 million barrels in 1985 to 456 million in 2009 (**Figure 1.5**).

Geologists are experts at discovering new resources. It should be kept in mind, however, that the assessment of resources is much less certain than the assessment of reserves. Any figure cited as representing the resources of a particular material is only an educated guess about how much will be available in the future.

Energy Resources

Energy is required do work, so it is fundamental to all aspects of human activity. A crisis in the supply of energy can bring a modern society to a halt. Wars have been fought over access to supplies of fuel; economic recession and destructive currency inflation have resulted from fluctuations in the price of oil. It is no surprise that the fossil-fuel industry is the world's biggest business!

Basic Resource Types Before the industrial revolution of the mid-nineteenth century, most of the energy used in the United States came from the burning of wood (**Figure 1.6**). A wood fire, in chemical terms, is the combustion of *biomass:* organic matter consisting of carbon and hydrogen compounds. The ultimate source of the energy is the sunlight green plants use to grow. In this capacity, the biomass acts as a short-term reservoir for storing solar energy. It is a **renewable resource** because the environment is constantly producing new biomass; for example, a forest chopped down for wood can be regrown and harvested again. Humans have used a variety of renewable energy sources to power mills and other machinery for

FIGURE 1.5 New technologies used aboard offshore platforms in the Gulf of Mexico can recover oil and gas from rock reservoirs below very deep waters. Concern over the environmental risks of this type of energy production has grown since the 2010 Deepwater Horizon disaster spilled 5 million barrels of oil into the Gulf of Mexico, causing severe ecological and economic damage. [*Larry Lee Photography/Corbis.*]

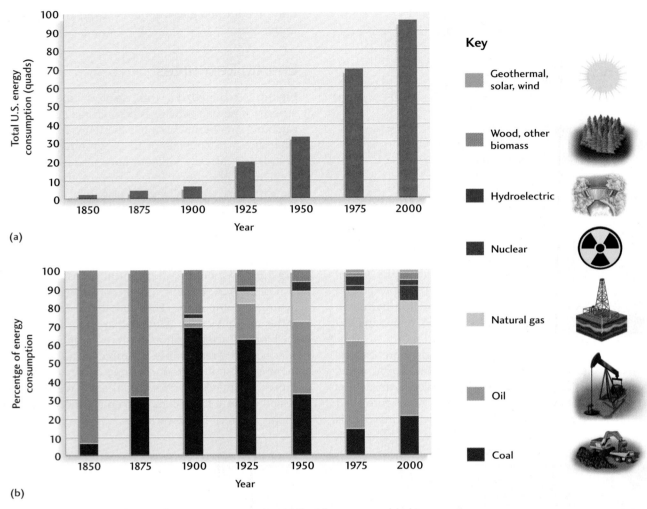

FIGURE 1.6 Human society's use of energy resources has shifted from renewable biomass to fossil fuels as the rate of energy consumption has grown. (a) Total energy consumption in the United States from 1850 to 2000, in quads (1 quad = 10^{15} Btu). (b) Percentages of various types of energy consumed in the United States from 1850 to 2000. [*U.S. Energy Information Agency.*]

thousands of years, including moving air, falling water, and the work of oxen, horses, and elephants.

Some of the biomass from plants that grew millions of years ago was buried beneath Earth's surface and transformed into the combustible rock called *coal* (**Figure 1.7**). When we burn coal, we are using "fossilized" energy from ancient sunlight stored by plants. Our other major fuels, crude oil (petroleum) and natural gas, were also created by the burial and heating of dead organic matter. These **fossil fuels** are considered to be **nonrenewable resources** because geologic processes produce them much more slowly than we are using them up. Our reserves of oil, natural gas, and coal will be exhausted long before Earth can replenish them.

Rise of the Fossil-Fuel Economy Industrialization increased the demand for energy beyond what the traditional renewable sources could supply. By the late eighteenth century, James Watt and others had developed coal-fired steam engines that could do the work of hundreds of horses. Steam technology lowered the price of energy dramatically, in part because it made coal mining possible on a very large scale—in Britain, from the coalfields of England and Wales; in continental Europe, from the coal basins of western Germany and bordering countries; and in North America, from the Appalachian coalfields of Pennsylvania and West Virginia. The availability of cheap energy sparked the industrial revolution. By the end of the nineteenth century, coal accounted for more than 60 percent of the U.S. energy supply (see Figure 1.6).

The first oil well was drilled in Pennsylvania by Colonel Edwin L. Drake in 1859. The idea that petroleum could be profitably mined like coal provoked skeptics to call the project "Drake's Folly" (**Figure 1.8**). The skeptics were

FIGURE 1.7 Coal is mined at Peabody Energy's North Antelope Rochelle coal mine, near Gillette, Wyoming. [*Matthew Staver/The New York Times/Redux.*]

wrong, of course; by the early twentieth century, oil and natural gas were beginning to displace coal as the fuels of choice. Not only did they burn more cleanly, producing no ash, but they could be transported by pipeline as well as by rail and ship. Moreover, gasoline and diesel fuels refined from crude oil were suitable for burning in the newly invented internal combustion engine. By 1950, the use of energy from oil and natural gas exceeded that from coal, and these fossil fuels taken together supplied almost 90 percent of U.S. energy needs.

Nuclear reactors that could produce electricity from radioactive elements such as uranium were developed

FIGURE 1.8 Edwin L. Drake (*right*) in front of the oil well that initiated the "age of petroleum." This photo was taken by John Mather in 1866 in Titusville, Pennsylvania. [*Bettmann/Corbis.*]

FIGURE 1.9 Japan's Kashiwazaki-Kariwa facility is the world's largest nuclear power plant, with seven reactors and a total generating capacity exceeding 8200 megawatts. It was damaged by a powerful earthquake (magnitude 6.6) that struck the region on July 16, 2007. The plant was shut down and required extensive repairs. [STR/AFP/Getty Images.]

after World War II (**Figure 1.9**). The early expectation that nuclear fuels would provide a large, low-cost, environmentally safe source of energy has not been realized, however. Safety concerns, the inability to dispose of nuclear wastes, and the escalating costs of stringent safety and security measures have slowed the construction of nuclear power plants. Nuclear power supplies a substantial fraction of the electrical energy used by some countries, such as France (76 percent) and Sweden (47 percent), but this proportion is much smaller in the United States (20 percent). Overall, nuclear power accounts for less than a tenth of the total U.S. energy budget.

Today the engine of civilization runs primarily on fossil fuels: coal, oil, and natural gas (methane: CH_4). Taken together, they account for 85 percent of the energy consumed by the United States and the world at large. The use of renewable energy resources, especially of biofuels such as ethanol (made from sugarcane and corn), is increasing, but the total U.S. energy production from all renewable resources was just under 7 percent in 2007.

> *A natural resource is renewable if Earth can replenish it as rapidly as humans consume it.*

Global Energy Consumption Energy use is often measured in units appropriate to the fuel; for example, barrels of oil, cubic feet of natural gas, and tons of coal. But the comparisons are easier if they are made in a standard unit of energy such as the British thermal unit (Btu). One Btu is the amount of energy needed to raise the temperature of 1 pound of water by 1°F (1054 joules). In 2007, the United States used 101 quadrillion Btu, or *quads,* of energy per year (1 quad = 10^{15} Btu), compared with a global total of about 480 quads (**Figure 1.10**). Thus, the United States, with less than 5 percent of the world's population, has been consuming about four and a half times more energy per person than the global average.

The total annual U.S. consumption of energy of all types actually declined by 7 percent between 2007 and 2009, the first multiyear drop in recent history. Much of this decrease can be attributed to the severe economic recession, but there are promising signs that new technologies and conservation efforts are beginning to lessen U.S. energy appetites.

The modest reductions in energy consumption by the United States, Japan, and Western Europe since 2007 have been more than offset by increases in the developing world, led by the world's two most populous countries, China (+8 percent per year) and India (+7 percent per year). In 2009, China's total energy consumption exceeded that of the United States for the first time. Still, China's average individual (or *per capita*) energy use remains almost eight times less, due to its much larger population. As it and other developing economies strive to improve their standards of living, global energy use per capita is bound to rise, accelerating overall energy consumption.

> *Energy use in the developing world is increasing rapidly, but the average consumption per person remains higher in the United States than in any other large nation.*

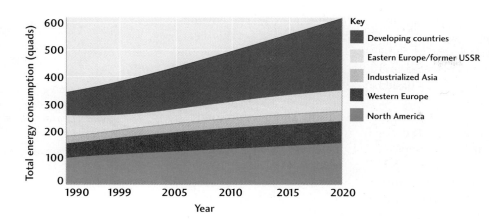

FIGURE 1.10 Actual and projected energy consumption, in quads, 1990–2020, by world regional groupings. [*U.S. Energy Information Agency.*]

Petroleum Exploration and Production Geologists are refining methods for finding new oil and gas resources in remote regions of the continents and on the continental margins beneath deep water (**Figure 1.11**). If discoveries in these regions significantly expand our petroleum resources, oil may stay relatively cheap for a longer time. The U.S. Geological Survey estimates that the world's total oil resources are at least 2 trillion barrels, equivalent to almost 12,000 quads of energy, and it has drawn the optimistic conclusion that any peak in petroleum production is decades away. The rate at which new petroleum resources are being discovered is not keeping up with rising oil production, however, leading some pessimists to estimate that global production will begin to decline in the next few years. We will revisit this debate in Chapter 14.

More petroleum will be available for future needs if reserves can be more efficiently extracted from deep rock formations. The subterranean reservoirs that hold oil and natural gas are complex geologic systems. Geologists

(a)   (b)

FIGURE 1.11 (a) The *Geco Topaz*, a vessel operated by WesternGeco Inc., conducting a three-dimensional seismic survey in the North Sea. The bubbles behind the ship are compressed-air explosions that send out sound waves; the reflections from the rocks below are recorded by the ship to produce an image of the subsurface structure. (b) Depiction of a three-dimensional seismic survey, showing acquisition system and a subsurface "cube" of seismic data. The colors show the layers of sediments beneath the seafloor, some trapping oil and natural gas. [*(a) Courtesy of Oil & Gas UK. (b) Courtesy of Satoil, Veritas, and BP.*]

can map the reservoir rocks in three dimensions using various techniques, such as seismic imaging (see Figure 1.11). The three-dimensional pictures they obtain show them where the bulk of the oil and gas is located and allow them to predict how it will flow out of holes drilled into the reservoir.

Drilling holes deep into Earth's rocky crust has become a very sophisticated and expensive business (**Figure 1.12**). Petroleum engineers use three-dimensional models to steer drill bits on swooping paths into the richest parts of a reservoir. To coax oil out of stubborn formations, they inject water and carbon dioxide down strategically positioned drill holes to push the oil into areas where it can be more efficiently pumped through other drill holes. These methods have increased the fraction of oil that can be extracted from known oil fields, increasing our reserves. The drilling from a single oil platform in the deep waters of the Gulf of Mexico (see Figure 1.5) can cost over $100 million.

The economic advantages of offshore drilling must be balanced against the considerable environmental risks. The blowout of the Deepwater Horizon well on April 20, 2010, released nearly 5 million barrels of oil into the Gulf of Mexico. The environmental damages of this disaster—the largest marine oil spill in history—are difficult to calculate (and may not be known for many years), but the short-term economic losses in the region are estimated to run into the tens of billions of dollars.

> Geologists are improving methods for finding new oil and gas resources and for extracting known resources more efficiently, but the environmental risks are increasing.

Our Energy Future Despite efforts to locate more reserves and improve production, the major petroleum reservoirs will become depleted sometime in the next few decades. Oil prices will rise to uneconomic levels, and alternative energy resources will have to take up more and more of the demand. How quickly will this transition to a post-petroleum economy occur? Which alternative sources of energy will replace oil? The predictions and choices, which are fiercely debated, will depend rather heavily on progress in Earth science and technology.

The financial costs of replacing oil (and eventually natural gas) with alternative energy sources must also be

FIGURE 1.12 Casing crew working on an oil rig drill floor. [Peter Bowater/Photo Researchers.]

balanced with the environmental costs. Nuclear power is an attractive option because it generates no greenhouse gases and because enough uranium resources are available as nuclear fuels to produce an estimated 240,000 quads of energy, more than twice the reserves of all fossil fuels combined. However, switching wholesale to nuclear power would create other environmental problems—people don't want nuclear power plants or nuclear waste dumps in their backyards—and would heighten concerns about the proliferation of nuclear weapons.

Developing solar power (**Figure 1.13**) and other renewable energy sources—such as wind, water, and geothermal power, as well as biofuels—will certainly help, but current projections indicate that the total energy production from these resources will fall short of our needs for many decades to come, unless there are unanticipated technological breakthroughs.

Coal, a plentiful and cheap alternative to oil and natural gas, is expected to fill the gap. Major deposits of coal are found on all continents of the Northern Hemisphere; the total resource amounts to over 3 trillion metric tons, which is enough to generate 67,000 quads of energy. But coal is a notoriously dirty fuel, polluting the atmosphere with gases and particles that cause smog and acid rain. Moreover, for a given amount of energy, coal produces 25 percent more carbon dioxide than oil and 70 percent more than natural gas. Other "unconventional" fossil fuels, such as the tar sands of Alberta and the oil shales of Colorado, are also abundant but polluting.

New technologies could help us to reduce coal pollution. If the carbon dioxide emitted from coal-fired power plants could be economically captured and stored underground, the world's abundant coal resources would become much more attractive as a replacement for petroleum. For example, "clean coal" energy could be used to generate hydrogen fuel for transportation. Turning carbon capture and storage into a feasible technology will be a great geological challenge.

Minerals and Other Raw Materials

Coal and oil are examples of raw materials we extract from Earth's crust. Many other raw materials are needed to meet modern society's demands, particularly *minerals*, the crystalline components of rocks. Gem minerals such as diamonds and rubies are prized for their beauty, but their economic value is tiny compared with that of the more mundane minerals and other raw materials that are mined in huge quantities. Industrial mining provides stone for buildings and roads, phosphates for fertilizers, cement

The depletion of petroleum reserves forces us to balance the availability, financial costs, and environmental costs of alternative energy sources. Coal is likely to reemerge as the world's primary energy resource.

In addition to energy resources, human activities, especially construction, require raw materials such as stone, minerals, and metals.

FIGURE 1.13 Solar cells convert sunlight, a renewable resource, into electrical energy at this utility in a remote village in Nepal.
[*Ned Gillette/Corbis.*]

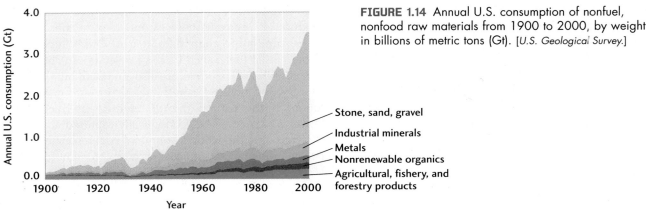

FIGURE 1.14 Annual U.S. consumption of nonfuel, nonfood raw materials from 1900 to 2000, by weight in billions of metric tons (Gt). [*U.S. Geological Survey.*]

for construction, clays for ceramics, sand for silicon chips and fiber-optic cables, and the metals used everywhere in everyday life.

Rising Economic Demand In 2000, the U.S. consumption of raw materials, exclusive of fuels and food, was about 3.6 Gt (gigatons, or billion metric tons). This use, almost 13,000 kg per capita, was double the amount in 1960 (**Figure 1.14**). Three-fourths was bulk material for construction, such as crushed stone, sand, and gravel. The remainder comprised industrial minerals, such as sand for making glass (10 percent), metals (5 percent), nonrenewable organics, such as asphalt and lubricants (4 percent), and agricultural, fishery, and forestry products (6 percent).

U.S. consumption of raw materials, about one-third of the global total, has been increasing at 2.8 percent per year, more than twice the rate of U.S. population growth. Global demand is rising even faster, due to the rapid expansion of the Asian economies. Worldwide, the average person is using more raw materials each year.

Growing consumption of raw materials has led to a boom in the mining industry, particularly in the mining of metals. After declining in the 1990s, the prices of important metals such as copper, zinc, lead, tin, and nickel more than tripled between 2002 and 2007 (**Figure 1.15**), pushed upward by the strong demand for industrial products in developing countries such as China and India. In the case of uranium minerals, which are refined to make fuels for nuclear reactors (and nuclear bombs), the price increased by more than a factor of 10. In late 2007, the global economy entered into a recession and the prices of metals dropped significantly. They are now climbing again as the economy improves and demand increases.

The need for metals can be accommodated in part by *recycling* the metal content of discarded goods and by substituting low-cost ceramics, composites, and plastics. For

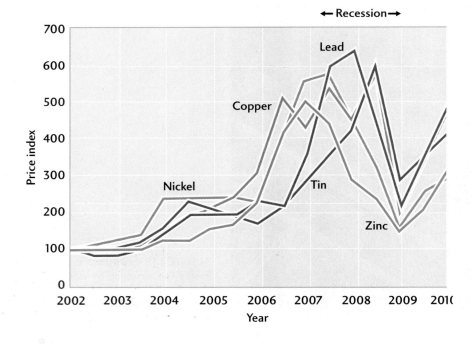

FIGURE 1.15 Metal prices rose substantially during the early 2000s, fueled by increasing global demand. Prices dropped considerably in 2007 because of the worldwide recession but are rising again. This graph shows the price indices of five important metals from 2002 (index = 100) to 2010. For example, $100 worth of nickel purchased in 2002 was worth $560 at the beginning of 2007, $160 at the beginning of 2009, and $300 at the beginning of 2010. [*London Metal Exchange.*]

example, over one-third of the total U.S. aluminum supply now comes from recycling. However, the stocks of copper, zinc, and other essential metals, even if they are recycled, may not meet the needs of the global population over the long haul. According to one recent study, the amount of copper that would be required to supply the current world population with the amount now used per capita by developed nations equals all the copper in known reserves, plus all the copper currently in use as electrical wiring and in computers, medicines, and a wide array of manufactured goods.

> *Despite recycling and substitution of lower-cost materials, the demand for useful metals is increasing rapidly as developing countries become more industrialized.*

Mineral Exploration Rising demand is driving an increasingly systematic search for new mineral resources. Geology provides us with knowledge about how mineral deposits are formed and where to look for more of them (**Figure 1.16**). It also guides us in exploiting these natural resources more efficiently and in minimizing the damage caused by mining to the environment.

Useful chemical elements, such as metals, are widely distributed in many common rocks and minerals. Theoretically, with enough money and energy, we could extract both abundant and rare elements from many types of rocks. A more practical concern, however, is the identification of new reserves, mineral deposits that are profitable to mine and purify. We can reasonably expect that new discoveries will add to current reserves but at an uncertain rate. When the richest deposits have been mined out, we will be forced to rely on lower-grade deposits that could be more expensive to recover.

The high-grade deposits of minerals from which we can extract metals are called *ores*. Ores can be found in a variety of geological environments, in young rocks and old rocks, on land and beneath the sea. Many of the most valuable ores were formed in regions of volcanic activity, where hot circulating groundwater leached the metals from nearby rocks and deposited them in a concentrated form. For this reason, information about volcanism in the geologic past is valuable in the search for new metal resources. We will examine this connection in Chapter 4.

> *Most metals are extracted from high-grade mineral deposits called ores. Many of the most valuable ores were deposited by hot groundwater circulating in volcanically active regions.*

Undiscovered mineral resources may lie in regions that are difficult to explore, such as the ocean floor. How important the deep sea will become as a source of minerals is an open question. The answer depends in part on the development of efficient marine technology for deep-sea exploration and mining, as well as on the resolution of legal issues regarding ownership of deep-sea deposits.

In 1982, the United Nations adopted an agreement known as the Convention on the Law of the Sea, which set up a legal and regulatory system for the development of deep-sea resources; the vote was 130 in favor, 4 opposed.

FIGURE 1.16 A geologist examines a rock containing lead at Greenside Mines, Helvellyn, Lake District, England. [*Ashley Cooper/Alamy.*]

FIGURE 1.17 Mine dumps near Johannesburg, South Africa, an impressive example of the wastes produced by mining. [*George Steinmetz.*]

The United States was one of the four nations opposed because it was concerned that the agreement might unfairly limit the commercialization of seafloor resources. A quarter of a century later, the United States still has not signed the agreement.

It is generally agreed that a nation has exclusive rights to mineral deposits in the offshore area within 200 nautical miles of its coast—its *exclusive economic zone.* Still in question is who owns the mineral deposits on the seafloor beyond this zone. Are they owned by all nations in common? By the discoverer and developer? By both? Because the economies of many mineral-exporting nations are threatened by competition from new sources on the seafloor, those nations would like to limit such development and obtain a share of the profits.

Crafting acceptable agreements on the exploitation of natural resources in international waters such as the Arctic Ocean, as well as on the continent of Antarctica (which is also regulated by a global treaty), will require good geological information about the abundance and distribution of mineral and energy resources.

Environmental Damage by Mining Despite its benefits, mining has been called the original dirty industry. Mining strips off more of Earth's surface each year than natural erosion by rivers. Annually, its waste products exceed the world's accumulation of municipal garbage (**Figure 1.17**). There has been progress in developing clean and environmentally benign mining operations, but mining and processing of minerals remain major contributors to environmental degradation and poor health in many countries. Geologists can provide the information needed to guide clean mining practices and the disposal of mining wastes.

Some people argue that searching for more ore deposits to mine is not the solution to shortages of raw materials because the extraction, processing, and waste disposal connected with exploiting the raw materials are so damaging to the environment and human health that they cannot be sustained. They propose a new "materials economy" based on recycling, greater efficiency in the use of raw materials, and substitution of advanced, environmentally benign materials for minerals. One example is the use of fiber-optic glass cables made from sand instead of copper wire.

In any case, society will have to find new mineral resources, and we will have to weigh the short-term economic gains of mining against its long-term environmental and health costs.

> *Mining has considerable negative consequences: it produces a tremendous amount of waste as well as environmental degradation and health problems.*

Water Resources

Water is vital to all life on Earth. Even the hardiest desert plants and animals require water, and humans cannot survive more than a few days without it. However, the amount of water consumed by modern society far exceeds what humans need for simple physical survival. A human can make do with a daily ration of about 2 liters of water, but today in the United States, the average individual uses about 250 liters per day. Industrial and agricultural uses increase the U.S. per capita total to 6000 liters per day. This per capita consumption is two to four times greater than in western Europe, where people pay up to three times as much for their water.

FIGURE 1.18 Irrigation in California's Imperial Valley, a natural desert. [*David McNew/Getty Images.*]

Fresh water is consumed in immense quantities by agriculture, industry, and sewerage systems. On average, U.S. agriculture takes about 35 percent of the water withdrawn from the nation's water supplies, and industry takes about 53 percent, primarily for electricity generation in power plants fired by fossil fuels. The remaining 12 percent accounts for all domestic and other commercial consumption, including sewerage.

Per capita water use in the western United States, mostly for agricultural irrigation, is 10 times greater than that in the eastern states, and the prices of water are lower, even though the arid western states receive only one-fourth of the average national rainfall. In California, for instance, 85 percent of the water is used for irrigation (**Figure 1.18**), 10 percent for municipalities, and 5 percent for industry. A 15 percent reduction in irrigation use would double the amount of water available for use by cities and industries. Such conversions may become necessary if water shortages accompany continuing population growth in the western states.

> *Per capita water use in the United States, primarily for agriculture, industry, and sewerage, is much higher than in other countries.*

Water Shortages Shortages due to increased demand for water can be compounded by *droughts,* dry periods of months or years that damage crops and cause significant reductions in water supplies. Droughts can occur in all climates, but arid regions are especially vulnerable. Lacking replenishment from rainfall, rivers may shrink and dry up, reservoirs may evaporate, and the soil may dry up and turn to dust, killing vegetation and other life.

From 2000 to 2008, the southwestern United States experienced its worst drought in more than a hundred years of recorded history. In the lower Colorado River basin, water reservoirs such as Lake Powell and Lake Mead had dropped to less than 60 percent of their capacity (**Figure 1.19**).

FIGURE 1.19 Lake Powell, Utah, in 2004 showing "bathtub rings" from recent drought. [*Christopher J. Morris/Redux.*]

Our climate history can give us perspectives on the severity of droughts. For example, during the 400-year period from 1500 to 1900, the Southwest was drier, on average, than it has been during the last century. Moreover, the geologic record shows droughts that were more severe and of longer duration than recent droughts have been (at least so far). By exploring the past, geologists and climate scientists are trying to predict the future. Are the recent droughts just short-term fluctuations in climate, or do they signal a return to an extended dry period? How will global climate change affect rainfall in the Southwest?

Despite its drought problems, the United States is relatively well endowed with water resources. Other countries are less fortunate. A severe drought from 1960 to 1980 destroyed farming and grazing along the southern border of the Sahara Desert, in a region called the Sahel, causing a famine that killed a million people.

According to the United Nations Environment Programme, more than one-third of the world's population is without a safe water supply, and one-fourth will experience chronic water shortages during the next decade. Within two decades, dozens of countries may suffer from water shortages, running up against a "water barrier" to further development.

These problems could be exacerbated by global climate change, which could bring more droughts to already dry regions. Some climate scientists think that global warming is permanently expanding the Sahara Desert southward into the Sahel and may further dry out the already parched southwestern United States.

As the demand on limited water supplies increases, the geological study of water resources, called *hydrology* (from *hydros,* the Greek word for "water"), is becoming an increasingly important field of Earth science.

Groundwater Resources The world's water supply, about 1.46 billion cubic kilometers, is truly enormous. But most of that water, about 96 percent, is salt water in the oceans, and another 3 percent is locked up in glaciers and the polar ice caps. Only 1 percent is fresh water, the hydrologic resource most needed for agriculture and civilization. A small fraction of this fresh water occupies lakes and rivers; most of it is **groundwater** contained in rocks beneath the land surface.

Water is continuously on the move, cycling among its natural reservoirs (**Figure 1.20**). The movement of water from the oceans into the atmosphere by evaporation and its subsequent precipitation as rain or snow on land convert salt water into fresh water. Precipitation recharges groundwater reservoirs, and the runoff flows through rivers and streams back into the oceans. Geologists work with meteorologists, oceanographers, and environmental engineers to investigate this *hydrologic cycle,* which we will discuss further in Chapter 11.

> *Most of Earth's fresh water is groundwater, contained in rocks beneath the land surface.*

FIGURE 1.20 Angel Falls, Venezuela, is the tallest waterfall on Earth. The falls plunge 914 m from a flat-topped mountain composed of 1.7-billion-year-old sandstones. The falls were named for pilot Jimmy Angel, who crash-landed atop the falls in the 1930s. [Robert Hildebrand.]

Large parts of North America rely on **aquifers**—reservoirs of groundwater contained in porous rock formations—for almost all their water needs. In the natural course of the hydrologic cycle, the groundwater depleted by evaporation and runoff is recharged by precipitation. In this sense, aquifers are a renewable resource. However, if rapid pumping of water out of the ground depletes the aquifer, precipitation may take hundreds or even thousands of years to replenish it. At our current high rates of water use, many aquifers are effectively a nonrenewable resource.

To illustrate this problem, we consider the huge Ogallala aquifer, which consists of porous formations of sand and gravel that lie beneath the southern Great Plains (**Figure 1.21**). For more than 100 years, the Ogallala aquifer has supplied fresh groundwater to the towns, ranches, and farms of the region, now home to more than a million people. Pumping of water from the aquifer, primarily for irrigation, has been so extensive—about 6 billion cubic meters of water per year from 170,000 wells—that recharge from rainfall

If groundwater is pumped from an aquifer more rapidly than it can be recharged by precipitation, the aquifer is effectively a nonrenewable resource.

cannot keep up. Pressure in the wells has declined steadily, and the water table has dropped by 30 m or more.

Because water use is outpacing recharge, the Ogallala aquifer will decline as a resource during the early decades of this century. If the water it produces cannot be replaced, about 5.1 million acres of irrigated land in western Texas and eastern New Mexico will dry up—and so will 12 percent of the country's supply of cotton, corn, sorghum, and wheat and a significant fraction of the feedlots for the nation's cattle.

Aquifers in the northern Great Plains and elsewhere in North America are also being rapidly depleted. Geologists will be helping communities assess and deal with these problems.

Improving Water Resources The traditional ways of increasing water supplies, such as building dams and reservoirs and drilling wells, have become extremely costly because most of the best (and therefore cheapest) sites have already been used. Furthermore, the building of more dams to hold larger reservoirs carries environmental costs, such as the flooding of inhabited areas, detrimental changes in river flows above and below the dams, and the disturbance of fish and other wildlife habitats. Factoring in these costs has led to delays in dam projects and rejection of proposals for new dams.

Informed decision making about the conservation and management of water resources requires knowledge of how water moves through the hydrologic cycle and how this flow responds to natural changes and human modifications. At what rates can we pump water from aquifers without depleting them? What will be the effects of climate change on water supplies?

Geological studies of rivers and aquifers allow us to improve the quality of our water resources as well as their quantity. Water supplies can be contaminated by a variety of human activities—sewage, industrial wastes, and mining wastes are particular concerns—but also through natural processes involving water-rock interactions. Geochemical techniques have been used, for example, to investigate elevated levels of arsenic in groundwater, which pose a serious health hazard in many parts of the world. The problem is particularly acute in Bangladesh, where groundwater provides 97 percent of the drinking water supply. Geologists are helping to guide the placement of new wells that draw water with acceptable concentrations of arsenic.

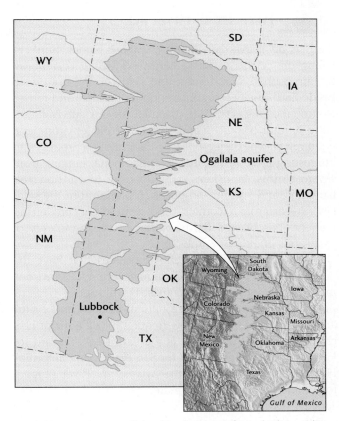

FIGURE 1.21 A map of the Ogallala aquifer, which supplies the southern Great Plains with much of the area's fresh water. [*U.S. Geological Survey.*]

Soils

Soils are an intricate combination of organic material and weathered rock, found in great variety on almost all land surfaces. They constitute an essential and immensely

FIGURE 1.22 Furrows of green beans growing in the rich soil of Monterey County, California. [*Corbis.*]

valuable natural resource (**Figure 1.22**). Soils are the primary reservoir of nutrients for agriculture and the ecological systems that produce renewable natural resources. They are also essential for waste disposal and water filtration. Clays and other soil materials are used for construction and in manufacturing activities.

Soils are a huge reservoir of carbon, containing twice as much as the atmosphere and three times more than all of the world's vegetation. They recycle the organic carbon from dead plants and animals and release it into the atmosphere. They also have a major influence on our water supplies. The quality of the fresh water most people use is determined largely by the soils it passes through.

Soil management is thus crucial to sustaining and improving the human environment, and issues related to soil quality now figure prominently in many policy decisions. Industrial-scale farming has worsened the environmental problems related to soil erosion and the accumulation of toxic elements and salt in soils, as well as downstream contamination of aquatic systems from agricultural runoff. In the United States, new agricultural practices are being adopted to counter these detrimental effects, but much has yet to be learned.

Soils are an important natural resource that must be carefully managed.

Good soil management will require increasing investments in the geological study of soils, including research on the fundamental physical, chemical, and biological processes involved in soil development.

NATURAL HAZARDS

Earth is an active, constantly changing planet. Sunlight falling on Earth's surface energizes swirling currents in the atmosphere and oceans, and heat coming from deep inside Earth melts rock, pushes up mountains, and moves continents. This activity is essential for our survival, creating natural resources and sustaining our environment, but it can also be dangerous. The Earth system produces an interesting array of hazardous phenomena, ranging from severe storms, landslides, and flooding to earthquakes, tsunamis, and volcanic eruptions. Material from the solar system—asteroids, comets, and chunks of broken-up planets—has blasted Earth's surface with the energy of thousands of nuclear bombs.

Events produced by natural processes that have the potential to kill people and damage our built environment are called **natural hazards.** Earth science gives us practical tools to characterize natural hazards and evaluate their risks to society. How frequently will a major U.S. city get hit by a hurricane? Is the earthquake hazard for Hilo, Hawaii, as high as that for San Francisco, California? How should towns along the Mississippi River prepare for floods? What poses a greater hazard for the Seattle-Tacoma area, an eruption of Mount Rainier—the active volcano that towers over the metropolis (see Figure 5.33) —or an earthquake caused by the geologic faults that lie beneath it?

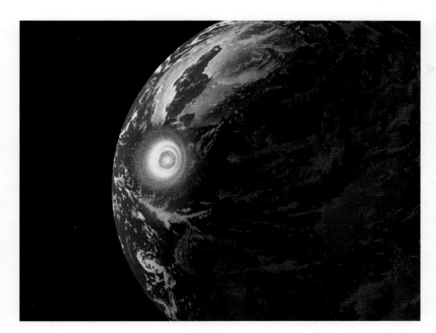

FIGURE 1.23 An artist's rendition of an asteroid impact on Mexico's Yucatán Peninsula 65 million years ago. Such extreme events are very rare but can have major geological and biological consequences. This impact wiped out the dinosaurs and accelerated the evolution of mammals. [*From "The Day the World Burned" by D. A. Kring and D. D. Durda, Scientific American, December 2003. Art by Chris Butler.*]

Answering questions such as these is complicated by the need to consider the sizes, locations, and frequencies of hazardous events. While bigger events are likely to cause more damage than smaller events, they do not occur as often. The very biggest events are very rare indeed; geologists call them *extreme events.*

About 65 million years ago, an asteroid the size of Manhattan smacked into Mexico's Yucatán Peninsula, blowing a huge amount of dust and smoke into the atmosphere (**Figure 1.23**). The environmental changes caused by this extreme event killed off the dinosaurs and many other species, allowing mammals (and eventually humans) to evolve. Geologists have found evidence of similar bad-news collisions at other places and times in Earth history. However, because the span of Earth history is so incredibly long—billions of years—the chances of an asteroid striking next year or even during the next century turn out to be extremely small. More common natural hazards, such as hurricanes and earthquakes, pose greater threats to society in the short term.

A Rogue's Gallery of Natural Hazards

Let's take a brief tour, introducing ourselves to the most important natural hazards and some of their effects on society. Later chapters will describe the natural phenomena that produce these hazards, as well as their geologic consequences, in more detail.

Hurricanes The greatest storms on Earth are **hurricanes,** swirling masses of dense clouds hundreds of kilometers across that suck their energy from the warm surface of tropical seas. In the Northern Hemisphere, they form during late summer when sea surface temperatures are highest; the resulting winds circulate counterclockwise around a calm, central "eye," where the atmospheric pressures are low (**Figure 1.24**). Their name originates from Huracan, a

FIGURE 1.24 Hurricane Katrina on August 28, 2005, a few hours before it struck New Orleans. It was one of the most powerful hurricanes on record, with maximum sustained winds of up to 280 km/hour (175 miles/hour) and gusts up to 360 km/hour (225 miles/hour). [*NASA/Jeff Schmaltz, MODIS Land Rapid Response Team.*]

god of storms to the Mayan people of Central America. In the western Pacific and the China Sea, hurricanes are known as *typhoons,* from a Cantonese word meaning "great wind." In Australia, Bangladesh, Pakistan, and India, they are called *cyclones.*

Whatever you call these intense storms, their high winds and torrential rains can wreak havoc as the storms move from the ocean onto land. Low pressures in and around the storm's eye raise water levels, sometimes by many meters, and the resulting "storm surge" can flood coastal areas, particularly when it coincides with normal high tides. The storm surge is the most deadly of a hurricane's hazards.

> *Hurricanes are intense storms that draw their energy from warm tropical waters.*

In 2005, when Hurricane Katrina struck Louisiana, more than 1800 people lost their lives, primarily by drowning in the storm surge. A catastrophic cyclone struck the coastal lowlands of Bangladesh in 1970, drowning as many as 500,000 people—perhaps the deadliest natural disaster of modern times. Another cyclone hit the same region in 1991, drowning at least 140,000 (**Figure 1.25**). The 1991 storm was more intense, but its death toll was lower due to better disaster preparations; 2 million people were evacuated.

Hurricanes pose tremendous economic risks to states along the Gulf Coast and Eastern Seaboard. The worst hurricane season on record was in 2005, when seven hurricanes caused over $130 billion in damages. Katrina alone accounted for more than $80 billion, making it the most costly disaster in U.S. history. In 2008, Hurricane Ike devastated the Texas coastline, killing 112 people and causing $30 billion in direct economic losses.

The tropical storms that grow into hurricanes can be tracked by satellites, and weather conditions inside the storms can be probed by airplanes. By feeding many types of data into computer models, meteorologists can do a pretty good job of predicting a storm's track and changes in its intensity up to several days in advance. The National Hurricane Center accurately predicted that Katrina would hit New Orleans as a severe hurricane 3 days before it actually did.

Floods When water flows into a region faster than it can flow out, the result is a **flood.** Floods can result from inundation by the sea (e.g., during storm surges), but the most frequent cause is water flowing into a river system more rapidly than the system can discharge it. Some rivers overflow their banks almost every year when the snows melt or rains arrive; others flood at irregular intervals. The floods deposit sediments containing minerals and organic nutrients on the floodplains of the river, making them fertile lands for agriculture (see Chapter 12). Because rivers also supply abundant water and ready-made transportation, floodplains have attracted human settlement since the beginning of civilization.

Small floods are common and usually cause little damage, but the larger episodes that happen every few decades or so can be destructive. In 1993, the Mississippi River and its tributaries went on a rampage with a flood that resulted in 487 deaths and more than $15 billion in property damage (**Figure 1.26**). At St. Louis, Missouri,

> *Rivers flood when water from rain or snowmelt flows into the river system more rapidly than it can be discharged from the system.*

FIGURE 1.25 Disaster in Chittagong, Bangladesh, caused by a cyclone on February 5, 1991. [*Pablo Bartholomew/Liaison Agency/Getty Images.*]

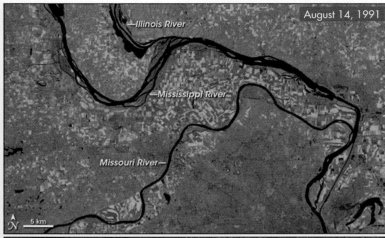

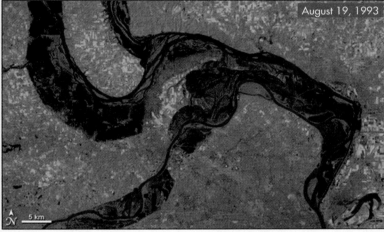

FIGURE 1.26 Satellite photo of the great flood of the Mississippi River, 1993. [NASA images created by Jesse Allen, Earth Observatory, from data courtesy of the Landsat Project Science Office.]

the Mississippi stayed above flood stage for 144 of the 183 days between April and September. In an unexpected secondary effect, the floodwaters leached agricultural chemicals from farmlands and deposited them in the flooded areas, causing widespread pollution.

Today, most large cities are protected by artificial levees and dams, which can help control the flooding, though they cannot entirely eliminate the risk. The flooding of New Orleans by the storm surge from Hurricane Katrina was primarily due to levee failures. In early May 2010, after exceptionally heavy rains, the Cumberland River crested at 52 feet above normal, flooding Nashville, Tennessee (**Figure 1.27**).

Figuring out how to protect society from floods presents some knotty problems in geological engineering. A river hemmed in by levees can no longer erode its banks and widen its channel to accommodate additional water during periods of high flow, and its floodplain no longer receives deposits of sediment. In the case of New Orleans, the floodplain has sunk below the level of the Mississippi River, making future flooding more likely.

In a dramatic step after the 1993 Mississippi flood, the citizens of Valmeyer, Illinois, voted to move the entire town to high ground several miles away. The new site was chosen with the help of a team of geologists from the Illinois Geological Survey. Yet some people who had lived on floodplains all their lives wanted to stay there and were prepared to live with the risk. The costs of protecting some floodplains are prohibitive, and these places will continue to pose public policy problems.

The prediction of river floods and their heights has become much more reliable as more rainfall and river

FIGURE 1.27 The Cumberland River overflows its banks in downtown Nashville, Tennessee, on Tuesday, May 4, 2010, after heavy rains during the week before. [AP Photo/Jeff Roberson.]

measurements have become available. Geologists can now forecast the rise and fall of river levels as much as several months in advance, and they can issue reliable flood warnings days in advance.

Earthquakes Shaking episodes caused by the breaking of rocks along geologic faults are called **earthquakes.** *Faults* are surfaces of weakness in Earth's crust where the rocks on one side slip past those on the other side (**Figure 1.28**). The rupture starts at some point on the fault and propagates quickly (at more than 6000 miles per hour), like a crack through glass. The vibrations generated by this cracking travel away from the fault and, if the crack gets big enough, these *seismic waves* can cause strong shaking tens or even hundreds of kilometers away.

Earthquakes can be incredibly destructive to the built environment, ripping apart urban infrastructure and causing buildings to collapse. In cities, most casualties during earthquakes are caused by falling buildings and their contents. The death tolls can be especially high in densely populated areas of developing countries, where buildings

are often constructed from bricks and mortar without adequate steel reinforcement. This unfortunate fact was tragically illustrated by the Haiti earthquake of January 12, 2010, which killed more than 220,000 people, making it the fifth-deadliest seismic disaster in recorded history. Though geologists expected high-intensity earthquakes to occur in this part of the Caribbean, buildings in the Haitian city of Port-au-Prince were not constructed to withstand such seismic shaking (**Figure 1.29**).

> *Earthquakes are generated by the breaking of rocks along surfaces of weakness in Earth's crust, called faults. The resulting shaking can knock down buildings and trigger many secondary hazards, including fires, landslides, and tsunamis.*

Secondary effects of earthquakes, such as landslides and fires, can also take their toll. The city of San Francisco was largely destroyed by the fires that followed the great earthquake of April 18, 1906; nearly 3000 of its inhabitants died. A great earthquake in China's Kansu Province in 1920 triggered an extensive landslide that covered a region larger than 100 km^2 and resulted in roughly 200,000 deaths.

When large earthquakes lift up the seafloor, they can generate water waves that propagate across the ocean and increase in size when they pile up along the shore, inundating coastal regions with destructive walls of water. These **tsunamis** (Japanese for "harbor waves") have been responsible for two great disasters in less than a decade. On December 26, 2004, an undersea fault broke west of the Indonesian island of Sumatra, lifting up the seafloor and sending a tsunami across the Indian Ocean. The giant waves drowned more than 300,000 people living on coastlines from Thailand to Africa. A similar earthquake east of Japan on March 11, 2011, destroyed many communities along the northeastern coastline of Honshu (**Figure 1.30**), killing more than 25,000 people and causing major radioactive leakage from the Fukushima nuclear power plant.

Earthquakes can be understood in terms of the basic geological machinery of plate tectonics, which moves continents and shapes the face of the planet (see Chapter 3). We can predict where and how frequently earthquakes will occur and how big they will be, but no one has yet figured out how to predict *when* large earthquakes will occur.

Even so, geologists can reduce earthquake damage by identifying dangerous faults and helping engineers to design buildings, dams, bridges, and other structures to withstand earthquake shaking. Sensitive instruments can pinpoint the location of the disturbance within seconds of a fault rupture, providing timely information to firefighters and medical personnel who must respond to a disaster.

FIGURE 1.28 This scarp was formed by slip (left side downward) along a dipping fault surface during the 1954 Fairview Peak earthquake in Nevada. [*Karl V. Steinbrugge Collection, Earthquake Engineering Research Center.*]

Volcanic Eruptions Like earthquakes, **volcanic eruptions** are a result of plate tectonics. These events deposit molten

FIGURE 1.29 Homes in Port-au-Prince destroyed by the Haiti earthquake of January 12, 2010. [*AP Photo/Michael Laughlin, Sun-Sentinel.*]

rock from Earth's interior onto its surface and spew gases into its atmosphere. Like many other geologic phenomena, volcanoes are a mixed blessing, providing society with mineral resources, fertile soils, and thermal energy, but also posing threats to the communities around them.

The rim of the Pacific Ocean is called the "Ring of Fire" (see Figure 3.6) because it contains many active volcanoes, including the majestic cones of Mount Rainier in the Pacific Northwest and Mount Fuji in Japan. These mountains can remain dormant for hundreds of years and then erupt in violent spasms, sometimes blowing their tops into the stratosphere and covering huge areas with hot rocks and ash.

Other types of volcanoes, like those found on the island of Hawaii in the middle of the Pacific, erupt more slowly and steadily, releasing hot, fluid lavas that fill up their craters and flood downhill in glowing rivers that bury everything in their path (**Figure 1.31**). Over hundreds of thousands of years, these flows can build up huge piles of lava. Measured from the seafloor, the Hawaiian volcanoes are the world's tallest mountains!

Like earthquakes, volcanic eruptions can trigger other hazards. The 1883 explosion of Krakatau, a volcano in Indonesia, generated a tsunami that reached 40 m in height and drowned 36,000 people on nearby coastlines. In 1985, the eruption of Nevado del Ruiz in the Colombian

FIGURE 1.30 Swath of destruction through the Japanese coastal town of Minamisanriku from the massive tsunami caused by the great Tohoku earthquake of March 11, 2011. [*Kyodo/Reuters.*]

FIGURE 1.31 A partly buried school bus in Kalapana, Hawaii. The village was buried by a lava flow from the Kilauea volcano. [Roger Ressmeyer/ Corbis.]

Andes melted ice and snow on its flanks, which rushed down the mountain in a huge mudflow that buried the town of Armero 50 km away, killing more than 25,000 people (**Figure 1.32**).

A less deadly but still costly hazard comes from the eruption of ash clouds that can damage the jet engines of airplanes that fly through them. The eruption of the Eyjafjallajökull, Iceland, volcano, which began in April of 2010, disrupted air traffic across the North Atlantic, resulting in over a billion dollars of losses to commercial airlines (**Figure 1.33**).

Volcanic eruptions deposit molten rocks from Earth's interior onto its surface and spew gases and ash into its atmosphere.

FIGURE 1.32 Armero, Colombia, submerged by mudflows, after an eruption of the long-dormant Nevado del Ruiz volcano in 1985. [STF/ASP/Getty Images.]

FIGURE 1.33 Iceland's Eyjafjallajökull volcano erupting on April 17, 2010. The ash cloud from this eruption disrupted air traffic across northern Europe.
[*Joanna Vestey/Corbis.*]

Geologists can identify dangerous volcanoes and characterize their hazards by studying the deposits laid down in earlier eruptions. Such investigations indicate that Mount Rainier, which looms over the heavily populated cities of Seattle and Tacoma, Washington, probably poses the greatest volcanic risk in the United States. Hundreds of thousands of people live in areas where mudflows have swept down from the volcano over the past 6000 years.

In many cases, volcanic eruptions can be predicted. Sensitive instruments can detect activity associated with the upward movement of molten rocks, such as earthquakes, swelling of the volcano, and gas emissions. People at risk can be evacuated if the authorities are organized and prepared. Prior to the May 1980 eruption of Mount St. Helens, a volcano in the Cascade Range in Washington State, the U.S. Geological Survey was issuing warnings of tremors. Finally, in April, as seismic tremors increased, the USGS issued a serious warning and ordered people out of the vicinity. The eruption of Mount St. Helens caused the most massive landslide ever recorded anywhere.

Landslides Downhill movements of rock, mud, and soil are called **landslides.** They occur when the friction that keeps the material from slipping downslope is overcome by the downward force of gravity, often due to water saturation. High seasonal rainfall in 2005 triggered many landslides in Southern California, including the deadly La Conchita slide that killed 10 people on the night of January 10 (**Figure 1.34**).

FIGURE 1.34 A massive mudslide buried homes in La Conchita, California, in 2005. [*AP Photo/KevorkDjansezian.*]

Earthquakes, hurricanes, and other natural hazards can also trigger landslides. In 1998, one of the most catastrophic storms of the twentieth century, Hurricane Mitch, dropped torrential rains on Central America; at least 9000 people were killed in the ensuing floods and landslides. One of the hardest-hit places was near the Nicaragua-Honduras border, where a series of landslides and mudflows buried at least 1500 people. Dozens of villages were simply obliterated, engulfed by a sea of mud. The flanks of a crater on the Casita volcano collapsed and started a series of slides and flows that were described as a moving wall of mud more than 7 m high. Those in the direct path of the landslides could not escape, and many were buried alive as they tried to outrun the fast-moving mud. An immense landslide (over 80 million cubic meters) triggered by a 1970 earthquake in Peru destroyed the mountain towns of Yungay and Ranrahirca, killing more than 66,000 people (**Figure 1.35**).

Geologists can map the scars and deposits from previous landslides and issue timely warnings about future earth movements. They can also identify regions that are susceptible to landslides due to steep terrain and weak rock types. Geologic maps showing where landslides are most likely to occur can help land-use planners and developers avoid this type of hazard.

> *A landslide is a downhill movement of rock, mud, and soil that is often due to the weakening of a slope by water saturation.*

Bolides Chunks of rock and other debris from outer space, called **bolides,** may explode as fireballs in Earth's atmosphere or strike its surface. Early in the history of the solar system, before about 3.9 billion years ago, large bolide impacts were much more common than they are today (most of the craters on the Moon date from this period). Collisions with planets have since swept the solar system fairly clean of debris.

Nevertheless, some 40,000 tons of extraterrestrial material continue to fall on Earth each year, mostly as dust and unnoticed small objects. Smaller pieces of debris heat up and vaporize in Earth's atmosphere before they reach the ground, but a few larger pieces make it through. In 1908, a bolide about 20 m in diameter exploded 5 to 10 km above the land surface near Tunguska, Siberia, flattening the forests across an area of 2000 km² (**Figure 1.36**). Collisions of this size, which release as much energy as 10 megatons of TNT (the equivalent of a large nuclear blast), probably take place every 500 to 1000 years. Larger chunks of matter 1 to 2 km in diameter impact Earth every few million years or so, releasing as much as 10,000 megatons of energy.

> *Large bolide impacts on Earth are infrequent, but space agencies are monitoring objects that might someday collide with Earth.*

Because an event this size would pose a global threat to civilization, NASA and other space agencies have as-

FIGURE 1.35 In 1970, an earthquake-induced landslide on Mount Huascarán, Peru, buried the towns of Yungay and Ranrahirca. The landslide traveled 17 km at a speed of up to 280 km/hour and is estimated to have consisted of up to 100 million cubic meters of water, mud, and rock. The death toll from the earthquake and landslide was 66,700 persons. (*left*) Towns of Yungay and Ranrahirca before the landslide. (*right*) Aftermath of the landslide. [*Both photos by Lloyd Cluff/Corbis.*]

FIGURE 1.36 Tree fall caused by the Tunguska explosion in 1908, photographed by the Kulik expedition of 1927.

signed telescopes to search the solar system for orbiting objects that might someday slam into Earth, and a systematic inventory of near-Earth objects is being conducted. Plans have even been formulated to use spacecraft to deflect such objects away from an Earth collision.

Reducing Risks from Natural Hazards

In discussing how Earth science deals with potential disasters, we need to differentiate between *hazard* and *risk.* The hazard depends on the intensity of the phenomenon, such as the wind speed during a hurricane or the intensity of shaking during an earthquake and how often it occurs. In other words, hazard describes the chances that nature might throw something big at us.

Risk, on the other hand, describes the damage that is likely to be caused when a disaster actually strikes, usually measured in lost lives and dollars. Unlike hazard, risk depends on the population and number of buildings in a region, as well as on how easily the structures might be damaged. The risk to society from natural hazards is increasing, primarily because our exposure to the hazards is going up through the process of *urbanization,* or city building.

Consider earthquakes as an example. The coastal regions of both California and Alaska experience lots of earthquakes and therefore have comparable earthquake

hazards. However, the population of California now exceeds 36 million, while fewer than 1 million people live in Alaska, so the risk that somebody will get hurt or something will be damaged by an earthquake is much greater in California.

The risk goes up as the hazard increases. The earthquake risk in Los Angeles is far greater than in Miami because destructive earthquakes are rare in southern Florida. The hurricane risk is reversed, since destructive hurricanes are common in Miami but not in Los Angeles.

It is an interesting fact, though an unfortunate one, that the most severe natural disasters—earthquakes, hurricanes, and volcanic eruptions—tend to be concentrated in coastal regions with warm climates, where otherwise favorable environments support booming populations and intense economic development. Consequently, the exposure of society to natural hazards is growing most rapidly where the hazards are highest.

Natural disasters cannot be prevented, but their catastrophic effects can be reduced by using Earth science in formulating public policies.

■ Land-use policy can discourage urbanization in regions of high hazard, such as in the likely path of lava flows down active volcanoes.

■ Regulations can be enacted to require builders and building owners to reinforce structures that could be

FIGURE 1.37 Emergency response personnel from Israel search for casualties in the rubble of a ruined building in the Haitian capital Port-au-Prince following the catastrophic earthquake of January 12, 2010. [*Xinhua News Agency/David de la Paz/eyevine/Redux.*]

damaged by high winds from hurricanes or intense ground shaking from earthquakes.

■ Cities can prepare for anticipated disasters by stockpiling supplies and training people how to respond to emergencies (**Figure 1.37**).

Characterizing natural hazards and developing techniques for predicting them are important jobs for Earth scientists. History can be a guide: a volcano that has erupted in the past is likely to erupt again. However, the largest disasters are rare enough that the last big one may have happened before anyone was recording history. Earth scientists must therefore extend the historical record by searching for clues to prehistoric extreme events. For example, they can map unusual layers of sand along coastlines that mark old storm surges and tsunamis.

Enlightened public policies informed by Earth science and improved prediction methods can reduce the risk of many natural hazards.

 ## NATURAL ENVIRONMENT

Life is everywhere on Earth's surface, from its equator to its poles, in its atmosphere, throughout its oceans, and even kilometers down into the rocky outer layers of its solid body. Life has been found in steaming volcanic vents at the bottom of the ocean, in lakes beneath the Antarctic ice, on the frozen peaks of the highest mountains, and in the deepest mines. All the living organisms everywhere, taken together, form the **biosphere,** *the part of Earth that is alive* (**Figure 1.38**).

Geologists use methods from both biology and geology to understand how the biosphere interacts with the physical environment. Life constitutes only a small fraction of the environment in terms of mass or volume, but it plays a very active role in maintaining hospitable conditions for future generations of life. When plants grow, they produce the oxygen that we animals breathe as well as much of the food we eat. Microorganisms—life we can see only under a microscope—are particularly important in sustaining the environment. Some types of microorganisms feed on rocks, mobilizing the mineral nutrients essential for plant growth. Others recycle dead organic material and return its components to the environment.

You will notice that people often discuss the environment from a local point of view, as when you see a report on "the environmental pollution of New York City" or read an "environmental impact statement" written to win approval for a shopping mall. We try to "save the environment" by recycling our trash and cleaning up toxic chemicals dumped into our soils and rivers. Geological knowledge helps us understand how local environments depend on natural resources and how we might adapt human activities to maintain local environmental quality.

More and more of our environmental concerns are global in scale, however. To address such concerns, we must consider the environment in its most general sense:

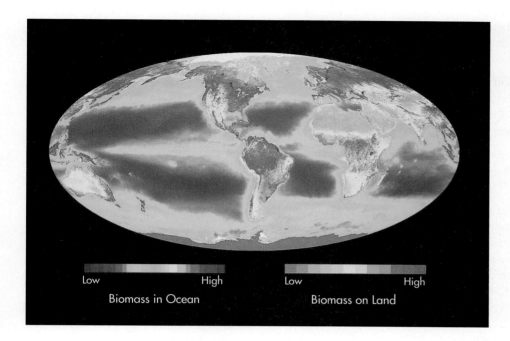

FIGURE 1.38 Earth's biosphere, represented by the global distribution of plant material on land and in the oceans, as mapped by NASA's SeaWiFS satellite. [*NASA/Goddard Space Flight Center.*]

the collection of all habitats of all living species—*the environment of the biosphere as a whole.*

The Human Factor

Humans have altered the environment by deforestation, agriculture, and other types of land uses since the first civilizations arose about 10,000 years ago (**Figure 1.39**). But human effects in earlier times were usually restricted to local or regional habitats. Energy production on an industrial scale now makes it possible for humans to compete with geologic processes in modifying the environment on a global scale. We are transforming the environment in fundamental ways, as illustrated by some startling observations:

- Dams and reservoirs built by humans now trap about 30 percent of the sediments transported down the world's rivers.

FIGURE 1.39 In Indonesia, deforestation by slash and burn. [*Charles O'Rear/Corbis.*]

- In most developed countries, construction workers move more tons of soil and rock each year than do all natural processes of erosion combined.

- Within 50 years after the invention of the artificial coolant freon, enough of it had leaked out of refrigerators and air conditioners and floated into the upper atmosphere to damage Earth's protective ozone layer.

- Humans have converted about one-third of the world's forested area to other land uses, primarily agriculture, in the last half-century.

- Since the industrial revolution began in the early nineteenth century, deforestation and the burning of fossil fuels have increased the concentration of carbon dioxide in the atmosphere by almost 40 percent. The atmospheric concentration of carbon dioxide is rising at the unprecedented rate of 4 percent per decade and is likely to cause significant global climate change during the lifetimes of most people now in college.

Modern industrialized society is capable of changing the environment on an unprecedented global scale.

The expression **global change** entered the world's vocabulary when it became clear that emissions from fossil-fuel burning and other human activities were beginning to alter the chemistry of the atmosphere. People became concerned about changes such as

- mass die-offs of plants and animals due to acid rain

- ozone depletion in the upper atmosphere resulting in increased exposure to the Sun's ultraviolet rays

- global warming and ocean acidification due to increased carbon dioxide and other greenhouse gases in the atmosphere

- extinction of plant and animal species by loss of natural habitat

These changes are all **anthropogenic,** or human-generated (from *anthropos*, the Greek word for "man"). However, they differ considerably in scope and magnitude. Acid rain is primarily a regional problem, although if left unabated, it could have global consequences. Ozone depletion in the upper atmosphere is clearly a global problem, but its cause is mainly chemicals of industrial origin, which are now strictly regulated through an international treaty—a major environmental success. In contrast, there appears to be little agreement over what we should do about global warming and related aspects of human-induced climate change.

We will discuss the problems of acid rain, stratospheric ozone depletion, and climate change to illustrate the problems of anthropogenic global change and some of their possible solutions.

Acid Rain

In many industrialized areas of the world, the air is severely polluted with sulfur-containing gases. These gases are emitted primarily from the smokestacks of power plants that burn coal containing large amounts of sulfur. Coals mined in the eastern and midwestern regions of the United States contain more sulfur than coals mined in the western states. Although volcanoes and coastal marshes also add sulfur gases to the atmosphere, more than 90 percent of the sulfur emissions in eastern North America are of anthropogenic origin.

Sulfur gases in the atmosphere react with oxygen and water to form sulfuric acid. Small amounts of sulfuric acid can turn harmless rainwater into **acid rain.** Although it is much too weak to sting human skin, acid rain causes widespread damage to delicate organisms and solid rock.

By acidifying sensitive lakes and streams, acid rain has caused massive kills of fish in many lakes in Canada, the northeastern United States, and Scandinavia. A survey of more than 1000 lakes and thousands of miles of streams in the northeastern United States showed that acid rain has affected 75 percent of the lakes and 50 percent of the streams. In some acidified lakes and streams, fish species such as the brook trout have been completely eradicated. The salmon habitat of eastern Canada has been greatly affected. Acid rain also damages mountain forests, particularly those at high elevations. Acid moisture in the air reduces visibility.

Acid rain causes noticeable damage to fabrics, paints, metals, and rocks, and it rapidly weathers stone monuments and outdoor sculptures (**Figure 1.40**). In Canada alone, acid rain causes about $1 billion in damage every year to buildings and monuments.

Concerned environmentalists have long recommended the restriction of sulfur emissions from coal-burning power plants. In 1990, after many years of wrangling, the U.S. Congress created the Acid Rain Program as an amendment to the 1970 Clean Air Act. The goal of the program was to limit, or "cap" sulfur gas emissions, beginning in 2000. As part of this program, coal-burning power plants have been allowed to trade sulfur emission allowances. The results

Sulfur-containing gases, emitted primarily from coal-burning power plants, react with oxygen and water in the atmosphere to produce acid rain.

FIGURE 1.40 A monument before and after deterioration caused by acid rain.
[*Westfälisches Amt für Denkmalpflege.*]

are impressive: sulfur gas emissions have been reduced from over 15 million tons in 1990 to less than 6 million tons in 2009.

Stratospheric Ozone Depletion

Ozone (O_3^+) is a reactive form of oxygen gas. Near Earth's surface, ozone is a major constituent of smog. It undermines health, damages crops, and corrodes materials. Low-lying ozone forms when sunlight interacts with chemical wastes from industrial processes and automobile exhausts.

Ozone also exists in Earth's upper atmosphere, or *stratosphere,* where it is concentrated in a layer 25 to 30 km above the surface. There, solar radiation transforms ordinary oxygen gas into ozone, forming a protective layer that shields Earth by absorbing cell-damaging ultraviolet (UV) radiation from the Sun. Skin cancer, cataracts, impaired immune systems, and reduced crop yields are attributable to excessive UV exposure.

In 1995, the Nobel Prize in chemistry was awarded to Mario Molina, Sherwood Rowland, and Paul Crutzen for the hypotheses that the protective ozone layer can be depleted by chemical reactions involving anthropogenic compounds. One class of compounds, chlorofluorocarbons (CFCs)—which were used as refrigerants, spray-can propellants, and cleaning solvents—raised special concerns. These compounds are stable and harmless—except when they migrate to the stratosphere. Molina and Rowland proposed that these chemicals undergo reactions that

destroy ozone in the stratosphere, thinning the protective ozone layer. Molina and Rowland's hypothesis was confirmed when a large hole in the ozone layer was discovered over Antarctica in 1985 (**Figure 1.41**). Subsequently, **stratospheric ozone depletion** was found to be a global phenomenon.

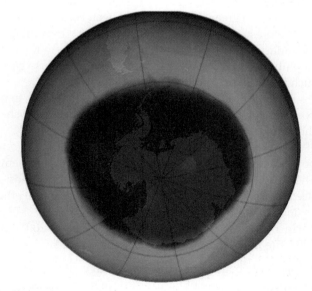

FIGURE 1.41 The Antarctic ozone hole at its annual maximum on September 12, 2008. On that day, the area depleted in ozone stretched over 27 million square kilometers, or 10.5 million square miles. This map was made using data from NASA's *Aura* satellite. [*NASA.*]

In the 1980s, when scientists were trying to convince government and industry officials that the ozone layer was being depleted due to CFCs, a senior government official remarked that the solution was for people to wear hats, sunscreen, and dark glasses. Fortunately, political wisdom prevailed. In 1987, a group of nations entered into a global treaty, called the Montreal Protocol, to protect the ozone layer. Today, all nations recognized by the United Nations have ratified the treaty. The Montreal Protocol required that CFC production be phased out by 1996, and it set up a fund, paid for by developed nations, to help developing nations switch to ozone-safe chemicals. There have been several amendments to the Montreal Protocol to increase the list of ozone-depleting chemicals that must be phased out and to postpone some of the phase-out dates. The manufacturing of safer alternatives has led to a successful phase-out of CFCs, and long-term projections indicate that the depletion of the ozone layer will diminish over the next several decades. The Montreal Protocol has become a model for how scientists, industrial leaders, and government officials can work together to head off an environmental disaster.

> The ozone layer in the stratosphere, which shields the biosphere from harmful ultraviolet radiation, has been damaged by chlorofluorocarbons. Use of these chemicals continues to be phased out.

Global Climate Change

Carbon dioxide (CO_2) is a **greenhouse gas;** it helps to trap solar heat in the atmosphere, somewhat like the frosted windows in a greenhouse trap solar heat. Fossil-fuel burning, deforestation, and agriculture since the beginning of the industrial era have caused a significant rise in the concentrations of carbon dioxide and other greenhouse gases, such as methane, in the atmosphere.

Recognizing the potential problems that this trend poses for global climate, the United Nations established the Intergovernmental Panel on Climate Change (IPCC) in 1988 to assess the risk of anthropogenic climate change, its potential effects, and possible solutions to the problems. The IPCC provides a continuing forum for hundreds of scientists, economists, and policy experts to work together to understand these issues.

In major assessment reports, published in 2001 and 2007, the IPCC drew the following conclusions:

- Since the beginning of the twentieth century, the average temperature of Earth's surface has risen, on average, by about 0.6°C (**Figure 1.42**).

- Much of this "twentieth-century warming" has been caused by anthropogenic increases in greenhouse gases.

- Anthropogenic warming is already causing changes in many natural systems, including reductions in Arctic sea

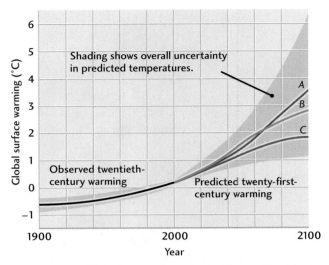

FIGURE 1.42 Global warming, past and future. The black line shows the average surface temperatures observed during the twentieth century (heavily smoothed). Colored lines show three possible scenarios for the twenty-first century. (A) Slow economic growth and continued reliance on fossil fuels. (B) Rapid economic growth and conversion to nonfossil fuels at a moderate rate. (C) Rapid economic growth and rapid conversion to nonfossil fuels. Not shown is a scenario with rapid economic growth and continued reliance on fossil fuels, which would result in global warming in excess of 4°C. The shaded region represents the overall uncertainties in the predicted temperatures from lack of knowledge of the climate system, as well as uncertainties in economic growth and energy technologies. [IPCC, Climate Change 2007: The Physical Scientific Basis. Cambridge University Press, 2007.]

ice and mountain glaciers, warming of lakes and rivers, earlier timing of spring events such as leaf unfolding and bird migrations, and longer growing seasons for plants.

- Levels of greenhouse gases will continue to increase throughout the twenty-first century, primarily because of human activities.

- The rate of increase will depend on a number of socioeconomic factors that will govern the rate of greenhouse gas emissions, including active steps by society to limit those emissions.

- About 30 percent of the carbon dioxide emitted from fossil-fuel burning is being absorbed into the oceans, causing the oceans to become more acidic. This *ocean acidification* will decrease the ability of shellfish and corals to produce their hard exteriors, which could severely harm ocean ecosystems.

- The increase in greenhouse gases will cause significant global warming during the twenty-first century. Projections of the amount of warming are highly uncertain because of doubt about the amount of future greenhouse gas emissions and an incomplete knowledge of how the climate system works. The range of likely temperature increases accepted by most experts is 1° to 6°C (see Figure 1.42).

According to the IPCC, global warming during the twenty-first century is likely to be accompanied by substantial global and regional changes. The warming will probably be greater over land than over the ocean, and regional climates will probably become more variable. Other likely effects include increasing rainfall in the tropics, decreasing rainfall and increasing droughts in temperate zones such as the United States, stronger hurricanes, greater flooding, more heat waves, melting of polar ice, and a rising sea level.

> *Anthropogenic global warming is caused primarily by rising concentrations of greenhouse gases from the burning of fossil fuels, deforestation, and agriculture.*

Some of the effects of global warming can already be seen and felt (**Figure 1.43**). January 2000 to December 2009 was the warmest decade on record, with 1998, 2005, and 2009 being the three warmest years since accurate records began in 1880. The average global temperature has increased about 0.8°C in the past three decades. The number of very strong hurricanes has almost doubled over the past three decades, and increases in tropical storm intensity have been observed around the world. The amount of sea ice in the Arctic Ocean is decreasing, and the melting of ice in Antarctica and Greenland is accelerating rapidly. Sea level is expected to rise by more than half a meter during the twenty-first century, creating serious problems for low-lying countries such as Bangladesh, as well as the Eastern Seaboard and Gulf Coast of the United States, where flooding during storm surges could become much worse.

> *Anthropogenic global warming is causing polar ice to melt and storm intensities to increase. It is likely to cause the sea level to rise and species to go extinct.*

Because anthropogenic warming will be rapid, many plant and animal species will have difficulty adjusting or migrating. Those that cannot cope with rapid warming may become extinct, resulting in a loss of biodiversity. Global warming is already being blamed for a variety of adverse ecological effects, such as the disruption of Arctic ecosystems as ice lines move northward and permafrost begins to melt, and the spread of malaria as more of the world experiences a tropical climate. Ocean acidification could have severe negative impacts on marine ecosystems in just a few decades.

The potentially dire consequences of anthropogenic global change are motivating politicians to work together in ways they never have before, as we all try to avoid the "tragedy of the commons"—the spoiling of our commonly held environmental resources. Neighboring nations are enacting mutually beneficial regulations to address regional environmental problems, and new multinational treaties are being formulated in attempts to manage

FIGURE 1.43 Global warming is reducing Arctic Sea ice, adversely affecting the habitat of Arctic animals such as polar bears. [*Thomas and Pat Leeson/Photo Researchers.*]

FIGURE 1.44 The authors at work. (*left*) Tom Jordan at a depth of 2800 m in the Mponeng gold mine, South Africa. (*right*) John Grotzinger on Namibia's Skeleton Coast.

anthropogenic effects on the global environment. Earth science gives us the knowledge to make rational choices about global environmental management.

OUR REASONS FOR STUDYING EARTH

As you have seen in this chapter, Earth science is a very practical business. It provides our society with the knowledge to manage our natural resources; strategies for characterizing natural hazards and reducing risk; and the ability to understand the environment and predict global change. The authors of this textbook have worked as geologists on many aspects of these practical problems, and we have been rewarded for our efforts (**Figure 1.44**).

Yet our satisfaction as scientists comes from a deeper, more abstract motivation: to understand the world around us, how it works, and how it has evolved. Although we use the tools of physics, chemistry, biology, and mathematics, our inspiration comes from the complexity and beauty of the Earth system, this great spaceship on which we sail.

When we descend into the deep gold mines of South Africa, we are aware of the riches being pulled from the excavations, but we also image the primeval conditions that deposited the gold in those rocks billions of years ago. The stories of those rocks are tales we try to unravel.

When we hike up the slopes of an active volcano, we recognize the danger it poses to the communities below, but we also realize that the molten rock and gases blowing out its top come from many kilometers below Earth's surface. To us, it is a window through which we can look into Earth's deep interior.

When we drive or fly across a landscape, we always ask the question, What are we seeing, and how did it get that way? Follow us in this textbook to sharpen your geological insight. We will indeed give you new eyes!

■ SUMMARY

What is geology? Geology is the science that deals with Earth—its history, its composition and internal structure, and its surface features. Geology provides the scientific knowledge needed to develop natural resources, deal with natural disasters, preserve the environment, and adapt to global change. Geophysics, geochemistry, and geobiology are subfields of geology that apply the methods of physics, chemistry, and biology to geologic problems.

What is the goal of sustainable development? Sustainable development is the idea that human activities can use natural resources in a way that satisfies the present needs of society without compromising the ability of future generations to meet their own needs. This goal will require managing natural resources to ensure that adequate energy, water, and raw materials are available; preparing society for inevitable natural disasters; repairing environmental damage; and maintaining a habitable environment over the long term.

How do we categorize our natural resources? Reserves are the known supplies of natural resources that can be exploited economically under current conditions. The most important resources available from the natural environment are energy, minerals and other mined raw materials, water, and soils. Natural resources can be classified as renewable and nonrenewable resources, depending on whether they are replenished at rates comparable to the rates at which we are consuming them. Fossil fuels are examples of nonrenewable energy resources; solar power and biofuels are examples of renewable energy resources.

How important are fossil fuels to our energy economy? Fossil fuels, which include oil, natural gas, and coal, dominate our present energy use. Taken together, these resources account for 86 percent of the energy consumed by the United States and the world at large. Despite efforts to locate more reserves and improve production, the major reservoirs of oil will become depleted sometime in the next few decades. Renewable energy sources, which include biofuels and solar, wind, water, and geothermal power, are likely to fall short of replacing oil unless there are unanticipated technological breakthroughs. Many observers expect that coal, which is geologically plentiful and economically cheap to develop, will fill the gap.

What resources come from mining? Mining provides the minerals and other raw materials that are used in huge quantities by our industrialized society. Demand for these materials, especially metals, is increasing due to the rapid expansion of the Asian economies. Ores, the high-grade deposits of minerals from which metals are extracted, can be found in a variety of geological environments. Many of the most valuable ores were formed in regions of volcanic activity, where hot circulating groundwater leached metals from rocks and deposited them in a concentrated form. Geologists are exploring for new mineral resources in remote areas, including polar regions and the ocean floor. In exploiting mineral resources, society must balance the economic gains of mining against its considerable environmental and health costs.

Where does society get its water resources? About 1 percent of Earth's water is fresh water. Only a small fraction of the fresh water occupies lakes and rivers; most is groundwater contained in rocks beneath the land surface. Many regions depend on reservoirs of groundwater in porous rock formations, called aquifers, for their water needs. Pumping water out of aquifers more rapidly than precipitation recharges them can deplete them. In these situations, groundwater becomes a nonrenewable resource.

Why are soils classified as a natural resource? Soils are the primary reservoir of nutrients for agriculture and the ecological systems that produce renewable natural resources. They serve as raw materials for construction and manufacturing. They contain twice as much carbon as the atmosphere and three times more than all the world's vegetation. They recycle the organic carbon from dead plants and animals and release it into the atmosphere. The quality of the fresh water most people use is determined largely by the soils it passes through.

What are the principal natural hazards? Natural hazards are events produced by natural processes that have the potential to kill people and damage our built environment. Hurricanes (also called typhoons and cyclones) are great storms that suck energy into the atmosphere from the warm surface of tropical seas. Floods can result from inundation by the sea, but the most frequent cause is river flooding, which occurs when water from rainfall and snowmelt flows into a river system faster than it can flow out. Earthquakes are shaking episodes caused by the breaking of rocks along surfaces of weakness in Earth's crust, called faults. Earthquakes beneath the seafloor can generate destructive tsunamis that inundate coastal regions. Volcanic eruptions deposit molten rock from Earth's interior on its surface. Landslides are downhill movements of rock, mud, and soil that occur when the friction that keeps the material from slipping is overcome by the downward force of gravity, often due to saturation by water. Bolides are chunks of rock and other debris from outer space that explode as fireballs in Earth's atmosphere or impact its surface.

What is the difference between hazard and risk? The hazard to society from a natural phenomenon depends on the intensity of the phenomenon and how often it occurs. Risk describes the damage that is likely to be caused when a disaster actually strikes; it is usually measured in lost lives and dollars. The risk depends on the population and structures exposed to the hazard, as well as how easily the structures might be damaged, whereas the hazard is independent of these factors. The risk to society from natural hazards is increasing, primarily because our exposure to natural hazards is increasing through the process of urbanization.

In what ways are humans affecting the environment? The environment is the region near Earth's surface inhabited by living organisms (which, taken together, constitute the biosphere). The effects of human activities on the environment are growing with the human population, which now exceeds 6 billion people. Humans have always affected their environment on a local scale, but the industrialization of society has harnessed enough energy to compete with geologic processes in causing changes on a global scale. Examples of anthropogenic global change include acid rain, stratospheric ozone depletion by chlorofluorocarbons, global warming due to the emission of greenhouse gases, and ocean acidification.

How might human activities change global climate? Fossil-fuel burning and deforestation release carbon dioxide, a greenhouse gas. The rising concentration of carbon dioxide and other greenhouse gases in the atmosphere is likely to cause significant global climate change during the twenty-first century. The nature of that climate change remains uncertain, but scientists estimate that the average global temperature will increase by 1° to 6°C before 2100. This warming is likely to have a number of adverse environmental effects, such as melting of polar ice, a rising sea level, the intensification of storms, prolonged droughts, and extinctions of plant and animal species both on land and in the sea.

■ KEY TERMS AND CONCEPTS

acid rain (p. 30)

anthropogenic (p. 30)

aquifer (p. 17)

biosphere (p. 28)

bolide (p. 26)

earthquake (p. 22)

environment (p. 3)

flood (p. 20)

fossil fuels (p. 6)

geology (p. 2)

global change (p. 30)

greenhouse gas (p. 32)

groundwater (p. 16)

hurricane (p. 19)

landslide (p. 25)

natural hazards (p. 18)

natural resources (p. 4)

nonrenewable resource (p. 6)

renewable resource (p. 5)

reserves (p. 4)

risk (p. 27)

soils (p. 17)

stratospheric ozone depletion (p. 31)

sustainable development (p. 4)

tsunami (p. 22)

volcanic eruption (p. 22)

■ EXERCISES

1. What are the main goals of sustainable development?

2. Approximately what percentage of the U.S. energy supply comes from renewable resources?

3. How do geologists contribute to the expansion of our oil reserves?

4. Why have the worldwide prices of many useful metals, such as iron, copper, zinc, and nickel, increased so rapidly in the last few years?

5. What human activity consumes the most water resources in the western United States?

6. Give an example of a situation in which water has become a nonrenewable resource.

7. Why are soils classified as a valuable natural resource? Are they a renewable or nonrenewable resource?

8. The earthquake hazards for both the cities of Anchorage, Alaska, and San Francisco, California, are high. Which place has the higher earthquake risk? Why?

9. Which natural hazard poses the greatest risk to your community?

10. A major bolide impact could destroy civilization, yet the risks from hurricanes and earthquakes are considered to be much greater than the risk from bolides. Why is this the case?

11. In what ways can geology help people living near volcanoes prepare for potentially disastrous eruptions?

12. What factors have made anthropogenic effects on the environment a global problem, rather just a regional one? What are some of those effects?

13. What causes acid rain, and what are its effects?

14. What is causing the depletion of ozone in the stratosphere, and what are its effects?

15. How does the burning of fossil fuels cause global climate change?

16. In one sentence or two, why do you think it is important to study Earth?

Visual Literacy Task

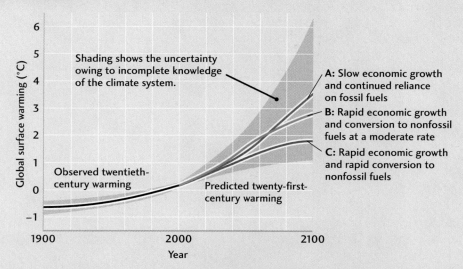

FIGURE 1.42 Global warming, past and future. The black line shows the average surface temperatures observed during the twentieth century (heavily smoothed). Colored lines show three possible scenarios for the twenty-first century. (A) S-1low economic growth and continued reliance on fossil fuels. (B) Rapid economic growth and conversion to non-fossil fuels at a moderate rate. (C) Rapid economic growth and rapid conversion to non-fossil fuels. Not shown is a scenario with rapid economic growth and continued reliance on fossil fuels, which would result in global warming in excess of 4° C. The shaded region represents the overall uncertainties in the predicted temperatures from lack of knowledge of the climate system, as well as uncertainties in economic growth and energy technologies. [*IPCC, Climate Change 2007: The Physical Scientific Basis. Cambridge University Press, 2007.*]

1. What does 0 degrees Celsius on the scale of the graph mean?

a. The temperature at which water freezes
b. The average global temperature today
c. The baseline from which the differences in temperature are measured

2. How does the temperature change from 1900 to 2000 compare to the predicted change from 2000 to 2100?

a. 1900–2000 had a larger increase.
b. 2000–2100 will have a larger increase.
c. The increase for both times are the same.

3. What does the dark gray area signify?

a. The range of predicted trends because we are still learning about Earth
b. The summer and winter extreme temperatures
c. The possible range of temperatures from year to year

4. Why does the amount of warming slow over time for Scenarios B and C?

a. They have rapid economic growth.
b. They have conversion to nonfossil fuels.
c. They have both rapid economic growth and conversion to nonfossil fuels.
d. They have neither rapid economic growth nor conversion to nonfossil fuels.

5. Do the scenarios allow that the temperature could be 5 degrees Celsius warmer by the year 2100?

a. Yes, it is within the uncertainty.
b. No, the maximum is 3.6 degrees Celsius.
c. No, the average increase will be 2.7 degrees Celsius.

Thought Question: How would you answer a fellow student who asks, "Does this graph mean that the average temperature of Earth is going to rise between 1.8 and 3.6 degrees Celsius in 2100?"

How We Study Earth

2

Measuring Mount Everest

ON FEBRUARY 6, 1800, COLONEL WILLIAM LAMBTON of the 33rd Foot Regiment received orders to begin the Great Trigonometrical Survey of India, the most ambitious scientific project of the nineteenth century. Over the next several decades, intrepid British explorers led by Lambton and his successor, George Everest, hauled bulky telescopes and heavy surveying equipment through the jungles of the Indian subcontinent, triangulating the positions of reference monuments established at high points in the terrain, from which they could accurately establish Earth's size and shape. Along the way, in 1852, they discovered that an obscure Himalayan peak, known on their maps only as "Peak XV," was the highest mountain on Earth. They promptly named it Mount Everest, in honor of their former boss.

On February 11, 2000, almost exactly 200 years after Lambton commenced his exploration, NASA launched another great survey, the Shuttle Radar Topography Mission (SRTM). The space shuttle *Endeavour* carried two large radar antennas into low Earth orbit, one in the cargo bay and the second mounted on a mast that could extend outward to 60 m. Working together like a pair of eyes, these antennas mapped the height of the land surface below the shuttle on a very dense geographic grid, rendering the terrain in unprecedented three-dimensional detail. Remarkably, the height of Mount Everest as confirmed by the SRTM (8850 m, or 29,035 ft) turned out to be only 10 m more than the original 1852 estimate.

> In just 11 days, the SRTM mapped 2.6 billion points covering 80 percent of Earth's land surface.

Though the accuracy of the Great Trigonometrical Survey was impressive, collecting data was a slow process. It took the British over 70 years to measure the positions of 2700 stations across the Indian subcontinent, an average of about one position every 3 months. In comparison, the SRTM collected about 3000 position measurements *each second.* In just 11 days, the SRTM mapped 2.6 billion points covering 80 percent of Earth's land surface, including many remote areas that had not been previously surveyed. And, unlike the British, the shuttle crew did not have to contend with malaria or tigers! ◆

Artist's depiction of the space shuttle *Endeavour* during the Shuttle Radar Topography Mission. The SRTM's dual-antenna radar was able to penetrate clouds and map 80 percent of Earth's land surface in just 11 days. [*Ball Aerospace & Technologies Corporation/NASA.*]

THIS CHAPTER PAINTS a broad picture of how we study Earth. It starts with the scientific method, the observational approach to the physical universe on which all scientific inquiry is based. Throughout this textbook, you will see the scientific method in action as we describe how Earth scientists gather and interpret information about our planet. In this chapter, we will illustrate how the scientific method has been applied to discover some of Earth's basic features—its shape and its internal layering. We will also introduce you to a geologist's view of time.

You may start to think about time differently as you begin to comprehend the immense span of geologic history. Earth and the other planets in our solar system formed about 4.5 billion years ago. More than 3 billion years ago, living cells developed on Earth's surface, and life has been evolving ever since. Yet our human origins date back only a few million years—a mere few hundredths of a percent of Earth's existence. The scales that measure individual lives in decades and mark off periods of human history in hundreds or thousands of years are inadequate for studying Earth history.

To explain features that are millions or even billions of years old, we look at what is happening on Earth today. We study our complex natural world as an Earth system involving many interrelated components. Some of these components, such as the atmosphere and oceans, are clearly visible above Earth's solid surface; others lie hidden deep within its interior. By observing the ways these components interact, scientists have built up an understanding of how the Earth system has changed through geologic time.

THE SCIENTIFIC METHOD

Scientists believe that physical events have physical explanations, even if they may be beyond our present capacity to understand them. The **scientific method,** on which all scientists rely, is the general procedure for discovering how the physical universe works through systematic observations and experiments. Using the scientific method to make new discoveries and to confirm old ones is the process of *scientific research* (**Figure 2.1**).

When scientists propose a **hypothesis**—a tentative explanation based on observational data and experiments—they present it to the community of scientists for criticism and repeated testing. A hypothesis is supported if it explains new data or predicts the outcome of new experiments. A hypothesis that is confirmed by other scientists gains credibility.

Here are four interesting scientific hypotheses we encountered in the first chapter of this textbook:

- Earth is billions of years old.

- Coal is a rock formed primarily from dead plants.

- Earthquakes are caused by the breaking of rocks along geologic faults.

- The burning of fossil fuels is causing global warming.

The first hypothesis agrees with the ages of thousands of ancient rocks as measured by precise laboratory techniques, and the next two hypotheses have also been confirmed by many independent observations. The fourth hypothesis has been more controversial, though so many new data support it that most scientists now accept it as

FIGURE 2.1 Research is the process of discovery and confirmation through observation of the natural world. These geologists are researching soil samples near a lake in Minnesota. [*U.S. Geological Survey.*]

true (see Chapters 10 and 14). Identifying other examples of hypotheses from Chapter 1 is a good exercise for you to try.

A coherent set of hypotheses that explains some aspect of nature constitutes a *theory*. Good theories are supported by substantial bodies of data and have survived repeated challenges. They usually obey *physical laws*, general principles about how the universe works that can be applied in every situation, such as Newton's law of gravity.

Some hypotheses and theories have been so extensively tested that all scientists accept them as true, at least to a good approximation. For instance, the theory that Earth is almost spherical, which follows from Newton's law of gravity, is supported by so much experience and direct evidence (ask any astronaut) that we take it to be a fact. The longer a theory holds up to all scientific challenges, the more confidently it is held.

Yet theories can never be considered completely proved. The essence of science is that no explanation, no matter how believable or appealing, is closed to questioning. If convincing new evidence indicates that a theory is wrong, scientists will discard it or modify it to account for the data. A theory, like a hypothesis, must always be testable; any proposal about the universe that cannot be evaluated by observing the natural world should not be called a scientific theory!

> *The scientific method is the procedure for discovering how the universe works through systematic observations and experiments.*

For scientists engaged in research, the most interesting hypotheses are often the most controversial rather than

the most widely accepted. The hypothesis that fossil-fuel burning causes global warming has been widely debated. Because the long-term predictions of this hypothesis are so important, many Earth scientists are now involved in testing it.

Knowledge based on many hypotheses and theories can be used to create a *scientific model*—a precise representation of how a natural process operates or how a natural system behaves. Scientists combine related ideas in a model to test the consistency of their knowledge and to make predictions. Like a good hypothesis or theory, a good model makes predictions that agree with observations.

A scientific model is often formulated as a computer program that simulates the behavior of a natural system through numerical calculations. The forecast of rain or shine you may see on TV tonight comes from a computer model of the weather. A computer can be programmed to simulate geologic phenomena that are too big to replicate in a laboratory or that operate over periods of time that are too long for humans to observe. For example, models used for predicting weather have been extended to predict climate changes decades into the future.

To encourage discussion of their ideas, scientists share their ideas and the data on which they are based. They present their findings at professional meetings (**Figure 2.2**), publish them in professional journals, and explain them in informal conversations with colleagues. Scientists learn from one another's work as well as from the discoveries of the past. Most of the great concepts of science, whether they emerge as a flash of insight or in the course of painstaking analysis, result from untold numbers of such

FIGURE 2.2 Presentations at scientific meetings, such as this one at a German university, are part of the scientific process. [*Technische Universität Berlin/Sabine Böck.*]

interactions. Albert Einstein put it this way: "In science . . . the work of the individual is so bound up with that of his scientific predecessors and contemporaries that it appears almost as an impersonal product of his generation."

Because such free intellectual exchange can be subject to abuses, a code of ethics has evolved among scientists. Scientists must acknowledge the contributions of all others on whose work they have drawn. They must not falsify data, use the work of others without recognizing them, or be otherwise deceitful in their work. They must also accept responsibility for training the next generation of researchers and teachers. These principles are supported by the basic values of scientific cooperation, which a former president of the National Academy of Sciences, Bruce Alberts, has aptly described as "honesty, generosity, a respect for evidence, openness to all ideas and opinions."

GEOLOGY AS A SCIENCE

In the popular media, scientists are often portrayed as people who do experiments wearing white coats. That stereotype is not inappropriate: many scientific problems are best investigated in the laboratory. What forces keep atoms together? How do chemicals react? Can viruses cause cancer? The phenomena that scientists observe to answer such questions are sufficiently small and happen quickly enough to be studied in the controlled environment of the laboratory.

FIGURE 2.4 Geologist Peter Gray setting up a Global Positioning System station in the crater of Mount St. Helens. A network of such stations monitors the changing shape of the land surface as molten rock moves upward within the volcano. [Lyn Topinka/USGS.]

The major questions of geology, however, involve processes that operate on much longer and larger scales. Controlled laboratory measurements yield critical data for testing hypotheses and theories—the ages and properties of rocks, for instance—but they are usually insufficient to solve major geologic problems. Almost all of the great discoveries described in this textbook were made by observing Earth processes in their uncontrolled, natural environment.

For this reason, geology is an outdoor science with its own particular style and outlook. Geologists "go into the field" to observe nature directly. They learn how mountains were formed by climbing up steep slopes and examining the exposed rocks. They discover how ocean basins have evolved by sailing rough seas to study the ocean floor (**Figure 2.3**).

Geology is also a science that requires very advanced technology. In the opening story of this chapter, we saw how instruments aboard spacecraft can amass huge quantities of information about Earth's surface. This type of *remote sensing* allows us to map the continents, chart motions of the atmosphere and oceans, and monitor how our environment is changing. Geologists also deploy sensitive instruments on land and on the seafloor to collect data on earthquakes, volcanic eruptions, and other activity within the solid Earth (**Figure 2.4**).

FIGURE 2.3 Marine scientists Craig Marquette and Will Ostrom, from the Woods Hole Oceanographic Institution, deploy a mooring to measure temperatures from the Research Vessel *Oceanus* during a gale off Cape Hatteras. [Chris Linder, Woods Hole Oceanographic Institution.]

The Geologic Record

A special aspect of geology is its ability to probe Earth's long history by reading what has been "written in stone."

FIGURE 2.5 The geologic record preserves evidence of Earth's long history. These multicolored layers of sandstone at Colorado National Monument were deposited more than 200 million years ago when this part of the western United States was a vast Sahara-like desert. They were subsequently overlain by other rocks, welded by pressure into sandstone, uplifted by mountain-building events, and eroded by wind and water into today's stunning landforms. [*Lonely Planet Images/ Mark Newman.*]

The **geologic record** is the information preserved in the rocks that have been formed at various times throughout a history more than 4 billion years long. Geologists decipher the geologic record by combining information from many kinds of work: examination of rocks in the field (**Figure 2.5**); careful mapping of their positions relative to older and younger rock formations; collection of representative samples; and determination of their ages using sensitive laboratory instruments (**Figure 2.6**).

> *The information preserved in rocks—the geologic record—can be deciphered to understand Earth's long history.*

In *Annals of the Former World*, a compendium of colorful stories about geologists, the popular writer John McPhee offers his view of how geologists bring field and laboratory observations together to visualize the big picture:

They look at mud and see mountains, in mountains oceans, in oceans mountains to be. They go up to some rock and figure out a story, another rock, another story, and as the stories compile through time they connect—and long case histories are constructed and written from interpreted patterns of clues. This is detective work on a scale unimaginable to most detectives, with the notable exception of Sherlock Holmes.

The geologic record tells us that, for the most part, the processes we see in action on Earth today have worked in much the same way throughout the geologic past. This important concept is known as the **principle of uniformitarianism.** It was stated as a scientific hypothesis in the late eighteenth century by a Scottish physician and geologist, James Hutton. In 1830, the British geologist Charles Lyell summarized the concept in a memorable line: "The present is the key to the past."

Millions of years are required for continents to drift apart, for mountains to be raised and eroded, and for river systems to deposit thick layers of sediments. Yet the principle of uniformitarianism does not mean that all geologic phenomena proceed at the same gradual pace. Geologic processes take place over a tremendous range of scales in both space and time (**Figure 2.7**). Some of the most important processes happen as sudden events. A large meteorite that impacts Earth can gouge out a vast crater in a matter of seconds. A volcano can blow its top and a fault can rupture the ground in an earthquake almost as quickly. Geology is the study of *extreme events* as well as gradual change.

> *The principle of uniformitarianism states that the geologic processes we see in action on Earth today have worked in much the same way throughout the geologic past.*

FIGURE 2.6 Sensitive instruments like this mass spectrometer can determine the age of a rock by measuring minute amounts of radioactive elements and how that radioactivity has decayed since the rock first formed. [*Rosenfeld Images Ltd./ Photo Researchers.*]

Over millions of years, layers of sediments built up over the oldest rocks. The most recent layer—the top—is about 250 million years old.

About 50,000 years ago, the explosive impact of a meteorite (perhaps weighing 300,000 tons) created this 1.2-km-wide crater in just a few seconds.

The rocks at the bottom of the Grand Canyon are 1.7–2.0 billion years old.

FIGURE 2.7 Some geologic processes take place over thousands of centuries or can occur with dazzling speed. (*left*) The Grand Canyon, Arizona. [*John Wang/PhotoDisc/ Getty Images.*] (*right*) Meteor Crater, Arizona. [*John Sanford/Photo Researchers.*]

Natural Hazards: Extreme Events in the Geologic Record

Extreme events leave their imprints on the geologic record, so we can determine how often they have happened in the past. The observed frequencies of extreme events allow us to infer their hazards to human society. Meteorites as big as the one that formed Meteor Crater in Arizona (see Figure 2.7) have hit Earth about once every 1000 years; hence, the chances that such an event might happen in the near future, say, during the next year, are fairly small (approximately 1 in 1000). Smaller impacts have been more frequent, whereas larger impacts have been less frequent. Yet we can discern no strict limits to the size of the impacting bodies. Looking across greater

spans of the geologic record, we see larger craters that were caused by bigger meteorites.

The same rules apply to other natural hazards such as earthquakes, volcanic eruptions, and floods: event frequency decreases with event size, but longer geologic intervals are marked by larger events. Consequently, the most extreme geologic events are exceptionally rare and have not been directly witnessed by humans. This does not contradict the principle of uniformitarianism. The geologic record tells us that Earth has been repeatedly hit by meteorites thousands of times more massive than the one that formed Meteor Crater, and we can surmise that such impacts are likely to happen again—with severe consequences for life on the planet. The same can be said for the vast volcanic outpourings that have covered areas bigger than Texas with lava and poisoned the global atmosphere with volcanic gases. The long history of Earth is punctuated by many such extreme though infrequent events, resulting in rapid changes to Earth's environment.

> Though extreme events are rare, they pose significant natural hazards to human society and can result in rapid changes to the natural environment.

Limitations to Uniformitarianism

From Hutton's day onward, geologists have observed nature at work and applied the principle of uniformitarianism to interpret features found in rock formations. Although this approach has been very successful, Hutton's principle is too confining for geologic science as it is now practiced. Modern geology must deal with the entire range of Earth's history, which began more than 4.5 billion years ago. As we will see in Chapter 9, the violent processes that shaped Earth's early history were distinctly different from the processes that operate today.

EARTH'S SHAPE AND SURFACE

How do we know Earth is round? No one had looked down on Earth from space before the early 1960s, but its shape was understood long before that time. In 1492, Columbus set a westward course for India because he believed in a theory that had been favored by Greek philosophers: *we live on a sphere.* His math was poor, however, so he badly underestimated Earth's circumference. Instead of a shortcut, he took the long way around, finding a New World instead of the Spice Islands! Had Columbus properly understood the ancient Greeks, he might not have made this fortuitous mistake, because they had accurately measured Earth's size more than 17 centuries earlier.

The credit for determining Earth's size goes to Eratosthenes, who was chief librarian at the Great Library of Alexandria, in Egypt. Sometime around 250 B.C., a traveler told him about a very interesting observation. At noon on the first day of summer (June 21), a deep well in the city of Syene, about 800 km south of Alexandria, was completely lit up by sunlight because the Sun was directly overhead. Acting on a hunch, Eratosthenes did an experiment. He set up a vertical pole in his own city, and at high noon on the summer solstice, the pole cast a shadow.

Because the Sun cast a shadow in Alexandria but was directly overhead at Syene, Eratosthenes concluded that the land surface must be curved. He hypothesized that Earth had a perfectly spherical shape (the Greeks admired geometrical perfection), and using the distance between the two cities and the size of the pole's shadow, he then deduced a circumference close to its modern value of 40,000 km (**Figure 2.8**). His experiment thus led to a simple model: *Earth is a sphere with a radius of about 6370 km.*

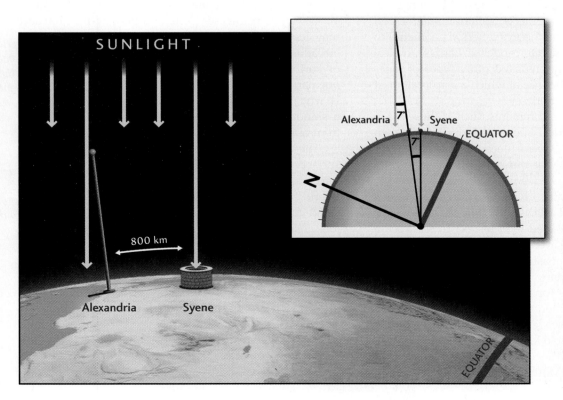

FIGURE 2.8 How Eratosthenes measured Earth's circumference. At noon on the summer solstice (June 21), a vertical well in the southern Egyptian town of Syene (near modern Aswan) was filled with light, indicating that the Sun was directly overhead, whereas a vertical pole in Eratosthenes' hometown of Alexandria cast a shadow. From the length of the shadow, Eratosthenes found that Syene and Alexandria were separated by about 1/50 of Earth's circumference, about 7° out of 360° (inset diagram). Since the measured distance between the two towns was approximately 800 km, he calculated a circumference close to its modern value of 40,000 km.

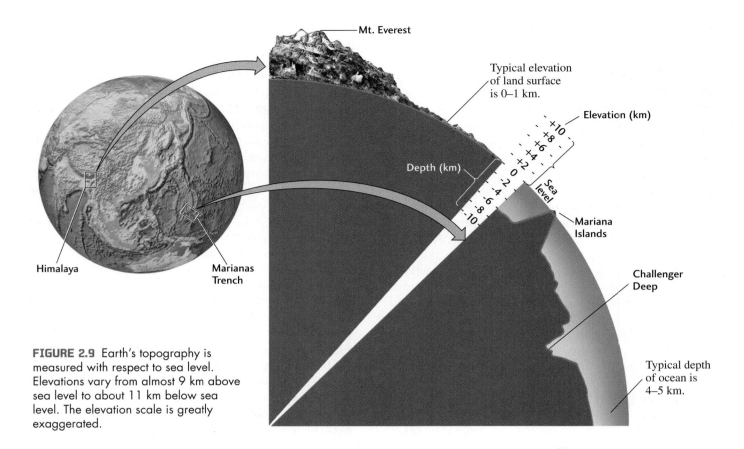

FIGURE 2.9 Earth's topography is measured with respect to sea level. Elevations vary from almost 9 km above sea level to about 11 km below sea level. The elevation scale is greatly exaggerated.

In this powerful demonstration of the scientific method, Eratosthenes made observations (the length of the shadow), formed a hypothesis (spherical shape), and applied some mathematical theory (spherical geometry) to propose a remarkably accurate model of Earth's physical form. The model correctly predicted other types of measurements, such as the distance at which a ship's tall mast would disappear over the horizon.

Much more precise measurements, such as those made by the Great Trigonometrical Survey of India, have shown that Earth is not a perfect sphere. Because of its daily rotation, it bulges out slightly at the equator and is slightly squashed at the poles.

In addition, the smooth curvature of Earth's surface is broken by mountains and valleys and other ups and downs. This **topography** is measured with respect to *sea level*, a smooth surface set at the average level of ocean water that conforms closely to the squashed spherical shape expected for the rotating Earth. Many features of geologic significance stand out in Earth's topography (**Figure 2.9**). Its two largest features are continents, which have typical elevations of 0 to 1 km above sea level, and ocean basins, which have typical depths of 4 to 5 km below sea level.

The elevation of Earth's surface varies by nearly 20 km from the highest point (Mount Everest, in the Himalaya, at 8850 m above sea level) to the lowest point (Challenger Deep, in the Marianas Trench in the Pacific Ocean, at 11,030 m below sea level). Although the Himalaya may loom large to us, their elevation is a small fraction of Earth's radius, only about one part in a thousand, which is why the globe looks like a smooth sphere when seen from outer space.

The topography of Earth's surface is created by great forces acting in Earth's deep interior. To begin to understand these forces, we must look inside the planet, which is layered like an onion.

Earth is almost spherical, but it bulges out slightly at the equator and is slightly squashed at the poles, due to its rotation.

Earth's topography varies by about 20 km from the highest point above sea level to the lowest point below sea level.

PEELING THE ONION: DISCOVERY OF A LAYERED EARTH

Ancient thinkers such as Eratosthenes divided the universe into two parts, the Heavens above and Hades below. The sky was transparent and full of light, and the ancients could directly observe its stars and track its wandering planets. But Earth's interior was dark and closed to human view. In some places, the ground quaked and erupted hot lava. Surely something terrible was going on down there!

So it remained until about a century ago, when geologists began to peer downward into Earth's interior, not with waves of light (which can't penetrate rock), but with waves produced by earthquakes. An earthquake occurs when geologic forces cause brittle rocks to fracture, sending out vibrations like the cracking of ice on a river. These **seismic waves** (from the Greek word for "earthquake," *seismos*) when recorded on sensitive instruments called *seismographs*, allow geologists to locate earthquakes and also to make pictures of Earth's inner workings, much as doctors use ultrasound and CAT scans to image the inside of your body.

> Seismic waves, which are produced by earthquakes, can be used to create images of Earth's interior.

When the first networks of seismographs were installed around the world at the end of the nineteenth century, geologists began to discover that Earth's interior was divided into concentric layers of different compositions, separated by sharp, nearly spherical boundaries (**Figure 2.10**).

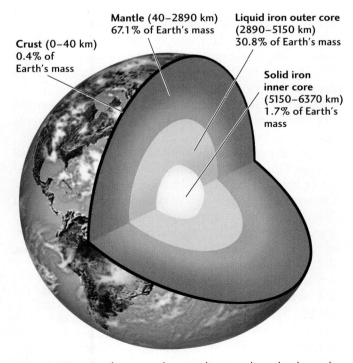

Crust (0–40 km) 0.4% of Earth's mass

Mantle (40–2890 km) 67.1% of Earth's mass

Liquid iron outer core (2890–5150 km) 30.8% of Earth's mass

Solid iron inner core (5150–6370 km) 1.7% of Earth's mass

FIGURE 2.10 Earth's major layers, showing their depths and their masses as a percentage of Earth's total mass.

Earth's Density

Layering of Earth's deep interior was first proposed by the German physicist Emil Wiechert at the end of the nineteenth century, before much seismic data had become available. He wanted to understand why our planet is so heavy, or more precisely, so *dense*. The density of a substance is easy to calculate: just measure its mass on a scale and divide by its volume. A typical rock, such as the granite used for tombstones, has a density of about 2.7 grams per cubic centimeter (2.7 g/cm³).

Estimating the density of an entire planet is a little harder, but not much. Eratosthenes had shown how to measure Earth's volume in 250 B.C., and sometime around 1680, the great English scientist Isaac Newton figured out how to calculate its mass from the gravitational force that pulls objects to its surface. The details, which involved careful laboratory experiments to calibrate Newton's law of gravity, were worked out by another Englishman, Henry Cavendish. In 1798, he calculated Earth's average density to be about 5.5 g/cm³, twice that of tombstone granite.

Wiechert was puzzled. He knew that a planet made entirely of common rocks could not have such a high density. Most common rocks, such as granite, contain a high proportion of silica (silicon plus oxygen, SiO_2) and have relatively low densities, below 3 g/cm³. Some iron-rich rocks brought to the surface by volcanoes have densities as high as 3.5 g/cm³, but no ordinary rock approached Cavendish's value. He also knew that, going downward into Earth's interior, the pressure on rock increases with the weight of the overlying mass. The pressure squeezes the rock into a smaller volume, making its density higher. But Wiechert found that even the effect of pressure was too small to account for the density Cavendish had calculated.

> Earth's average density is higher than would be expected if it were made entirely of common rocks, implying that it has a dense core.

The Mantle and Core

In thinking about what lay beneath his feet, Wiechert turned outward to the solar system, in particular to meteorites, which are pieces of the solar system that have fallen to Earth. He knew that some meteorites are made of an *alloy* (a mixture) of two heavy metals, iron and nickel, and thus have densities as high as 8 g/cm³ (**Figure 2.11**). He also knew that these elements are relatively abundant throughout our solar system. So, in 1896, he proposed a grand hypothesis: sometime in Earth's past, most of the iron and nickel in its interior had dropped inward to its center under the force of gravity. This movement created a dense **core,** which was surrounded by a shell of silicate-rich rock that he called the **mantle** (using the German word for "coat").

(a)

(b)

FIGURE 2.11 Two common types of meteorites. (a) This stony meteorite, which is similar in composition to Earth's silicate mantle, has a density of about 3 g/cm³. (b) This iron-nickel meteorite, which is similar in composition to Earth's core, has a density of about 8 g/cm³.
[*John Grotzinger/Ramón Rivera-Moret/Harvard Mineralogical Museum.*]

With this hypothesis, he could come up with a two-layer Earth model that agreed with Cavendish's value for Earth's average density. He could also explain the existence of iron-nickel meteorites: they were chunks from the core of an Earthlike planet (or planets) that had broken apart, most likely by collision with other planets.

Wiechert got busy testing his hypothesis using seismic waves recorded by seismographs located around the globe (he designed one himself). The first results showed a shadowy inner mass that he took to be the core, but he had problems identifying some of the seismic waves. These waves come in two basic types: *compressional waves*, which expand and compress the material as they travel through a solid, liquid, or gas, and *shear waves*, which move the material from side to side (shearing motion). Shear waves can propagate only through solids, which resist shearing, and not through fluids (liquids or gases) such as air and water, which have no resistance to this type of motion.

In 1906, a British seismologist, Robert Oldham, was able to sort out the paths traveled by these two types of seismic waves and to show that shear waves did not propagate through the core. The core, at least in its outer part, is liquid! This turns out to be not too surprising. Iron melts at a lower temperature than silicates, which is why metallurgists can use containers made of ceramics (which are silicate materials) to hold molten iron. Earth's deep interior is hot enough to melt iron and nickel but not silicate rock. Beno Gutenberg, one of Wiechert's students, confirmed Oldham's observations and, in 1914, determined that the depth of the *core-mantle boundary* is just shy of 2890 km (see Figure 2.10).

The Crust

Five years earlier, a Croatian scientist had detected another boundary at the relatively shallow depth of 40 km beneath the European continent. This boundary, named the *Mohorovičić discontinuity* (Moho for short) after its discoverer, separates a **crust** composed of low-density silicates, which are rich in aluminum and potassium, from the higher-density silicates of the mantle, which contain more magnesium and iron.

Like the core-mantle boundary, the Moho is a global feature. However, it was found to be substantially shallower beneath the oceans than beneath the continents. On average, the thickness of oceanic crust is only about 7 km, compared with almost 40 km for continental crust. Moreover, rocks in the oceanic crust contain more iron and are therefore denser than continental rocks. Because the continental crust is thicker but less dense than the oceanic crust, the continents ride higher by floating like buoyant rafts on the denser mantle **Figure 2.12**), much as icebergs float on the ocean. Continental buoyancy

Observations of seismic waves indicate that Earth's outer core is liquid whereas its mantle is solid.

The continental crust is less dense than either the oceanic crust or the mantle and so rides higher on the mantle than the oceanic crust.

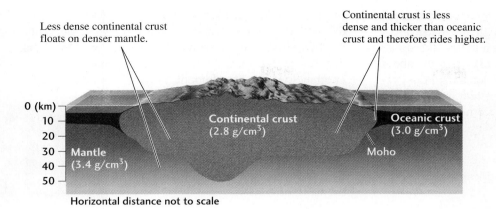

Less dense continental crust floats on denser mantle.

Continental crust is less dense and thicker than oceanic crust and therefore rides higher.

0 (km)
10
20
30
40
50

Continental crust (2.8 g/cm³)

Oceanic crust (3.0 g/cm³)

Mantle (3.4 g/cm³)

Moho

Horizontal distance not to scale

FIGURE 2.12 Because crustal rocks are less dense than mantle rocks, Earth's crust floats on the mantle. Continental crust is thicker and has a lower density than oceanic crust, which causes it to ride higher, explaining the elevation difference between continents and the deep seafloor.

explains the most striking feature of Earth's surface topography: why the elevations shown in Figure 2.9 fall into two main groups, 0 to 1 km above sea level for much of the land surface and 4 to 5 km below sea level for much of the deep sea.

Shear waves travel well through the mantle and crust, so we know that both are solid rock. How can continents float on solid rock? Rock can be solid and strong over the short term (seconds to years) but weak over the long term (thousands to millions of years). The mantle below a depth of about 100 km has little strength, and over very long periods, it flows as it adjusts to support the weight of continents and mountains.

The Inner Core

Because the mantle is solid and the outer part of the core is liquid, the core-mantle boundary reflects seismic waves just as a mirror reflects light waves. In 1936, Danish seismologist Inge Lehmann discovered another sharp spherical boundary at the much greater depth of 5150 km, indicating a central mass with a higher density than the liquid core. Studies following her pioneering research showed that the inner core can transmit both shear waves and compressional waves. The **inner core** is therefore a solid metallic sphere suspended within the liquid **outer core**—a "planet within a planet." The radius of the inner core is 1220 km, about two-thirds the size of the Moon (see Figure 2.10).

Geologists were puzzled by the existence of this "frozen" inner core. They knew that temperatures inside Earth should increase with depth. According to the best current estimates, Earth's temperature rises from about 3500°C at the core-mantle boundary to almost 5000°C at its center. If the inner core is hotter, how could it be solid while the outer core

Earth's inner core is solid despite its high temperature because of the extremely high pressures at Earth's center.

is molten? The mystery was eventually solved by laboratory experiments on iron-nickel alloys, which showed that the "freezing" was due to higher pressures rather than lower temperatures at Earth's center.

Chemical Composition of Earth's Major Layers

By the mid-twentieth century, geologists had discovered all of Earth's major layers—crust, mantle, outer core, and inner core—plus a number of more subtle features in its interior. They found, for example, that the mantle itself is layered into an *upper mantle* and a *lower mantle*, separated by a *transition zone* where the rock density increases in a series of steps. These density steps are caused not by changes in the rock's chemical composition but rather by changes in its compactness due to the increasing pressure with depth. The two largest density jumps in the transition zone are located at depths of about 400 km and 650 km, but they are smaller than the density increases across the Moho and core-mantle boundary, which are due to changes in chemical composition (see Figure 2.10).

Geologists were also able to show that Earth's outer core cannot be made of a pure iron-nickel alloy because the densities of these metals are higher than the observed density of the outer core. About 10 percent of the outer core's mass must be made of lighter elements, such as oxygen and sulfur. On the other hand, the density of the solid inner core is slightly higher than that of the outer core and is consistent with a nearly pure iron-nickel alloy.

The boundaries between the inner core, outer core, mantle, and crust mark changes in Earth's chemical composition.

By bringing together many lines of evidence, geologists have put together a model of the composition of Earth and its various layers. In addition to the seismic data, the evidence includes the compositions of crustal and mantle rocks as well as the compositions of meteorites, thought to be samples of the cosmic material from which planets like Earth were originally made.

Only 8 elements, out of more than 100, make up 99 percent of Earth's mass (**Figure 2.13**). In fact, about 90 percent of Earth consists of only 4 elements: iron, oxygen, silicon, and magnesium. The first two are the most abundant elements, each accounting for nearly a third of the planet's overall mass, but they are distributed very differently. Iron, the densest of these common elements, is concentrated in the core, whereas oxygen, the least dense, is concentrated in the crust and mantle. Silicon, another low-density element, is more abundant in the crust than in the mantle, and it is nearly absent in the core. These relationships confirm Wiechert's hypothesis that the different compositions of Earth's layers are primarily the work of gravity.

Ninety percent of Earth's mass is composed of only four elements: iron, oxygen, silicon, and magnesium.

Natural Resources: Concentration of Valuable Elements in the Continental Crust

Though Earth's composition is dominated by just a handful of elements, the processes that led to its chemical layering have also concentrated many rarer but useful elements into the continental crust. Sodium and potassium, two important nutrients for life, are prime examples. The continental crust contains 8 times more sodium and almost 50 times more potassium than the mantle.

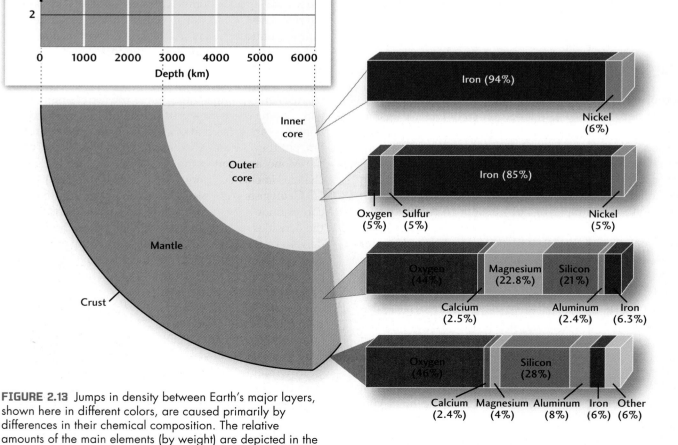

FIGURE 2.13 Jumps in density between Earth's major layers, shown here in different colors, are caused primarily by differences in their chemical composition. The relative amounts of the main elements (by weight) are depicted in the bars on the right.

As we will learn in future chapters, this concentration resulted from the formation of the crust by partial melting of the mantle. Molten rock, or *magma,* is less dense than solid rock and therefore rises toward the surface, where it cools and solidifies. When the mantle begin to melt, many of the rarer elements, including economically valuable metals such as aluminum, copper, zinc, silver, and gold, move from the solid rock into the liquid magma and are then transported by rise of the magma into the crust, where additional melting and transport lead to further concentration. Over the long history of the planet, these processes have endowed the continental crust with a rich stockpile of materials important to human society.

> The melting of mantle rocks has concentrated many rare but valuable elements into the continental crust.

EARTH AS A SYSTEM OF INTERACTING COMPONENTS

Earth is a restless planet, continually changing through geologic activity manifested in volcanism, earthquakes, and glaciation. This activity is powered by two heat engines: one internal, the other external (**Figure 2.14**). A *heat* engine—for example, the gasoline engine of an automobile—transforms heat into mechanical motion or work. Earth's *internal heat engine* is powered by the heat energy trapped in its deep interior during its violent origin and released inside the planet by radioactivity. This internal heat drives the motions in the mantle and core and supplies the energy that melts rock, moves continents, and lifts up mountains. Earth's *external heat engine* is driven by solar energy—heat supplied to Earth's surface by the Sun. Heat from the Sun energizes the atmosphere and oceans and is responsible for Earth's climate and weather. Rain, wind, and ice erode mountains and shape the landscape, and the shape of the landscape, in turn, influences the climate.

> Earth's activity is powered by two energy sources: solar energy and the heat energy of its deep interior.

All the parts of our planet and all their interactions, taken together, constitute the **Earth system.** Although Earth scientists have long thought in terms of natural systems, it was not until the late twentieth century that they had the tools to investigate how the Earth system actually works. Networks of instruments and Earth-orbiting satellites now collect information about the Earth system on a global scale, and computers are now powerful enough to calculate the mass and energy transfers within the system.

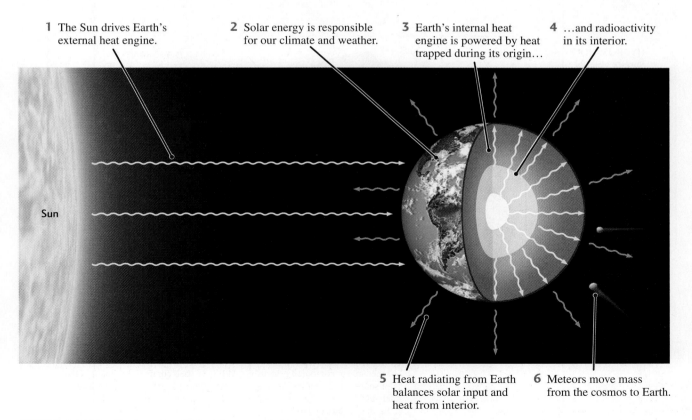

1 The Sun drives Earth's external heat engine.

2 Solar energy is responsible for our climate and weather.

3 Earth's internal heat engine is powered by heat trapped during its origin…

4 …and radioactivity in its interior.

Sun

5 Heat radiating from Earth balances solar input and heat from interior.

6 Meteors move mass from the cosmos to Earth.

FIGURE 2.14 The Earth system is an open system that exchanges energy and mass with its surroundings.

The major components of the Earth system can be represented as a set of domains, or "spheres" (**Figure 2.15**). In our discussion of the environment in Chapter 1, we encountered some of these components: the atmosphere, the hydrosphere (all the liquid water in the oceans, lakes, and rivers), the cryosphere (all the ice in the polar caps, mountain glaciers, and snowfields), and the biosphere (all living organisms). We will define the others shortly.

We will talk about the Earth system throughout this textbook. Let's get started by looking at some of its basic features. The Earth system is an *open system* that exchanges energy and mass with its surroundings. Radiant energy from the Sun energizes the weathering and erosion of Earth's surface, as well as the growth of plants, which feed almost all living things. Earth's climate is controlled by the balance between the solar energy coming into the Earth system and the heat energy Earth radiates back into space. These days, the exchange of mass between Earth and space is relatively small; only about 40,000 tons of meteorites—equivalent to a cube 24 m on a side—fall to Earth each year. Mass transfer was much greater during the early life of the solar system.

Although we think of Earth as a single system, it is a challenge to study the whole thing all at once. Instead, we will focus our attention on the particular parts of the Earth system (subsystems) we are trying to understand.

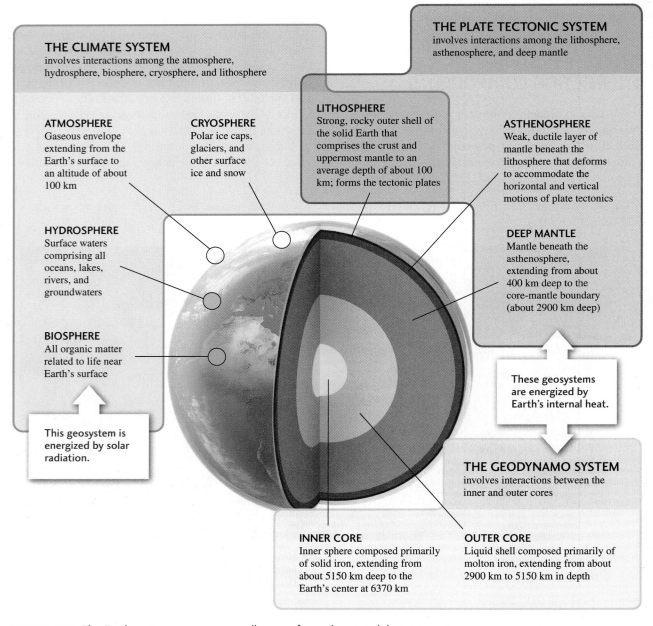

THE PLATE TECTONIC SYSTEM
involves interactions among the lithosphere, asthenosphere, and deep mantle

THE CLIMATE SYSTEM
involves interactions among the atmosphere, hydrosphere, biosphere, cryosphere, and lithosphere

ATMOSPHERE
Gaseous envelope extending from the Earth's surface to an altitude of about 100 km

CRYOSPHERE
Polar ice caps, glaciers, and other surface ice and snow

LITHOSPHERE
Strong, rocky outer shell of the solid Earth that comprises the crust and uppermost mantle to an average depth of about 100 km; forms the tectonic plates

ASTHENOSPHERE
Weak, ductile layer of mantle beneath the lithosphere that deforms to accommodate the horizontal and vertical motions of plate tectonics

HYDROSPHERE
Surface waters comprising all oceans, lakes, rivers, and groundwaters

DEEP MANTLE
Mantle beneath the asthenosphere, extending from about 400 km deep to the core-mantle boundary (about 2900 km deep)

BIOSPHERE
All organic matter related to life near Earth's surface

These geosystems are energized by Earth's internal heat.

This geosystem is energized by solar radiation.

THE GEODYNAMO SYSTEM
involves interactions between the inner and outer cores

INNER CORE
Inner sphere composed primarily of solid iron, extending from about 5150 km deep to the Earth's center at 6370 km

OUTER CORE
Liquid shell composed primarily of molton iron, extending from about 2900 km to 5150 km in depth

FIGURE 2.15 The Earth system encompasses all parts of our planet and their interactions.

For instance, in our discussion of global climate change, we will primarily consider interactions among the atmosphere, hydrosphere, cryosphere, and biosphere that are driven by solar energy. Our coverage of how the continents are deformed to make mountains will focus on interactions between the crust and mantle that are driven by Earth's internal energy.

Specialized subsystems that produce specific types of behaviors, such as climate change or mountain building, are called **geosystems.** The Earth system can be thought of as the collection of many open, interacting (and often overlapping) geosystems. In this section, we will introduce three important geosystems that operate on a global scale: the climate system, the plate tectonic system, and the geodynamo system. Later in this textbook, we will have occasion to discuss a number of smaller geosystems, such as volcanoes that erupt hot lava (Chapter 5), hydrologic systems that give us our drinking water (Chapter 11), and petroleum reservoirs that produce oil and gas (Chapter 14).

The Climate System

Weather is the term we use to describe the temperature, precipitation, cloud cover, and winds observed at a particular location and time on Earth's surface. We all know how variable the weather can be—hot and rainy one day, cool and dry the next—depending on the movements of storm systems, warm and cold fronts, and other atmospheric disturbances. Because the atmosphere is so complex, even the best forecasters have a hard time predicting the weather more than 4 or 5 days in advance. However, we can guess in rough terms what our weather will be much further into the future because weather is governed primarily by the changes in solar energy input on seasonal and daily cycles: summers are hot, winters cold; days are warmer, nights cooler. The **climate** produced by these weather cycles can be described by averaging temperature and other variables over many years of observation. A complete description of climate also includes measures of how variable the weather has been, such as the highest and lowest temperatures ever recorded on a given day of the year.

The **climate system** includes all the Earth system components that determine climate on a global scale and how climate changes with time. In other words, the climate system describes not only the behavior of the atmosphere, but also the influences of the hydrosphere, cryosphere, biosphere, and lithosphere (see Figure 2.15).

When the Sun warms Earth's surface, some of the heat is trapped by water vapor, carbon dioxide, and other gases in the atmosphere, much as heat is trapped by frosted glass in a greenhouse. This *greenhouse effect* explains why Earth has a climate that makes life possible. If its atmosphere contained no greenhouse gases, Earth's surface would be frozen solid! Therefore, greenhouse gases, particularly carbon dioxide, play an essential role in regulating climate. As we will learn in later chapters, the concentration of carbon dioxide in the atmosphere is a balance between the amount spewed out of Earth's interior in volcanic eruptions and the amount withdrawn during the weathering of silicate rocks. In this way, the behavior of the atmosphere is regulated by interactions with the lithosphere.

> *Solar radiation energizes the climate system, which involves the atmosphere, hydrosphere, cryosphere, biosphere, and lithosphere.*

To understand these types of interactions, scientists build numerical models—virtual climate systems—on large computers. They improve the models by matching them to climate observations, and they use them to simulate future climate change. A particularly urgent problem is to understand the global warming that is being caused by *anthropogenic* (human-generated) emissions of carbon dioxide and other greenhouse gases. In Chapter 10 we will discuss some aspects of how the climate system works and in Chapter 14 the practical problems posed by anthropogenic climate change.

The Plate Tectonic System

Some of Earth's most dramatic geologic events—volcanic eruptions and earthquakes, for example—also result from interactions within the Earth system. These phenomena are driven by Earth's internal heat, which is transferred upward through the circulation of material in Earth's mantle.

We have seen that Earth is zoned by chemistry: its crust, mantle, and core are chemically distinct layers. Earth is also zoned by *strength*, a property that measures how much an Earth material can resist being deformed. Material strength depends on both chemical composition (bricks are strong, soap bars are weak) and temperature (cold wax is strong, hot wax is weak).

In some ways, the outer part of the solid Earth behaves like a ball of hot wax. Cooling of the surface forms a strong outer shell, or **lithosphere** (from the Greek *lithos*, meaning "stone"), which encases a hot, weak **asthenosphere** (from the Greek *asthenes*, meaning "weak"). The lithosphere includes the crust and the top part of the mantle down to an average depth of about 100 km. The asthenosphere is the portion of mantle, perhaps 300 km thick, immediately below the lithosphere. When subjected to force, the lithosphere tends to behave like a nearly rigid and brittle shell, whereas the underlying asthenosphere flows like a moldable, or *ductile*, solid.

> *Earth's internal heat energizes the plate tectonic system, which involves the lithosphere, asthenosphere, and deep mantle.*

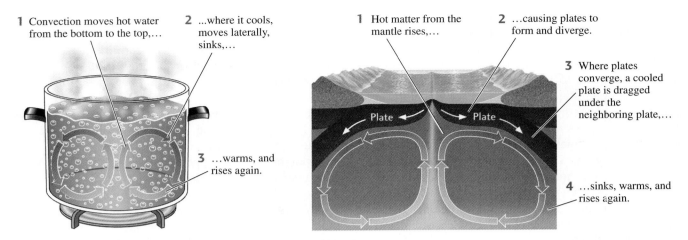

1 Convection moves hot water from the bottom to the top,...

2 ...where it cools, moves laterally, sinks,...

3 ...warms, and rises again.

1 Hot matter from the mantle rises,...

2 ...causing plates to form and diverge.

Plate   Plate

3 Where plates converge, a cooled plate is dragged under the neighboring plate,...

4 ...sinks, warms, and rises again.

FIGURE 2.16 Convection in Earth's mantle can be compared to the pattern of movement in a pot of boiling water. Both processes carry heat upward through the movement of matter.

According to the remarkable theory of **plate tectonics,** the lithosphere is not a continuous shell; it is broken into about a dozen large plates that move over Earth's surface at rates of a few centimeters per year. Each lithospheric plate is a rigid unit that rides on the asthenosphere, which is also in motion. The lithosphere that forms a plate may vary from just a few kilometers thick in volcanically active areas to more than 200 km thick beneath the older, colder parts of continents. The discovery of plate tectonics in the 1960s led to the first unified theory that explained the worldwide distribution of earthquakes and volcanoes, continental drift, mountain building, and many other geologic phenomena. Chapter 3 is devoted to a description of plate tectonics.

Why do the plates move across Earth's surface instead of locking up into a completely rigid shell? The forces that push and pull the plates come from the mantle. Driven by Earth's internal heat engine, hot mantle material rises at boundaries where plates separate, forming new lithosphere. The lithosphere cools and becomes more rigid as it moves away from these boundaries, eventually sinking back into the mantle under the pull of gravity at other boundaries where plates converge. This general process, in which hotter material rises and cooler material sinks, is called *convection* (**Figure 2.16**).

> The lithosphere is composed of rigid plates that ride on a ductile asthenosphere. The plates are pushed and pulled by convection as hot mantle material rises from Earth's interior and cooler material sinks back into the interior.

The Geodynamo System

The third global geosystem involves interactions that produce a **magnetic field** deep inside Earth, in its liquid outer core. This magnetic field reaches far into outer space,

causing compass needles to point north and shielding the biosphere from the Sun's harmful radiation. When rocks form, they become slightly magnetized by this magnetic field, so geologists can study how the field behaved in the past and use it to help them decipher the geologic record.

Earth rotates about an axis that goes through its North and South Poles. Earth's magnetic field behaves as if a powerful bar magnet were located at Earth's center and inclined about 11° from the rotational axis. The magnetic force points into Earth at the north magnetic pole and outward from Earth at the south magnetic pole (**Figure 2.17**). At any place on Earth (except near the magnetic poles), a compass needle that is free to swing under the influence of the magnetic field will rotate into a position parallel to the local line of force, approximately in the north-south direction.

Although a permanent magnet at Earth's center could explain the dipole ("two-pole") nature of the observed magnetic field, this hypothesis can be easily rejected. Laboratory experiments have demonstrated that the field of a permanent magnet is destroyed when the magnet is heated above about 500°C. We know that the temperatures in Earth's deep interior are much higher than that—thousands of degrees at its center—so, unless the magnetism were constantly regenerated, it could not be maintained.

Scientists theorize that heat flowing out of Earth's core causes convection that generates and maintains the magnetic field. Why is a magnetic field created by convection in the outer core but not by convection in the mantle? First, the outer core is made primarily of iron, which is a very good electrical conductor, whereas the silicate rocks of the mantle are poor electrical conductors. Second, the convective flow is a million times more rapid in the

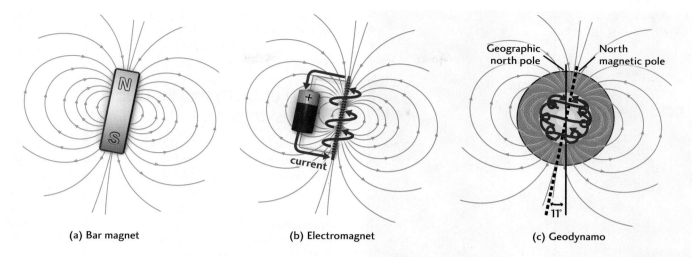

(a) Bar magnet   (b) Electromagnet   (c) Geodynamo

FIGURE 2.17 (a) A bar magnet creates a dipole field with north and south poles. The direction of the magnetic force is indicated by the blue lines. (b) A dipole field can also be produced by electric currents flowing through a coil of metallic wire, as shown for this battery-powered electromagnet. (c) Earth's magnetic field, which is approximately dipole, is produced by electric currents flowing in the liquid-metal outer core, which are powered by convection.

> *Convection of liquid iron in Earth's outer core creates a geodynamo, which generates Earth's magnetic field.*

liquid outer core than in the solid mantle. The rapid flow stirs up electric currents in the iron-nickel alloy to produce the magnetic field. Thus, this **geodynamo** is more like an electromagnet than a bar magnet (see Figure 2.17).

For some 400 years, scientists have known that a compass needle points north because of Earth's magnetic field. Imagine how stunned they were when they found geologic evidence that the direction of the magnetic force can be reversed. Over about half of geologic time, a compass needle would have pointed south!

Magnetic reversals occur at irregular intervals ranging from tens of thousands to millions of years. The processes that cause them are not well understood, but computer models of the geodynamo show sporadic reversals occurring in the absence of any external factors—that is, purely through interactions within Earth's core. As we will see in the next chapter, magnetic reversals, which leave their imprint on the geologic record, have helped geologists figure out the movements of the tectonic plates.

NATURAL ENVIRONMENT:
Interactions Among Geosystems Support Life

The natural environment—the habitat of life—is largely controlled by the climate system. The biosphere partici-

pates as an active component of this geosystem, regulating, for example, the amount of carbon dioxide, methane, and other greenhouse gases in the atmosphere, which in turn determines the planet's surface temperature. As we shall see in Chapter 9, the evolution of the biosphere and atmosphere have gone hand-in-hand throughout the last 3.5 billion years of climate-system history.

Perhaps less obvious is the coupling of the natural environment to the other two global geosystems. Plate tectonics produces volcanoes that resupply the atmosphere and oceans with water and gases from Earth's deep interior, and it is responsible for the tectonic processes that raise mountains. The interactions of the atmosphere, hydrosphere, and cryosphere with the surface topography create a variety of habitats that enrich the biosphere and, through the erosion of rock and dissolution of minerals, provide life with essential nutrients.

Unlike the convective motions of plate tectonics, the swirling currents in Earth's outer core are too deep to deform the crust or alter its chemistry. However, the magnetic field produced by this geodynamo reaches outward into space far beyond Earth's atmosphere (see Figure 2.17). There it forms a barrier to highly energetic particles that stream outward from the Sun at speeds of more than 400 km/s—the *solar wind* (**Figure 2.18**). Without this shield, Earth's surface would be bombarded by harmful solar radiation, which would kill many forms of life that now prosper in its biosphere.

> *All three global geosystems— climate, plate tectonics, and the geodynamo—are important in sustaining our natural environment.*

FIGURE 2.18 Earth's magnetic field protects life by shielding Earth's surface from harmful solar radiation. This solar wind contains highly energetic charged particles ejected from the Sun, which distort Earth's magnetic field lines, shown here in light blue. The distances in this picture are not to scale. [SOHO (ESA and NASA).]

AN OVERVIEW OF GEOLOGIC TIME

So far, we have discussed Earth's size and shape, its internal layering and composition, and the operation of its three major geosystems. How did Earth get its layered structure in the first place? How have the global geosystems evolved through geologic time? To begin to answer these questions, we present a brief overview of geologic time from the birth of the planet to the present. Later chapters will fill in the details.

Comprehending the immensity of geologic time is a challenge. John McPhee has noted that geologists look into the "deep time" of Earth's early history (measured in billions of years), just as astronomers look into the "deep space" of the outer universe (measured in billions of light-years). **Figure 2.19** presents the geologic time line marked with some major events and transitions.

Origin of Earth and Its Global Geosystems

Using evidence from meteorites, geologists have been able to show that Earth and the other planets of the solar system formed about 4.56 billion years ago through the rapid condensation of a dust cloud that circulated around the young Sun. This violent process, which involved the aggregation and collision of progressively larger clumps of matter, will be described in more detail in Chapter 9. Within just 100 million years (a relatively short period of time, geologically speaking), the Moon had formed and Earth's core had separated from its mantle. Exactly what happened during the next several hundred million years is hard to figure out. Very little of the rock record survived intense bombardment by the large meteorites that were smashing into Earth. This early period of Earth's history is appropriately called the geologic "dark ages."

The oldest rocks now found on Earth's surface are nearly 4.3 billion years old. Rocks as ancient as 3.8 billion years show evidence of erosion by water, indicating the existence of a hydrosphere and the operation of a climate system not too different from that of the present. Rocks only slightly younger, 3.5 billion years old, record a magnetic field about as strong as the one we see today, showing that the geodynamo was operating by that time. By 2.5 billion years ago, enough low-density crust had collected at Earth's surface to form large continental masses. The geologic processes that

> *Earth formed about 4.56 billion years ago. Rocks almost 4.3 billion years old have been found at Earth's surface.*

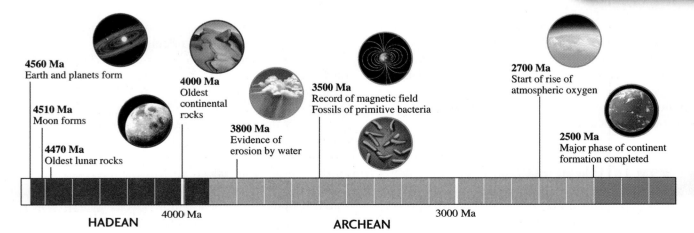

4560 Ma Earth and planets form

4510 Ma Moon forms

4470 Ma Oldest lunar rocks

4000 Ma Oldest continental rocks

3800 Ma Evidence of erosion by water

3500 Ma Record of magnetic field Fossils of primitive bacteria

2700 Ma Start of rise of atmospheric oxygen

2500 Ma Major phase of continent formation completed

HADEAN   4000 Ma   ARCHEAN   3000 Ma

FIGURE 2.19 This geologic time line shows some of the major events observed in the geologic record, beginning with the formation of the planets. (Ma: million years ago.)

then modified those continents were very similar to the plate tectonic processes we see operating today.

The Evolution of Life

Life also began very early in Earth's history, as we can tell from the study of **fossils,** traces of organisms preserved in the geologic record. Fossils of primitive bacteria have been found in rocks dated at 3.5 billion years ago. A key event was the evolution of organisms that release oxygen into the atmosphere and oceans. The buildup of oxygen in the atmosphere was under way by 2.7 billion years ago. Atmospheric oxygen concentrations probably rose to modern levels in a series of steps over a period as long as 2 billion years.

Life on early Earth was simple, consisting mostly of small, single-celled organisms that floated near the surface of the oceans or lived on the seafloor. Between 1 billion and 2 billion years ago, more complex life-forms such as algae and seaweeds evolved. The first animals appeared about 600 million years ago, evolving in a series of waves. In a period starting 542 million years ago and probably lasting less than 10 million years, eight entirely new branches of the animal kingdom were established, including the ancestors of nearly all animals inhabiting Earth today. It was during this evolutionary explosion, sometimes called biology's Big Bang, that animals with shells first left their shelly fossils in the geologic record.

Primitive fossils indicate that life on Earth began before 3.5 billion years ago.

Although biological evolution is often viewed as a very slow process, it is punctuated by brief periods of rapid change. Spectacular examples are *mass extinctions*, during which many types of organisms suddenly disappeared from the geologic record. Five of these huge turnovers are marked on the time ribbon in Figure 2.19. The most recent one was caused by a large bolide impact 65 million years ago. The meteorite, about 10 km in diameter, caused the extinction of half of Earth's species, including all the dinosaurs (see Figure 1.23).

The causes of the other mass extinctions are still being debated. In addition to bolide impacts, scientists have proposed other kinds of extreme events, such as rapid climate changes brought on by glaciations and massive eruptions of volcanic material. The evidence is often ambiguous or inconsistent, however. For example, the largest mass extinction of all time took place about 251 million years ago, wiping out nearly 95 percent of all species. A bolide impact has been proposed by some investigators as the cause, but the geologic record shows that ice sheets expanded and seawater chemistry changed at this time, a finding consistent with a major climate change. At the same time, an enormous volcanic eruption covered an area in Siberia almost half the size of the United States with 2 million to 3 million cubic kilometers of lava. This mass extinction has been dubbed "Murder on the Orient Express" because there are so many suspects!

Mass extinctions reduce the number of species competing for space in the biosphere. By "thinning out the crowd," these extreme events can promote the evolution of new species. After the demise of the dinosaurs 65 million years ago, mammals became the dominant class of animals. The rapid evolution of mammals into species with bigger brains and more dexterity led to the emergence of humanlike species (*hominids*) around 5 million years ago and our own species, *Homo sapiens* (Latin for "knowing human"), about 200,000 years ago. As newcomers to the biosphere, we're just beginning to leave our mark on the geologic record. Indeed, our short history as a species can be appreciated by noting that it spans less than a vertical line's width on the geologic time scale in Figure 2.19.

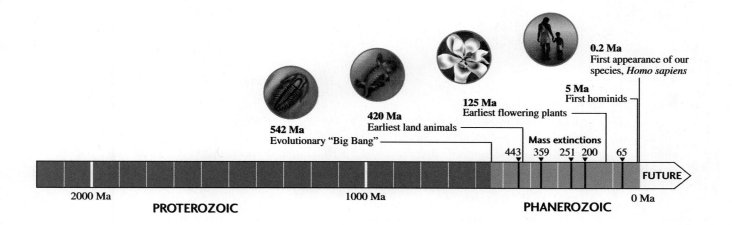

0.2 Ma
First appearance of our species, *Homo sapiens*

5 Ma
First hominids

125 Ma
Earliest flowering plants

420 Ma
Earliest land animals

542 Ma
Evolutionary "Big Bang"

Mass extinctions
443 359 251 200 65

FUTURE

2000 Ma   1000 Ma   0 Ma
PROTEROZOIC   PHANEROZOIC

GOOGLE EARTH PROJECT

Earth is a dynamic and complex system of interrelated components. A great many factors work to shape Earth's surface, and they are brought together by the overarching theory of plate tectonics. In this exercise, we will use Google Earth (GE) to explore the topographic extremes of our planet; in subsequent exercises in later chapters we will explore the origin of those features. Let's start at the roof of the world: the Himalaya.

LOCATION   Topographic exploration from the Himalaya, in central Asia, to the Challenger Deep, off the southern coast of Guam in the Pacific Ocean

GOAL   Demonstrate the topographic variation of our planet and introduce the tools of Google Earth

LINKED   Figure 2.9

1. Enter "Mount Everest" into your GE search engine and use the cursor to find its highest point. What is its approximate elevation above sea level? It may be helpful to tilt your frame of view to the north in order to pick out the highest point.

   a. 10,400 m
   b. 7380 m
   c. 8850 m
   d. 9230 m

2. Zoom out from Mount Everest proper and take a look at the shape of the Himalaya as a whole (try an eye altitude of 4400 km). Which of the following descriptions best captures what you see?

   a. Triangular mountain range composed of a single high peak
   b. East-west–oriented mountain range composed of dozens of high peaks
   c. North-south–oriented mountain range composed of high peaks in the middle and lower peaks around the edges
   d. Circular mountain range closed around a central broad dome

3. From the Himalaya, move to one of the deepest places on Earth's surface by typing "Challenger Deep" into the search panel. GE should take you immediately out to sea, off the coast of the Philippines.

Use the GE "line" measurement tool to determine the approximate horizontal surface distance between the two locations. What is the distance?

   a. 6300 km
   b. 2200 km
   c. 185,000 km
   d. 75,500 km

4. Zoom out from Challenger Deep to an eye altitude of 4200 km. Notice the unique surface feature that links Challenger Deep to deep regions of the ocean here. How would you describe this large-scale feature?

   a. Challenger Deep is part of an undersea mountain range with a roughly north-south orientation.
   b. Challenger Deep is part of an arcuate deep-sea trench in the Pacific Ocean that trends almost east-west at this location.
   c. Challenger Deep is the deepest part of a broad, almost flat, plain near the middle of the Pacific Ocean.
   d. Challenger Deep is at the top of an undersea volcano that extends high above the Pacific Ocean floor.

Optional Challenge Question

5. Using the answer to question 1 and using your cursor to note the maximum depth of Challenger Deep below mean sea level, calculate the approximate total difference in elevation of the two locations. Which of the following numbers is closest to the difference?

   a. 14,000 m
   b. 20,000 m
   c. 18,000 m
   d. 26,000 m

◼ SUMMARY

How do geologists study Earth? Geologists, like other scientists, use the scientific method. They develop and test hypotheses, which are tentative explanations for natural phenomena based on observations and experiments. A coherent set of hypotheses that have survived repeated challenges constitutes a theory. Hypotheses and theories can be combined into a scientific model that represents a natural system or process. Confidence grows in those hypotheses, theories, and models that withstand repeated tests and are able to predict the results of new observations and experiments.

What is Earth's shape? Earth's overall shape is a sphere with an average radius of 6370 km that bulges slightly at the equator and is slightly squashed at the poles due to the planet's rotation. Its topography varies by about 20 km from the highest point to the lowest point. Elevations fall into two main groups: 0 to 1 km above sea level for much of the land surface and 4 to 5 km below sea level for much of the deep sea.

What are Earth's major layers? Earth's interior is divided into concentric layers of different compositions, separated by sharp, nearly spherical boundaries. The outer layer is the crust, made up of mainly silicate rock, which varies in thickness from about 40 km in the case of continental crust to about 7 km for oceanic crust. Below the crust is the mantle, a thick shell of denser silicate rock that extends to the core-mantle boundary at a depth of about 2890 km. The core, which is composed primarily of iron and nickel, is divided into two layers—a liquid outer core and a solid inner core—separated by a boundary at a depth of 5150 km. Jumps in density between these layers are caused by differences in their chemical composition.

How do we study Earth as a system of interacting components? When we try to understand a complex system such as the Earth system, we find that it is often easier to focus on subsystems (which we call geosystems). This textbook discusses three major global geosystems: the climate system, which involves interactions among the atmosphere, hydrosphere, cryosphere, biosphere, and lithosphere; the plate tectonic system, which involves interactions among Earth's solid components (lithosphere, asthenosphere, and deep mantle); and the geodynamo, which involves interactions within Earth's core. The climate system is driven by heat from the Sun, whereas the plate tectonic and geodynamo are driven by Earth's internal heat engine.

What are the basic elements of plate tectonics? The lithosphere is broken into about a dozen large plates. Driven by convection in the mantle, these plates move over Earth's surface at rates of a few centimeters per year. Each plate acts as a rigid unit, riding on the ductile asthenosphere, which also is in motion. Hot mantle material rises at boundaries where plates form and separate, cooling and becoming more rigid as it moves away. Eventually, most of it sinks back into the mantle at boundaries where plates converge.

What are some major events in Earth's history? Earth formed 4.56 billion years ago. Rocks as old as 4.3 billion years have survived in Earth's crust. Liquid water existed on Earth's surface by 3.8 billion years ago. Rocks about 3.5 billion years old show evidence of a magnetic field, and the earliest evidence of life has been found in rocks of the same age. By 2.7 billion years ago, the oxygen content of the atmosphere was rising because of oxygen production by early organisms, and by 2.5 billion years ago, large continental masses had formed. Animals appeared suddenly about 600 million years ago, diversifying rapidly in a great evolutionary explosion. The subsequent evolution of life was marked by a series of mass extinctions, the most recent of which was caused by a large bolide impact 65 million years ago. Our species, *Homo sapiens,* first appeared about 200,000 years ago.

■ KEY TERMS AND CONCEPTS

asthenosphere (p. 53)	inner core (p. 49)
climate (p. 53)	lithosphere (p. 53)
climate system (p. 53)	magnetic field (p. 54)
core (p. 49)	mantle (p. 43)
crust (p. 48)	outer core (p. 49)
Earth system (p. 51)	plate tectonics (p. 54)
fossil (p. 57)	principle of
geodynamo (p. 55)	uniformitarianism (p. 43)
geologic record (p. 42)	scientific method (p. 40)
geosystem (p. 53)	seismic wave (p. 47)
hypothesis (p. 40)	topography (p. 46)

■ EXERCISES

1. Illustrate the differences between a hypothesis, a theory, and a model with some examples drawn from this chapter.

2. How does science differ from religion as a way to understand the world?

3. Give an example of how the model of Earth's spherical shape developed by Eratosthenes could be experimentally tested.

4. Give two reasons why Earth's shape is not a perfect sphere.

5. If you made a model of Earth that was 10 cm in radius, how high would Mount Everest rise above sea level?

6. It is thought that a large bolide impact 65 million years ago caused the extinction of three-quarters of Earth's species, including all the dinosaurs. Does this event disprove the principle of uniformitarianism? Explain your answer.

7. Imagine you are a tour guide on a journey from Earth's surface to its center. How would you describe the material that your tour group encounters on the way down? Why is the density of the material always increasing as you go deeper?

8. How does the chemical composition of Earth's crust differ from the composition of its deeper interior? From the composition of its core?

9. Explain how Earth's outer core can be a liquid while the deep mantle is a solid.

10. How does viewing Earth as a system of interacting components help us to understand our planet? Give an example of an interaction between two or more geosystems that could affect the geologic record.

11. How do the terms *weather* and *climate* differ? Express the relationship between climate and weather using examples from your experience.

12. Earth's mantle is solid, but it undergoes convection as part of the plate tectonic system. Explain why these statements are not contradictory.

13. In what general ways are the climate system, the plate tectonic system, and the geodynamo system similar? In what ways are they different?

14. Not every planet has a geodynamo. Why not? If Earth did not have a magnetic field, what might be different about our planet?

15. Based on the material presented in this chapter, what can we say about how long ago the three major global geosystems began to operate?

16. If no theory can be proved true, why do almost all geologists believe strongly in Darwin's theory of evolution?

Visual Literacy Task

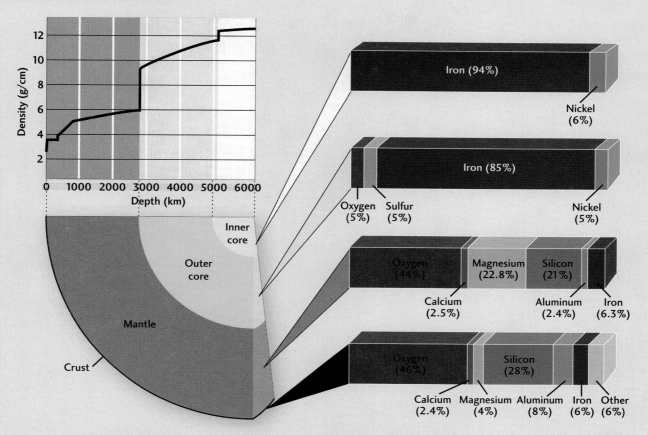

FIGURE 2.13 Jumps in density between Earth's major layers, shown above in different colors, are caused primarily by differences in their chemical composition. The relative amounts of the main elements (by weight) are depicted in the bars on the right.

1. What is the thinnest layer?

a. Inner core   c. Mantle
b. Outer core   d. Crust

2. How does density change with increasing depth?

a. It increases at a rate that stays the same.
b. It increases at a rate that varies.
c. It decreases at a rate that stays the same.
d. It decreases at a rate that varies.

3. What is the major difference between the crust, mantle, and core layers?

a. The layers are different colors.
b. The layers have different compositions.
c. The layers have different behaviors.
d. The layers have different phases (solid, liquid, gas).

4. Which layer has the highest percentage of magnesium?

a. Inner core   c. Mantle
b. Outer core   d. Crust

5. Which combination of elements is most dense?

a. Silicon and oxygen
b. Silicon, oxygen and magnesium
c. Iron and nickel

Thought Question: Explain what you would tell a fellow student who said, "The density of Earth increases the deeper you go primarily because there is more pressure, which makes the materials more compact."

Plate Tectonics: The Unifying Theory of Geology

3

Geology at Sea

AN OCEANOGRAPHIC VESSEL, like most working ships, does not provide posh accommodations. As the lyrics of Jimmy Buffett's song "Landfall" describe it, "you live half your life in an eight-by-five room, just cruisin' to the sound of the big diesel boom." During the other half, you stand watch in the lab, attending to the hull-mounted sonar array and magnetometers hung over the side to map the seafloor as the ship steams along, or out on deck, washed by spray, using thick steel cables to lower instruments into the water or dredge rocks from craggy outcrops several kilometers below the rough sea surface. Sometimes you light the fuse on a kilogram or two of high explosives and toss it off the fantail, making a bang big enough to send sound waves deep into the crust, which then bounce back and tell you something interesting about the properties of the submerged rocks.

> Marine geology is a tedious and sometimes dangerous business.

Marine geology is a tedious and sometimes dangerous business. Maurice "Doc" Ewing, who was responsible for collecting much of the data that led to the discoveries highlighted in this chapter, was once washed overboard with two mates and his brother John, also a marine geologist, by a freak wave in a storm north of Bermuda. Skillful maneuvering by the ship's captain saved the brothers and one mate from drowning, but Doc Ewing walked with a limp for the rest of his life. Bruce Heezen, another marine geologist important to our story, died aboard a research submarine while exploring the Mid-Atlantic Ridge southwest of Iceland.

Seventy percent of Earth's solid surface is hidden beneath the oceans. The first global oceanographic expedition sailed on the HMS *Challenger* (namesake of the space shuttle) from Portsmouth, England, in 1872. During its 3-year cruise, scientists discovered the symmetrical undersea mountains of the Mid-Atlantic Ridge and the Challenger Deep of the Marianas Trench, the deepest spot in all the world's oceans. Nearly a century of further exploration and many untold adventures aboard ships at sea were required to explain these peculiar features and reveal the secrets of plate tectonics—the grand, unifying theory of geology. ◆

The research vessel *Oceanus*, owned by the National Science Foundation and operated by the Woods Hole Oceanographic Institution, battles the high seas and rough weather of the North Atlantic. [*Chris Linder, Woods Hole Oceanographic Institution.*]

CHAPTER OUTLINE

THE LITHOSPHERE—Earth's strong, rigid outer shell of rock—is broken into about a dozen plates, which slide past, converge with, or separate from one another as they move over the weaker, ductile asthenosphere. Plates are created where they separate and are recycled where they converge in a continuous process of creation and destruction. Continents, embedded in the lithosphere, drift along with the moving plates. The theory of **plate tectonics** describes the movements of plates and the forces acting on them. It also explains volcanoes, earthquakes, and the distribution of mountain chains, rock assemblages, and structures on the seafloor—all of which result from events at plate boundaries. Plate tectonics provides a conceptual framework for a large part of this textbook and, indeed, for much of geology.

This chapter lays out the theory of plate tectonics and how it was discovered, describes plate movements today and in the geologic past, and examines how the forces that drive these movements arise from the mantle convection system.

DISCOVERY OF PLATE TECTONICS

In the 1960s, a great revolution in thinking shook the world of geology. For almost 200 years, geologists had been developing various theories of *tectonics* (from the Greek *tekton*, meaning "builder")—the general term used to describe mountain building, volcanism, earthquakes, and other processes that construct geologic features on Earth's surface. It was not until the discovery of plate tectonics, however, that a single theory could satisfactorily explain the whole range of geologic processes. Physics had a comparable revolution at the beginning of the twentieth century, when the theory of relativity revised the physical laws that govern space, time, mass, and motion. Biology had a comparable revolution in the middle of the twentieth century, when the discovery of DNA allowed biologists to explain how organisms transmit the information that controls their growth and functioning from generation to generation.

In Chapter 2, we discussed how the scientific method guides the work of geologists. In the context of the scientific method, plate tectonics is a confirmed theory whose strength lies in its simplicity, its generality, and its consistency with many types of observations.

The basic ideas of plate tectonics were put together as a unified theory of geology during the mid-1960s. The scientific synthesis that led to the theory of plate tectonics, however, really began much earlier in the twentieth century, with the recognition of evidence for continental drift.

CONTINENTAL DRIFT

Such changes in the superficial parts of the globe seemed to me unlikely to happen if the earth were solid to the center. I therefore imagined that the internal parts might be a fluid more dense, and of greater specific gravity than any of the solids we are acquainted with, which therefore might swim in or upon that fluid. Thus the surface of the earth would be a shell, capable of being broken and disordered by the violent movements of the fluid on which it rested.

Benjamin Franklin, 1782, in a letter to the French geologist Abbé J. L. Giraud-Soulavie

The concept of **continental drift**—large-scale movements of continents—has been around for a long time.

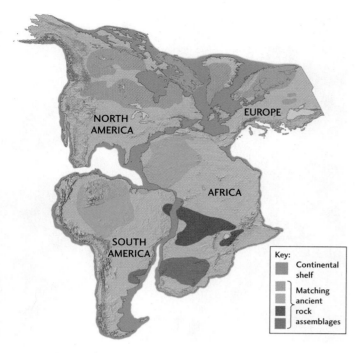

FIGURE 3.1 The jigsaw-puzzle fit of the continents bordering the Atlantic Ocean formed the basis of Alfred Wegener's theory of continental drift. In his book *The Origin of Continents and Oceans,* Wegener cited as additional evidence the similarity of geologic features on opposite sides of the Atlantic. The matches between ancient crystalline rocks in adjacent regions of South America and Africa and of North America and Europe are shown here. [*Geographic fit from data of E. C. Bullard; geologic data from P. M. Hurley.*]

In the late sixteenth century and the seventeenth century, European scientists noticed the jigsaw-puzzle fit of the coasts on both sides of the Atlantic Ocean, as if the Americas, Europe, and Africa had once been part of a single continent and had subsequently drifted apart. By the close of the nineteenth century, the Austrian geologist Eduard Suess had put together some of the pieces of the puzzle. He postulated that the present-day southern continents had once formed a single giant continent called Gondwana (or *Gondwanaland*). In 1915, Alfred Wegener, a German meteorologist who was recovering from wounds suffered in World War I, wrote a book on the breakup and drift of continents, in which he laid out the remarkable similarity of geologic features on opposite sides of the Atlantic (**Figure 3.1**). In the years that followed, Wegener postulated a supercontinent, which he called **Pangaea** (Greek for "all lands"), that broke up into the continents as we know them today.

Although Wegener was correct in asserting that the continents had drifted apart, his hypotheses about how fast they were moving and what forces were pushing them across Earth's surface turned out to be wrong, as we shall see, and these errors reduced his credibility among other scientists. After about a decade of spirited debate,

physicists convinced geologists that Earth's outer layers were too rigid for continental drift to occur, and Wegener's ideas were rejected by all but a few geologists.

Wegener and other advocates of the drift hypothesis pointed not only to the geographic matching of geologic features, but also to similarities in rock ages and trends in geologic structures on opposite sides of the Atlantic. They also offered arguments, accepted now as good evidence of drift, based on fossil and climate data. Identical 300-million-year-old fossils of the reptile *Mesosaurus,* for example, have been found in Africa and South America but nowhere else, suggesting that the two continents were joined when the *Mesosaurus* was alive (**Figure 3.2**). The animals and plants on the different continents showed similarities in their evolution until the postulated breakup time. After that, they followed different evolutionary paths because of the isolation and changing environments of the separating continents. In addition, rocks deposited by glaciers that existed 300 million years ago are now distributed across South America, Africa, India, and Australia. If the southern continents had once been part of Gondwana near the South Pole, a single continental glacier could account for all of these glacial deposits.

> *The jigsaw-puzzle fit of the continents on both sides of the Atlantic Ocean, as well as rock ages, fossil evidence, and glacial deposits, indicate that those continents were once part of the supercontinent Pangaea.*

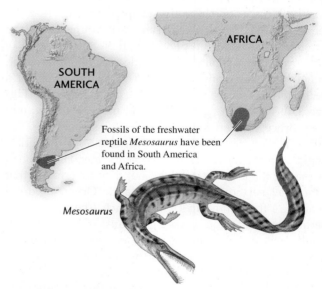

Fossils of the freshwater reptile *Mesosaurus* have been found in South America and Africa.

Mesosaurus

FIGURE 3.2 Fossils of the freshwater reptile *Mesosaurus,* 300 million years old, are found in South America and Africa and nowhere else in the world. If *Mesosaurus* were able to swim across the South Atlantic Ocean, it should have been able to cross other oceans and should have spread more widely. The observation that it did not suggests that South America and Africa must have been joined 300 million years ago. [*After A. Hallam, "Continental Drift and the Fossil Record," Scientific American (November 1972): 57–66.*]

Seafloor Spreading

The geologic evidence did not convince the skeptics, who maintained that continental drift was physically impossible. No one had yet come up with a plausible driving force that could have split Pangaea and moved the continents apart. Wegener, for example, thought the continents floated like boats across the solid oceanic crust, dragged along by the tidal forces of the Sun and Moon. His hypothesis was quickly rejected, however, because it could be shown that tidal forces are much too weak to move continents.

The breakthrough came when scientists realized that convection in Earth's mantle (described in Chapter 2) could push and pull continents apart, creating new oceanic crust through the process of **seafloor spreading.** In 1928, the British geologist Arthur Holmes proposed that convection currents "dragged the two halves of the original continent apart, with consequent mountain building in the front where the currents are descending, and the ocean floor development on the site of the gap, where the currents are ascending." Many still argued, however, that Earth's crust and mantle are rigid and immobile, and Holmes conceded that "purely speculative ideas of this kind, specially invented to match the requirements, can have no scientific value until they acquire support from independent evidence."

Evidence emerged from exploration of the seafloor after World War II. Doc Ewing, the marine geologist mentioned at the opening of this chapter, showed that the seafloor of the Atlantic Ocean is made of young basalt, not old granite, as had been previously thought (**Figure 3.3**). Moreover, the mapping of an undersea mountain chain called the Mid-Atlantic Ridge led to the discovery of a deep, cracklike valley, or *rift*, running down its center (**Figure 3.4**). Two of the geologists who mapped this feature were Bruce Heezen and Marie Tharp, assistants to Doc Ewing at Columbia University (**Figure 3.5**). "I thought it might be a rift valley," Tharp said years later. Heezen had dismissed the idea as "girl talk," but they soon found that almost all earthquakes in the Atlantic Ocean occurred near the rift, confirming Tharp's hunch (see Figure 3.4). Because most earthquakes are generated by faulting, their results indicated that the rift was a tectonically active feature. Other mid-ocean ridges with similar rifts and earthquake activity were found in the Pacific and Indian oceans.

In the early 1960s, Harry Hess of Princeton University and Robert Dietz of the Scripps Institution of Oceanography proposed that Earth's crust separates along the rifts in mid-ocean ridges and that new crust is formed by the upwelling of hot molten rock into these cracks. The

> Seafloor spreading occurs as plates separate along a rift and hot molten rock wells up to create new crust, which moves away from the mid-ocean ridge as more crust is continually created.

FIGURE 3.3 This photo, taken in the summer of 1947, shows Maurice "Doc" Ewing (*center*) beaming as he looks at a piece of young basalt dredged from the depths of the Atlantic Ocean by the research vessel *Atlantis I*. On the near left is Frank Press, who initiated the series of geology textbooks that includes this one. [*Photo courtesy of Lamont-Doherty Earth Observatory, The Earth Institute at Columbia University.*]

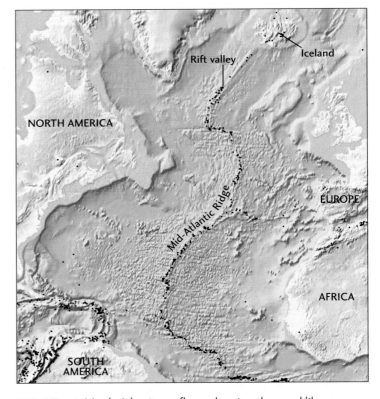

FIGURE 3.4 North Atlantic seafloor, showing the cracklike rift valley running down the center of the Mid-Atlantic Ridge and the locations of earthquakes (black dots).

FIGURE 3.5 Marie Tharp and Bruce Heezen inspect a map of the seafloor. Their discovery of active rifts on the mid-ocean ridges provided important evidence for seafloor spreading. [*The Earth Institute at Columbia University.*]

new seafloor—actually the surface of newly created lithosphere—spreads laterally away from the rift and is replaced by even newer crust in a continuing process of plate creation.

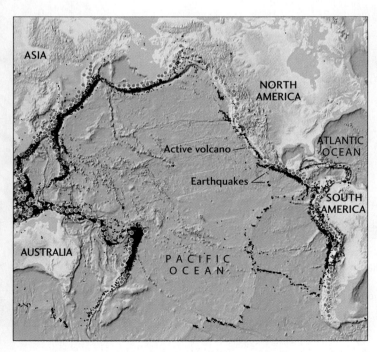

FIGURE 3.6 The Pacific Ring of Fire, with its active volcanoes (large red circles) and frequent earthquakes (small black dots), marks plate boundaries where oceanic lithosphere is being recycled.

The Great Synthesis: 1963–1968

The seafloor spreading hypothesis put forward by Hess and Dietz in 1962 explained how the continents could drift apart through the creation of new lithosphere at mid-ocean ridges. Could the seafloor and its underlying lithosphere be destroyed by recycling back into Earth's interior? If not, Earth's surface area would have to increase over time. For a time in the early 1960s, some physicists and geologists, including Heezen, actually believed in this idea of an expanding Earth. Other geologists recognized that the seafloor was instead being recycled. They were convinced this was happening in regions of intense volcanic and earthquake activity around the margins of the Pacific Ocean basin, known collectively as the Ring of Fire (**Figure 3.6**). The details of the process, however, remained unclear.

In 1965, the Canadian geologist J. Tuzo Wilson first described tectonics around the globe in terms of rigid plates moving over Earth's surface. He characterized three basic types of boundaries where plates move apart, come together, or slide past each other. Soon after, other scientists showed that almost all contemporary tectonic deformation—the process by which rocks are folded, faulted, sheared, or compressed by tectonic forces—is concentrated at these boundaries. They measured the rates and directions of crustal movements and demonstrated that these movements are mathematically consistent with a system of rigid plates moving over the planet's spherical surface. The basic elements of the new theory of plate tectonics were established by the end of 1968. By 1970, the evidence for plate tectonics had become so persuasive that almost all Earth scientists embraced the theory. Textbooks were revised, and specialists began to consider the implications of the new concept for their own fields.

THE PLATES AND THEIR BOUNDARIES

According to the theory of plate tectonics, the rigid lithosphere is not a continuous shell but is broken into a mosaic of 13 large, rigid plates that move over Earth's surface (**Figure 3.7**). Each plate travels as a distinct unit, riding on the asthenosphere, which is also in motion. The largest is the Pacific Plate, which comprises much (though not all) of the Pacific Ocean basin. Some of the plates are named after the continents they include, but in no case is a plate identical with a continent. The North American Plate, for instance, extends from the Pacific coast of North America to the middle of the Atlantic Ocean, where it meets the Eurasian and African plates.

In addition to the major plates, there are a number of smaller ones. An example is the Juan de Fuca Plate, a small piece of oceanic lithosphere trapped between the giant Pacific and North American plates just offshore of the

(*Continued on page 70.*)

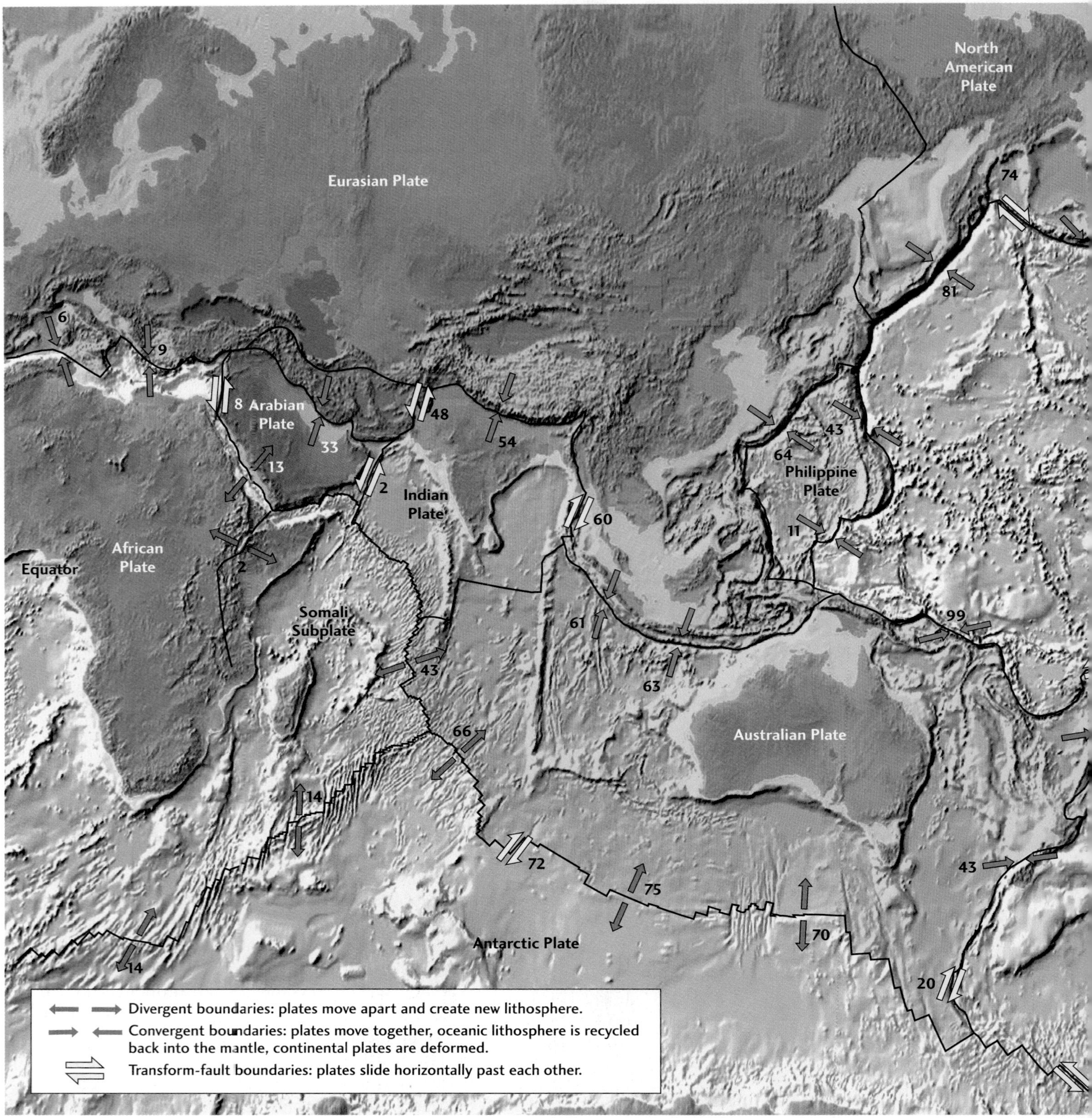

FIGURE 3.7 Earth's surface is a mosaic of 13 major plates of rigid lithosphere, as well as a number of smaller plates, that move slowly over the ductile asthenosphere. Only one of the smaller plates, the Juan de Fuca Plate, off the west coast of North America, is shown on this map. The arrows show the relative motion of two plates at a point on their boundary. The numbers next to the arrows give the relative plate speeds in millimeters per year. [*Plate boundaries by Peter Bird, UCLA.*]

Eurasian Plate

North American Plate

Juan de Fuca Plate

African Plate

Caribbean Plate

Cocos Plate

Pacific Plate

Nazca Plate

South American Plate

Antarctic Plate

18

47

63

73

34

50

22

24

11

64

89

12

27

118

31

50

138

73

84

79

72

150

35

55

80

92

34

81

18

18

50

64

18

DIVERGENT BOUNDARIES

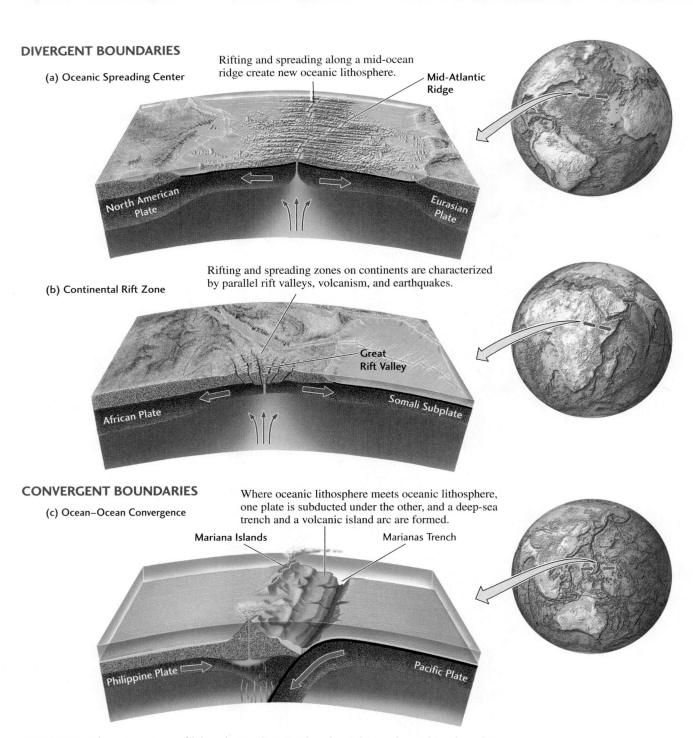

(a) Oceanic Spreading Center

Rifting and spreading along a mid-ocean ridge create new oceanic lithosphere.

Mid-Atlantic Ridge

North American Plate

Eurasian Plate

(b) Continental Rift Zone

Rifting and spreading zones on continents are characterized by parallel rift valleys, volcanism, and earthquakes.

Great Rift Valley

African Plate

Somali Subplate

CONVERGENT BOUNDARIES

(c) Ocean–Ocean Convergence

Where oceanic lithosphere meets oceanic lithosphere, one plate is subducted under the other, and a deep-sea trench and a volcanic island arc are formed.

Mariana Islands

Marianas Trench

Philippine Plate

Pacific Plate

FIGURE 3.8 The interactions of lithospheric plates at their boundaries depend on the relative direction of plate movement and the type of lithosphere involved.

The lithosphere is broken into 13 large plates plus a number of smaller ones, all in constant motion.

northwestern United States. Others are continental fragments, such as the small Anatolian Plate, which includes much of Turkey.

To see plate tectonics in action, go to a plate boundary. Depending on which boundary you visit, you may find earthquakes, volcanoes, rising mountains, long, narrow rifts, folding, or faulting. Many

geologic features develop through the interactions of plates at their boundaries.

There are three basic types of plate boundaries (**Figure 3.8**), all defined by the direction of movement of the plates relative to each other:

■ At **divergent boundaries,** plates move apart and new lithosphere is created (plate area increases).

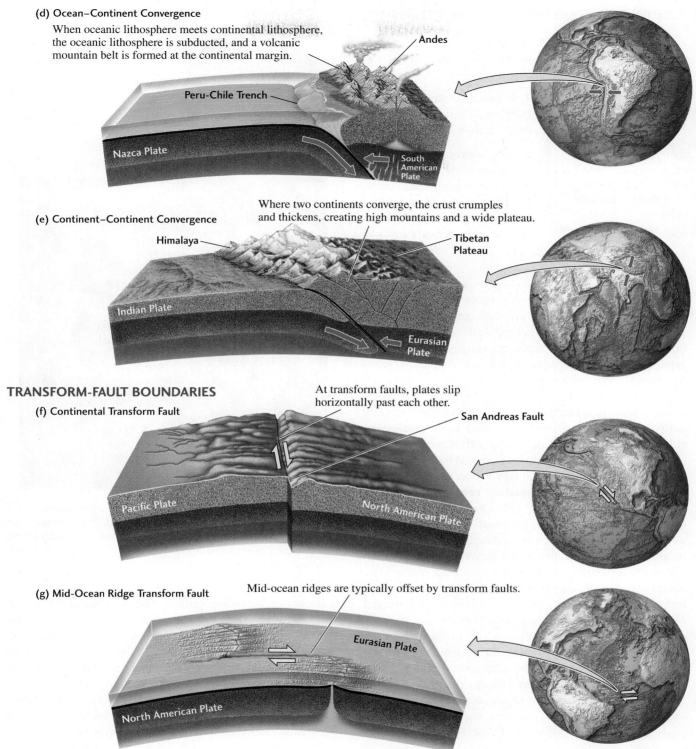

(d) Ocean–Continent Convergence

When oceanic lithosphere meets continental lithosphere, the oceanic lithosphere is subducted, and a volcanic mountain belt is formed at the continental margin.

Andes

Peru-Chile Trench

Nazca Plate

South American Plate

(e) Continent–Continent Convergence

Where two continents converge, the crust crumples and thickens, creating high mountains and a wide plateau.

Himalaya

Tibetan Plateau

Indian Plate

Eurasian Plate

TRANSFORM-FAULT BOUNDARIES

(f) Continental Transform Fault

At transform faults, plates slip horizontally past each other.

San Andreas Fault

Pacific Plate

North American Plate

(g) Mid-Ocean Ridge Transform Fault

Mid-ocean ridges are typically offset by transform faults.

Eurasian Plate

North American Plate

■ At **convergent boundaries,** plates come together and one plate is recycled into the mantle (plate area decreases).

■ At **transform faults,** plates slide horizontally past each other (plate area does not change).

Like many models of nature, these three basic types are idealized. In addition, there are "oblique" boundaries that combine divergence or convergence with some amount of transform faulting. Moreover, what actually goes on at a plate boundary depends on the type of lithosphere involved, because continental and oceanic lithosphere behave differently. The continental crust is made of rocks that are both lighter and weaker than the rocks in either the oceanic crust or the mantle beneath the crust (see Figure 2.12). Later chapters will examine these differences in more detail; for now, you need to keep in mind only two of their consequences:

> There are three basic types of plate boundaries: divergent boundaries, where plates are moving apart; convergent boundaries, where plates are moving together; and transform faults, where plates are sliding past each other.

1. Because continental crust is lighter, it is not as easily recycled as oceanic crust.

2. Because continental crust is weaker, plate boundaries that involve continental crust tend to be more spread out and more complicated than oceanic plate boundaries.

Divergent Boundaries

At divergent boundaries, plates move apart. Divergent boundaries within ocean basins are narrow rifts that approximate the idealization of plate tectonics. Divergent boundaries within continents are usually more complicated and distributed over a wider area. This difference is illustrated in Figures 3.8a and b.

Oceanic Spreading Centers On the seafloor, the boundary between separating plates is marked by a **mid-ocean ridge,** an undersea mountain chain that exhibits earthquakes, volcanism, and rifting, all caused by the tensional (stretching) forces of mantle convection that are pulling the two plates apart. At **spreading centers,** the seafloor spreads as hot molten rock, called *magma*, wells up into the rifts to form new oceanic crust. Figure 3.8a shows what happens at one such spreading center on the Mid-Atlantic Ridge, where the North American and Eurasian plates are separating. (A more detailed map of the Mid-Atlantic Ridge is shown in Figure 3.4.) The island of Iceland exposes a segment of the otherwise submerged Mid-Atlantic Ridge, allowing geologists to view the processes of plate separation and seafloor spreading directly (**Figure 3.9**). The Mid-Atlantic Ridge continues in the Arctic Ocean north of Iceland and connects to a nearly globe-encircling system of mid-ocean ridges that winds through the Indian and Pacific oceans, ending along the western coast of North America. These spreading centers have created the millions of square kilometers of oceanic crust that now form the floors of the world's oceans.

> At spreading centers on mid-ocean ridges, plates separate and new oceanic crust is formed.

Continental Rifting Early stages of plate separation, such as the Great Rift Valley of East Africa (see Figure 3.8b), can be found on some continents. These divergent boundaries are characterized by rift valleys, volcanism, and earthquakes distributed over a wider zone than is found

Iceland

FIGURE 3.9 The Mid-Atlantic Ridge, a divergent plate boundary, rises above sea level in Iceland. This cracklike rift valley, filled with newly formed volcanic rocks, indicates that plates are being pulled apart. [*Gudmundur E. Sigvaldason, Nordic Volcanological Institute.*]

at oceanic spreading centers. The Red Sea and the Gulf of California are rifts that are further along in the spreading process (**Figure 3.10**). In these cases, the continents have separated enough for new seafloor to form along the spreading axis, and the ocean has flooded the rift valleys.

Sometimes continental rifting slows or stops before the continent actually splits apart. The Rhine Valley, along the border of Germany and France in western Europe, is a weakly active continental rift that may be this type of "failed" spreading center. Will the East African Rift continue to open, causing the Somali Subplate to split away from Africa completely and form a new ocean basin, as happened between Africa and the island of Madagascar? Or will the spreading slow and eventually stop, as appears to be happening in western Europe? Geologists don't know the answers.

> Where continents split apart, rift valleys, volcanic activity, and earthquakes are distributed over a wider zone than at oceanic spreading centers.

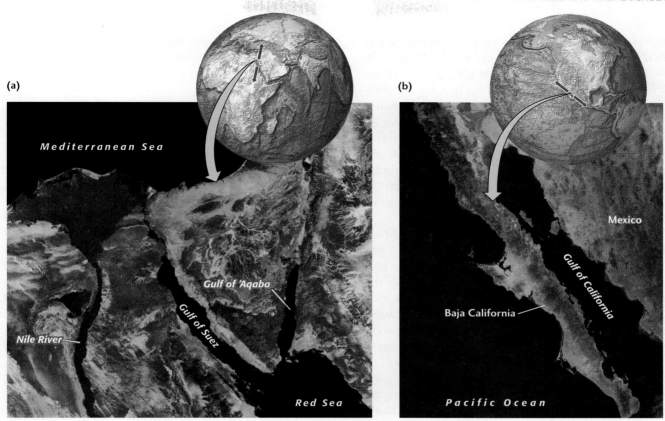

FIGURE 3.10 Rifting of continental crust. (a) The Arabian Plate, on the right, is moving northeastward relative to the African Plate, on the left, opening the Red Sea (lower right). The Gulf of Suez is a failed rift that became inactive about 5 million years ago. North of the Red Sea, most of the plate motion is now taken up by rifting and transform faulting along the Gulf of 'Aqaba and its northward extension. (b) Baja California, on the Pacific Plate, is moving northwestward relative to the North American Plate, opening the Gulf of California between Baja and the Mexican mainland. [*(a) Earth Satellite Corporation. (b) Jeff Schmaltz, MODIS Rapid Response Team, NASA/GSFC.*]

Convergent Boundaries

Lithospheric plates cover the globe, so if they separate in one place, they must converge somewhere else, if Earth's surface area is to remain the same. (As far as we can tell, our planet is not expanding!) Thus, where plates come together, they form convergent boundaries. The variety of geologic events resulting from plate convergence makes these boundaries the most complex type observed.

Ocean-Ocean Convergence If the lithosphere of both converging plates is oceanic, one plate descends beneath the other in a process known as **subduction** (see Figure 3.8c). The lithosphere of the subducting plate sinks into the asthenosphere and is eventually recycled by the mantle convection system. This sinking produces a long, narrow deep-sea trench. In the Marianas Trench of the western Pacific, the ocean reaches its greatest depth, about 11 km—deeper than the height of Mount Everest.

As the cold slab of lithosphere descends deeper into Earth's interior, the pressure on it increases. Water trapped in the rocks is squeezed out and rises into the asthenosphere above the slab. This fluid causes the mantle material above it to melt. The resulting magma produces a chain of volcanoes, called an **island arc**, behind the trench. The subduction of the Pacific Plate has formed the volcanically active Aleutian Islands west of Alaska as well as the Mariana Island and other island arcs in the western Pacific. The lithospheric slabs descending into the mantle cause earthquakes as deep as 690 km beneath some island arcs.

Where two oceanic plates converge, one plate is subducted beneath the other and sinks into the asthenosphere.

Ocean-Continent Convergence If one plate has a continental edge, it overrides the oceanic crust of the other plate, because continental crust is lighter and less easily subducted (see Figure 3.8d). The submerged margin of the continent is crumpled by the compressive (squeezing) forces of convergence, which can cause huge earthquakes that generate tsunamis. The resulting deformation of the continental crust uplifts rocks into a mountain

> *Continental crust overrides oceanic crust at a convergent boundary because it is lighter and less easily subducted.*

chain roughly parallel to the deep-sea trench. Over time, materials are scraped off the descending slab and incorporated into the adjacent mountain belt, leaving geologists with a complex (and often confusing) record of the subduction process. As in the case of ocean-ocean convergence, the water carried downward by the subducting oceanic lithosphere causes mantle material to melt and forms volcanoes in the mountain belt behind the trench.

The west coast of South America, where the South American Plate converges with the oceanic Nazca Plate, is a subduction zone of this type. A great chain of high mountains, the Andes, rises on the continental side of this convergent boundary, and a deep-sea trench lies just off the coast. The volcanoes here are active and deadly. One of them, Nevado del Ruiz in Colombia, killed 25,000 people when it erupted in 1985. Some of the world's greatest earthquakes have been recorded along this boundary.

Another example is the Cascadia subduction zone, where the small Juan de Fuca Plate converges with the North American Plate off the coast of western North America. This convergent boundary gives rise to the dangerous volcanoes of the Cascade Range, such as Mount St. Helens, which had a major eruption in 1980 and a minor one in 2004. There is increasing worry that a large earthquake will occur in the Cascadia subduction zone along the coasts of Oregon, Washington, and British Columbia. Such an earthquake could generate a disastrous tsunami like the one caused by the great Tohoku earthquake of March 11, 2011, off the northeast coast of Honshu, Japan.

Continent-Continent Convergence Where two continents converge (see Figure 3.8e), the kind of subduction seen at other convergent plate boundaries cannot occur. The geologic consequences of such a continent-continent collision are impressive. The collision of the Indian and Eurasian plates, both with continents at their leading edges, provides the best example. The Eurasian Plate overrides the Indian Plate, but India and Asia remain afloat on the mantle. The collision creates a double thickness of crust, forming the highest mountain range in the world, the Himalaya, as well as the vast high Tibetan Plateau. Severe earthquakes occur in the crumpling crust of this and other continent-continent collision zones.

Many episodes of mountain building throughout Earth's history were caused by continent-continent collisions. The Appalachian Mountains, which run along the eastern coast of North America, were uplifted when

> *Where two continents converge, the collision deforms and thickens the crust, raising high mountains.*

North America, Eurasia, and Africa collided to form the supercontinent Pangaea about 300 million years ago.

Transform Faults

At boundaries where plates slide past each other, lithosphere is neither created nor destroyed. Such boundaries are transform faults: fractures along which the plates slip horizontally past each other (see Figures 3.8f, g).

The San Andreas fault in California, where the Pacific Plate slides past the North American Plate, is a prime example of a continental transform fault. Because the plates have been sliding past each other for millions of years, the rocks facing each other on the two sides of the fault are of different types and ages (**Figure 3.11**). Large earthquakes, such as the one that destroyed San Francisco in 1906, can occur on transform faults. There is much concern that, within the next several decades, a sudden rupture of the San Andreas fault or of related faults near Los Angeles and San Francisco will result in an extremely destructive earthquake.

Transform-fault boundaries are typically found along mid-ocean ridges where the continuity of a spreading

FIGURE 3.11 A view southeast along the San Andreas fault in the Carrizo Plain of central California. The San Andreas is a transform fault, forming a portion of the sliding boundary between the Pacific Plate (*right*) and the North American Plate *left*). [*Kevin Schafer/Peter Arnold/Alamy.*]

zone is broken and the boundary is offset in a steplike pattern (see Figure 3.8g). An example can be found along the boundary between the African Plate and the South American Plate in the central Atlantic Ocean. Transform faults can also connect divergent plate boundaries with convergent plate boundaries and convergent boundaries with other convergent boundaries. Can you find examples of these types of transform-fault boundaries in Figure 3.7?

> *Plates slide past each other along transform faults.*

Combination of Plate Boundaries

Each plate is bordered by some combination of divergent, convergent, and transform-fault boundaries. For example, the Nazca Plate in the Pacific is bounded on three sides by spreading centers, offset in a steplike pattern by transform faults and on one side by the Peru-Chile subduction zone (see Figure 3.7). The North American Plate is bounded on the east by the Mid-Atlantic Ridge, a spreading center, and on the west by subduction zones and transform-fault boundaries.

Natural Hazards: The Pacific Ring of Fire

The intense geologic activity along plate boundaries can make them very dangerous places. To observe the close relationship between geologic hazards and plate movements, notice how the volcanoes marking the Pacific Ring of Fire (red dots in Figure 3.6) correspond to the convergent plate boundaries (blue arrows in Figure 3.7). Here, subducting oceanic lithosphere carries water-rich rocks beneath the overriding plates, causing the mantle to melt. The magmas rise toward the surface, where they can produce explosive eruptions like the one that blew the top off Mount St. Helens in 1980. Almost all the active volcanoes in the United States are located within the Ring of Fire. (Interesting exceptions are Hawaii's active volcanoes in the middle of the Pacific Plate, which are discussed later in this chapter.)

The Ring of Fire also contains much of the world's seismic activity. The great earthquake and tsunami that devastated the northeastern coast of Japan on March 11, 2011, were caused by a huge fault break along the boundary where the Pacific Plate subducts westward beneath the Japanese island of Honshu. Earthquakes in subduction zones can be exceptionally large because the faults that form the boundary between the two plates slope beneath the overriding plate at shallow angles, which makes them flatter and much wider than other plate-boundary faults

> *Geologic activity at plate boundaries is responsible for the geologic hazards most dangerous to human society.*

(see Figure 3.8). Like bigger springs, they can accumulate more energy before breaking.

Natural Resources: Mineral Concentrations at Plate Boundaries

Although volcanism on plate boundaries can be hazardous, it is also beneficial in providing society with valuable mineral resources. Melting of mantle rocks at spreading centers generates new oceanic crust, and melting in subduction zones produces new continental crust. Both types of melting concentrate metals into liquid magmas, which rise into the crust, where they interact with water near Earth's surface to form metal deposits rich enough to be mined.

Giant deposits of copper, silver, gold, and other metals have been found within the island arcs and continental mountain belts around the Pacific Ocean, and all of them are associated with the subduction-zone volcanism of the Ring of Fire. The discoveries include the Mother Lode of California's Sierra Nevada, which led to the Gold Rush of 1849, and the famous Comstock Lode of western Nevada.

Modern exploration for these resources now relies on principles of plate tectonics. In 1988, a giant deposit was discovered in Irian Jaya, Indonesia (**Figure 3.12**). The

FIGURE 3.12 Aerial view of the Grasberg copper and gold mine, the world's richest, which lies at an altitude of approximately 4100 meters (14,000 ft) in the Indonesian province of Papua. [*Rob Huibers/Hollandse Hoogte/Redux.*]

> *Volcanism on plate boundaries is an important process in creating mineral resources.*

Grasberg mine developed at this site is now the world's richest mine; its reserves have been estimated to contain more than 25 million tons of copper, 10,000 tons of silver, and 2,200 tons of gold!

RATES AND HISTORY OF PLATE MOVEMENTS

How fast do plates move? Do some plates move faster than others, and if so, why? Is the velocity of plate movements today the same as it was in the geologic past? Geologists have developed ingenious methods to answer these questions and thereby to gain a better understanding of plate tectonics. In this section, we will examine three of these methods.

The Seafloor as a Magnetic Tape Recorder

During World War II, extremely sensitive magnetometers were developed to detect submarines by the magnetic fields emanating from their steel hulls. Geologists modified these instruments slightly and towed them behind research ships to measure the local magnetic field created by magnetized rocks on the seafloor. Steaming back and forth across the ocean, seagoing scientists discovered regular patterns in the strength of the local magnetic field that completely surprised them. In many areas, the intensity of the magnetic field alternated between high and low values in long, narrow parallel bands, called **magnetic anomalies,** that were almost perfectly symmetrical with respect to the crest of a mid-ocean ridge (**Figure 3.13**). The detection of these patterns was one of the great discoveries that confirmed the seafloor spreading hypothesis and led to the theory of plate tectonics. It also allowed geologists to trace plate movements far back into geologic time. To understand these advances, we need to look more closely at how rocks become magnetized.

The Rock Record of Magnetic Reversals on Land Magnetic anomalies are evidence that Earth's magnetic field does not remain constant over time. At present, the north magnetic pole is closely aligned with the geographic north pole (see Figure 2.17), but small changes in the geodynamo can flip the orientation of the north and south magnetic poles by 180°, causing a *magnetic reversal.*

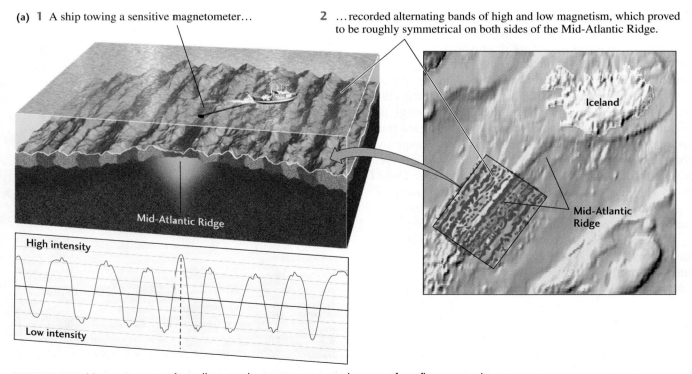

(a) **1** A ship towing a sensitive magnetometer…

2 …recorded alternating bands of high and low magnetism, which proved to be roughly symmetrical on both sides of the Mid-Atlantic Ridge.

Mid-Atlantic Ridge

High intensity

Low intensity

Iceland

Mid-Atlantic Ridge

FIGURE 3.13 Magnetic anomalies allow geologists to measure the rate of seafloor spreading. (a) An oceanographic survey over the Mid-Atlantic Ridge just southwest of Iceland revealed a banded pattern of magnetic field strength. (b) Geologists found and dated similar magnetic anomalies in volcanic lavas on land to construct a magnetic time scale. (c) That magnetic time scale was used to date the magnetic anomalies on the seafloor worldwide.

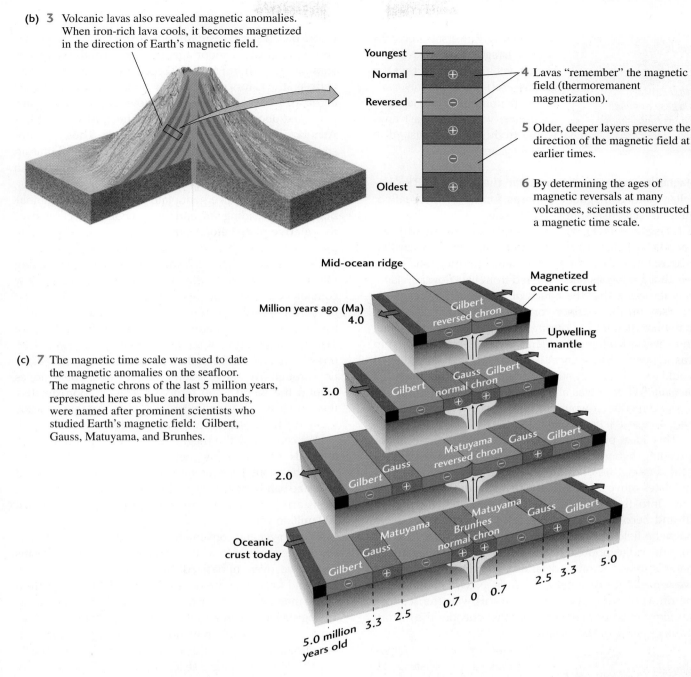

(b) **3** Volcanic lavas also revealed magnetic anomalies. When iron-rich lava cools, it becomes magnetized in the direction of Earth's magnetic field.

Youngest

Normal

Reversed

Oldest

4 Lavas "remember" the magnetic field (thermoremanent magnetization).

5 Older, deeper layers preserve the direction of the magnetic field at earlier times.

6 By determining the ages of magnetic reversals at many volcanoes, scientists constructed a magnetic time scale.

Mid-ocean ridge

Magnetized oceanic crust

Million years ago (Ma)
4.0

Gilbert reversed chron

Upwelling mantle

(c) **7** The magnetic time scale was used to date the magnetic anomalies on the seafloor. The magnetic chrons of the last 5 million years, represented here as blue and brown bands, were named after prominent scientists who studied Earth's magnetic field: Gilbert, Gauss, Matuyama, and Brunhes.

3.0

Gilbert

Gauss Gilbert
normal chron

2.0

Gilbert Gauss

Matuyama reversed chron

Gauss Gilbert

Oceanic crust today

Gilbert

Matuyama
Gauss

Brunhes normal chron

Matuyama Gauss Gilbert

5.0 million years old   3.3   2.5   0.7   0   0.7   2.5   3.3   5.0

In the early 1960s, geologists discovered that a precise record of this peculiar behavior can be obtained from layered flows of volcanic lava. When iron-rich lavas cool, they become slightly but permanently magnetized in the direction of Earth's magnetic field. This phenomenon is called *thermoremanent magnetization* because the rock "remembers" the magnetization long after the magnetic field has changed.

In layered lava flows, such as those in a volcanic cone, the rocks at the top represent the most recent layer, while layers deeper in the cone are older. The age of each layer can be determined by various dating methods (described in Chapter 8). The direction of magnetization of rock samples from each layer then reveals the direction of Earth's magnetic field at the time when that layer cooled (Figure 3.13b). By repeating these measurements at hundreds of places around the world, geologists have worked out the detailed history of magnetic reversals going back in geologic time. The **magnetic time scale** of the past 5 million years is shown in Figure 3.13c.

About half of all volcanic rocks studied have been found to be magnetized in a direction opposite to that of Earth's present magnetic field. Apparently, the field has flipped frequently over geologic time, so normal fields (same as now) and reversed fields (opposite to now) are equally likely. Major periods during which the field is normal or reversed are called *magnetic chrons*; they last about half a million years, on average, although

> *By measuring the magnetization of volcanic lavas, geologists have determined that Earth's magnetic field has reversed itself many times throughout Earth's history.*

the pattern of reversals becomes highly irregular as we move back in geologic time. Within the chrons are short-lived reversals of the field, known as magnetic *subchrons,* which may last anywhere from several thousand to 200,000 years.

Magnetic Anomaly Patterns on the Seafloor The peculiar banded magnetic patterns found on the seafloor puzzled scientists until 1963, when two Englishmen, F. J. Vine and D. H. Mathews—and, independently, two Canadians, L. Morley and A. Larochelle—made a startling proposal. Based on the new evidence for magnetic reversals that geologists had collected from lava flows on land, they reasoned that the bands of high and low magnetic intensity on the seafloor corresponded to bands of rock that were magnetized during ancient episodes of normal and reversed magnetism. That is, when a research ship was above rocks magnetized in the normal direction, it would record a locally stronger field, or a *positive magnetic anomaly.* When it was above rocks magnetized in the reversed direction, it would record a locally weaker field, or a *negative magnetic anomaly.*

This idea provided a powerful test of the seafloor spreading hypothesis, which states that new seafloor is created along the rift at the crest of a mid-ocean ridge as the plates move apart. Magma rising from Earth's interior flows into the rift, where it cools and solidifies in the rift and becomes magnetized in the direction of Earth's magnetic field at the time. As the seafloor spreads away from the ridge, approximately half the newly magnetized material moves to one side and half to the other, forming two symmetrical magnetized bands. Newer material fills the rift, continuing the process. In this way, the seafloor acts like a magnetic tape recorder that encodes the history of the opening of the oceans.

> *Bands of high and low magnetic intensity on the seafloor provided evidence that new seafloor is continually being created at and spreading away from mid-ocean ridges.*

The seafloor spreading hypothesis provides a consistent explanation for the symmetrical patterns of magnetic anomalies found at mid-ocean ridges around the world. Moreover, it gives us a precise tool for measuring rates of seafloor spreading now and in the geologic past.

Inferring Seafloor Ages and Relative Plate Velocities By using the ages of magnetic reversals that had been worked out from magnetized lavas on land, geologists were able to assign ages to the bands of magnetized rocks on the seafloor. They could calculate how fast the seafloor was spreading by using the formula *speed = distance / time,*

where distance is measured from the mid-ocean ridge axis and time equals seafloor age. For instance, the magnetic anomaly pattern in Figure 3.12c shows that the boundary between the Gauss normal chron and the Gilbert reversed chron, which was dated from lava flows at 3.3 million years, is located about 30 km away from the crest of the Mid-Atlantic Ridge just southwest of Iceland. Thus, seafloor spreading has moved the North American and Eurasian plates apart by about 60 km in 3.3 million years, giving a spreading rate of 18 km per million years or, equivalently, 18 mm/year. On a divergent plate boundary, the combination of the spreading rate and the spreading direction gives the **relative plate velocity,** the velocity at which one plate moves relative to the other.

If you look at Figure 3.7, you will see that the spreading rate at the Mid-Atlantic Ridge south of Iceland is fairly low compared with the rate at many other places. The speed record for spreading can be found on the East Pacific Rise just south of the equator, where the Pacific and Nazca plates are separating at a rate of about 150 mm/year—an order of magnitude faster than the rate in the North Atlantic. A rough average spreading rate for mid-ocean ridges around the world is 50 mm/year. This is approximately the rate at which your fingernails grow—so, geologically speaking, the plates move very fast indeed!

We can follow the magnetic time scale through many reversals of Earth's magnetic field. The corresponding magnetic anomalies on the seafloor, which can be thought of as age bands, have been mapped in detail from the ridge crests across the ocean basins over a time span of almost 200 million years.

The power and convenience of using magnetic anomalies on the seafloor to work out the history of ocean basins cannot be overemphasized. Simply by steaming a ship back and forth over the ocean and correlating the pattern of the magnetic anomalies with the magnetic time scales, geologists have been able to determine the ages of various regions of the seafloor without directly examining rock samples. In effect, they learned how to "replay the tape."

Although measuring seafloor magnetization is a very effective technique, it is an indirect, or remote, sensing method in that rocks are not recovered from the seafloor and their ages are not directly determined in the laboratory. The few geologists who remained skeptical of plate tectonics demanded direct evidence of seafloor spreading and plate movement. Deep-sea drilling supplied it.

Deep-Sea Drilling

In 1968, a program of drilling into the seafloor was launched as a joint project of several major oceanographic institutions and the National Science Foundation. Later, many nations joined the effort (**Figure 3.14**). Using hollow drills, scientists brought up cores containing sections of

FIGURE 3.14 The deep-sea drilling vessel *JOIDES Resolution* is 143 m long and carries a drilling derrick 61 m high that is capable of drilling into the seafloor beneath the deepest ocean. Samples from cores of sediment recovered by drilling have confirmed the age of the seafloor deduced from magnetic anomalies. Such samples have also shed new light on the history of ocean basins and ancient climatic conditions. [*Courtesy of Ocean Drilling Program/Texas A&M University.*]

seafloor rocks from many locations in the oceans. In some cases, the drilling penetrated thousands of meters below the seafloor surface. This program gave geologists an opportunity to work out the history of the ocean basins from direct evidence.

One of the most important facts geologists sought was the age of each rock sample. Small particles falling through the ocean waters—dust from the atmosphere, organic material from marine plants and animals—begin to accumulate as seafloor sediments as soon as new oceanic crust forms. Therefore, the age of the oldest sediments in a core—those immediately on top of the crust—tells the geologist how old the seafloor is at that spot. The ages of sediments can be obtained primarily from the fossil skeletons of tiny single-celled planktonic organisms that live at the surface of the open ocean and sink to the bottom when they die. Geologists found that the ages of the sediments in the cores increased with distance from mid-ocean ridges and that the age of the seafloor at any one place agreed almost perfectly with the age determined from magnetic reversal data. This agreement validated the magnetic time scale and provided strong evidence for seafloor spreading.

> *Direct measurements of rock age obtained by drilling into the seafloor validated the magnetic time scale and confirmed the hypothesis of seafloor spreading.*

GPS Measurements of Plate Movements

In his publications advocating the continental drift hypothesis, Alfred Wegener made a big mistake: he proposed that North America and Europe were drifting apart at a rate of nearly 30 meters per year—a thousand times faster than the North Atlantic seafloor is actually spreading! This unbelievably high speed was one of the reasons that many scientists roundly rejected his notions of continental drift. Wegener made his estimate by incorrectly assuming that the continents were joined together as Pangaea as recently as the last ice age (which occurred only about 20,000 years ago).

Wegener imagined that he could measure continental drift in the following way: Two observers, one in Europe and the other in North America, would simultaneously determine their positions relative to the fixed stars. From those positions, they would calculate the distance between their two observation posts at that instant. They would then repeat this distance measurement from the same observation posts sometime later—say, after 1 year. If the continents were drifting apart, then the distance would increase, and the value of the increase would determine the speed of the drift. Unfortunately, in Wegener's day, the accuracy of astronomical positioning was poor. We now know that the spreading of the Mid-Atlantic Ridge from one year to the next is only a couple of centimeters, a thousand times too small to be observed by the techniques that were then available.

Since the mid-1980s, geologists have used a constellation of 24 Earth-orbiting satellites, called the Global Positioning System (GPS), to make much more accurate positioning measurements, and they can now track the movements of the continents using inexpensive portable radio receivers

> *The Global Positioning System provides geologists with a way to observe plate movements as they are occurring.*

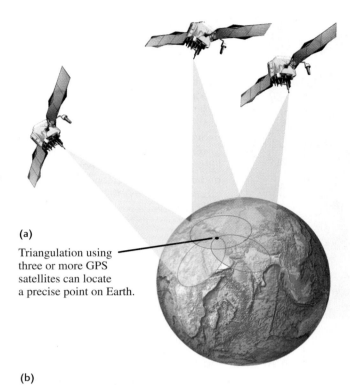

(a)

Triangulation using three or more GPS satellites can locate a precise point on Earth.

(b)

A GPS station

FIGURE 3.15 The Global Positioning System allows geologists to monitor plate movements. (a) GPS satellites provide a fixed frame of reference outside Earth. (b) Small GPS receivers can be easily placed anywhere on Earth. Displacements of receiver locations over a period of years can be used to measure plate movements. [*Photo courtesy of Southern California Earthquake Center.*]

smaller than this textbook (**Figure 3.15**). This technology is similar to the GPS systems that are now used in automobiles and by hikers, though much more precise.

The changes in distance between land-based GPS receivers recorded over several years agree in both magnitude and direction with the changes in distance calculated from magnetic anomalies on the seafloor, indicating that plate movements are remarkably steady over periods ranging from just a few years up to millions of years. Geologists are now using GPS to measure plate movements on a regular basis at many locations around the globe.

THE GRAND RECONSTRUCTION

The supercontinent Pangaea was the only major landmass that existed 250 million years ago. One of the great triumphs of modern geology is the reconstruction of the events that led to the assembly of Pangaea and to its later fragmentation into the continents we know today. Let's use what we have learned about plate tectonics to see how this feat was accomplished.

Seafloor Isochrons

The color map in **Figure 3.16** shows the ages of the rocks on the seafloor as determined by magnetic anomaly data and fossils from deep-sea drilling. Each colored band represents the span of time when the rocks within that band formed. Notice how the seafloor becomes progressively older on both sides of the mid-ocean ridges. The boundaries between bands, called **isochrons,** are contours that connect rocks of equal age.

Isochrons tell us the time that has elapsed since the rocks were injected as magma into a spreading zone and, therefore, the amount of spreading that has occurred since they formed. For example, the distance from a ridge axis to a 140-million-year isochron (boundary between green and blue bands) indicates the extent of new seafloor created over that time span. The more widely spaced isochrons of the eastern Pacific signify faster spreading rates than those in the Atlantic.

In 1990, after a 20-year search, geologists found the oldest oceanic rocks by drilling into the seafloor of the western Pacific. These rocks turned out to be about 200 million years old, only about 4 percent of Earth's age. This finding indicates how geologically young the seafloor is compared with the continents. Over a period of 100 million to 200 million years in some places and only tens of millions of years in others, oceanic lithosphere forms by seafloor spreading, cools, and is recycled into the underlying mantle. In contrast, the oldest known continental rocks are over 4 billion years old.

Seafloor isochrons have been used to reconstruct the movements of plates since the breakup of the supercontinent Pangaea.

Reconstructing the History of Plate Movements

Earth's plates behave as rigid bodies. That is, the distances between three points on the same rigid plate—say,

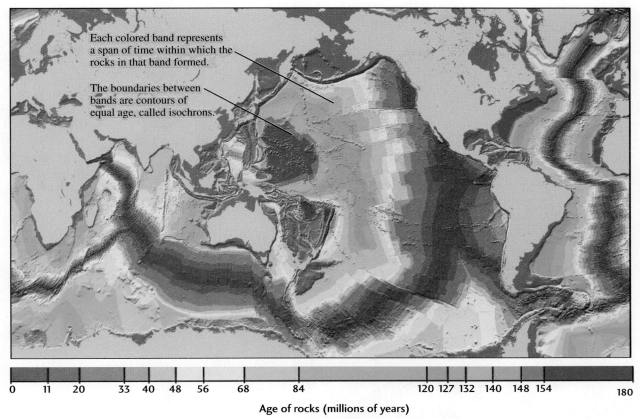

Each colored band represents a span of time within which the rocks in that band formed.

The boundaries between bands are contours of equal age, called isochrons.

Age of rocks (millions of years)

0　11　20　33　40　48　56　68　84　120 127 132 140 148 154　180

FIGURE 3.16 This global isochron map shows the ages of rocks on the seafloor. The time scale at the bottom gives the age of the seafloor in millions of years since its creation at mid-ocean ridges. Light gray indicates land; dark gray indicates shallow water over continental shelves. Mid-ocean ridges, along which new seafloor is extruded, coincide with the youngest rocks (red). [*Journal of Geophysical Research* 102 (1997): 3211–3214. Courtesy of R. Dietmar Müller.]

New York, Miami, and Bermuda on the North American Plate—do not change very much, no matter how far the plate moves. But the distance between, say, New York and Lisbon increases over time because those two cities are on two different plates that are separating along the Mid-Atlantic Ridge. The direction of the movement of one plate in relation to another depends on two geometric principles that govern the behavior of rigid plates on a sphere:

- *Transform-fault boundaries indicate the directions of relative plate movement.* With few exceptions, no overlap, buckling, or separation occurs along typical transform-fault boundaries in the oceans. The two plates merely slide past each other without creating or destroying plate material. Look for a transform-fault boundary if you want to deduce the directions of relative plate movement, because the orientation of the fault is the direction in which one plate slides with respect to the other (see Figures 3.8f, g).

- *Seafloor isochrons reveal the positions of divergent boundaries in earlier times.* Isochrons on the seafloor are

roughly parallel and symmetrical to the ridge axis along which they were created (see Figure 3.16). Because each isochron was at the divergent boundary at an earlier time, isochrons that are of the same age but on opposite sides of a mid-ocean ridge can be brought together to show the positions of the plates and the configuration of the continents embedded in them as they were in that earlier time.

The Breakup of Pangaea

Using these principles, geologists have reconstructed the opening of the Atlantic Ocean and the breakup of Pangaea (**Figure 3.17**). Figure 3.17e shows the supercontinent Pangaea as it existed 240 million years ago. It began to break apart when North America rifted away from Europe about 200 million years ago (Figure 3.17f). The opening of the North Atlantic was accompanied by the separation of the northern continents (referred to as Laurasia) from the southern continents (Gondwana) and the rifting of Gondwana along what is now the eastern coast of

(Continued on page 84.)

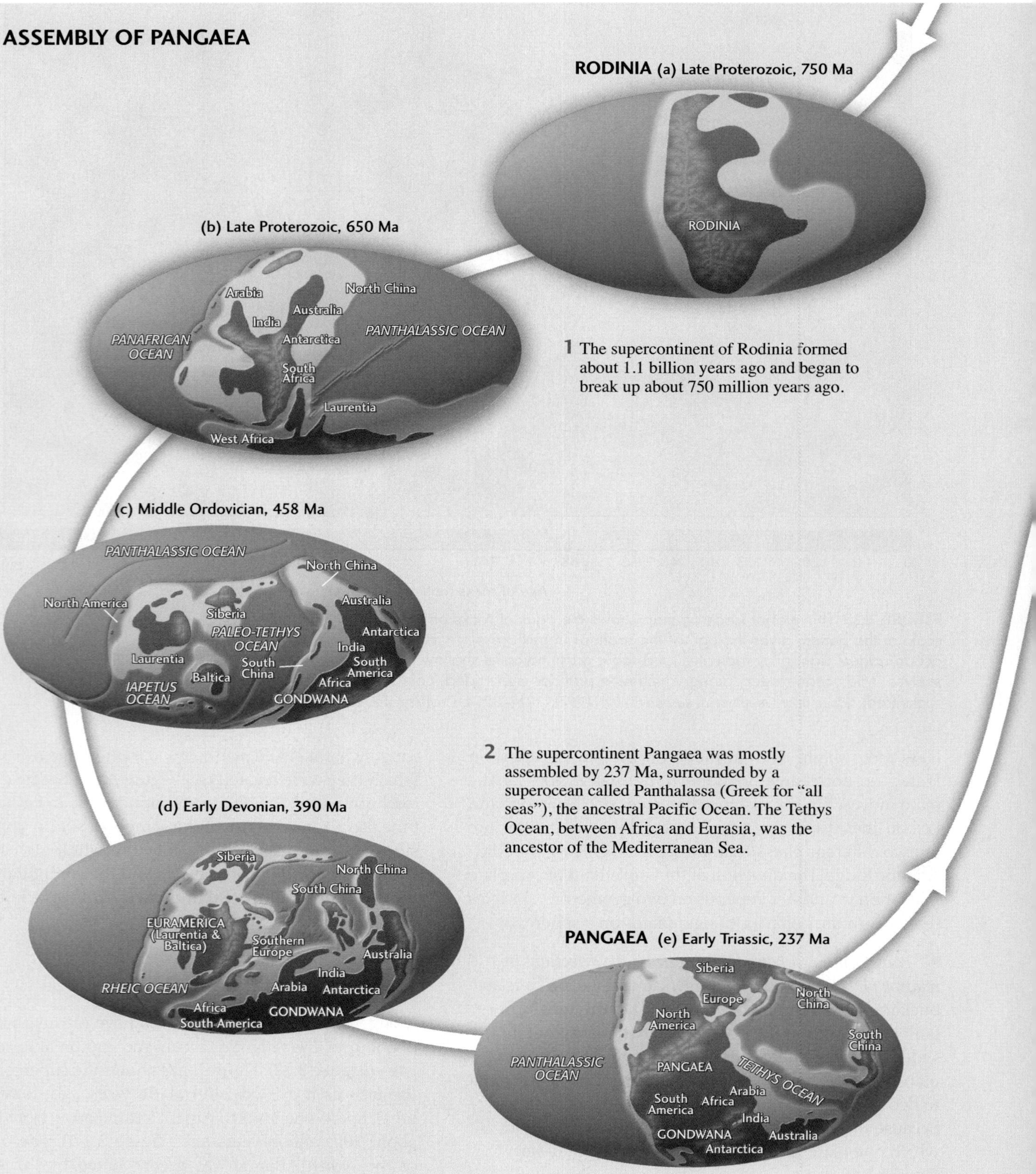

ASSEMBLY OF PANGAEA

RODINIA (a) Late Proterozoic, 750 Ma

RODINIA

(b) Late Proterozoic, 650 Ma

Arabia
North China
Australia
India
PANAFRICAN
OCEAN
Antarctica
PANTHALASSIC OCEAN
South
Africa
Laurentia
West Africa

1 The supercontinent of Rodinia formed about 1.1 billion years ago and began to break up about 750 million years ago.

(c) Middle Ordovician, 458 Ma

PANTHALASSIC OCEAN
North China
North America
Australia
Siberia
Antarctica
PALEO-TETHYS
OCEAN
India
Laurentia
South
China
South
America
IAPETUS
Baltica
Africa
OCEAN
GONDWANA

2 The supercontinent Pangaea was mostly assembled by 237 Ma, surrounded by a superocean called Panthalassa (Greek for "all seas"), the ancestral Pacific Ocean. The Tethys Ocean, between Africa and Eurasia, was the ancestor of the Mediterranean Sea.

(d) Early Devonian, 390 Ma

Siberia
North China
South China
EURAMERICA
(Laurentia &
Baltica)
Southern
Europe
Australia
India
RHEIC OCEAN
Arabia
Antarctica
Africa
South America
GONDWANA

PANGAEA (e) Early Triassic, 237 Ma

Siberia
Europe
North
China
North
America
South
China
PANTHALASSIC
OCEAN
PANGAEA
TETHYS OCEAN
South
America
Africa
Arabia
India
GONDWANA
Australia
Antarctica

FIGURE 3.17 Continental rifting, drifting, and collisions assembled and then disassembled the supercontinent Pangaea. [*Paleogeographic maps by Christopher R. Scotese, 2003 PALEOMAP Project (www.scotese.com).*]

BREAKUP OF PANGAEA

(f) Early Jurassic, 195 Ma

3 The breakup of Pangaea was signaled by the opening of rifts from which lava poured. Rock assemblages that are relics of this great event can be found today in 200-million-year-old volcanic rocks from Nova Scotia to North Carolina.

(g) Late Jurassic, 152 Ma

4 By about 150 million years ago, Pangaea was in the early stages of breakup. The Atlantic Ocean had partially opened, the Tethys Ocean had contracted, and the northern continents (Laurasia) had all but split away from the southern continents (Gondwana). India, Antarctica, and Australia began to split away from Africa.

5 By 66 million years ago, the South Atlantic had opened and widened. India was well on its way northward toward Asia, and the Tethys Ocean was closing to form the Mediterranean.

(h) Late Cretaceous, 66 Ma

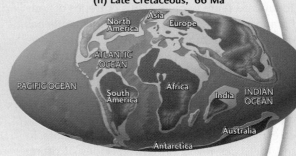

THE PRESENT-DAY AND FUTURE WORLD

6 The modern world has been produced over the past 65 million years. India collided with Asia, ending its trip across the ocean, and is still pushing northward into Asia. Australia has separated from Antarctica.

(i) PRESENT-DAY WORLD

(j) 50 million years in the future

Africa (Figure 3.17g). The breakup of Gondwana separated South America, Africa, India, and Antarctica, narrowing the Tethys Ocean and creating the South Atlantic and Southern Ocean (Figure 3.17h). The separation of Australia from Antarctica and the ramming of India into Eurasia closed the Tethys Ocean, giving us the world we see today (Figure 3.17i).

The plate movements have not ceased, of course, so the configuration of the continents will continue to evolve. A plausible scenario for the distribution of continents and plate boundaries 50 million years in the future is displayed in Figure 3.17j.

The Assembly of Pangaea

The isochron map in Figure 3.16 tells us that all of the seafloor on Earth's surface today has been created since the breakup of Pangaea. We know from the geologic record in older continental mountain belts, however, that plate tectonics had been operating for billions of years before this breakup. Evidently, seafloor spreading took place just as it does today, and there were previous episodes of continental drift and collision. Subduction into the mantle has destroyed the seafloor created in those earlier times, however, so we must rely on the older evidence preserved on continents to identify and chart the movements of ancient continents (*paleocontinents*).

Old mountain belts, such as the Appalachians of North America and the Urals, which separate Europe from Asia, help us locate ancient collisions of the paleocontinents. In many places, the rocks reveal ancient episodes of rifting and subduction. Rock types and fossils also indicate the distribution of ancient seas, glaciers, lowlands, mountains, and climates. Knowledge of ancient climates enables geologists to locate the latitudes at which continental rocks formed, which in turn helps them to assemble the jigsaw puzzle of paleocontinents. When volcanism or mountain building produces new continental rocks, these rocks also record the direction of Earth's magnetic field, just as oceanic rocks do when they are created by seafloor spreading. Like a compass frozen in time, the thermoremanent magnetization of a continental fragment records its ancient orientation and magnetic latitude.

The left side of Figure 3.17 shows one of the latest efforts to depict the pre-Pangaean configuration of continents. It is truly impressive that modern science can recover the geography of this strange world of hundreds of millions of years ago. The evidence from rock types, fossils, climate, and rock magnetization has allowed scientists to reconstruct an earlier supercontinent, called **Rodinia,** that formed about 1.1 billion years ago and began to break up about 750 million years ago (Figure 3.17a). They have been able to chart its fragments over the subsequent 500 million years as those fragments drifted and reassembled into the supercontinent Pangaea. Geologists continue to sort out the details of this complex jigsaw puzzle, whose individual pieces have changed shape over geologic time.

Natural Environment: How Drifting Continents Alter the Climate System

Movements of the tectonic plates change the locations of continents and oceans. Over millions of years, these changes have affected the climate system in profound ways. In the present arrangement, the waters of the Southern Ocean are able to circulate all the way around Antarctica, forming a "circumpolar seaway" that isolates the continent from the warmer water and air of tropical latitudes. This isolation keeps the southern polar regions colder than they might otherwise be, maintaining a massive ice sheet across the entire Antarctic continent.

The situation was rather different 66 million years ago, as shown in Figure 3.17h. Australia was still connected to Antarctica, allowing currents of warmer water to flow southward and heat the polar continent. Also at this time, the North and South American continents were separated, so water could flow between the Atlantic and Pacific oceans. The circumpolar seaway did not form until Australia broke away from Antarctica around 40 million years ago. Somewhat later, only about 5 million years ago, subduction in the eastern Pacific Ocean formed the Isthmus of Panama, connecting North and South America and isolating the Atlantic from the Pacific.

These changes, combined with the collision of India with Asia, which formed the high plateau of Tibet (see Figure 3.17i), cooled the entire planet enough to create ice sheets of Antarctica in the southern hemisphere and Greenland in the northern hemisphere. The resulting modification of the climate system is thought to have initiated oscillations of climate between very cold periods (ice ages) and somewhat warmer periods, such as the one we now enjoy.

> *The rearrangement of continents and oceans by plate tectonics explains many of the climate changes observed in the geologic record.*

MANTLE CONVECTION: THE ENGINE OF PLATE TECTONICS

Everything discussed in this chapter so far might be called descriptive plate tectonics. But a description is not an explanation. We need a comprehensive theory that explains *why* plates move.

As Arthur Holmes and other early advocates of continental drift realized, mantle convection is the "engine" that drives the large-scale tectonic processes operating on Earth's surface. In Chapter 2, we described the mantle as a moldable, or ductile, solid. Hot mantle material is capable of flowing like a sticky, or viscous, fluid. Heat escaping from Earth's deep interior causes this material to undergo convection (circulate upward and downward) at speeds of a few tens of millimeters per year. The details of the mantle convection system have not been completely worked out, but we can gain some insights by trying to answer three important questions.

Where Do the Plate-Driving Forces Originate?

The lithospheric plates are not just dragged along Earth's surface by the mantle convection currents; they are active participants in the flow. In convergence zones, the gravitational pull exerted by the cold (and thus heavy) slabs of old subducting lithosphere drags the plates downward into the mantle. This hypothesis explains why plates being subducted along a large fraction of their boundaries (e.g., the Pacific, Nazca, Cocos, and Indian plates) move faster, while those that do not have significant fractions of subducting slabs (e.g., the North American, South American, African, Eurasian, and Antarctic plates) move more slowly. In other words, the plates "fall back" into the mantle under their own weight.

There is also evidence that overriding plates, as well as subducting plates, are pulled toward convergent boundaries by gravitational forces. For example, as the Nazca Plate is subducted beneath South America, it causes the convergent boundary at the Peru-Chile Trench to retreat toward the Pacific, "sucking" the South American Plate to the west.

Other forces are evident from the history of plate movements. When the continents came together to form Pangaea, they acted as an insulating blanket, preventing heat from getting out of Earth's mantle (as it otherwise would through the process of seafloor spreading). The heat in the mantle built up over time, forming hot bulges in the mantle beneath the supercontinent. These bulges raised Pangaea slightly and caused it to rift apart in a kind of "landslide" off the top of the bulges. Gravitational forces continue to drive subsequent seafloor spreading as the plates "slide downhill" off the crest of the Mid-Atlantic Ridge. Earthquakes that sometimes occur in plate interiors provide direct evidence of the compression of plates by these ridge-related gravitational forces.

> *Oceanic lithosphere slides down and away from mid-ocean ridges and is pulled into subduction zones by its own weight.*

How Deep Does Plate Recycling Occur?

For plate tectonics to work, the lithospheric material that descends into the mantle at subduction zones must be recycled through the mantle and eventually return to the surface as new lithosphere created at spreading centers. How deep into the mantle does this recycling process extend? That is, where is the lower boundary of the mantle convection system?

The deepest the lower boundary could be is about 2900 km below Earth's surface, where a sharp boundary separates the mantle from the core (**Figure 3.18**). As we

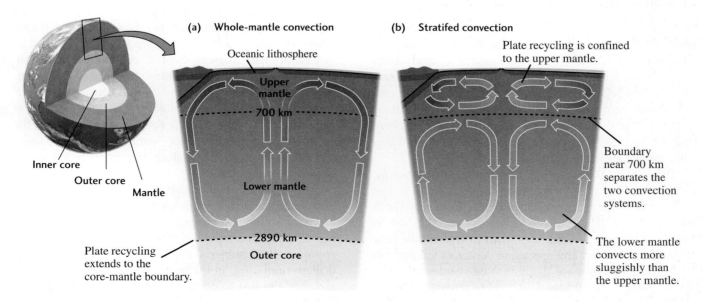

FIGURE 3.18 Two competing hypotheses on the extent of the mantle convection system.

saw in Chapter 2, the iron-rich liquid below this core-mantle boundary is much denser than the solid rock of the mantle, preventing any significant exchange of material between the two layers. We can thus imagine a system of *whole-mantle convection* in which the material from the plates circulates all the way through the mantle, down as far as the core-mantle boundary (Figure 3.18a).

However, some scientists think that the mantle is divided into two layers: an upper mantle system above 700 km, where the recycling of lithosphere takes place, and a lower mantle system from a depth of 700 km to the core-mantle boundary, where convection is much more sluggish. According to this hypothesis, called *stratified convection,* the separation of the two systems is maintained because the upper system consists of lighter rock than the lower system and thus floats on top, in the same way the mantle floats on the core (Figure 3.18b).

To test these two competing hypotheses, scientists have looked for "lithospheric graveyards" below convergent boundaries where plates have been subducted. Old subducted lithosphere is colder than the surrounding mantle and can therefore be "seen" using seismic waves. Sure enough, scientists have found regions of colder material in the deep mantle under North and South America, eastern Asia, and other sites adjacent to convergent boundaries. These zones occur as extensions of descending lithospheric slabs, and some appear to go down as far as the core-mantle boundary. From this evidence, most scientists have concluded that plate recycling takes place through whole-mantle convection rather than stratified convection.

What Is the Nature of Rising Convection Currents?

What about the rising currents of hot mantle material needed to balance subduction? Are there concentrated, sheetlike upwellings directly beneath the mid-ocean ridges? Most scientists who study the problem think not. Instead, they believe that the rising currents are slower and spread out over broader regions. This view is consistent with the idea that seafloor spreading is a rather passive process: pull the plates apart almost anywhere, and you will generate a spreading center.

There is one big exception, however: a type of narrow, jetlike upwelling called a *mantle plume* (**Figure 3.19**). The best evidence for mantle plumes comes from regions of intense, localized volcanism (called *hot spots*), such as Hawaii, where huge volcanoes form in the middle of a plate, far away from any spreading center or subduction zone. Mantle plumes are thought to be slender cylinders of fast-rising material, less than 100 km across, that come

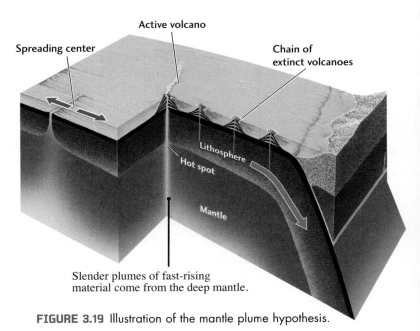

FIGURE 3.19 Illustration of the mantle plume hypothesis.

from the deep mantle. Mantle plumes are so intense that they can literally burn holes in the plates and erupt tremendous volumes of lava. Plumes may be responsible for outpourings of lava so massive that they may have changed Earth's climate and killed off many life-forms in mass extinction events (see Chapter 9). We will describe mantle plumes in more detail in Chapter 5.

THE THEORY OF PLATE TECTONICS AND THE SCIENTIFIC METHOD

In Chapter 1, we considered the scientific method and how it guides the work of geologists. In the context of the scientific method, plate tectonics is a confirmed theory whose strength lies in its simplicity, its generality, and its consistency with many types of observations. Theories can always be overturned or modified. As we have seen, competing hypotheses about how mantle convection generates plate tectonics have been advanced. But the theory of plate tectonics itself—like the theories of Earth's age, the evolution of life, and genetics—explains so much so well and has survived so many efforts to prove it false that geologists treat it as fact.

The question remains, Why wasn't plate tectonics discovered earlier? Why did it take the scientific establishment so long to move from skepticism about continental drift to acceptance of plate tectonics? Scientists approach their subjects differently. Scientists with particularly inquiring, uninhibited, and synthesizing minds are often the first to perceive great truths. Although their perceptions frequently turn out to be false (think of the mistakes Wegener made in proposing continental drift), these visionary people are often the first to see the great generalizations of science. Deservedly, they are the ones history remembers.

Most scientists, however, proceed more cautiously and wait out the slow process of gathering supporting evidence. Continental drift and seafloor spreading were slow to be accepted largely because these audacious ideas came far ahead of any firm evidence. Scientists had to explore the oceans, develop new instruments, and drill the seafloor before the majority could be convinced. Today, many scientists are still waiting for more evidence of how the mantle convection system really works.

GOOGLE EARTH PROJECT

This chapter focuses on the fundamental theory of plate tectonics and how the theory integrated previously independent geologic observations into a unified whole. To appreciate plate tectonics on a global scale, we'll begin by viewing Earth at an eye altitude of 11,000 km. Rotate the globe by using the virtual joystick in the upper right corner of the screen. Notice that this global view eliminates the distortion of a Mercator map at high latitudes and allows you to see the polar regions not shown on that type of map.

LOCATION   Antarctica, Mid-Atlantic Ridge, and South Pacific Ocean

GOAL   Investigate divergent plate boundaries

LINKED   Figures 3.1, 3.7, and 3.16

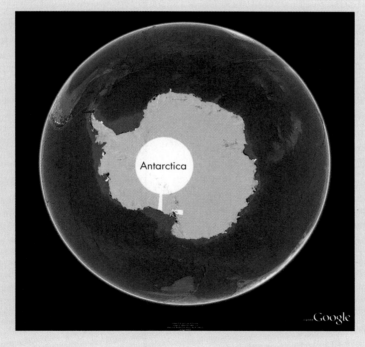

1. Navigate to the South Pole and view the continent of Antarctica. From an eye altitude of about 7000 km, examine the continent and the plate boundaries that surround it. Use Figure 2.7 to help you identify the types of plate boundaries around the continent. Based on those types, how is the surface area of the Antarctic Plate changing over time?

   a. The surface area of the plate is decreasing.
   b. The surface area of the plate is increasing.
   c. There is no net change in surface area of the plate.
   d. Not enough information is available to say how the surface area of the plate is changing over time.

(Continued.)

2. Navigate northward into the Atlantic Ocean basin. Find the conspicuous undersea mountain range that runs from south to north through the middle of the ocean basin—the Mid-Atlantic Ridge—and consider its relationship to the continents on both sides. Notice that the submerged edges of eastern South America and western Africa would fit nicely together—evidence of continental drift. As you move north, focus on the section of the ridge between 15°N and 30°N from an eye altitude of about 2200 km. You may want to activate the "grid" option from the "View" tab along the top of the GE browser to find this location easily. From your observations, what is the best description of the plate boundary along this portion of the mid-Atlantic Ridge?

   a. A continuous divergent boundary

   b. A continuous convergent boundary

   c. A steplike pattern of spreading centers separated by perpendicular transform faults

   d. A steplike pattern of subduction zones separated by perpendicular transform faults

3. Now that you have familiarized yourself with the Mid-Atlantic Ridge system, let's use Google Earth to measure the average speed of continental drift between North America and Africa. From the reconstruction of the supercontinent Pangaea in Figure 3.1, you can see that the North American margin just east of Charleston, South Carolina, once fit against the African margin just west of Dakar, Senegal. From the isochron map in Figure 3.16, you can estimate that the two continents began to rift apart about 200 million to 180 million years ago (see also Figure 3.17). Using the GE ruler tool to measure the ocean width at these locations, estimate the average rate at which the Atlantic has opened. What is that rate, and how does it compare with the present-day rate of continental drift derived from Figure 3.7?

   a. 5–10 mm/year, much slower than the present-day rate

   b. 15–20 mm/year, slower than the present-day rate

   c. 20–25 mm/year, comparable to the present-day rate

   d. 30–35 mm/year, faster than the present-day rate

4. Use the GE search window to locate Easter Island off the west coast of South America (it belongs to the country of Chile). By zooming out to an eye altitude of 5250 km, you can appreciate how small and remote this island is. By comparing topographic features on the seafloor with those in Figure 3.7, you should be able to locate Easter Island in the figure (it's not labeled). Which plate boundary is Easter Island nearest, and what is the present-day rate of seafloor spreading at that boundary?

   a. North American Plate–Pacific Plate boundary; 63 mm/year

   b. Pacific Plate–Nazca Plate boundary; 150 mm/year

   c. North American Plate–African Plate boundary; 24 mm/year

   d. Nazca Plate–South American Plate boundary; 79 mm/year

Optional Challenge Question

5. Locate Isla San Ambrosio, another tiny island off the west coast of Chile, at 26°20′34″S, 79°53′19″ W, and measure its distance from Easter Island using the ruler tool. From the isochron map in Figure 3.16, you can see that the seafloor near Isla San Ambrosio is approximately 35 million years old. What is the average rate of seafloor spreading over those 35 million years, and how does it compare with the present-day rate near Easter Island? (Hint: Assume that seafloor spreading has been symmetrical across the Pacific Plate–Nazca Plate boundary over the last 35 million years.)

   a. 70–90 mm/year, much slower than the present-day rate

   b. 140–160 mm/year, comparable to the present-day rate

   c. 160–180 mm/year, slightly faster than the present-day rate

   d. 200–220 mm/year, much faster than the present-day rate

SUMMARY

What is the theory of plate tectonics? According to the theory of plate tectonics, the lithosphere is broken into about a dozen rigid plates that move over Earth's surface. Three types of plate boundaries are defined by the direction of the movements of plates in relation to each other: divergent, convergent, and transform-fault boundaries. Earth's surface area does not change over time; therefore, the area of new lithosphere created at divergent boundaries equals the area of lithosphere recycled at convergent boundaries by subduction into the mantle.

What are some of the geologic characteristics of plate boundaries? Many geologic features develop at plate boundaries. Divergent boundaries are typically marked by volcanism and earthquakes at the crest of a mid-ocean ridge. Convergent boundaries are marked by deep-sea trenches, earthquakes, mountain building, and volcanoes. Transform faults, along which plates slide past each other, can be recognized by earthquake activity and offsets in geologic features.

How can the age of the seafloor be determined? We can measure the age of the seafloor by using thermoremanent magnetization. Magnetic anomaly patterns mapped on the seafloor can be compared with a magnetic time scale that was established by measuring the magnetic anomalies of lavas of known ages on land. Seafloor ages have been verified through dating of rock samples obtained by deep-sea drilling. Geologists can now draw isochron maps for most of the world's oceans, which allow them to reconstruct the history of seafloor spreading over the past 200 million years. Using this method and other geologic data, geologists have developed a detailed model of how Pangaea broke apart and the continents drifted into their present configuration.

What is the engine that drives plate tectonics? The plate tectonic system is driven by mantle convection, the energy for which comes from Earth's internal heat. Gravitational forces act on the cooling lithosphere as it slides downhill from spreading centers and sinks into the mantle at subduction zones. Subducted lithosphere extends as deep as the core-mantle boundary, indicating that the whole mantle is involved in the convection system that recycles the plates. Rising convection currents may include mantle plumes, intense upwellings from the deep mantle that cause localized volcanism at hot spots in the middle of a plate.

KEY TERMS AND CONCEPTS

continental drift (p. 64)
convergent boundary (p. 71)
divergent boundary (p. 70)
island arc (p. 73)
isochron (p. 80)
magnetic anomaly (p. 76)
magnetic time scale (p. 77)
mid-ocean ridge (p. 72)
Pangaea (p. 65)
plate tectonics (p. 64)
relative plate velocity (p. 78)
Rodinia (p. 84)
seafloor spreading (p. 66)
spreading center (p. 72)
subduction (p. 73)
transform fault (p. 71)

EXERCISES

1. Why are there active volcanoes along the Pacific coast in Washington and Oregon but not along the east coast of the United States?

2. What mistakes did Wegener make in formulating his theory of continental drift? Do you think the geologists of his era were justified in rejecting his theory?

3. Would you characterize plate tectonics as a hypothesis, a theory, or a fact? Why?

4. How do the differences between continental and oceanic crust affect the way plates interact?

5. From Figure 3.7, trace the boundaries of the South American Plate on a sheet of paper and identify segments that are divergent, convergent, and transform-fault boundaries. Approximately what fraction of the plate area is occupied by the South American continent? Is the fraction of the South American Plate occupied by oceanic crust increasing or decreasing over time? Explain your answer using the principles of plate tectonics.

6. In Figure 3.16, the isochrons are symmetrically distributed in the Atlantic Ocean but not in the Pacific Ocean. For example, seafloor as much as 180 million years old (in darkest blue) is found in the western Pacific but not in the eastern Pacific. Why?

7. From the isochron map in Figure 3.16, estimate how long ago the continents of Australia and Antarctica were separated by seafloor spreading. Did this happen before or after South America separated from Africa?

8. Name three mountain belts formed by continent-continent collisions that are occurring now or have occurred in the past.

9. Most active volcanoes are located on or near plate boundaries. Give an example of a volcano that is not on a plate boundary and describe a hypothesis consistent with plate tectonics that can explain it.

Visual Literacy Task

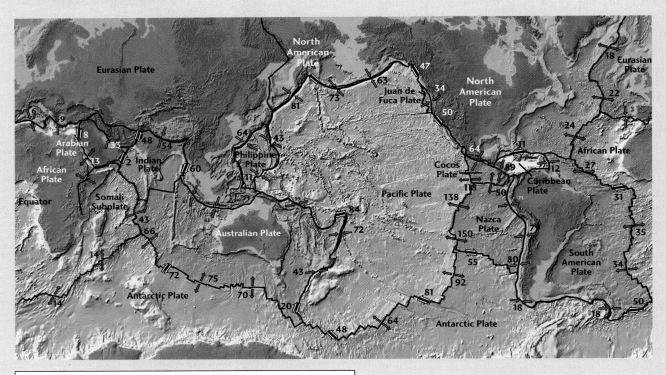

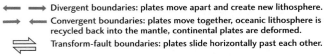

Divergent boundaries: plates move apart and create new lithosphere.

Convergent boundaries: plates move together, oceanic lithosphere is recycled back into the mantle, continental plates are deformed.

Transform-fault boundaries: plates slide horizontally past each other.

FIGURE 3.7 Earth's surface is a mosaic of 13 plates of rigid lithosphere, as well as a number of smaller plates, that move slowly over the ductile asthenosphere. Only one of the smaller plates – the Juan de Fuca Plate, off the west coast of North America – is shown on this map. The arrows show the relative movement of two plates at a point on their boundary. The numbers next to the arrows give the relative plate velocities, in millimeters per year. [*Plate boundaries by Peter Bird, UCLA.*]

1. What do the pastel colors, such as yellow or green, represent?

a. Continents

b. Tectonic plates

c. Tectonic plate boundaries

2. What types of plate boundaries surround the South American Plate?

a. Divergent and convergent boundaries

b. Divergent and transform boundaries

c. Transform and convergent boundaries

d. Divergent, convergent, and transform boundaries

3. Which of the following does the North American Plate include?

a. The North American continent

b. The North American continent and part of the Atlantic Ocean

c. The North American continent and all of the Atlantic Ocean

d. The North American continent and the Australian continent

4. How does the spreading rate of the East Pacific Rise (EPR) compare to the Mid-Atlantic Ridge (MAR)?

a. The East Pacific Rise is spreading faster than the Mid-Atlantic Ridge.

b. The East Pacific Rise is spreading slower than the Mid-Atlantic Ridge.

c. The East Pacific Rise is spreading at the same rate as the Mid-Atlantic Ridge.

5. In what approximate direction is the Nazca Plate moving relative to the Antarctic Plate?

a. Northward

b. Southward

c. Eastward

d. Westward

Thought Question: How would you answer a fellow student who asked, "Are continents the same thing as plates?"

Earth Materials: Minerals and Rocks

The Diamond Invention

The French were bred to die for love
They delight in fighting duels
But I prefer a man who lives
And gives expensive jewels
A kiss on the hand may be quite continental
But diamonds are a girl's best friend
A kiss may be grand but it won't pay the rental
On your humble flat, or help you at the automat
Men grow cold as girls grow old
And we all lose our charms in the end
But square cut or pear shaped
These rocks don't lose their shape
Diamonds are a girl's best friend

"DIAMONDS ARE A GIRL'S BEST FRIEND" is a song most famously performed by Marilyn Monroe in the 1953 movie of the Broadway musical *Gentlemen Prefer Blondes*. The song is a testimony to a brilliant marketing campaign, known as the "diamond invention." At its root is De Beers Consolidated Mines, Ltd., whose plan it was to romanticize diamonds by subtly changing the public's perception of the way a man courts—and wins—a woman. The marketing agency recruited movie idols, the symbols of romance for

> That "Diamonds are Forever" is not true, geologically speaking...

the general public, to advertise diamonds as symbols of enduring love. Magazine stories stressed the large diamonds that celebrities presented to their lovers, and photographs showed prominent women conspicuously flaunting the sparkling stone on their hands.

The diamond invention was a cunning scheme for sustaining the value of diamonds for the few people who produced them. It involved gaining complete control over the production and distribution of all of the world's diamonds. A system was created for allocating the gems to a select number of diamond cutters who all agreed to follow a strict set of rules intended to ensure that the quantity of cut diamonds available never exceeded the public's demand for them.

The most important ingredient in the campaign's success was to instill in the buying public the idea that they must never resell their diamonds. That "Diamonds Are Forever" is not true, geologically speaking, but it creates an illusion in the mass mindset that diamonds are "forever" in the sense that they must never be resold. You own them for life and then pass them on through generations to come. ◆

Marilyn Monroe, with diamonds. [*LGI Stock/Corbis.*]

CHAPTER OUTLINE

THE DIAMOND INVENTION has resulted in a staggering stockpile of uncut diamonds in a well-protected vault in London; a billion-dollar cash reserve in European banks; an intelligence network operating out of Antwerp, Tel Aviv, Johannesburg, and London; a global array of advertising agencies, brokers, and distributors; corporate fronts in Africa for concealing massive diamond purchases; and private treaties with nations establishing quotas for annual production. For example, until recently, diamonds surpassed even oil as the way Russia raised most of its cash to underpin its economy—and only through cooperation with De Beers in price fixing.

But the diamond invention goes far beyond being a monopoly whose goal is to fix diamond prices. It is a process for converting tiny crystals of a mineral called diamond, made up of colorless carbon—judged by many to be less beautiful than rubies or sapphires—into universally recognized tokens of power and romance.

The diamond invention is a relatively recent development in the history of social power. Up until the late 1800s, diamonds were genuinely rare gemstones, found only in riverbeds in India and in the tropical rain forests of Brazil. The entire world production amounted to only a few pounds each year. All this changed when "kimberlites" were discovered in South Africa in 1870. Kimberlites are rocks formed by volcanic eruptions containing rocks transported from depths of 100 to 200 km—all the way from Earth's upper mantle. All diamonds are formed in the upper mantle, where extreme pressures allow carbon to form its distinctive diamond structure.

The kimberlites of South Africa meant that instead of the rare chance of finding diamonds in a riverbed, diamonds could be gathered by the bucketful. Suddenly the market was swamped by an ever-expanding flood of diamonds. The British capitalists who had financed the operation of the kimberlite mines quickly learned that their investment would be threatened. In fact, diamonds were so plentiful as to be common. Their price had been completely dependent on their scarcity; produced at their new, true abundance, diamonds might now be even less valuable than semiprecious stones such as tourmaline and garnet.

The investors in the new mines had no choice but to merge their interests into a single company that would be powerful enough to control the production of diamonds and perpetuate the illusion of their scarcity. Thus, De Beers Consolidated, a company incorporated in South Africa, was born. While other commodities fluctuate wildly in price due to various global economic influences, diamonds continue to appreciate in value each year. However, the diamond invention is not forever. If De Beers either loses its control of diamond production or fails to continue to convince the general public never to sell their diamonds, the diamond invention could fail.

We begin this chapter with a study of minerals, often using diamond as a model of how minerals form. Minerals are the building blocks of rocks, which are, in turn, the records of geologic history.

To tell Earth's story, geologists often adopt a "Sherlock Holmes" approach: they use current evidence to deduce the tectonic processes and events that occurred in the past at some particular place. The kinds of minerals found in volcanic rocks, for example, provide evidence of eruptions that brought molten rock to Earth's surface, while the minerals in granite reveal that it crystallized deep in the crust under the very high temperatures and pressures produced when two continental plates collide. Understanding the geology of a region also allows us to make informed guesses about where undiscovered deposits of economically important mineral resources might lie.

The focus of this chapter is **mineralogy,** the branch of geology that studies the composition, structure, appearance, stability, occurrence, and associations of minerals.

WHAT ARE MINERALS?

Minerals are the building blocks of rocks. With the proper tools, most rocks can be separated into their constituent minerals. A few rocks, such as limestone, contain only a single mineral (in this case, calcite). Other rocks, such as granite, are made of several different minerals. To identify and classify the many kinds of rocks that compose Earth and understand how they formed, we must know about minerals.

Geologists define a **mineral** as a *naturally occurring, solid crystalline substance, usually inorganic, with a specific chemical composition.* Minerals are homogeneous: they cannot be divided mechanically into smaller components.

Let's examine each part of our definition of a mineral in a little more detail.

Naturally occurring   To qualify as a mineral, a substance must be found in nature. The diamonds mined in South

Africa are minerals. The synthetic versions produced in industrial laboratories are not minerals, nor are the thousands of laboratory products invented by chemists.

Minerals are the building blocks of rocks. They are solid, usually inorganic, naturally occurring, crystalline structures that have a specific chemical composition.

Solid crystalline substance
Minerals are solid substances—they are neither liquids nor gases. When we say that a mineral is *crystalline*, we mean that the tiny particles of matter, or atoms, that compose it are arranged in an orderly, repeating, three-dimensional array. Solid materials that have no such orderly arrangement are referred to as *glassy* or *amorphous* (without form) and are not conventionally called minerals. Windowpane glass is amorphous, as are some natural glasses formed during volcanic eruptions. Later in this chapter, we will explore in detail the process by which crystalline materials form.

Usually inorganic   Minerals are defined as inorganic substances and so exclude the organic materials that make up plant and animal bodies. Organic matter is composed of organic carbon, the form of carbon found in all organisms, living or dead. Decaying vegetation in a swamp may be geologically transformed into coal, which is also made of organic carbon, but although it is found in naturally occurring deposits, coal is not considered a mineral. Many minerals, however, are secreted by organisms. One such mineral, calcite (**Figure 4.1**), which forms the shells of oysters and many other organisms, contains inorganic carbon. The calcite of these shells, which constitute the bulk of many limestones, fits the definition of a mineral because it is inorganic and crystalline.

With a specific chemical composition   The key to understanding the composition of Earth's materials lies in knowing how the chemical elements are organized into minerals. What makes each mineral unique is its chemical composition and the arrangement of its atoms in an internal structure. A mineral's chemical composition either is fixed or varies within defined limits. The mineral quartz, for example, has a fixed ratio of two atoms of oxygen to one atom of silicon. This ratio never varies, even though quartz is found in many different kinds of rocks. Similarly, the chemical elements that make up the mineral olivine—iron, magnesium, oxygen, and silicon—always have a fixed ratio. Although the numbers of iron and magnesium atoms may vary, the sum of the two atoms in relation to the number of silicon atoms always forms a fixed ratio.

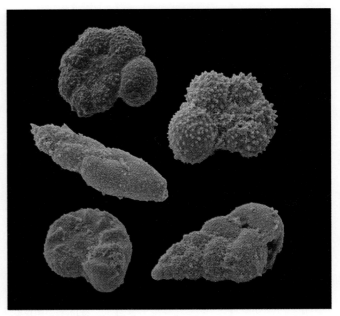

FIGURE 4.1 The mineral calcite is found in the shells of many organisms, such as foraminifers.
[*(left)* John Grotzinger/Ramón Rivera-Moret/Harvard Mineralogical Museum. *(right)* Andrew Syred/Photo Researchers.]

THE STRUCTURE OF MATTER

In 1805, John Dalton hypothesized that each of the various chemical elements consists of a different kind of atom, that all atoms of any given element are identical, and that chemical compounds are formed by various combinations of atoms of different elements in definite proportions. By the early twentieth century, physicists, chemists, and mineralogists, building on Dalton's ideas, had come to understand the structure of matter much as we understand it today. We now know that an **atom** is the smallest unit of an element that retains the physical and chemical properties of that element. We also know that atoms are the small units of matter that combine in chemical reactions and that atoms themselves are divisible into even smaller units.

The Structure of Atoms

Understanding the structure of atoms allows us to predict how chemical elements will react with one another and form new crystal structures. The structure of an atom is defined by a nucleus, which contains protons and neutrons and which is surrounded by electrons. (For a more detailed review of the structure of atoms, see Appendix 3.)

At the center of every atom is a dense *nucleus* containing virtually all the mass of the atom in two kinds of particles: protons and neutrons (**Figure 4.2**). A *proton* has a positive electrical charge of +1. A *neutron* is electrically neutral—that is, uncharged. Atoms of the same chemical

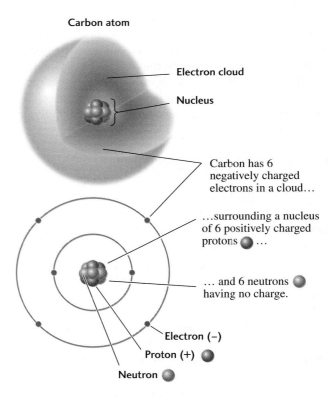

Carbon atom

Electron cloud

Nucleus

Carbon has 6 negatively charged electrons in a cloud…

…surrounding a nucleus of 6 positively charged protons …

… and 6 neutrons having no charge.

Electron (−)

Proton (+)

Neutron

FIGURE 4.2 Structure of the carbon atom (carbon-12). The six electrons, each with a charge of −1, are represented as a negatively charged cloud surrounding the nucleus, which contains six protons, each with a charge of +1, and six neutrons, each with no charge. The size of the nucleus is greatly exaggerated in these drawings; in actuality, it is much too small to show at a true scale.

element may have different numbers of neutrons, but the number of protons does not vary. For instance, all carbon atoms have six protons.

Surrounding the nucleus is a cloud of moving particles called *electrons*, each with a mass so small that it is conventionally taken to be zero. Each electron carries a negative electrical charge of –1. The number of protons in the nucleus of any atom is balanced by the same number of electrons in the cloud surrounding the nucleus, so the atom is electrically neutral. Thus, the nucleus of a carbon atom is surrounded by six electrons (see Figure 4.2).

> *Atoms are the smallest units of matter. The structure of an atom is defined by a nucleus, which contains protons and neutrons and is surrounded by electrons.*

Atomic Number and Atomic Mass

The number of protons in the nucleus of an atom is called its *atomic number*. Because all atoms of the same element have the same number of protons, they also have the same atomic number. All atoms with six protons, for example, are carbon atoms (atomic number 6). In fact, the atomic number of an element can tell us so much about the element's behavior that the periodic table organizes elements according to their atomic number (see Appendix 3). Elements in the same vertical column of the periodic table, such as carbon and silicon, tend to react similarly.

The *atomic mass* of an element is the sum of the masses of its protons and its neutrons. (Electrons, because they have so little mass, are not included in this sum.) Atoms of the same chemical element always have the same number of protons but may have different numbers of neutrons and therefore different atomic masses. Atoms of the same element with different numbers of neutrons are called **isotopes.** Isotopes of the element carbon, for example, all have six protons but may have six, seven, or eight neutrons, giving atomic masses of 12, 13, and 14.

In nature, the chemical elements exist as mixtures of isotopes, so their atomic masses are never whole numbers. Carbon's atomic mass, for example, is 12.011. It is close to 12 because the isotope carbon-12 is by far the most abundant. The relative abundances of the various isotopes of an element on Earth are determined by processes that enhance the abundances of some isotopes over others. Carbon-12, for example, is favored by some chemical reactions, such as photosynthesis, in which organic carbon compounds are produced from inorganic carbon compounds.

Chemical Reactions

The structure of an atom determines its chemical reactions with other atoms. **Chemical reactions** are interactions of the atoms of two or more chemical elements in certain fixed proportions that produce chemical compounds. For example, when two hydrogen atoms combine with one oxygen atom, they form a new chemical compound, water (H_2O). The properties of a chemical compound may be entirely different from those of its constituent elements. For example, when an atom of sodium, a metal, combines with an atom of chlorine, a noxious gas, they form the chemical compound sodium chloride, better known as table salt. We represent this compound by the chemical formula NaCl, in which the symbol Na stands for the element sodium and the symbol Cl for the element chlorine. (Every chemical element has been assigned its own symbol, which we use as a kind of shorthand for writing chemical formulas and equations; these symbols are given in the periodic table in Appendix 3.)

Chemical compounds, such as minerals, are formed either by *electron sharing* or by *electron transfer* between the reacting atoms. Carbon and silicon tend to form compounds by electron sharing. Diamond is a compound composed entirely of carbon atoms sharing electrons (**Figure 4.3**).

> *In chemical reactions, new chemical compounds are formed.*

In the reaction between sodium (Na) and chlorine (Cl) atoms to form sodium chloride (NaCl), electrons are transferred. The sodium atom loses one electron, which the chlorine atom gains (**Figure 4.4**). An atom or group of atoms that has an electrical charge, either positive or negative, because of the loss or gain of one or more electrons is called an **ion.** Because the chlorine atom has gained a negatively charged electron, it is now a negatively charged ion, Cl^-. Likewise, the loss of an electron gives sodium a positive charge, Na^+. The compound NaCl itself remains electrically neutral because the positive charge on Na^+ is exactly balanced by the negative charge on Cl^-. A positively charged ion is a *cation*, and a negatively charged ion is an *anion*.

The carbon atoms in diamond are arranged in regular tetrahedra...

...in which each atom shares an electron with each of its four neighbors.

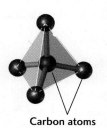

Carbon atoms

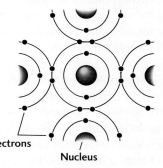

Electrons

Nucleus

FIGURE 4.3 Some atoms, such as the carbon atoms in diamond, share electrons to form covalent bonds.

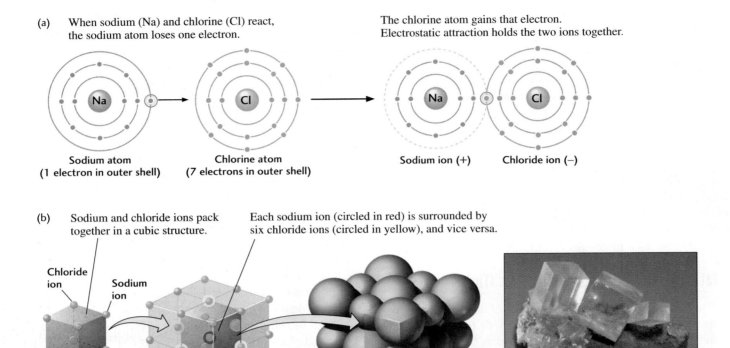

(a) When sodium (Na) and chlorine (Cl) react, the sodium atom loses one electron.

The chlorine atom gains that electron.
Electrostatic attraction holds the two ions together.

Sodium atom
(1 electron in outer shell)

Chlorine atom
(7 electrons in outer shell)

Sodium ion (+)   Chloride ion (−)

(b) Sodium and chloride ions pack together in a cubic structure.

Each sodium ion (circled in red) is surrounded by six chloride ions (circled in yellow), and vice versa.

Chloride ion

Sodium ion

NaCl crystal

Halite (table salt)

FIGURE 4.4 Some atoms, such as those in sodium chloride, transfer electrons to form ionic bonds. [*John Grotzinger/Ramón Rivera-Moret/Harvard Mineralogical Museum.*]

Chemical Bonds

When a chemical compound is formed either by electron sharing or by electron transfer, the ions or atoms that make up the compound are held together by electrostatic attraction between negatively charged electrons and positively charged protons. These attractions, or **chemical bonds,** between shared electrons or between gained and lost electrons may be strong or weak. Strong bonds keep a substance from decomposing into its elements or into other compounds. They also make minerals hard and keep them from cracking or splitting. Two major types of bonds are found in most rock-forming minerals: ionic bonds and covalent bonds.

■ The simplest form of chemical bond is the *ionic bond.* Bonds of this type form by electrostatic attraction between ions of opposite charge, such as Na^+ and Cl^- in sodium chloride (see Figure 4.4a), when electrons are transferred. This attraction is of exactly the same nature as the static electricity that can make nylon or silk clothing cling to the body. The strength of an ionic bond

decreases greatly as the distance between ions increases, and it increases as the electrical charges of the ions increase. Ionic bonds are the dominant type of chemical bonds in mineral structures: *about 90 percent of all minerals are essentially ionic compounds.*

■ Elements that do not readily gain or lose electrons to form ions and instead form compounds by sharing electrons are held together by *covalent bonds.* These bonds are generally stronger than ionic bonds. One mineral with a covalently bonded crystal structure is diamond, which consists of a single element, carbon. Each carbon atom can share four of its electrons with other carbon atoms and can acquire another four electrons by sharing with other carbon atoms. In diamond, every carbon atom is surrounded by four others arranged in a regular *tetrahedron*, a four-sided pyramidal form, each side a triangle (see Figure 4.3). In this configuration, each carbon atom shares an electron with each of its four neighbors, resulting in a very stable configuration. Figure 4.10 (page 102) shows a network of carbon tetrahedra linked together.

Atoms of metallic elements, which have strong tendencies to lose electrons, pack together as cations, and the freely mobile electrons are shared and dispersed among the ions. This free electron sharing results in a kind of covalent bond that we call a *metallic bond.* It is found in a small number of minerals, among them the metal copper and some sulfides.

> *In chemical reactions, new bonds are formed. Some electrons are shared in a covalent bond or transferred from one atom to another to form an ionic bond.*

The chemical bonds of some minerals are intermediate between pure ionic and pure covalent bonds because some electrons are exchanged and others are shared.

THE FORMATION OF MINERALS

Let's look more closely at the orderly forms that result from the chemical bonds we have just described. In this section, we examine the crystal structures of minerals and the conditions under which minerals form. Later in this chapter, we will see how the crystal structures affect the physical properties of minerals.

Minerals form by the process of **crystallization,** in which the atoms of a gas or liquid come together in the proper chemical proportions and crystalline arrangement to form a solid substance. (Remember that the atoms in a mineral are arranged in an orderly three-dimensional array.) The bonding of carbon atoms in diamond, a covalently bonded mineral, is one example of crystallization. Under the very high pressures and temperatures in Earth's mantle, carbon atoms bond together in tetrahedra, and each tetrahedron attaches to another, building up a regular three-dimensional structure from a great many atoms. As a diamond crystal grows, it extends its tetrahedral structure in all directions, always adding new atoms in the proper geometric arrangement. Diamonds can be artificially synthesized under very high pressures and temperatures that mimic the conditions in Earth's mantle, where natural diamonds form.

The Crystal Structure of Minerals

The sodium and chloride ions that make up sodium chloride, an ionically bonded mineral, also crystallize in an orderly three-dimensional array. In Figure 4.4b, we can see the geometry of their arrangement, with each ion of one kind surrounded by six ions of the other kind in a series of *cubic* structures extending in three directions. We can think of ions as solid spheres, packed together in close-fitting structural units. Figure 4.4b also shows the relative sizes of the ions in NaCl. There are six neighboring ions in NaCl's basic structural unit. The relative sizes of the sodium and chloride ions allow them to fit together in a closely packed arrangement.

Many of the cations of abundant minerals are relatively small, while most anions are large (**Figure 4.5**). This is the case with the most common anion on Earth, oxygen (O^{2-}). Because anions tend to be larger than cations, most of the space of a crystal is occupied by the anions, and the cations fit into the spaces between them. As a result, crystal structures are determined largely by how the anions are arranged and how the cations fit between them.

> *The crystal structure of a mineral is defined by an orderly three-dimensional arrangement of atoms in specific proportions.*

Cations of similar sizes and charges tend to substitute for one another and to form compounds having the same crystal structure but differing in chemical composition. *Cation substitution* is common in minerals that contain the silicate ion (SiO_4^{4-}), such as olivine, which is abundant in many volcanic rocks. Iron (Fe^{2+}) and magnesium (Mg^{2+}) ions are similar in size, and both have two positive charges, so they easily substitute for one another in the

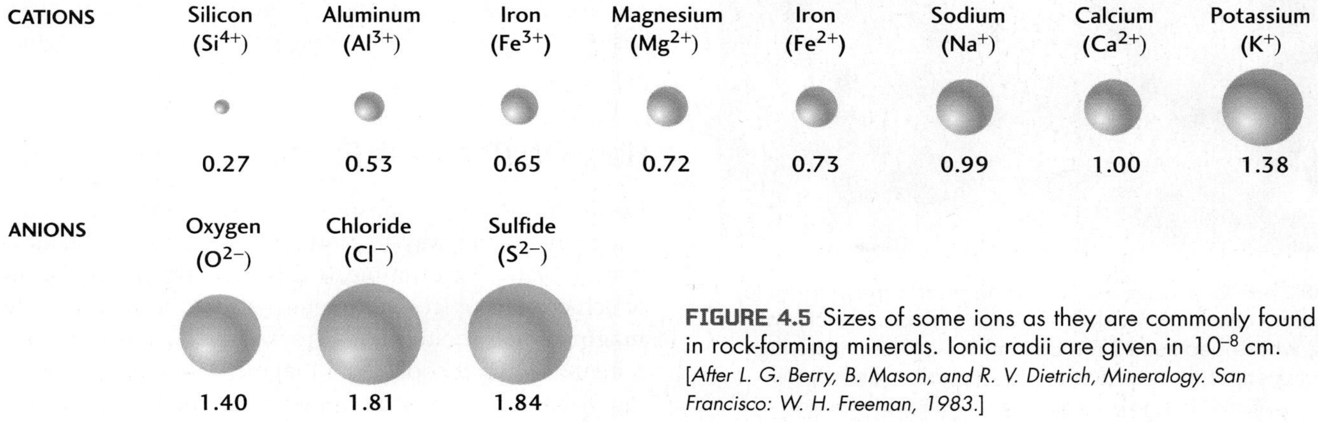

CATIONS	Silicon (Si^{4+})	Aluminum (Al^{3+})	Iron (Fe^{3+})	Magnesium (Mg^{2+})	Iron (Fe^{2+})	Sodium (Na^+)	Calcium (Ca^{2+})	Potassium (K^+)
	0.27	0.53	0.65	0.72	0.73	0.99	1.00	1.38

ANIONS	Oxygen (O^{2-})	Chloride (Cl^-)	Sulfide (S^{2-})
	1.40	1.81	1.84

FIGURE 4.5 Sizes of some ions as they are commonly found in rock-forming minerals. Ionic radii are given in 10^{-8} cm. [After L. G. Berry, B. Mason, and R. V. Dietrich, *Mineralogy. San Francisco: W. H. Freeman, 1983.*]

structure of olivine. The composition of pure magnesium olivine is Mg_2SiO_4; that of pure iron olivine is Fe_2SiO_4. The composition of olivine containing both iron and magnesium is given by the formula $(Mg,Fe)_2SiO_4$, which simply means that the number of iron and magnesium cations may vary, but their combined total (expressed as a subscript 2) does not vary in relation to each SiO_4^{4-} ion. The proportion of iron to magnesium is determined by the relative abundances of the two elements in the molten material from which the olivine crystallized. In many silicate minerals, aluminum (Al^{3+}) substitutes for silicon (Si^{4+}). Aluminum and silicon ions are so similar in size that aluminum can take the place of silicon in many crystal structures. The difference in charge between aluminum (3) and silicon (4) ions is balanced by an increase in the number of other cations, such as sodium (1+).

The Crystallization of Minerals

Crystallization starts with the formation of microscopic single **crystals,** orderly three-dimensional arrays of atoms in which the basic arrangement is repeated in all directions. The boundaries of crystals are natural flat (*planar*) surfaces called *crystal faces* (**Figure 4.6**). The crystal faces of a mineral are the external expression of the mineral's internal atomic structure. **Figure 4.7** pairs a drawing of a

Crystal faces

A perfect quartz crystal   A natural quartz crystal

FIGURE 4.7 A perfect crystal is rare in nature, but no matter how irregular the shapes of the faces may be, the angles are always exactly the same. [Breck P. Kent.]

perfect quartz crystal with a photograph of the actual mineral. The six-sided (hexagonal) shape of the quartz crystal corresponds to its hexagonal internal atomic structure.

During crystallization, the initially microscopic crystals grow larger, maintaining their crystal faces as long as they are free to grow. Large crystals with well-defined faces form when growth is slow and steady and space is adequate to allow growth without interference from other crystals nearby (**Figure 4.8**). For this reason, most large mineral crystals form in open spaces in rocks, such as fractures or cavities.

Often, however, the spaces between growing crystals fill in, or crystallization proceeds rapidly. Crystals then grow over one another and coalesce to become a solid mass of crystalline particles, or *grains*. In this case, few or no grains show crystal faces. Large crystals that can be seen with the naked eye are relatively unusual, but many minerals in rocks display crystal faces that can be seen under a microscope.

> As a crystal grows, the basic arrangement of its atoms is repeated in all directions. The crystal faces of a mineral reflect this basic arrangement.

Unlike crystalline minerals, glassy materials—which solidify from liquids so quickly that they lack any internal atomic order—do not form crystals with planar faces. Instead, they are found as masses with curved, irregular surfaces. The most common glass is volcanic glass.

How Do Minerals Form?

Lowering the temperature of a liquid below its freezing point is one way to start the process of crystallization. In water, for example, 0°C is the temperature below which crystals of ice—a mineral—start to form. Similarly, **magma**—hot, molten liquid rock—crystallizes into solid minerals when it cools. As a magma falls below its melting point, which may be higher than 1000°C, crystals of

FIGURE 4.6 Crystals of amethyst and quartz growing on top of epidote crystals (green). The planar surfaces are crystal faces, whose geometries are determined by the underlying arrangement of the atoms that make up the crystals. [John Grotzinger/Ramón Rivera-Moret/Harvard Mineralogical Museum.]

FIGURE 4.8 Giant crystals are sometimes found in caves where they have room to grow. These selenite crystals are a gem-quality form of gypsum (calcium sulfate). [*Javier Trueba/MSF/Photo Researchers.*]

silicate minerals such as olivine or feldspar begin to form. (Geologists usually refer to melting points of magmas rather than freezing points because freezing implies cold.)

Crystallization can also occur as liquids evaporate from a solution. A *solution* is a homogeneous mixture of one chemical substance with another, such as salt and water. As the water evaporates from a salt solution, the concentration of salt eventually gets so high that the solution can hold no more salt and is said to be *saturated*. If evaporation continues, the salt starts to *precipitate*, or drop out of solution as crystals. Deposits of table salt, or halite, form under just these conditions when seawater evaporates to the point of saturation in some hot, arid bays or arms of the ocean (**Figure 4.9**).

> *Crystallization occurs when a liquid is cooled below its melting point or when a liquid precipitates from an evaporating solution.*

FIGURE 4.9 Halite crystals precipitating within a modern hypersaline lagoon, San Salvador Island, Bahamas. Note the cubic shape of the crystals. [*John Grotzinger.*]

Diamond and graphite (the material used as the "lead" in pencils) exemplify the dramatic effects that temperature and pressure can have on mineral formation. These two minerals are **polymorphs,** alternative structures formed from a single chemical compound (**Figure 4.10**). They are both formed from carbon but have different crystal structures and very different appearances. From experimentation and geological observation, we know that diamond forms and remains stable at the very high pressures and temperatures found in Earth's mantle. High pressures force the atoms in diamond into a closely

Graphite is formed at lower pressures and temperatures than diamond. Its carbon forms sheets whose atoms are more loosely packed than those in diamond.

Within its sheets, carbon atoms are joined by strong bonds.

Weak bonds connect carbon atoms in alternating sheets.

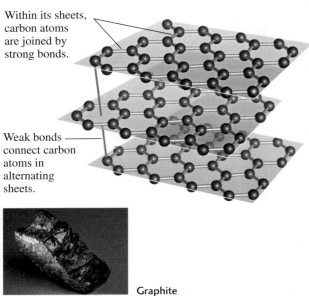

Graphite

Natural **diamond** is formed by very high pressures and temperatures in Earth's mantle. Its carbon atoms are closely packed.

All carbon atoms in diamond are closely packed, and all the bonds are very strong.

Diamond

FIGURE 4.10 Graphite and diamond are polymorphs, alternative structures formed from a single chemical compound, carbon. [John Grotzinger/Ramón Rivera-Moret/Harvard Mineralogical Museum.]

packed structure. Diamond therefore has a higher **density** (mass per unit volume, usually expressed in grams per cubic centimeter, g/cm^3) than graphite, which is less closely packed: diamond has a density of 3.5 g/cm^3, while the density of graphite is only 2.1 g/cm^3. Graphite forms and is stable at moderate pressures and temperatures, such as those in Earth's crust.

> *Polymorphs are alternative mineral structures with the same chemical composition that are formed under different temperatures and pressures.*

Low temperatures can also produce close packing. Quartz and cristobalite are polymorphs of silica (SiO_2). Quartz forms at low temperatures and is relatively dense (2.7 g/cm^3). Cristobalite, which forms at higher temperatures, has a more open structure and is therefore less dense (2.3 g/cm^3).

ROCK-FORMING MINERALS

All minerals on Earth have been grouped into eight classes according to their chemical composition (**Table 4.1**). Some minerals, such as copper, occur naturally as un-ionized pure elements, and they are classified as *native elements.* Most others are classified by their anions.

Although many thousands of minerals are known, geologists commonly encounter only about 30 of them. These minerals are the building blocks of most crustal rocks and are called *rock-forming minerals.* Their relatively small number corresponds to the small number of elements that are abundant in Earth's crust. As we learned in Chapter 2, 99 percent of the crust is made up of only eight elements.

> *The most common classes of rock-forming minerals are the silicates, carbonates, oxides, sulfides, and sulfates.*

In the following pages, we consider the five most common classes of rock-forming minerals:

■ *Silicates*, the most abundant class of minerals in Earth's crust, are composed of oxygen (O) and silicon (Si)—the two most abundant elements in the crust—mostly in combination with cations of other elements.

■ *Carbonates* are minerals composed of carbon and oxygen in the form of the carbonate anion (CO_3^{2-}) in combination with calcium and magnesium. Calcite ($CaCO_3$) is one such mineral.

■ *Oxides* are compounds of the oxygen anion (O^{2-}) and metallic cations; an example is the mineral hematite (Fe_2O_3).

TABLE 4.1	Some Chemical Classes of Minerals	
Class	**Defining Anions**	**Example**
Native elements	None: no charged ions	Copper metal (Cu)
Oxides	Oxygen ion (O^{2-})	Hematite (Fe_2O_3)
Hydroxides	Hydroxyl ion (OH^-)	Brucite ($Mg[OH]_2$)
Halides	Chloride (Cl^-), fluoride (F^-), bromide (Br^-), iodide (I^-)	Halite (NaCl)
Carbonates	Carbonate ion (CO_3^{2-})	Calcite ($CaCO_3$)
Sulfates	Sulfate ion (SO_4^{2-})	Anhydrite ($CaSO_4$)
Silicates	Silicate ion (SiO_4^{4-})	Olivine (Mg_2SiO_4)
Sulfides	Sulfide ion (S^{2-})	Pyrite (FeS^2)

- *Sulfides* are compounds of the sulfide anion (S^{2-}) and metallic cations; an example is the mineral sphalerite (ZnS).

- *Sulfates* are compounds of the sulfate anion (SO_4^{2-}) and metallic cations; an example is the mineral anhydrite ($CaSO_4$).

The other chemical classes of minerals, including native elements and halides, are not as common as the rock-forming minerals.

Silicates

The basic building block of all silicate mineral structures is the silicate ion. It is a tetrahedron—a pyramidal structure with four sides—composed of a central silicon ion (Si^{4+}) surrounded by four oxygen ions (O^{2-}), giving the formula SiO_4^{4-} (**Figure 4.11**). Because the silicate ion has a negative charge, it often bonds to cations to form electrically neutral minerals. The silicate ion typically bonds with sodium (Na^+), potassium (K^+), calcium (Ca^{2+}), magnesium (Mg^{2+}), and iron (Fe^{2+}). Alternatively, it can share oxygen ions with other silicon-oxygen tetrahedra. Tetrahedra may be isolated (linked only to cations), or they may be linked to other silicate tetrahedra in rings, single chains, double chains, sheets, or frameworks. Some of these structures are shown in Figure 4.11.

Silicates are the most abundant class of minerals in Earth's crust. Their basic structure is a tetrahedron, which links to other silica tetrahedra or to cations to form a variety of structures.

Isolated Tetrahedra Isolated tetrahedra are linked by the bonding of each oxygen ion of the tetrahedron to a cation (see Figure 4.11a). The cations, in turn, bond to the oxygen ions of other tetrahedra. The tetrahedra are thus isolated from one another by cations on all sides. Olivine is a rock-forming mineral with this structure.

Single-Chain Structures Single chains form by the sharing of oxygen ions. Two oxygen ions of each tetrahedron bond to adjacent tetrahedra in an open-ended chain (see Figure 4.11b). Single chains are linked to other chains by cations. Minerals of the pyroxene group are single-chain silicate minerals. Enstatite, a pyroxene, is composed of iron or magnesium ions, or both, and is limited to a chain of tetrahedra in which the two cations may substitute for each other, as in olivine. The formula $(Mg,Fe)SiO_3$ represents this structure.

Double-Chain Structures Two single chains may combine to form double chains linked to each other by shared oxygen ions (see Figure 4.11c). Adjacent double chains linked by cations form the structure of minerals in the amphibole group. Hornblende, a member of this group, is an extremely common mineral in both igneous and metamorphic rocks. It has a complex composition that includes calcium (Ca^{2+}), sodium (Na^+), magnesium (Mg^{2+}), iron (Fe^{2+}), and aluminum (Al^{3+}).

Sheet Structures In sheet structures, each tetrahedron shares three of its oxygen ions with adjacent tetrahedra to build stacked sheets of tetrahedra (see Figure 4.11d). Cations may be interlayered with tetrahedral sheets. The

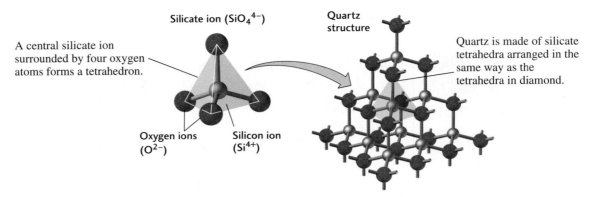

Silicate ion (SiO_4^{4-})

A central silicate ion surrounded by four oxygen atoms forms a tetrahedron.

Oxygen ions (O^{2-})

Silicon ion (Si^{4+})

Quartz structure

Quartz is made of silicate tetrahedra arranged in the same way as the tetrahedra in diamond.

Silicate tetrahedra can be arranged into a number of different structures.

Mineral	Chemical formula	Cleavage planes	Structure	Specimen
Olivine (a)	$(Mg,Fe)_2SiO_4$	1 plane	Isolated tetrahedra	
Pyroxene (b)	$(Mg,Fe)SiO_3$	2 planes at 90°	Single chains	
Amphibole (c)	$Ca_2(Mg,Fe)_5Si_8O_{22}(OH)_2$	2 planes at 60° and 120°	Double chains	
Mica (d)	Muscovite: $KAl_2(AlSi_3O_{10})(OH)_2$	1 plane	Sheets	
Feldspar (e)	Orthoclase feldspar: $KAlSi_3O_8$ Plagioclase feldspar: $(Ca,Na)AlSi_3O_8$	2 planes at 90°	Three-dimensional frameworks	

FIGURE 4.11 The silicate ion is the basic building block of all silicate mineral structures.

[*John Grotzinger/Ramón Rivera-Moret/Harvard Mineralogical Museum.*]

micas and clay minerals are the most abundant sheet silicates. Muscovite, $KAl_2(AlSi_3O_{10})(OH)_2$, is one of the most common sheet silicates and is found in many types of rocks. It can be separated into extremely thin, transparent sheets. Kaolinite, $Al_2Si_2O_5(OH)_4$, which also has this structure, is a common clay mineral found in sediments and is the basic raw material for pottery.

Frameworks Three-dimensional frameworks may form as each tetrahedron shares all its oxygen ions with other tetrahedra. Feldspars, the most abundant minerals in Earth's crust, are framework silicates (see Figure 4.11e), as is another of the most common minerals, quartz (SiO_2).

Silicate Compositions Chemically, the simplest silicate is silicon dioxide, also called silica (SiO_2), which is found most often as the mineral quartz. When the silicate tetrahedra of quartz are linked, sharing two oxygen ions for each silicon ion, the total formula adds up to SiO_2.

In other silicate minerals, the basic structural units—rings, chains, sheets, and frameworks—are bonded to cations such as sodium (Na^+), potassium (K^+), calcium (Ca^{2+}), magnesium (Mg^{2+}), and iron (Fe^{2+}). As noted in the discussion of cation substitution, aluminum (Al^{3+}) substitutes for silicon in many silicate minerals.

Carbonates

The mineral calcite (calcium carbonate, $CaCO_3$) is one of the most abundant minerals in Earth's crust and is the chief constituent of a group of rocks called limestones (**Figure 4.12**). Its basic building block, the carbonate ion (CO_3^{2-}), consists of a carbon ion surrounded by three oxygen ions in a triangle, as shown in Figure 4.12b. The carbon atom shares electrons with the oxygen atoms. Groups of carbonate ions are arranged in sheets somewhat like those of the sheet silicates, which are bonded together by layers of cations (see Figure 4.12c). The sheets of carbonate ions in calcite are separated by layers of calcium ions. Dolomite ($CaMg[CO_3]_2$), another major mineral of crustal rocks, is made up of the same carbonate sheets separated by alternating layers of calcium ions and magnesium ions.

Oxides

Oxide minerals are compounds in which oxygen is bonded to atoms or cations of other elements, usually metallic cations such as iron (Fe^{2+} or Fe^{3+}). Most oxide minerals are ionically bonded, and their structures vary with the size of the metallic cations. This class of minerals has great economic importance because it includes the ores containing many of the metals, such as chromium and titanium, used in the industrial and technological manufacture of metallic materials and devices. Hematite (Fe_2O_3), shown in **Figure 4.13**, is a chief ore of iron.

Another of the abundant minerals in this class, spinel, is an oxide of two metals, magnesium and aluminum ($MgAl_2O_4$). Spinel (see Figure 4.13) has a closely packed cubic structure and a high density (3.6 g/cm^3), reflecting the conditions of high pressure and temperature under which it forms. Transparent gem-quality spinel resembles ruby and sapphire and is found in the crown jewels of England and Russia.

Sulfides

The chief ores of many valuable minerals—such as copper, zinc, and nickel—are members of the sulfide class. This class includes compounds in which the sulfide ion (S^{2-}) is bonded to metallic cations. In the sulfide ion, a sulfur atom has gained two electrons in its outer shell.

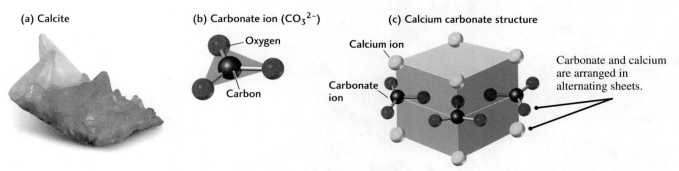

(a) Calcite   (b) Carbonate ion (CO_3^{2-}) — Oxygen, Carbon   (c) Calcium carbonate structure — Calcium ion, Carbonate ion, Carbonate and calcium are arranged in alternating sheets.

FIGURE 4.12 Carbonate minerals, such as calcite (calcium carbonate, $CaCO_3$), have a layered structure. (a) Calcite. [*John Grotzinger/Ramón Rivera-Moret/Harvard Mineralogical Museum.*] (b) Top view of the carbonate ion, composed of a carbon ion surrounded in a triangle by three oxygen ions, with a net charge of –2. (c) View of the alternating layers of calcium and carbonate ions in calcite.

FIGURE 4.13 Oxides include many economically valuable minerals. (*left*) Hematite. (*right*) Spinel. [*John Grotzinger/Ramón Rivera-Moret/Harvard Mineralogical Museum.*]

Most sulfide minerals look like metals, and almost all are opaque. The most common sulfide mineral is pyrite (FeS_2), often called "fool's gold" because of its yellowish metallic appearance (**Figure 4.14**).

Sulfates

The basic building block of all sulfates is the sulfate ion (SO_4^{2-}). It is a tetrahedron made up of a central sulfur atom surrounded by four oxygen ions (O^{2-}). One of the most abundant minerals of this class is gypsum (**Figure 4.15**), the primary component of plaster. Gypsum, a calcium sulfate, forms when seawater evaporates. During evaporation, Ca^{2+} and SO_4^{2-}, two ions that are abundant in seawater, combine and precipitate as layers of sediment, forming calcium sulfate ($CaSO_4 \cdot 2H_2O$). (The dot in this formula signifies that two water molecules are bonded to the calcium and sulfate ions.)

Another calcium sulfate, anhydrite ($CaSO_4$), differs from gypsum in that it contains no water. Its name is derived from the word *anhydrous*, meaning "free from water." Gypsum is stable at the low temperatures and

FIGURE 4.14 Pyrite, a sulfide mineral, is also known as "fool's gold." [*John Grotzinger/Ramón Rivera-Moret/Harvard Mineralogical Museum.*]

FIGURE 4.15 Gypsum is a sulfate formed when seawater evaporates. [*John Grotzinger/Ramón Rivera-Moret/Harvard Mineralogical Museum.*]

pressures found at Earth's surface, whereas anhydrite is stable at the higher temperatures and pressures where sedimentary rocks are buried.

As we discovered in 2004, sulfate minerals were precipitated from water early in the history of Mars. Much as on Earth, these minerals were precipitated when lakes and shallow seas dried up and formed sedimentary layers. However, many of these sulfate minerals are quite different from the sulfate mineral precipitates commonly found on Earth and include strange iron-bearing sulfates that precipitated from very harsh, acidic waters.

PHYSICAL PROPERTIES OF MINERALS

Geologists use their knowledge of mineral composition and structure to understand the origins of rocks. First they must identify the minerals that make up a rock. To do so, they rely greatly on chemical and physical properties that can be observed relatively easily. In the nineteenth and early twentieth centuries, geologists carried field kits for rough chemical analyses of minerals that would help in their identification. One such test is the origin of the phrase "the acid test." It consists of dropping diluted hydrochloric acid (HCl) on a mineral to see if it fizzes (**Figure 4.16**). The fizzing indicates that carbon dioxide (CO_2) is escaping, which means that the mineral is likely to be calcite, a carbonate.

In this section, we review the physical properties of minerals, many of which contribute to their practical and decorative value.

> Geologists identify minerals by measuring key physical properties that include hardness, cleavage, fracture, luster, color, density, and crystal habit.

Hardness

Hardness is a measure of the ease with which the surface of a mineral can be scratched. Just as diamond, the hardest mineral known, scratches glass, a quartz crystal, which is harder than feldspar, scratches a feldspar crystal. In 1822, Friedrich Mohs, an Austrian mineralogist, devised a scale (now known as the **Mohs scale of hardness**) based on the ability of one mineral to scratch another. At one extreme is the softest mineral (talc); at the other, the hardest

FIGURE 4.16 The acid test. One easy but effective way to identify certain minerals is to drop diluted hydrochloric acid (HCl) on the substance of interest. If it fizzes, indicating the escape of carbon dioxide, the mineral is likely to be calcite. [*Chip Clark.*]

TABLE 4.2	Mohs Scale of Hardness	
Mineral	**Scale Number**	**Common Objects**
Talc	1	
Gypsum	2	Fingernail
Calcite	3	Copper coin
Fluorite	4	
Apatite	5	Knife blade
Orthoclase	6	Window glass
Quartz	7	Steel file
Topaz	8	
Corundum	9	
Diamond	10	

(diamond) (**Table 4.2**). The Mohs scale is still one of the best practical tools for identifying an unknown mineral. With a knife blade and a few of the minerals on the hardness scale, a field geologist can gauge an unknown mineral's position on the scale. If the unknown mineral is scratched by a piece of quartz but not by the knife, for example, it lies between 5 and 7 on the scale.

Recall that covalent bonds are generally stronger than ionic bonds. The hardness of any mineral depends on the strength of its chemical bonds: the stronger the bonds, the harder the mineral. Hardness varies with crystal structure within the silicate class of minerals. For example, hardness varies from 1 in talc, a sheet silicate, to 8 in topaz, a silicate with isolated tetrahedra. Most silicates fall in the 5 to 7 range on the Mohs scale. Only sheet silicates are relatively soft, with hardnesses between 1 and 3.

Within groups of minerals having similar crystal structures, increasing hardness is related to other factors that also increase bond strength:

■ *Size* The smaller the atoms or ions, the smaller the distance between them and the greater the electrical attraction—and thus the stronger the bond.

■ *Charge* The larger the charge of ions, the greater the attraction between them and thus the stronger the bond.

■ *Packing of atoms or ions* The closer the packing of atoms or ions, the smaller the distance between them and thus the stronger the bond.

Size is an especially important factor for most metallic oxides and for most sulfides of metals with high atomic numbers—such as gold, silver, copper, and lead. Minerals of these groups are soft, with hardnesses of less than 3, because their metallic cations are so large. Carbonates and sulfates, whose structures are not closely packed, are also soft, with hardnesses of less than 5.

Cleavage

Cleavage is the tendency of a crystal to split along planar surfaces (**Figure 4.17**). The term *cleavage* is also used to describe the geometric pattern produced by such breakage. Cleavage varies inversely with bond strength: high bond strength produces poor cleavage, while low bond strength produces good cleavage. Because of their strength, covalent bonds generally give poor or no cleavage. Ionic bonds are relatively weak, so they give good cleavage. However, bond strength also varies along the different

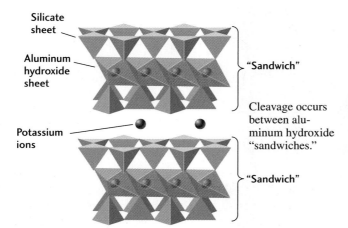

FIGURE 4.17 Cleavage of mica. The diagram shows the cleavage plane in the crystal structure, oriented perpendicular to the plane of the page. Horizontal lines mark the interfaces of silica-oxygen tetrahedral sheets and the sheets of aluminum hydroxide bonding the two tetrahedral sheets into a sandwich. Cleavage takes place between tetrahedral–aluminum hydroxide sandwiches. The photograph shows thin sheets of mica separating along the cleavage planes. [*Chip Clark.*]

planes depending on whether a mineral has only covalent bonds or only ionic bonds. For example, all the bonds in diamond are covalent bonds, which are very strong, but some planes are more weakly bonded than others. Thus, diamond, the hardest mineral of all, can be cleaved along these weaker planes to produce perfect planar surfaces.

If the bonds between some of the planes of atoms or ions in a crystal are relatively weak, the mineral can be made to split along those planes. Muscovite, a mica sheet silicate, splits along smooth, lustrous, flat, parallel surfaces, forming transparent sheets less than a millimeter thick. Mica's excellent cleavage results from the relative weakness of the bonds between the sandwiched layers of cations and tetrahedral silica sheets (see Figure 4.17).

Cleavage is classified according to two primary sets of characteristics: the number of planes and pattern of cleavage, and the quality of surfaces and ease of cleaving.

Number of Planes and Pattern of Cleavage The number of planes and pattern of cleavage are identifying hallmarks of many rock-forming minerals. Muscovite, for example, has only one plane of cleavage, whereas calcite and dolomite crystals have three excellent cleavage planes that give them a rhomboidal shape (**Figure 4.18**).

A crystal's structure determines its cleavage planes and its crystal faces. Crystals have fewer cleavage planes than possible crystal faces. Faces may be formed along any of numerous planes defined by rows of atoms or ions. Cleavage occurs along any of those planes across which the bonding is weak. All crystals of a mineral exhibit its characteristic cleavage planes, whereas only some crystals display particular faces.

Galena (lead sulfide, PbS) and halite (sodium chloride, NaCl) cleave along three planes, forming perfect cubes.

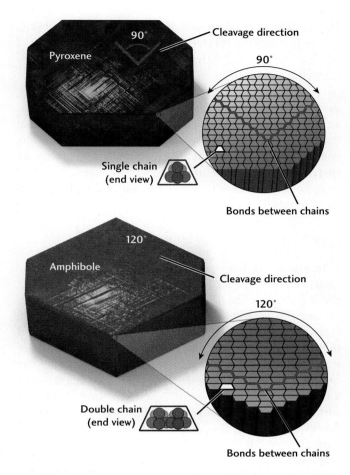

FIGURE 4.19 Pyroxene and amphibole often look very much alike, but their differing angles of cleavage can be used to identify and classify them.

Distinctive angles of cleavage help identify two important groups of silicates, the pyroxenes and amphiboles, that otherwise often look alike (**Figure 4.19**). Pyroxenes have a single-chain structure and are bonded so that their cleavage planes are almost at right angles (about 90°) to each other. In cross section, the cleavage pattern of pyroxene is nearly a square. In contrast, amphiboles, which have a double-chain structure, are bonded so as to give two cleavage planes that intersect to produce angles at 60° and 120° to each other. They produce a diamond-shaped cross section.

Quality of Surfaces and Ease of Cleaving A mineral's cleavage is assessed as perfect, good, or fair, according to the quality of surfaces produced and the ease of cleaving. Muscovite can be cleaved easily, producing extremely high-quality, smooth surfaces; its cleavage is *perfect*. The single-chain and double-chain silicates (the pyroxenes and amphiboles, respectively) show *good* cleavage. Although these minerals split easily along the cleavage plane, they also break across it, producing cleavage surfaces that

FIGURE 4.18 Example of rhomboidal cleavage in calcite.
[*Charles D. Winters/Photo Researchers.*]

are not as smooth as those of mica. *Fair* cleavage is shown by the ring silicate beryl. Beryl's cleavage is irregular, and the mineral breaks relatively easily along directions other than cleavage planes.

Many minerals are so strongly bonded that they lack even fair cleavage. Quartz, a framework silicate, is so strongly bonded in all directions that it breaks only along irregular surfaces. Garnet, an isolated tetrahedral silicate, is also bonded strongly in all directions and so shows no cleavage. This absence of a tendency to cleave is found in most framework silicates and in silicates with isolated tetrahedra.

Fracture

Fracture is the tendency of a crystal to break along irregular surfaces other than cleavage planes. All minerals show fracture, either across cleavage planes or—in such minerals as quartz—with no cleavage in any direction. Fracture is related to how bond strengths are distributed in directions that cut across cleavage planes. Breakage of these bonds results in irregular fractures. Fractures may be *conchoidal*, showing smooth, curved surfaces like those of a thick piece of broken glass. Another common fracture surface with an appearance like split wood is described as *fibrous* or *splintery*. The shapes and appearances of many kinds of irregular fractures depend on the particular structure and composition of the mineral.

Luster

The way the surface of a mineral reflects light gives it a characteristic **luster.** Mineral lusters are described by the terms listed in **Table 4.3**. Luster is controlled by the kinds of atoms present and their bonding, both of which affect the way light passes through or is reflected by the mineral. Ionically bonded crystals tend to have a glassy, or vitreous, luster, but covalently bonded materials are more variable. Many tend to have an adamantine luster, like that of diamond. Metallic luster is shown by pure metals, such as gold, and by many sulfides, such as galena (lead sulfide, PbS). A pearly luster results from multiple reflections of light from planes beneath the surfaces of translucent minerals, such as the mother-of-pearl inner surfaces of many clamshells, which are made of the mineral aragonite. Luster, although an important criterion for field classification, depends heavily on the visual perception of reflected light. Textbook descriptions fall short of the actual experience of holding the mineral in your hand.

Color

The **color** of a mineral is imparted by light, either transmitted through or reflected by crystals or irregular masses

TABLE 4.3	Mineral Luster
Luster	**Characteristics**
Metallic	Strong reflections produced by opaque substances
Vitreous	Bright, as in glass
Resinous	Characteristic of resins, such as amber
Greasy	The appearance of being coated with an oily substance
Pearly	The whitish iridescence of such materials as pearl
Silky	The sheen of fibrous materials such as silk
Adamantine	The brilliant luster of diamond and similar minerals

of the mineral or a streak of mineral powder. **Streak** refers to the color of the fine deposit of mineral powder left on an abrasive surface, such as a tile of unglazed porcelain, when a mineral is scraped across it. Such a tile, called a *streak plate* (**Figure 4.20**), is a good diagnostic tool because the uniform small grains of mineral that are present in the powder are revealed on the plate. Hematite (Fe_2O_3), for example, may be black, red, or brown, but this mineral will always leave a trail of reddish brown powder on a streak plate.

Color is a complex and not yet fully understood property of minerals. It is determined both by the kinds of ions found in the pure mineral and by trace impurities.

FIGURE 4.20 Hematite may be black, red, or brown, but it always leaves a reddish brown streak when scraped along a ceramic streak plate. [*Breck P. Kent.*]

Ions and Mineral Color The color of pure minerals depends on the presence of certain ions, such as iron or chromium, that strongly absorb portions of the light spectrum. Olivine that contains iron, for example, absorbs all colors except green, which it reflects, so we see this type of olivine as green. We see pure magnesium olivine as white (transparent and colorless).

Trace Impurities and Mineral Color All minerals contain impurities. Instruments can now measure even very small quantities of some elements—as little as a billionth of a gram in some cases. Elements that make up much less than 0.1 percent of a mineral are reported as "traces," and many of them are called *trace elements*.

Some trace elements can be used to interpret the origins of the minerals in which they are found. Others, such as the traces of uranium in some granites, contribute to local natural radioactivity. Still others, such as small dispersed flakes of hematite that color a feldspar crystal brownish or reddish, are notable because they give a general color to an otherwise colorless mineral. Many of the gem varieties of minerals, such as emerald (green beryl) and sapphire (blue corundum), get their color from trace elements dissolved in the solid crystal (**Figure 4.21**). Emerald derives its color from chromium; the sources of sapphire's blue color are iron and titanium.

The color of a mineral may be distinctive, but it is not the most reliable clue to its identity. Some minerals always show the same color; others may have a range of colors. Many minerals show a characteristic color only on freshly broken surfaces or only on weathered surfaces. Some—precious opals, for example—show a stunning display of colors on reflecting surfaces. Others change color slightly with a change in the angle of the light shining on their surfaces.

FIGURE 4.21 Trace elements give gems their colors. Sapphire (*left*) and ruby (*center*) are formed of the same common mineral, corundum (aluminum oxide). Small amounts of impurities produce the intense colors that we value. Ruby, for example, is red because of small amounts of chromium, the same element that gives emerald (*right*) its green color.
[*John Grotzinger/Ramón Rivera-Moret/Harvard Mineralogical Museum.*]

Density

You can easily feel the difference in weight between a piece of hematite iron ore and a piece of sulfur of the same size by lifting the two pieces. A great many common rock-forming minerals, however, are too similar in density for such a simple test. Scientists therefore need some easy method to measure this property of minerals. A standard measure of density is **specific gravity,** which is the weight of a mineral divided by the weight of an equal volume of pure water at 4°C.

Density depends on the atomic mass of a mineral's ions and how closely they are packed in its crystal structure. Consider the iron oxide magnetite, with a density of 5.2 g/cm^3. This high density results partly from the high atomic mass of iron and partly from the closely packed structure that magnetite has in common with the other members of the spinel group of oxides (see page 105). The density of the iron silicate olivine, at 4.4 g/cm^3, is lower than that of magnetite for two reasons. First, the atomic mass of silicon, one of the elements that make up olivines, is lower than that of iron. Second, iron olivine has a more openly packed structure than minerals of the spinel group. The density of magnesium olivine is even lower, 3.32 g/cm^3, because magnesium's atomic mass is much lower than that of iron.

Increases in density caused by increases in pressure affect the way minerals transmit light, heat, and seismic waves. Experiments at extremely high pressures have shown that the structure of olivine converts into the denser structure of the spinel group at pressures corresponding to a depth of 400 km. At a greater depth, 670 km, mantle materials are further transformed into silicate minerals with the even more densely packed structure of perovskite (CaTiO$_3$). Because of the huge volume of the lower mantle, perovskite is probably the most abundant mineral in Earth as a whole. Some perovskite minerals have been synthesized to be used as high-temperature semiconductors, which conduct electricity without loss of current and may have great commercial potential. Mineralogists experienced with natural perovskites helped unravel the structure of these newly created materials. Temperature also affects density: the higher the temperature, the more open and expanded the structure of the mineral, and thus the lower its density.

Crystal Habit

A mineral's **crystal habit** is the shape in which individual crystals or aggregates of crystals grow. Crystal habits are often named after common geometric shapes, such as blades, plates, and needles. Some minerals have such a distinctive crystal habit that they are easily recognizable. An example is quartz, with its six-sided column topped by

FIGURE 4.22 Chrysotile, a type of asbestos. Fibers are readily combed from the solid mineral. [*Runk/Schoenberger/ Grant Heilman Photography.*]

a pyramid-like set of faces (see Figure 4.7). These shapes indicate not only the planes of atoms or ions in the mineral's crystal structure but also the typical speed and direction of crystal growth. Thus, a needlelike crystal is one that grows very quickly in one direction and very slowly in all other directions. In contrast, a plate-shaped crystal (often referred to as *platy*) grows fast in all directions that

are perpendicular to its single direction of slow growth. *Fibrous* crystals take shape as multiple long, narrow fibers, essentially aggregates of long needles (**Figure 4.22**).

Asbestos is a generic name for a group of silicates with a more or less fibrous habit that allows the crystals to become embedded in the lungs if they are inhaled. Other minerals with deleterious effects include arsenic-containing pyrites, some of which are poisonous when ingested and others that release toxic fumes when heated. Mineral dust diseases affect many miners, who may face large occupational exposures. An example is silicosis, a disease of the lungs caused by inhaling quartz dust.

Table 4.4 summarizes the physical properties of minerals that we discussed in this section.

WHAT ARE ROCKS?

A geologist's primary aim is to understand the properties of rocks and to deduce their geologic origins from those properties. Such deductions further our understanding of our planet, and they also provide important information about fuel reserves. For example, knowing that oil forms in certain kinds of sedimentary rocks that are rich in organic matter allows us to explore for new oil reserves more intelligently. Similarly, our knowledge of the properties of rocks helps us find new reserves of other useful and economically valuable mineral and energy resources,

TABLE 4.4	Physical Properties of Minerals
Property	**Relation to Composition and Crystal Structure**
Hardness	Strong chemical bonds result in hard minerals. Covalently bonded minerals are generally harder than ionically bonded minerals.
Cleavage	Cleavage is poor if bonds in crystal structure are strong, good if bonds are weak. Covalent bonds generally give poor or no cleavage; ionic bonds are weaker and so give good cleavage.
Fracture	Related to distribution of bond strengths across irregular surfaces other than cleavage planes.
Luster	Tends to be glassy for ionically bonded crystals, more variable for covalently bonded crystals.
Color	Determined by kinds of atoms or ions and trace impurities. Many ionically bonded crystals are colorless. Iron tends to color strongly.
Streak	Color of fine mineral powder is more characteristic than that of massive mineral because of uniformly small size of grains.
Density	Depends on atomic weight of atoms or ions and their closeness of packing in crystal structure. Iron minerals and metals have high density; covalently bonded minerals have more open packing and so have lower density.
Crystal habit	Depends on planes of atoms or ions in a mineral's crystal structure and the typical speed and direction of crystal growth

such as natural gas, coal, and metallic ores. Understanding how rocks form also guides us in solving environmental problems. For example, the underground storage of radioactive and other wastes depends on analysis of the rock to be used as a repository: Will this rock be prone to earthquake-triggered landslides? How might it transmit polluted waters in the ground?

A **rock** is a naturally occurring solid aggregate of minerals or, in some cases, nonmineral solid matter. Some rocks, such as white marble, are composed of just one mineral (in this case, calcite). A few rocks are composed of nonmineral matter. These include the noncrystalline, glassy volcanic rocks obsidian and pumice as well as coal, which is made up of compacted plant remains. In an *aggregate*, minerals are joined in such a way that they retain their individual identity (**Figure 4.23**).

> *A rock is a naturally occurring solid aggregate of minerals or, in some cases, nonmineral solid matter.*

What determines the physical appearance of a rock? Rocks vary in color, in the sizes of their crystals or grains, and in the kinds of minerals that compose them. Along a road cut, for example, we might find a rough, speckled white and pink rock composed of interlocking crystals large enough to be seen with the naked eye. Nearby, we might see a grayish rock containing many large glittering crystals of mica and some grains of quartz and feldspar. Overlying both the white and pink rock and the gray one, we might see horizontal layers of a striped white and mauve rock that appear to be made up of sand grains cemented together. And these rocks might all be overlain by a dark fine-grained rock with tiny white dots in it.

The identity of a rock is determined partly by its mineralogy and partly by its texture. Here, the term *mineralogy* refers to the relative proportions of a rock's constituent minerals. **Texture** describes the sizes and shapes of a rock's mineral crystals or grains and the way they are

Constituent minerals

Orthoclase feldspar   Quartz   Biotite   Plagioclase feldspar

Plagioclase feldspar
Orthoclase feldspar
Biotite
Quartz

Rock (granite)

FIGURE 4.23 Rocks are naturally occurring aggregates of minerals. [John Grotzinger/ Ramón Rivera-Moret/Harvard Mineralogical Museum.]

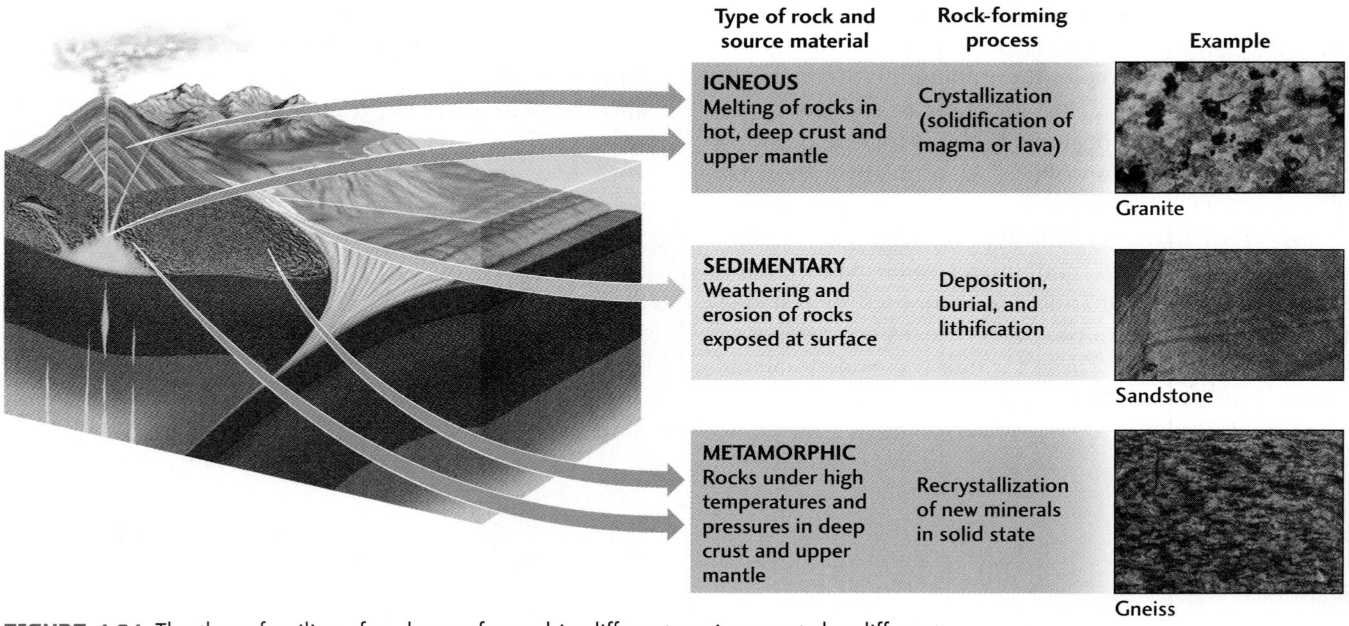

Type of rock and source material	Rock-forming process	Example
IGNEOUS Melting of rocks in hot, deep crust and upper mantle	Crystallization (solidification of magma or lava)	Granite
SEDIMENTARY Weathering and erosion of rocks exposed at surface	Deposition, burial, and lithification	Sandstone
METAMORPHIC Rocks under high temperatures and pressures in deep crust and upper mantle	Recrystallization of new minerals in solid state	Gneiss

FIGURE 4.24 The three families of rocks are formed in different environments by different geologic processes. [*Granite and gneiss: John Grotzinger/Ramón Rivera-Moret/Harvard Mineralogical Museum. Sandstone: John Grotzinger/Ramón Rivera-Moret/MIT.*]

put together. If the crystals or grains, which are only a few millimeters in diameter in most rocks, are large enough to be seen with the unaided eye, the rock is categorized as *coarse-grained*. If they are not large enough to be seen, the rock is categorized as *fine-grained*.

The mineralogy and texture that determine a rock's appearance are themselves determined by the rock's geologic origin—where and how it formed (**Figure 4.24**). The dark rock that caps the sequence of rocks in our road cut, called basalt, was formed by a volcanic eruption. Its mineralogy and texture depend on the chemical composition of rocks that were melted deep within Earth. All rocks formed by the solidification of molten rock are called **igneous rocks.**

> *The physical appearance of a rock is defined primarily by its color, its mineralogy, and the sizes of its crystals or grains.*

The striped white and mauve layers in the road cut are sandstone, formed as sand particles accumulated, perhaps on an ancient beach, and eventually were covered over, buried, and cemented together. All rocks formed as the burial products of layers of sediments (such as sand, mud, or calcium carbonate shells), whether they were laid down on land or under the sea, are called **sedimentary rocks.**

The grayish rock of our road cut, a schist, contains crystals of mica, quartz, and feldspar. It formed deep in Earth's crust as high temperatures and pressures transformed the mineralogy and texture of a buried sedimentary rock. All rocks formed by the transformation of preexisting solid rocks under the influence of high pressure and temperature are called **metamorphic rocks.**

The three types of rocks seen in our road cut represent the three great families of rock: igneous, sedimentary, and metamorphic. Let's take a closer look at each of these families and at the geologic processes that form them.

> *The mineralogy and texture of rocks are determined by the geologic processes by which they form.*

IGNEOUS ROCKS

Igneous rocks (from the Latin *ignis*, meaning "fire") form by crystallization from a magma, a mass of melted rock that originates deep in the crust or upper mantle, where temperatures reach the 700°C or more needed to melt most rocks. When a magma cools slowly in Earth's interior, microscopic crystals start to form. As the magma cools below its melting point, some of these crystals have time to grow to several millimeters in diameter or larger before the whole mass crystallizes as a coarse-grained igneous rock. But when a magma erupts from a volcano onto Earth's surface, it cools and solidifies so rapidly that individual crystals have no time to grow gradually. In that case, many tiny crystals form simultaneously, and the result is a fine-grained igneous rock. Geologists distinguish two major types of igneous

> *Igneous rocks are formed by solidification of molten rock.*

rocks—intrusive and extrusive—on the basis of the sizes of their crystals.

Intrusive and Extrusive Igneous Rocks

Intrusive igneous rocks crystallize when magma intrudes into unmelted rock masses deep in Earth's crust. Large crystals grow as the magma slowly cools, producing coarse-grained rocks (**Figure 4.25**). Intrusive igneous rocks can be recognized by their large, interlocking crystals. Granite is an intrusive igneous rock.

Extrusive igneous rocks form from magmas that erupt at the surface through volcanoes and cool rapidly. Extrusive igneous rocks, such as basalt, are easily recognized by their glassy or fine-grained texture (see Figure 4.25).

Common Minerals in Igneous Rocks

Most of the minerals of igneous rocks are silicates, partly because silicon is so abundant and partly because many silicate minerals melt at the high temperatures and pressures reached in deeper parts of the crust and in the mantle. The common silicate minerals found in igneous rocks include quartz, feldspar, mica, pyroxene, amphibole, and olivine (**Table 4.5**).

SEDIMENTARY ROCKS

Sediments, the precursors of sedimentary rocks, are found on Earth's surface as layers of loose particles, such as sand, silt, and the shells of organisms. These particles originate in the processes of weathering and erosion. **Weathering**

TABLE 4.5	Some Common Minerals of Igneous, Sedimentary, and Metamorphic Rocks	
Igneous Rocks	**Sedimentary Rocks**	**Metamorphic Rocks**
Quartz	Quartz	Quartz
Feldspar	Clay minerals	Feldspar
Mica	Feldspar	Mica
Pyroxene	* Calcite	Garnet
Amphibole	* Dolomite	Pyroxene
Olivine	* Gypsum	Staurolite
	* Halite	Kyanite

An asterisk indicates a nonsilicate mineral.

refers to all the chemical and physical processes that break up and decay rocks into fragments and dissolved substances of various sizes. These particles are then transported by **erosion,** the set of processes that loosen soil and rock and move them downhill or downstream to the spot where they are deposited as layers of sediment (**Figure 4.26**). Sediments are deposited in two ways:

- *Siliciclastic sediments* are made up of physically deposited particles, such as grains of quartz and feldspar derived from weathered granite. (*Clastic* is derived from the Greek word *klastos*, meaning "broken.") These sediments are laid down by running water, wind, and ice and form layers of sand, silt, and gravel.

Extrusive igneous rocks form when lava erupts at the surface and cools rapidly.

The resulting rocks are fine-grained, like this basalt, or have a glassy texture.

Intrusive igneous rocks form when magma intrudes into unmelted rock and cools slowly.

The slow cooling allows large crystals to grow. The resulting rocks are coarse-grained, like this granite.

FIGURE 4.25 Igneous rocks are formed by the crystallization of magma. [John Grotzinger/Ramón Rivera-Moret/Harvard Mineralogical Museum.]

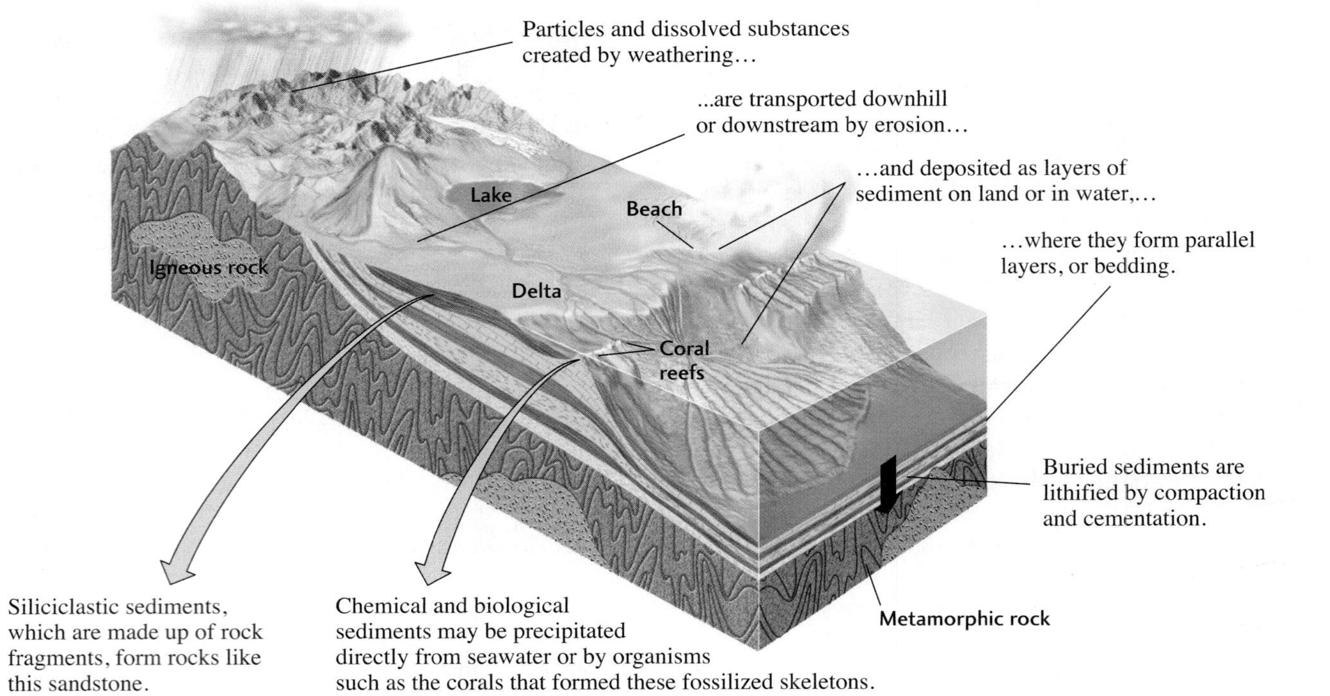

Particles and dissolved substances created by weathering…

…are transported downhill or downstream by erosion…

…and deposited as layers of sediment on land or in water,…

…where they form parallel layers, or bedding.

Lake

Beach

Igneous rock

Delta

Coral reefs

Buried sediments are lithified by compaction and cementation.

Metamorphic rock

Siliciclastic sediments, which are made up of rock fragments, form rocks like this sandstone.

Chemical and biological sediments may be precipitated directly from seawater or by organisms such as the corals that formed these fossilized skeletons.

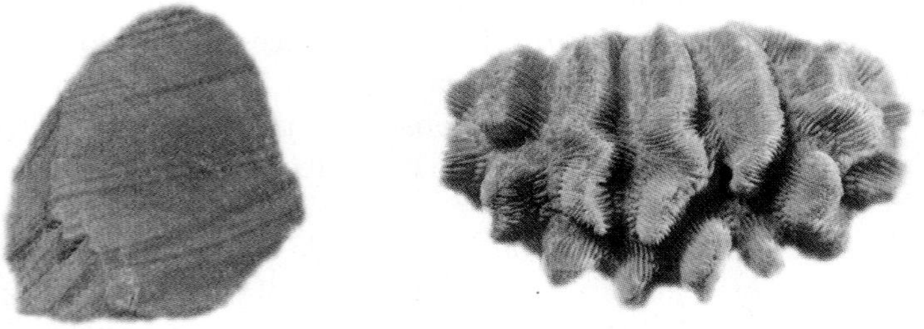

FIGURE 4.26 Sedimentary rocks are formed from the particles of other rocks. Weathering breaks down rocks into fragments and dissolved materials that are then carried downhill and downstream by erosion to be deposited as layers of sediment. (*left*) Sandstone. (*right*) Fossilized coral. [John Grotzinger/Ramón Rivera-Moret/MIT.]

■ *Chemical and biological sediments* are new chemical substances that form by precipitation when some of a rock's components dissolve during weathering and are carried in river waters to the sea. These sediments include layers of such minerals as halite (sodium chloride) and calcite (calcium carbonate, most often found in the form of reefs and shells).

From Sediment to Solid Rock

Lithification is the process that converts sediments into solid rock. It occurs in one of two ways:

■ In *compaction*, particles are squeezed together by the weight of overlying sediment into a mass denser than the original.

■ In *cementation*, minerals precipitate around deposited particles and bind them together.

Sediments are compacted and cemented after they are buried under additional layers of sediment. Sandstone forms by the lithification of sand particles, and limestone forms by the lithification of shells and other particles of calcium carbonate.

Layers of Sediment

Sediments and sedimentary rocks are characterized by **bedding,** the formation of parallel layers of sediment as particles settle to the bottom of a water body or on a land surface. Because sedimentary rocks are formed by surface processes, they cover much of Earth's land surface and seafloor. In terms of surface area, most rocks found at Earth's surface are sedimentary, but these rocks weather easily, so their volume is small compared with that of the igneous and metamorphic rocks that make up the main volume of the crust.

Common Minerals

The common minerals of siliciclastic sediments are silicates, because silicate minerals predominate in the rocks that weather to form sedimentary particles (see Table 4.5). The most abundant silicate minerals in siliciclastic

> *Sedimentary rocks are the result of surface processes on Earth, including weathering, erosion, and the precipitation of dissolved minerals.*

sedimentary rocks are quartz, feldspar, and clay minerals. Clay minerals form by weathering and alteration of preexisting silicate minerals, such as feldspar.

The most abundant minerals of chemically or biologically precipitated sediments are carbonates, such as calcite, the main constituent of limestone. Dolomite, also found in limestone, is a calcium-magnesium carbonate formed by precipitation during lithification. Two other chemical sediments—gypsum and halite—form by precipitation as seawater evaporates.

METAMORPHIC ROCKS

Metamorphic rocks take their name from the Greek words for "change" (*meta*) and "form" (*morphe*). These rocks are produced when high temperatures and pressures deep within Earth cause any kind of preexisting rock—igneous, sedimentary, or other metamorphic rock—to change its mineralogy, texture, or chemical composition while maintaining its solid form. The temperatures of metamorphism are below the melting points of the rocks (about 700°C), but high enough (above 250°C) for the rocks to be changed by recrystallization and chemical reactions.

> Metamorphic rocks are produced by modification of preexisting rocks under conditions of high temperature and pressure deep within Earth.

Regional and Contact Metamorphism

Metamorphism may take place over a widespread area or a limited one (**Figure 4.27**). **Regional metamorphism** occurs where high pressures and temperatures extend over large regions, as happens where plates collide. Regional metamorphism accompanies plate collisions that result in mountain building and the folding and breaking of sedimentary layers that were once horizontal. Where

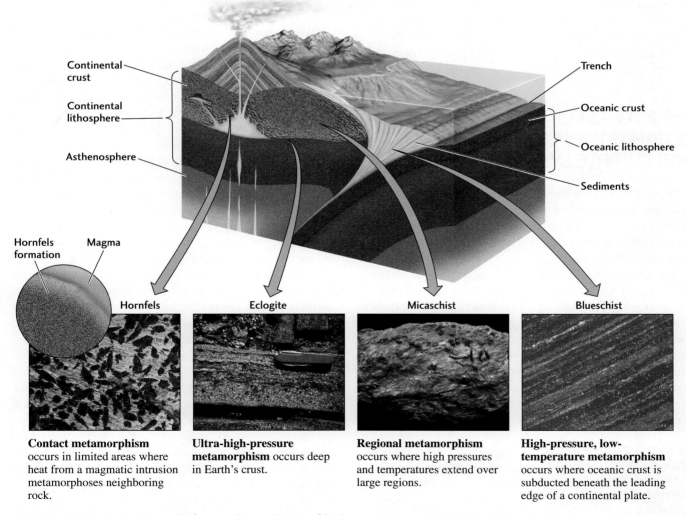

Contact metamorphism occurs in limited areas where heat from a magmatic intrusion metamorphoses neighboring rock.

Ultra-high-pressure metamorphism occurs deep in Earth's crust.

Regional metamorphism occurs where high pressures and temperatures extend over large regions.

High-pressure, low-temperature metamorphism occurs where oceanic crust is subducted beneath the leading edge of a continental plate.

FIGURE 4.27 Metamorphic rocks form under conditions of high temperatures and pressures. [*hornfels: Biophoto Associates/Photo Researchers; eclogite: Julie Baldwin; micaschist: John Grotzinger; blueschist: Mark Cloos.*]

high temperatures are restricted to smaller areas, as in the rocks near and in contact with a magmatic intrusion, rocks are transformed by **contact metamorphism.**

Many regionally metamorphosed rocks, such as schists, have characteristic **foliation,** wavy or flat planes produced when the rock was structurally deformed into folds. Granular textures are more typical of most contact metamorphic rocks and of some regional metamorphic rocks formed by very high pressures and temperatures.

Common Minerals of Metamorphic Rocks

Silicates are the most abundant minerals of metamorphic rocks because the parent rocks from which they are formed are also rich in silicates (see Table 4.5). Typical minerals of metamorphic rocks are quartz, feldspar, mica, pyroxene, and amphibole—the same kinds of silicates characteristic of igneous rocks. Several other silicates—kyanite, staurolite, and some varieties of garnet—are characteristic of metamorphic rocks alone. These minerals form under conditions of high pressure and temperature in the crust and are not characteristic of igneous rocks. They are therefore good indicators of metamorphism. Calcite is the main mineral of marbles, which are metamorphosed limestones.

THE ROCK CYCLE: INTERACTIONS BETWEEN THE PLATE TECTONIC AND CLIMATE SYSTEMS

Earth scientists have known for over 200 years that the three families of rocks—igneous, metamorphic, and sedimentary—all can evolve from one to another. The **rock cycle** explains how each type of rock is converted into one of the other two types.

We also now know that the cycling of rock is the result of interactions between plate tectonics and climate. For example, the melting of subducting lithospheric slabs and the formation of magma result from processes operating within the plate tectonic system. When these molten rocks erupt to the land surface, the newly formed igneous rocks are subject to weathering by the climate system. The same process injects volcanic ash and carbon dioxide gas high into the atmosphere, where they may affect global climate. As global climate changes, perhaps becoming warmer or cooler, the rate of rock weathering changes, which in turn influences the rate at which rock material (sediment) is returned to Earth's interior.

> The rock cycle is a set of processes that can transform each type of rock into the other two types.

Let's trace one turn of the rock cycle, beginning with the creation of new oceanic lithosphere at a mid-ocean ridge spreading center as two continents drift apart (**Figure 4.28**). The ocean gets wider and wider, until at some point the process reverses itself and the ocean closes. As the ocean basin closes, igneous rocks created at the mid-ocean ridge eventually descend into a subduction zone beneath a continental plate. Sediments that were formed on the continent and transported to its edge may also be dragged down into the subduction zone. Ultimately, the two continents, which were once drifting apart, may now move closer together and collide. As the igneous rocks and sediments that descend into the subduction zone go deeper and deeper into Earth's interior, they begin to melt to form a new generation of igneous rocks. The great heat associated with the intrusion of these igneous rocks, coupled with the heat and pressure that come with being pushed to levels deep within Earth, transforms these igneous rocks—and other surrounding rocks—into metamorphic rocks. When the continents collide, these igneous and metamorphic rocks are uplifted into a high mountain chain as a section of Earth's crust crumples and deforms.

Weathering of these uplifted igneous and sedimentary rocks results in the formation of loose materials that erosion then strips away. Water and wind transport some of these materials across the continents and eventually to the edges of the continents, where they are deposited as sediments. The sediments laid down where the land meets the ocean are buried under successive layers of sediments, where they slowly lithify into sedimentary rock. These oceans, like those mentioned at the beginning of the cycle, may also have formed by spreading along mid-ocean ridges, thus completing the rock cycle.

The particular pathway illustrated here—that of a continent breaking apart, forming a new ocean basin, then closing back up again—is only one variation among the many that may take place in the rock cycle. Any type of rock—igneous, sedimentary, or metamorphic—can be uplifted during a mountain-building event and then weathered and eroded to form new sediments. Some stages may be omitted: as a sedimentary rock is uplifted and eroded, for example, metamorphism and melting are skipped. In some cases, the rock cycle may proceed very slowly. For example, we know that some igneous and metamorphic rocks many kilometers deep in the crust may be uplifted or exposed to weathering and erosion only after billions of years have passed.

The essence of the rock cycle is that plate tectonic processes drive rocks at the surface deep into Earth, where they may become metamorphosed or even melt, and then return these rocks to the surface. Weathering at the surface decomposes these rocks to produce loose materials that may accumulate as thick sediments, only to be driven down deep into Earth again by plate tectonics.

1 The cycle begins with rifting within a continent. Sediments erode from the continental interior and are deposited in rift basins, where they are buried to form sedimentary rocks.

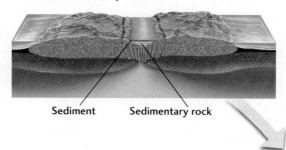

Sediment   Sedimentary rock

2 Rifting and spreading continue, and a new ocean basin develops. Magma rises from the asthenosphere at mid-ocean ridges and chills to form basalt, an igneous rock.

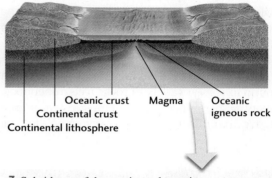

Oceanic crust   Magma   Oceanic
Continental crust   igneous rock
Continental lithosphere

6 Streams transport sediment away from collision zones to oceans, where it is deposited as layers of sand and silt. Layers of sediment are buried and lithify to form sedimentary rock.

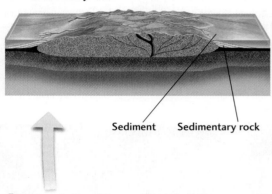

Sediment   Sedimentary rock

3 Subsidence of the continental margin—sinking of Earth's lithosphere—leads to accumulation of sediment and formation of sedimentary rock during burial.

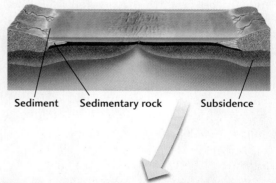

Sediment   Sedimentary rock   Subsidence

5 Further closing of the ocean basin leads to continental collision, forming high mountain ranges. Where continents collide, rocks are buried deeper or modified by heat and pressure, forming metamorphic rocks. Uplifted mountains force moisture-laden air to rise, cool, and release its moisture as precipitation. Weathering creates loose material—soils and sediment—that erosion strips away.

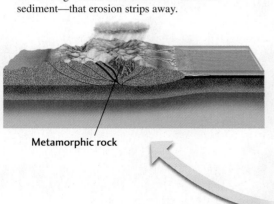

Metamorphic rock

4 Oceanic crust subducts beneath a continent, building a volcanic mountain chain. The subducting plate melts as it descends. Magma rises from the melting plate and mantle and cools to make granitic igneous rocks.

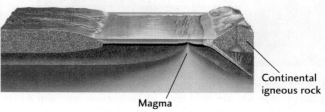

Continental
igneous rock

Magma

FIGURE 4.28 The rock cycle results from the interaction of the plate tectonic and climate systems.

The rock cycle never ends. It is always operating at different stages in various parts of the world, forming and eroding mountains in one place and laying down and burying sediments in another. The rocks that make up the solid Earth are recycled continuously, but we can see only the surface parts of the cycle. We must deduce the recycling of the deep crust and the mantle from indirect evidence.

NATURAL RESOURCES: Concentrations of Valuable Minerals

The rock cycle turns out to be crucial in creating economically important concentrations of valuable minerals in Earth's crust. Finding these minerals and extracting them is a vital job for Earth scientists, so we now turn our attention to how and where some of these geological prizes are formed.

The chemical elements of Earth's crust are widely distributed in many kinds of minerals, and those minerals are found in a great variety of rocks (see Appendix 4 for a summary of the properties of the most common minerals of Earth's crust). In most places, any given element will be found homogenized with other elements in amounts close to its average concentration in the crust. An ordinary granitic rock, for example, may contain a small percentage of iron, close to the average concentration of iron in Earth's crust.

When an element is present in higher concentrations, it means that the rock underwent some geologic process that concentrated larger quantities of that element than normal. The *concentration factor* of an element in a mineral deposit is the ratio of the element's abundance in the deposit to its average abundance in the crust. High concentrations of elements are found in a limited number of specific geological settings. These settings are of economic interest because the higher the concentration of a resource in a given deposit, the lower the cost to recover it.

Ore Minerals

Ores are rich deposits of minerals from which valuable metals can be recovered profitably. The minerals containing these metals are *ore minerals*. Ore minerals include sulfides (the largest group), oxides, and silicates. The ore minerals in each of these groups are compounds of metallic elements with sulfur, oxygen, and silicon oxide,

FIGURE 4.29 Some metals are found in their native state. (*left*) A geologist examines rock samples in an underground gold mine in Zimbabwe, southern Africa. [*Peter Bowater/Photo Researchers.*] (*right*) Native gold on a quartz crystal. [*Chip Clark.*]

respectively. The copper ore mineral covelite, for example, is a copper sulfide (CuS). The iron ore mineral hematite (Fe_2O_3) is an iron oxide. The nickel ore mineral garnierite is a nickel silicate, $Ni_3Si_2O_5(OH)_4$. In addition, some metals, such as gold, are found in their native state—that is, uncombined with other elements (**Figure 4.29**).

Recall in our discussion of the rock cycle that continental margins where subduction occurs may be associated with melting of oceanic lithosphere to form igneous rocks. Very large ore deposits can be formed in such a tectonic setting when hot water solutions—also known as **hydrothermal solutions**—are formed around bodies of molten rock. This happens when circulating groundwater or seawater comes into contact with a magmatic intrusion, reacts with it, and carries off significant quantities of elements and ions released by the reaction.

These elements and their ions then interact with one another to deposit ore minerals, usually as the solution cools.

Exploration for ore minerals is an important and challenging activity that employs many geologists. Finding a promising deposit is only the first step toward extracting useful materials, however. The shape of the deposit and the distribution and concentration of the ore must be estimated before mining begins. This is done by drilling closely spaced holes and obtaining continuous cores through the ore deposit and the surrounding rock (**Figure 4.30**). Information from the cores is used to create a three-dimensional model of the ore deposit. That model is then used to evaluate whether or not the deposit is large enough, and has a high enough concentration of minerals to justify opening a mine.

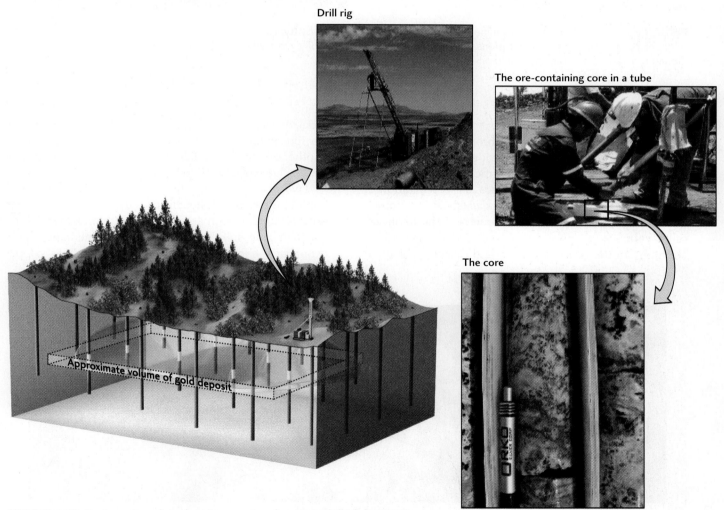

Drill rig

The ore-containing core in a tube

The core

FIGURE 4.30 During mineral exploration, an ore deposit is drilled to provide core samples for geochemical and mineral analysis. A rotating metal tube, studded with diamond teeth, cuts into the deposit. The hollow space in the tube becomes filled with solid rock, which is then extracted when the tube is pulled out of the rock. The core has the shape of a cylinder. [*Photos by Ben Whiting, P.Geo.*]

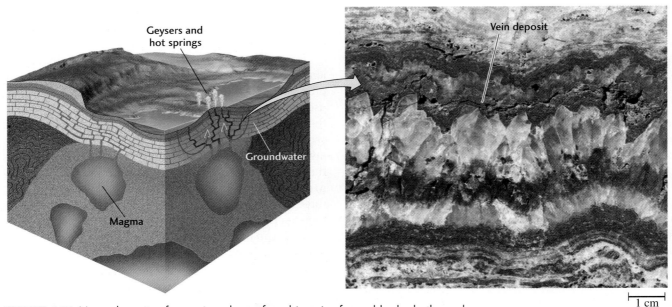

FIGURE 4.31 Many deposits of ore minerals are found in veins formed by hydrothermal solutions rising from magmatic intrusions. This quartz vein deposit (about 1 cm thick) in Oatman, Arizona, contains gold and silver ores. [*Peter Kresan.*]

Veins

Hydrothermal solutions moving through fractured rocks often deposit ore minerals. These fluids flow easily through the fractures and joints of the rocks, cooling rapidly in the process. Quick cooling causes fast precipitation of the ore minerals. The resulting *tabular* (sheetlike) deposits of precipitated minerals in the fractures and joints are called **veins.** Some ore minerals are found in veins; others are found in the rocks surrounding the veins, which are altered when the hot solutions heat and infiltrate those rocks. As the solutions react with surrounding rocks, they may precipitate ore minerals together with quartz, calcite, or other common vein-filling minerals. Vein deposits are a major source of gold (**Figure 4.31**).

Hydrothermal vein deposits are among the most important sources of metallic ores. Typically, metallic ores exist as sulfides, such as iron sulfide (pyrite), lead sulfide (galena), zinc sulfide (sphalerite), and mercury sulfide (cinnabar)—shown in **Figure 4.32**—as well as copper

Galena (lead sulfide)   Cinnabar (mercury sulfide)   Pyrite (iron sulfide)   Sphalerite (zinc sulfide)

FIGURE 4.32 Metalic sulfide ores. Sulfides are the most common types of metallic ores. [*Chip Clark.*]

sulfide (covelite and chalcocite). Hydrothermal solutions reach the surface as hot springs and geysers, many of which precipitate metallic ores—including ores of lead, zinc, and mercury—as they cool.

Disseminated Deposits

Mineral deposits that are scattered through volumes of rock much larger than veins are called *disseminated deposits*. In both igneous and sedimentary rocks, minerals are disseminated along abundant cracks and fractures. Among the economically important disseminated deposits are the copper deposits of Chile and the southwestern United States. These deposits develop in geologic provinces with abundant igneous rocks, usually emplaced as large intrusive bodies. In Chile, these intrusive igneous rocks are related to the subduction of oceanic lithosphere beneath the Andes, an event very similar to what was described in our example of the rock cycle. The most common copper mineral in these deposits is chalcopyrite, a copper sulfide (**Figure 4.33**). The copper was deposited when ore-forming minerals were introduced into a great number of tiny fractures in granitic intrusive rocks and in the rocks surrounding the higher parts of the igneous intrusions. Some unknown process associated with the magmatic intrusion or its aftermath broke these rocks

into millions of pieces. Hydrothermal solutions penetrated and recemented the rocks by precipitating ore minerals throughout the extensive network of tiny fractures. This widespread dispersal produced a low-grade but very large resource of many millions of tons of ore, which can be mined economically by large-scale methods (**Figure 4.34**).

Disseminated hydrothermal deposits are also present in sedimentary rocks. This is the case in the lead-zinc province of the Upper Mississippi Valley, which extends from southwestern Wisconsin to Kansas and Oklahoma. The ores in this province are not associated with a known magmatic intrusion that could have been a source of hydrothermal fluids, so their origin must be very different. Some geologists speculate that the ores were deposited by groundwater that was driven out of the ancestral Appalachian Mountains when they were much higher. A continental collision between North America and Africa may have created a continental-scale squeegee that pushed fluids from deep within the collision zone all the way into the continental interior of North America. Groundwater may have penetrated hot crustal rocks at great depths and dissolved soluble ore minerals, then moved upward into the overlying sedimentary rocks, where it precipitated the minerals as fillings in cavities. In some cases, it appears that these solutions infiltrated limestone formations and

Chalcopyrite
(a copper sulfide)

Malachite
(a copper carbonate)

Chalcocite
(a copper sulfide)

FIGURE 4.33 Copper ores. Chalcopyrite and chalcocite are copper sulfide ores. Malachite is a carbonate of copper found in association with sulfides of copper. [*Chip Clark.*]

FIGURE 4.34 Kennecott Copper Mine, Utah, an open-pit mine. Open-pit mining is typical of the large-scale methods used to exploit widely disseminated ore deposits. [*David R. Frazier/The Image Works.*]

dissolved some carbonates, then replaced the carbonates with equal volumes of new sulfide crystals. The major minerals of the hydrothermal deposits in this province are lead sulfide (galena) and zinc sulfide (sphalerite).

Igneous Deposits

The most important *igneous ore deposits*—deposits of ore in igneous rocks—are found as segregations of ore minerals near the bottoms of intrusions. These deposits form when minerals crystallize from molten magma, settle, and accumulate on the floor of a magma chamber. Most of the chromium and platinum ores in the world, such as the deposits in South Africa and Montana, are found as layered accumulations of minerals that formed in this way (**Figure 4.35**). One of the richest ore bodies ever found, at Sudbury, Ontario, is a large mafic igneous intrusion containing great quantities of layered nickel, copper, and iron sulfides near its base. Geologists believe that these sulfide deposits formed from crystallization of

a dense, sulfide-rich liquid that separated from the rest of the cooling magma and sank to the bottom of the chamber before it congealed.

Pegmatites are extremely coarse-grained intrusive rocks of granitic composition. As the magma in a large granitic intrusion cools, the last material to crystallize forms pegmatites, in which minerals present in only trace amounts in the parent rock are concentrated. Pegmatites may contain rare mineral deposits rich in such elements as boron, lithium, fluorine, niobium, and uranium and in such gem minerals as tourmaline.

Sedimentary Deposits

Sedimentary mineral deposits include some of the world's most valuable mineral sources. Many economically important minerals segregate as an ordinary result of sedimentary processes. Sedimentary mineral deposits are also important sources of copper, iron, and other metals. These deposits are chemically precipitated in sedimentary

FIGURE 4.35 Chromite (chromium ore, visible as dark layers) in a layered igneous intrusion in the Bushveldt Complex, South Africa. [*Spence Titley.*]

environments to which large quantities of metals are transported in solution. Some of the important sedimentary copper ores, such as those of the Permian Kupferschiefer (German for "copper slate") beds of Germany, may have precipitated from hydrothermal solutions rich in metal sulfides that interacted with sediments on the seafloor. The tectonic setting of these deposits may have been something like the mid-ocean ridge described in our example of the rock cycle, except that it developed within a continent. Here, rifting of the continental crust led to development of a deep trough, where sediments and ore minerals were deposited in a very still, narrow sea.

Many rich deposits of gold, diamonds, and other heavy minerals such as magnetite and chromite are found in *placers*, sedimentary ore deposits that have been concentrated by the mechanical sorting action of river currents. These ore deposits originate where uplifted rocks weather to form grains of sediment, which are then sorted by weight when currents of water flow over them. Because heavy minerals settle out of a current more quickly than lighter minerals such as quartz and feldspar, the heavy minerals tend to accumulate on river bottoms and sandbars. Here the current is strong enough to keep the lighter minerals suspended and in transport, but too weak to move the heavier minerals. Similarly, ocean waves preferentially deposit heavy minerals on beaches or shallow offshore bars. A gold panner accomplishes the same thing: the shaking of a water-filled pan allows the lighter minerals to be washed away, leaving the heavier gold in the bottom of the pan (**Figure 4.36**). Some placers can be traced upstream to the location of the original mineral deposit, usually of igneous origin, from which the minerals were eroded. Erosion of the Mother Lode, an extensive gold-bearing vein

> *Ore deposits are formed in geological settings where elements accumulate in concentrations that far exceed their normal abundances. Precipitation from hydrothermal solutions, crystallization and settlement in magmas, and selective transport of heavy minerals in streams are a few examples of the processes by which elements can be concentrated.*

FIGURE 4.36 Panning for gold was popularized by "forty-niners" during the California gold rush and is still popular in the San Gabriel River today. [*(left) Bo Zaunders/Corbis; (right) David Butow/Corbis SABA.*]

system lying along the western flanks of the Sierra Nevada, produced the placers that were discovered in 1848 and led to the California gold rush. The placers were found before their source was discovered. Placers also led to the discovery of the Kimberley diamond mines of South Africa two decades later.

■ SUMMARY

What is a mineral? Minerals, the building blocks of rocks, are naturally occurring, inorganic solids with specific crystal structures and chemical compositions that either are fixed or vary within a defined range. A mineral is constructed of atoms, the small units of matter that combine in chemical reactions. An atom is composed of a nucleus made up of protons and neutrons and surrounded by electrons. The atomic number of an element is the number of protons in its nucleus, and its atomic mass is the sum of the masses of its protons and neutrons.

How do atoms combine to form the crystal structures of minerals? Chemical elements react with one another to form compounds by either gaining or losing electrons to become ions or by sharing electrons. The ions in some chemical compounds are held together by ionic bonds, which form by electrostatic attraction between positive ions (cations) and negative ions (anions). Atoms that share electrons to form a compound are held together by covalent bonds. When a mineral crystallizes, atoms or ions come together in the proper proportions to form a crystal structure—an orderly three-dimensional array in which the basic arrangement of the atoms is repeated in all directions.

What are the major rock-forming minerals? Silicate minerals, the most abundant minerals in Earth's crust, are crystal structures built of silicate tetrahedra linked in various ways. Tetrahedra may be isolated (olivines) or in single chains (pyroxenes), double chains (amphiboles), sheets (micas), or frameworks (feldspars). Carbonate minerals are made of carbonate ions bonded to calcium or magnesium or both. Oxide minerals are compounds of oxygen and metallic elements. Sulfide and sulfate minerals are composed of sulfur atoms in combination with metallic elements.

What are the physical properties of minerals? A mineral's physical properties, which indicate its composition and structure, include hardness—the ease with which its surface is scratched; cleavage—its ability to split or break along flat surfaces; fracture—the way it breaks along irregular surfaces; luster—the way it reflects light; color—imparted by transmitted or reflected light to crystals or irregular masses, or visible as streak (the color of a fine powder); density—mass per unit volume; and crystal habit—the shapes of individual crystals or aggregates.

What determines the properties of the various kinds of rocks that form in and on Earth's surface? Mineralogy (the kinds and proportions of minerals that make up a rock) and texture (the sizes, shapes, and spatial arrangement of its crystals or grains) define a rock. The mineralogy and texture of a rock are determined by the geologic processes by which it formed.

What are the three families of rocks, and how do they form? Igneous rocks form by the crystallization of magmas as they cool. Intrusive igneous rocks form in Earth's interior and have large crystals. Extrusive igneous rocks, which form at Earth's surface where lavas erupt from volcanoes, have a glassy or fine-grained texture. Sedimentary rocks form by the lithification of sediments after burial. Sediments are derived from the weathering and erosion of rocks exposed at Earth's surface. Metamorphic rocks form by changes in igneous, sedimentary, or other metamorphic rocks that are subjected to high temperatures and pressures in Earth's interior.

How does the rock cycle explain the transformation of rocks from one type into another? The rock cycle relates geologic processes to the formation of the three types of rocks from one another. We can view these processes by starting at any point in the cycle, such as the creation of new oceanic lithosphere at a mid-ocean ridge as two continents drift apart. The ocean basin gets wider until at some point the process reverses itself. As the basin closes, igneous rocks and sediments are subducted beneath a continental plate. As they descend into Earth's interior, they begin to melt to form a new generation of igneous rocks. The heat associated with the intrusion of these igneous rocks, coupled with the heat and pressure that come with being pushed to levels deep within Earth, transforms these igneous rocks—and other surrounding rocks—into metamorphic rocks. Ultimately,

the two continents may collide, and these igneous and metamorphic rocks may be uplifted into a high mountain chain. The uplifted rocks slowly weather, and their fragments are deposited as sediments. Plate tectonics and climate are the mechanisms by which the rock cycle operates.

What is hydrothermal mineral deposition? Hydrothermal deposits, which are some of the most important ore mineral deposits, are formed when circulating groundwater or seawater reacts with a magmatic intrusion to form a hydrothermal solution. The heated water leaches soluble minerals in its path and transports them to cooler rocks, where they are precipitated in fractures, joints, and voids. The resulting ores may be found in veins or in disseminated deposits.

How do igneous ore deposits form? Igneous ore deposits typically form when minerals crystallize from molten magma, settle, and accumulate on the floor of a magma chamber. They are often found as layered accumulations of minerals.

■ KEY TERMS AND CONCEPTS

atom (p. 96)

bedding (p. 116)

chemical bond (p. 98)

chemical reaction (p. 97)

cleavage (p. 108)

color (p. 110)

contact metamorphism (p. 118)

crystal (p. 100)

crystal habit (p. 111)

crystallization (p. 99)

density (p. 102)

erosion (p. 115)

extrusive igneous rock (p. 115)

foliation (p. 118)

fracture (p. 110)

hardness (p. 107)

hydrothermal solution (p. 121)

igneous rock (p. 114)

intrusive igneous rock (p. 115)

ion (p. 97)

isotope (p. 97)

lithification (p. 116)

luster (p. 110)

magma (p. 100)

metamorphic rock (p. 114)

mineral (p. 95)

mineralogy (p. 95)

Mohs scale of hardness (p. 107)

ore (p. 120)

polymorph (p. 102)

regional metamorphism (p. 117)

rock (p. 113)

rock cycle (p. 118)

sediment (p. 115)

sedimentary rock (p. 114)

specific gravity (p. 111)

streak (p. 110)

texture (p. 113)

vein (p. 122)

weathering (p. 115)

■ EXERCISES

1. Define a mineral.

2. What is the difference between an atom and an ion?

3. Name two types of chemical bonds.

4. List the basic crystal structures of silicate minerals.

5. Name three classes of minerals other than silicates.

6. How would a field geologist measure hardness?

7. What are the differences between extrusive and intrusive igneous rocks?

8. What are the differences between regional and contact metamorphism?

9. What are the differences between siliciclastic and chemical or biological sediments?

10. List three common silicate minerals found in each of the three families of rocks: igneous, sedimentary, and metamorphic.

11. Of the three families of rocks, which form at Earth's surface and which in the interior of the crust?

12. What are the characteristics of an economically important ore deposit?

13. Choose two minerals from Appendix 4 that you think might make good abrasive or grinding stones for sharpening steel, and describe the physical property that causes you to believe they would be suitable for that purpose.

14. Aragonite, with a density of 2.9 g/cm^3, has exactly the same chemical composition as calcite, with a density of 2.7 g/cm^3. Other things being equal, which of these two minerals is more likely to have formed under high pressure?

15. There are at least seven physical properties that can be used to identify an unknown mineral. Which ones are most useful in discriminating between minerals that look similar? Describe a strategy that would allow you to prove that an unknown clear calcite crystal is not the same as a known clear crystal of quartz.

16. Using the rock cycle, trace the path from a magma to a granite intrusion to a metamorphic gneiss to a sandstone. Be sure to include the role of tectonics and the specific processes that create the rocks.

17. Where are igneous rocks most likely to be found? How could you be certain that the rocks were igneous and not sedimentary or metamorphic?

18. Back in the late 1800s, gold miners used to pan for gold by placing sediment from rivers in a pan and filtering water through the pan while swirling the pan's contents. The miners wanted to be certain that they had found real gold, not pyrite ("fool's gold"). Why did this method work? What mineral property does the process of panning for gold use? What is another possible method for distinguishing between gold and pyrite?

Visual Literacy Task

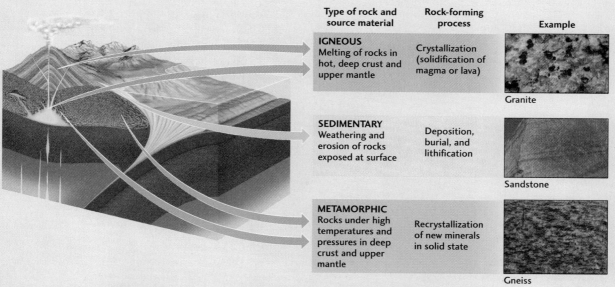

Type of rock and source material	Rock-forming process	Example
IGNEOUS Melting of rocks in hot, deep crust and upper mantle	Crystallization (solidification of magma or lava)	Granite
SEDIMENTARY Weathering and erosion of rocks exposed at surface	Deposition, burial, and lithification	Sandstone
METAMORPHIC Rocks under high temperatures and pressures in deep crust and upper mantle	Recrystallization of new minerals in solid state	Gneiss

FIGURE 4.24 The three families of rocks are formed in different geologic environments by different geologic processes. [*Granite and gneiss: John Grotzinger/Ramón Rivera-Moret/ Harvard Mineralogical Museum. Sandstone: John Grotzinger/Ramón Rivera-Moret/MIT.*]

1. Which rock type *cannot* form deep within Earth's crust?

a. Igneous
b. Sedimentary
c. Metamorphic

2. What is the primary type of rock you would expect to see at the center of Location A?

a. Igneous
b. Sedimentary
c. Metamorphic

3. Gabbro forms from cooling magma. Based on how it forms, which of the pictured rocks is gabbro most similar to?
a. Granite
b. Sandstone
c. Gneiss

4. What rock-forming process results in the formation of minerals large enough that you can see them individually?

a. Deposition
b. Weathering
c. Melting
d. Melting

5. What rock type can form from broken-up pieces of other rock?

a. Igneous
b. Sedimentary
c. Metamorphic

Thought Question: How would you use this illustration to explain the rock cycle to a fellow student?

Igneous Processes and Volcanism

5

A Volcanic Ash Cloud over Europe

ON APRIL 14, 2010, Eyjafjallajökull volcano in Iceland began a series of eruptions that caused tremendous disruption to air travel over western and northern Europe for a period of 6 days. (Legend has it that the word "Eyjafjallajökull" in the native Icelandic language means "name that no one can pronounce." Actually, it is pronounced aye'-ya-fyah'-dla-jow-kudl, and it means "island-mountain glacier.") These eruptions led to the closure of most of Europe's larger airports which caused the cancellation of many flights to and from Europe, resulting in the highest level of air traffic disruption since World War II. Many people were stranded for days with little comfort as flights were sequentially canceled, leaving people in many cities struggling to find alternative means of transportation or hotels. In the first 6 days following the eruption, it is estimated that 250,000 British, French, and Irish citizens were stranded abroad, the European economy may have lost almost $2 billion, and the aviation industry lost up to $200 million per day.

The eruptions were predicted well in advance. Seismic activity in and around Eyjafjallajökull began in late 2009 and increased in intensity and frequency until March 20, 2010, when a small eruption occurred. A second, much larger eruption occurred on April 14, resulting in an estimated 250 million m^3 of ejected volcanic ash. The ash cloud rose to elevations of 9000 m and, though not as large as the 1980 Mount St. Helens eruption in Oregon, it was high enough to enter the jet stream that flowed directly over Iceland at the time. In the northern hemisphere the jet stream flows from west to east, so the ash was transported directly to Europe, where it spread out over a large part of the continent. Without the push of the jet stream, the ash would have mostly settled in the Atlantic Ocean.

> The ash cloud rose to elevations of 9000 m, high enough to enter the jet stream that created a conveyor belt direct to Europe.

Much of this volcanic ash was very fine-grained, its particles less than 2 mm in size. At this size the particles are fine enough to get caught up in jet engines, which produce temperatures of up to 2000°C, hot enough to remelt the ash. When this happens the ash is essentially turned back into lava, which then sticks to engine parts, possibly causing engine failure. In extreme examples, planes have had to literally glide their way out of the ash cloud before engines could be restarted.

Volcanoes have tremendous potential to use Earth's heat for alternative power production, and they also form beautiful landscapes. The eruption of Eyjafjallajökull is a reminder that volcanoes can also negatively influence our lives, even if we live a long way away. ◆

In this April 16, 2010, photo, the volcano in southern Iceland's Eyjafjallajökull glacier sends ash into the air just prior to sunset.
[AP Photo/Brynjar Gauti.]

CHAPTER OUTLINE

AS WE LEARNED IN CHAPTER 4, much of Earth's crust is composed of igneous rock. Understanding the processes that melt and resolidify rocks is a key to understanding how Earth's crust forms.

We also learned that plate tectonic processes create a wide variety of igneous rocks. Specifically, igneous rocks form at spreading centers where plates move apart, along convergent boundaries where one plate descends beneath another, and at hot spots where magmas ascend from great depths in the mantle (**Figure 5.1**).

In this chapter, we will look in more detail at the igneous processes associated with specific plate tectonic settings. We still have much to learn about the exact mechanisms of melting and solidification: How do igneous rocks differ from one another? Where do igneous rocks form? How do rocks solidify from magma? Where do magmas form? We will give special attention to volcanoes, which allow us to see into Earth's deep interior to understand the igneous processes that have generated Earth's oceanic and continental crust.

We will examine how magma rises through the crust, emerges onto the surface as lava, and cools into hard volcanic rock. We will see how tectonic processes produce volcanism at plate boundaries. We will also consider the destructive power of volcanoes and the hazards they pose to human society. The vast active volcano that stretches across the wilderness of Yellowstone National Park, for example, expels more energy in the form of heat than all the electric power consumed in the three surrounding states of Wyoming, Idaho, and Montana combined. That energy is not released steadily, however; some of it builds up in hot magma chambers until the volcano blows its top. A cataclysmic eruption of the Yellowstone volcano 630,000 years ago ejected 1000 km^3 of material into the air, covering regions as far away as Texas and California with a layer of volcanic ash.

The geologic record shows that volcanic explosions nearly as big as or bigger than the Yellowstone eruption have occurred in the western United States at least six times during the last 2 million years, so we can be fairly certain that such an eruption will happen again. We can only imagine what it might do to civilization. Hot ash will snuff out all life within 100 km or more, and cooler but choking ash will blanket the ground more than 1000 km away. Dust thrown high into the stratosphere will dim the Sun for several years, dropping temperatures and plunging the Northern Hemisphere into an extended volcanic winter.

HOW DO IGNEOUS ROCKS DIFFER FROM ONE ANOTHER?

Today we classify igneous rock samples in the same way some geologists did in the late nineteenth century: by their texture and by their chemical and mineral composition.

Texture

Two hundred years ago, the first division of igneous rocks was made on the basis of texture, which largely reflects

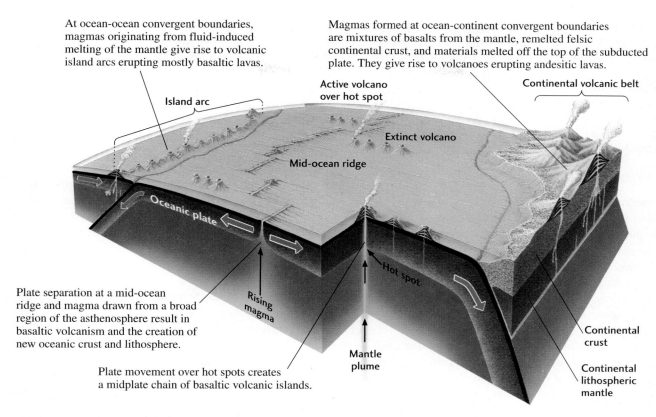

At ocean-ocean convergent boundaries, magmas originating from fluid-induced melting of the mantle give rise to volcanic island arcs erupting mostly basaltic lavas.

Magmas formed at ocean-continent convergent boundaries are mixtures of basalts from the mantle, remelted felsic continental crust, and materials melted off the top of the subducted plate. They give rise to volcanoes erupting andesitic lavas.

Island arc

Active volcano over hot spot

Continental volcanic belt

Extinct volcano

Mid-ocean ridge

Oceanic plate

Plate separation at a mid-ocean ridge and magma drawn from a broad region of the asthenosphere result in basaltic volcanism and the creation of new oceanic crust and lithosphere.

Rising magma

Hot spot

Continental crust

Mantle plume

Continental lithospheric mantle

Plate movement over hot spots creates a midplate chain of basaltic volcanic islands.

FIGURE 5.1 Plate tectonic processes explain the global pattern of igneous processes.

differences in mineral *crystal size.* Geologists classified rocks as either coarse-grained or fine-grained (see Chapter 4). Crystal size is a simple characteristic that geologists can easily see in the field. A coarse-grained rock such as granite has distinct crystals that are easily visible to the unaided eye. In contrast, the crystals of fine-grained rocks such as basalt are too small to be seen, even with a magnifying glass. **Figure 5.2** shows samples of granite and

Granite

Basalt

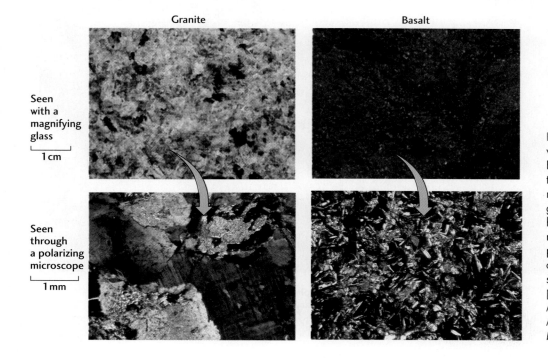

Seen with a magnifying glass

1 cm

Seen through a polarizing microscope

1 mm

FIGURE 5.2 Igneous rocks were first classified by texture. Early geologists assessed texture with a small hand-held magnifying glass. Modern geologists have access to high-powered polarizing microscopes, which can produce photomicrographs of thin, transparent rock slices like those shown here. [*John Grotzinger/Ramón Rivera-Moret/Harvard Mineralogical Museum. Photomicrographs by Steven Chemtob.*]

basalt, accompanied by photomicrographs of very thin, transparent slices of each rock. *Photomicrographs*, which are simply photographs taken through a microscope, give us an enlarged view of minerals and their textures. Textural differences were clear to early geologists, but several more clues were needed to unravel the meaning of the differences.

> The variation among igneous rocks is due to variations in their texture, which largely reflects crystal size, and variations in their mineral composition.

First Clue: Volcanic Rocks Early geologists observed volcanic rocks forming from lava during volcanic eruptions. (**Lava** is the term we apply to magma flowing out onto Earth's surface.) They noted that when lava cooled rapidly, it formed either a fine-grained rock or a glassy one in which no crystals could be distinguished. Where lava cooled more slowly, as in the middle of a thick flow many meters high, somewhat larger crystals were present.

Second Clue: Laboratory Studies of Crystallization Just over a hundred years ago, experimental scientists began to understand the nature of crystallization. Anyone who has frozen a tray of ice cubes knows that water solidifies to ice in a few hours as its temperature drops below the freezing point. If you have ever attempted to retrieve your ice cubes before they were completely solid, you may have seen thin ice crystals forming at the surface and along the sides of the tray. During crystallization, the water molecules take up fixed positions in the solidifying crystal structure, and they are no longer able to move freely, as they did when the water was liquid. All other liquids, including magmas, crystallize in this way.

The first tiny crystals form a pattern. Other atoms or ions in the crystallizing liquid then attach themselves in such a way that the tiny crystals grow larger. It takes some time for the atoms or ions to "find" their correct places on a growing crystal, so large crystals form only if they have time to grow slowly. If a liquid solidifies very quickly, as a magma does when it erupts onto the cool surface of Earth, the crystals have no time to grow. Instead, a large number of tiny crystals form simultaneously as the liquid cools and solidifies.

Third Clue: Granite—Evidence of Slow Cooling By studying volcanoes, early geologists determined that finely crystalline textures indicate quick cooling at Earth's surface and that finely crystalline igneous rocks are evidence of former volcanism. But in the absence of direct observation, how could geologists deduce that coarse-grained rocks form by *slow* cooling deep in Earth's interior? Granite—one of the commonest rocks of the continents—turned out to be the crucial clue (**Figure 5.3**). James Hutton, one of geology's founding fathers, saw granite cutting across and disrupting

Granitic intrusion

Metamorphosed sedimentary rock

FIGURE 5.3 Granitic intrusions (light-colored) cutting across metamorphosed sedimentary rock. [*Tom Bean/DRK PHOTO.*]

layers of sedimentary rocks as he worked in the field in Scotland. He noticed that the granite had somehow fractured and invaded the sedimentary rocks, as though the granite had been forced into the fractures as a liquid.

As Hutton looked at more and more granites, he began to focus on the sedimentary rocks bordering them. He observed that the minerals of the sedimentary rocks in contact with the granite were different from those found in sedimentary rocks at some distance from the granite. He concluded that the changes in the sedimentary rocks must have resulted from great heat and that the heat must have come from the granite. Hutton also noted that the granite was composed of interlocked crystals (see Figure 5.2). By this time, chemists had established that a slow crystallization process produces this pattern.

> *The texture of igneous rocks depends on crystal size. Crystal size depends on how fast magma cools—the more slowly it cools, the larger the crystal size in the resulting rock.*

With these three lines of evidence, Hutton proposed that granite forms from a hot molten material that solidifies deep within Earth. The evidence was conclusive, because no other explanation could accommodate all the facts. Other geologists, who saw the same characteristics of granites in widely separated places in the world, came to recognize that granite and many similar coarsely crystalline rocks were the products of magma that had crystallized slowly in the interior of Earth.

Intrusive Igneous Rocks   The full significance of an igneous rock's texture is now clear: its texture is linked to the rate and therefore the place of cooling. Slow cooling of magma in Earth's interior allows adequate time for the growth of the interlocking large, coarse crystals that characterize intrusive igneous rocks (**Figure 5.4**). An *intrusive igneous rock* is one that has forced its way into surrounding rock. That surrounding rock is called **country rock.**

> *Intrusive igneous rocks form by the cooling of magma in Earth's interior. Extrusive igneous rocks form by the cooling of magma erupted onto Earth's surface.*

A **porphyry** is an igneous rock that has a mixed texture in which large crystals "float" in a predominantly fine-grained matrix (see Figure 5.4). The large crystals,

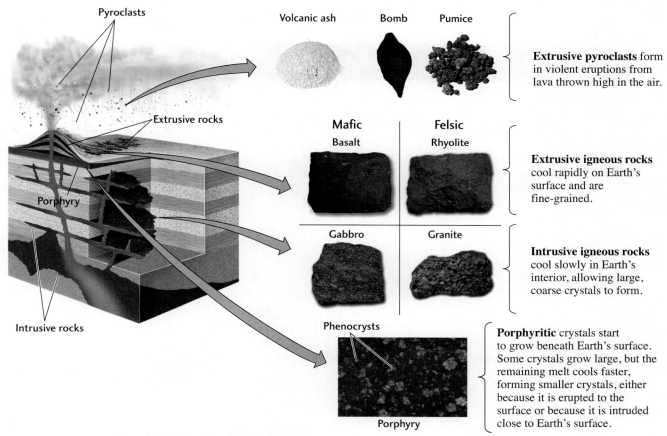

FIGURE 5.4 Igneous rock types can be identified by texture.
[*John Grotzinger/Ramón Rivera-Moret/Harvard Mineralogical Museum.*]

called *phenocrysts*, form while the magma is still below Earth's surface. Then, before other crystals can grow, a volcanic eruption brings the magma to the surface, where it cools quickly to a finely crystalline mass. In some cases, porphyries form as intrusive igneous rocks; for example, they may form where magmas cool quickly at very shallow levels in the crust. Porphyry textures are important to geologists because they indicate that different minerals crystallize at different rates.

> A porphyry is an igneous rock consisting of large crystals dispersed within a finer-grained matrix.

Extrusive Igneous Rocks Rapid cooling at Earth's surface produces the fine-grained texture or glassy appearance of the *extrusive igneous rocks* (see Figure 5.4). These rocks, composed partly or largely of volcanic glass, form when lava or other material erupts from volcanoes. For this reason, they are also known as *volcanic rocks*. They fall into two major categories based on the type of erupted material from which they are formed:

■ *Lavas* Volcanic rocks formed from lavas range in appearance from smooth and ropy to sharp, spiky, and jagged, depending on the conditions under which the rocks formed.

■ *Pyroclasts* In more violent eruptions, **pyroclasts** form when fragments of lava are thrown high into the air. Pyroclasts range in size from fine particles of volcanic ash to huge boulders. As they fall to the ground and cool, these fragments of volcanic debris may stick together to form rocks.

> Extrusive igneous rocks are formed from two kinds of materials, lavas and pyroclasts.

Later in this chapter, we will examine some special forms of intrusive igneous rocks, and we will look more closely at how extrusive igneous rocks form during volcanism. Now, however, we turn to the second way in which the family of igneous rocks is subdivided.

Chemical and Mineral Composition

We have seen that igneous rocks can be subdivided according to their texture. They can also be classified by their chemical and mineral composition. Volcanic glass, which is formless even under a microscope, is often classified by chemical analysis. One of the earliest classifications of igneous rocks was based on a simple chemical analysis of their silica (SiO_2) content. Silica is abundant in most

TABLE 5.1	Common Minerals of Igneous Rocks		
Compositional Group	**Mineral**	**Chemical Composition**	**Silicate Structure**
FELSIC	Quartz	SiO_2	Frameworks
	Potassium feldspar	$KAlSi_3O_8$	
	Plagioclase feldspar	$NaAlSi_3O_8$; $CaAl_2Si_2O_8$	
	Muscovite (mica)	$KAl_3Si_3O_{10}(OH)_2$	Sheets
MAFIC	Biotite (mica)	K Mg Fe Al $\Big\}$ $Si_3O_{10}(OH)_2$	
	Amphibole group	Mg Fe Ca Na $\Big\}$ $Si_8O_{22}(OH)_2$	Double chains
	Pyroxene group	Mg Fe Ca Al $\Big\}$ SiO_3	Single chains
	Olivine	$(Mg,Fe)_2SiO_4$	Isolated tetrahedra

igneous rocks, as noted in Chapter 4, and accounts for 40 to 70 percent of their total weight. We still refer to rocks rich in silica, such as granite, as *silicic*.

Modern classifications now group igneous rocks according to their relative proportions of silicate minerals (**Table 5.1**; Appendix 5). The silicate minerals—quartz, feldspar (both orthoclase and plagioclase), muscovite and biotite micas, the amphibole and pyroxene groups, and olivine—form a systematic series. *Felsic* minerals are the highest in silica; *mafic* minerals are the lowest in silica. The adjectives *felsic* (from *fel*dspar and *si*lica) and *mafic* (from *ma*gnesium and *f*erric, from the Latin *ferrum*, "iron") are applied to both the minerals and the rocks that have high contents of these minerals. Mafic minerals crystallize at higher temperatures—that is, earlier in the cooling of a magma—than felsic minerals.

As the mineral and chemical compositions of

> *Igneous rocks can be classified by the proportions of silica and silicate minerals they contain.*

igneous rocks became known, geologists soon noticed that some extrusive and intrusive rocks were identical in composition and differed only in texture. Basalt, for example, is an extrusive rock formed from lava. Gabbro has exactly the same mineral composition as basalt but forms deep in Earth's crust (see Figure 5.4). Similarly, rhyolite and granite are identical in composition but differ in texture. Thus, extrusive and intrusive rocks form two chemically and mineralogically parallel sets of igneous rocks. Conversely, most chemical and mineral compositions can appear in either extrusive or intrusive rocks. The only exceptions are very highly mafic rocks, which rarely appear as extrusive igneous rocks.

> *Extrusive and intrusive igneous rocks may have the same chemical composition and differ only in texture.*

Figure 5.5 is a model that portrays these relationships. The horizontal axis plots silica content as a percentage of a given rock's weight. The percentages given—from high silica content at 70 percent to low silica content

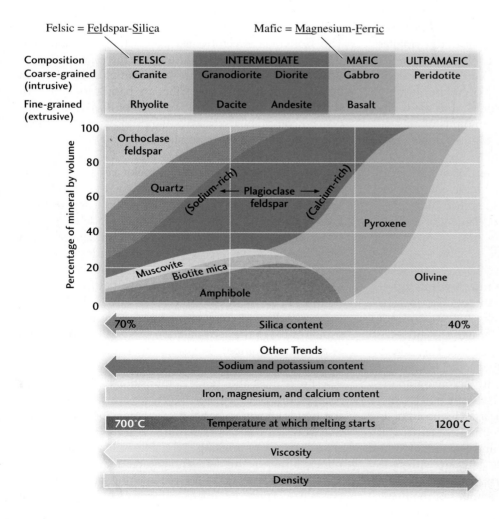

Felsic = <u>Fel</u>dspar-<u>Si</u>lica   Mafic = <u>Ma</u>gnesium-<u>F</u>erri<u>c</u>

Composition	FELSIC	INTERMEDIATE		MAFIC	ULTRAMAFIC
Coarse-grained (intrusive)	Granite	Granodiorite	Diorite	Gabbro	Peridotite
Fine-grained (extrusive)	Rhyolite	Dacite	Andesite	Basalt	

Other Trends

Sodium and potassium content

Iron, magnesium, and calcium content

700°C Temperature at which melting starts 1200°C

Viscosity

Density

FIGURE 5.5 Classification model of igneous rocks. The vertical axis shows the minerals contained in a given rock as a percentage of its volume. The horizontal axis shows the silica content of a given rock as a percentage of its weight. Thus, if you knew by chemical analysis that a coarsely textured rock sample was about 70 percent silica, you could deduce that its composition was about 6 percent amphibole, 3 percent biotite, 5 percent muscovite, 14 percent plagioclase feldspar, 22 percent quartz, and 50 percent orthoclase feldspar. Your rock would be granite. Although rhyolite has the same mineral composition, its fine texture would eliminate it from consideration.

at 40 percent—cover the range found in igneous rocks. The vertical axis displays a scale measuring the minerals contained in a given rock as a percentage of its volume. If you know the silica content of a rock sample, you can determine its mineral composition and, from that, the type of rock.

We can use Figure 5.5 to guide our discussion of the intrusive and extrusive igneous rocks. We begin with the felsic rocks, on the extreme left of the model.

Felsic Rocks **Felsic rocks** are poor in iron and magnesium and rich in minerals that are high in silica. Such minerals include quartz, orthoclase feldspar, and plagioclase feldspar. Orthoclase feldspar is more abundant than plagioclase feldspar. Plagioclase feldspars contain both calcium and sodium. As Figure 5.5 indicates, they are richer in sodium near the felsic end and richer in calcium near the mafic end. Thus, just as mafic minerals crystallize at higher temperatures than felsic minerals, calcium-rich plagioclases crystallize at higher temperatures than sodium-rich plagioclases.

Felsic rocks tend to be light in color. **Granite,** one of the most abundant intrusive igneous rocks, contains about 70 percent silica. Its mineral composition includes abundant quartz and orthoclase feldspar and a smaller amount of plagioclase feldspar. These light-colored felsic minerals give granite its pink or gray color. Granite also contains small amounts of muscovite and biotite micas and amphibole.

Felsic rocks are rich in minerals with a high silica content. They include granite, an intrusive rock, and rhyolite, an extrusive rock.

Rhyolite is the extrusive equivalent of granite. This light brown to gray rock has the same felsic composition and light coloration as granite, but it is much more fine-grained. Many rhyolites are composed largely or entirely of volcanic glass.

Intermediate Igneous Rocks Midway between the felsic and mafic ends of the series are the **intermediate igneous rocks.** As their name indicates, these rocks are neither as rich in silica as the felsic rocks nor as poor in it as the mafic rocks. We find the intermediate intrusive rocks to the right of granite in Figure 5.5. The first is **granodiorite,** a light-colored rock that looks something like granite. It is also similar to granite in having abundant quartz, but its predominant feldspar is plagioclase, not orthoclase. To its right is **diorite,** which contains still less silica and is dominated by plagioclase feldspar, with little or no quartz. Diorites contain a moderate amount of the mafic minerals biotite, amphibole,

Intermediate igneous rocks, which fall between the felsic and mafic rocks in silica content, include intrusive rocks called granodiorite and diorite and extrusive rocks called dacite and andesite.

and pyroxene. They tend to be darker than granite or granodiorite.

The volcanic equivalent of granodiorite is **dacite.** To its right in the extrusive series is **andesite,** the volcanic equivalent of diorite. Andesite derives its name from the Andes, the volcanic mountain chain in South America.

Mafic Rocks **Mafic rocks** contain large proportions of pyroxenes and olivines. These minerals are relatively poor in silica but rich in magnesium and iron, from which they get their characteristic dark colors. **Gabbro,** with even less silica than is found in the intermediate igneous rocks, is a coarse-grained, dark gray, intrusive igneous rock. Gabbro has an abundance of mafic minerals, especially pyroxene. It contains no quartz and only moderate amounts of calcium-rich plagioclase feldspar.

Mafic rocks have a low silica content. They include gabbro, an intrusive rock, and basalt, an extrusive rock.

Basalt is dark gray to black and is the fine-grained extrusive equivalent of gabbro. Basalt is the most abundant igneous rock of the crust, and it underlies virtually the entire seafloor. In some places on the continents, extensive thick sheets of basalt make up large plateaus. The Columbia River basalts of Washington State and the remarkable formation known as the Giant's Causeway in Northern Ireland are two examples. The Deccan basalts of India and the Siberian basalts of northern Russia are enormous outpourings of basalt that appear to coincide closely with two of the greatest periods of mass extinction in the fossil record.

Ultramafic Rocks **Ultramafic rocks** consist primarily of mafic minerals and contain less than 10 percent feldspar. At the far right of Figure 5.5, we find **peridotite,** a coarse-grained, dark greenish gray rock made up primarily of olivine with smaller amounts of pyroxene; it has a silica content of only about 45 percent. Peridotites are the dominant rocks in Earth's mantle, and as we will see, they are the source of the basaltic magmas that form rocks at mid-ocean ridges. Ultramafic rocks are rarely found as extrusives. Because they form at such high temperatures, usually through the accumulation of crystals at the bottom of a magma chamber, they are rarely liquid and hence do not form typical lavas.

The names and exact compositions of the various rocks in the felsic-to-mafic series are less important than the trends shown in **Table 5.2.** There is a strong correlation between a rock's mineralogy and its temperature of crystallization or melting. As Table 5.2 indicates, mafic minerals melt at higher temperatures than felsic minerals. At temperatures below the melting point, minerals crystallize. Therefore, mafic minerals also crystallize at higher temperatures than felsic minerals. We can also see

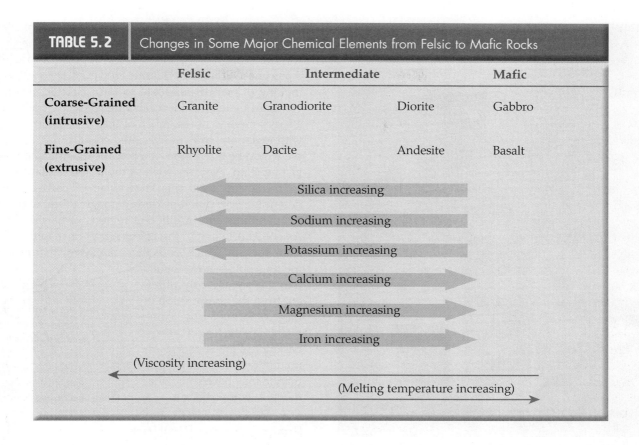

TABLE 5.2	Changes in Some Major Chemical Elements from Felsic to Mafic Rocks			
	Felsic	**Intermediate**		**Mafic**
Coarse-Grained (intrusive)	Granite	Granodiorite	Diorite	Gabbro
Fine-Grained (extrusive)	Rhyolite	Dacite	Andesite	Basalt

Silica increasing ←
Sodium increasing ←
Potassium increasing ←
Calcium increasing →
Magnesium increasing →
Iron increasing →

(Viscosity increasing) ←

(Melting temperature increasing) →

from the table that silica content increases as we move from the mafic group to the felsic group. Increasing silica content results in increasingly complex silicate structures (see Table 5.1), which interfere with a melted rock's ability to flow. As a structure grows more complex, its ability to flow decreases. Thus, **viscosity**—the measure of a liquid's *resistance* to flow—increases as silica content increases. Viscosity is an important factor in the behavior of lavas.

It is clear that a rock's minerals are an indication of the conditions under which the rock's parent magma formed and crystallized. To interpret this information accurately, however, we must understand more about igneous processes. We turn to that topic next.

> *Ultramafic rocks have a very low silica content and include peridotite, an intrusive rock. Extrusive ultramafic rocks are very rare.*

HOW DO MAGMAS FORM?

We know from the way Earth transmits seismic waves that the bulk of the planet is solid for thousands of kilometers down to the core-mantle boundary (see Chapter 2). The evidence of volcanic eruptions, however, tells us that there must also be liquid regions where magmas originate. How do we resolve this apparent contradiction? The answer lies in the processes that melt rocks and create magmas.

Although we do not yet understand the exact mechanisms of melting and solidification within Earth, geologists have learned a great deal from laboratory experiments using high-temperature furnaces (**Figure 5.6**). From these experiments, we know that a rock's melting point depends on its composition and on conditions of temperature and pressure (**Table 5.3**).

> *Magmas form where the temperature is high enough to melt rock.*

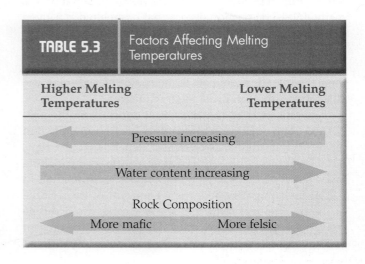

TABLE 5.3	Factors Affecting Melting Temperatures
Higher Melting Temperatures	**Lower Melting Temperatures**

Pressure increasing ←

Water content increasing →

Rock Composition
More mafic ←   More felsic →

FIGURE 5.6 Experimental device used to melt rocks in the laboratory. [*Sally Newman.*]

Temperature and Melting

A hundred years ago, geologists discovered that a rock does not melt completely at a given temperature. Instead, rocks undergo **partial melting** because the minerals that compose them melt at different temperatures. As temperatures rise, some minerals melt and others remain solid. If the same conditions are maintained at any given temperature, the same mixture of solid and melted rock is maintained. The fraction of rock that has melted at a given temperature is called a *partial melt*. To visualize a partial melt, think of how a chocolate chip cookie would look if you heated it to the point at which the chocolate chips melted while the main part of the cookie stayed solid. The chips represent the partial melt, or magma.

The ratio of solid to partial melt depends on the proportions and melting temperatures of the minerals that make up the original rock. It also depends on the temperature at the depth in the crust or mantle where melting takes place. At the lower end of a rock's melting range, a partial melt might be less than 1 percent of the volume of the original rock. Much of the hot rock would still be solid, but appreciable amounts of liquid would be present as small droplets in the tiny spaces between crystals throughout the mass. In the upper mantle, for example, some basaltic partial melts are produced by only 1 to 2 percent melting of peridotite. However, 15 to 20 percent melting of mantle peridotite to form basaltic magmas is common beneath mid-ocean ridges. At the high end of a rock's melting range, much of the rock would be liquid, with lesser amounts of unmelted crystals in it. An example would be a reservoir of a basaltic magma and crystals just beneath a volcano such as the island of Hawaii.

Geologists used this knowledge of partial melts to determine how different kinds of magma form at different temperatures and in different regions of Earth's interior. As you can imagine, the composition of magma formed from rock in which only the minerals with the lowest melting points have melted may be significantly different from the composition of magma formed from completely melted rock. Thus, basaltic magmas that form in different regions of the mantle may have somewhat different compositions.

> *Melting is not complete at a given temperature. Instead, rocks undergo partial melting because the minerals within the rocks melt at different temperatures.*

Pressure and Melting

To get the whole story on melting, we must consider pressure as well as temperature. Pressure increases with depth as a result of the increased weight of overlying rock. Geologists found that as they melted rocks under various pressures in the laboratory, higher pressures led to higher melting temperatures. Thus, rocks that would melt at a given temperature at Earth's surface would remain solid at the same temperature in Earth's interior. For example, a rock that melts at 1000°C at Earth's surface might have a much higher melting temperature, perhaps 1300°C, deep in the interior, where pressures are many thousands of times greater than those at the surface. It is the effect of pressure that explains why the rocks in most of the crust and mantle do not melt. Rock can melt only when both the temperature and pressure conditions are right.

Just as an increase in pressure can keep rock solid, a decrease in pressure can make rock melt, given a suitably high temperature. Because of convection currents, mantle material rises to Earth's surface at mid-ocean ridges at a more or less constant temperature. As the material rises and the pressure decreases below a critical point, solid rock melts spontaneously, without the

> *Higher pressures result in higher melting temperatures. If pressure is decreased at a given temperature, decompression melting occurs.*

introduction of any additional heat. This process, known as **decompression melting,** produces the greatest volume of magma anywhere on Earth. It is the process by which most basalts form on the seafloor.

Water and Melting

The many experiments on melting temperatures and partial melting paid other dividends as well. One of them was a better understanding of the role of water in rock melting. Geologists working on natural lavas in the field determined that water was present in some magmas. This gave them the idea to add water to their experimental melts back in the laboratory. By adding small but differing amounts of water, they discovered that the compositions of partial melts varied not only with temperature and pressure, but also with the amount of water present.

Consider, for example, the effect of dissolved water on pure albite, a sodium-rich plagioclase feldspar, at the low pressures at Earth's surface. If only a small amount of water is present in the rock, it will remain solid at temperatures just over 1000°C, hundreds of degrees above the boiling point of water. At these temperatures, the water in the albite is present as a vapor (gas). If large amounts of water are present, however, the melting temperature of the albite will decrease, dropping to as low as 800°C. This behavior follows the general rule that dissolving one substance (in this case, water vapor) in another (in this case, albite) lowers the melting temperature of the solution. If you live in a cold climate, you are probably familiar with this principle because you know that salt sprinkled on icy roads lowers the melting temperature of the ice. By the same principle, the melting temperature of albite—and of all the feldspars and other silicate minerals—drops considerably in the presence of large amounts of water. Their melting points decrease in proportion to the amount of water dissolved in the molten silicate.

Water content is a significant factor in determining the melting temperatures of mixtures of sedimentary and other rocks. Sedimentary rocks contain an especially large volume of water in their pore spaces, more than is found in igneous or metamorphic rocks. As we will see later in this chapter, the water in sedimentary rocks plays an important role in melting in Earth's interior.

The presence of water in rock lowers its melting temperature.

The Formation of Magma Chambers

Most substances are less dense in their liquid form than in their solid form. The density of a melted rock is lower than the density of a solid rock of the same composition.

With this knowledge, geologists reasoned that large bodies of magma could form in the following way. If the less dense melt were given a chance to move, it would move upward—just as oil, which is less dense than water, rises to the surface of a mixture of oil and water. Being liquid, the partial melt could move slowly upward through pores and along the boundaries between crystals of the surrounding rock. As the hot drops of molten rock moved upward, they would mix with other drops, gradually forming larger pools of magma within Earth's solid interior.

The rise of magmas through the mantle and crust may be slow or rapid. Magmas rise at rates of 0.3 m/year to almost 50 m/year, over periods of tens of thousands or even hundreds of thousands of years. As they ascend, magmas may mix with other melts and may also melt portions of the crust. We now know that large pools of molten rock form **magma chambers**—magma-filled cavities in the lithosphere that form as rising pools of melted rock push aside surrounding solid rock. A magma chamber may encompass a volume as large as several cubic kilometers. We cannot yet say exactly how magma chambers form, nor what they look like in three dimensions. We think of them as large, liquid-filled cavities in solid rock, which expand as more of the surrounding rock melts or as magma migrates through cracks and other small openings. Magma chambers contract as they expel magma to the surface in eruptions. We know that magma chambers exist because seismic waves have shown us the depth, size, and general outlines of the magma chambers underlying some active volcanoes.

A magma chamber is a large cavity in the lithosphere that is filled with molten rock.

Temperature and Depth

Our understanding of igneous processes is based mainly on data from two sources. The first is volcanoes, which give us information about where magmas are located. The second source of data is the record of temperatures measured in deep drill holes and mine shafts. This record shows that the temperature of Earth's interior increases with depth. Using these measurements, scientists have been able to estimate the rate at which temperature rises as depth increases.

The temperatures recorded at a given depth in some locations are much higher than the temperatures recorded at the same depth in other locations. These results indicate that some parts of Earth's mantle and crust are hotter than others. For example, the Great Basin of the western United States is an area where the North American continent is being stretched and thinned, with the result that temperature increases with depth at an exceptionally rapid rate, reaching 1000°C at 40 km, not far below the base of the crust. This temperature is almost high enough to melt basalt. By contrast, in tectonically stable regions, such as the central parts of continents, temperature increases much more slowly, reaching only 500°C at the same depth.

We now know that magmas formed by partial melting can vary in their composition. And we know that temperatures in Earth's interior are sufficient to create magmas. Let's turn now to the question of why there are so many different types of igneous rocks.

MAGMATIC DIFFERENTIATION

The processes we've discussed so far account for the melting of rocks to form magmas. But what accounts for the variety of igneous rocks? Are magmas of different chemical compositions made by the melting of different kinds of rocks? Or do magmatic processes produce a variety of rocks from an originally uniform parent material?

Again, the answers to these questions came from laboratory experiments. Geologists mixed chemical elements in the same proportions that are found in natural igneous rocks and then melted these mixtures. As the melts cooled and solidified, the experimenters recorded the temperatures at which crystals formed as well as the chemical compositions of those crystals. This research gave rise to the theory of **magmatic differentiation,** a process by which rocks of varying composition can arise from a uniform parent magma. Magmatic differentiation occurs because different minerals crystallize at different temperatures. During crystallization, the composition of the magma changes as it is depleted of the chemical elements that form the crystallized minerals.

In a kind of mirror image of partial melting, the last minerals to melt are the first minerals to crystallize from a cooling magma. This initial crystallization withdraws chemical elements from the melt, changing the magma's composition. Continued cooling crystallizes the minerals that melted at the next lower temperature range. Again, the magma's chemical composition changes as various elements are withdrawn. Finally, as the magma solidifies completely, the last minerals to crystallize are the ones that melted first. Thus, the same parent magma, because of its changing chemical composition throughout the crystallization process, can give rise to different igneous rocks.

> *Because different minerals crystallize at different temperatures, the composition of a magma changes progressively as various elements are withdrawn to form crystallized minerals.*

Fractional Crystallization

Fractional crystallization is the process by which the crystals formed in a cooling magma are segregated from the remaining liquid. This segregation happens in several ways, following a sequence commonly described as *Bowen's reaction series* (**Figure 5.7**). In the simplest scenario, crystals formed in a magma chamber settle to the chamber's floor and are thus removed from further reaction with the remaining liquid. The magma then migrates to new locations, forming new chambers. Thus, crystals

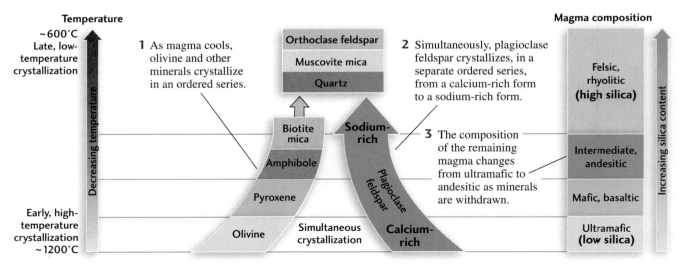

FIGURE 5.7 Bowen's reaction series provides a model of fractional crystallization.

> *Fractional crystallization is the process by which different minerals in a magma crystallize with progressively decreasing temperatures.*

that form early are segregated from the remaining magma, which continues to crystallize as it cools.

The effects of fractional crystallization can be seen in the Palisades, a line of imposing cliffs that faces the city of New York on the west bank of the Hudson River (**Figure 5.8**). This igneous formation is about 80 km long and, in places, more than 300 m high. It formed as a magma of basaltic composition intruded into almost horizontal sedimentary rocks. It contains abundant olivine near the bottom, pyroxene and plagioclase feldspar in the middle, and mostly plagioclase feldspar near the top. These variations in mineral composition from bottom to top made the Palisades a perfect site for testing the theory of fractional crystallization.

Geologists melted rocks with about the same mineral compositions as those found in the Palisades intrusion and determined that the temperature of the magma had to have been about 1200°C. The parts of the magma within a few meters of the relatively cold sedimentary rocks above and below it cooled quickly. The quick cooling formed a fine-grained basalt and preserved the chemical composition of the original melt. The hot interior cooled more slowly, as evidenced by the slightly larger crystals found in the intrusion's interior.

The theory of fractional crystallization leads us to expect that the first mineral to crystallize from the slowly cooling interior of the Palisades intrusion would have been olivine, as this heavy mineral would sink through

the melt to the bottom of the intrusion. It can be found today as a coarse-grained, olivine-rich layer just above the chilled, fine-grained basaltic layer along the bottom zone of contact with the underlying sedimentary rock. Continued cooling would have produced pyroxene crystals, followed almost immediately by calcium-rich plagioclase feldspar. These minerals, too, would have settled out through the magma and accumulated in the lower third of the Palisades intrusion. The abundance of plagioclase feldspar in the upper parts of the intrusion is evidence that the magma continued to change composition until successive layers of settled crystals were topped off by a layer of mostly sodium-rich plagioclase feldspar crystals.

Granite from Basalt: Complexities of Magmatic Differentiation

Studies of the lavas of volcanoes have shown that basaltic magmas are common—far more common than the rhyolitic magmas that correspond in composition to granites. How, then, could the granites have become so abundant in the crust?

The original theory of magmatic differentiation suggested that a basaltic magma would gradually cool and differentiate into a cooler, more silicic melt. The early stages of this differentiation would produce andesitic magma, which might erupt to form andesitic lavas or solidify by slow crystallization to form dioritic intrusives. Intermediate stages would make magmas of granodiorite composition. If this process were carried far enough, its

As magma cools, minerals crystallize at different temperatures and settle out of the magma in a particular order.

Basaltic intrusion 245–275 m (800–900 ft)

Sandstone
Basalt

Mostly sodium-rich plagioclase feldspar; no olivine

Calcium-rich plagioclase feldspar and pyroxene; no olivine

Olivine
Basalt

Sandstone

FIGURE 5.8 Fractional crystallization explains the composition of the basaltic intrusion that forms the Palisades. [*Zehdreh Allen-Lafayette.*]

late stages would form rhyolitic lavas and granitic intrusions (see Figure 5.7).

We have since learned that magmatic differentiation is a more complex process than had been originally described. For instance, so much time would be needed for small crystals of olivine to settle through a dense, viscous magma that they might never reach the bottom of a magma chamber. In fact, there are many layered intrusions—similar to but much larger than the Palisades—that do not show the simple progression of layers predicted by the original theory.

The greatest sticking point in the original theory, however, was the source of granite. The great volume of granite found on Earth could not have formed from basaltic magmas by magmatic differentiation because large quantities of liquid volume would be lost by crystallization during successive stages of differentiation. To produce a given amount of granite, an initial volume of basaltic magma 10 times the size of the granitic intrusion would be required! But geologists could not find anything like that amount of basalt. Even where great volumes of basalt are found—at mid-ocean ridges—there is no wholesale conversion into granite through magmatic differentiation.

Most in question is the original idea that all granitic rocks are formed from the differentiation of a single type of magma, a basaltic melt. Instead, geologists now believe that the melting of varied source rocks of the upper mantle and crust is responsible for much of the observed variation in the composition of magmas:

1. Rocks in the upper mantle might undergo partial melting to produce basaltic magmas.

2. Mixtures of sedimentary rocks and basaltic oceanic rocks, such as those found in subduction zones, might melt to form andesitic magmas.

3. Mixtures of sedimentary, igneous, and metamorphic continental crustal rocks might melt to produce granitic magmas.

Thus, the mechanisms of magmatic differentiation are much more complex than first recognized:

■ Magmatic differentiation can be achieved by the partial melting of mantle and crustal rocks over a range of temperatures and water contents.

■ Magmas do not cool uniformly; they may exist transiently at a range of temperatures within a magma chamber. Differences in temperature within and among magma chambers may cause the chemical composition of the magma to vary from one region to another.

■ A few magmas are *immiscible*—they do not mix with one another, just as oil and water do not mix. When such magmas coexist in one magma chamber, each forms its own crystallization products. Magmas that are *miscible*—that *do* mix—may follow a crystallization path different from that followed by any one magma alone.

Forms of Igneous Intrusion

As noted earlier, we cannot directly observe the shapes of the igneous intrusions formed when magmas intrude into the crust. We can deduce their shapes and distributions only by observing them today where intrusive rocks have been uplifted and exposed by erosion, millions of years after the magma intruded and cooled.

> *The most basic forms of igneous intrusions are plutons, sills and dikes, and veins.*

We do have indirect evidence of current magmatic activity. Seismic waves, for example, show us the general outlines of the magma chambers that underlie some active volcanoes. In some nonvolcanic but tectonically active regions, such as an area near the Salton Sea in Southern California, measurements of temperatures in deep drill holes reveal a crust much hotter than normal, which may be evidence of an intrusion nearby. But these methods cannot reveal the detailed shapes or sizes of intrusions.

Most of what we know about igneous intrusions is based on the work of field geologists who have examined and compared a wide variety of outcrops and have reconstructed their history. In the following pages, we consider some of these forms: plutons, sills and dikes, and veins. **Figure 5.9** illustrates a variety of extrusive and intrusive structures.

Plutons

Plutons are large igneous bodies formed deep in Earth's crust. They range in size from a cubic kilometer to hundreds of cubic kilometers. We can study these large bodies when uplift and erosion uncover them or when mines or drill holes cut into them. Plutons are highly variable, not only in size but also in shape and in their relationship to the country rock.

> *Plutons are large igneous bodies formed deep within Earth's crust. They create space for themselves by wedging open, melting, and breaking off pieces of country rock.*

This wide variation is due in part to the different ways in which magma makes space for itself as it rises through the crust. Most plutons intrude at depths greater than 8 to 10 km. At these depths, few holes or openings exist because the pressure of the overlying rock would close

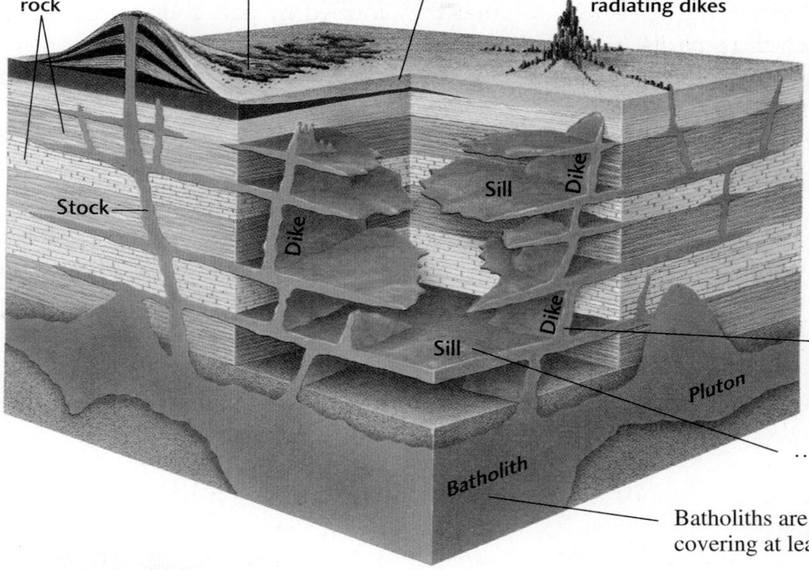

Country rock   Volcano   Lava flow   Ash falls and pyroclastics   Eroded volcano with radiating dikes

Stock

Sill   Dike

Dike

Dike

Sill   Dike

Pluton

Batholith

Dikes cut across layers of country rock…

…but sills run parallel to them.

Batholiths are the largest forms of plutons, covering at least 100 km^2.

FIGURE 5.9 The basic forms of extrusive and intrusive igneous structures.

them. But the upwelling magma overcomes even that great pressure.

Magma rising through the crust makes space for itself in three ways (**Figure 5.10**) that may be referred to collectively as *magmatic stoping:*

1. *Wedging open the overlying rock.* As the rising magma lifts the great weight of the overlying rock, it fractures the rock, penetrates the cracks, wedges them open, and so flows into the rock. Overlying rocks may bow upward during this process.

Rising magma wedges open and fractures overlying country rock.

The overlying rock may bow up.

The magma melts surrounding rock.

The melted country rock mixes with the magma and changes its composition.

The magma also breaks off blocks of overlying rock— xenoliths— that sink into the magma and melt.

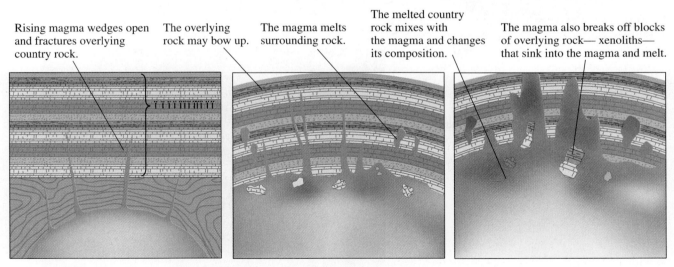

FIGURE 5.10 Magmas make their way into country rock in three basic ways: by invading cracks and wedging open overlying rock, by breaking off pieces of rock, and by melting country rock. Pieces of broken-off country rock, called xenoliths, can become completely dissolved in the magma. If many xenoliths are dissolved and the country rock differs in composition from the magma, the composition of the magma will change.

2. *Breaking off large blocks of rock.* Magma can push its way upward by breaking off blocks of the invaded crust. These blocks, known as *xenoliths,* sink into the magma, melt, and blend into the liquid, in some places changing the composition of the magma.

3. *Melting surrounding rock.* Magma also makes its way by melting country rock.

Most plutons show sharp zones of contact with country rock and other evidence of the intrusion of a liquid magma into solid rock. Other plutons grade into country rock and contain structures vaguely resembling those of sedimentary rocks. The features of these plutons suggest that they formed by partial or complete melting of preexisting sedimentary rocks.

Batholiths, the largest plutons, are great irregular masses of coarse-grained igneous rock that by definition cover at least 100 km² (see Figure 5.9). Batholiths are thick, horizontal, sheetlike or lobe-shaped bodies extending from a funnel-shaped central region. Their bottoms may extend 10 to 15 km deep, and a few are estimated to go even deeper. The coarse grains of batholiths result from slow cooling at great depths. Other, smaller plutons are called **stocks.** Both batholiths and stocks cut across the layers of the country rock that they intrude.

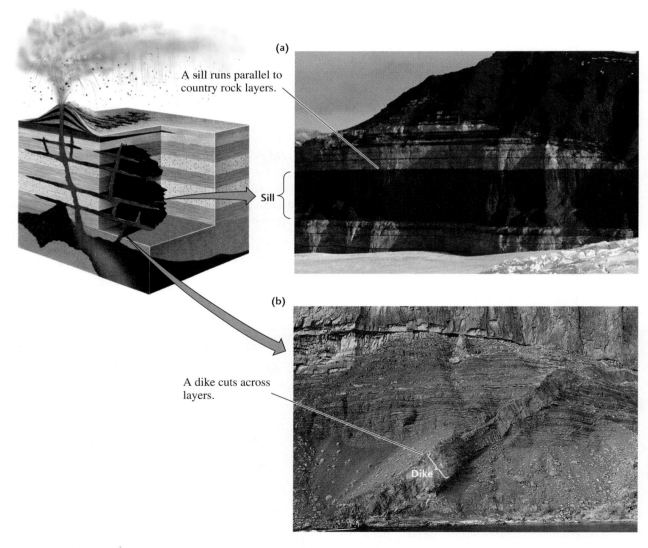

(a)

A sill runs parallel to country rock layers.

Sill

(b)

A dike cuts across layers.

Dike

FIGURE 5.11 (a) At Finger Mountain, situated in the Dry Valleys of Antarctica, sandstone beds are split by sills parallel to the bedding. (b) A dike of igneous rock (dark) intrudes into shaley sedimentary rock (reddish brown) in Grand Canyon National Park, Arizona. [(a) Colin Monteath/AUSCAPE; (b) Tom Bean/DRK PHOTO.]

Sills and Dikes

Sills and dikes are similar to plutons in many ways, but they are smaller and have a different relationship to the layering of the country rock. A **sill** is a sheetlike body formed by the injection of magma between parallel layers of preexisting bedded rock (**Figure 5.11**). The boundaries of sills lie parallel to those layers, whether or not the layers are horizontal. Sills range in thickness from a single centimeter to hundreds of meters, and they can extend over considerable areas. Figure 5.11a shows a large sill at Finger Mountain, Antarctica. The 300-m-thick Palisades intrusion (see Figure 5.8) is another large sill.

Sills may superficially resemble layers of lava flows and pyroclastic deposits, but they differ from those layers in four ways:

1. They lack the ropy, blocky, and vesicle-filled structures that characterize many volcanic rocks.

2. They are more coarse-grained than volcanics because they have cooled more slowly.

3. Rocks above and below sills show the effects of heating: their color may have been changed, or their mineral composition may have been altered by contact metamorphism.

4. Many lava flows overlie weathered older flows or soils formed between successive flows; sills do not.

Dikes are the major route of magma transport in the crust. They are like sills in being sheetlike igneous bodies, but dikes cut across the layers in bedded country rock (see Figure 5.11b). Dikes sometimes form by forcing open existing fractures, but more often they create channels through new cracks opened by the pressure of rising magma. Some individual dikes can be followed in the field for tens of kilometers. Their thicknesses vary from many meters to a few centimeters. In some dikes, xenoliths provide evidence of disruption of the surrounding rock during the intrusion process. Dikes rarely exist alone; more typically, swarms of hundreds or thousands of dikes are found in a region that has been deformed by a large igneous intrusion.

> Sills are small intrusive bodies formed by the injection of magma between layers of bedded country rock. Dikes are small intrusive bodies that cut across layers of country rock.

The texture of dikes and sills varies. Many are coarse-grained, with an appearance typical of intrusive rocks. Many others are fine-grained and look much more like volcanic rocks. Because we know that texture corresponds to the rate of cooling, we can conclude that the fine-grained dikes and sills invaded country rock nearer Earth's surface, where the country rock was cold compared with the intrusions. Their fine texture is the result of fast cooling. The coarse-grained ones formed at depths of many kilometers and invaded warmer rocks whose temperatures were much closer to their own.

NATURAL RESOURCES: Fractional Crystallization, Porphyries, and Veins

Igneous rocks are often the host for accumulations of economically important minerals. Three types are discussed here, and the discussions build on material introduced so far in this chapter.

In some places *fractional crystallization* has led to concentration of economically important volumes of minerals. So-called large layered igneous complexes are host to very unusual suites of minerals that yield "platinum-group" metals such as platinum, palladium, osmium, iridium, and rhodium in addition to more ordinary metals such as iron, tin, chromium and titanium. The Bushveld intrusion is a gigantic mafic intrusion that covers about 66,000 km^2—about the size of Ireland—and is about 9000 meters thick (**Figure 5.12**). It took all this magma to generate one layer only 30 to 90 cm thick that contains approximately 90% of the world's platinum-group metal reserves! The minerals in which these metals occur simply crystallized at the right time according to their temperature of crystallization, settling to the bottom of the magma chamber to form the thin layer.

FIGURE 5.12 Rock samples from the Bushveld intrusion, South Africa, show lustrous metals of platinum and palladium. [*Dirk Wiersma/Photo Researchers.*]

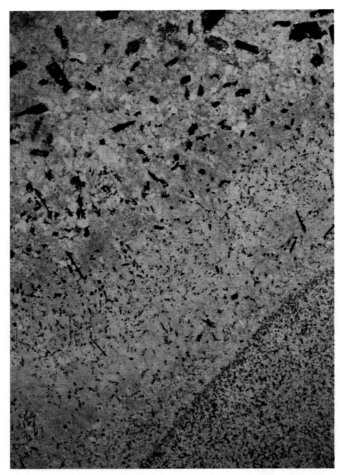

FIGURE 5.13 A granite pegmatite vein. The center of the intrusion (*upper right*) cooled more slowly and developed coarser crystals. The margin of the intrusion (*lower left*) has finer crystals due to more rapid cooling. [*John Grotzinger/ Ramón Rivera-Moret/Harvard Mineralogical Museum.*]

Porphyries can also form economically important mineral deposits. Circulating groundwater interacts with the intrusive melts causing alteration of the surrounding country rock in which fractures are filled with minerals. Copper is one of the most important metals to form in the alteration zone, along with molybdenum, silver, and gold. Porphyry copper deposits are the world's largest known source of copper and almost all mines are run as enormous open pits, such as the Bingham Canyon mine in Utah. Most porphyry mineral deposits are associated with the subduction-related volcanoes that flank the Pacific Ocean basin.

Veins are deposits of minerals found within a rock fracture that are foreign to the country rock. Irregular pencil- or sheet-shaped veins branch off from the tops and sides of many intrusive bodies. They may be a few millimeters to several meters across, and they tend to be

tens of meters to kilometers long or wide. The well-known Mother Lode of the gold rush of 1849 in California is a vein of gold-bearing quartz crystals. An example of gold in a quartz-rich vein is shown in Figure 4.29. Veins of extremely coarse-grained granite cutting across a much finer grained country rock are called **pegmatites** (**Figure 5.13**). Pegmatites crystallize from a water-rich magma in the late stages of solidification.

IGNEOUS PROCESSES AND PLATE TECTONICS

The facts and theories of igneous rock formation fit nicely into a framework based on plate tectonic theory. Batholiths, for example, are found in the cores of many mountain ranges formed by the convergence of two plates. This indicates a connection between pluton formation, mountain-building processes, and plate tectonics. The geometry of plate motions is the link we need to tie tectonic activity to magmatic processes (see Figure 5.1).

> *Magma forms most abundantly at two types of plate boundaries: mid-ocean ridges and subduction zones.*

Magma forms most abundantly at two types of plate boundaries: mid-ocean ridges, where two plates diverge and the seafloor spreads, and subduction zones, where one plate dives beneath another. Mantle plumes, though not associated with plate boundaries, are also the result of partial melting and form near the core-mantle boundary deep within Earth's interior.

Spreading Centers as Magma Factories

Most igneous rocks are formed by the globally extensive network of mid-ocean ridges. Each year, approximately 3 km³ of basaltic lava erupts along the mid-ocean ridges in the process of seafloor spreading. This is a truly enormous volume. In comparison, all the active volcanoes along convergent plate boundaries (about 400) generate volcanic rock at a rate of less than 1 km³/year. Enough magma has erupted in seafloor spreading over the past 200 million years to create the entire present-day seafloor, which covers nearly two-thirds of Earth's surface. Throughout the mid-ocean ridge network, decompression melting of the mantle creates magmas that well up along rising convection currents and are then erupted onto the seafloor.

Ophiolite suites (**Figure 5.14**) are pieces of oceanic crust that are found on land. These rock assemblages consist of deep-sea sediments, submarine basaltic lavas, and

OPHIOLITE SUITE

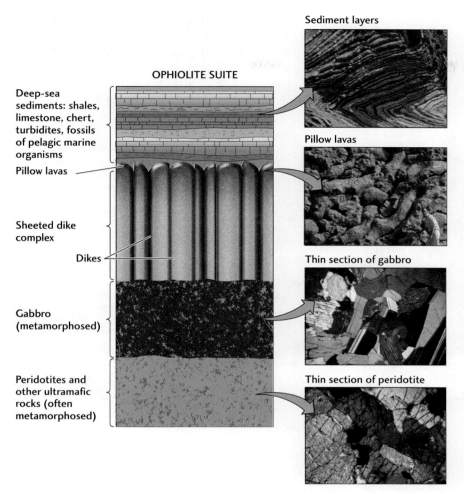

Deep-sea
sediments: shales,
limestone, chert,
turbidites, fossils
of pelagic marine
organisms

Pillow lavas

Sheeted dike
complex

Dikes

Gabbro
(metamorphosed)

Peridotites and
other ultramafic
rocks (often
metamorphosed)

Sediment layers

Pillow lavas

Thin section of gabbro

Thin section of peridotite

FIGURE 5.14 Idealized section of an ophiolite suite. The combination of deep-sea sediments, submarine pillow lavas, sheeted basaltic dikes, and mafic igneous intrusions indicates a deep-sea origin. [*Photos courtesy of John Grotzinger. Thin sections courtesy of T. L. Grove.*]

mafic igneous intrusions. They were created by seafloor spreading and then raised above sea level and thrust onto a continent in a later episode of plate collision. Ophiolite suites, along with information gleaned from ocean drilling and sound-wave profiling, tell us what may be happening in sub-seafloor magma chambers.

How does seafloor spreading work? We can think of a spreading center as a huge factory that processes mantle material to produce oceanic crust. **Figure 5.15** is a highly simplified representation of a seafloor spreading center. The raw material fed into this factory is peridotite that comes from the asthenosphere. The mineral composition of the average peridotite in the mantle is chiefly olivine, with smaller amounts of pyroxene and garnet. Temperatures in the asthenosphere are hot enough to melt a small

Seafloor spreading at mid-ocean ridges generates tremendous volumes of basaltic magma. Peridotite rising from the mantle undergoes decompression, leading to partial melting and the formation of basaltic magmas. These magmas then crystallize to form oceanic crust.

fraction of this peridotite (less than 1 percent), but not hot enough to generate substantial volumes of magma.

Decompression melting is the process that generates great volumes of magma from peridotite. Recall from earlier in this chapter that the melting temperature of a mineral depends on the pressure at which it melts as well as its composition: decreasing the pressure generally lowers a mineral's melting temperature. Consequently, if a mineral is near its melting point and the pressure on it is lowered while its temperature is kept constant, the mineral will melt. As the plates pull apart, the partially molten peridotite is sucked inward and upward toward the spreading center; the decrease in pressure as the peridotite rises melts a large fraction of the rock (up to 15 percent).

The peridotites subjected to this process do not melt evenly; the garnet and pyroxene minerals they contain melt at a lower temperature than the olivine. For this reason, the magma generated by decompression melting is not peridotitic in composition; rather, it is enriched in silica and iron and has the same composition as basalt

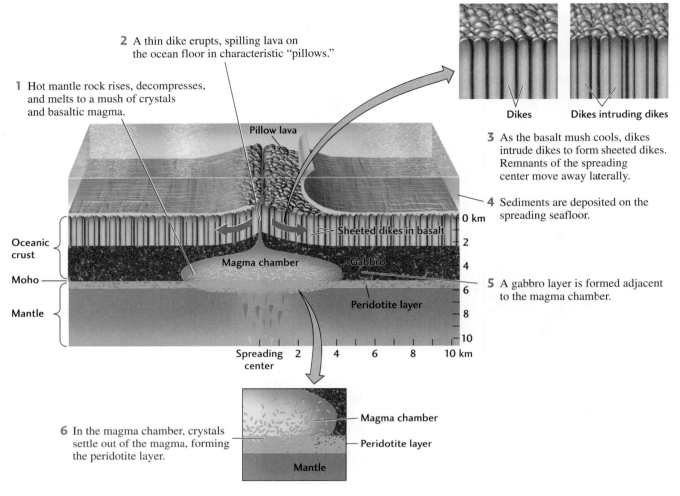

2 A thin dike erupts, spilling lava on the ocean floor in characteristic "pillows."

1 Hot mantle rock rises, decompresses, and melts to a mush of crystals and basaltic magma.

Pillow lava

Oceanic crust

Moho

Mantle

Magma chamber

Sheeted dikes in basalt

Gabbro

Peridotite layer

Spreading center

Dikes

Dikes intruding dikes

3 As the basalt mush cools, dikes intrude dikes to form sheeted dikes. Remnants of the spreading center move away laterally.

4 Sediments are deposited on the spreading seafloor.

5 A gabbro layer is formed adjacent to the magma chamber.

6 In the magma chamber, crystals settle out of the magma, forming the peridotite layer.

Magma chamber

Peridotite layer

Mantle

FIGURE 5.15 Decompression melting creates magma at seafloor spreading centers.

(see Figure 5.15). This basaltic melt forms a magma chamber below the mid-ocean ridge crest, from which it separates into three layers:

1. Some magma rises through the narrow cracks that open where the plates are separating and erupts into the ocean, forming the basaltic pillow lavas that cover the seafloor (see Figure 5.15).

2. Some magma freezes in the cracks as vertical, sheeted dikes of gabbro.

3. The remaining magma solidifies as massive intrusions of gabbro as the underlying magma chamber is pulled apart by seafloor spreading.

These igneous units—pillow lavas, sheeted dikes, and massive gabbros—are the basic layers of the crust that geologists have found throughout the world's oceans. Seafloor spreading results in another layer beneath this oceanic crust: the residual peridotite from which the

basaltic magma was originally derived. Above the pillow lavas, a thin blanket of deep-sea sediments begins to cover the newly formed oceanic crust. As the seafloor spreads, these layers of sediments, lavas, dikes, and gabbros are transported away from the mid-ocean ridge where this characteristic sequence of rocks is assembled— almost like a production line.

Subduction Zones as Magma Factories

Other kinds of magmas underlie regions in which volcanoes are highly concentrated, such as the Andes mountain range of South America and the Aleutian Islands of Alaska. Both of these regions lie over subduction zones, where one plate sinks under another. Subduction zones are also major magma factories (**Figure 5.16**). They generate magmas of varying composition, depending on how much and what kinds of materials are subducted. Where subduction takes place beneath a continent, the resulting

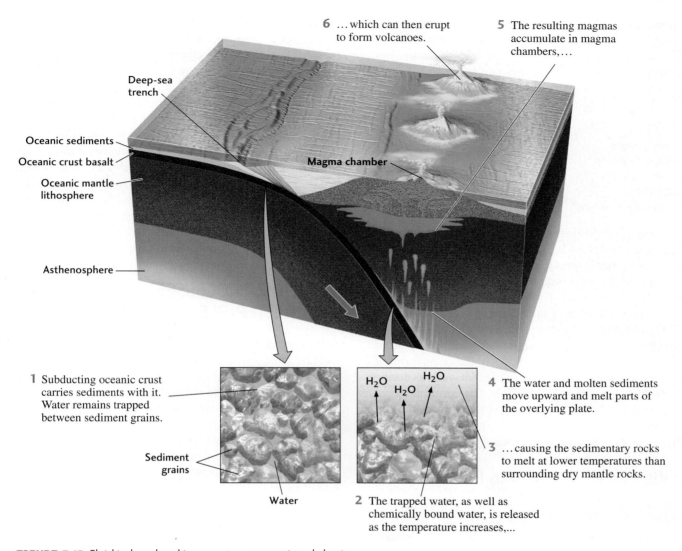

6 ...which can then erupt to form volcanoes.

5 The resulting magmas accumulate in magma chambers,...

Deep-sea trench

Oceanic sediments

Oceanic crust basalt

Magma chamber

Oceanic mantle lithosphere

Asthenosphere

1 Subducting oceanic crust carries sediments with it. Water remains trapped between sediment grains.

Sediment grains

Water

H_2O H_2O H_2O

4 The water and molten sediments move upward and melt parts of the overlying plate.

3 ...causing the sedimentary rocks to melt at lower temperatures than surrounding dry mantle rocks.

2 The trapped water, as well as chemically bound water, is released as the temperature increases,...

FIGURE 5.16 Fluid-induced melting creates magma in subduction zones.

volcanoes and volcanic rocks form a mountainous volcanic belt on land, such as the Andes, which mark the subduction of the oceanic Nazca Plate beneath continental South America. Similarly, subduction of the small Juan de Fuca Plate beneath western North America has generated the Cascade Range, with its active volcanoes, in northern California, Oregon, and Washington. Where one oceanic plate is subducted under another, a deep-sea trench and a volcanic island arc are formed.

Subduction zones generate magmas of varying composition.

When an oceanic plate collides with and dives beneath another oceanic plate, several processes that cause **fluid-induced melting** are set in motion. The fluid involved is primarily water, which, as we have seen, lowers the melting temperature of rock. By the time oceanic lithosphere

is subducted at a convergent plate boundary, a lot of water has been incorporated into its outer layers. Some of the seawater that circulates through the crust near a spreading center reacts with basalt to form new minerals with water bound into their structures. In addition, as the lithosphere ages and moves across the ocean basin, sediments containing water are deposited on its surface. Some of these sediments get scraped off at the deep-sea trench where the plate subducts, but much of this water-laden material is carried downward into the subduction zone.

As the lithosphere moves downward, it is subjected to increasing pressure. Water is squeezed out of the minerals in the outer layers of the descending crust and rises buoyantly into the mantle wedge above the crust. At depths greater than about 5 km, the temperature increases, and the remaining water is released by metamorphic chemical reactions. All this water then causes melting of the descending basalt-rich oceanic crust and the overlying peridotite-rich mantle wedge. Most of the resulting mafic magma accumulates at the base of the crust of the overriding plate, and some of it intrudes into the crust to form magma chambers, which results in the formation of a volcanic island arc. When an oceanic plate is subducted beneath a continental plate, a similar process gives rise to a belt of volcanoes.

> Subduction of water-rich oceanic lithosphere leads to fluid-induced melting of the subducting lithosphere.

The magmas produced by this type of fluid-induced melting are essentially basaltic, although their composition is more variable than that of mid-ocean ridge basalts. The composition of the magmas is further altered during their residence in the crust. Within the magma chambers, the process of fractional crystallization increases the magma's silica content, producing eruptions of andesitic lavas. Where the overlying plate is continental, the heat from the magmas can melt the felsic rocks in the crust, forming magmas with an even higher silica content, such as dacitic and rhyolitic magmas (see Table 5.2). The contribution of lithospheric fluids to these magmas is inferred because trace elements known to be present in oceanic crust and sediments are found in the magmas.

Mantle Plumes as Magma Factories

We learned earlier that the magmas at mid-ocean ridges are formed as a result of decompression melting. This process is also important in the formation of **mantle plumes.** Mantle plumes, like spreading centers, are sites of decompression, but they differ from spreading centers by forming within lithospheric plates rather than along the margins of plates. Basalts similar to those produced at mid-ocean ridges are found in thick accumulations over some parts of continents distant from plate boundaries. In the states of Washington, Oregon, and Idaho, the Columbia and Snake rivers flow over a large area covered by this kind of basalt, which solidified from lavas that flowed out millions of years ago. Large quantities of basalt are also erupted in isolated volcanic islands far from plate boundaries, such as the Hawaiian Islands. In such places, slender, pencil-like plumes of hot mantle rise from deep within Earth, perhaps as deep as the core-mantle boundary.

> Mantle plumes produce hot spots where basaltic magmas form by decompression melting. The movement of plates over these hot spots may produce a trail of volcanoes.

Mantle plumes that reach Earth's surface and form **hot spots** are responsible for these outpourings of basalt. The basalt is produced by decompression melting of the mantle. When the peridotites brought up toward the surface in a mantle plume reach low pressures at shallow depths, they begin to melt, producing basaltic magma. The magma eventually penetrates the lithosphere and erupts at the surface. The current position of the plate over the hot spot is marked by an active volcano, which becomes inactive as plate movement pushes it away from the hot spot. Plate motion thus generates a trail of extinct, progressively older volcanoes. As shown in **Figure 5.17**, the Hawaiian Islands fit this pattern well, revealing a rate of movement of the Pacific Plate over the Hawaiian hot spot of about 100 mm/year. The Yellowstone volcano is the product of a hot spot beneath a continental plate.

VOLCANIC PROCESSES AND MATERIALS

Volcanoes and volcanic rocks deserve special attention because they provide geologists with portals into the igneous processes that operate deep within Earth, which might not otherwise be visible. The natural processes that give rise to volcanoes and volcanic rocks are known collectively as *volcanism*.

A **volcano** consists of a magma chamber, to which magma is transported from deep within Earth; a plumbing system that transports the magma to Earth's surface; one or more vents at the surface of Earth; and often a hill or mountain constructed from the accumulation of lava and pyroclasts erupted from the vents. Volcanoes can be viewed as factories that process raw materials (magmas) and transport the end products (lavas and pyroclasts) to the surface.

As we have seen, the magmas that feed volcanoes vary in their chemical composition. Some volcanoes are fed by multiple magma chambers with different compositions, leading to volcanoes made up of stacked deposits of different compositions. The shape of a volcano is controlled both by magma composition and by the eruption process itself. For example, basaltic magmas tend to produce volcanoes with broad, gently sloping forms, whereas andesitic and rhyolitic magmas tend to form volcanoes that have narrower, more conical forms. Volcanoes produced by andesitic and rhyolitic magmas are also more likely to explode, which can dramatically modify their shapes.

> Volcanoes are formed by the transport of magmas from deep within Earth to magma chambers closer to the surface. Eruptions then transport the magmas to Earth's surface.

(a)

The Pacific Plate has moved north-west over the Hawaiian hot spot,...

...resulting in a chain of volcanic islands and seamounts.

The ages of the volcanoes are consistent with plate movement of about 100 mm/year.

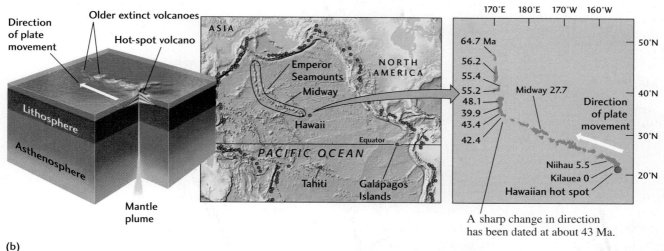

A sharp change in direction has been dated at about 43 Ma.

(b)

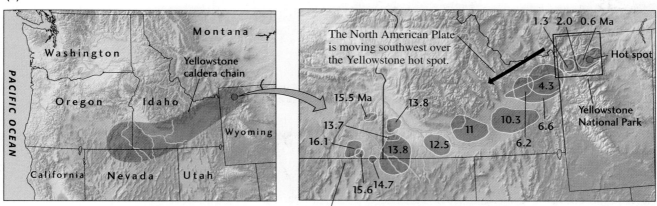

The ages of the calderas are consistent with plate movement of about 25 mm/year.

FIGURE 5.17 Plate motion generates a trail of progressively older volcanoes. (a) The Hawaiian Island chain and its extension into the northwestern Pacific reveal the northwestward movement of the Pacific Plate. (b) The Yellowstone volcanic track marks the movement of the North American Plate over a hot spot during the past 16 million years. (Ma: million years ago) [*Wheeling Jesuit University/NASA Classroom of the Future.*]

Volcanic Processes

In most cases, the magmas that rise to Earth's surface through volcanoes have been generated by partial melting in the asthenosphere, but in some cases they result from partial melting of the lithosphere. The starting composition of the magma has an important effect on the final composition of the lava that is erupted at Earth's surface. However, chemical processing in magma chambers also has an important effect on lava composition.

Transport of Magmas to Magma Chambers A volcano can be thought of as a plumbing system through which

magma is transported to the surface (**Figure 5.18**). Magmas rising buoyantly through the lithosphere pool together in a magma chamber, usually at a shallow depth in the crust. This chamber periodically empties to the surface through a pipelike central vent in repeated cycles of eruptions. Lava can also erupt from other vents on the flanks of a volcano or from vertical cracks (fissures).

Chemical Modification of Magmas Volcanoes can also be thought of as chemical factories that process magmas to produce lavas. As we have seen, a magma gains chemical components as it melts the surrounding rocks while rising through the lithosphere. It loses other components when

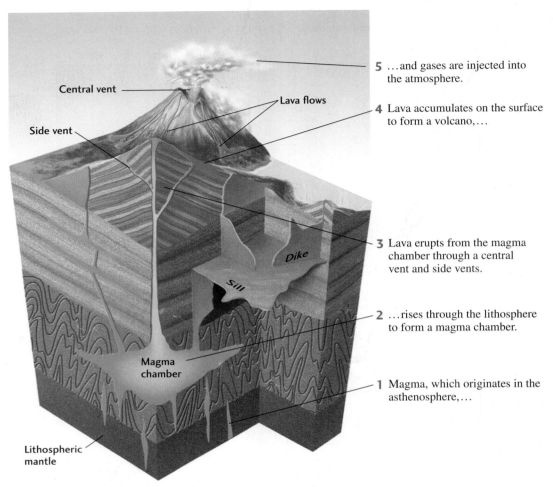

5 ...and gases are injected into the atmosphere.

4 Lava accumulates on the surface to form a volcano,...

3 Lava erupts from the magma chamber through a central vent and side vents.

2 ...rises through the lithosphere to form a magma chamber.

1 Magma, which originates in the asthenosphere,...

Central vent

Side vent

Lava flows

Dike

Sill

Magma chamber

Lithospheric mantle

FIGURE 5.18 Volcanoes transport magma from Earth's interior to its surface, where rocks are formed and gases are emitted.

crystals settle out along the way or sink to the bottom of shallow magma chambers. And its gaseous constituents escape into the atmosphere or ocean as it erupts at the surface. These changes in composition result in changes in the physical properties of the lava, such as its viscosity, which in turn create the variation in the types of volcanoes that we see on Earth.

Eruption of Lavas and Formation of Volcanic Rocks Volcanic rocks form when magmas erupt at Earth's surface as lavas. As we will learn, lavas of different composition produce different landforms. The differences depend on the chemical composition, gas content, and temperature of lavas. The higher the silica content and the lower the temperature, for example, the more viscous the lava is and the more slowly it moves. The more gas a lava contains, the more violent its eruption is likely to be.

Figure 5.19 shows the locations of the world's active volcanoes that occur on land or above the ocean's surface.

About 80 percent are found at convergent plate boundaries, 15 percent at divergent plate boundaries, and the remaining few within plate interiors. There are many more active volcanoes than shown on this map, however. Most of the lava erupted on Earth's surface comes from vents beneath the sea, located on the spreading centers of the mid-ocean ridges.

Types of Lava

Erupted lavas usually solidify into one of three major types of igneous rock: basalts, andesites, and rhyolites.

> *Lavas can be classified into three major types: basaltic, andesitic, and rhyolitic.*

Basaltic Lavas Basalt is an extrusive igneous rock of mafic composition (high in iron, magnesium, and calcium) and the lowest silica content of the three igneous rock types. Basaltic magmas are the most common magma type. They

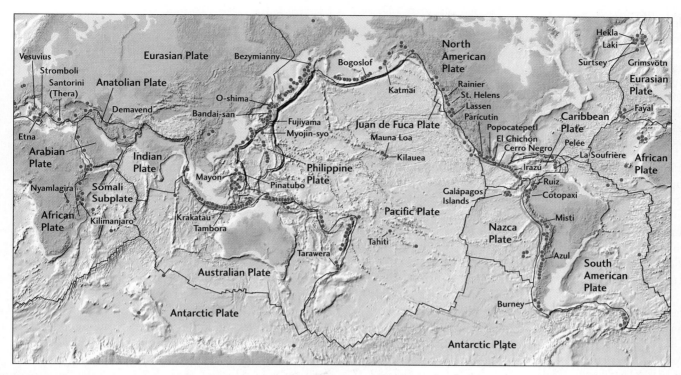

FIGURE 5.19 The active volcanoes of the world with vents on land or above the ocean surface are represented by red dots. Black lines represent plate boundaries. Not shown on this map are the numerous volcanoes of the mid-ocean ridge system below the water's surface.

are produced along mid-ocean ridges and at hot spots, as well as in continental rift valleys. The volcanic island of Hawaii, which is made up primarily of basaltic lava, lies above a hot spot.

When cool, *basaltic lavas* are black or dark gray, but at their high eruption temperatures (1000° to 1200°C), they glow in reds and yellows (**Figure 5.20**). Because their temperatures are high and their silica content low, basaltic lavas are extremely fluid and can flow downhill fast and far. Lava streams flowing as fast as 100 km/hour have been observed, although velocities of a few kilometers per hour are more common. In 1938, two daring Russian

FIGURE 5.20 A central vent eruption from Kilauea, a shield volcano on the island of Hawaii, produces a river of hot, fast-flowing basaltic lava. [*J. D. Griggs/USGS.*]

volcanologists measured temperatures and collected gas samples while floating down a river of molten basalt on a raft of colder solidified lava. The surface temperature of the raft was 300°C, and the river temperature was 870°C. Lava streams have been observed to travel more than 50 km from their sources.

> Basaltic lavas, which are the most common type, have the lowest silica content of the three types of lava.

Basaltic eruptions are rarely explosive. Basaltic lavas erupt when hot, fluid magmas fill up the volcano's plumbing system and overflow, sending lava down the flanks of the volcano in great streams that can engulf everything in their path. Basaltic lava flows take on different forms depending on how they cool. On land, they solidify as pahoehoe (pronounced pa-ho'-ee-ho'u-ee) or aa (ah-ah) (**Figure 5.21**).

Pahoehoe (the word is Hawaiian for "ropy") forms when a highly fluid lava spreads in sheets and a thin, glassy, elastic skin congeals on its surface as it cools. As the molten liquid continues to flow below the surface, the skin is dragged and twisted into coiled folds that resemble rope.

> Basaltic lava flows take on different forms depending on how they cool.

"*Aa*" is what the unwary exclaim after venturing barefoot onto lava that looks like clumps of moist, freshly plowed earth. Aa forms when lava loses its gases and consequently flows more slowly than pahoehoe, allowing a thick skin to form. As the flow continues to move, the thick skin breaks into rough, jagged blocks. The blocks pile up in a steep front of angular boulders that advances like a tractor tread. Aa is truly treacherous to cross. A good

pair of boots may last about a week on it, and the traveler can count on cut knees and elbows.

A single downhill basaltic flow commonly has the features of pahoehoe near its source, where the lava is still fluid and hot, and of aa farther downstream, where the flow's surface—having been exposed to cool air longer—has developed a thicker outer skin.

Basaltic magma that cools under water forms *pillow lavas*—piles of ellipsoidal, pillowlike blocks of basalt about a meter wide (see Figure 5.14). Pillow lavas are an important indicator that a region on dry land was once under water. Scuba-diving geologists have actually observed pillow lavas forming on the ocean floor off Hawaii. Tongues of molten basaltic lava develop a tough, plastic skin on contact with the cold ocean water. Because the lava inside the skin cools more slowly, the pillow's interior develops a crystalline texture, whereas the quickly chilled skin solidifies to a crystal-less glass.

Andesitic Lavas Andesite is an extrusive igneous rock with an intermediate silica content that makes andesitic lavas more viscous than basaltic lavas. Andesitic lavas are erupted mainly in the volcanic belts above subduction zones. The name comes from a prime example—the Andes of South America.

> Andesitic lavas have intermediate silica content and are more viscous than basaltic lavas.

The temperatures of *andesitic lavas* are lower than those of basalts, and their silica content is higher, so they flow more slowly and lump up in sticky masses. If one of these sticky masses plugs up the central vent of the volcano, gases can build up beneath the plug and eventually

FIGURE 5.21 Two forms of basaltic lava, ropy pahoehoe (*left*) and jagged blocks of aa (*right*), produced by Mauna Loa on the island of Hawaii. [*John Grotzinger.*]

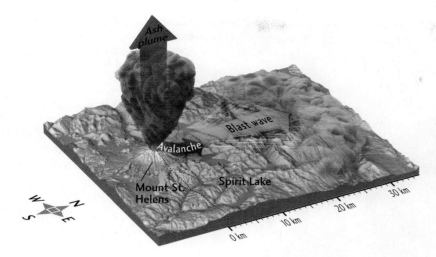

The eruption of Mount St. Helens on May 18, 1980, sent an ash plume into the stratosphere and an avalanche and blast wave toward the north.

(May 17, 3 P.M.) View of Mount St. Helens the day before its eruption. The north side of the volcano has bulged outward from magma intruded at shallow levels during the previous two months.

(May 18, 8:33 A.M.) An earthquake and massive landslide "uncork" the volcano, releasing an ash plume and a powerful lateral blast wave.

FIGURE 5.22 Mount St. Helens, an andesitic volcano in southwestern Washington State, before and after its cataclysmic eruption in May 1980, which ejected about 1 km³ of pyroclastic material. The collapsed northern flank can be seen in the photo on the right. [*Photos by Keith Ronnholm.*]

blow off the top of the volcano. The explosive eruption of Mount St. Helens in 1980 (**Figure 5.22**) is a famous example.

Some of the most destructive volcanic eruptions in history have been *phreatic*, or steam, explosions, which occur when hot, gas-charged magma encounters groundwater or seawater, generating vast quantities of superheated steam (**Figure 5.23**). The island of Krakatau, an andesitic volcano in Indonesia, was destroyed by a phreatic explosion in 1883. This legendary eruption was heard thousands of kilometers away, and it generated a tsunami that killed more than 40,000 people.

Rhyolitic Lavas Rhyolite is an extrusive igneous rock of felsic composition (high in sodium and potassium) with a silica content greater than 68 percent. Rhyolitic magmas are produced in zones where heat from the mantle has melted large volumes of continental crust. Today, the Yellowstone volcano is producing huge amounts of rhyolitic magmas that are building up in shallow chambers.

Rhyolitic lavas have the highest silica content of the three lava types and therefore are the most viscous and potentially explosive. These lavas are produced where melting of continental crust occurs.

FIGURE 5.23 A phreatic eruption of an island arc volcano spews out plumes of steam. The volcano, about 6 miles off the Tongan island of Tongatapu, is one of about 36 in the area. [*Dana Stephenson/Getty Images.*]

Rhyolite is light in color, often a pretty pink. It has a lower melting point than andesite, becoming liquid at temperatures of only 600° to 800°C. Because rhyolitic lavas are richer in silica than any other lava type, they are the most viscous lava type. A rhyolitic flow typically moves more than 10 times more slowly than a basaltic flow, and it tends to pile up in thick, bulbous deposits. Gases are easily trapped beneath rhyolitic lavas, and large rhyolitic volcanoes such as Yellowstone produce the most explosive of all volcanic eruptions.

Textures of Volcanic Rocks

The textures of volcanic rocks, like the surfaces of solidified lava flows, reflect the conditions under which they solidified. Coarse-grained textures with visible crystals can result if lavas cool slowly. Lavas that cool quickly tend to have fine-grained textures. If they are silica-rich, rapidly cooled lavas can form **obsidian,** a glassy volcanic rock that is solid and dense. Chipped or fragmented obsidian produces very sharp edges, and Native Americans and many other hunting groups have used it for arrowheads and a variety of cutting tools.

Volcanic rock often contains little bubbles, created as gases are released during an eruption. Magma is typically charged with gas, like soda in an unopened bottle. When magma rises, the pressure on it decreases, just as the pressure on the soda drops when the bottle cap is removed.

And just as the carbon dioxide in the soda forms bubbles when the pressure is released, the water vapor and other dissolved gases escaping from lava as it erupts create gas cavities, or *vesicles* (**Figure 5.24**). **Pumice** is an extremely vesicular volcanic rock, usually rhyolitic in composition. Some pumice has so many vesicles that it is light enough to float on water.

FIGURE 5.24 Vesicular basalt sample. [*John Grotzinger.*]

Pyroclastic Deposits

Water and gases in magmas can have even more dramatic effects. Before a magma erupts, the confining pressure of the overlying rock keeps these volatiles from escaping. When the magma rises close to the surface and the pressure drops, the volatiles may be released with explosive force, shattering the lava and any overlying solidified rock into pyroclasts of various sizes, shapes, and textures (**Figure 5.25**). Such

> *Pyroclastic deposits include fragments thrown from volcanoes during explosive eruptions (pyroclasts) and flows of hot gases mixed with dust and ash (pyroclastic flows).*

explosive eruptions are particularly likely with gas-rich, viscous rhyolitic and andesitic lavas.

Pyroclasts are classified according to their size. The finest pyroclasts are fragments less than 2 mm in diameter, usually of glass, that form when escaping gases force a fine spray of magma from a volcano. These fragments are classified as **volcanic ash.** Fragments ejected as blobs of lava that cool in flight and become rounded, or as chunks torn loose from previously solidified volcanic rock, can be much larger. These fragments are called volcanic **bombs** (**Figure 5.26**). Volcanic bombs as large as houses have been thrown more than 10 km by explosive eruptions. Volcanic ash fine enough to stay aloft can be carried great distances. Within 2 weeks of the 1991 eruption of

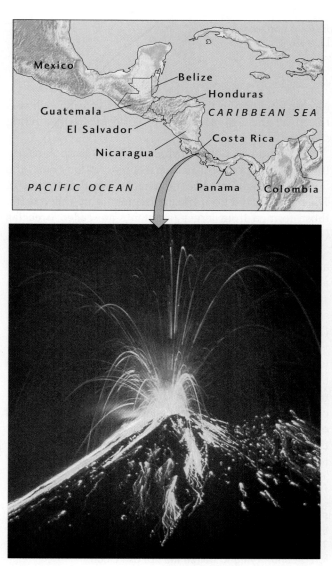

FIGURE 5.25 Explosive eruption at Arenal volcano, Costa Rica. [*Gregory G. Dimijian/Photo Researchers.*]

FIGURE 5.26 A volcanologist examines a volcanic bomb ejected from Asama volcano, Japan. [*Science Source/Photo Researchers.*]

(a)

(b)

~0.3 m

FIGURE 5.27 (a) Welded tuff from an ash-flow deposit in the Great Basin of northern Nevada. (b) Volcanic breccia. [*(a) John Grotzinger; (b) Doug Sokell/Visual Unlimited.*]

Mount Pinatubo in the Philippines, its volcanic ash was traced all the way around the world by Earth-orbiting satellites.

Sooner or later, these pyroclasts fall to Earth, building the largest deposits near their source. As they cool, the hot, sticky fragments become welded together (lithified). The rocks created from smaller fragments are called **tuffs;** those formed from larger fragments are called **breccias** (**Figure 5.27**).

Pyroclastic flows, which are particularly spectacular and often deadly, occur when hot ash, dust, and gases are ejected in a glowing cloud that rolls downhill at high speeds. The solid particles are buoyed up by the hot gases, so there is little frictional resistance to their movement (**Figure 5.28**).

In 1902, with very little warning, a pyroclastic flow with an internal temperature of 800°C exploded from the side of Mont Pelée, on the Caribbean island of Martinique. The avalanche of choking hot gas and glowing volcanic ash plunged down the slopes at a hurricane speed of 160 km/hour. In 1 minute and with hardly a sound, the searing emulsion of gas, ash, and dust enveloped the town of St. Pierre and killed 29,000 people.

FIGURE 5.28 This pyroclastic flow plunged down the slopes of Mount Unzen, in Japan, in June 1991. Note the fireman and fire engine in the foreground, trying to outrun the hot ash cloud descending on them. Three scientists who were studying this volcano died when they were engulfed by a similar flow. [*AP/Wide World Photos.*]

ERUPTIVE STYLES AND LANDFORMS

The shape a volcano takes varies with the properties of the magma, especially its chemical composition and gas content, the type of material (lava versus pyroclasts) erupted, and the environmental conditions under which it erupts—on land or under the sea. Volcanic landforms also depend on the rate at which lava is produced and the plumbing system that gets it to the surface (**Figure 5.29**).

> *The shape of a volcano depends on the composition, amount, and type of material erupted as well as on where it erupts.*

Central Eruptions

Central eruptions discharge lava or pyroclasts from a *central vent*. The vent consists of an opening atop a pipe-like feeder channel rising from the magma chamber. The magma ascends through this channel to erupt at Earth's surface. Central eruptions create the most familiar of all volcanic features: the volcanic mountain, shaped like a cone.

Shield Volcanoes A lava cone is built by successive flows of lava from a central vent. If the lava is basaltic, it flows easily and spreads widely. If flows are copious and frequent, they create a broad, shield-shaped volcano 2 or more kilometers high and many tens of kilometers in circumference with relatively gentle slopes. Mauna Loa, on the island of Hawaii, is the classic example of a **shield volcano** (Figure 5.29a). Although it rises only 4 km above sea level, it is actually the world's tallest mountain: measured from its base on the seafloor, Mauna Loa is 10 km high, higher than Mount Everest! It has a base diameter of 120 km, covering three times the area of Rhode Island. It grew to this enormous size by the accumulation of thousands of lava flows, each only a few meters thick, over a period of about a million years. The island of Hawaii actually consists of the overlapping tops of active shield volcanoes emerging from the ocean.

Volcanic Domes In contrast to basaltic lavas, andesitic and rhyolitic lavas are so viscous that they can barely flow. They often produce a *volcanic dome*, a rounded, steep-sided mass of rock. Domes look as though lava had been squeezed out of a vent like toothpaste, with very little lateral spreading. Domes often plug vents, trapping gases beneath them. Pressure increases until an explosion occurs, blasting the dome into fragments. Mount St. Helens produced such an explosion when it erupted in 1980 (Figure 5.29b).

Cinder Cones When volcanic vents discharge pyroclasts, these solid fragments can build up to create *cinder cones*.

The profile of a cinder cone is determined by the maximum angle at which the fragments remain stable rather than sliding downhill. The larger fragments, which fall near the vent, form very steep but stable slopes. Finer particles are carried farther from the vent and form gentle slopes at the base of the cone. The classic concave-shaped volcanic cone with its summit vent develops in this way (Figure 5.29c).

Stratovolcanoes When a volcano emits lava as well as pyroclasts, alternating lava flows and beds of pyroclasts build a concave-shaped composite volcano, or **stratovolcano.** Stratovolcanoes are common above subduction zones. Famous examples are Mount Fuji in Japan (Figure 5.29d), Mount Vesuvius and Mount Etna in Italy, and Mount Rainier in Washington State. Mount St. Helens used to have a near-perfect stratovolcano shape until its eruption in 1980 destroyed its northern flank (see Figure 5.22).

Craters A bowl-shaped pit, or **crater,** is found at the summit of most volcanoes, surrounding the central vent. During an eruption, the upwelling lava overflows the crater walls. When the eruption ceases, the lava that remains in the crater often sinks back into the vent and solidifies. When the next eruption occurs, that material may be blasted out of the crater. The crater later becomes partly filled by rock fragments that fall back into it. Because a crater's walls are steep, they may cave in or become eroded over time. In this way, a crater can grow to several times the diameter of the vent and hundreds of meters deep. The crater of Mount Etna in Sicily, for example, is currently 300 m in diameter.

Calderas When great volumes of magma are discharged rapidly, a large magma chamber cannot support its roof. In such cases, the overlying volcanic structure can collapse catastrophically, leaving a large, steep-walled, basin-shaped depression much larger than the crater, called a **caldera** (Figure 5.29e). Calderas can be impressive features, ranging in size from a few kilometers to 50 km or more in diameter. The Yellowstone volcano, which has an area greater than Rhode Island, is the largest active caldera in the United States.

After some hundreds of thousands of years, enough fresh magma may reenter the collapsed magma chamber to reinflate it, forcing the caldera floor to dome upward again and creating a *resurgent caldera*. The cycle of eruption, collapse, and resurgence may occur repeatedly over geologic time. Three times over the last 2 million years, Yellowstone Caldera has erupted catastrophically, in each instance ejecting hundreds or thousands of times more material than the 1980 Mount St. Helens eruption and

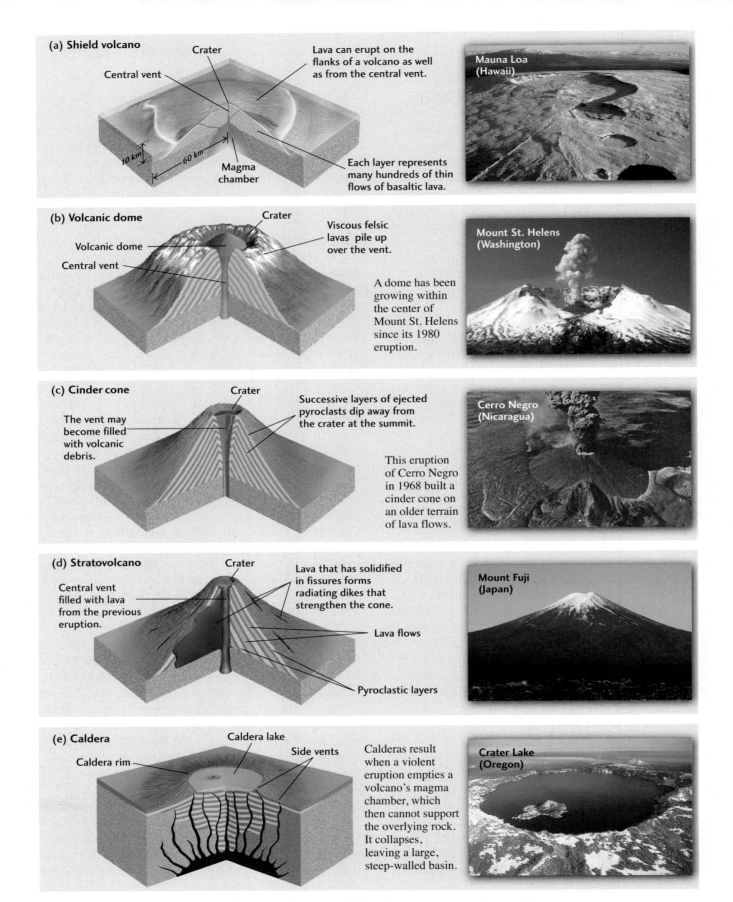

(a) Shield volcano

Crater

Central vent

10 km

60 km

Magma chamber

Lava can erupt on the flanks of a volcano as well as from the central vent.

Each layer represents many hundreds of thin flows of basaltic lava.

Mauna Loa (Hawaii)

(b) Volcanic dome

Crater

Volcanic dome

Central vent

Viscous felsic lavas pile up over the vent.

A dome has been growing within the center of Mount St. Helens since its 1980 eruption.

Mount St. Helens (Washington)

(c) Cinder cone

Crater

The vent may become filled with volcanic debris.

Successive layers of ejected pyroclasts dip away from the crater at the summit.

This eruption of Cerro Negro in 1968 built a cinder cone on an older terrain of lava flows.

Cerro Negro (Nicaragua)

(d) Stratovolcano

Crater

Central vent filled with lava from the previous eruption.

Lava that has solidified in fissures forms radiating dikes that strengthen the cone.

Lava flows

Pyroclastic layers

Mount Fuji (Japan)

(e) Caldera

Caldera lake

Caldera rim

Side vents

Calderas result when a violent eruption empties a volcano's magma chamber, which then cannot support the overlying rock. It collapses, leaving a large, steep-walled basin.

Crater Lake (Oregon)

FIGURE 5.29 The eruptive styles of volcanoes and the landforms they create are determined by the composition of magma. [(a) U.S. Geological Survey. (b) Lyn Topinka/USGS Cascades Volcano Observatory. (c) Mark Hurd Aerial Surveys. (d) Corbis. (e) Greg Vaughn/Tom Stack & Associates.]

depositing ash over much of what is now the western United States. Other resurgent calderas are Valles Caldera in New Mexico and Long Valley Caldera in California, which last erupted about 1.2 million and 760,000 years ago, respectively.

Fissure Eruptions

The largest volcanic eruptions come not from a central vent but through large, nearly vertical cracks in Earth's surface, sometimes tens of kilometers long (**Figure 5.30**). Such *fissure eruptions* are the main style of volcanism along mid-ocean ridges, where new oceanic crust is formed. A moderate-sized fissure eruption occurred in 1783 on a segment of the Mid-Atlantic Ridge that comes ashore in Iceland (**Figure 5.31**). One-fifth of Iceland's population perished from starvation as a result. A fissure 32 km long opened and spewed out some 12 km³ of basalt, enough to cover Manhattan to a height about halfway up the Empire State Building. Fissure eruptions continue on Iceland, although on a smaller scale.

FIGURE 5.30 A fissure eruption generates a "curtain of fire" on Kilauea, Hawaii, in 1992. [*U.S. Geological Survey.*]

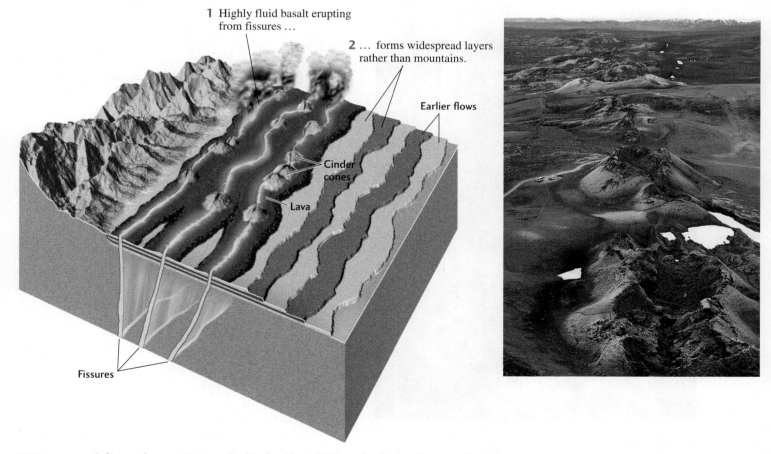

1 Highly fluid basalt erupting from fissures …

2 … forms widespread layers rather than mountains.

Earlier flows

Cinder cones

Lava

Fissures

FIGURE 5.31 (*left*) In a fissure eruption, highly fluid basaltic lava flows rapidly away from fissures and forms widespread layers, rather than building up into a volcanic mountain. [*After R. S. Fiske/ USGS.*] (*right*) These volcanic cones lie along the Laki fissure in Iceland, which opened in 1783 and erupted the largest flow of lava on land in recorded history. [*Tony Waltham.*]

400 km

FIGURE 5.32 The Columbia River basalts. Successive flows of flood basalts piled up to build this immense plateau, which covers a large area of Washington and Oregon. [*Dave Schiefelbein.*]

Flood Basalts Highly fluid basaltic lavas erupting from fissures on continental plates can spread out in thin sheets over flat terrain. Successive flows often pile up into immense basaltic lava plateaus, called *flood basalts*, rather than forming a shield volcano as they do when the eruption is confined to a central vent. A huge eruption of flood basalts about 16 million years ago buried 160,000 km^2 of preexisting topography in Washington, Oregon, and Idaho (**Figure 5.32**). Individual flows were more than 100 m thick, and some were so fluid that they traveled more than 500 km from their source. An entirely new landscape with new river valleys has since evolved atop the lava that buried the old surface. Plateaus made by flood basalts are found on every continent.

Ash-Flow Deposits Eruptions of pyroclasts on continents have produced extensive sheets of hard volcanic tuffs called *ash-flow deposits*. A succession of forests in Yellowstone National Park have been buried under ash flows. Some of the largest pyroclastic deposits on the planet are the ash flows erupted in the mid-Cenozoic era, 45 million to 30 million years ago, through fissures in what is now the Basin and Range province of the western United States. The amount of material released during this pyroclastic flare-up was a staggering 500,000 km^3—enough to cover the entire state of Nevada with a layer of rock nearly 2 km thick! As far as we know, humans have never witnessed one of these spectacular events.

NATURAL HAZARDS: The Effect of Volcanoes

Volcanic eruptions have a prominent place in human history and mythology. Ancient philosophers were awed by volcanoes and their fearsome eruptions of molten rock. The myth of the lost continent of Atlantis may have its source in the explosion of Thera (also known as Santorini), a volcanic island in the Aegean Sea. The eruption, which has been dated at 1623 B.C., formed a caldera 7 km by 10 km in diameter, visible today as a lagoon as much as 500 m deep with two small active volcanoes in the center. The eruption and the tsunami that followed it destroyed dozens of coastal settlements over a large part of the eastern Mediterranean. Some scientists have attributed the mysterious demise of the Minoan civilization to this ancient catastrophe.

Of Earth's 500 to 600 active volcanoes, at least one in six is known to have claimed human lives. In the past 500 years alone, more than 250,000 people have been killed by volcanic eruptions. Statistics on these fatalities and their causes are shown in **Figure 5.33**.

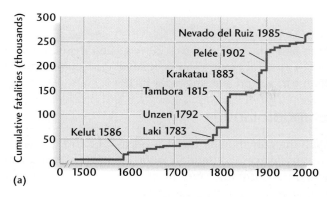

(a)

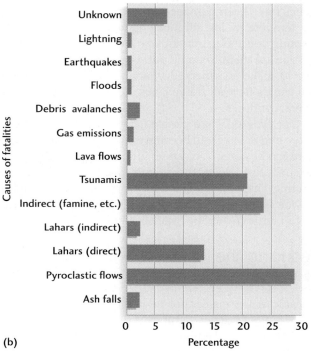

(b)

FIGURE 5.33 (a) Cumulative statistics on fatalities caused by volcanoes since A.D. 1500. The seven eruptions that dominate the record, each of which claimed 10,000 or more victims, are named. These eruptions account for two-thirds of the total deaths. (b) Causes of volcano fatalities since A.D. 1500. [*After T. Simkin, L. Siebert, and R. Blong, Science 291 (2001): 255.*]

Volcanoes can kill people and damage property in many ways, some of which are listed in Figure 5.33 and depicted in **Figure 5.34**. We have already mentioned some of these hazards, including pyroclastic flows and tsunamis. Several additional volcanic hazards are of special concern.

Lahars

Among the most dangerous volcanic events are the torrential mudflows of wet volcanic debris called **lahars.** They can occur when a pyroclastic flow meets a river or a snowbank, when the wall of a water-filled crater breaks, when a lava flow melts glacial ice, or when heavy rainfall transforms new ash deposits into mud. One extensive layer of volcanic debris in the Sierra Nevada of California contains 8000 km^3 of material of lahar origin, enough to cover all of Delaware with a deposit more than a kilometer thick. Lahars have been known to carry huge boulders for tens of kilometers. When Nevado del Ruiz in the Colombian Andes erupted in 1985, lahars triggered by the melting of glacial ice near the summit plunged down the slopes and buried the town of Armero 50 km away, killing more than 25,000 people.

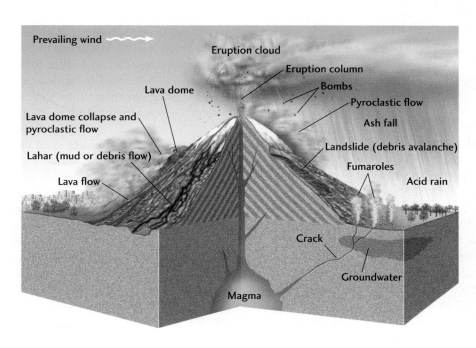

FIGURE 5.34 Some of the volcanic hazards that can kill people and destroy property. [*B. Meyers et al./USGS.*]

Flank Collapse

A volcano is constructed from thousands of deposits of lava or pyroclasts or both, which is not the best way to build a stable structure. A volcano's sides may become too steep and break or slip off. In recent years, volcanologists have discovered many prehistoric examples of catastrophic structural failures in which a big piece of a volcano broke off, perhaps because of an earthquake, and slid downhill in a massive, destructive landslide. On a worldwide basis, such *flank collapses* occur at an average rate of about four times per century. The collapse of one side of Mount St. Helens was the most damaging part of its 1980 eruption (see Figure 5.22).

Surveys of the seafloor off the Hawaiian Islands have revealed many giant landslides on the underwater flanks of the Hawaiian Ridge. When they occurred, these massive movements of earth probably triggered huge tsunamis. In fact, coral-bearing marine sediments have been found some 300 m above sea level on one of the Hawaiian Islands. These sediments were probably deposited by a giant tsunami that was caused by a prehistoric flank collapse.

The southern flank of Kilauea, on the island of Hawaii, is advancing toward the sea at a rate of 250 mm/year, which is relatively fast, geologically speaking. This advance became even more worrisome when it suddenly accelerated by a factor of several hundred on November 8, 2000, probably as a result of heavy rainfall a few days earlier. A network of motion sensors detected an ominous surge in velocity of about 50 mm/day. The surge lasted for 36 hours, after which the normal motion was reestablished. Someday, maybe thousands of years from now but perhaps sooner, the southern flank of the volcano is likely to break off and slide into the ocean. This catastrophic event would trigger a tsunami that could prove disastrous for Hawaii, California, and other Pacific coastal areas.

Caldera Collapse

Though infrequent, collapses of large calderas are one of the most destructive natural phenomena on Earth. Monitoring the activity of calderas is very important because of their long-term potential for widespread destruction. Fortunately, no catastrophic collapses have occurred in North America during recorded history, but geologists are concerned about an increase in small earthquakes in Yellowstone and Long Valley calderas as well as other indications of activity in their underlying magma chambers. For example, carbon dioxide leaking into the soil from magma in the crust has been killing trees since 1992 on Mammoth Mountain, a volcano on the boundary of Long Valley Caldera. Other indications of resurgence in this caldera have been the uplifting of the center of the caldera by more than half a meter in the past 20 years and the occurrence of nearly continuous swarms of small earthquakes (more than a thousand in a single day in 1997).

Eruption Clouds

With the growth in air travel, a volcanic hazard that is attracting increased attention is the clouds of volcanic ash lofted into air traffic lanes by erupting volcanoes. Over a period of 25 years, more than 60 commercial jet passenger planes have been damaged by such clouds. One Boeing 747 temporarily lost all four engines when ash from an erupting volcano in Alaska was sucked into the engines and caused them to flame out. Fortunately, the pilot was able to make an emergency landing. Warnings of eruption clouds near air traffic lanes are now being issued by several countries.

Our lives today are more reliant on air travel than ever before and are likely to become even more so. At any time, there can be several volcanoes erupting around the world. We are likely to see further examples of volcanic ash clouds entering airspace, and new measures will be needed to provide solutions from the lessons learned from the volcanic ash crisis created by Eyjafjallajökull in 2010.

Reducing the Risks of Volcanic Hazards

Volcanic eruptions cannot be prevented, but their catastrophic effects can be significantly reduced by a combination of science and enlightened public policy. Volcanology has progressed to the point that we can identify the world's dangerous volcanoes and characterize their potential hazards by studying deposits laid down in earlier eruptions. Some potentially dangerous volcanoes in the United States and Canada are identified in **Figure 5.35**. Assessments of their hazards can be used to guide zoning regulations to restrict land use in areas prone to volcano risk—the most effective measure to reduce property losses and casualties.

Studies indicate that Mount Rainier, because of its proximity to the heavily populated cities of Seattle and Tacoma, probably poses the greatest volcanic risk in the United States (**Figure 5.36**). Some 150,000 people live in areas where the geologic record shows evidence of floods and lahars that have swept down from the volcano over the past 6000 years. An eruption could kill thousands of people and cripple the economy of the Pacific Northwest.

Predicting Eruptions

These potential risks raise an important question: Can volcanic eruptions be predicted? In many cases, the answer is yes. Instrumented monitoring can detect signals such as

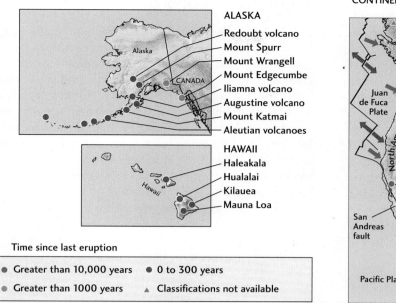

ALASKA
- Redoubt volcano
- Mount Spurr
- Mount Wrangell
- Mount Edgecumbe
- Iliamna volcano
- Augustine volcano
- Mount Katmai
- Aleutian volcanoes

HAWAII
- Haleakala
- Hualalai
- Kilauea
- Mauna Loa

Time since last eruption

- Greater than 10,000 years
- Greater than 1000 years
- 0 to 300 years
- ▲ Classifications not available

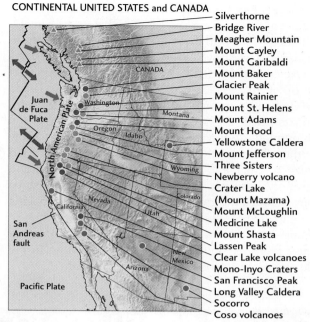

CONTINENTAL UNITED STATES and CANADA
- Silverthorne
- Bridge River
- Meagher Mountain
- Mount Cayley
- Mount Garibaldi
- Mount Baker
- Glacier Peak
- Mount Rainier
- Mount St. Helens
- Mount Adams
- Mount Hood
- Yellowstone Caldera
- Mount Jefferson
- Three Sisters
- Newberry volcano
- Crater Lake (Mount Mazama)
- Mount McLoughlin
- Medicine Lake
- Mount Shasta
- Lassen Peak
- Clear Lake volcanoes
- Mono-Inyo Craters
- San Francisco Peak
- Long Valley Caldera
- Socorro
- Coso volcanoes

FIGURE 5.35 Locations of potentially hazardous volcanoes in the United States and Canada. Volcanoes within each U.S. group are color-coded by time since their last eruption; those that have erupted most recently are thought to present the greatest cause for concern. (These classifications are subject to revision as studies progress and are not available for Canadian volcanoes.) Note the relationship between the volcanoes extending from northern California to British Columbia and the convergent boundary between the North American Plate and the Juan de Fuca Plate. [*After R. A. Bailey, P. R. Beauchemin, F. P. Kapinos, and D. W. Klick/USGS.*]

earthquakes, swelling of the volcano, and gas emissions that warn of impending eruptions. People at risk can be evacuated if the authorities are organized and prepared. Scientists monitoring Mount St. Helens were able to warn people before its eruption in 1980, as described at the opening of this chapter. Government infrastructure was in place to evaluate the warnings and to enforce evacuation orders, so very few people were killed.

Another successful warning was issued a few days before the cataclysmic eruption of Mount Pinatubo in the Philippines on June 15, 1991. A quarter of a million people were evacuated, including some 16,000 residents

FIGURE 5.36 Mount Rainier, seen from Tacoma, Washington. [*Joe Becker/Age fotostock/Photolibrary.*]

of the nearby U.S. Clark Air Force Base (which was heavily damaged by the eruption and has since been permanently abandoned). Tens of thousands of lives were saved from the lahars that destroyed everything in their paths. Casualties were limited to the few who disregarded the evacuation order. In 1994, 30,000 residents of Rabaul, Papua New Guinea, were successfully evacuated by land and sea hours before the two volcanoes on either side of the town erupted, destroying or damaging most of it. Many owe their lives to the government, which conducted evacuation drills, and to scientists at the local volcano observatory, who issued a warning when their seismographs recorded the ground tremors that signaled magma moving toward the surface.

Improving our ability to predict eruptions is important because there are about 100 high-risk volcanoes in the world and some 50 volcanic eruptions per year.

Controlling Eruptions

Can we go further by actually controlling volcanic eruptions? Not likely, because large volcanoes release energy on a scale that dwarfs our capabilities for control. However, under special circumstances and on a small scale, the damage can be reduced. Perhaps the most successful attempt to manage volcanic activity was made on the Icelandic island of Heimaey in January 1973. By spraying advancing lava with seawater, residents cooled and slowed the flow, preventing the lava from blocking the port entrance and saving some homes from destruction.

In the years ahead, the best policy for protecting the public will be the establishment of more warning and evacuation systems and more rigorous restriction of settlements in potentially dangerous locations.

GOOGLE EARTH PROJECT

Some of the most spectacular and dangerous volcanoes occur in the island arcs and volcanic mountain belts above subduction zones. Google Earth is a good tool for observing the sizes and shapes of these volcanoes. We will use it to investigate a famous example, Mount Fuji, on the Japanese island of Honshu.

LOCATION	Mount Fuji, Japan, and Sarychev Peak, Kurile Islands
GOAL	Observe the sizes and shapes of active stratovolcanoes
LINKED	Figure 5.29

Image © 2009 DigitalGlobe
Image © 2009 TerraMetrics
Image © 2009 Digital Earth Technology
Image © 2009 GeoEye

Google Earth view of Mount Fuji, Japan.

1. Type "Mt. Fuji, Japan" into the GE search window. Once you arrive there, tilt your frame of view to the north and observe the topography of the mountain from an eye altitude of several kilometers. Use the cursor to measure the peak height above sea level. Which of the following answers best describes the general shape of Mount Fuji?

   a. Large linear fissure in Earth's surface

   b. Low-relief, very broad volcano

   c. Steep-sided, low-elevation cinder cone

   d. High-elevation, steep-sided stratovolcano

2. From your observations of Mount Fuji and the surrounding area, what single feature convinces you that you are looking at a volcano?

   a. Number of trees and the amount of snow present on the mountainside

   b. Presence of a crater at the top of the mountain

   c. Steepness of the mountain slopes and the large landslide on the south slope

   d. Proximity of the mountain to the coastline of Japan and its distance from China

3. After considering the visible characteristics of Mount Fuji from various angles, how would you classify its level of volcanic activity at the time the satellite photo was taken?

   a. The eroded shapes of the landscape around the volcano indicate that it is now extinct, a conclusion further supported by the presence of snow.

   b. The steep slopes, circular shape, and well-defined crater indicate recent volcanic activity, but the presence of abundant snow near the crater rim suggests that the volcano is not currently erupting.

   c. The steep slopes, circular shape, and well-defined crater, combined with fresh lava on the snowfields of the main summit, indicate that the volcano is active and currently erupting.

   d. The circular shape and well-defined crater suggest a once-active volcano, but the lack of fresh lava and the presence of snow indicate that the volcano is extinct.

4. Tokyo, Japan, one of the largest cities on Earth, is home to more than 12 million people. To assess the hazard to Tokyo from Mt. Fuji, consider that the prevailing winds are expected to blow the cloud from a major eruption to the east, dumping up to a meter of ash more than 100 km from the volcano. Measure the distance and direction from the volcano to the urban center of Tokyo. Which of the following statements is most consistent with this information?

   a. Mount Fuji is too far away from Tokyo to pose a significant hazard.

   b. The volcano poses a significant hazard to Tokyo because it is close to the city and because the prevailing winds are likely to blow an eruption cloud in its direction.

   c. The volcano poses only a moderate hazard to Tokyo; it is close enough, but the prevailing winds are likely to blow any eruption cloud away from the city.

   d. The volcano is not a hazard to Tokyo because it is extinct and not expected to erupt.

Optional Challenge Question

5. Zoom out to an eye altitude of 3000 km. Look for the deep-sea trench that marks a subduction zone east of Mount Fuji. Move along the subduction zone to the northeast until you encounter Matua Island in the Kurile Islands chain, which belongs to Russia. This island is dominated by Sarychev Peak, one of the most active volcanoes of the Kurile Islands. Measure the height of the volcano and observe its features. Which of the following statements best describes your observations?

   a. Sarychev Peak is an island arc volcano, smaller but currently more active than Mount Fuji.

   b. Sarychev Peak is located in a continental volcanic mountain belt; it is smaller and currently less active than Mount Fuji.

   c. Sarychev Peak is a mid-ocean ridge volcano, larger and currently more active than Mount Fuji.

   d. Sarychev Peak is a hot-spot shield volcano, smaller but currently more active than Mount Fuji.

SUMMARY

How are igneous rocks classified? All igneous rocks can be divided into two broad textural classes: (1) coarse-grained rocks, which are intrusive and therefore cooled slowly; and (2) fine-grained rocks, which are extrusive and cooled rapidly. Within each of these broad categories, igneous rocks are classified as felsic, intermediate, or mafic on the basis of their silica content, or mineralogically on the basis of their proportions of lighter-colored felsic minerals and darker mafic minerals.

How and where do magmas form? Magmas form at places in the lower crust and mantle where temperatures are high enough for at least partial melting of rock. Because the minerals within a rock melt at different temperatures, the composition of magmas varies with melting temperature. Pressure raises the melting temperature of rock, and water lowers it. Because melted rock is less dense than solid rock, magma rises through the surrounding rock, and drops of magma come together to form magma chambers.

How does magmatic differentiation account for the variety of igneous rocks? Because different minerals crystallize at different temperatures, the composition of magma changes as it cools as various elements are withdrawn to form crystallized minerals. When a melt undergoes such fractional crystallization, the last rocks to form may be more felsic than the earlier, more mafic rocks. Fractional crystallization does not adequately explain the abundance of granite in Earth's crust, however, nor does differentiation of basaltic magma explain the observed compositions and abundances of igneous rocks. Different kinds of igneous rocks may be produced by variations in the compositions of magmas caused by the melting of different mixtures of sedimentary and other rocks and by the mixing of magmas.

What are the forms of igneous intrusions? Large intrusive igneous bodies are called plutons. The largest plutons are batholiths, which are thick tabular masses extending from a central funnel. Stocks are smaller plutons. Less massive than plutons are sills, which lie parallel to the layers of bedded country rock, and dikes, which cut across the layers.

How do tectonic processes affect magma production? Magmas are produced at mid-ocean ridges, where material rises from the upper mantle and undergoes decompression melting to form basaltic magma, and at subduction zones, where subducting oceanic lithosphere undergoes fluid-enhanced melting to generate magmas of varying composition. Mantle plumes penetrate the interiors of lithospheric plates as "hot spots," resulting in great outpourings of basaltic lavas.

What are the three major types of lavas? Lavas are classified as basaltic (mafic), andesitic (intermediate), and rhyolitic (felsic) on the basis of their silica content. Basaltic lavas are relatively fluid and flow freely; andesitic and rhyolitic lavas are more viscous. Lavas differ from pyroclasts, which are formed by explosive eruptions and vary in size from fine ash particles to house-sized bombs.

How are volcanoes shaped? The chemical composition and gas content of magma are important factors in a volcano's eruptive style and in the shape of the landforms it creates. A shield volcano grows from repeated eruptions of basaltic lava from a central vent. Andesitic and rhyolitic lavas are more viscous and tend to erupt explosively. The erupted pyroclasts may pile up into a cinder cone or cover an extensive area with ash-flow deposits. A stratovolcano is built of alternating layers of lava flows and pyroclastic deposits. The rapid ejection of magma from a large magma chamber, followed by collapse of the chamber's roof, results in a large depression, or caldera. Basaltic lavas can erupt from fissures along mid-ocean ridges as well as on continents, where they flow over the landscape in thin sheets to form flood basalts.

What hazards does volcanism present for human society? Volcanic hazards that can kill people and damage property include pyroclastic flows, tsunamis, lahars, flank collapses, caldera collapses, eruption clouds, and ash falls. Volcanic eruptions have killed about 250,000 people in the past 500 years.

KEY TERMS AND CONCEPTS

andesite (p. 138)
basalt (p. 138)
batholith (p. 146)
bomb (p. 159)
breccia (p. 160)
caldera (p. 161)
country rock (p. 135)
crater (p. 161)
dacite (p. 138)
decompression melting (p. 141)
dike (p. 147)
diorite (p. 138)
felsic rock (p. 138)
fluid-induced melting (p. 151)
fractional crystallization (p. 142)
gabbro (p. 138)
granite (p. 138)
granodiorite (p. 138)
hot spot (p. 152)
intermediate igneous rock (p. 138)

lahar (p. 165)

lava (p. 134)

mafic rock (p. 138)

magma chamber
(p. 141)

magmatic differentiation
(p. 142)

mantle plume (p. 152)

obsidian (p. 158)

partial melting (p. 140)

pegmatite (p. 148)

peridotite (p. 138)

pluton (p. 144)

porphyry (p. 135)

pumice (p. 158)

pyroclast (p. 136)

pyroclastic flow (p. 160)

rhyolite (p. 138)

shield volcano (p. 161)

sill (p. 147)

stock (p. 146)

stratovolcano (p. 161)

tuff (p. 160)

ultramafic rock (p. 138)

vein (p. 148)

viscosity (p. 139)

volcanic ash (p. 159)

volcano (p. 152)

▨ EXERCISES

1. Why are intrusive igneous rocks coarse-grained and extrusive rocks fine-grained?

2. What kinds of minerals would you find in a mafic igneous rock?

3. What kinds of igneous rocks contain quartz?

4. Name two intrusive igneous rocks with a higher silica content than that of gabbro.

5. How does fractional crystallization lead to magmatic differentiation?

6. Where in the crust, mantle, or core would you find a partial melt of basaltic composition?

7. Why do melts migrate upward?

8. Where on the ocean floor would you find basaltic magmas being extruded?

9. Much of Earth's crustal area and nearly its entire mantle are composed of basaltic or ultramafic rocks. Why are granitic and dioritic rocks as plentiful as they are on Earth? Where do the materials that constitute these rocks come from?

10. What is the difference between magma and lava? Describe a geologic situation in which a magma does not form a lava.

11. What are the three major types of volcanic rocks and their intrusive counterparts?

12. Most volcanism occurs near plate boundaries. What type of plate boundary can produce large amounts of rhyolitic lavas?

13. How do scientists predict volcanic eruptions?

14. How would you classify a coarse-grained igneous rock that contains about 50 percent pyroxene and 50 percent olivine?

15. What kind of rock would contain some plagioclase feldspar crystals about 5 mm long "floating" in a dark gray matrix of crystals of less than 1 mm?

16. What differences in crystal size might you expect to find between two sills, one intruded at a depth of about 12 km, where the country rock is very hot, and the other at a depth of 0.5 km, where the country rock is moderately warm?

17. If you were to drill a hole through the crust of a mid-ocean ridge, what intrusive or extrusive igneous rocks might you expect to encounter at or near the surface? What intrusive or extrusive igneous rocks might you expect at the base of the crust?

18. What observations might you make to show that a pluton solidified during fractional crystallization?

19. What processes create the unequal sizes of crystals in porphyries?

20. Water is abundant in the sedimentary rocks and oceanic crust of subduction zones. How would that water affect melting in these zones?

21. Why are eruptions of stratovolcanoes generally more explosive than eruptions of shield volcanoes?

22. Give a few examples of what geologists have learned about Earth's interior by studying volcanoes and volcanic rocks.

23. While on a field trip, you come across a volcanic formation that resembles a field of sandbags. The individual ellipsoidal forms have a smooth, glassy surface texture. What type of lava is this, and what information does the type give you about its history?

24. What might be the effects on civilization of a Yellowstone-type caldera eruption, such as the one described in this chapter?

Visual Literacy Task

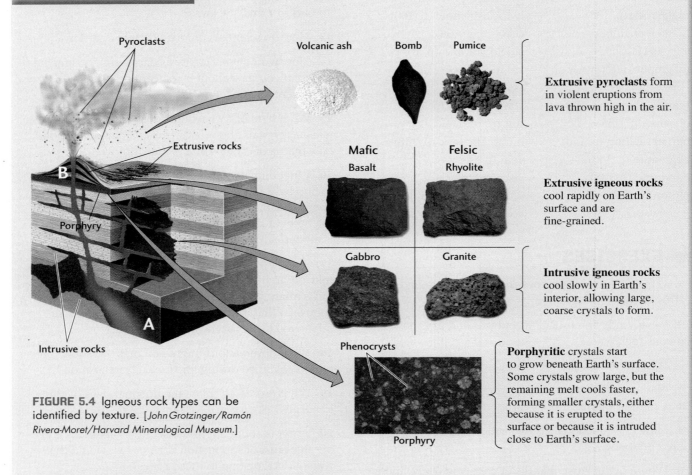

Pyroclasts

Volcanic ash   Bomb   Pumice

Extrusive pyroclasts form in violent eruptions from lava thrown high in the air.

Extrusive rocks

B

Porphyry

Intrusive rocks

A

Mafic
Basalt

Felsic
Rhyolite

Extrusive igneous rocks cool rapidly on Earth's surface and are fine-grained.

Gabbro   Granite

Intrusive igneous rocks cool slowly in Earth's interior, allowing large, coarse crystals to form.

Phenocrysts

Porphyry

Porphyritic crystals start to grow beneath Earth's surface. Some crystals grow large, but the remaining melt cools faster, forming smaller crystals, either because it is erupted to the surface or because it is intruded close to Earth's surface.

FIGURE 5.4 Igneous rock types can be identified by texture. [*John Grotzinger/Ramón Rivera-Moret/Harvard Mineralogical Museum.*]

1. What makes granite and gabbro different?

a. Cooling rate (fast vs. slow)

b. Size of minerals (large vs. small)

c. Location of formation (at the surface vs. deep in the crust)

d. Composition (mafic vs. felsic)

2. Which rocks could form at Location A and Location B, respectively?

a. A: Basalt and B: Gabbro

b. A: Gabbro and B: Granite

c. A: Granite and B: Rhyolite

d. A: Rhyolite and B: Basalt

3. The early part of the formation of a porphyry is similar to the formation of what type of rock?

a. Extrusive igneous pyroclasts

b. Extrusive igneous rocks

c. Intrusive igneous rocks

4. Which rocks cool the fastest?

a. Extrusive pyroclasts

b. Extrusive igneous rocks

c. Intrusive igneous rocks

d. Porphyritic igneous rocks

5. All igneous rocks that formed at Location B would have what characteristic?

a. Dark colored

b. Light colored

c. Fine-grained (small minerals)

d. Coarse-grained (large minerals)

Thought Question: Consider the generalization: "Fine-grained igneous rocks form during rapid cooling near the surface, whereas coarse-grained igneous rocks form during slow cooling at depth." Does the existence of porphyritic rocks imply that this generalization is incorrect?

Sedimentation: Rocks Formed by Surface Processes

6

Deep-Water Oil Reserves

ON APRIL 20, 2010, the world woke up to news that one of largest oil spills in history was under way in the Gulf of Mexico. An explosion had occurred on the Deepwater Horizon drilling platform poised above one of the deepest and most difficult places on Earth to explore for oil. This explosion—or "blowout"—introduced many people for the first time to the hazards of drilling in such difficult situations. The accident resulted in 11 deaths and oil flowed uncontrollably for 3 months as the well operator struggled to stem the flow. When the well was finally shut down on July 15 approximately 4 million barrels (about 170 million gallons) of oil had been leaked into the Gulf; it had washed up on shore from Mississippi to Texas and formed great pods of tar suspended in the deep waters of the Gulf of Mexico. It will take years to understand the full extent of this environmental catastrophe. Given the risks and consequences of such activity, one may responsibly ask what exactly is the prize that lurks in such deep waters?

> As oil reserves dwindle their value will go up, ensuring the commercial viability of drilling in very deep water.

As we will learn in this chapter, there are vast tracts of sediments and sedimentary rocks that lie along the margins of Earth's continents. Some date back to the time when the supercontinent Pangaea broke apart, creating rift basins in the Earth's crust that later became filled with sedimentary rocks. These sedimentary rocks can be observed using geological exploration techniques discussed in Chapter 1 (see Figure 1.11). In some cases the sediments were deposited in lakes and rivers that supplied both organic matter to create the oil and the reservoir rock that stores the oil. Then, as these rift basins became connected to the world ocean, they became flooded with seawater that evaporated to form a continuous salt layer. This salt layer blanketed the reservoir rocks to very effectively "seal" the system and prevent the escape of hydrocarbons out of the trough. In other cases, because salt is less dense than overlying rocks, it will sometimes rise up through those rocks and create "salt diapirs" that seal off oil and gas reservoirs. This was likely the case when the Deepwater Horizon blowout occurred.

The key to successful exploration involves identifying both the reservoir rock and the sealing rock; the Gulf of Mexico as well as much of the Atlantic Ocean along the coasts of Brazil and central Africa have well-developed troughs that are sealed with salt and full of oil. Estimated reserves are enormous, perhaps rivaling their well-known cousins in the Middle East. However, unlike in the Middle East, the cost to produce these reserves is much higher because operations involve drilling and production in water as deep as 3000 meters and up to hundreds of kilometers offshore where small floating towns must be built out at sea to support the drilling and production operations.

As oil reserves dwindle their value will go up, ensuring the commercial viability of drilling in very deep water. The reserves in deep water have proven to be a very large prize. ◆

An aerial view of the Ocean Confidence, a deep-water drilling rig, as it works in the Gulf of Mexico. [AP Photo/Bill Haber.]

CHAPTER OUTLINE

THE PRINCIPLES OF EXPLORATION for oil and gas are simple enough. Four basic factors are important: First, sedimentary rocks that lie close to the organic matter from which oil is created—so-called "source rocks"—must be present. Second, the source rocks must lie close to "reservoir rocks," porous sedimentary rocks into which oil can migrate. Third, "seal rocks"—sedimentary rocks that are impermeable to oil (or natural gas) and will not let it escape—must be present. Finally, "the trap"—very often the simple folding of the reservoir and seal rocks—forces the oil to accumulate as a pool. In this manner, the largest oil fields on Earth, such as those in Saudi Arabia, are created. All known oil originates in sedimentary rocks.

As supplies of oil and natural gas begin to diminish, the challenges for geologists increase. These challenges include finding new oil fields as well as squeezing out what is left behind in existing fields. Ultimately, it is the availability of source rocks that limits how much oil can be found. Source rocks were more abundant in some periods of Earth history, and they formed more easily in certain parts of the world. So there are geologic constraints that we must learn to accept. But we can learn to be smarter about how we explore for what little oil is left, and the need for well-trained geologists has never been greater

Much of Earth's surface, including its seafloor, is covered with sediments. These layers of loose particles have diverse origins. Most sediments are created by weathering of the continental crust. Some are the remains of mineral shells secreted by organisms. Yet others consist of inorganic crystals that precipitated when dissolved chemicals in oceans and lakes combined to form new minerals.

Sedimentary rocks were once sediments, so they are records of the conditions at Earth's surface when and where the sediments were deposited. Geologists can work backward to infer the sources of the sediments from which the rocks were formed and the kinds of places in which the sediments were originally deposited. The top of Mount Everest, for example, is composed of fossiliferous (fossil-containing) limestones. Because such limestones are formed from carbonate minerals in seawater, Mount Everest must once have been part of an ocean floor!

The analysis used to determine the history of rock formations at the top of Mount Everest applies just as well to ancient shorelines, mountains, plains, deserts, and swamps. In one area, for example, sandstone may record an earlier time when beach sands accumulated along a shoreline that no longer exists. In a bordering area, carbonate reefs may have been laid down along the perimeter of a tropical island. Beyond, there may

FIGURE 6.1 Sedimentary rocks can reveal former plate tectonic events and processes. These sedimentary rock layers, exposed in a road cut in Guatemala, originally were deposited as horizontal layers of sand and gravel by streams and landslides. After their deposition, the sediments were broken by a fault that has dropped the layers to the left of the fault down about 3 meters relative to the layers to the right. [*Fletcher & Baylis, Photo Researchers.*]

have been a nearshore area in which the sediments were shallow marine carbonate muds that later became thin-bedded limestone. By reconstructing such environments, we can map the continents and oceans of long ago.

Sedimentary rocks also reveal former plate tectonic events and processes by their presence within or adjacent to volcanic arcs, continental rift valleys, or collisional mountain belts (**Figure 6.1**). In cases in which sediments and sedimentary rocks are derived from the weathering of preexisting rocks, we can form hypotheses about ancient climates and environments. We can also use sedimentary rocks formed by precipitation from seawater to read the history of changes in Earth's climate and seawater chemistry.

The study of sediments and sedimentary rocks has great practical value as well. Oil and gas, our most valuable sources of energy, are found in sedimentary rocks. These precious resources are becoming increasingly difficult to find, so it is more important than ever to understand how sedimentary rocks form. As oil and gas decline in abundance, coal—which is a distinct type of sedimentary rock—will be used increasingly to generate energy. Another important energy source that may accumulate in sedimentary rocks is uranium, which is used for nuclear power. Phosphate rock used for fertilizer is also sedimentary, as is much of the world's iron ore. Knowing how these kinds of sediments form helps us to find and use these limited resources.

Finally, because virtually all sedimentary processes take place at or near Earth's surface where we humans live, we can study sedimentary rocks to improve our understanding of our environment.

In this chapter, we will see how geologic surface processes produce sediments and sedimentary rocks. We will describe the compositions, textures, and structures of sediments and sedimentary rocks and examine how they correlate with the kinds of environments in which those sediments and rocks were laid down.

WEATHERING AND SURFACE PROCESSES OF THE ROCK CYCLE

Sediments and the sedimentary rocks formed from them are produced by weathering and other surface processes of the rock cycle (discussed in Chapter 4). These processes act on rocks after the rocks have been moved from Earth's interior to its surface and uplifted by mountain building and before the rocks are returned to Earth's interior by subduction.

These processes move materials from a *source area*, where sediment particles are created, to a *sink area*, where they are deposited in layers. The path that sediment particles follow from source to sink may be a very long journey—one that involves several important processes resulting from interactions between the plate tectonic and climate systems.

> *Sediments and sedimentary rocks are produced by the surface processes of the rock cycle.*

Let's look at the role of the Mississippi River in a typical sedimentary process. Tectonic plate movement lifts up rocks in the Rocky Mountains. Rainfall in the

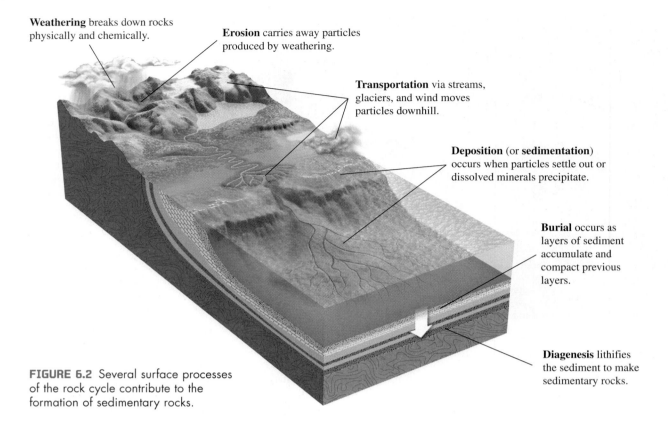

Weathering breaks down rocks physically and chemically.

Erosion carries away particles produced by weathering.

Transportation via streams, glaciers, and wind moves particles downhill.

Deposition (or **sedimentation**) occurs when particles settle out or dissolved minerals precipitate.

Burial occurs as layers of sediment accumulate and compact previous layers.

Diagenesis lithifies the sediment to make sedimentary rocks.

FIGURE 6.2 Several surface processes of the rock cycle contribute to the formation of sedimentary rocks.

mountains—a source area—causes weathering of the rocks there. If precipitation increases in the mountains, weathering also increases. Faster weathering produces more sediment to be released into the river and transported downhill and downstream. At the same time, if the flow in the river also increases because of the higher rainfall, transportation of the sediment down the length of the river will increase. This will increase the volume of sediment delivered to sink areas—sites of deposition, also known as sedimentary basins—in the Mississippi delta and the Gulf of Mexico. And in these sedimentary basins, the sediments will pile up on top of one another—layer after layer—and be buried deep in Earth's crust, to depths where they may become filled with valuable oil and natural gas.

The surface processes of the rock cycle that are important in the formation of sedimentary rocks are reviewed in **Figure 6.2** and summarized here.

■ **Weathering** is the general process by which rocks are broken down at Earth's surface to produce sediment particles. There are two types of weathering. **Physical weathering** takes place when solid rock is fragmented by mechanical processes, such as freezing and thawing or wedging by tree roots (**Figure 6.3**), that do not change its chemical composition. The rubble of broken stone often seen at the tops of mountains and hills is primarily

FIGURE 6.3 Plant roots contribute to physical weathering by penetrating fractures and wedging rocks apart. [*David R. Frazier/Photo Researchers.*]

Physical weathering refers to processes that fragment rock without changing its chemical composition. Chemical weathering refers to processes that change the chemical composition of rock and dissolve it.

the result of physical weathering. **Chemical weathering** refers to processes by which the minerals in a rock are chemically altered or dissolved. The blurring or disappearance of lettering on old gravestones and monuments is caused mainly by chemical weathering (**Figure 6.4**).

▪ *Erosion* refers to processes that dislodge particles of rock produced by weathering and move them away from the source area. Erosion occurs most commonly when rainwater runs downhill.

▪ *Transportation* refers to processes by which sedimentary particles are moved to sink areas. It occurs when water currents, wind, or the moving ice of glaciers transport particles to new locations downhill or downstream.

▪ *Deposition* (also called *sedimentation*) refers to processes by which sedimentary particles settle out as water currents slow, winds die down, or glacier edges melt to form layers of sediment in sedimentary basins. In aquatic environments, particles settle out, chemical precipitates form and are deposited, and the bodies and shells of dead organisms are broken up and deposited.

▪ *Burial* occurs as layers of sediment accumulate in sedimentary basins on top of older, previously deposited sediments, which are compacted and progressively buried deep within the basin. These sediments remain at depth, as part of Earth's crust, until they are either uplifted again or subducted by plate tectonic processes.

▪ *Diagenesis* refers to the physical and chemical changes—caused by pressure, heat, and chemical reactions—by which sediments buried within sedimentary basins are *lithified*, or converted into sedimentary rocks.

The Sources of Sediment: Weathering and Erosion

Chemical and physical weathering reinforce one another. Chemical weathering weakens rocks and makes them more susceptible to fragmentation. The smaller the fragments produced by physical weathering, the greater the surface area exposed to chemical weathering. Together, chemical weathering and physical weathering of rock create both solid particles and dissolved products, and erosion carries them away. The end products can be classified either as siliciclastic sediments or as chemical and biological sediments.

Siliciclastic Sediments The physical and chemical weathering of preexisting rocks forms *clastic particles* that are transported and deposited as sediments. Clastic particles range in size from boulders and pebbles to particles of sand, silt, and clay. They also vary widely in shape. Natural breakage along bedding planes and fractures in the parent rock determine the shapes of boulders, cobbles, and pebbles. Sand grains are the remnants of individual crystals formerly interlocked in the parent rock, and their shapes tend to reflect the shapes of those crystals.

FIGURE 6.4 These early-nineteenth-century gravestones at Wellfleet, Massachusetts, show the results of chemical weathering. The stone on the right is limestone and has become so weathered that it is unreadable. The stone on the left is slate and has remained legible under the same conditions. [*Courtesy of Raymond Siever.*]

TABLE 6.1	Minerals Present in Sediments Derived from a Granite Outcrop Under Varying Intensities of Weathering		
Low	**Medium**	**High**
Quartz	Quartz	Quartz
Feldspar	Feldspar	Clay minerals
Mica	Mica	
Pyroxene	Clay minerals	
Amphibole		

Most clastic particles are produced by the weathering of rocks composed largely of silicate minerals, so sediments formed from these particles are called **siliciclastic sediments.** The mixture of minerals in siliciclastic sediments varies. Minerals such as quartz resist weathering and thus are found chemically unaltered in siliciclastic sediments. These sediments may also contain partly altered fragments of minerals, such as feldspar, that are less resistant to weathering and therefore less stable. Still other minerals in siliciclastic sediments, such as clay minerals, are newly formed by chemical weathering. Varying intensities of weathering can produce different sets of minerals in sediments derived from the same parent rock.

Where weathering is intense, the sediment contains only clastic particles made of chemically stable minerals mixed with clay minerals. Where weathering is slight, many minerals that are unstable under land surface conditions will survive as clastic particles in the sediment. **Table 6.1** shows three possible sets of minerals in sediments derived from a typical granite outcrop.

> *Siliciclastic sediments are formed by the weathering of silicate rocks.*

Chemical and Biological Sediments Chemical weathering produces dissolved ions and molecules that accumulate in the waters of soils, rivers, lakes, and oceans. Chemical and biological reactions then precipitate these substances to form chemical and biological sediments. **Chemical sediments** form at or near their place of deposition. The evaporation of seawater, for example, often leads to the precipitation of gypsum or halite (**Figure 6.5**).

> *Chemical sediments are formed by the precipitation of products of chemical weathering, usually at or near their place of deposition.*

Biological sediments also form near their place of deposition, but they are the result of mineral precipitation by organisms. The abundance of biological sediments depends strongly on climate. Most are restricted to the subtropics and tropics, where carbonate-secreting

FIGURE 6.5 Salts precipitate when water containing dissolved minerals evaporates, which has occurred here in Death Valley, California. [*John G. Wilbanks/Agefoto.*]

FIGURE 6.6 One kind of sedimentary rock of biological origin is formed entirely of shell fragments. [*John Grotzinger.*]

organisms grow well. After the organisms die, their mineral remains, such as shells or skeletons, accumulate as sediments (**Figure 6.6**).

In shallow marine environments, biological sediments consist of layers of particles, such as whole or fragmented shells of marine organisms. Many different types of organisms, ranging from corals to clams to algae, may contribute their shells. Sometimes the shells are transported, further broken up, and deposited as **bioclastic sediments.** These shallow-water sediments consist predominantly of two calcium carbonate minerals, calcite and aragonite, in variable proportions. Other minerals, such as phosphates and sulfates, are locally abundant only in bioclasitc sediments.

> *Biological sediments are formed from the remains of organisms that precipitate minerals to form shells or skeletons.*

In the deep sea, biological sediments are made of the shells of only a few kinds of organisms. These shells are composed predominantly of the calcium carbonate mineral calcite, but silica is precipitated broadly over some parts of the deep sea. Because these biological particles accumulate in very deep water, where agitation by sediment-transporting currents is uncommon, they rarely form bioclastic sediments.

We distinguish between chemical and biological sediments for convenience only; in practice, many chemical and biological sediments overlap. In most of the world,

much more rock is fragmented by physical weathering than is dissolved by chemical weathering. Thus, clastic sediments are about 10 times more abundant in Earth's crust than chemical and biological sediments.

Transportation and Deposition: The Downhill Journey to Sedimentary Basins

After clastic particles and dissolved ions have been formed by weathering and dislodged by erosion, they start their journey to a sedimentary basin. This journey may be very long; for example, it might span thousands of kilometers from the tributaries of the Mississippi River in the Rocky Mountains to the wetlands of the Mississippi delta.

Most transport agents carry sediment on a one-way trip downhill. Rocks falling from a cliff, sand carried by a river flowing to the sea, and glacial ice slowly creeping downhill are all responses to gravity. Although winds may blow material from a low elevation to a higher one, in the long run the effects of gravity prevail. When a wind-blown particle drops into the ocean and settles through the water, it is trapped. It can be picked up again only by an ocean current, which can transport it to and deposit it in another site on the seafloor. Ocean currents transport sediments over a shorter distance than do big rivers on land, and the short transportation distance for chemical and biological sediments contrasts with the much greater distances over which siliciclastic sediments are transported. But eventually, all sediment transportation paths, as simple or complicated as they may be, lead downhill into a sedimentary basin.

Currents as Transport Agents Most sediments are transported by air and water currents. The enormous quantities of all kinds of sediments found in the oceans result primarily from the transporting capacities of rivers, which annually carry a solid and dissolved sediment load of about 25 billion tons (25×10^{15} g) (**Figure 6.7**).

Air currents—winds—move sediments, too, but in far smaller quantities than rivers or ocean currents. As particles are lifted into the air or water, the current carries them downwind or downstream. The stronger the current—that is, the faster it flows—the larger the particles it can transport.

Current Strength, Particle Size, and Sorting Deposition starts where transportation stops. For clastic particles, gravity is the driving force of deposition. The tendency of particles to settle under the pull of gravity works against a current's ability to carry them. A particle's settling velocity is proportional to its density and to its size. Because

all clastic particles have roughly the same density, we use particle size as the best indicator of how quickly a particle will settle. (We will take a more specific look at categories of particle size later in this chapter.) In water, large particles settle faster than small ones. This is also true in air, but the difference is much smaller.

Current strength, which is directly related to current velocity, determines the size of the particles deposited in a particular place. As a wind or water current begins to slow, it can no longer keep the largest particles suspended, and those particles settle. As the current slows even more, smaller particles settle. When the current stops completely, even the smallest particles settle. Currents segregate particles in the following ways:

■ *Strong currents* (faster than 50 cm/s) carry gravel (which includes boulders, cobbles, and pebbles), along with an abundant supply of smaller particles. Such currents are common in swiftly flowing streams in mountainous terrain, where erosion is rapid. Beach gravels are deposited where ocean waves erode rocky shores.

■ *Moderately strong currents* (20–50 cm/s) lay down sand beds. Currents of moderate strength are common in most rivers, which carry and deposit sand in their channels. Rapidly flowing floodwaters may spread sand

FIGURE 6.7 Sediments are easily transported by flowing water. In this photo, small ripples of sand in the channel are evidence of sediment transport. [*John Grotzinger.*]

Channel flow

Ripples of sand

over the width of a river valley. Waves and currents deposit sand on beaches and in the ocean. Winds also blow and deposit sand, especially in deserts. However, because air is much less dense than water, much higher current velocities are required for it to move sediments of the same size and density.

- *Weak currents* (slower than 20 cm/s) carry muds composed of the finest clastic particles (silt and clay). Weak currents are found on the floor of a river valley when floodwaters recede slowly or stop flowing entirely. In the ocean, muds are generally deposited some distance from shore, where currents are too slow to keep even fine particles in suspension. Much of the floor of the open ocean is covered with mud particles originally transported by surface waves and currents or by the wind. These particles slowly settle to depths where currents and waves are stilled and, ultimately, all the way to the bottom of the ocean.

As you can see, currents may begin by carrying particles of widely varying sizes, which then become separated as the strength of the current changes. A strong, fast current may lay down a bed of gravel, while keeping sand and mud in suspension. If the current weakens and slows, it will lay down a bed of sand on top of the gravel. If the current then stops altogether, it will deposit a layer of mud on top of the sand bed. This tendency for variations in current velocity to segregate sediments according to size is called **sorting.** A well-sorted sediment consists mostly of particles of a uniform size. A poorly sorted sediment contains particles of many sizes (**Figure 6.8**).

As cobbles, pebbles, and large sand grains are being transported by water or air currents, they tumble and strike one another or rub against the underlying rock. The resulting *abrasion* affects particles in two ways: it reduces their size, and it rounds off their rough edges (**Figure 6.9**). These effects apply mostly to the larger particles; smaller sand grains and silt undergo little abrasion.

Particles are generally transported intermittently rather than steadily. A river may transport large quantities of sand and gravel when it floods, then drop them as the flood recedes, only to pick them up again and carry them even farther in the next flood. Likewise, strong winds may carry large amounts of dust for a few days, then die down and deposit the dust as a layer of sediment. The strong tidal currents along some ocean margins may transport broken shell fragments to places farther offshore and drop them there.

The total time it takes for clastic particles to be transported may be many hundreds or thousands of years, depending on the distance to the final sedimentary basin and the number of stops along the way. Clastic particles eroded by the headwaters of the Missouri River in the mountains of western Montana, for example, take hundreds of years to travel the 3200 km down the Missouri and Mississippi rivers to the Gulf of Mexico.

> *Oceans contain abundant dissolved substances produced by chemical weathering, which can precipitate to form chemical or biological sediments.*

Oceans as Chemical Mixing Vats The driving force of chemical and biological sedimentation is precipitation

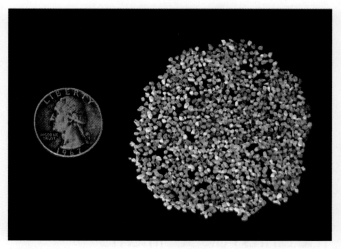 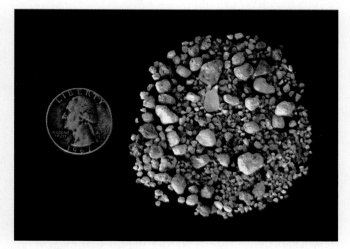

FIGURE 6.8 As the strength of currents changes, sediments are sorted according to particle size. The relatively homogeneous group of sand grains on the left is well sorted; the group on the right is poorly sorted. [*Bill Lyons.*]

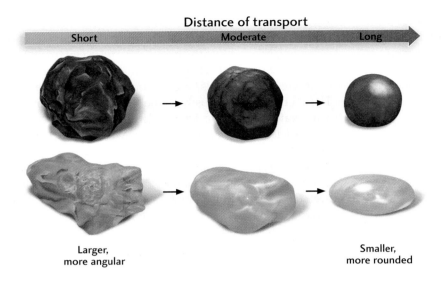

Distance of transport
Short · Moderate · Long

Larger,
more angular

Smaller,
more rounded

FIGURE 6.9 Abrasion during transportation reduces the size and angularity of clastic particles. Particles become more rounded and slightly smaller as they are transported, although the general shape of the particle may not change significantly.

rather than gravity. Substances dissolved in water by chemical weathering are carried along with the water. These materials are part of the water solution itself, so gravity cannot cause them to settle out. As the dissolved materials are carried down rivers, they ultimately enter the ocean.

Oceans may be thought of as huge chemical mixing vats. Rivers, rain, wind, and glaciers constantly bring in dissolved materials. Smaller quantities of dissolved materials enter the oceans through hydrothermal chemical reactions between seawater and hot basalt at mid-ocean ridges. The oceans lose water continuously by evaporation at the surface. The inflow and outflow of water to and from the oceans are so exactly balanced that the amount of water in the oceans remains constant over such geologically short times as years, decades, or even centuries. Over a time scale of thousands to millions of years, however, the balance may shift. During the most recent ice age, for example, significant quantities of seawater were converted into glacial ice, and sea level was drawn down by more than 100 m.

The entry and exit of dissolved materials, too, are balanced. Each of the many dissolved components of seawater participates in some chemical or biological reaction that eventually precipitates it out of the water and onto the seafloor. As a result, the ocean's **salinity**—the total amount of dissolved substances in a given volume of water—remains constant. Totaled over all the oceans of the world, mineral precipitation balances the total inflow of dissolved material from continental weathering processes and from hydrothermal activity at mid-ocean ridges—yet another way in which the Earth system maintains balance.

We can better understand this chemical balance by considering the element calcium. Calcium is a component of the most abundant biological precipitate formed in the oceans, calcium carbonate ($CaCO_3$). On land, calcium dissolves when limestone and calcium-containing silicate minerals, such as some feldspars and pyroxenes, are weathered. The calcium is then transported to the ocean as dissolved calcium ions (Ca^{2+}). There, many different marine organisms take up the calcium ions and combine them with carbonate ions (CO_3^{2}), also present in seawater, to form their calcium carbonate shells. Thus, the calcium that entered the ocean as dissolved ions leaves it as solid sediment particles when the organisms die and their shells settle to the seafloor and accumulate there as calcium carbonate sediments. Ultimately, the calcium carbonate sediments are buried and transformed into limestone. The chemical balance that keeps the levels of calcium dissolved in the ocean constant is thus controlled in part by the activities of organisms.

Nonbiological mechanisms also maintain chemical balance in the oceans. For example, sodium ions (Na^+) transported to the oceans react chemically with chloride ions (Cl^-) to form the precipitate sodium chloride ($NaCl$). This happens when evaporation raises the concentrations of sodium and chloride ions past the point of saturation. As we saw in Chapter 4, minerals precipitate when solutions become so saturated with dissolved materials that they can hold no more. The intense evaporation required to crystallize sodium chloride may take place in warm, shallow arms of the sea or in saline lakes.

SEDIMENTARY BASINS: THE SINKS FOR SEDIMENTS

As we have seen, the currents that move sediments across Earth's surface generally flow downhill. Therefore, sediments tend to accumulate in depressions in Earth's crust. Such depressions are formed by *subsidence*, in which a broad area of the crust sinks (subsides) relative to the surrounding crust. Subsidence is induced partly by the weight of sediments on the crust but is caused mainly by tectonic processes.

Sedimentary basins are depressions in the Earth's crust where the combination of subsidence and sedimentation has formed thick accumulations of sediments and sedimentary rock. Sedimentary basins are Earth's primary sources of oil and natural gas. Commercial exploration for these resources has helped us better understand the deep structure of sedimentary basins and of the continental lithosphere.

Rift Basins and Thermal Subsidence Basins

When plate separation begins within a continent, subsidence results from the stretching, thinning, and heating of the underlying lithosphere by the tectonic processes that are causing the separation (**Figure 6.10**). A long, narrow rift develops, bounded by great downdropped blocks of crustal rock. Hot, ductile mantle material rises, melts,

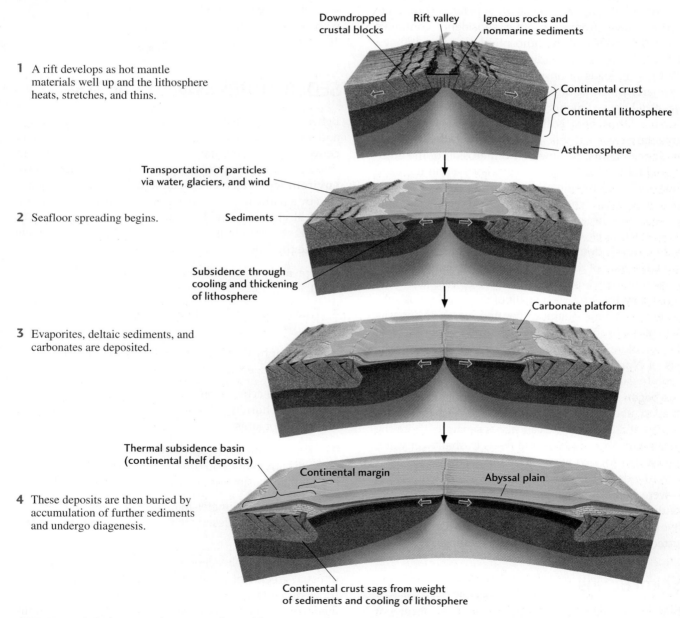

1 A rift develops as hot mantle materials well up and the lithosphere heats, stretches, and thins.

Downdropped crustal blocks | Rift valley | Igneous rocks and nonmarine sediments

Continental crust
Continental lithosphere
Asthenosphere

Transportation of particles via water, glaciers, and wind

2 Seafloor spreading begins.

Sediments

Subsidence through cooling and thickening of lithosphere

3 Evaporites, deltaic sediments, and carbonates are deposited.

Carbonate platform

Thermal subsidence basin (continental shelf deposits)

Continental margin | Abyssal plain

4 These deposits are then buried by accumulation of further sediments and undergo diagenesis.

Continental crust sags from weight of sediments and cooling of lithosphere

FIGURE 6.10 Sedimentary basins are formed by tectonic plate separation.

and fills the space created by the thinned lithosphere and crust, initiating the eruption of basaltic lavas in the rift zone. Such **rift basins** are deep, narrow, and long, with thick successions of sedimentary rocks and extrusive and intrusive igneous rocks.

The rift valleys of East Africa, the Rio Grande Valley, and the Jordan Valley in the Middle East are examples of rift basins. In places like the ancient Gulf of Mexico and the proto-Atlantic Ocean, rift basins may have become filled with sediments that created prolific hydrocarbon systems, like that discussed at the beginning of the chapter.

At later stages of plate separation, when rifting has led to seafloor spreading and the newly separated continents are drifting away from each other, subsidence continues through the cooling of the lithosphere that was thinned and heated during the earlier rifting stage (see Figure 6.10). Cooling leads to an increase in the density of the lithosphere, which in turn leads to its subsidence below sea level, where sediments can accumulate. Because cooling of the lithosphere is the main process creating the sedimentary basins at this stage, they are called **thermal subsidence basins.** Sediments from erosion of the adjacent land fill the basin nearly to sea level along the edge of the continent, creating a **continental shelf.**

The continental shelf continues to receive sediments for a long time because the trailing edge of the drifting continent subsides slowly and because the continents provide a tremendous area from which sediments can be derived. The load of the growing mass of sediment further depresses the crust, so the basins can receive still more material from the land. As a result of this continuous subsidence and sediment supply, continental shelf deposits can accumulate in an orderly fashion to thicknesses of 10 km or more. The continental shelves off the Atlantic coasts of North and South America, Europe, and Africa are good examples of thermal subsidence basins. These basins began to form when the supercontinent Pangaea split apart about 200 million years ago and the North American and South American plates separated from the Eurasian and African plates. Oil fields in shallower water in the Gulf of Mexico were formed in thermal sag basins that overlie deeper and older rift basins. Petroleum reservoirs were created when large volumes of sand were swept off of the continental shelf into deep water. The rocks that were being drilled when the Deepwater Horizon blowout occurred were probably formed in this setting.

Flexural Basins

A third type of sedimentary basin develops at convergent plate boundaries where one plate pushes up over the

other. The weight of the overriding plate causes the underlying plate to bend or flex down, producing a **flexural basin.** The Mesopotamian Basin in Iraq is a flexural basin formed when the Arabian Plate collided with and was subducted beneath the Eurasian Plate. The enormous oil reserves in Iraq (second only to the reserves of Saudi Arabia) owe their size to having the right ingredients in this important flexural basin. In effect, oil that had formed in the rocks now beneath the Zagros Mountains in Iran was squeezed out, forming several great pools of oil with volumes larger than 10 billion barrels.

> *Plate separation produces rift basins. As separation continues, plate cooling forms thermal subsidence basins. Flexural basins are generated where plates converge and one plate bends down under the weight of the other.*

SEDIMENTARY ENVIRONMENTS

Between the source area where sediments are formed and the sedimentary basin where they are buried and converted to sedimentary rocks, sediments travel a path through many sedimentary environments. A **sedimentary environment** is an area of sediment deposition characterized by a particular combination of climate conditions and physical, chemical, and biological processes (**Figure 6.11**). Important characteristics of sedimentary environments include the following:

- Type and amount of water (ocean, lake, river, arid land)

- Type and strength of transport agents (water, wind, ice)

- Topography (lowland, mountain, coastal plain, shallow sea, deep sea)

- Biological activity (precipitation of shells, growth of coral reefs, churning of sediments by worms and other burrowing organisms)

- Plate tectonic settings of sediment source areas (volcanic arc, continent-continent collision zone) and sedimentary basins (rift, thermal subsidence, flexural)

- Climate (cold climates may form glaciers; arid climates may form deserts where minerals precipitate by evaporation)

Consider the beaches of Hawaii, famous for their unusual green sands, which are a result of their distinctive sedimentary environment. Hawaii is a volcanic island made of olivine-bearing basalt, which is released during weathering. Rivers transport the olivine to the beach,

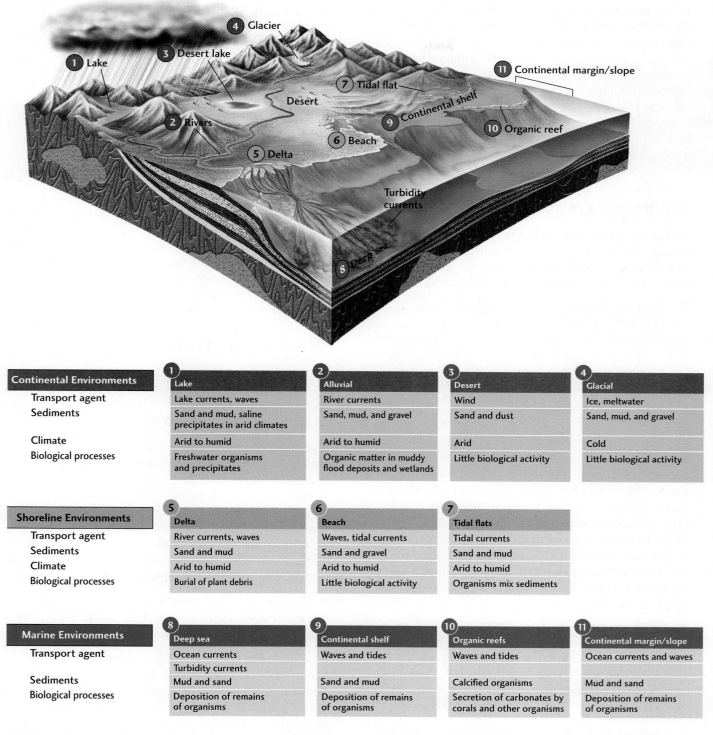

FIGURE 6.11 Multiple factors interact to create sedimentary environments.

Continental Environments	1 Lake	2 Alluvial	3 Desert	4 Glacial
Transport agent	Lake currents, waves	River currents	Wind	Ice, meltwater
Sediments	Sand and mud, saline precipitates in arid climates	Sand, mud, and gravel	Sand and dust	Sand, mud, and gravel
Climate	Arid to humid	Arid to humid	Arid	Cold
Biological processes	Freshwater organisms and precipitates	Organic matter in muddy flood deposits and wetlands	Little biological activity	Little biological activity

Shoreline Environments	5 Delta	6 Beach	7 Tidal flats
Transport agent	River currents, waves	Waves, tidal currents	Tidal currents
Sediments	Sand and mud	Sand and gravel	Sand and mud
Climate	Arid to humid	Arid to humid	Arid to humid
Biological processes	Burial of plant debris	Little biological activity	Organisms mix sediments

Marine Environments	8 Deep sea	9 Continental shelf	10 Organic reefs	11 Continental margin/slope
Transport agent	Ocean currents Turbidity currents	Waves and tides	Waves and tides	Ocean currents and waves
Sediments	Mud and sand	Sand and mud	Calcified organisms	Mud and sand
Biological processes	Deposition of remains of organisms	Deposition of remains of organisms	Secretion of carbonates by corals and other organisms	Deposition of remains of organisms

where waves and wave-producing currents concentrate the olivine and remove fragments of basalt to form olivine-rich sand deposits.

> *A sedimentary environment is a geographic location characterized by a particular combination of climate conditions and physical, chemical, and biological processes.*

Sedimentary environments are often grouped by location: on continents, near shorelines, or in the ocean. This very general subdivision highlights the processes that give sedimentary environments their distinct identities.

Continental Environments

Sedimentary environments on continents are diverse due to the wide variation in temperature and rainfall on the surface of the land. These environments are built around lakes, rivers, deserts, and glaciers (see Figure 6.11).

■ *Lake environments* include inland bodies of fresh or saline water in which the transport agents are relatively small waves and moderate currents. Chemical sedimentation of organic matter and carbonates may occur in freshwater lakes. Saline lakes, such as those found in deserts, evaporate and precipitate a variety of *evaporite* minerals, such as halite. The Great Salt Lake in Utah is an example.

■ *Alluvial environments* include the channel of a river, its borders and associated wetlands, and the flat valley floor on either side of the channel that is covered by water when the river floods (the floodplain). Rivers are present on all the continents except Antarctica, so alluvial deposits are widespread. Organisms are abundant in the muddy flood deposits and produce organic sediments that accumulate in swamplands adjacent to river channels. Climates vary from arid to humid. An example is the Mississippi River and its floodplains.

■ *Desert environments* are arid. Wind and the rivers that flow intermittently through deserts transport sand and dust. The dry climate inhibits abundant organic growth, so organisms have little effect on the sediments. Desert dune fields are an example of such an environment.

■ *Glacial environments* are dominated by the dynamics of moving masses of ice and are characterized by a

> *Continents offer diverse sedimentary environments because factors such as temperature and rainfall vary widely.*

cold climate. Vegetation is present but has little effect on sediments. At the melting border of a glacier, meltwater streams form a transitional alluvial environment.

Shoreline Sedimentary Environments

The dynamics of waves, tides, and river currents on sandy shores dominate shoreline environments (see Figure 6.11). Shoreline environments include the following:

■ *Deltas*, where rivers enter lakes or oceans

■ *Tidal flats*, where extensive areas exposed at low tide are dominated by tidal currents

■ *Beaches*, where the strong waves approaching and breaking on the shore distribute sediments on the beach, depositing strips of sand or gravel

In most cases, the sediments that accumulate in shoreline environments are siliciclastic. Organisms affect these sediments mostly by burrowing into and mixing them. However, in some tropical and subtropical settings, sediment particles, particularly carbonate sediments, may be of biological origin. These biological sediments are also subject to transport by waves and tidal currents.

Marine Environments

Marine sedimentary environments are usually classified by water depth, which determines the kinds of currents that are present (see Figure 6.11). Alternatively, they can be classified on the basis of distance from land.

■ *Continental shelf environments* are located in the shallow waters off continental shores, where sedimentation is controlled by relatively gentle currents. Sediments may be composed of either siliciclastic particles or biological carbonate particles, depending on how much siliciclastic sediment is supplied by rivers and on the abundance of carbonate-producing organisms. Sedimentation may also be chemical if the climate is arid and an arm of the sea becomes isolated and evaporates.

■ *Organic reefs* are carbonate structures, formed by carbonate-secreting organisms, built up on continental shelves or on oceanic volcanic islands.

■ *Continental margin and slope environments* are found in the deeper waters at and off the edges of the continents, where sediments are deposited by turbidity currents. A *turbidity current* is a turbulent submarine avalanche of sediment and water that moves downslope. Most sediments deposited by turbidity currents are siliciclastic, except at sites where organisms produce a lot of carbonate sediments. In this case, continental margin and slope sediments may be rich in carbonates.

■ *Deep-sea environments* are found far from continents, where the waters are much deeper than the reach of waves and tidal currents. These environments include the lower portion of the continental slope, which is built up by turbidity currents traveling far from continental margins; the abyssal plains of the deep sea, which accumulate carbonate sediments provided mostly by the shells of plankton; and the mid-ocean ridges.

> Marine sedimentary environments vary with water depth and distance from land.

Siliciclastic versus Chemical and Biological Sedimentary Environments

Sedimentary environments can be grouped not only by their location but also according to the kinds of sediments found in them, or according to the dominant sediment formation process. Grouping in this manner produces two broad classes: siliciclastic sedimentary environments and chemical and biological sedimentary environments.

Siliciclastic sedimentary environments are those dominated by siliciclastic sediments. They include all the continental sedimentary environments as well as the shoreline environments that serve as transitional zones between continental and marine environments. They also include the marine environments of the continental shelf, continental margin, and deep seafloor where siliciclastic sands and muds are deposited (**Figure 6.12**). The sediments of these siliclastic environments are often called *terrigenous sediments* to indicate their origin on land.

Chemical and biological sedimentary environments are characterized principally by chemical and biological precipitation (**Table 6.2**). By far the most abundant of these environments are *carbonate environments*, marine settings where calcium carbonate, mostly secreted by organisms, is the main sediment. Hundreds of species of mollusks and other invertebrate organisms, as well as calcareous (calcium-containing) algae, secrete carbonate shell materials. Various populations of these organisms live at different depths, both in quiet areas and in places where waves and currents are strong. As they die, their shells accumulate to form carbonate sediments.

Carbonate environments, except those of the deep sea, are found mostly in the warmer tropical and subtropical regions of the oceans, where carbonate-secreting organisms flourish. These environments include organic

FIGURE 6.12 Sedimentary rocks exposed at El Capitan, in the Guadalupe Mountains of Texas, were formed in an ancient ocean about 260 million years ago. The lower slopes of the mountains contain siliciclastic sedimentary rocks formed in deep-sea environments. The overlying cliffs of El Capitan are limestone and dolostone, which formed in a shallow sea when carbonate-secreting organisms died, leaving their shells to form a reef. [*John Grotzinger.*]

TABLE 6.2	Major Chemical and Biological Sedimentary Environments	
Environment	**Agent of Precipitation**	**Sediments**
SHORELINE AND MARINE		
Carbonate (reefs, platforms, deep sea, etc.)	Shelled organisms, some algae; inorganic muds, reefs	Carbonate sands and precipitation from seawater
Evaporite	Evaporation of seawater	Gypsum, halite, other salts
Siliceous (deep sea)	Shelled organisms	Silica
CONTINENTAL		
Evaporite	Evaporation of lake water	Halite, borates, nitrates, carbonates, other salts
Swamp	Vegetation	Peat

reefs, carbonate sand beaches, tidal flats, and shallow carbonate platforms. In a few places, carbonate sediments form in cooler waters that are supersaturated with carbonate ions—waters that are generally below 20°C, such as some regions of the Antarctic Ocean south of Australia. These carbonate sediments are formed by a very limited group of tiny organisms that mainly secrete calcite shell materials.

Siliceous environments are unique deep-sea sedimentary environments named for the remains of silica shells deposited in them. The planktonic organisms that secrete these tiny silica shells grow in surface waters where nutrients are abundant. When they die, their shells settle to the deep seafloor and accumulate as layers of siliceous sediments.

An *evaporite environment* is created when the warm seawater of an arid inlet or arm of the sea evaporates more rapidly than it can mix with seawater from the open ocean. The degree of evaporation and the length of time it has proceeded control the salinity of the evaporating seawater and thus the kinds of chemical sediments formed. Evaporite environments also form in lakes lacking river outlets. Such lakes may produce sediments of halite, borate, nitrates, and other salts.

SEDIMENTARY STRUCTURES

Sedimentary structures include all kinds of features formed at the time of deposition. Sediments and sedimentary rocks are characterized by **bedding,** or *stratification*, which occurs when layers of sediment, or *beds*, with different particle sizes or compositions are deposited on top of one another. These beds range from only millimeters or centimeters thick to meters or even many meters thick. Most bedding is horizontal, or nearly so, at the time of deposition. Some types of bedding, however, form at a high angle relative to the horizontal.

> Bedding is produced when layers of sediment with different particle sizes or compositions are deposited on top of one another.

Cross-Bedding

Cross-bedding consists of beds deposited by wind or water and inclined at angles as much as 35° from the horizontal (**Figure 6.13**). Cross-beds form when sediment particles are deposited on the steeper, downcurrent (leeward) slopes of sand dunes on land or of sandbars in rivers and on the seafloor (**Figure 6.14**). Cross-bedding in wind-deposited sand dunes may be complex as a result of rapidly changing wind directions. Cross-bedding is common in sandstones and is also found in gravels and some carbonate sediments. Cross-bedding is easier to see in sandstones than in sands, which must be excavated to see a cross section.

> Cross-bedding consists of layers of sediments deposited at an angle to the horizontal.

Graded Bedding

Graded bedding is most abundant in continental slope and deep-sea sediments deposited by dense, muddy

FIGURE 6.13 Cross-bedding in a desert environment: Navajo sandstone, Zion National Park, southwestern Utah. The varying angles of cross-bedding in this sandstone are due to changes in wind direction at the time the sand dunes were deposited. [*Peter Kresan.*]

turbidity currents, which hug the bottom of the ocean as they move downhill. Each bed progresses from large particles at the bottom to small particles at the top. As

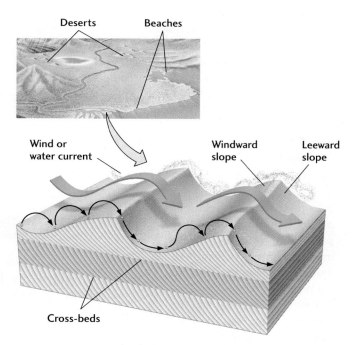

FIGURE 6.14 Cross-beds form when sediment particles are deposited on the steeper, downcurrent (leeward) slope of a sand dune or sandbar.

the current progressively slows, it drops progressively smaller particles. The grading indicates a weakening of the current that deposited the particles. A graded bed consists of a set of sediment particles, normally ranging from a few centimeters to several meters thick, that formed a horizontal or nearly horizontal layer at the time of deposition. Accumulations of many individual graded beds can reach a total thickness of hundreds of meters. A graded bed formed as a result of deposition by a turbidity current is called a *turbidite*.

> *Graded bedding is formed when particle sizes within a bed progress from large at the bottom of the bed to small at the top.*

Ripples

Ripples are very small ridges of sand or silt whose long dimension is at right angles to the current. They form low, narrow ridges, usually only a centimeter or two high, separated by wider troughs. These sedimentary structures are common in both modern sands and ancient sandstones (**Figure 6.15**). Ripples can be seen on the surfaces of windswept dunes, on underwater sandbars in shallow streams, and under the waves at beaches. Geologists can distinguish the symmetrical ripples made by waves moving back and forth on a beach from the asymmetrical ripples

> *Ripples are very small ridges of sand or silt formed by currents of water or wind.*

FIGURE 6.15 Ripples. (*left*) Ripples in modern sand on a beach. [*John Grotzinger.*] (*right*) Ancient ripple-marked sandstone. [*Photos by John Grotzinger.*]

formed by currents moving in a single direction over river sandbars or windswept dunes (**Figure 6.16**).

Bioturbation Structures

In many sedimentary rocks, the bedding is broken or disrupted by roughly cylindrical tubes a few centimeters in diameter that extend vertically through several beds. These sedimentary structures are remnants of burrows and tunnels excavated by clams, worms, and other marine organisms that live on the ocean floor. These organisms churn and burrow through muds and sands—a process called **bioturbation.** They ingest the sediment, digest the bits of organic matter it contains, and leave behind the reworked sediment, which fills the burrow (**Figure 6.17**). From bioturbation structures, geologists can determine the behavior of the organisms that burrowed in the sediment. Since the behavior of burrowing organisms is controlled partly by environmental factors, such as the strength of currents or the availability of nutrients, bioturbation structures help us reconstruct past sedimentary environments.

> *Bioturbation is the disruption of sediments by burrowing organisms.*

Bedding Sequences

Bedding sequences are built of interbedded and vertically stacked layers of different sedimentary rock types. A bedding sequence might consist of cross-bedded sandstone, overlain by bioturbated siltstone, overlain in turn by rippled sandstone—in any combination of thicknesses for each rock type in the sequence.

Bedding sequences help geologists reconstruct how sediments were deposited and so provide insight into the history of geologic processes and events that occurred at Earth's surface long ago. **Figure 6.18** shows a bedding sequence typically formed in alluvial sedimentary environments. A river lays down sediments as its channel meanders back and forth across the valley floor. Thus, the lower part of each sequence contains the beds deposited in the deepest part of the river channel, where the current was strongest. The middle part contains the beds deposited in the shallower parts of the channel, where the current was weaker, and the upper part contains the beds deposited on the floodplain. Typically, a bedding sequence formed in this manner consists of sediment particles that grade upward from large to small. This sequence may be repeated a number of times as the river meanders back and forth.

> *Bedding sequences consist of vertically stacked beds of different sedimentary rock types. They provide a history of geologic processes and events that occurred when the beds were formed.*

Most bedding sequences consist of a number of small-scale subdivisions. In the example shown in Figure 6.18,

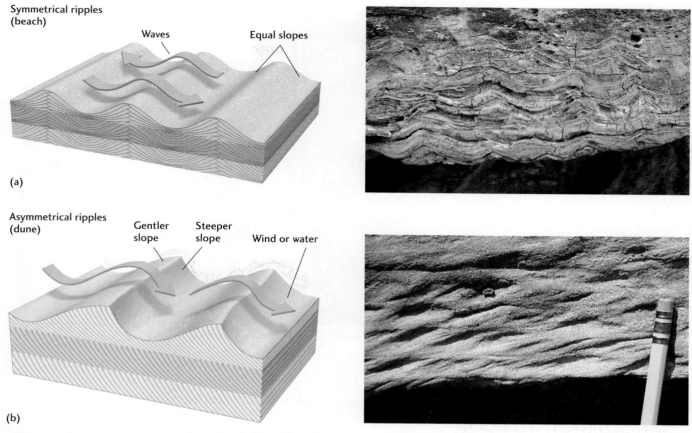

Symmetrical ripples (beach)

Waves   Equal slopes

(a)

Asymmetrical ripples (dune)

Gentler slope   Steeper slope   Wind or water

(b)

FIGURE 6.16 (a) The shapes of ripples on beach sand, produced by the back-and-forth movements of waves, are symmetrical. (b) Ripples on dunes and river sandbars, produced by the movement of a current in one direction, are asymmetrical. [*Photos by John Grotzinger.*]

the basal layers of the bedding contain cross-bedding. These layers are overlain by more cross-bedded layers, but the cross-beds are smaller in scale. Horizontal bedding occurs at the top of the bedding sequence. Today, computer models are used to analyze how bedding sequences of sands were deposited in alluvial environments.

FIGURE 6.17 Bioturbation structures. This rock is crisscrossed with fossilized tunnels originally made by organisms burrowing through the mud. [*John Grotzinger/Ramón Rivera-Moret/Harvard Mineralogical Museum.*]

FIGURE 6.18 A typical bedding sequence formed by a meandering river. [*USDA-NRCS photo by Jim R. Fortner.*]

BURIAL AND DIAGENESIS: FROM SEDIMENT TO ROCK

Most of the siliciclastic particles produced by weathering and erosion on land end up deposited in various sedimentary basins in the oceans. A smaller amount of siliciclastic sediment is deposited in sedimentary environments on land. Most chemical and biological sediments are also deposited in ocean basins, although some are deposited in continental basins containing lakes and wetlands.

Burial

Once sediments reach the ocean floor, they are trapped there. The deep seafloor is the ultimate sedimentary basin and for most sediments their final resting place. As the sediments are buried under new layers of sediments, they are subjected to increasingly high temperatures and pressures as well as chemical changes.

Diagenesis

After sediments are deposited and buried, they are subject to **diagenesis**—the many physical and chemical changes that result from the increasing temperatures and pressures as they are buried ever deeper in Earth's crust (**Figure 6.19**). These changes continue until the sediment or sedimentary rock is either exposed to weathering or metamorphosed by more extreme heat and pressure.

Temperature increases with depth in Earth's crust at an average rate of 30°C for each kilometer of depth. Thus, at a depth of 4 km, buried sediments may reach 120°C or more, the temperature at which certain types of organic matter may be converted to oil and natural gas. Pressure also increases with depth—on average, about 1 atmosphere for each 4.4 meters of depth. This increased pressure is responsible for the compaction of buried sediments.

> *Diagenesis is the set of physical and chemical processes that change sediments after they are deposited and buried.*

Buried sediments are also continuously bathed in groundwater full of dissolved minerals, which can precipitate in the pores between the sediment particles and bind them together—a chemical change called **cementation.** Cementation decreases **porosity,** the percentage of a rock's volume consisting of open pores between particles. In some sands, for example, calcium carbonate is precipitated as calcite, which acts as a cement that binds the grains and hardens the resulting mass into sandstone (**Figure 6.20**). Other minerals, such as quartz, may cement sands, muds, and gravels into sandstone, mudstone, and conglomerate.

The major physical diagenetic change is **compaction,** a decrease in the volume and porosity of a sediment. Compaction occurs as sediment particles are squeezed closer together by the weight of overlying sediment. Sands are fairly well packed during deposition, so they

1 Sediments are deposited and buried at shallow depths in Earth's crust.

2 Diagenesis includes the physical and chemical processes that change sediments to sedimentary rocks.

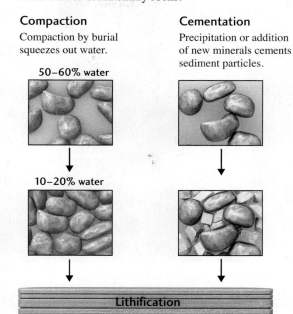

Compaction

Compaction by burial squeezes out water.

50–60% water

10–20% water

Cementation

Precipitation or addition of new minerals cements sediment particles.

Lithification

3 Different sediments result in different sedimentary rocks.

Fine ⟶ Coarse

Mud   Sand   Gravel

Shale   Sandstone   Conglomerate

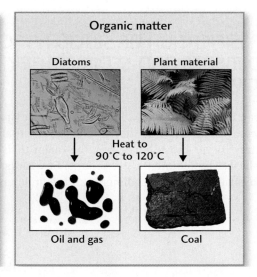

Organic matter

Diatoms   Plant material

Heat to 90°C to 120°C

Oil and gas   Coal

FIGURE 6.19 Diagenesis is the set of physical and chemical changes that convert sediments into sedimentary rocks. [*mud, sand, gravel: John Grotzinger; shale: John Grotzinger/Ramón Rivera-Moret/Harvard Mineralogical Museum; sandstone, conglomerate, coal: John Grotzinger/Ramón Rivera-Moret/MIT; diatoms: Mark B. Edlund/National Science Foundation; plant material: Kevin Rosseel; oil and gas: Wasabi/Alamy.*]

do not compact much. However, newly deposited muds, including carbonate muds, are highly porous. Often, more than 60 percent of these sediments consists of water in pore spaces. As a result, muds compact greatly after burial, losing more than half their water.

Both cementation and compaction result in **lithification,** the hardening of soft sediment into rock.

Cementation, the precipitation of minerals that fill in the pores between sediment particles and bind them together, and compaction, the reduction of pore space by squeezing particles together, result in lithification, the transformation of sediment into sedimentary rock.

FIGURE 6.20 This photomicrograph of sandstone shows quartz grains (white and gray) cemented by calcite (brightly colored and variegated) introduced after deposition. [Peter Kresan.]

CLASSIFICATION OF SILICICLASTIC SEDIMENTS AND SEDIMENTARY ROCKS

We can now use our knowledge of sedimentary processes to classify sediments and their lithified counterparts, sedimentary rocks. As we have seen, the major divisions are the siliciclastic sediments and sedimentary rocks and the chemical and biological sediments and sedimentary rocks.

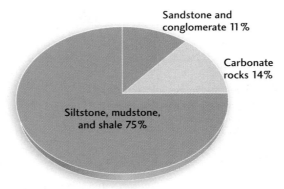

FIGURE 6.21 The relative abundances of the major sedimentary rock types. In comparison with these three types, all other sedimentary rock types—including evaporites, cherts, and other chemical sedimentary rocks—exist in only minor amounts.

Siliciclastic sediments and sedimentary rocks constitute more than three-fourths of the total mass of all types of sediments and sedimentary rocks in Earth's crust (**Figure 6.21**). We therefore begin with them.

Classification by Particle Size

Siliciclastic sediments and rocks are categorized primarily by particle size (**Table 6.3**):

- *Coarse-grained:* gravel and conglomerate

- *Medium-grained:* sand and sandstone

- *Fine-grained:* silt and siltstone; mud, mudstone, and shale; clay and claystone

We classify siliciclastic sediments and rocks on the basis of their particle size because it distinguishes them by one of the most important conditions of sedimentation: current strength. As we have seen, the larger the particle, the stronger the current needed to transport and deposit it. This relationship between current strength and particle size is the reason like-sized particles tend to accumulate in sorted beds. In other words, most sand beds do not contain pebbles or mud, and most muds consist only of particles finer than sand.

Of the various types of siliciclastic sediments and sedimentary rocks, the fine-grained clastics are by far the most abundant—about three times more common than the coarser-grained siliciclastics. The abundance of the fine-grained siliciclastics, which contain large amounts of clay minerals, is due to the chemical weathering of the large quantities of feldspar and other silicate minerals in Earth's crust into clay minerals. We turn now to a consideration of each of the three types of siliciclastic sediments and sedimentary rocks in more detail (**Figure 6.22**).

Coarse-Grained Siliciclastics: Gravel and Conglomerate

Gravel is the coarsest siliciclastic sediment, consisting of particles larger than 2 mm in diameter and including pebbles, cobbles, and boulders (see Table 6.3). **Conglomerate** is the lithified equivalent of gravel (Figure 6.22a). Pebbles, cobbles, and boulders are easy to study and identify because of their large size, which can tell us the strength of the currents that transported them. In addition, their composition can tell us about the nature of the distant terrain where they were produced.

There are relatively few sedimentary environments—mountain streams, rocky beaches with high waves, and glacier meltwaters—in which currents are strong enough to transport gravel. Strong currents also carry sand,

TABLE 6.3	Major Classes of Siliciclastic Sediments and Sedimentary Rocks		
Particle Size		**Sediment**	**Rock**
COARSE		GRAVEL	
Larger than 256 mm		Boulder ⎫	
256–64 mm		Cobble ⎬	Conglomerate
64–2 mm		Pebble ⎭	
MEDIUM			
2–0.062 mm		SAND	Sandstone
FINE		MUD	
0.062–0.0039 mm		Silt	Siltstone
Finer than 0.0039 mm		Clay	⎧ Mudstone (blocky fracture) ⎨ Shale (breaks along bedding) ⎩ Claystone

> *Conglomerates are sedimentary rocks formed from lithified coarse-grained sediments, equivalent in particle size to gravel.*

and we almost always find sand between gravel particles. Some of it is deposited with the gravel, and some infiltrates the spaces between particles after the gravel is deposited.

Medium-Grained Siliciclastics: Sand and Sandstone

Sand consists of medium-sized particles, ranging from 0.062 to 2 mm in diameter (see Table 6.3). These particles can be moved by moderate currents, such as those of rivers, waves at shorelines, and the winds that blow sand into dunes. Sand particles are large enough to be seen with the unaided eye, and many of their features are easily discerned with a low-power magnifying glass. The lithified equivalent of sand is **sandstone** (Figure 6.22b).

> *Sandstones are sedimentary rocks formed from lithified sand.*

Sizes and Shapes of Sand Grains Medium-sized siliciclastic particles—sand grains—are subdivided into fine, medium, and coarse grains. The average size of the grains in any one sandstone can be an important clue to both the strength of the current that carried it and the sizes

(a) Conglomerate

(b) Sandstone

(c) Shale

FIGURE 6.22 Clastic sedimentary rocks. [(a), (b) John Grotzinger/Ramón Rivera-Moret/MIT; (c) John Grotzinger/ Ramón Rivera-Moret/Harvard Mineralogical Museum.]

of the crystals eroded from the parent rock. The range of grain sizes and their relative abundances are also significant. If all the grains are close to the average size, the sand is well sorted. If many grains are much larger or smaller than the average, the sand is poorly sorted. The degree of sorting can help us distinguish, for example, between sands deposited on beaches (which tend to be well sorted) and sands deposited by glaciers (which tend to be muddy and poorly sorted). The shapes of sand grains can also be important clues to their origin. Sand grains, like pebbles and cobbles, are abraded and rounded during transport. Angular grains imply short transport distances; rounded ones indicate long journeys down a large river system.

Mineralogy of Sands and Sandstones Siliciclastics can be further subdivided by mineralogy, which can help identify the parent rocks. Thus, there are quartz-rich and feldspar-rich sandstones. Some sands are bioclastics, rather than siliciclastics; they are formed from material such as carbonate minerals that were originally precipitated as shells and then broken up and transported by currents. Thus, the mineralogy of sands and sandstones indicates the source areas and materials that were eroded to produce the sand grains. Sodium- and potassium-rich feldspars with abundant quartz, for example, might indicate that the sediments were eroded from a granitic terrain. Other minerals, as we will see in Chapter 7, might indicate metamorphic parent rocks.

The mineral content of sands and sandstones also indicates the plate tectonic setting of the parent rock. Sandstones containing abundant fragments of mafic volcanic rock, for example, might indicate that the sand grains were derived from a volcanic arc at a subduction zone.

> *In addition to particle size, siliciclastic sedimentary rocks can be classified by the sorting, shape, and mineralogy of the component particles.*

Fine-Grained Siliciclastics

The finest-grained siliciclastic sediments and sedimentary rocks are the silts and siltstones; the muds, mudstones, and shales; and the clays and claystones. All of them consist of particles that are less than 0.062 mm in diameter, but they vary widely in their range of grain sizes and in their mineral compositions. Fine-grained sediments are deposited by the gentlest currents, which allow the finest sediment particles to settle slowly to the bottom in quiet waves.

Silt and Siltstone Siltstone is the lithified equivalent of **silt,** a siliciclastic sediment in which most of the grains are between 0.0039 and 0.062 mm in diameter. Siltstone looks similar to mudstone or very fine grained sandstone.

Mud, Mudstone, and Shale Mud is a siliciclastic sediment containing water in which most of the particles are less than 0.062 mm in diameter (see Table 6.3). Thus, mud can be made of silt- or clay-sized sediments or varying quantities of both. The general term "mud" is very useful in fieldwork because it is often difficult to distinguish between silt- and clay-sized particles without a microscope.

Muds are deposited by rivers and tides. As a river recedes after flooding, the current slows, and mud, some of it containing abundant organic matter, settles on the floodplain. This mud contributes to the fertility of river floodplains. Muds are also left behind by ebbing tides along many tidal flats where wave action is mild. Much of the deep seafloor, where currents are weak or absent, is blanketed by muds.

The fine-grained rock equivalents of muds are mudstones and shales. **Mudstones** are blocky and show poor or no bedding. Distinct beds may have been present when the sediments were first deposited but then lost through bioturbation. **Shales** (Figure 6.22c) are composed of silt plus a significant component of clay, which causes them to break readily along bedding planes. Many muds contain more than 10 percent calcium carbonate sediments, forming calcareous mudstones and shales. Black, or organic, shales contain abundant organic matter. Some, called oil shales, contain large quantities of oily organic material, which makes them a potentially important source of oil. (We consider the oil shales in more detail in Chapter 14.)

Clay and Claystone Clay is the most abundant component of fine-grained sediments and sedimentary rocks and consists largely of clay minerals. Clay-sized particles are less than 0.0039 mm in diameter (see Table 6.3). Rocks made up exclusively of clay-sized particles are called **claystones.**

> *Fine-grained siliciclastic sedimentary rocks include siltstones, mudstones, shales, and claystones.*

NATURAL RESOURCES: Sandstone as a Source of Water, Oil and Natural Gas, and Uranium

Groundwater geologists, petroleum geologists, and mining geologists all have a special interest in sandstones. Groundwater geologists study the origin of sandstones to predict possible supplies of water in areas of porous sandstone, such as those found in the western plains of North America. In regions where tectonic uplift of mountains has occurred, sandstones may form the only type

FIGURE 6.23 Some uranium deposits form in sedimentary rocks when the chemistry of uranium-enriched fluids changes from reducing to oxidizing, causing precipitation of uranium. Here, the change of color from maroon to greenish gray represents that change in chemistry, and the mine is developed along the contact, where uranium is present. [*Matt Affolter/Wikimedia.*]

of aquifer in an entire region. For example, in the Great Valley of central California, groundwater is pumped from sandstone aquifers to provide water for irrigation of some of the country's largest vegetable and fruit farms.

Petroleum geologists must understand the porosity and cementation of sandstones because much of the oil and natural gas discovered in the last 150 years has been found in buried sandstones. For example, the prolific petroleum reservoir that was being drilled when the April 20, 2010, Deepwater Horizon blowout occurred is made out of sandstone. Sandstones are important reservoirs because their cementation may proceed more slowly than the cementation of carbonate rocks, which can lose much of their primary porosity by early cementation—perhaps even on the seafloor where they form. In contrast, sandstones may stay porous and permeable to greater burial depths thus allowing more time for petroleum migration into the reservoir.

Finally, much of the uranium used for nuclear power plants and weapons has come from uranium deposits precipitated in sandstones (**Figure 6.23**). These deposits formed when ancient groundwaters percolated through porous sandstones. Uranium can be dissolved in the groundwater if it lacks oxygen, and it can be transported for significant distances. However, if during this transport the oxygen-depleted groundwater intersects a different groundwater that contains higher levels of oxygen, then uranium will precipitate as uranium oxide minerals, potentially creating an ore deposit. These ore deposits often have a geometry that "freezes in" the intersection surface between the two groundwater masses.

CLASSIFICATION OF CHEMICAL AND BIOLOGICAL SEDIMENTS AND SEDIMENTARY ROCKS

Chemical and biological sediments and sedimentary rocks can be classified by their chemical composition (**Table 6.4**). Geologists distinguish between chemical and biological sediments not only for convenience, but also to emphasize the importance of organisms as the chief mediators of biological sedimentation. Both kinds of sediments can tell us about chemical conditions in the ocean, the predominant environment of deposition.

Carbonate Sediments and Rocks

Most **carbonate sediments** and **carbonate rocks** are formed by the accumulation and lithification of carbonate minerals that are secreted by organisms. The most abundant of these carbonate minerals is calcite (calcium carbonate—$CaCO_3$); in addition, most carbonate sediments contain aragonite, a less stable form of calcium carbonate. Some organisms precipitate calcite, some precipitate aragonite, and some precipitate both.

Most carbonate sediments of shallow marine environments are bioclastic sediments originally secreted as shells by organisms living near the surface or on the bottom. After the organisms die, they break apart, producing shells or fragments of shells that constitute individual particles, or *clasts*, of carbonate sediment. These sediments are found in tropical and subtropical environments

TABLE 6.4	Classification of Biological and Chemical Sediments and Sedimentary Rocks		
Sediment	**Rock**	**Chemical Composition**	**Minerals**
BIOLOGICAL			
Sand and mud (primarily bioclastic)	Limestone	Calcium carbonate ($CaCO_3$)	Calcite, aragonite
Siliceous sediment	Chert	Silica (SiO_2)	Opal, chalcedony, quartz
Peat, organic matter	Organics	Carbon compounds; carbon compounded with oxygen and hydrogen	(Coal, oil, gas)
No primary sediment (formed by diagenesis)	Phosphorite	Calcium phosphate ($Ca_3[PO_4]_2$)	Apatite
CHEMICAL			
No primary sediment (formed by diagenesis)	Dolostone	Calcium-magnesium carbonate ($CaMg[CO_3]_2$)	Dolomite
Iron oxide sediment	Iron formation	Iron silicate; oxide (Fe_2O_3); limonite, carbonate	Hematite, siderite
Evaporite sediment	Evaporite	Calcium sulfate ($CaSO_4$); sodium chloride (NaCl)	Gypsum, anhydrite, halite, other salts

Carbonate sediments and sedimentary rocks form from the accumulation and lithification of carbonate minerals that are secreted by organisms.

from the Pacific islands to the Caribbean and the Bahamas. Carbonate sediments are most accessible for study in these spectacular vacation spots, but the deep sea is where most carbonate sediments are deposited today.

Most of the carbonate sediments deposited on the abyssal plains of the deep sea are derived from the tiny calcite shells of *foraminifers* and other planktonic organisms that live in surface waters and secrete calcium carbonate. When the organisms die, their shells settle to the seafloor and accumulate there as sediments.

Reefs are moundlike or ridgelike organic structures composed of the carbonate skeletons and shells of millions of organisms. In the warm seas of the present, reefs are built mainly by corals, but hundreds of other organisms, such as algae, clams, and snails, also contribute. In contrast to the soft, loose sediments produced in other carbonate environments, the reef forms a rigid, wave-resistant structure of solid calcite and aragonite that is built up to and slightly above sea level (**Figure 6.24**). The solid calcite and aragonite of the reef are produced directly by the carbonate-cementing action of the organisms; there is no loose sediment stage.

Carbonate rocks are abundant because of the large amounts of calcium and carbonate minerals present in seawater, which organisms can convert directly into shells. Calcium is supplied by the weathering of feldspars and other minerals in igneous and metamorphic rocks. Carbonate minerals are derived from carbon dioxide in the atmosphere. Calcium and carbonate minerals also come from the easily weathered limestone on the continents. During burial and diagenesis, carbonate sediments react with water to form a new suite of carbonate minerals. In all these processes, the minerals precipitated are either calcium carbonates (calcite or aragonite) or calcium-magnesium carbonate (dolomite).

The dominant biological sedimentary rock lithified from carbonate sediments is **limestone,** which is composed mainly of calcite (calcium carbonate, $CaCO_3$) (**Figure 6.25a**; see Table 6.4). Limestone is formed from carbonate sands and muds and, in some cases, ancient reefs (see Figure 6.12). Although reefs are constructed mainly by corals today, they were constructed by other carbonate-secreting organisms—such as a now-extinct variety of mollusk—at earlier times in Earth's history (**Figure 6.26**).

Limestones are sedimentary rocks lithified from carbonate sediments. They are composed mainly of calcite.

Another abundant carbonate rock is **dolostone,** made up of the mineral dolomite, which is composed of calcium-magnesium carbonate (see Table 6.4). Dolostones are diagenetically altered carbonate sediments and limestones. The mineral dolomite does not form as a primary precipitate from ordinary seawater, and no organisms secrete

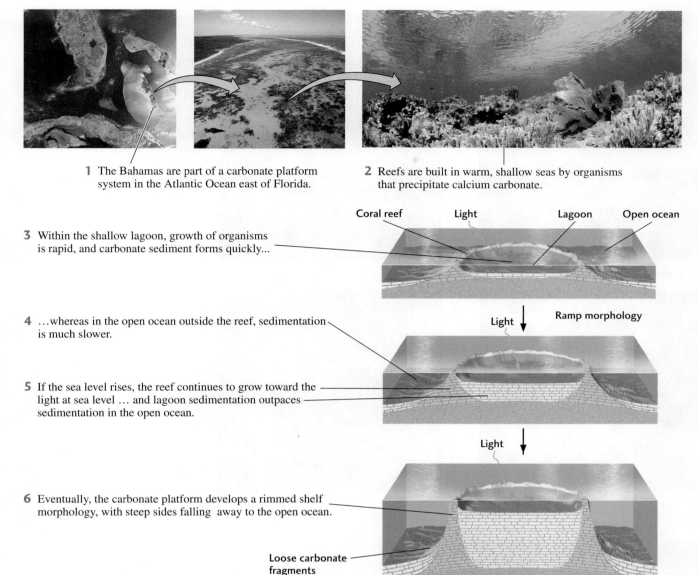

1 The Bahamas are part of a carbonate platform system in the Atlantic Ocean east of Florida.

2 Reefs are built in warm, shallow seas by organisms that precipitate calcium carbonate.

Coral reef   Light   Lagoon   Open ocean

3 Within the shallow lagoon, growth of organisms is rapid, and carbonate sediment forms quickly...

4 ...whereas in the open ocean outside the reef, sedimentation is much slower.

Light   Ramp morphology

5 If the sea level rises, the reef continues to grow toward the light at sea level ... and lagoon sedimentation outpaces sedimentation in the open ocean.

Light

6 Eventually, the carbonate platform develops a rimmed shelf morphology, with steep sides falling away to the open ocean.

Loose carbonate fragments

FIGURE 6.24 Marine organisms create carbonate platform systems. [*left: NASA; middle: Joseph R. Melanson/Aerials Only Photographs; right: Stephen Frink/Index Stock Imagery.*]

shells of dolomite. Instead, some calcium ions in the calcite or aragonite are exchanged for magnesium ions from seawater (or magnesium-rich groundwater) slowly passing through the pores of the sediment. This exchange converts calcium carbonate, $CaCO_3$, into dolomite, $CaMg(CO_3)_{2-}$.

> *Dolostone is a sedimentary rock made of the mineral dolomite, which forms by diagenetic alteration of carbonate sediments or limestones.*

Evaporite Sediments and Rocks: Products of Evaporation

Evaporite sediments and **evaporite rocks** are chemically precipitated from evaporating seawater and, in some cases, lake water.

Marine Evaporites Marine evaporites are chemical sediments and sedimentary rocks formed by the evaporation of seawater. These sediments and rocks contain minerals formed by the crystallization of sodium chloride (halite), calcium sulfate (gypsum and anhydrite), and other combinations of ions commonly found in seawater. As evaporation proceeds and the ions in the seawater become more concentrated, the minerals crystallize in a set sequence. As dissolved ions precipitate to form each mineral, the composition of the evaporating seawater changes.

Seawater has the same composition in all the oceans, which explains why marine evaporites are so similar the world over. No matter where seawater evaporates, the same sequence of minerals

> *Marine evaporites contain minerals that are precipitated from evaporating seawater*

(a) Limestone

(b) Gypsum

FIGURE 6.25 Chemical and biological sedimentary rocks: (a) limestone, lithified from carbonate sediments; (b) gypsum and (c) halite, marine evaporites that crystallize out of shallow seawater basins; and (d) chert, made up of siliceous sediments. [*John Grotzinger/Ramón Rivera-Moret/ Harvard Mineralogical Museum.*]

(c) Halite

(d) Chert

FIGURE 6.26 Limestone made from a reef constructed by now-extinct clams (rudists) in the Cretaceous Shuiba formation, Sultanate of Oman. [*John Grotzinger.*]

always forms. The study of evaporite sediments also shows us that the composition of the world's oceans has stayed more or less constant over the past 1.8 billion years. Before that time, however, the precipitation sequence may have been different, indicating that the composition of seawater may also have been different.

The great volume of many marine evaporites, some of which are hundreds of meters thick, shows that they could not have formed from the small amount of water that could be held in a small, shallow bay or pond. A huge amount of seawater must have evaporated to form them. The way in which such large quantities of seawater evaporate is very clear in bays or arms of the sea that meet the following conditions (**Figure 6.27**):

- The freshwater supply from rivers is small.

- Connections to the open sea are constricted.

- The climate is arid.

In such locations, water evaporates steadily, but the connections allow seawater to flow in to replenish the evaporating waters of the bay. As a result, those waters

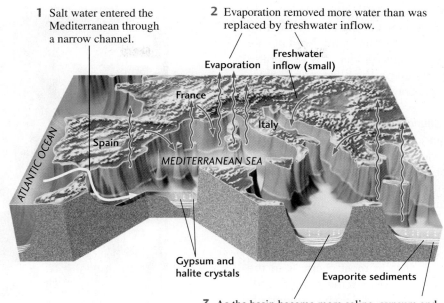

1 Salt water entered the Mediterranean through a narrow channel.

2 Evaporation removed more water than was replaced by freshwater inflow.

Evaporation

Freshwater inflow (small)

France

Italy

Spain

ATLANTIC OCEAN

MEDITERRANEAN SEA

Gypsum and halite crystals

Evaporite sediments

3 As the basin became more saline, gypsum and halite precipitated, forming evaporite sediments.

FIGURE 6.27 A marine evaporite environment of the past. The drier climate of the Miocene epoch made the Mediterranean Sea shallower than it is today, and its restricted connection to the open ocean created conditions suitable for evaporite formation. As seawater evaporated, gypsum precipitated to form evaporite sediments. A further increase in salinity led to the crystallization of halite. (The basin depth is greatly exaggerated in this drawing.)

stay at a constant volume but become more saline than the open ocean. The evaporating bay waters remain more or less constantly supersaturated and steadily deposit evaporite minerals on the floor of the bay.

As seawater evaporates, the first precipitates to form are the carbonates. Continued evaporation leads to the precipitation of gypsum, or calcium sulfate—$CaSO_4 \cdot 2H_2O$ (Figure 6.25b). By the time gypsum precipitates, almost no carbonate ions are left in the water. Gypsum is the principal component of plaster of Paris and is used in the manufacture of wallboard, which lines the walls of most new houses.

After still further evaporation, the mineral halite, or sodium chloride (NaCl)—one of the most common chemical sediments precipitated from evaporating seawater—starts to form (Figure 6.25c). Halite, as you may remember from Chapter 4, is table salt. Deep under the city of Detroit, Michigan, beds of salt laid down by an evaporating arm of an ancient ocean are commercially mined.

In the final stages of evaporation, after the sodium chloride is gone, magnesium and potassium chlorides and sulfates precipitate from the water. The salt mines near Carlsbad, New Mexico, contain commercial quantities of potassium chloride. Potassium chloride is often used as

> *Gypsum and halite are common marine evaporites.*

a substitute for table salt by people with certain dietary restrictions.

This sequence of mineral precipitation from seawater has been studied in the laboratory and is matched by the bedding sequences found in certain natural salt formations. Most of the world's evaporites consist of thick sequences of dolomite, gypsum, and halite and do not contain the final-stage precipitates. Many do not go even as far as halite. The absence of the final stages indicates that the water did not evaporate completely, but was replenished by normal seawater as evaporation continued.

Nonmarine Evaporites Evaporite sediments also form in arid-region lakes that typically have few or no river outlets. In such lakes, evaporation controls the lake level, and incoming minerals derived from chemical weathering accumulate as sediments. The Great Salt Lake is one of the best known of these lakes. In the dry climate of Utah, evaporation has more than balanced the inflow of fresh water from rivers and rain. As a result, the concentrations of dissolved ions in the lake make it one of the saltiest bodies of water in the world—eight times more saline than seawater. Sediments form when these ions precipitate.

> *Minerals precipitate and accumulate as sediments in nonmarine environments such as lakes as well as in oceans.*

Small lakes in arid regions may precipitate unusual salts, such as borates (compounds of the element boron), and some become alkaline. The water in this kind of lake is poisonous. Economically valuable deposits of borates and nitrates (minerals containing the element nitrogen) are found in the sediments beneath some of these lakes.

Siliceous Sediments: Sources of Chert There are several less common biological and chemical sediments that are locally abundant. One of the first sedimentary rocks to be used for practical purposes by our prehistoric ancestors was **chert,** which is made up of silica (SiO_2) (Figure 6.25d). Early hunters used it for arrowheads and other tools because it could be chipped and shaped to form hard, sharp implements. A common name for chert is *flint*; the two terms are virtually interchangeable. The silica in most cherts is in the form of extremely fine crystalline quartz. Some geologically young cherts consist of opal, a less well crystallized form of silica.

> *Chert is a sedimentary rock formed from siliceous sediments, which are generally precipitated biologically.*

Like calcium carbonate sediments, many siliceous sediments are precipitated biologically, as silica shells secreted by planktonic organisms that settle to the deep seafloor and accumulate as layers of sediment. After these sediments are buried by later sediments, they are cemented into chert. Chert may also form as nodules and irregular masses replacing carbonate in limestones and dolostones.

NATURAL RESOURCES: Coal, Oil, and Natural Gas

Coal is a biological sedimentary rock composed almost entirely of organic carbon formed by the diagenesis of swamp vegetation. Vegetation may be preserved from decay and accumulate as a rich organic material called *peat*, which contains more than 50 percent carbon. If peat is ultimately buried, it may be transformed into coal. Coal is classified as an *organic sedimentary rock*, a class that consists entirely or partly of organic carbon-rich deposits formed by the diagenesis of once-living material that has been buried.

> *Coal is a sedimentary rock formed from organic sediments. Oil and natural gas are fluids that are also created by diagenesis of organic matter.*

In both lake and ocean waters, the remains of algae, bacteria, and other microscopic organisms may accumulate in fine-grained sediments as organic matter that can be transformed by diagenesis into oil and natural gas.

FIGURE 6.28 Petroleum reservoir rock from a core obtained 4 kilometers deep in the Earth. This carbonate rock contains voids filled with oil. The rock was formed by microorganisms that built a porous reeflike structure that was then buried and filled with migrating hydrocarbons. Reservoir rocks of this type may be very important to future deep offshore drilling in the South Atlantic region. Scale subdivided in centimeters. [*John Grotzinger.*]

Crude oil (petroleum) and *natural gas* are fluids that are not normally classed with sedimentary rocks. They can be considered organic sediments, however, because they form by the diagenesis of organic material in the pores of sedimentary rocks. Deep burial changes the organic matter originally deposited along with inorganic sediments into a fluid that then escapes to porous rock formations and becomes trapped there (**Figure 6.28**). As noted earlier in this chapter, oil and natural gas are found mainly in sandstones and limestones.

As supplies of oil and natural gas begin to diminish, the challenges for geologists increase. These challenges include finding new oil fields as well as squeezing out what is left behind in existing fields. Ultimately, it is the availability of organic sediments that limits how much oil and gas can be found. These sediments were more abundant in some period of Earth history, and they were formed more easily in certain parts of the world. So there are geologic constraints that we must learn to accept. But we can learn to be smarter about how we explore for what little oil is left and the need for well-trained geologists has never been greater.

GOOGLE EARTH PROJECT

Sediments are deposited in specific geologic environments where the formation of sedimentary rock takes place. Specifically, the processes that transform sediment into sedimentary rock happen near Earth's surface and always involve the presence of liquid water in some form. Thus, Google Earth is an ideal tool for interpreting and appreciating the spectrum of environments in which sedimentary rocks form.

Among the unique sedimentary environments are carbonate platforms. Carbonate platforms form when ions dissolved in seawater precipitate to form carbonate sediments. This process is often controlled by organisms. Consider the Great Barrier Reef off the northeastern coast of Australia. Here you will see a sedimentary environment driven by the life cycles of small marine organisms and the calcite-enriched water they live in. What causes the distinctive blue-green color of the water here, and how are soluble minerals such as $CaCO_3$ precipitated as sediment on the ocean floor? Appreciate the geometry of the reef feature and its geographic limits to the north and south. Now compare the geometry of the Great Barrier Reef with a slightly different carbonate environment. Travel down to the equatorial waters of Bora Bora atoll in the South Pacific Ocean and see how the carbonate deposits there differ. Why does the reef take its circular shape, and what inspired it to begin growing here? These questions and many more can be explored through the GE interface

LOCATION	Great Barrier Reef, northeastern Australian coast, and Bora Bora atoll, South Pacific Ocean
GOAL	Explore an area of significant sediment deposition in a modern sedimentary environment
LINKED	Figure 6.24

Southern limit of Great Barrier Reef

Data SIO, NOAA, U.S. Navy, NGA, GEBCO
©2009 Cnes/Spot Image
Image ©2009 TerraMetrics
Image ©2009 DigitalGlobe

Google

1. Navigate to the Great Barrier Reef on the northeastern coast of Australia and zoom in to an eye altitude of 20 km on Pipon Island, Queensland, Australia. Notice the white material around the island itself and along the coast of the peninsula just to the south (on mainland Australia). From your investigation of the images here (feel free to zoom in and out), how would you best characterize this white material? Be sure to

(Continued.)

consider any patterns you see in the distribution of this coastal material when choosing an answer.

   a. Unconsolidated carbonate sediment

   b. Large boulder deposits

   c. Cemented olivine sand

   d. Deltaic siliciclastic mudstone

2. From Pipon Island, zoom out to an eye altitude of 2300 km and appreciate the length of the offshore reef features paralleling the coast. Use the path measurement tool in GE to determine the approximate length of this reef system.

   a. 2000 km

   b. 1400 km

   c. 2800 km

   d. 750 km

3. Notice that the Great Barrier Reef provides some protection to the coastal environment where it is present. As you follow the reef to the south, it becomes less distinct and provides less coastal protection. At the southern end of the reef, waves from the open South Pacific are free to break on the Australian coast, and some of the best surfing in the world results. At approximately what southerly latitude does the reef system end?

   a. 10°30′23″S; 143°30′06″E

   b. 17°56′25″S; 146°42′57″E

   c. 24°39′49″S; 153°15′18″E

   d. 21°06′31″S; 151°38′53″E

4. Following up on questions 2 and 3, the Great Barrier Reef provides protection to the Australian coastline and allows for sedimentary processes to occur there. As one moves farther from the equator to latitudes of 25° S, it is clear that reef formation stops. Consider the conditions in which reef-building organisms precipitate calcium carbonate. What might be the primary climate-related factor controlling the southern limit of the Great Barrier Reef?

   a. Sea surface temperatures of less than 18°C

   b. Depth of ocean water along the coast to the south

   c. Amount of sediment on the beaches near Brisbane

   d. Color of the seawater along the coast south of 25°

Optional Challenge Question

5. Now let's travel to warmer climes by typing "Bora Bora atoll" into the GE search window and zooming in to an eye altitude of 20 km once you arrive there. In contrast to the Great Barrier Reef, this island in the South Pacific has a very limited reef system, yet the reef system has a unique geometry. The formation of an atoll like this one involves a unique relationship between biotic and geologic factors. From your observation and exploration of the atoll, which pair of biotic and abiotic factors properly reflects the relationship present here?

   a. Birds and quartz sand beaches

   b. Coral reefs and volcanic islands

   c. Foraminifera and outcrops of marine shale

   d. Whales and carbonate platforms

■ SUMMARY

What are the major processes that form sedimentary rock? Weathering produces the particles that compose siliciclastic sediments and the dissolved ions that precipitate to form biological and chemical sediments. Currents of water and air and the movement of glaciers transport the sediments to their ultimate resting place in a sedimentary basin. Sedimentation (also called deposition) is the settling out of particles or precipitation of minerals to form layers of sediments. Lithification and diagenesis harden the sediments into sedimentary rock.

What are the two major types of sediments? Sediments and the sedimentary rocks that form from them can be classified as siliciclastic or chemical and biological. Siliciclastic sediments form from fragments of parent rock produced by physical and chemical weathering and are transported to sedimentary basins by water, wind, or ice. Chemical and biological sediments originate from ions dissolved in and transported by water. Through chemical and biological reactions, these ions are precipitated from solution to form sediments.

How are the major kinds of siliciclastic sediments and chemical and biological sediments classified? Siliciclastic sediments and sedimentary rocks are classified by the size of their particles. The major classes, in descending particle size order, are gravels and conglomerates; sands and sandstones; silts and siltstones; muds, mudstones, and shales; and clays and claystones. This classification method emphasizes the importance of the strength of the current that transported the sediments. Chemical and biological sediments and sedimentary rocks are classified on the basis of their chemical composition. The most abundant of these rocks are the carbonate rocks: limestone and dolostone. Limestone is made up largely of biologically precipitated calcite. Dolostone is formed by the diagenetic alteration of limestone. Other chemical and biological sediments include evaporites, siliceous sediments such as chert, and peat and other organic matter that is transformed into coal, oil, and gas.

KEY TERMS AND CONCEPTS

bedding (p. 190)

bedding sequence (p. 192)

bioclastic sediment (p. 181)

biological sediment (p. 180)

bioturbation (p. 192)

carbonate rock (p. 199)

carbonate sediment (p. 199)

cementation (p. 194)

chemical and biological sedimentary environment (p. 189)

chemical sediment (p. 180)

chemical weathering (p. 179)

chert (p. 204)

clay (p. 198)

claystone (p. 198)

coal (p. 204)

compaction (p. 194)

conglomerate (p. 196)

continental shelf (p. 186)

cross-bedding (p. 190)

diagenesis (p. 194)

dolostone (p. 200)

evaporite rock (p. 201)

evaporite sediment (p. 201)

flexural basin (p. 186)

graded bedding (p. 190)

gravel (p. 196)

limestone (p. 200)

lithification (p. 195)

mud (p. 198)

mudstone (p. 198)

physical weathering (p. 178)

porosity (p. 194)

reef (p. 200)

rift basin (p. 186)

ripple (p. 191)

salinity (p. 184)

sand (p. 197)

sandstone (p. 197)

sedimentary basin (p. 185)

sedimentary environment (p. 186)

sedimentary structure (p. 190)

shale (p. 198)

siliciclastic sediment (p. 180)

siliciclastic sedimentary environment (p. 189)

silt (p. 198)

siltstone (p. 198)

sorting (p. 183)

thermal subsidence basin (p. 186)

weathering (p. 178)

EXERCISES

1. What processes change sediment into sedimentary rock?

2. How do siliciclastic sedimentary rocks differ from chemical and biological sedimentary rocks?

3. How and on what basis are the siliciclastic sedimentary rocks classified?

4. What kinds of sedimentary rocks are formed by the evaporation of seawater?

5. Define a sedimentary environment, and name three siliciclastic environments.

6. Explain how plate tectonic processes control the development of sedimentary basins.

7. Name two kinds of carbonate rocks and explain how they differ.

8. How do organisms produce or modify sediments?

9. In what two kinds of sedimentary rocks are most oil and gas found?

10. If you drilled into the bottom of a sedimentary basin an oil well that is 1 km deep and another that is 5 km deep, which would have the higher pressures and temperatures? Oil turns into natural gas at high basin temperatures. In which well would you expect to find more gas?

11. You are looking at a cross section of a rippled sandstone. What sedimentary structure would tell you the direction of the current that deposited the sand?

12. You discover a bedding sequence that has a conglomerate at the base; grades upward to a sandstone and then to a shale; and finally, at the top, grades to a limestone of cemented carbonate sand. What changes in the sediment's source area or in the sedimentary environment could have been responsible for this sequence?

13. How can you use the size and sorting of sediment particles to distinguish between sediments deposited in a glacial environment and those deposited in a desert?

14. Where are reefs likely to be found?

Visual Literacy Task

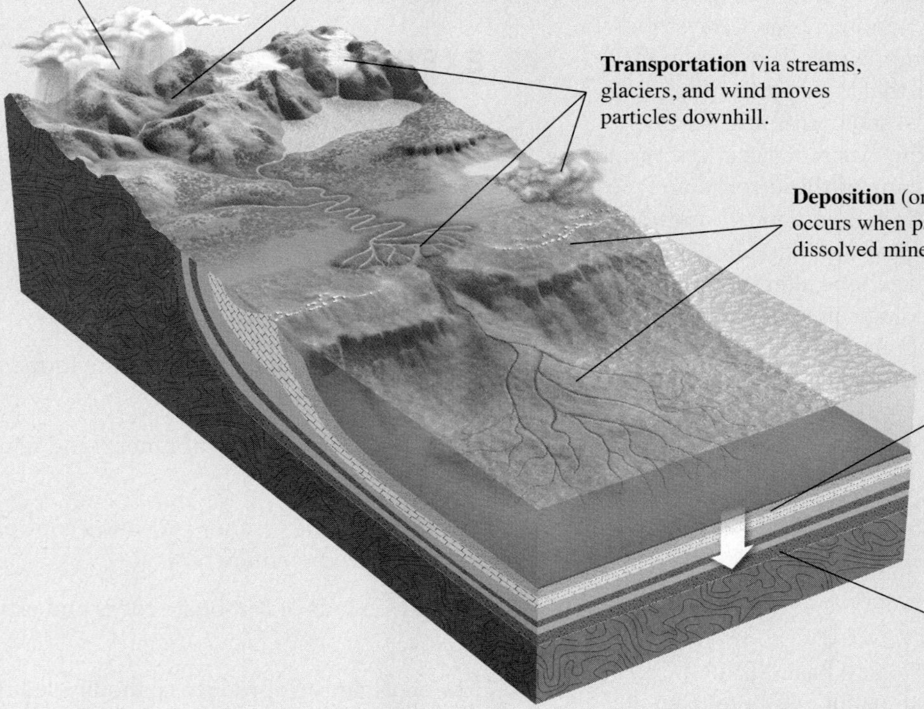

Weathering breaks down rocks physically and chemically.

Erosion carries away particles produced by weathering.

Transportation via streams, glaciers, and wind moves particles downhill.

Deposition (or **sedimentation**) occurs when particles settle out or dissolved minerals precipitate.

Burial occurs as layers of sediment accumulate and compact previous layers.

Diagenesis lithifies the sediment to make sedimentary rocks.

FIGURE 6.2 Several surface processes of the rock cycle contribute to the formation of sedimentary rocks.

1. What process happens primarily underground?

a. Weathering
b. Transportation
c. Deposition (sedimentation)
d. Diagenesis

2. What are the primary processes that take place in glaciers, deltas, and deserts?

a. Transportation and deposition
b. Weathering and burial
c. Erosion and diagenesis

3. Sediments form at the surface of Earth. Where do sedimentary rocks form?

a. At the surface only on land
b. At the surface on the land or ocean floor
c. Beneath the surface of land only
d. Beneath the surface of land or ocean floor

4. During what process do sediment particles not move?

a. During deposition (sedimentation)
b. During transportation
c. During erosion
d. During diagenesis

5. Which process can happen only after deposition?

a. Weathering
b. Burial
c. Erosion
d. Transportation

Thought Question: Construct a list of the basic sedimentary processes in the order in which they typically occur.

Deformation and Metamorphism

7

The Inspiration of Marble

THE USE OF MARBLE IN ARCHITECTURE dates back thousands of years to ancient Egyptian and Mesopotamian cultures. The Parthenon, built in the sixth century B.C., was one of the first buildings constructed entirely of marble. From miles away, incoming ships sailing through the Mediterranean could see the white marble columns of the most famous temple in Athens shining like a beacon of light. The Parthenon still stands today, a testament to the resilience of one of the world's most luxurious building materials. Capable of bearing immense weight, marble was ideally suited for monolithic columns and supporting structures in public, private, and religious buildings.

> The word *marble* originates from the Greek *marmaros*, meaning "a snow white and spotless stone."

The word *marble* originates from the Greek *marmaros*, meaning "a snow white and spotless stone" (although not all marble is white). And because no other stone offers the opportunity for refinement quite like marble, it has been the medium of choice for some of the world's greatest sculptors. The most famous, Michelangelo, carved *David* from a single block of marble. Michelangelo preferred to use Carrara marble, which is the stone still preferred by the best sculptors today. Stone from the Carrara region, in the Apennines of Italy, was employed in Rome for architectural purposes in the time of Augustus, but the finer varieties, best suited to the needs of sculptors, were not discovered until later. The purest varieties of Carrara marble have a snow-white color.

In the United States, marble production has been a valued enterprise since George Yule, a mining engineer, discovered marble deposits and recognized their value. The town of Marble, Colorado, was incorporated in 1899 and began its history by producing the finest pure white marble in the world. The quarry, operated until 1941, is most famous for producing the blocks of marble used for the Lincoln Memorial and the Tomb of the Unknowns. Marble was a typical "boom-and-bust" mining town, and the quarry was closed in 1941 at the advent of World War II. At the time it closed, it had the largest marble finishing mill of its kind in the world, employing over 1000 people. With the quarry gone, Marble's population dropped to just 50 residents. ◆

The Lincoln Memorial, in Washington, D.C. [Cpenler/Dreamstime.com.]

CHAPTER OUTLINE

TODAY, MARBLE IS INCREASINGLY a part of our daily lives. Modern technologies have enabled home designers to take advantage of its beauty and elegance at a fraction of what was paid even 20 years ago. Hewn from Earth's crust, cut with diamond saws, and polished like glass, thin slabs of marble are routinely installed as kitchen and bathroom countertops.

If you call on your neighborhood marble dealer, you will discover that not all the "marble" being sold is really marble. Most of what is on display may actually be various types of granite, an igneous intrusive rock, or gneiss, a metamorphic rock formed from granite or sedimentary rocks. Granite and gneiss are extremely popular because, unlike true marble, they are formed of silicate minerals, which are much harder and better resist scratching and staining. Other types of "marble" include onyx, a sedimentary rock formed in ancient caves, and travertine, a sedimentary rock formed in thermal springs. These types of "marble" are softer than true marble and require special care, but they add a touch of elegance in the right place.

What is it that gives marble its strength and beauty? To find out, we must look at the transformations deep in Earth's crust that produce marble and other metamorphic rocks.

During the rock cycle, rocks may be subjected to tectonic forces great enough to tilt, bend, or fracture them. They may also be subjected to temperatures and pres-sures great enough to cause changes in their mineralogy, texture, or chemical composition. **This chapter examines deformation and metamorphism, the processes that cause these kinds of changes in continental crust.**

Deformation and metamorphism are both intimately associated with plate tectonic processes. The steady, relative motion between two rigid lithospheric plates causes deformation along the plate boundary. Within continents, however, the deformation can be "smeared out" across a plate boundary in zones hundreds or even thousands of kilometers wide. Within these broad zones, rocks are deformed by **folding** and **faulting** (**Figure 7.1**).

Folds in rocks are like folds in clothing. Just as cloth pushed together from opposite sides bunches up in folds, layers of rock slowly compressed by tectonic forces in the crust can be pushed into folds (Figure 7.1a). Tectonic forces can also cause a rock formation to break and slip on both sides of the fracture, forming a fault (Figure 7.1b). When such a break occurs suddenly, the result is an earthquake. Active zones of continental deformation are marked by frequent earthquakes.

Patterns of deformation also fit into the larger framework of the plate tectonic movements involved in **orogeny,** or "mountain making." Tectonic forces build mountains through the folding and faulting of rock layers, often with accompanying volcanism, especially at ocean-continent convergences. These enormous forces transform the tex-

(a)

(b)

FIGURE 7.1 (a) An outcrop of originally horizontal rock layers bent into folds by compressive tectonic forces. [*Phil Dombrowski.*] (b) An outcrop of once-continuous rock layers displaced along small faults by tensional tectonic forces. [*Tom Bean.*]

ture and composition of the rocks themselves. Great belts of metamorphic rock stretching across a region are often found where continents collided and built mountains. In the cores of the major mountain chains of the world, from the Appalachians to the Alps, we find long belts of regionally metamorphosed and deformed rocks that parallel the lines of folds and faults in the mountains.

At each outcrop within a region, there are distinct **formations,** groups of rock layers that can be identified throughout the region by their physical properties. Some formations consist of a single rock type, such as limestone. Others are made up of thin, interlayered beds of different kinds of sediments, such as sandstone and shale. However they vary, each formation is composed of a distinctive set of rock layers that can be recognized and mapped as a unit.

MAPPING GEOLOGIC STRUCTURES

Faults and folds are examples of the basic features geologists observe and map to understand crustal deformation. To better understand this process, we need information about the geometry of faults and folds. The best place to find this information is an *outcrop,* where the solid rock that underlies the ground surface—the *bedrock*—is exposed (not obscured by soil or loose boulders). Figure 7.1a is a picture of an outcrop showing sedimentary beds bent into a fold. Often, however, folded rocks are only partly

exposed in an outcrop and can be seen only as an inclined layer (**Figure 7.2**). The orientation of the layer is an important clue that we can use to piece together a picture of the overall geologic structure. Two measurements describe the orientation of a rock layer exposed at an outcrop: the strike and the dip of the layer's surface.

FIGURE 7.2 Dipping limestone and shale beds on the coast of Somerset, England. Children are walking along the strike of beds that dip to the left at an angle of about 15°. [*Chris Pellant.*]

Measuring Strike and Dip

The **strike** is the compass direction of a rock layer as it intersects with a horizontal surface. The **dip,** which is measured at right angles to the strike, is simply the amount of tilting—the angle at which the rock layer inclines from the horizontal. **Figure 7.3** shows how the strike and the dip are measured in the field. A geologist might describe the outcrop in this figure as "a bed of coarse-grained sandstone striking east-west and dipping 45° south." Strike and dip can also be used to map the orientation of other types of geologic surfaces, such as fault planes or ancient erosional surfaces.

Measurements of strike and dip provide information on the orientation of rock layers, fault planes, and other geologic surfaces.

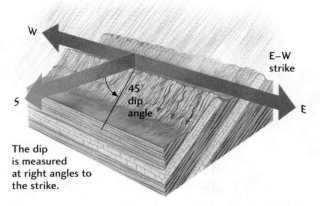

FIGURE 7.3 The strike and dip of a rock layer define its orientation at a particular place. Strike is the compass direction of a rock layer as it intersects with a horizontal surface; dip is the angle of steepest descent of the rock layer from the horizontal, measured at right angles to the strike. Here the strike is east-west and the dip is 45° to the south. [*After A. Maltman, Geological Maps: An introduction. New York: Van Nostrand Reinhold, 1990, p. 37.*]

Geologic Maps

Geologic maps are two-dimensional representations of the rock formations exposed at Earth's surface. To create a map, geologists must choose an appropriate scale—the ratio of the distance on the map to true surface distance.

A common scale for geologic field mapping is 1:24,000 (pronounced "one to twenty-four thousand"), which means that 1 inch on the map corresponds to 24,000 inches (2000 feet) on Earth's surface.

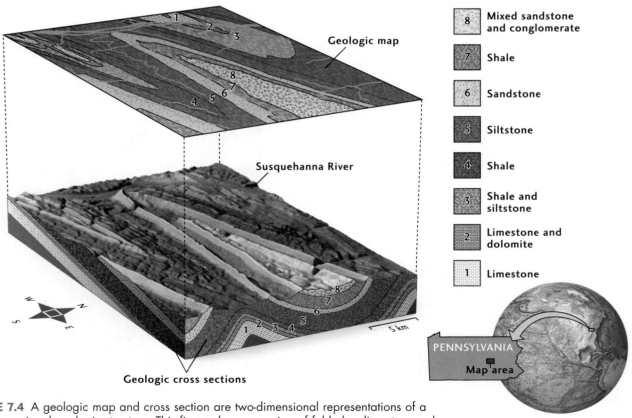

FIGURE 7.4 A geologic map and cross section are two-dimensional representations of a three-dimensional geologic structure. This figure shows a region of folded sedimentary rock in central Pennsylvania east of the Susquehanna River. The rock formations exposed at the surface are labeled from the oldest (formation 1) to the youngest (formation 8).

To keep track of different rock formations, each rock formation is assigned a particular color on the map, usually keyed to the rock's type and age (**Figure 7.4**). Many different rock formations may be exposed in highly deformed regions, so geologic maps can be very colorful! Softer rocks, such as mudstones, are more easily eroded than harder rocks, such as limestones. Consequently, rock types can exert a strong influence on the topography of the land surface and the exposure of rock formations. The important relationships between geology and topography can be made clear by plotting the contours of the land surface on a geologic map (see Figure 7.4).

Geologic Cross Sections

Once a region has been mapped, the two-dimensional geologic map must be interpreted in terms of the underlying three-dimensional geologic structure. How can the shapes of the rock layers be reconstructed, especially when erosion has removed parts of a formation? The process is like putting together a three-dimensional jigsaw puzzle with some of the pieces missing. Common sense and intuition play important roles, as do basic geologic principles.

To piece together the puzzle, geologists construct **geologic cross sections**—diagrams showing the features that would be visible if vertical slices were made through part of the crust. Some small cross sections can actually be seen in the vertical faces of cliffs, quarries, and road cuts (**Figure 7.5**). Cross sections spanning much larger areas can be constructed from the information on a geologic map, including the strikes and dips observed at outcrops. The accuracy of cross sections based on surface mapping can be improved by drilling boreholes to collect rock samples as well as by seismic imaging. But drilling and seismic imaging are expensive, so data collected by these methods are usually available only for areas that have been explored for oil, water, or other valuable natural resources.

Figure 7.4 shows a geologic map and geologic cross sections of an area where sedimentary rocks, originally horizontal, were bent into a series of folds and eroded into a set of linear ridges and valleys. We will explore some of the geologic relationships seen in this map later in this chapter. But first we will investigate the basic processes by which rocks deform.

> Geologic maps and geologic cross sections can be constructed to illustrate the three-dimensional geologic structure of a region.

FIGURE 7.5 Geologic cross sections can sometimes be observed directly in the field. This road cut, where Interstate 70 crosses Dinosaur Ridge west of Denver, Colorado, provides a near-vertical cross section through a sequence of sedimentary beds tilted by the uplift of the Rocky Mountains. [*Courtesy of Mark McNaught, Mount Union College.*]

HOW ROCKS DEFORM

Rocks deform in response to the tectonic forces acting on them. Whether they respond to tectonic forces by folding, faulting, or some combination of the two depends on the orientation of the forces, the rock type, and the physical conditions (such as temperature and pressure) during deformation.

Brittle and Ductile Behavior of Rocks in the Laboratory

In the mid-1900s, geologists began to explore the forces of deformation by using powerful hydraulic rams to bend and break small samples of rock. Engineers had invented such machines to measure the strength of concrete and other building materials, but geologists modified them to discover how rocks deform at pressures and temperatures high enough to simulate physical conditions deep in Earth's crust.

In one such experiment, the researchers applied compressive (squeezing) force by pushing down with a hydraulic ram on one end of a small cylinder of marble, while at the same time maintaining the force of confining pressure on the cylinder (**Figure 7.6**). Under low confining pressures, equivalent to those found at shallow depths in Earth's crust, the marble sample deformed only

a small amount until the compressive force on its end was increased to the point that the entire sample suddenly broke (see Figure 7.6, left side). This experiment showed that marble behaves as a **brittle** material at low confining pressures. Repeating the experiment under high confining pressures equivalent to pressures that often accompany metamorphism produced a different result: the marble sample slowly and steadily deformed into a shortened, bulging shape without fracturing (see Figure 7.6, right side). Marble thus behaves as a pliable or **ductile** material at the high confining pressures found deep in the crust.

Other experiments showed that when marble was heated to temperatures as high as those that accompany metamorphism, it acted as a ductile material at a lower confining pressure—just as heating wax changes it from a hard material that can break into a soft material that flows. The researchers concluded that the particular marble they were working with would deform by faulting at depths shallower than a few kilometers but would deform by folding at the greater crustal depths where metamorphism generally occurs.

> *Rocks behaving as brittle materials respond to deformation by fracturing, whereas rocks behaving as ductile materials respond to deformation by deforming without fracturing.*

Brittle and Ductile Behavior in Earth's Crust

Natural conditions in Earth's crust cannot be reproduced exactly in the laboratory. Tectonic forces are applied over millions of years, whereas laboratory experiments are rarely conducted for more than a few hours or at most a few weeks. Nevertheless, the results of laboratory experiments can help us interpret what we see in the field. Geologists keep the following points in mind as they map crustal folds and faults:

■ The same rock can be brittle at shallow depths (where temperatures and pressures are relatively low) and ductile deep in the crust (where temperatures and pressures are higher). Metamorphism is usually accompanied by ductile deformation.

■ Rock type affects deformation. In particular, the hard igneous and metamorphic rocks that form the crystalline *basement* of a continent (the crust beneath the layers of sediments) often behave as brittle materials, fracturing along fault planes in

> *Brittle behavior is more likely with lower temperatures and pressures, harder rock types, more rapid deformation, and tensional rather than compressive forces.*

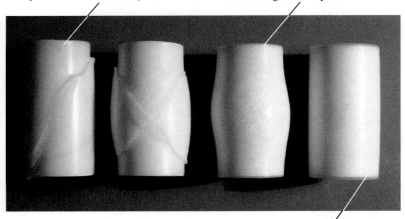

This sample was compressed under conditions representative of the shallow crust. The fractures indicate that the marble is brittle at the laboratory equivalent of shallow depth.

This sample was compressed under conditions representative of the deeper crust. It deformed by bulging smoothly, indicating that marble is ductile at greater depth.

Undeformed sample

FIGURE 7.6 Results of laboratory experiments conducted to discover how a rock—in this case, marble—is deformed by compressive forces. The marble samples are encased in translucent plastic jackets, which explains their shiny appearance. [*Fred and Judith Chester/John Handin Rock Deformation Laboratory of the Center for Tectonophysics.*]

earthquakes, while softer sedimentary rocks that overlie them often behave as ductile materials, folding gradually.

■ A rock formation that would behave as a ductile material if deformed slowly may behave as a brittle material if deformed more rapidly. (Think of Silly Putty, which flows as a ductile clay when you squeeze it slowly but breaks into pieces when you pull it apart very quickly.)

■ Rocks break more easily when subjected to tensional (pulling and stretching) forces than when subjected to compressive forces. Sedimentary rock formations that will deform by folding during compression will often break along faults when subjected to tension.

BASIC DEFORMATION STRUCTURES

The simple geometric concepts and measurements described earlier in this chapter are used to classify features such as faults and folds into different types of deformation structures.

Tectonic Forces Determine Types of Deformation Structures

The types of faults and folds we see depend on the main tectonic forces in the upper crust, which are produced mostly by horizontal movements of the plates. The tectonic forces that act at plate boundaries are thus predominantly *horizontally directed.* Horizontal plate movements exert three types of tectonic forces:

■ **Tensional forces,** which stretch and pull rock formations apart, dominate at divergent boundaries, where plates move away from each other.

■ **Compressive forces,** which squeeze and shorten rock formations, dominate at convergent boundaries, where plates move toward each other.

Horizontal plate movements exert three types of tectonic forces: tensional forces (which stretch and pull rocks apart), compressive forces (which squeeze and shorten rocks), and shearing forces (which push rocks in opposite directions).

■ **Shearing forces,** which push two sides of a rock formation in opposite directions, dominate at transform-fault boundaries, where plates slide past each other.

Faults

As we saw in Chapter 1, a **fault** is a fracture surface across which rock formations have been displaced. We can measure the orientation of a fault surface by its strike and dip, just as we do for other geologic surfaces (see Figure 7.3).

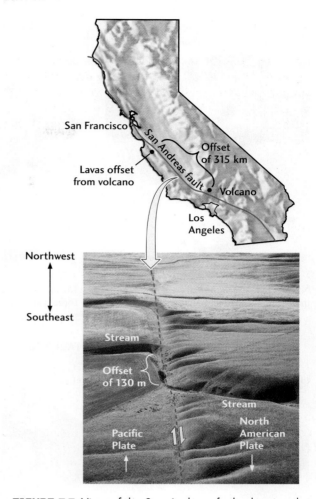

FIGURE 7.7 View of the San Andreas fault, showing the northwestward movement of the Pacific Plate with respect to the North American Plate. The map shows a formation of volcanic rocks 23 million years old that has been displaced by 315 km. The fault runs from top to bottom (dashed line) near the middle of the photograph. Note the offset of the stream (Wallace Creek) by 130 m as it crosses the fault. [*John S. Shelton.*]

The movement of the block of rock on one side of the fault with respect to that on the other side can be described by a *slip direction* and by the total displacement, or *offset.* For a small fault, such as the one pictured in Figure 7.1b, the offset might be only a couple of meters, whereas the offset along a major transform fault, such as the San Andreas fault, can amount to hundreds of kilometers (**Figure 7.7**).

Faults are classified by their slip direction (**Figure 7.8**). A **dip-slip fault** is one on which there has been relative movement of blocks of rock up or down the dip of the fault plane. A **strike-slip fault** is a fault on which the movement has been horizontal, parallel to the strike of the fault plane. When blocks of rock move along the strike and simultaneously up or down the dip, the result is an *oblique-slip fault.* Dip-slip faults are caused

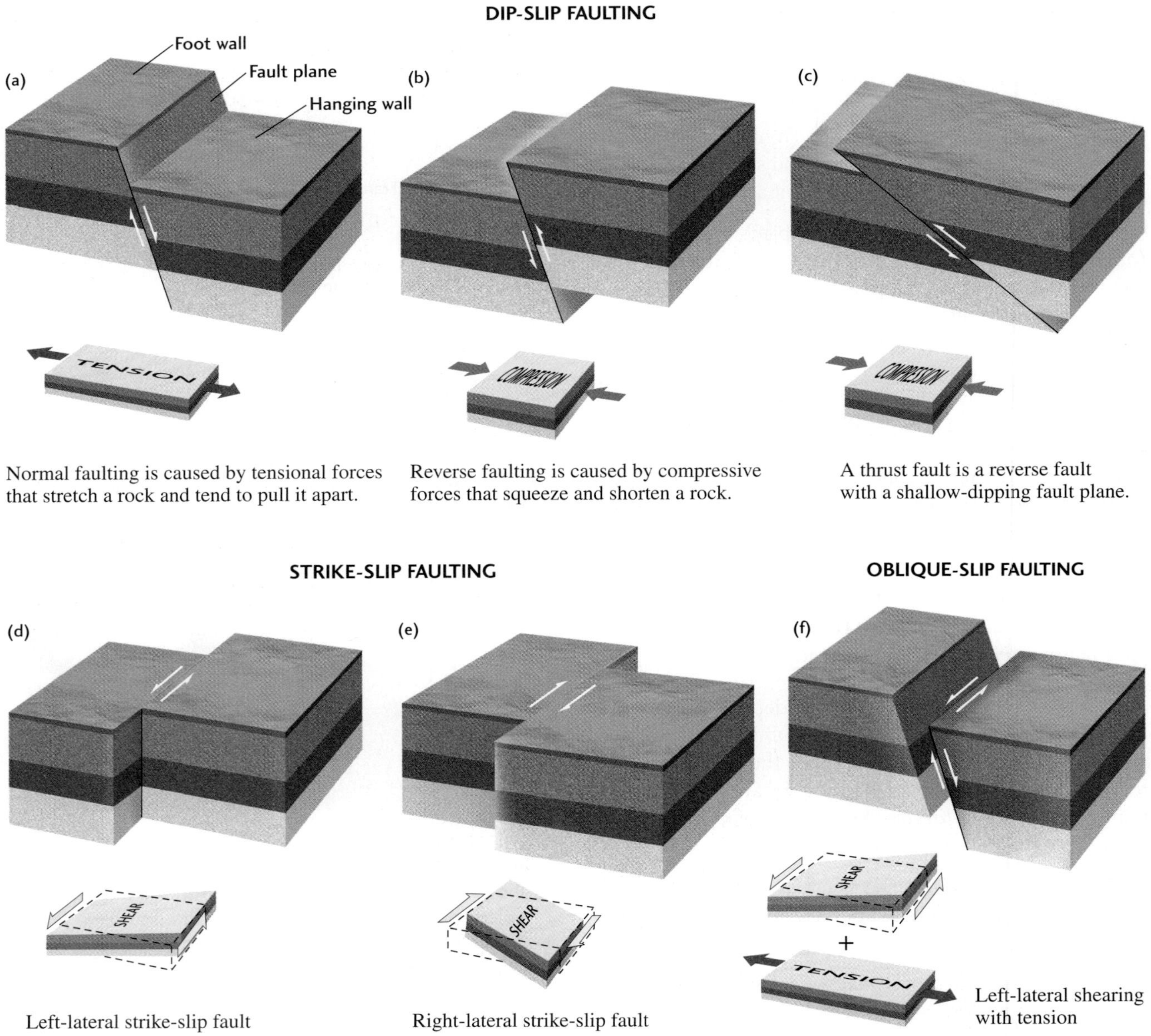

DIP-SLIP FAULTING

(a) Foot wall / Fault plane / Hanging wall

(b)

(c)

TENSION

COMPRESSION

COMPRESSION

Normal faulting is caused by tensional forces that stretch a rock and tend to pull it apart.

Reverse faulting is caused by compressive forces that squeeze and shorten a rock.

A thrust fault is a reverse fault with a shallow-dipping fault plane.

STRIKE-SLIP FAULTING

OBLIQUE-SLIP FAULTING

(d)

(e)

(f)

SHEAR

SHEAR

SHEAR

+

TENSION

Left-lateral strike-slip fault

Right-lateral strike-slip fault

Left-lateral shearing with tension

FIGURE 7.8 The orientation of tectonic forces determines the style of faulting. Dip-slip faulting (parts a–c) is caused by tensional and compressive forces. Strike-slip faulting (parts d–e) is caused by shearing forces. Oblique-slip faulting (part f) is caused by a combination of tensional, compressive, and shearing forces.

by compressive or tensional forces, whereas strike-slip faults are the work of shearing forces. An oblique-slip fault results from shearing in combination with either compression or tension.

These fault types require further classification because the movement can be up or down or right or left. To describe these movements, geologists borrow some terminology used by miners, calling the block of rock above a dipping fault plane the **hanging wall** and the block of rock below it the foot wall. A dip-slip fault is called a

normal fault if the rock's hanging wall moves downward relative to the foot wall, extending the structure horizontally (Figure 7.8a). A dip-slip fault is called a *reverse fault* if the hanging wall moves upward relative to the foot wall, causing a shortening of the structure (Figure 7.8b)—the reverse of what geologists have (somewhat arbitrarily) chosen as "normal." A **thrust fault** is a low-angled reverse fault, that is, one with a dip of less than 45°, so the movement is more horizontal than vertical (Figure 7.8c). When subjected to horizontal compression, brittle rocks of the continental crust usually break along thrust faults with dips of about 30° or less, rather than along more steeply dipping reverse faults.

A strike-slip fault is a *left-lateral fault* if an observer on one side of the fault sees that the block on the opposite side has moved to the left (Figure 7.8d). It is a *right-lateral fault* if the block on the opposite side appears to have moved to the right (Figure 7.8e). As you can tell from the stream offset in Figure 7.7, the San Andreas fault is a right-lateral transform fault. Other faults show a combination of motions, perhaps one part strike-slip and one part dip-slip. These faults are known as **oblique slip faults** (Figure 7.8f).

Geologists can recognize faults in the field in several ways. A fault may form a *scarp* (a cliff) that marks the position where the fault surface intersects the ground surface (**Figure 7.9**). If the offset has been large, as it is for

A fault is a surface across which rock formations have been displaced. Dip-slip faults show movement of blocks up (reverse fault) and down (normal fault) the dip of the fault plane. Strike-slip faults show horizontal movement of blocks parallel to the strike of the fault plane.

transform faults such as the San Andreas, the rock formations currently facing each other across the fault usually differ in type and age. When movements are smaller, offset features can be observed and measured. (As an exercise, try to match up the beds offset by the small faults in Figure 7.1b.) In establishing the time of faulting, geologists apply a simple rule: a fault must be younger than the youngest rocks that it cuts (the rocks had to be there before they could break!) and older than the oldest undisrupted formation that covers it.

Folds

Folding is a common form of deformation observed in layered rocks (as in Figure 7.1a). Folds occur when an originally planar structure, such as a sedimentary bed, is bent into a curved structure. The bending can be produced by either horizontally or vertically directed forces in the crust, just as either pushing together the opposite edges of a piece of paper or pushing up or down on one side or the other can fold it.

Like faults, folds come in all sizes. In many mountain belts, majestic, sweeping folds can be traced over many

FIGURE 7.9 This fault scarp is a fresh surface feature that formed by normal faulting during the 1954 Fairview Peak earthquake in Nevada. [*Karl V. Steinbrugge Collection, Earthquake Engineering Research Center.*]

FIGURE 7.10 Large-scale folds in the sedimentary rocks that form Mount Kidd, Peter Lougheed Provincial Park, Alberta, Canada. [*Peter French/DRK Photos.*]

FIGURE 7.11 Small-scale folds in sedimentary beds of anhydrite (light) and shale (dark) in West Texas. [*John Grotzinger/Ramón Rivera-Moret/Harvard Mineralogical Museum.*]

kilometers (**Figure 7.10**). On a much smaller scale, very thin sedimentary beds can be crumpled into folds a few centimeters long (**Figure 7.11**). The bending can be gentle or severe, depending on the magnitude of the applied forces, the length of time that they were applied, and the resistance of the rocks to deformation.

Folds in which layered rocks are bent upward into arches are called **anticlines;** those in which rocks are bent downward into troughs are called **synclines** (**Figure 7.12**). The two sides of a fold are its *limbs*. The *axial plane* is an imaginary surface that divides a fold as symmetrically as possible, with one limb on either side of the plane.

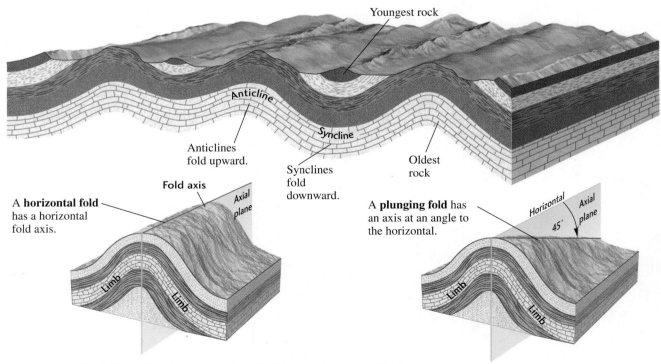

FIGURE 7.12 The folding of rock layers is described by the direction of folding (upward or downward) and by the orientation of the fold axis and the axial plane.

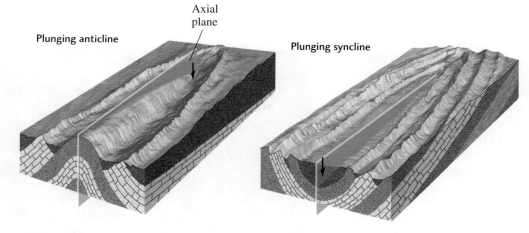

Plunging anticline

Axial plane

Plunging syncline

FIGURE 7.13 The geometry of plunging folds. Note the converging pattern of the layers of rock where they intersect the land surface.

The line made by the lengthwise intersection of the axial plane with the rock layers is the *fold axis*. A symmetrical horizontal fold has a horizontal fold axis and a vertical axial plane with limbs dipping symmetrically away from the axis.

Folds rarely stay horizontal, however. Follow the axis of any fold in the field, and sooner or later the fold dies out or appears to plunge into the ground. If a fold's axis is not horizontal, it is called a *plunging fold*. **Figure 7.13** diagrams the geometry of plunging anticlines and plunging synclines. In eroded mountain belts, a zigzag pattern of outcrops may appear in the field after erosion has removed much of the surface rock from the folds. The geologic map of Figure 7.4 shows this characteristic pattern.

Nor do folds remain symmetrical. With increasing amounts of deformation folds can be pushed into asymmetrical shapes, with one limb dipping more steeply than the other (**Figure 7.14**). This kind of deformation occurs

> *Deformed layers of rock may be folded upward into anticlines or downward into synclines.*

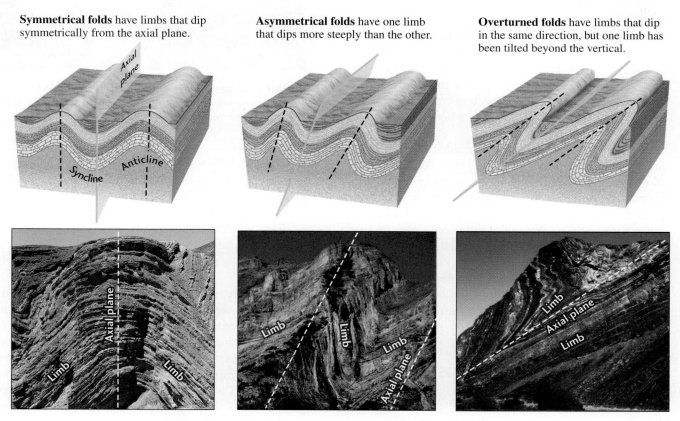

Symmetrical folds have limbs that dip symmetrically from the axial plane.

Axial plane

Syncline   Anticline

Asymmetrical folds have one limb that dips more steeply than the other.

Overturned folds have limbs that dip in the same direction, but one limb has been tilted beyond the vertical.

Limb   Axial plane   Limb

Limb   Limb   Limb   Axial plane

Limb   Axial plane   Limb

FIGURE 7.14 As deformation increases, folds are pushed into asymmetrical shapes. [(left) *Courtesy of Cleet Carlton/Golden Gate Photo.* (center) *Courtesy of Mark McNaught.* (right) *John Grotzinger.*]

when the direction of the deformation force is oblique to the layering of the beds. Such *asymmetrical folds* are common. When the deformation is so intense that one limb has been tilted beyond the vertical, the fold is called an *overturned fold*. Both limbs of an overturned fold dip in the same direction, but the order of the layers in the bottom limb is precisely the reverse of their original sequence— that is, older rocks are on top of younger rocks.

Observations in the field seldom provide geologists with complete information. Bedrock may be obscured by overlying soils, or erosion may have removed much of the evidence of former structures. So geologists search for clues they can use to work out the relationship of one bed to another. For example, in the field or on a geologic map, an eroded anticline might be recognized by a strip of older rocks forming a core bordered on both sides by younger rocks dipping away from the core. These relationships are illustrated in Figures 7.4 and 7.13. An eroded syncline would show as a core of younger rocks bordered on both sides by older rocks dipping toward the core. Determining the subsurface structure of folds by surface mapping has been an important method for finding oil, as discussed in the next section.

Joints

We have seen how rock deformation depends on tectonic forces and the conditions under which tectonic forces are applied. Some layers crumple into folds, and some fracture. A fracture that has displaced the rock formations on either side is called a fault. A second type of fracture is a **joint**—a crack along which there has been no appreciable movement.

Joints are found in almost every outcrop, and they create surfaces that are vulnerable to weathering. If two or more sets of joints intersect, weathering may cause a rock formation to break into large columns or blocks (**Figure 7.15**). Some joints are caused by tectonic forces. Like any other easily broken material, brittle rocks fracture more easily at flaws or weak spots when they are subjected to pressure. These flaws can be tiny cracks, fragments of other materials, or even fossils. Regional forces—compressive, tensional, or shearing—may leave a set of joints as their imprint long after they have vanished.

 ## natural Resources: Using Geologic Maps to find Oil and Gas

Crude oil, or *petroleum* (from the Latin words for "rock oil"), has been collected from natural seeps at Earth's surface since ancient times. The foul-smelling, tarry

FIGURE 7.15 Intersecting sets of joints have made this rock vulnerable to weathering. (The backpack is shown for scale.) [*Courtesy of Peter Kaufman.*]

substance was used as boat caulking, wheel grease, and medicine, but not commonly as a fuel until the process of oil refining was developed in the 1850s. Demand skyrocketed at that time, primarily because oil from whale blubber, the best fuel then available for lamps, had become terribly expensive ($60 per gallon in today's dollars!) as overfishing decimated whale populations.

The ability to refine clean lamp oil from petroleum set off North America's first oil boom. The mining of "black gold" was centered in areas around Lake Erie, where major petroleum seeps had been discovered—in northwestern Pennsylvania, northeastern Ohio, and southern Ontario. Early petroleum explorers, such as self-proclaimed "Colonel" Edwin Drake of Pennsylvania, simply drilled into the seeps, but this straightforward approach soon proved inadequate as a strategy for satisfying the new thirst for oil. Could geologic knowledge be used to locate large petroleum reservoirs hidden underground—

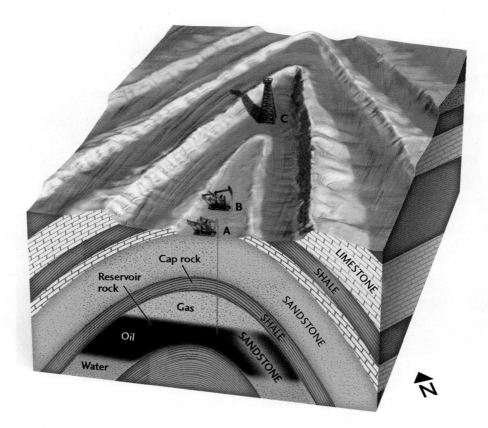

FIGURE 7.16 Oil is often found in large reservoirs along the fold of axes of anticlines. This figure illustrates a typical anticlinal trap.

that is, in regions where no oil seeped to the surface? An affirmative answer was provided in 1861 by T. Sterry Hunt, a Connecticut-born geochemist. As a member of the Geological Survey of Canada, Hunt had been active in the new science of mapping natural resources. He documented the petroleum seeps of southern Ontario in 1850, and he noticed that seeps and successful wells tended to be aligned along the crests of geologic folds.

Petroleum is lighter than water; because of this buoyancy, it tends to rise toward the surface. Hunt hypothesized that the rising petroleum could accumulate in porous "reservoir rocks," such as sandstones, if such rocks were overlain by impermeable "cap rocks," such as shales, that prevented the petroleum from rising farther. Moreover, the most likely place to find large reservoirs would be along the fold axes of anticlines, where substantial amounts of petroleum could be trapped without escaping to the surface.

Figure 7.16 illustrates a typical anticlinal trap, for which we can imagine the following narrative of geologic discovery. Erosion of the fold has exposed a sequence of sandstones, limestones, and shales. Mapping by an enterprising geologist shows that the axis of the anticline strikes to the northeast. Drilling at point A on the axis

of the anticline first penetrates a thick sandstone layer exposed at the surface and then a thinner shale layer. Immediately below the shale the drilling crew encounters another sandstone layer containing gas and, below the gas, significant quantities of oil. The geologist infers that the shale is capping a major petroleum reservoir in the deeper sandstone layer, so the crew is instructed to move along the strike of the anticline and drill at point B. Another successful oil well! The results of drilling anticlines have been impressive: most of the trillion barrels of crude oil produced since 1861 have come from anticlinal oil traps.

STYLES OF CONTINENTAL DEFORMATION

If we look closely enough, we can find all the basic deformation structures in any zone of continental deformation. But when we view continental deformation at a regional scale, we see distinctive patterns of faulting and folding that relate directly to the tectonic forces causing the deformation. **Figure 7.17** depicts the deformation styles typical of the three main types of tectonic forces.

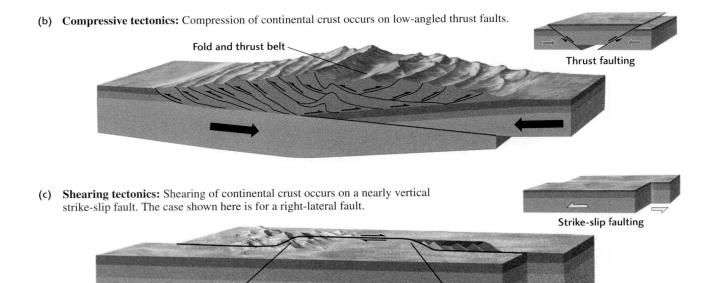

(a) Tensional tectonics: Extension of continental crust produces normal faults with high dip angles in the upper crust that flatten with depth, forming curved fault surfaces.

Brittle upper crust

Normal faulting

Ductile lower crust

(b) Compressive tectonics: Compression of continental crust occurs on low-angled thrust faults.

Fold and thrust belt

Thrust faulting

(c) Shearing tectonics: Shearing of continental crust occurs on a nearly vertical strike-slip fault. The case shown here is for a right-lateral fault.

Strike-slip faulting

A left bend in the fault results in local compression.

A right bend in the fault results in local extension.

FIGURE 7.17 The orientation of tectonic forces—(a) tensional, (b) compressive, and (c) shearing— determines the style of continental deformation. On a regional scale, the basic types of faulting shown in the inset figures can lead to distinctive, complex patterns of deformation. [*After John Suppe, Principles of Structural Geology. Upper Saddle River, N.J.: Prentice Hall, 1985.*]

Tensional Tectonics

In brittle crust, the tensional forces that produce normal faulting may split a plate apart, resulting in a *rift valley*— a long, narrow trough formed when a block of rock has dropped downward relative to its two flanking blocks along nearly parallel, steeply dipping normal faults (**Figure 7.18**). The rift valleys of East Africa, the rifts of mid-ocean ridges, the Rhine Valley, and the Red Sea are well-known examples. As we saw in Chapter 6, these structures form basins that fill with sediments eroded from the mountains of the rift walls as well as with volcanic rocks extruded from tensional cracks in the crust.

The Basin and Range province, which is centered on the Great Basin of Nevada and Utah, is a good example of a region defined by many adjacent rift valleys. This region, which is now more than 800 km wide, has been stretched and extended in a northwest-southeast direction by a factor of two during the last 15 million years. Here normal faulting has created an immense landscape of eroded, rugged fault-block mountains and smooth, sediment-filled valleys, some covered with recent volcanic rocks. This tensional deformation, which appears to be caused

Tensional deformation of the lithosphere produces rift valleys bounded by normal faults.

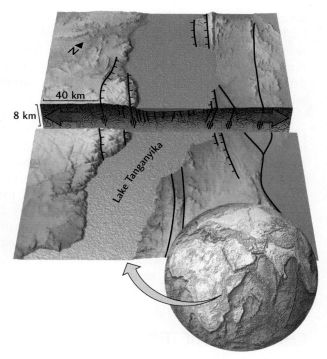

FIGURE 7.18 In East Africa, tensional forces are pulling the Somali subplate away from the African Plate creating rift valleys bounded by normal faults. The rift valley shown here is filled by sediments and Lake Tanganyika, on the boundary between Tanzania and the Democratic Republic of Congo. The cross section has been vertically exaggerated by a factor of 2.5:1, which exaggerates the fault dips; the actual dips of the normal faults are about 60°.

by upwelling convection currents beneath the Basin and Range province, continues today.

Compressive Tectonics

In subduction zones (see Figure 3.8c, d), oceanic lithosphere slips beneath an overriding plate along a huge thrust fault, or *megathrust*. The world's largest earthquakes, such as the great Sumatra earthquake of December 26, 2004, which generated a disastrous tsunami that killed more than 300,000 people, are caused by sudden slips on megathrusts. Thrust faulting is also the most common type of faulting within continents undergoing tectonic compression. Sheets of crust may glide over one another for tens of kilometers along nearly horizontal thrust faults, forming *overthrust* structures (**Figure 7.19**).

When two continents collide, the crust can be compressed across a wide zone, resulting in spectacular episodes of mountain building. During such collisions, the brittle basement rocks ride over one another by thrust faulting while the more ductile overlying sedimentary rocks compress into a series of great folds, forming a *fold*

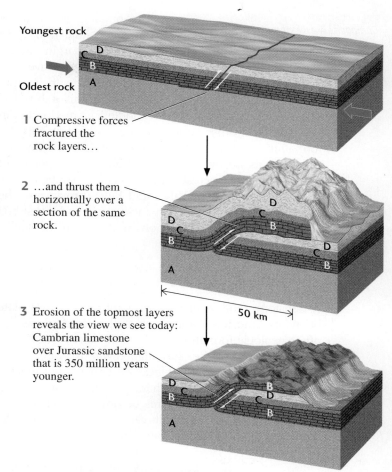

1 Compressive forces fractured the rock layers…

2 …and thrust them horizontally over a section of the same rock.

3 Erosion of the topmost layers reveals the view we see today: Cambrian limestone over Jurassic sandstone that is 350 million years younger.

Keystone thrust fault, southern Nevada

FIGURE 7.19 The Keystone thrust fault of southern Nevada is a large-scale overthrust structure of a kind formed during episodes of continental compression. Compressive forces have detached a sheet of rock layers (D, C, B) and thrust it a great distance horizontally over a section of the same rock layers (D, C, B, A). [Photo by Marli Miller/Visuals Unlimited.]

and thrust belt (Figure 7.17b). Large earthquakes are common in fold and thrust belts; a recent example is the great Wenchuan earthquake that hit Sichuan, China, on May 12, 2008, killing more than 80,000 people.

Compressive deformation of continental crust produces thrust faults and results in orogeny.

The ongoing collisions of Africa, Arabia, and India with the southern margin of the Eurasian continent have created fold and thrust belts from the Alps to the Himalaya, many of which are still active. The great oil reservoirs of the Middle East are located in anticlines and other structural traps formed by this deformation. Compression across western North America, caused by its westward movement during the opening of the Atlantic Ocean, created the fold and thrust belt of the Canadian Rockies. The Valley and Ridge province of the Appalachian Mountains is an ancient fold and thrust belt that dates back to the collisions that created the supercontinent Pangaea.

Shearing Tectonics

A transform fault is a strike-slip fault that forms a plate boundary. Transform faults such as the San Andreas can offset geologic formations by long distances (see Figure 7.7), but as long as they stay aligned with the direction of relative plate motion, the blocks on either side can slide past each other without much internal deformation. Long transform faults are rarely straight, however, so deformation patterns along these faults can be much more complicated. The faults may have bends and jogs that change the tectonic forces acting across portions of the plate boundary from shearing to compression or tension. Those forces, in turn, cause secondary faulting and folding (see Figure 7.17c).

Shearing deformation produces strike-slip faults, but bends in these faults produce more complicated deformation patterns.

A good example of how shearing works can be found in Southern California, where the right-lateral San Andreas fault bends first to the left and then to the right as one moves along its trace from south to north (see Figure 7.7). The segments of the fault on both sides of this "Big Bend" are aligned with the direction of relative plate movement, so the blocks slip past each other in simple strike-slip faulting. Within the Big Bend, however, the change in the fault orientation causes the blocks to push against each other, which causes thrust faulting to the south of the fault. This thrusting has raised the San Gabriel and San Bernadino mountains to elevations exceeding 3000 m and, during the last half-century, has produced a series of destructive ruptures, including the 1994 Northridge earthquake, which caused more than $40 billion in damage to Los Angeles (see Chapter 13).

At the southern end of the San Andreas between the Gulf of California and the Salton Sea, the boundary between the Pacific and North American plates jogs to the right in a series of steps. Within these jogs, the plate boundary is subjected to tensional forces, and normal faulting has formed rift valleys that are volcanically active, rapidly subsiding, and filling with sediments. The extension occurs within 200 km of the Big Bend compression, demonstrating how variable the tectonics along continental transform faults can be!

We have now described the major types of structural changes that occur when a rock is deformed. When temperatures get high enough, however, rocks undergo changes in their mineral composition, chemical composition, and texture as well as in their structure. These are the changes that are characteristic of metamorphic rocks. Although metamorphism is usually accompanied by ductile deformation, it can also occur without deformation, such as when hot magmas are injected into surrounding rocks. Let's turn next to the process of metamorphism and the types of changes that occur when rocks are subjected to high temperatures and pressures.

CAUSES OF METAMORPHISM

We are all familiar with some of the ways in which heat and pressure can transform materials. Cooking batter in a waffle iron not only heats up the batter but also puts pressure on it, transforming it into a rigid solid. In similar ways, rocks change as they encounter high temperatures and pressures. Tens of kilometers below the surface, temperatures and pressures are high enough to transform rock without being high enough to melt it. Increases in temperature and pressure and changes in the chemical environment can alter the mineral composition and crystalline textures of rock, *even though it remains solid all the while.* The result is the third large class of rocks: **metamorphic rock,** which has undergone changes in mineralogy, texture, chemical composition, or all three.

Changes due to metamorphism are short by geologic standards but usually require a million years or more. Examples of metamorphism include a limestone filled with fossils that may be transformed into a white marble in which no trace of fossils remains. The mineral and chemical composition of the rock may be unaltered, but its texture may have changed drastically, from small calcite crystals to large, interlocked calcite crystals that erase such former features as fossils. Shale, a well-bedded sedimentary rock so fine-grained that no individual crystal can be seen with the naked eye, may become a schist, in which the original bedding is hidden and the texture is dominated by large crystals of mica. In this case, both mineralogy and texture have changed, but the overall chemical composition of the rock has remained the same.

Metamorphic rocks are formed when rocks are subjected to significant changes in temperature or pressure.

Sediments and sedimentary rocks are products of Earth's surface environments, whereas igneous rocks are products of the magmas that originate in the lower crust and mantle. Metamorphic rocks, even those exposed at Earth's surface, are mainly the products of processes acting on rocks at depths ranging from the upper to the lower crust. Most of these rocks are formed at depths of 10 to 30 km, in the middle to lower half of the crust. But metamorphism can also occur at Earth's surface. We can see metamorphic changes, for example, in the baked surfaces of soils and sediments just beneath volcanic lava flows.

The heat and pressure in Earth's interior and its fluid composition are the three principal factors that drive metamorphism. In much of Earth's crust, temperature increases at a rate of 30°C per kilometer of depth, although that rate varies considerably among different regions (**Figure 7.20**). Thus, at a depth of 15 km, the temperature will be about 450°C—much higher than the average temperature at Earth's surface, which ranges from 10° to 20°C in most regions. The contribution of pressure is the result of vertically oriented forces exerted by the weight of overlying rocks as well as horizontally oriented forces developed as the rocks are deformed. The average

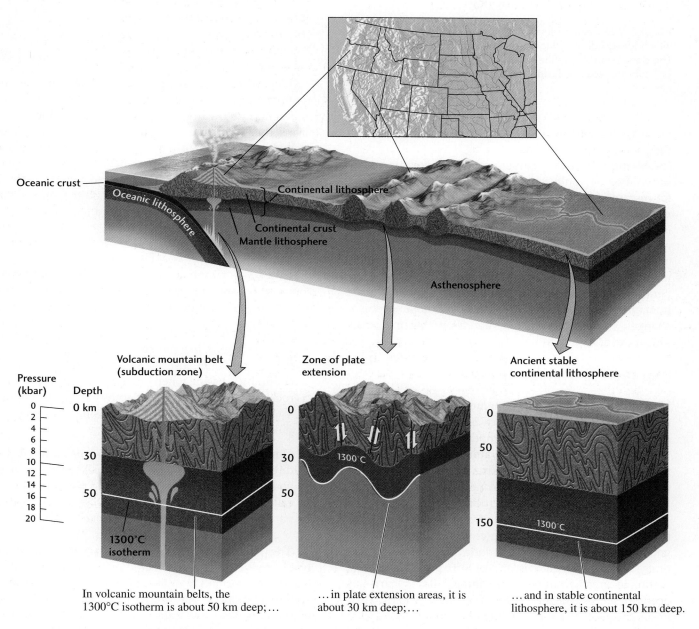

FIGURE 7.20 Pressure and temperature increase with depth in all regions, as shown in these cross sections of a volcanic mountain belt, a region of continental extension, and a region of ancient stable continental lithosphere.

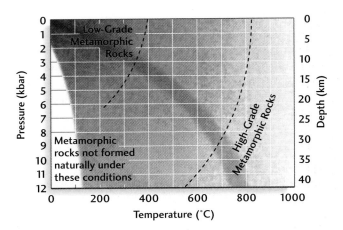

FIGURE 7.21 Temperatures, pressures, and depths at which low-grade and high-grade metamorphic rocks form. The dark band shows the rates at which temperature and pressure increase with depth over much of the continental lithosphere.

pressure at a depth of 15 km amounts to about 4000 times the pressure at the surface.

As high as these temperatures and pressures may seem, they are only in the middle range of conditions for metamorphism, as **Figure 7.21** shows. A rock's *metamorphic grade* reflects the temperatures and pressures it was subjected to during metamorphism. We refer to metamorphic rocks formed under the lower temperatures and pressures of shallower crustal regions as *low-grade* metamorphic rocks and those formed at the higher temperatures and pressures of deeper zones as *high-grade* metamorphic rocks. As the grade of metamorphism changes, the assemblages of minerals within metamorphic rocks also change.

The Role of Temperature

Heat greatly affects a rock's chemical composition, mineralogy, and texture. When rock is moved from Earth's surface to its interior, where temperatures are higher, the rock adjusts to the new temperature. Its atoms and ions recrystallize, linking up in new arrangements and creating new mineral assemblages. Many new crystals grow larger than the crystals in the original rock.

The increase in temperature with increasing depth in Earth's interior is called the *geothermal gradient*. The geothermal gradient varies among plate tectonic settings, but on average it is about 30°C per kilometer of depth. In areas where the continental lithosphere has been stretched and thinned, such as in Nevada's Great Basin, the geothermal gradient is *steep* (for example, 50°C per kilometer of depth). In areas where the continental

lithosphere is old and thick, such as beneath central North America, the geothermal gradient is *shallow* (for example, 20°C per kilometer of depth) (see Figure 7.20).

Because different minerals crystallize and remain stable at different temperatures, we can use a rock's mineral composition as a kind of *geothermometer* to gauge the temperature at which it formed. For example, as sedimentary rocks containing clay minerals are buried deeper and deeper, the clay minerals begin to recrystallize and form new minerals, such as micas. With additional burial at greater depths and temperatures, the micas become unstable and begin to recrystallize into new minerals, such as garnet.

> The geothermal gradient is the rate at which temperature increases with depth in Earth's crust; it varies depending on plate tectonic setting.

Plate tectonic processes such as subduction and continent-continent collision, which transport rocks and sediments into the hot depths of the crust, are the mechanisms that form most metamorphic rocks. In addition, limited metamorphism may occur where rocks are subjected to elevated temperatures near igneous intrusions. The heat is locally intense but does not penetrate deeply; thus, the intrusions can metamorphose the surrounding country rock, but the effect is localized.

The Role of Pressure

Pressure, like temperature, changes a rock's chemical composition, mineralogy, and texture. Solid rock is subjected to two basic kinds of pressure, also called *stress:*

1. *Confining pressure* is a general force applied equally in all directions, like the pressure a swimmer feels under water. Just as a swimmer feels greater confining pressure when diving to greater depths, a rock descending to greater depths in Earth's interior is subjected to progressively increasing confining pressure.

2. *Directed pressure* is force exerted in a particular direction, as when you squeeze a ball of clay between your thumb and forefinger. Directed pressure, also called *differential stress* is usually concentrated within particular zones or along discrete planes.

The compressive force exerted where lithospheric plates converge is a form of directed pressure, and it results in deformation of the rocks near the plate boundary. Heat reduces the strength of a rock, so directed pressure is likely to cause severe folding and other forms of ductile deformation, as well as metamorphism,

FIGURE 7.22 These rocks in Sequoia National Forest, California, show both the banding and the folding characteristic of sedimentary rocks metamorphosed into marble, schist, and gneiss. [*Gregory G. Dimijian/Photo Researchers.*]

in mountain belts where temperatures are high. Rocks subjected to differential stress may be severely distorted, becoming flattened in the direction the force is applied and elongated in the direction perpendicular to the force (**Figure 7.22**).

The minerals in a rock under pressure may be compressed, elongated, or rotated to line up in a particular direction, depending on the kind of stress applied to the rock. Thus, directed pressure guides the shape and orientation of the new crystals formed as minerals recrystallize under the influence of both heat and pressure. During the recrystallization of micas, for example, the crystals grow with the planes of their sheet silicate structures aligned perpendicular to the directed stress. The rock may develop a banded pattern as minerals of different compositions are segregated into separate planes.

> *Confining pressure increases with depth in Earth's interior. Confining pressure is constant in all directions, whereas directed pressure is force exerted in a particular direction.*

Marble, which we discussed at the opening of this chapter, owes its remarkable strength to this recrystallization process. When limestone, a sedimentary rock, is heated to the very high temperatures that cause it to recrystallize, the original minerals and crystals

> *Metamorphic recrystallization may align minerals in a particular direction depending on the kind of stress applied to the rock.*

become reoriented and tightly interlocked to form a very strong structure with no planes of weakness.

The pressure to which rock is subjected deep in Earth's interior is related to both the thickness and the density of the overlying rocks. Pressure is usually recorded in *kilobars* (1000 bars, abbreviated kbar) and increases at a rate of 0.3 to 0.4 kbar per kilometer of depth (see Figure 7.20). One bar is approximately equivalent to the pressure of air at Earth's surface. A diver touring the deeper part of a coral reef at a depth of 10 m would experience an additional bar of pressure.

The Role of Fluids

Metamorphic processes can alter a rock's mineralogy by introducing or removing chemical components that are soluble in heated water. Hydrothermal fluids carry dissolved carbon dioxide as well as chemical substances—such as sodium, potassium, silica, copper, and zinc—that are soluble in hot water under pressure. As hydrothermal solutions percolate up to the shallower parts of the crust, they react with the rocks they penetrate, changing their chemical and mineral compositions and sometimes completely replacing one mineral with another without changing the rock's texture. This kind of change in a rock's composition by fluid transport of chemical substances into or out of the rock is called **metasomatism.** Many valuable deposits of copper, zinc, lead, and other metallic ores are formed by this kind of chemical substitution, as we saw in Chapter 4.

Where do these chemically reactive fluids originate? Although most rocks appear to be completely dry and to have extremely low porosity, they characteristically contain water in minute pores (the spaces between grains). This water comes not from the pores of sedimentary rocks—from which most of the water is expelled during diagenesis—but rather from chemically bound water in clays. Water forms part of the crystal structures of metamorphic minerals such as micas and amphiboles. The carbon dioxide dissolved in these high-temperature fluids is derived largely from sedimentary carbonates: limestones and dolostones.

TYPES OF METAMORPHISM

Geologists can duplicate metamorphic conditions in the laboratory and determine the precise combinations of pressure, temperature, and parent rock composition under which transformations might take place. But to understand when, where, and how these conditions came about in Earth's interior, we must categorize metamorphic rocks on the basis of their geologic settings (**Figure 7.23**).

Regional Metamorphism

Regional metamorphism, the most widespread type, takes place where both high temperatures and high pressures are imposed over large parts of the crust. We use this term to distinguish this type of metamorphism from more localized transformations near igneous intrusions or faults. Regional metamorphism is a characteristic feature of convergent plate boundaries. It occurs in volcanic mountain belts, such as the Andes of South America, and in the cores of mountain chains produced by continent-continent collisions, such as the Himalaya of central Asia. These mountain chains are often linear features, so zones of regional metamorphism are often linear in their distribution. In fact, geologists usually interpret regionally extensive belts of metamorphic rocks as representing

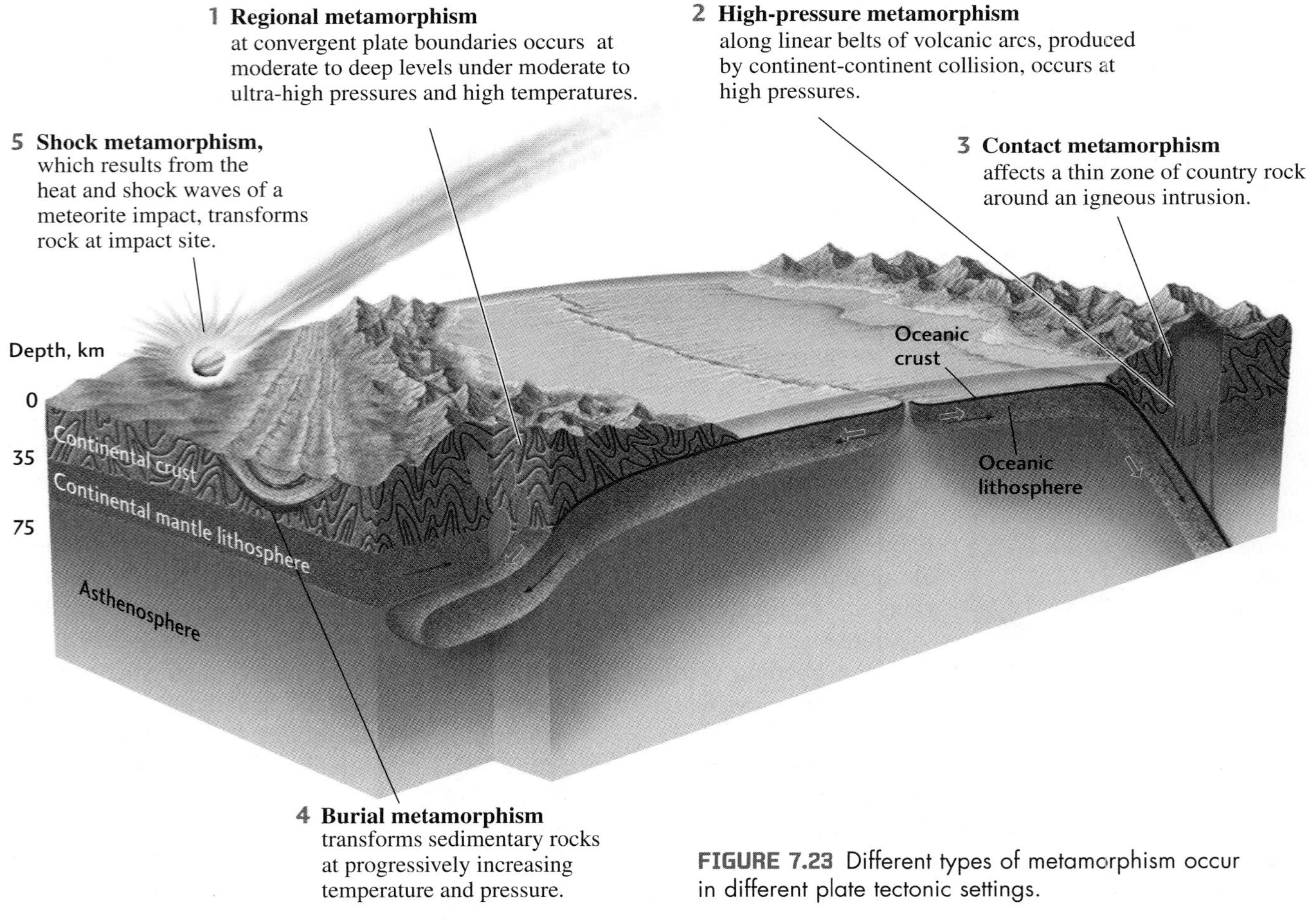

1 Regional metamorphism at convergent plate boundaries occurs at moderate to deep levels under moderate to ultra-high pressures and high temperatures.

2 High-pressure metamorphism along linear belts of volcanic arcs, produced by continent-continent collision, occurs at high pressures.

5 Shock metamorphism, which results from the heat and shock waves of a meteorite impact, transforms rock at impact site.

3 Contact metamorphism affects a thin zone of country rock around an igneous intrusion.

4 Burial metamorphism transforms sedimentary rocks at progressively increasing temperature and pressure.

FIGURE 7.23 Different types of metamorphism occur in different plate tectonic settings.

> *Regional metamorphism takes place where both high temperatures and high pressures are imposed over large parts of the crust. It is characteristic of convergent plate boundaries.*

> *Contact metamorphism is the localized transformation of rock adjacent to igneous intrusions, mainly as a result of high temperatures.*

sites of former mountain chains that were eroded over millions of years, exposing the rocks in their interiors.

Some regional metamorphic belts are created by the high temperatures and moderate to high pressures near the volcanic mountain belts formed where subducted plates sink deep into the mantle. Others are formed under the very high pressures and temperatures found deeper in the crust along boundaries where colliding continents deform rock and raise high mountain chains. In both cases, the metamorphosed rocks are typically transported to great depths in Earth's crust, then uplifted and eroded at Earth's surface.

Contact Metamorphism

In **contact metamorphism,** the heat from an igneous intrusion metamorphoses the immediately surrounding rock. This type of localized transformation normally affects only a thin zone of country rock along the zone of contact. In many contact metamorphic rocks, especially at the margins of shallow intrusions, the mineral and chemical transformations are largely related to the high temperature of the intruding magma. Pressure effects are important only where the magma is intruded at great depths. Here, the pressure results not from the intrusion forcing its way into the country rock, but from the presence of regional confining pressure. Contact metamorphism by volcanic deposits is limited to very thin zones because lavas cool quickly at the surface and their heat has little time to penetrate deep into the surrounding rocks and cause metamorphic changes.

Other Types of Metamorphism

There are several other types of metamorphism that produce smaller amounts of metamorphic rock. Some of these types are extremely important in helping geologists understand conditions deep within Earth.

Burial Metamorphism. Recall from Chapter 6 that sedimentary rocks are transformed by diagenesis as they are gradually buried. Diagenesis grades into *burial metamorphism,* or low-grade metamorphism, that is caused by the progressive increase in pressure exerted by the growing

layers of overlying sediments and sedimentary rocks and by the increase in heat associated with increased depth of burial.

> *Burial metamorphism is caused by the increase in temperatures and pressures as sediments are gradually buried deeper in the crust.*

Depending on the local geothermal gradient, burial metamorphism typically begins at depths of 6 to 10 km, where temperatures range between 100° and 200°C and pressures are less than 3 kbar. This fact is of great importance to the oil and gas industry, which defines its "economic basement" as the depth where low-grade metamorphism begins. Oil and gas wells are rarely drilled below this depth because temperatures above 130°C convert organic matter trapped in sedimentary rocks into carbon dioxide rather than crude oil and natural gas.

High-Pressure and Ultra-High-Pressure Metamorphism

Metamorphic rocks formed by *high-pressure metamorphism* (at 8 to 12 kbar) and *ultra-high-pressure metamorphism* (at pressures greater than 28 kbar) are rarely exposed at Earth's surface for us to study. These rocks are rare because they form at such great depths that it takes a very long time for them to be recycled back to the surface. Most high-pressure metamorphic rocks form in subduction zones as sediments scraped from subducting oceanic plates are plunged to depths of over 30 km, where they experience pressures of up to 12 kbar.

> *Rocks produced by high-pressure and ultra-high-pressure metamorphism are rare at Earth's surface because they form at great depths.*

Unusual metamorphic rocks once located at the base of Earth's crust can sometimes be found at Earth's surface. These rocks—called *eclogites*—indicate pressures of greater than 28 kbar, suggesting depths of over 80 km. Such rocks form at moderate to high temperatures, ranging from 800° to 1000°C. In a few cases, these rocks contain *microscopic diamonds*, indicative of pressures greater than 40 kbar and depths greater than 120 km!

Shock Metamorphism

Shock metamorphism occurs when a meteorite collides with Earth. Upon impact, the energy represented by the meteorite's mass and velocity is transformed into heat and shock waves that pass through the impacted country rock. The country rock can be shattered and even partially melted to produce *tektites*. The smallest tektites look like droplets of glass. In some cases, quartz is transformed into *coesite* and *stishovite*, two of its high-pressure forms. Earth's dense atmosphere causes most meteorites to burn up before they strike its surface, so shock metamorphism is rare on Earth.

METAMORPHIC TEXTURES

Metamorphism imprints new textures on the rocks it alters (**Figure 7.24**). The texture of a metamorphic rock is determined by the sizes, shapes, and arrangement of its constituent crystals. Some metamorphic rock textures depend on the particular kinds of minerals formed under metamorphic conditions. Variation in grain size is also important. In general, grain size increases as metamorphic grade increases. Each textural variety of metamorphic rock tells us something about the metamorphic process that created it. In this section, we examine two major textural classes of metamorphic rocks: foliated rocks and granoblastic rocks.

Foliated Rocks

The most prominent textural feature of regionally metamorphosed rocks is **foliation,** a set of flat or wavy parallel cleavage planes produced by deformation under directed

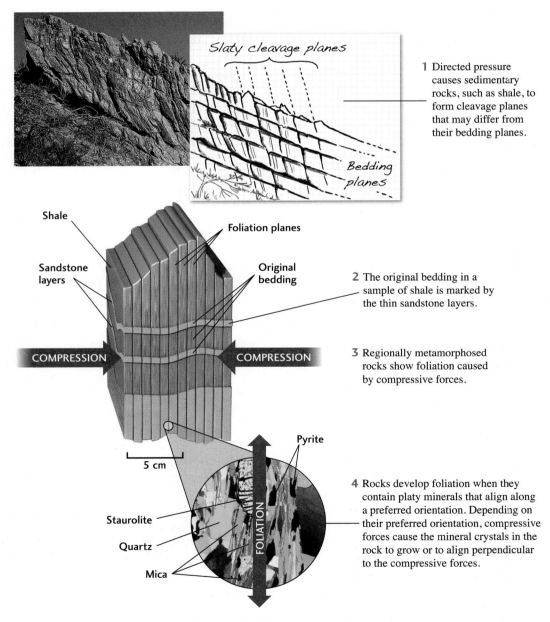

Slaty cleavage planes

1 Directed pressure causes sedimentary rocks, such as shale, to form cleavage planes that may differ from their bedding planes.

Bedding planes

Shale

Foliation planes

Sandstone layers

Original bedding

2 The original bedding in a sample of shale is marked by the thin sandstone layers.

COMPRESSION   COMPRESSION

3 Regionally metamorphosed rocks show foliation caused by compressive forces.

5 cm

Pyrite

FOLIATION

Staurolite

Quartz

Mica

4 Rocks develop foliation when they contain platy minerals that align along a preferred orientation. Depending on their preferred orientation, compressive forces cause the mineral crystals in the rock to grow or to align perpendicular to the compressive forces.

FIGURE 7.24 Directed pressure on rocks containing platy minerals causes foliation. [*(top) Marli Miller. (bottom) S. Dobos.*]

pressure. Foliation planes may cut through the bedding of the original sedimentary rock at any angle or be parallel to the bedding (see Figure 7.24). In general, as the grade of regional metamorphism increases, foliation becomes more pronounced.

A major cause of foliation is the formation of minerals with a platy crystal habit, chiefly the micas and chlorite. The planes of all the platy crystals are aligned parallel to the foliation, an alignment called the *preferred orientation* of the minerals (see Figure 7.24). As platy minerals crystallize, their preferred orientation is usually perpendicular to the main direction of the forces squeezing the rock during metamorphism. Crystals of preexisting minerals may contribute to the foliation by rotating until they also lie parallel to the developing foliation plane.

Minerals with an elongate, needlelike crystal habit also tend to assume a preferred orientation during metamorphism: these crystals, too, normally line up parallel to the foliation plane. Rocks that contain abundant amphiboles, typically metamorphosed mafic volcanics, have this kind of texture.

> **Foliation is created by the alignment of platy minerals under directed pressure.**

The **foliated rocks** are classified according to four main criteria:

1. Metamorphic grade
2. Grain (crystal) size
3. Type of foliation
4. Banding

Figure 7.25 shows examples of the major types of foliated rocks. In general, foliation progresses from one texture to another with increasing metamorphic grade. In this progression, as temperature and pressure increase, a shale may metamorphose first to a slate, then to a phyllite, then to a schist, then to a gneiss, and finally to a migmatite.

Slate **Slates** are the lowest grade of foliated rocks. These rocks are so fine-grained that their individual minerals cannot be seen easily without a microscope. They are commonly produced by the metamorphism of shales or, less frequently, of volcanic ash deposits. Slates usually range from dark gray to black, colored by small amounts of organic material originally present in the parent shale. Slate splitters learned long ago to recognize foliation planes and use them to make thick or thin slates for roofing tiles and blackboards. We still use flat slabs of slate for flagstone walks in parts of the country where slate is abundant.

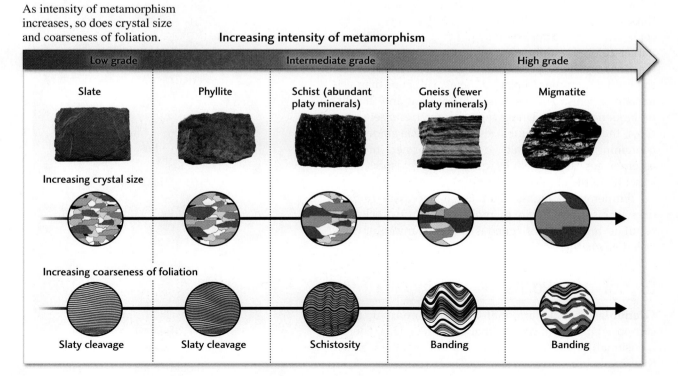

As intensity of metamorphism increases, so does crystal size and coarseness of foliation.

Increasing intensity of metamorphism

| Low grade | Intermediate grade | | High grade |

| Slate | Phyllite | Schist (abundant platy minerals) | Gneiss (fewer platy minerals) | Migmatite |

Increasing crystal size

Increasing coarseness of foliation

| Slaty cleavage | Slaty cleavage | Schistosity | Banding | Banding |

FIGURE 7.25 Foliated rocks are classified by metamorphic grade, grain size, type of foliation, and banding. [*slate, phyllite, schist, gneiss: John Grotzinger/Ramón Rivera-Moret/Harvard Mineralogical Museum; migmatite: Kip Hodges.*]

Phyllite The **phyllites** are rocks of a slightly higher grade than the slates but are similar in character and origin. They tend to have a more or less glossy sheen resulting from crystals of mica and chlorite that have grown a little larger than those of slates. Phyllites, like slates, tend to split into thin sheets, but phyllites split less perfectly than slates.

Schist At low grades of metamorphism, the crystals of platy minerals are generally too small to be seen, and foliation planes are closely spaced. As rocks are subjected to higher temperatures and pressures, however, the platy crystals grow large enough to be visible to the naked eye, and the minerals may tend to segregate into lighter and darker bands. This parallel arrangement of platy minerals produces the coarse, wavy foliation known as *schistosity*, which characterizes **schists.** Schists, which are intermediate-grade rocks, are among the most abundant metamorphic rock types. They contain more than 50 percent platy minerals, mainly the micas muscovite and biotite. Schists may contain thin layers of quartz, feldspar, or both, depending on the quartz content of the parent shale.

> *Slates, phyllites, schists, and gneisses are the principal types of foliated metamorphic rocks formed at progressively higher temperatures and pressures.*

Gneiss Even coarser foliation is shown by **gneisses,** light-colored rocks with coarse bands of light and dark minerals throughout the rock. This *gneiss foliation* results from the segregation of lighter-colored quartz and feldspar from darker-colored amphiboles and other mafic minerals. Gneisses are high-grade, coarse-grained rocks in which the ratio of granular to platy minerals is higher than it is in slate or schist. The result is poor foliation and thus little tendency to split. Under high pressures and temperatures, the mineral assemblages of the lower-grade rocks containing micas and chlorite are transformed into new assemblages dominated by quartz and feldspars, with lesser amounts of micas and amphiboles.

Temperatures higher than those necessary to produce gneiss may begin to melt the country rock. The result is a mixture of igneous and metamorphic rock called *migmatite.*

Granoblastic Rocks

Granoblastic rocks are nonfoliated metamorphic rocks composed mainly of crystals that grow in equant (equidimensional) shapes, such as cubes and spheres, rather than in platy or elongate shapes. These rocks result from metamorphic processes, such as contact metamorphism, in which directed pressure is absent, so foliation does not occur. Granoblastic rocks include hornfels, quartzite, marble, greenstone, amphibolite, and granulite (**Figure 7.26**). All granoblastic rocks, except hornfels, are defined by their mineralogy rather than their texture because they all have a homogeneous granular texture.

> *Granoblastic rocks are nonfoliated metamorphic rocks composed mainly of crystals that grow in equant shapes.*

Hornfels is a high-temperature contact metamorphic rock of uniform grain size that has undergone little or no deformation. It is formed from fine-grained sedimentary rock and other types of rock containing an abundance of

(a)

Quartzite

(b)

Marble

FIGURE 7.26 Granoblastic (nonfoliated) metamorphic rocks: (a) quartzite [*Breck P. Kent*]; (b) marble [*Diego Lezama Orezzoli/Corbis.*]

FIGURE 7.27 Michelangelo's *David*. [*Alinari Archives/Corbis.*]

covered with basalts that have been slightly or extensively altered in this way at mid-ocean ridges. An abundance of chlorite gives these rocks their greenish cast.

Amphibolite is a rock made up of amphibole and plagioclase feldspar. Amphibolites are typically the product of medium- to high-grade metamorphism of mafic volcanic rocks. Foliated amphibolites can be produced by directed pressure.

Granulite, a high-grade metamorphic rock that is also referred to as *granofels*, has a homogeneous granular texture. It is a medium- to coarse-grained rock in which the crystals are equant and show only faint foliation. It is formed by the metamorphism of shale, impure sandstone, and many kinds of igneous rock.

> *Granoblastic rocks include hornfels, quartzite, marble, greenstone, amphibolite, and granulite.*

Porphyroblasts are newly formed metamorphic minerals that may grow into large crystals surrounded by a much finer grained matrix of other minerals (**Figure 7.28**). These large crystals are found in rocks formed both by contact and by regional metamorphism. They grow as the chemical components of the matrix are reorganized and thus replace parts of the matrix. Porphyroblasts vary in size, ranging from a few millimeters to several centimeters in

silicate minerals. Hornfels has a granular texture overall, even though it commonly contains pyroxene, which makes elongate crystals, and some micas. It is not foliated, and its platy or elongate crystals are oriented randomly.

Quartzites are very hard, white rocks derived from quartz-rich sandstones. Some quartzites are homogeneous, unbroken by preserved bedding or foliation (see Figure 7.22). Others contain thin bands of slate or schist, relics of former interbedded layers of clay or shale.

Marbles are the metamorphic products of heat and pressure acting on limestones and dolomites. Some white, pure marbles, such as the famous Italian Carrara marbles prized by sculptors (**Figure 7.27**), show a smooth, even texture of interlocked calcite crystals of uniform size. Other marbles show irregular banding or mottling from silicate and other mineral impurities in the original limestone (see Figure 7.22). The Lincoln Memorial in Washington, D.C., samples the variety of marbles that exist in nature: white Colorado marble is used for the exterior, pink Tennessee marble for the floor, and Alabama marble for the ceiling.

Greenstones are metamorphosed mafic volcanic rocks. Many of these low-grade metamorphic rocks form when mafic lavas and ash deposits react with percolating seawater or other solutions. Large areas of the seafloor are

FIGURE 7.28 Garnet porphyroblasts in a schist matrix. The minerals in the matrix are continuously recrystallized as pressure and temperature change and therefore grow to only a small size. In contrast, porphyroblasts grow to a large size because they are stable over a broad range of pressures and temperatures. [*Chip Clark.*]

TABLE 7.1	Classification of Metamorphic Rocks by Texture		
Classification	**Characteristics**	**Rock Name**	**Typical Parent Rock**
Foliated	Distinguished by slaty cleavage, schistosity, or gneissic foliation; mineral grains show preferred orientation	Slate Phyllite Schist Gneiss	Shale, sandstone
Granoblastic (nonfoliated)	Granular, characterized by coarse or fine interlocking grains; little or no preferred orientation	Hornfels Quartzite Marble Argillite Greenstone Amphibolite[a] Granulite[b]	Shale, volcanics Quartz-rich sandstone Limestone, dolomite Shale Basalt Shale, basalt Shale, basalt
Porphyroblastic	Large crystals set in fine matrix	Slate to gneiss	Shale

[a]*Typically contains much amphibole, which may show alignment of long, needle-like crystals.*
[b]*High-temperature, high-pressure rock.*

diameter. Garnet and staurolite are two common minerals that form porphyroblasts, but many others are also found.

Table 7.1 summarizes the textural classes of metamorphic rocks and their main characteristics.

REGIONAL METAMORPHISM AND METAMORPHIC GRADE

As we have seen, metamorphic rocks form under a wide range of conditions, and their mineralogies and textures are clues to the pressures and temperatures in the crust where and when they formed. Geologists who study the formation of metamorphic rocks constantly seek to determine the intensity and character of metamorphism more precisely than is indicated by a designation of "low grade" or "high grade." To make these finer distinctions, geologists "read" minerals as though they were pressure gauges and thermometers. These techniques are best illustrated by their application to regional metamorphism.

Mineral Isograds: Mapping Zones of Change

When we study a broad belt of regional metamorphism, we can see many outcrops, some showing one set of minerals, some showing others. Different zones within the belt may be distinguished by *index minerals*, abundant minerals that each form under a limited range of temperatures and pressures (**Figure 7.29**). For example, a zone of

unmetamorphosed shales may lie next to a zone of weakly metamorphosed slates (Figure 7.29a). As we move from the shale zone into the slate zone, a new index mineral—chlorite—appears. Chlorite is an index mineral marking the point at which we move into a new zone with a higher metamorphic grade. If laboratory studies have determined the temperature and pressure at which the index mineral forms, we can draw conclusions about the conditions that existed when the rocks in the zone were formed.

We can use the occurrences of index minerals to make a map of the boundaries between metamorphic zones. Geologists define these boundaries by drawing lines called *isograds* that plot the transition from one zone to the next. Isograds are used in Figure 7.29a to show a series of mineral assemblages produced by the regional metamorphism of shale in New England. A pattern of isograds tends to follow the deformation features of a region, as outlined by folds and faults. An isograd based on a single index mineral, such as the chlorite isograd shown in the figure, is a good approximate measure of metamorphic pressure and temperature.

Because isograds reveal the pressures and temperatures at which the minerals in a regional metamorphic belt were formed, the isograd sequence in one belt may differ from that in another. The reason for this difference is that pressure and temperature do not increase at the same rate in all plate tectonic settings. As we discussed earlier in this chapter, pressure increases more rapidly than temperature in some settings and more slowly in others (see Figures 7.20 and 7.21).

Isograds show where one mineral zone, containing rocks of one metamorphic grade, changes to another.

(a) Isograds based on index minerals can be used to plot metamorphic grades over a regional metamorphic belt.

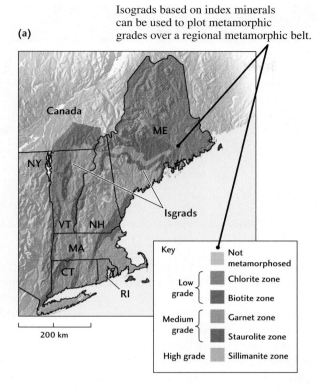

FIGURE 7.29 Index minerals define the different metamorphic zones within a belt of regional metamorphism. (a) Map of New England, showing mineral zones based on index minerals found in rocks metamorphosed from slate. (b) Rocks produced by the metamorphism of slate at different temperatures and pressures. [slate, phyllite, schist, gneiss: John Grotzinger/Ramón Rivera-Moret/Harvard Mineralogical Museum; blueschist: courtesy of Mark Cloos; migmatite: Kip Hodges.] (c) Changes in the mineral composition of metamorphic rock derived from slate define its metamorphic facies and are used to indicate its metamorphic grade.

(b) As a parent rock is metamorphosed, it progresses from low-grade rock to high-grade rock.

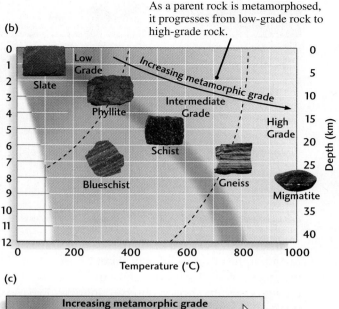

(c)

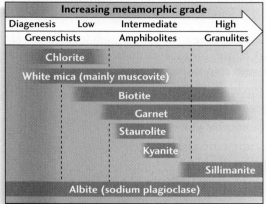

With increasing metamorphic grade, mineral composition changes.

Metamorphic Facies

We can plot this information about the grades of metamorphic rock in a regional metamorphic belt—derived from parent rocks of many different chemical compositions—on a graph of temperature and pressure (**Figure 7.30**).

Metamorphic facies are groupings of rocks of various mineral compositions formed under different grades of

FIGURE 7.30 Metamorphic facies correspond to particular combinations of temperature and pressure, which correspond to particular plate tectonic settings. The dashed lines indicate the overlapping nature of the boundaries between metamorphic rocks.

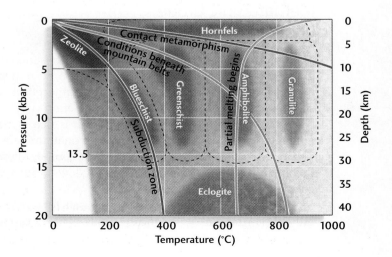

TABLE 7.2	Major Minerals of Metamorphic Facies Produced from Parent Rocks of Different Composition	
Facies	**Minerals Produced from Shale Parent**	**Minerals Produced from Basalt Parent**
Greenschist	Muscovite, chlorite, quartz, albite	Albite, epidote, chlorite
Amphibolite	Muscovite, biotite, garnet, quartz, albite, staurolite, kyanite, sillimanite	Amphibole, plagioclase feldspar
Granulite	Garnet, sillimanite, albite, orthoclase, quartz, biotite	Calcium-rich pyroxene, calcium-rich plagioclase feldspar
Eclogite	Garnet, sodium-rich pyroxene, quartz/coesite, kyanite	Sodium-rich pyroxene, garnet

metamorphism from different parent rocks. By delineating metamorphic facies, we can be more specific about the grades of metamorphism observed in rocks. Two essential points characterize the concept of metamorphic facies:

1. Given the same metamorphic grade, different kinds of metamorphic rocks form from parent rocks of different composition.

2. Given parent rocks of the same composition, different kinds of metamorphic rocks form at different metamorphic grades.

> *Metamorphic facies are groupings of rocks of various mineral compositions formed under different grades of metamorphism from different parent rocks.*

Table 7.2 lists the major minerals of the metamorphic facies produced from shale and basalt. Because parent rocks vary so greatly in composition, there are no sharp boundaries between metamorphic facies (see Figure 7.30). Perhaps the most important reason for analyzing metamorphic facies is that they give us clues to the tectonic processes responsible for metamorphism.

Metamorphic Grade and Parent Rock Composition

The kind of metamorphic rock that results from a given grade of metamorphism depends partly on the mineral composition of the parent rock. The metamorphism of slate, as shown in Figure 7.29c, reveals the effects of pressure and temperature on parent rock that is rich in clay minerals, quartz, and perhaps some carbonate minerals. The metamorphism of mafic volcanic rock, composed predominantly of feldspars and pyroxene, follows a different course (**Figure 7.31**).

In the regional metamorphism of a basalt, for example, the lowest-grade rocks characteristically contain various *zeolite* minerals. The silicate minerals in the zeolite class contain water in cavities within their crystal structure. Zeolite minerals form at very low temperatures and pressures. Rocks that include this group of minerals are thus placed in the zeolite grade.

Overlapping with the zeolite grade is a higher grade of metamorphosed mafic volcanic rocks, the **greenschists,** whose abundant minerals include chlorite. Next are the amphibolites, which contain large amounts of amphiboles. The granulites, coarse-grained rocks containing pyroxene and calcium plagioclase, constitute the highest grade of metamorphosed mafic volcanics. Rocks of the greenschist, amphibolite, and granulite grades are also formed during metamorphism of sedimentary rocks, as shown in Figure 7.29c.

> *The kind of metamorphic rock that results from a given grade of metamorphism depends partly on the mineral composition of the parent rock.*

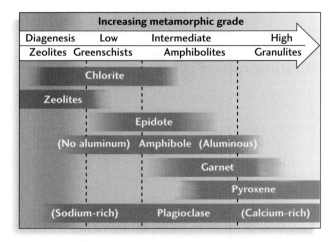

FIGURE 7.31 Changes in the mineral composition of mafic rock during metamorphism indicate its metamorphic grade.

The pyroxene-bearing granulites are the products of high-grade metamorphism in which the temperature is high and the pressure is moderate. The opposite conditions, in which the pressure is high and the temperature moderate, produce rocks of the **blueschist** grade from parent rocks of various starting compositions, from mafic volcanic rocks to shaley sedimentary rocks. The name comes from the abundance of glaucophane, a blue amphibole, in these rocks. Still another metamorphic rock, formed at extremely high pressures and moderate to high temperatures, is eclogite, which is rich in garnet and pyroxene.

PLATE TECTONICS AND METAMORPHISM

The concept of metamorphic grade can inform us of the maximum pressure or temperature to which a metamorphic rock has been subjected, but it says nothing about where the rock encountered those conditions. Nor does it tell us anything about how the rock was **exhumed,** or transported back to Earth's surface. It is important to understand that most metamorphism is a dynamic pro-

cess, not a static event. Plate tectonics provides the framework within which we can understand the dynamic processes of metamorphism.

Each metamorphic rock has a distinctive history of changing conditions of temperature and pressure that is reflected in its texture and mineral composition. This history is called a metamorphic *P-T path.* The P-T path can be a sensitive recorder of many important factors that influence metamorphism—such as sources of heat, which change temperatures, and rates of tectonic transport (burial and exhumation), which change pressures. Thus, P-T paths are characteristic of particular plate tectonic settings.

> *P-T paths are distinctive histories of temperatures and pressures that are characteristic of particular plate tectonic settings.*

Ocean-Continent Convergence

The P-T path of the rock assemblages that form when oceanic lithosphere is subducted beneath a plate carrying a continent on its leading edge is shown in **Figure 7.32a.** Thick sediments eroded from the continent rapidly fill the deep-sea trench that forms at the subduction zone. As it

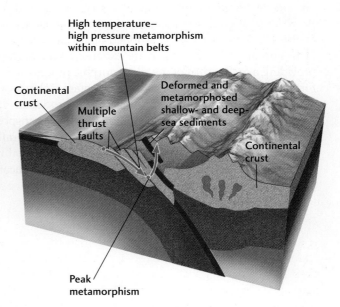

(a) Ocean-continent convergence

Low temperature–high pressure metamorphism at subduction zones
Deep-sea trench
Volcanic mountain belt
Deep-sea sediments
Oceanic crust
Ophiolites
Continental crust
Peak metamorphism

(b) Continent-continent convergence

High temperature–high pressure metamorphism within mountain belts
Continental crust
Multiple thrust faults
Deformed and metamorphosed shallow- and deep-sea sediments
Continental crust
Peak metamorphism

FIGURE 7.32 P-T paths indicate the trajectory of rocks during metamorphism at (a) an ocean-continent convergence and (b) a continent-continent convergence. The different P-T paths of rocks formed in these different plate tectonic settings indicate differences in geothermal gradients. Rocks transported to similar depths—and pressures—beneath mountain belts become much hotter than rocks at an equivalent depth in subduction zones.

descends, the cold slab of oceanic lithosphere stuffs the region below the inner wall of the trench (the wall closer to the continent) with these sediments, as well as with deep-sea sediments and shreds of ophiolite suites (see Figure 5.12) scraped off the descending plate. Regions of this sort, located between the volcanic arc on the continent and the trench offshore, are enormously complex and variable. The rocks formed there are all highly folded, intricately faulted, and metamorphosed. Blueschist-grade metamorphism, which is characteristic of high pressures and low temperatures, is found in these regions because the material may be carried relatively rapidly to depths as great as 30 km, where recrystallization occurs in the environment of the still-cool subducting slab. Typical P-T paths for blueschist-grade metamorphic rocks reveal this initial downward transport in the subduction zone. During exhumation, the trajectory reverses itself.

Continent–Continent Convergence

Because continental crust is buoyant, when a continent collides with another continent, both continents resist subduction and stay afloat on the mantle. As a result, a wide zone of intense deformation develops at the convergent boundary where the continents grind together, as shown in Figure 7.32b. The intense deformation results in a much-thickened continental crust in the collision zone, which often produces high mountains. As the lithosphere thickens, the deep parts of the continental crust heat up and undergo varying grades of metamorphism. Millions of years afterward, when erosion has stripped off the surface layers of the mountains, their cores are exposed at the surface, providing a rock record of the metamorphic processes that formed the schists, gneisses, and other metamorphic rocks.

P-T paths for metamorphic rocks produced by continent-continent collision indicate initial transport deep within the collision zone. Geologists generally interpret this initial trajectory as indicating the burial of rocks beneath high mountains. P-T paths then indicate transport of metamorphic rocks back to Earth's surface as

the buried rocks are uplifted and exhumed. P-T paths of metamorphic rocks produced by continent-continent collision have a different shape from those of rocks produced by subduction alone. Continent-continent collision generates higher temperatures than subduction; therefore, as rock is pushed to greater depths, the temperature that corresponds to a given pressure will be higher.

Exhumation: The Link Between Metamorphism and Climate

Forty years ago, plate tectonic theory provided a ready explanation for how metamorphic rocks could be produced by seafloor spreading, subduction, and continent-continent collision. By the mid-1980s, the study of P-T paths provided a clearer picture of the specific tectonic mechanisms involved in the deep burial of rocks. At the same time, it surprised geologists by providing an equally clear picture of the subsequent and often very rapid uplift and exhumation of these deeply buried rocks. Since the time of this discovery, geologists have been searching for exclusively tectonic mechanisms that could bring these rocks back to Earth's surface so quickly.

A relatively new idea is that climate and tectonic processes, not tectonic processes alone, drive the flow of rocks from the deep crust to the shallow crust through the process of rapid erosion. Thus, plate tectonics—which acts through mountain building—and climate processes—which act through weathering and erosion—*interact* to control the flow of metamorphic rocks to Earth's surface. After decades of emphasis solely on tectonic explanations for regional and global geologic processes, it now seems that two apparently unrelated processes—metamorphism and erosion—may be linked in an elegant way. As one geologist exclaimed: "Savor the irony should the metamorphic muscles that push mountains to the sky be driven by the pitter patter of tiny raindrops."

By increasing erosion rates, climate may help drive the flow of metamorphic rocks from deep in Earth's crust to Earth's surface.

GOOGLE EARTH PROJECT

When rocks deform, they may form folds. An anticline is a fold that is shaped like an arch, while a syncline is shaped like a trough. When examining a rock outcrop, we can recognize an anticline by a sequence of beds dipping, or tilting, away from the center of the fold. For a syncline, the beds would dip toward the center of the fold. Let's look at an example from Wyoming.

LOCATION   Sheep Mountain/Ribbon Canyon near Big Horn, Wyoming

GOAL   Learn to recognize simple fold structures

LINKED   Figures 7.3, 7.4, and 7.10

(a) Google Earth satellite image showing the northwestern United States and the location of Ribbon Canyon, Wyoming (marked with a red dot).

(b) Google Earth three-dimensional satellite image of Ribbon Canyon, Wyoming, showing beds dipping steeply in opposite directions

1. Type "Ribbon Canyon, Big Horn, Wyoming" into the GE search window. At the bottom left, under the "Layers" tab, click on "Terrain" to turn on the 3D viewing capability. Zoom in to an eye altitude of 8.75 km. What is the latitude and longitude of Ribbon Canyon, Wyoming?

   a. 44° 39′ 12.26″ N, 108° 11′ 59.32″ W

   b. 44° 31′ 30.16″ N, 107° 57′ 24.39″ W

   c. 45° 15′ 17.18″ N, 108° 59′ 03.12″ W

   d. 45° 29′ 07.45″ N, 107° 24′ 18.94″ W

(Continued.)

2. In this chapter, we've discussed various features that a geologist in the field might map. The map shown in part (b) is centered at 44°38′ 21.52″ N, 108° 10′ 50.64″ W, at an eye altitude of approximately 2.75 km. We have used the "Add path" tool (an icon at the top of the Google Earth screen) to trace out some of the beds on both sides of the mound. Try mapping (tracing) these beds yourself. You can change the colors of your paths by clicking on "Style/Color" in the "Add path" window. Which way do the beds on each side of the mound appear to be dipping?

   a. The blue beds are dipping to the southwest, while the red beds are dipping to the northeast.

   b. The blue beds and the red beds are dipping to the southwest.

   c. The blue beds are dipping to the northeast, while the red beds are dipping to the southwest.

   d. The blue beds and the red beds are dipping to the northwest.

3. In which direction do the beds on each side of the mound appear to be tilted in relation to the fold? (You may need to rotate the perspective view to see the beds from different sides. To do so, at the top right of the Google Earth screen, drag the "N" around the circle to change direction. You can also scroll farther to the southeast to see a larger extent of this feature.) Given this information, do you think this area represents an anticline or a syncline?

   a. The beds are tilted away from the center of the fold, making it a syncline.

   b. The beds are tilted toward the center of the fold, making it a syncline.

   c. The beds are tilted away from the center of the fold, making it an anticline.

   d. The beds are tilted toward the center of the fold, making it an anticline.

4. In the same area, what is the approximate highest elevation at the top of the mound?

   a. 1275 m

   b. 1320 m

   c. 1530 m

   d. 1610 m

Optional Challenge Question

5. Think about how anticlines and synclines are formed. Do you think the oldest beds are on the inside or the outside of the fold? (Hint: Imagine a stack of papers in which the papers on the bottom of the stack represent the oldest beds and the papers on top represent the youngest. If you fold the stack into a U shape [a syncline], are the "oldest" papers located on the inside or the outside of the fold? If you invert the stack into an upside-down U shape [an anticline], where are the oldest layers now?)

   a. At Ribbon Canyon, the oldest beds are on the outside of the fold.

   b. At Ribbon Canyon, the oldest beds are on the inside of the fold.

   c. At Ribbon Canyon, the oldest beds are all the same age.

   d. At Ribbon Canyon, the beds are faulted, not folded.

▨ SUMMARY

What are geologic maps and cross sections? A geologic map is a model of the rock formations exposed at Earth's surface. A geologic cross section is a diagram representing the geologic features that would be visible if a vertical slice were made through part of the crust.

What do laboratory experiments tell us about the way rocks deform? Laboratory studies show that rocks subjected to tectonic forces may behave as brittle materials or as ductile materials. These behaviors depend on temperature and pressure, the type of rock, the speed of deformation, and the orientation of tectonic forces.

What are the basic deformation structures that geologists observe in the field? The most important geologic structures in rock formations that result from deformation are folds and faults. Fractures are known as faults if rocks are displaced across the fracture surface and as joints if no displacement is observed.

What kinds of forces produce these deformation structures? Faults and folds are produced primarily by horizontally directed forces at plate boundaries. Horizontal tensional forces at divergent boundaries produce normal faults, horizontal compressive forces at convergent boundaries produce thrust faults, and horizontal shearing forces at transform-fault boundaries produce strike-slip faults.

Folds are usually formed in layered rock by compressive forces, especially in regions where continents collide.

What are the causes of metamorphism? Metamorphism is an alteration in the chemical composition, mineralogy, or texture of solid rock. It is caused by increases in pressure and temperature and by reactions with chemical components introduced by hydrothermal fluids. As rocks are pushed deep within the crust by plate tectonic processes and exposed to increasing pressures and temperatures, the chemical components of the parent rock rearrange themselves into a new set of minerals that are stable under the new conditions. Metamorphic rocks that form at relatively low temperatures and pressures are referred to as low-grade metamorphic rocks; those that form at high temperatures and pressures are high-grade metamorphic rocks. Chemical components may be added to or removed from a rock during metamorphism, usually by high temperature fluids.

What are the various types of metamorphism? The major types of metamorphism are regional metamorphism, during which large areas are metamorphosed by high pressures and temperatures generated during mountain building, and contact metamorphism, during which country rock close to an igneous intrusion is transformed primarily by the heat of the intruding magma. Less common types are burial metamorphism, during which sedimentary rocks are altered by increases in pressure and temperature as they are buried deeper in the crust; high-pressure and ultra-high-pressure metamorphism, which occur at great depths, as when sediments are subducted; and shock metamorphism, which results from meteorite impacts.

What are the chief kinds of metamorphic rocks? Metamorphic rocks fall into two major textural classes: foliated rocks (displaying foliation, a pattern of parallel cleavage planes resulting from a preferred orientation of crystals) and granoblastic, or nonfoliated, rocks. The kinds of rocks produced depend on the composition of the parent rock and the grade of metamorphism. Regional metamorphism of a shale leads to zones of foliated rock of progressively higher grade, from slate to phyllite, schist, gneiss, and finally migmatite. Among granoblastic rocks, marble is derived from the metamorphism of limestone, quartzite from quartz-rich sandstone, and greenstone from basalt. Hornfels is the product of contact metamorphism of fine-grained sedimentary rocks and other types of rock containing an abundance of silicate minerals.

What do metamorphic rocks reveal about the conditions under which they were formed? Zones of metamorphism can be mapped with isograds defined by the first appearance of an index mineral. The presence of an index mineral can indicate the temperature and pressure under which the rocks in the zone were formed. According to the concept of metamorphic facies, rocks of the same metamorphic grade may differ because of variations in the chemical composition of the parent rock, whereas rocks metamorphosed from the same parent rock may vary because they were subjected to different grades of metamorphism.

How are metamorphic rocks related to plate tectonic processes? During both subduction and continent-continent collision at convergent plate boundaries, rocks and sediments are pushed to great depths in Earth's crust, where they are subjected to increasing pressures and temperatures that result in metamorphism. Metamorphic P-T paths provide insight into the plate tectonic settings where these rocks are metamorphosed. In the case of ocean-continent convergences, P-T paths indicate subduction of rocks and sediments to sites with high pressures and relatively low temperatures. In continent-continent collision zones, rocks are pushed down to depths where pressure and temperature are both high. In both settings, the P-T paths show that after the rocks experience the maximum pressures and temperatures, they return to shallow depths. This process of exhumation may be driven by weathering and erosion at Earth's surface as well as by plate tectonic processes.

KEY TERMS AND CONCEPTS

anticline (p. 220)
blueschist (p. 239)
brittle (p. 216)
compressive force (p. 217)
contact metamorphism (p. 231)
dip (p. 214)
dip-slip fault (p. 217)
ductile (p. 216)
exhumation (p. 239)
fault (p. 217)
faulting (p. 212)
folding (p. 212)
foliated rocks (p. 233)
foliation (p. 232)
formation (p. 213)
geologic cross section (p. 215)
geologic map (p. 214)
gneiss (p. 234)
granoblastic rocks (p. 234)

greenschist (p. 238)
hanging wall (p. 218)
joint (p. 222)
metamorphic facies (p. 237)
metamorphic rock (p. 226)
metasomatism (p. 229)
normal fault (p. 218)
oblique slip fault (p. 219)
orogeny (p. 212)
phyllite (p. 234)
regional metamorphism (p. 230)
schist (p. 234)
shearing force (p. 217)
slate (p. 233)
strike (p. 214)
strike-slip fault (p. 217)
syncline (p. 220)
tensional force (p. 217)
thrust fault (p. 218)

■ EXERCISES

1. Is the small fault on the left side of Figure 7.1b a normal fault or a thrust fault? Estimate the fault's offset, expressing your answer in meters.

2. The motion of the San Andreas fault has offset the stream channel in Figure 7.7 by 130 m. Geologists have determined that this channel is 3800 years old. What is the slip rate along the San Andreas fault at this site, expressed in millimeters per year?

3. Show that a left jog in a right-lateral strike-slip fault will produce compression, whereas a right jog in a right-lateral strike-slip fault will produce extension. Write a similar rule for left-lateral strike-slip faults.

4. At what depths in Earth do metamorphic rocks form? What happens if temperatures get too high?

5. What does preferred orientation refer to in a metamorphic rock? Think about how the alignment of minerals relates to deformation.

6. You have mapped an area of metamorphic rocks, such as the region in Figure 7.29, and have observed mineral zones marked by north-south isograd lines running from sillimanite in the east to chlorite in the west. Were metamorphic temperatures higher in the east or in the west?

7. Subduction zones are generally characterized by high pressure–low temperature metamorphism. In contrast, continent-continent collision zones are marked by moderate pressure–high temperature metamorphism. Which type of plate boundary has a higher geothermal gradient? Explain.

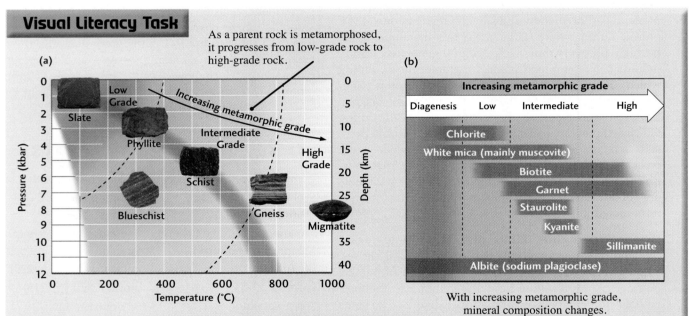

Visual Literacy Task

As a parent rock is metamorphosed, it progresses from low-grade rock to high-grade rock.

With increasing metamorphic grade, mineral composition changes.

FIGURE 7.29 Index minerals define the different metamorphic zones within a belt of regional metamorphism. (a) Rocks produced by the metamorphism of slate at different temperatures and pressures. [slate, phyllite, schist, gneiss: John Grotzinger/ Ramón Rivera-Moret/Harvard Mineralogical Museum; blueschist: courtesy of Mark Cloos; migmatite: Kip Hodges.] (b) Changes in the mineral composition of metamorphic rock derived from slate define its metamorphic facies and are used to indicate its metamorphic grade.

1. What set of conditions could be possible during high-grade metamorphism?

a. 3 kbar (~10 km deep) and 300°C
b. 6 kbar (~20 km deep) and 500°C
c. 7 kbar (~25 km deep) and 900°C

2. What mineral tells the least amount of information about metamorphic grade?

a. Chlorite   d. Sillimanite
b. Staurolite   e. Albite
c. Kyanite

3. Why does some rock metamorphose into blueschist while other rock might metamorphose into gneiss?

a. They reach different temperatures.
b. They reach different pressures.
c. They reach the same metamorphic conditions but started as different rocks.

4. What mineral would least likely be part of phyllite?

a. Biotite
b. White mica (muscovite)
c. Sillimanite
d. Albite

5. If you find staurolite in a rock, what rock would it likely be?

a. Slate
b. Schist
c. Migmatite

Thought Question: Why are most metamorphic rocks formed at convergent plate boundaries?

Timing the Geologic Record

8

The Abyss of Time

ON A CLEAR JUNE DAY IN 1788, the geologist James Hutton sailed along the Berwick-shire coast of southeastern Scotland with a mathematician friend, John Playfair, and Sir James Hall, looking for evidence of Earth's ancient past. They rounded Siccar Point and landed at the base of an impressive rock formation, soon to become one of the most famous sites in the study of geology. Nearly vertical layers of gray slate (schistus) rose up in a sheer cliff face, only to terminate abruptly against nearly horizontal beds of red sandstone. In the words of Playfair,

We felt ourselves necessarily carried back to the time when the schistus on which we stood was yet at the bottom of the sea, and when the sandstone before us was only beginning to be deposited, in the shape of sand or mud, from the waters of a superincumbent ocean. An epoch still more remote presented itself, when even the most ancient of these rocks, instead of standing upright in vertical beds, lay in horizontal planes at the bottom of the sea, and was not yet disturbed by that immeasurable force which has burst asunder the solid pavement of the globe. . . . The mind seemed to grow giddy by looking so far into the abyss of time.

> **"The mind seemed to grow giddy by looking so far into the abyss of time."**

How deep is that abyss? Hutton was a gentleman farmer and sometime philosopher who had carefully observed how water and ice, year by year, wore down the highlands a few rocks at a time and rivers washed away, grain by grain, the soils of his farmlands. He had realized that these erosional processes, which laid sediments down in the sea, must be balanced by the deformation and uplift of sedimentary layers to rejuvenate the landscape in a continuing rock cycle.

It was obvious to Hutton—though not to many others of his day—that the contorted geology of the British Isles could not have been created by such slow processes in the mere thousands of years of recorded human history. After all, Hadrian's Wall, built to protect the Roman emperor's English settlements from rowdy Scottish tribes, hardly looked the worse for wear despite 16 centuries of exposure to the elements. What much greater expanse of time must nature have taken to create the skewed, cross-cutting formations at Siccar Point? ◆

Siccar Point, southeastern Scotland, where gently dipping beds of red sandstone rest on near-vertical beds of gray slate. [*Dave Souza/Wikipedia.*]

IN HIS GREAT TREATISE *THEORY OF THE EARTH,* published in the same year as his expedition with Playfair, Hutton presented the evidence for his principle of uniformitarianism (introduced in Chapter 2). He described the indefinite span of geologic time in a memorable line: "We find no vestige of a beginning, no prospect of an end."

Spurred by Hutton's vision, geologists set out to explore Earth's history through ever more careful and precise readings of the rock record. They eventually discovered that Earth did indeed have a beginning—a violent one at that. The methods they invented to measure geologic time are described in this chapter; what they found is the subject of the next.

In this chapter, we will learn how geologists have plumbed the abyss of time by finding order in the **geologic record and using "clocks in rocks" to date the events that have occurred throughout Earth's 4.56-billion-year history.**

Geologic processes occur on time scales that range from seconds (meteorite impacts, volcanic explosions, earthquakes) to tens of millions of years (the recycling of oceanic lithosphere) and even billions of years (the tectonic evolution of continents). If we are careful enough, we can measure the rates of short-term processes, such as beach erosion or the seasonal variations in the transport of sediments by rivers, in a few years. Precise surveying can monitor the slow movements of glaciers (meters per year), and with the new GPS instruments depicted in Figure 3.14, we can track the even slower movements of tectonic plates (centimeters per year). Historical documents can provide certain types of geologic data, such

1871

1968

FIGURE 8.1 Two photographs of Bowknot Bend on the Green River in Utah, taken nearly 100 years apart, show that the configuration of rocks and geologic structures has changed very little in that time interval. [*left: E.O. Beaman/USGS; right: H.G. Stevens/USGS.*]

as the dates of major earthquakes or volcanic eruptions, from hundreds or, in some cases, thousands of years ago.

However, the record of human observation is far too short for the study of many slow geologic processes (**Figure 8.1**). In fact, it's not even long enough to capture some types of rapid but infrequent events; for example, we have never witnessed a meteorite impact as big as the one that left the crater shown in Figure 2.7. We must rely instead on the geologic record—the information preserved in rocks that have survived erosion and subduction. Almost all oceanic crust older than 200 million years has been subducted back into the mantle, so most of Earth's history is documented only in the older rocks of the continents. Geologists can reconstruct subsidence from the record of sedimentation; uplift from the erosion of rock layers; and deformation from faults, folds, and metamorphic rocks. But to measure the pace of these processes and understand their common causes, we must be able to date the geologic record.

Geologists speak carefully about time. "To date" means to measure the **absolute age** of an event in the geologic record—the number of years elapsed from the event until now. Before the twentieth century, no one knew much about absolute ages; geologists could determine only whether one event was older or younger than another—their **relative ages.** They could say, for instance, that fish bones were first deposited in marine sediments before mammal bones appeared in land sediments, but they couldn't tell how many millions of years ago the first fish or mammals appeared. As we will see in this chapter, powerful methods have been invented for estimating absolute ages and developing a precise and detailed geologic time scale.

RECONSTRUCTING GEOLOGIC HISTORY FROM THE STRATIGRAPHIC RECORD

The first geological observations pertaining to the question of deep time came in the mid-seventeenth century from the study of fossils. A *fossil* is an artifact of life preserved in the geologic record (**Figure 8.2**). However, few

(a)
(b)

FIGURE 8.2 Fossils are traces of living organisms preserved in the geologic record. (a) Ammonite fossils, ancient examples of a large group of invertebrate organisms that are now largely extinct. Their sole representative in the modern world is the chambered nautilus. (b) Petrified Forest, Arizona. These ancient logs are millions of years old. Their substance has been completely replaced by silica, which preserved all the original details of form. [*(a) Chip Clark; (b) Tom Bean.*]

(a)

(b)

(c)

FIGURE 8.3 Nicolaus Steno was the first to demonstrate that fossils are the remains of ancient life. (a) Portrait of Nicolaus Steno (1638–1686). (b) "Tongue stones" of the type found in sedimentary rocks in the Mediterranean region, where Steno worked. (c) This diagram is from Steno's 1667 book, which demonstrated that tongue stones are the fossilized teeth of ancient sharks. [*(a) Conserved at St. Anna's parsonage, Schwerin, Germany/Wikimedia; (b) Corbis RF/Alamy; (c) Wikipedia Commons.*]

people living in the seventeenth century would have understood this definition. Most thought that the seashells and other lifelike forms found in rocks dated from Earth's beginnings about 6000 years earlier, or grew there spontaneously.

In 1667, the Danish scientist Nicolaus Steno, who was working for the royal court in Florence, Italy, demonstrated that the peculiar "tongue stones" found in certain Mediterranean sedimentary rocks were essentially identical to the teeth of modern sharks (**Figure 8.3**). He concluded that tongue stones *really were* ancient shark teeth preserved in the rocks, and more generally, that fossils were the remains of ancient life deposited with sediments. To convince people of his ideas, Steno wrote a short but brilliant book about the geology of Tuscany in which he laid the foundation for the modern science of **stratigraphy**—the study of *strata* (layers) in rocks.

Stratigraphy is the study of layers, or strata, of sedimentary rock.

Principles of Stratigraphy

Geologists still use Steno's principles to interpret sedimentary strata. Two of his basic rules are so simple they seem obvious to us today (**Figure 8.4**):

1. The **principle of original horizontality** states that sediments are deposited under the influence of gravity as nearly horizontal beds. Observations in a wide variety of sedimentary environments support this principle. If we find folded or faulted strata, we know that the layers were deformed by tectonic forces after the sediments were deposited.

2. The **principle of superposition** states that each layer of an undeformed sedimentary sequence is younger than the one beneath it and older than the one above it. A new layer cannot be deposited beneath an existing layer. Thus, strata can be vertically ordered in time from the lowest (oldest) bed to the uppermost (youngest) bed. A chronologically ordered set of strata is known as a **stratigraphic succession.**

Sediments are deposited sequentially in horizontal layers with the oldest layers at the bottom and the youngest layers at the top.

We can apply Steno's principles in the field to determine whether one sedimentary formation is older than another by piecing together the formations exposed in different outcrops; we can sort them into chronological order and thus construct the stratigraphic succession of a region—at least in principle. In practice, there were two problems with this strategy. First, geologists almost always found gaps in a region's stratigraphic succession, indicating time intervals that had gone entirely unrecorded. Some of these intervals were short, such as periods of drought between floods; others lasted for millions of years—for example, periods of regional tectonic uplift when thick sequences of sedimentary rocks were removed by erosion. Second, it was difficult to

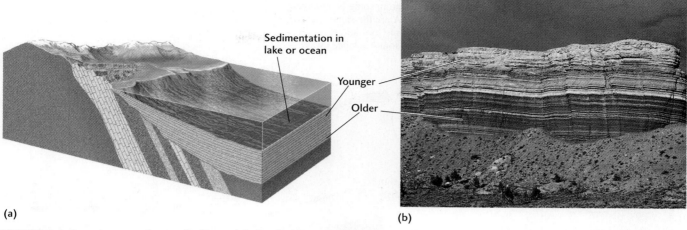

Sedimentation in lake or ocean

Younger

Older

(a)

(b)

FIGURE 8.4 Steno's principles guide the study of sedimentary strata. (a) Sediments are deposited in horizontal layers before being slowly transformed into sedimentary rock. If left undisturbed by tectonic processes, the youngest layers remain on the top and the oldest on the bottom. (b) Marble Canyon, part of the Grand Canyon, was cut by the Colorado River through what is now northern Arizona, revealing these undisturbed strata, which record millions of years of geologic history. [*Photo by Fletcher and Baylis/Photo Researchers.*]

determine the relative ages of two formations that were widely separated in space; stratigraphy alone couldn't determine whether a sequence of mudstones in, say, Tuscany was older, younger, or the same age as a similar sequence in England. It was necessary to expand Steno's ideas about the biological origin of fossils to solve these problems.

Fossils as Recorders of Geologic Time

In 1793, only 5 years after Hutton visited Siccar Point, William Smith, a surveyor working on the construction of canals in southern England, recognized that fossils could help geologists determine the relative ages of sedimentary rocks. Smith was fascinated by the variety of fossils that he collected from the strata exposed along canal cuts. He observed that different layers contained different sets of fossils, and he was able to tell one layer from another by the characteristic fossils in each. He established a general order for the sequence of fossil assemblages and strata, from lowest (oldest) to uppermost (youngest) rock layers. Regardless of its location, Smith could predict the stratigraphic position of any particular layer or formation in any outcrop in southern England based on its fossil assemblages. This stratigraphic ordering of the fossils of animal species (fauna) produces a sequence known as a *faunal succession.*

Smith's **principle of faunal succession** states that the sedimentary strata in an outcrop contain fossils in a definite sequence. The same sequence can be found in

outcrops at other locations, so strata in one location can be matched to strata in another location.

Using faunal successions, Smith was able to identify formations of the same age in different outcrops. By noting the vertical order in which the formations were found in each place, he compiled a composite stratigraphic succession for the entire region. His compilation showed how the complete succession would have looked if the formations at different levels in all the various outcrops could have been brought together at a single spot. **Figure 8.5** shows a composite stratigraphic succession for two outcrops.

> *Characteristic assemblages of fossils can be used to identify strata of the same age in different locations.*

Smith kept track of his work by mapping outcrops using colors assigned to specific formations, thus inventing the geologic map (see Figure 7.4). In 1815, he summarized his lifelong research by publishing his "General Map of Strata in England and Wales," a hand-colored masterpiece 8 feet tall and 6 feet wide—the first geologic map of an entire country. The original still hangs in the office of the Geologic Society of London.

The geologists who followed in Steno's and Smith's footsteps described and catalogued hundreds of fossils and their relationships to modern organisms, establishing the new science of paleontology—the historical study of ancient life-forms. The most common fossils they found were the shells of invertebrate animals. Some were similar to clams, oysters, and other living shellfish; others

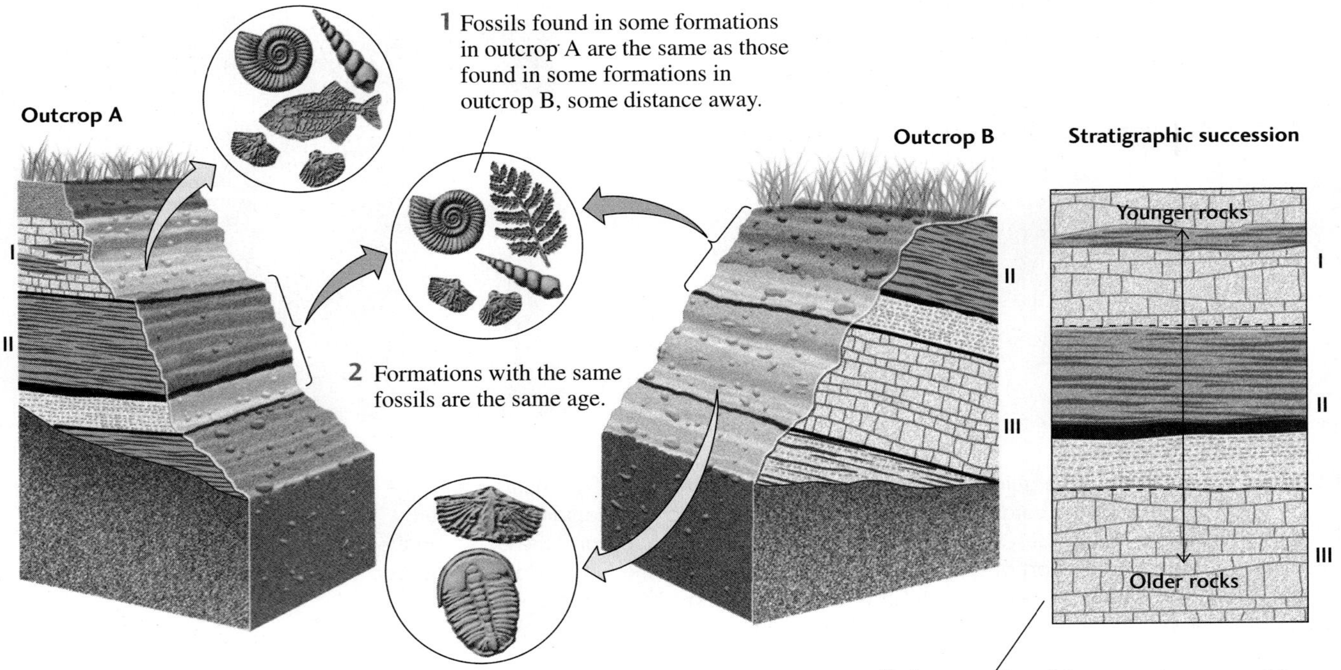

Outcrop A

1 Fossils found in some formations in outcrop A are the same as those found in some formations in outcrop B, some distance away.

Outcrop B

Stratigraphic succession

Younger rocks

I

II

III

Older rocks

2 Formations with the same fossils are the same age.

I

II

III

3 A composite of the two outcrops would show formations I and II both overlying formation III.

FIGURE 8.5 The principle of faunal succession can be used to correlate rock layers in different outcrops.

represented strange species with no living examples, such as the trilobites (**Figure 8.6**). Less common were the bones of vertebrates, such as mammals, birds, and the huge extinct reptiles they called dinosaurs. Plant fossils were found to be abundant in some rocks, particularly in coal beds, where leaves, twigs, branches, and even whole tree trunks could be recognized. Fossils were not found in intrusive igneous rocks—no surprise, since any biological material would have been destroyed when the rocks melted—nor were they found in high-grade metamorphic rocks, where any remains of organisms would have been distorted beyond recognition.

By the beginning of the nineteenth century, paleontology had become the single most important source of information about geologic history. The systematic study of fossils affected science far beyond geology, however.

FIGURE 8.6 Trilobites, an extinct life-form, preserved as fossils in rocks about 365 million years old found in Ontario, Canada. [*William E. Ferguson.*]

Charles Darwin studied paleontology as a young scientist, and he collected many unusual fossils on his famous voyage aboard the *Beagle* (1831–1836). During this world-circling tour, he also studied many unfamiliar animal and plant species in their native habitats. Darwin pondered what he had seen until 1859, when he proposed his theory of evolution by natural selection. His theory revolutionized the science of biology and provided a sound theoretical framework for paleontology: if organisms evolve progressively over time, then the fossils in each sedimentary bed must represent the organisms living when that bed was deposited.

Unconformities: Gaps in the Geologic Record

In compiling the stratigraphic succession of a region, geologists often find places in the geologic record where a formation is missing. Either no rock was ever deposited, or it was eroded away before the next strata were laid down. The surface between two beds that were laid down with a time gap between them—the boundary along which the two existing formations meet—is called an **unconformity** (**Figure 8.7**). A series of beds bounded above and below by unconformities is referred to as a *sedimentary sequence*. An unconformity, like a sedimentary sequence, represents the passage of time.

An unconformity may imply that tectonic forces raised the rock above sea level, where erosion removed some rock layers. Alternatively, the unconformity could have been produced by the erosion of newly exposed land as the sea level fell. As we will see in Chapter 10, global sea level can be lowered by hundreds of meters during ice ages, when water is withdrawn from the oceans to form continental ice sheets.

> *An unconformity indicates a gap in time between layers in a stratigraphic succession.*

Unconformities are classified according to the relationships between the layers above and below them. An unconformity in which the upper sedimentary sequence overlies an erosional surface developed on an undeformed, still horizontal lower sedimentary sequence is a *disconformity* (see Figure 8.7). Sea level drops and broad tectonic uplifts often create disconformities. In a *nonconformity,* the upper sedimentary beds overlie metamorphic or igneous rocks (see Figure 8.12 for an example). An *angular unconformity* is one in which the upper beds overlie lower beds that have been folded by tectonic processes and then eroded to a more or less even plane. In an angular unconformity, the two sequences have bedding planes that are not parallel. The photograph at the opening of this chapter shows the famous angular unconformity that Hutton discovered at Siccar Point.

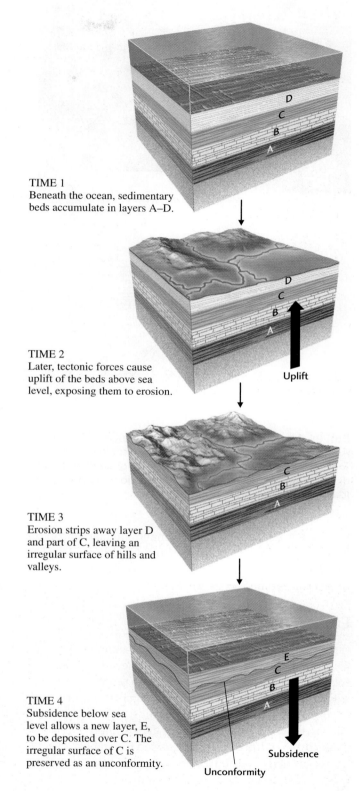

TIME 1
Beneath the ocean, sedimentary beds accumulate in layers A–D.

TIME 2
Later, tectonic forces cause uplift of the beds above sea level, exposing them to erosion.

Uplift

TIME 3
Erosion strips away layer D and part of C, leaving an irregular surface of hills and valleys.

TIME 4
Subsidence below sea level allows a new layer, E, to be deposited over C. The irregular surface of C is preserved as an unconformity.

Subsidence

Unconformity

FIGURE 8.7 An unconformity is a surface between two rock layers representing a layer that never formed or was eroded away. The type of unconformity shown here, created through uplift and erosion, followed by subsidence and another round of sedimentation, is called a disconformity.

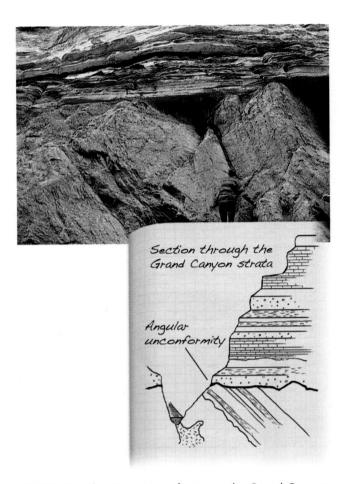

FIGURE 8.8 The Great Unconformity in the Grand Canyon, Arizona, is an angular unconformity between the horizontal Tapeats sandstone of Cambrian age (above) and the steeply dipping Grand Canyon beds of Precambrian age (below). [*GeoScience Features Picture Library.*]

Figure 8.8 depicts another dramatic angular unconformity found near the bottom of the Grand Canyon. The formation of an angular unconformity by tectonic processes is illustrated in **Figure 8.9**.

Cross-Cutting Relationships

Other disturbances of the layering of sedimentary strata also provide clues for determining the relative ages of rocks. Recall that dikes can cut through sedimentary beds; sills can be intruded parallel to bedding planes (see Chapter 6); and faults can displace bedding planes, dikes, and sills as they shift blocks of rock (see Chapter 7). These *cross-cutting relationships* can be used to establish the relative ages of igneous intrusions or faults within the stratigraphic succession. Because the deformation or intrusion events must have taken place after the affected sedimentary beds were deposited, those structures must be younger than the rocks

Cross-cutting relationships can be used to determine the relative ages of igneous intrusions, faults, and sedimentary strata.

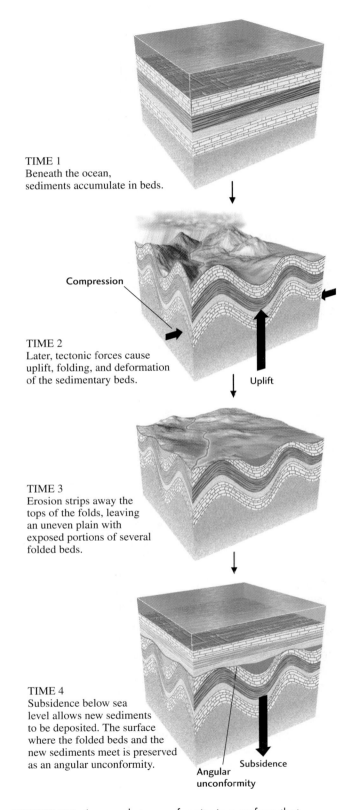

TIME 1
Beneath the ocean, sediments accumulate in beds.

TIME 2
Later, tectonic forces cause uplift, folding, and deformation of the sedimentary beds.

Compression

Uplift

TIME 3
Erosion strips away the tops of the folds, leaving an uneven plain with exposed portions of several folded beds.

TIME 4
Subsidence below sea level allows new sediments to be deposited. The surface where the folded beds and the new sediments meet is preserved as an angular unconformity.

Subsidence

Angular unconformity

FIGURE 8.9 An angular unconformity is a surface that separates two sedimentary sequences whose bedding planes are not parallel. This series of drawings shows how an angular conformity can form.

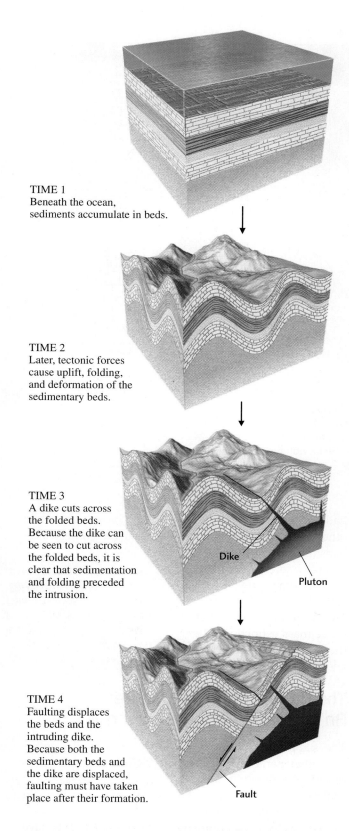

TIME 1
Beneath the ocean, sediments accumulate in beds.

TIME 2
Later, tectonic forces cause uplift, folding, and deformation of the sedimentary beds.

TIME 3
A dike cuts across the folded beds. Because the dike can be seen to cut across the folded beds, it is clear that sedimentation and folding preceded the intrusion.

Dike

Pluton

TIME 4
Faulting displaces the beds and the intruding dike. Because both the sedimentary beds and the dike are displaced, faulting must have taken place after their formation.

Fault

FIGURE 8.10 Cross-cutting relationships allow geologists to establish the relative ages of igneous intrusions or faults within a stratigraphic succession.

they cut (**Figure 8.10**). If the intrusions or fault displacements are eroded and planed off at an unconformity and then overlaid by younger sedimentary beds, we know that those structures are older than the younger strata.

THE GEOLOGIC TIME SCALE: RELATIVE AGES

Early in the nineteenth century, geologists began to apply Steno's and Smith's stratigraphic principles to outcrops all over the world. The same distinctive fossils were discovered in similar formations on many continents. Moreover, faunal successions from different continents often displayed the same changes in fossil assemblages. By matching up faunal successions and using cross-cutting relationships, geologists could determine the relative ages of rock formations on a global basis. By the end of the century, they had pieced together a worldwide history of geologic events—a **geologic time scale.**

> *Distinctive fossil assemblages mark the intervals of the geologic time scale.*

Intervals of Geologic Time

The geologic time scale divides Earth's history into intervals marked by distinctive sets of fossils, and it places the boundaries of those intervals at times when those sets of fossils changed abruptly (**Figure 8.11**). The basic intervals of this time scale are the **eras:** the Paleozoic (from the Greek *paleo*, meaning "old," and *zoi*, meaning "life"), the Mesozoic ("middle life"), and the Cenozoic ("new life").

The eras are subdivided into **periods,** usually named for the geographic locality in which the formations representing them were first or best described, or for some distinguishing characteristic of the formations. The Jurassic period, for example, is named for the Jura mountain range of France and Switzerland, and the Carboniferous period is named for the coal-bearing sedimentary rocks of Europe and North America. The Paleogene and Neogene periods of the Cenozoic are two exceptions; these Greek names mean "old origin" and "new origin," respectively.

Some periods are further subdivided into **epochs,** such as the Miocene, Pliocene, and Pleistocene epochs of the Neogene period (see Figure 8.11). Today we are living in the Holocene ("completely new") epoch of the Neogene period in the Cenozoic era.

Interval Boundaries Mark Mass Extinctions

Many of the major boundaries in the geologic time scale represent **mass extinctions**—short intervals during

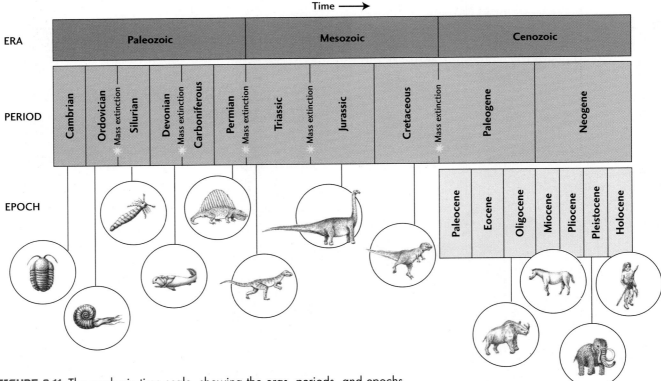

FIGURE 8.11 The geologic time scale, showing the eras, periods, and epochs distinguished by assemblages of fossils. The boundaries of these intervals are marked by the abrupt disappearance of some life-forms and the appearance of new ones. The five most dramatic mass extinctions are indicated by yellow stars. Note that this diagram shows only the relative ages of the intervals.

which a large proportion of the species living at the time simply disappeared from the fossil record, followed by the blossoming of many new species. These abrupt changes in the faunal succession were a great mystery to the geologists who discovered them. Darwin's theory of evolution explained how new species could evolve, but what had caused the mass extinctions?

In some cases, we think we know. The mass extinction at the end of the Cretaceous, which killed off 75 percent of the living species, including all the dinosaurs, was almost certainly the result of a large meteorite impact that darkened and poisoned the atmosphere and plunged Earth's climate into many years of bitter cold. This disaster marks the end of the Mesozoic era and the beginning of the Cenozoic.

In other cases, we are still not sure. The largest mass extinction, at the end of the Permian period, which defines the Paleozoic-Mesozoic boundary, eliminated nearly 95 percent of all living species, but the cause of this event is the subject of debate, as we will see in Chapter 9. The timing of this mass extinction precisely corresponds to a huge eruption of flood basalts in Siberia 251 million years ago, but why this particular eruption

had such a devastating effect on the biosphere (while others almost as large did not) remains a mystery. Perhaps the eruption injected enough material into the atmosphere to trigger catastrophic climate change. The extreme events that cause mass extinctions are the subject of very active geologic research.

Mass extinctions, in which a large proportion of living species disappeared from the fossil record, mark many of the major boundaries in the geologic time scale.

The Grand Canyon Sequence: An Exercise in Relative Dating

The sequence of rocks exposed in the Grand Canyon and other parts of the Colorado Plateau illustrates how relative dating works. These beds record a long history of sedimentation in a variety of environments, sometimes on land and sometimes under the sea. By matching the rock formations exposed at different localities, geologists have constructed a stratigraphic succession over a billion years long that spans both the Paleozoic and Mesozoic eras (**Figure 8.12**).

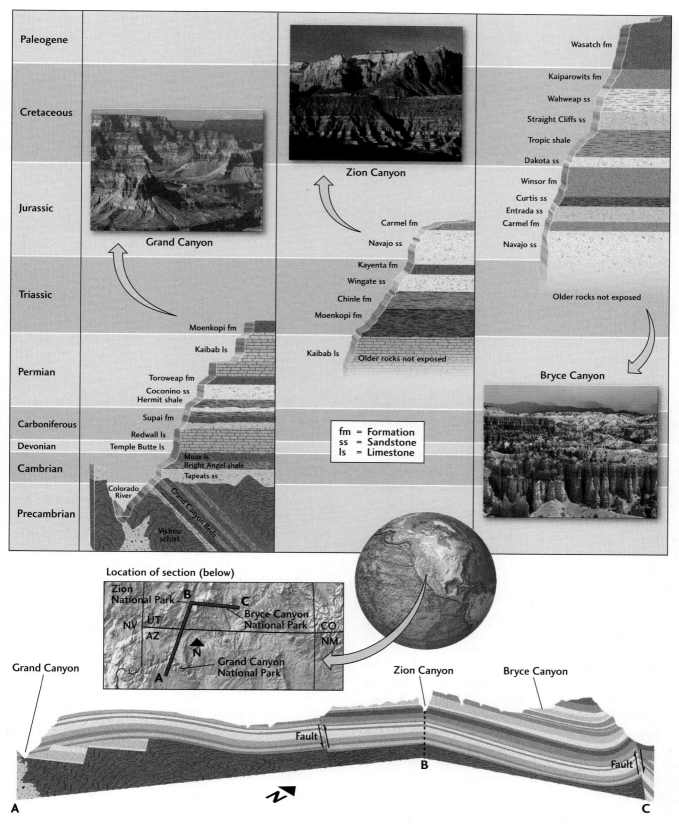

FIGURE 8.12 Stratigraphic sequence of the Colorado Plateau, reconstructed from strata exposed in Grand Canyon, Zion Canyon, and Bryce Canyon National Parks. [*Grand Canyon: John Wang/ Photo Disc/Getty Images. Zion Canyon: David Muench/Corbis. Bryce Canyon: Tim Davis/Photo Researchers.*]

The lowest—and therefore oldest—rocks exposed in the Grand Canyon are dark igneous and metamorphic rocks that make up the Vishnu schist, a group of formations that have been shown to be about 1.8 billion years old. Above the Vishnu schist are the younger Grand Canyon Beds. Although these sedimentary rocks contain fossils of single-celled microorganism that provide evidence of early life, they do not contain the shelly fossils distinctive of the Cambrian and later periods, so they are categorized as *Precambrian* rocks.

A nonconformity separates the Vishnu schist and the Grand Canyon Beds. This structure indicates a period of deformation accompanying metamorphism of the Vishnu schist and then a period of erosion before the deposition of the Grand Canyon Beds. The tilting of the Grand Canyon Beds from their originally horizontal position shows that they, too, were folded after deposition and burial.

An angular unconformity divides the Grand Canyon Beds from the overlying horizontal Tapeats sandstone (see Figure 8.8). This unconformity indicates a long period of erosion after the lower rocks had been tilted. The Tapeats sandstone and Bright Angel shale can be dated as Cambrian by their fossils, many of which are trilobites.

Above the Bright Angel shale is a group of horizontal limestone and shale formations (Muav limestone, Temple Butte limestone, Redwall limestone) that represent a long history of marine sedimentation from the late Cambrian period to the Carboniferous. There are so many time gaps represented by disconformities in these rocks that less than 40 percent of the Paleozoic is actually represented by rock strata.

The next set of strata, high up on the canyon wall, is the Supai formation (Carboniferous and Permian), which contain fossils of land plants like those found in the coal beds of North America and other continents. Overlying the Supai formation is the Hermit shale, a sandy red shale.

Continuing up the canyon wall, we find another continental deposit, the Coconino sandstone. This formation contains vertebrate animal tracks, which suggests that the Coconino was formed in a terrestrial environment during the Permian period. At the top of the cliffs at the canyon rim are two more formations of Permian age: the Toroweap, made mostly of limestone, overlain by the Kaibab, a massive layer of sandy and cherty limestone. These two formations record subsidence below sea level and the deposition of marine sediments. Above the Kaibab limestone and the canyon rim itself, but exposed within Grand Canyon National Park, is the Moenkopi formation, a red sandstone of Triassic age—the first appearance of rocks from the Mesozoic era in this stratigraphic succession.

The stratigraphic succession at the Grand Canyon, though picturesque and informative, represents an incomplete picture of Earth's history. Younger intervals of geologic time are not preserved there, and we must travel to nearby locations in Utah, such as Zion Canyon and Bryce Canyon, to fill in this more recent history. At Zion Canyon, we find equivalents of the Kaibab limestone and Moenkopi formation, which allow us to correlate this stratigraphic succession with the one at the Grand Canyon and establish a link. Unlike the Grand Canyon, however, the Zion strata extend upward to Jurassic time, including ancient sand dunes represented by the Navajo sandstone (see Figure 6.13). In Bryce Canyon, to the east of Zion, we again find the Navajo sandstone, as well as the strata that extend still farther upward, to the Wasatch formation of the Paleogene period.

The correlation of strata among these three areas of the Colorado Plateau shows how widely separated localities—each with an incomplete record of geologic time—can be pieced together to build a composite record of Earth's history.

The geologic history of the Colorado Plateau can be reconstructed from overlapping stratigraphic sequences exposed in the Grand Canyon, Zion, and Bryce Canyon National Parks.

NATURAL RESOURCES: Ages of Petroleum Source Rocks

Oil and natural gas come from organic matter that was buried in sedimentary rock formations at some time in the geologic past. The relative ages of these "petroleum source rocks" provide important clues about where to look for new oil and gas resources. Global surveys have shown that very little petroleum has come from Precambrian rocks, which makes sense, because the primitive organisms that existed before to the Cambrian period generated little organic matter.

Petroleum source rocks were deposited during all three of the geologic eras following the Cambrian, although certain periods of geologic time have produced much more of this resource than others (**Figure 8.13**). The clear winners are the Jurassic and Cretaceous periods of the Mesozoic era, which together have accounted for almost 60 percent of the world's petroleum production. Sedimentary formations of Jurassic and Cretaceous age were the source rocks for the great oil fields of the Middle East, the Gulf of Mexico, Venezuela, and the North Slope of Alaska.

If you examine Figure 3.16, you can see that, during these periods of geologic time, the supercontinent of Pangaea was breaking up into the modern continents. This tectonic activity formed many marine sedimentary basins and increased the rate at which sediments were

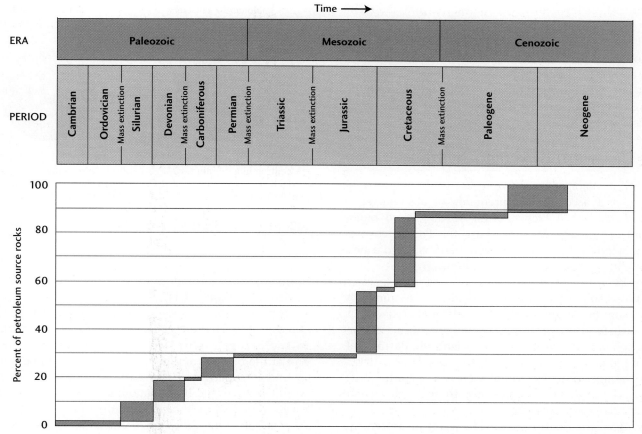

FIGURE 8.13 Relative ages and amounts of sedimentary rocks that contained the organic matter now found as oil and natural gas. Bars in the lower graph show the percentage of these petroleum source rocks found worldwide (height of the bar) within a given age range (width of the bar). Almost 60 percent of the total inventory was deposited during the Jurassic and Cretaceous periods of the Mesozoic era.

deposited into these basins. During the Jurassic and Cretaceous periods—the Age of Dinosaurs—marine life was abundant, providing much of the organic matter that was buried in the sediments. This carbon-rich material has since been "cooked up" and transported into the oil reservoirs, where we find it today.

> *Relative ages of petroleum source rocks provide important clues about where to look for new oil and gas resources.*

MEASURING ABSOLUTE TIME WITH ISOTOPIC CLOCKS

The geologic time scale based on studies of stratigraphy and faunal successions is a relative time scale. It tells us whether one formation or fossil assemblage is older than another, but not how long the eras, periods, and epochs were in actual years. Estimates of how long it takes mountains to erode and sediments to accumulate suggested that most geologic periods had lasted for millions of years, but geologists of the nineteenth century did not know whether the duration of a specific period was 10 million years, 100 million years, or even longer.

They did know that the geologic time scale was incomplete. The earliest period of geologic history recorded by faunal successions was the Cambrian, when animal life, in the form of shelly fossils, suddenly appeared in the geologic record. Many rock formations were clearly older than the Cambrian period because they occurred below Cambrian rocks in the stratigraphic successions. But these formations contained no recognizable fossils, so there was no way to determine their relative ages, and all such rocks were lumped into the general category "Precambrian." What fraction of Earth's history was locked up in these cryptic rocks? How old was the oldest Precambrian rock? How old was Earth itself?

These questions sparked a huge debate in the latter half of the nineteenth century. Physicists and astronomers argued for a maximum age of less than 100 million years, but most geologists regarded this age as much too young, even though they had no precise data to back themselves up.

The Discovery of Radioactivity

In 1896, a major advance in physics paved the way for reliable and accurate measurements of absolute ages. Henri Becquerel, a French physicist, discovered radioactivity in uranium. Within a year, the French chemist Marie Sklodowska-Curie discovered and isolated a new and highly radioactive element, radium.

In 1905, the physicist Ernest Rutherford suggested that the absolute age of a rock could be determined by measuring the decay of radioactive elements found in it. He calculated the age of one rock by measuring its uranium content. This was the start of **isotopic dating,** the use of naturally occurring radioactive elements to determine the ages of rocks. Within a decade of Rutherford's first attempt, geologists were able to show that some Precambrian rocks were billions of years old.

In 1956, the geologist Clare Paterson measured the decay of uranium in meteorites and terrestrial rocks to determine that the solar system—and, by implication, Earth—was formed 4.56 billion years ago. That age has been modified by less than 10 million years since Paterson's original measurement, so we might say that he completed the discovery of geologic time.

Radioactive Isotopes: The Clocks in Rocks

How do geologists use radioactivity to determine the age of a rock? Recall that the nucleus of an atom consists of protons and neutrons (see Chapter 4). For a given element, the number of protons is constant, but the number of neutrons can vary among different *isotopes* of the same element. Most isotopes are stable, but the nucleus of a *radioactive* isotope can spontaneously disintegrate, or *decay*, emitting particles and transforming the atom into an atom of a different element. We call the original atom the *parent* and the product of decay its *daughter*.

One useful element for isotopic dating is rubidium, which has 37 protons and two naturally occurring isotopes: rubidium-85, which has 48 neutrons and is stable, and rubidium-87, which has 50 neutrons and is radioactive. A neutron in the nucleus of a rubidium-87 atom can spontaneously eject an electron, changing to a proton that stays in the nucleus. The parent rubidium atom thus

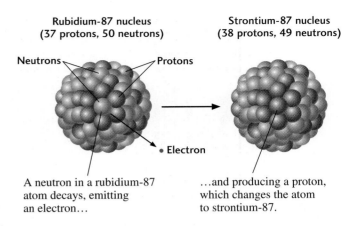

Rubidium-87 nucleus
(37 protons, 50 neutrons)

Strontium-87 nucleus
(38 protons, 49 neutrons)

Neutrons — Protons

• Electron

A neutron in a rubidium-87 atom decays, emitting an electron…

…and producing a proton, which changes the atom to strontium-87.

FIGURE 8.14 Radioactive decay of rubidium to strontium.

forms a daughter strontium-87 atom, with 38 protons and 49 neutrons (**Figure 8.14**).

A parent isotope decays into a daughter isotope at a constant rate. The rate of radioactive decay is measured by the isotope's **half-life**—the time required for one-half of the original number of parent atoms to be transformed into daughter atoms (**Figure 8.15**). Of course, we must know the initial amount of the daughter that was present in a rock to calculate the isotopic age. In the case of rubidium-strontium dating, the initial amount of strontium-87 can be estimated from the amount of strontium-86, a stable isotope that is not the product of radioactive decay and therefore does not change as the mineral ages.

Radioactive isotopes make good clocks because half-life does not vary with

Radioactive isotopes decay at a constant rate to form daughter isotopes. The age of a rock sample can be determined by measuring how much decay has occurred.

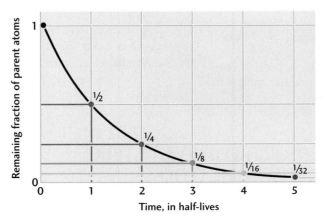

FIGURE 8.15 The fraction of atoms of a radioactive isotope in any mineral declines at a fixed rate over time. This rate of decay is measured by the half-life of the isotope.

the changes in temperature, pressure, chemical environment, or other factors that can accompany geologic processes on Earth or other planets. So when atoms of a radioactive isotope are created anywhere in the universe, they start to act like a ticking clock, steadily transforming from one type of atom to another at a fixed rate.

We can measure the ratio of parent and daughter atoms in a mineral with a mass spectrometer, which can detect even minute quantities of isotopes, and determine how much of the daughter has been produced from the parent. Knowing the half-life, we can then calculate the time elapsed since the radioactive clock began to tick.

The isotopic age of a rock corresponds to the time since the isotopic clock was "reset" when the isotopes were locked into the minerals of the rock. This locking usually occurs when a mineral crystallizes from a magma or recrystallizes during metamorphism. During crystallization, however, the number of daughter atoms in a mineral is not necessarily reset to zero, so the initial amount must be taken into account when calculating isotopic age.

Many other complications make isotopic dating a tricky business. A mineral can lose daughter isotopes by weathering or can be contaminated by fluids circulating in the rock. Metamorphism of igneous rocks can reset the isotopic age of minerals in those rocks to a date much later than the crystallization age.

Isotopic Dating Methods

Isotopic dating is possible only if a measurable number of parent and daughter atoms remain in the sample being dated. For example, if a rock is very old and the decay rate of an isotope is fast, almost all the parent atoms will already have been transformed. In that case, we could determine that the isotopic clock had run down, but we would not be able to say when. Thus, isotopes that decay slowly over billions of years, such as rubidium-87, are most useful in measuring the ages of older rocks, whereas those that decay rapidly, such as carbon-14, can be used to date only younger rocks (**Table 8.1**).

Carbon-14, which has a half-life of 5700 years, is especially useful for dating bone, shell, wood, and other organic materials in sediments less than a few tens of thousands of years old. Carbon is an essential element in the living cells of all organisms. As green plants grow, they continuously incorporate carbon into their tissues from carbon dioxide in the atmosphere. When a plant dies, however, it stops absorbing carbon dioxide. At the moment of death, the ratio of carbon-14 to the stable carbon isotopes in the plant is identical to that in the atmosphere. Thereafter, the ratio of carbon-14 in the dead tissue decays. Nitrogen-14, the daughter isotope of carbon-14, is a gas and thus leaks from the material, so it cannot be measured to determine the time that has elapsed since the plant died. We can, however, estimate the absolute age by comparing the ratio of carbon-14 left in the plant material with the ratio in the atmosphere at the time the plant died.

One of the most precise dating methods for old rocks is based on the decay of two related isotopes: the decay of uranium-238 to lead-206 and the decay of uranium-235 to lead-207. Isotopes of the same element behave similarly in the chemical reactions that alter rocks because the chemistry of an element depends mainly on its atomic number, not its atomic mass. The two uranium isotopes have different half-lives, however (see Table 8.1), so together they provide a consistency check that helps geologists compensate for the problems of weathering, contamination, and metamorphism discussed above.

Crustal minerals more than 4 billion years old have been dated using uranium isotopes.

The lead isotopes from single crystals of zircon—a crustal mineral with a relatively high concentration of

TABLE 8.1	Major Radioactive Elements Used in Isotopic Dating			
Isotopes		**Half-Life of Parent (years)**	**Effective Dating Range (years)**	**Examples of Minerals and Materials That Can Be Dated**
Parent	**Daughter**			
Uranium-238	Lead-206	4.4 billion	10 million–4.6 billion	Zircon, apatite
Uranium-235	Lead-207	0.7 billion	10 million–4.6 billion	Zircon, apatite
Potassium-40	Argon-40	1.3 billion	50,000–4.6 billion	Muscovite, biotite, hornblende
Rubidium-87	Strontium-87	47 billion	10 million–4.6 billion	Muscovite, biotite, potassium feldspar
Carbon-14	Nitrogen-14	5730	100–70,000	Wood, charcoal, peat, bone and tissue, shells and other calcium carbonates

uranium—can be used to date the oldest rocks on Earth with an uncertainty of less than 1 percent. These formations turn out to be more than 4 billion years old.

THE GEOLOGIC TIME SCALE: ABSOLUTE AGES

Armed with isotopic dating techniques, geologists of the twentieth century were able to nail down the absolute ages of the key events on which their predecessors had based the geologic time scale. More important, they were able to explore the early history of the planet recorded in Precambrian rocks. **Figure 8.16** presents the results of this century-long effort.

The assignment of absolute ages to the geologic time scale reveals great differences in the time spans of the geologic periods. The Cretaceous period (80 million years) turns out to be more than three time longer than the Neogene (only 23 million years), and the Paleozoic era (291 million years) lasted longer than the Mesozoic and Cenozoic eras combined. The biggest surprise is the Precambrian, which had a duration of over 400 million years—almost nine-tenths of Earth's history!

Eons: The Longest Intervals of Geologic Time

To represent the rich history of the Precambrian, a division of the geologic time scale longer than the era, called the **eon,** was introduced. Four eons, based on the isotopic ages of terrestrial rocks and meteorites, are now recognized.

Hadean Eon The earliest eon, whose name comes from *Hades* (the Greek word for "hell"), began with the formation of Earth 4.56 billion years ago and ended about 3.9 billion years ago. During its first 660 million years, Earth was bombarded by chunks of material from the early solar system. Although very few rock formations survived this violent period, individual zircon grains 4.4 billion years old have been found, indicating that Earth had a felsic crust within 160 million years of its formation. There is also evidence that some liquid water existed on Earth's surface at about this time, suggesting that the planet cooled rapidly. In Chapter 9, we will explore this early phase of Earth's history in more detail.

Archean Eon The name of the next eon comes from *archaios* (the Greek word for "ancient"). Rocks of Archean

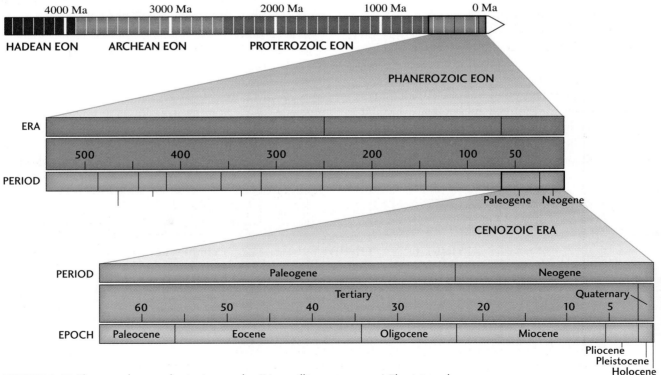

FIGURE 8.16 The complete geologic time scale. (Ma: million years ago.) The intervals labeled "Tertiary" and "Quaternary" are older divisions that have been largely replaced by the Paleogene and Neogene periods but are still sometimes used by geologists.

age range from 3.9 billion to 2.5 billion years old. The geo-dynamo and the climate system were established during the Archean eon, and felsic crust accumulated to form the first stable continental masses, as we will see in the next chapter. The processes of plate tectonics were probably operating, although perhaps substantially differently from the way they did later in Earth's history. Life, in the form of primitive single-celled microorganisms, became established, as indicated by the fossils found in sedimentary rocks of this age.

Proterozoic Eon The last part of the Precambrian is the Proterozoic eon (from the Greek words *proteros* and *zoi*, meaning "earlier life"), which spans the time interval from 2.5 billion to 542 million years ago. By the beginning of this eon, the plate tectonic and climate systems were operating much as they do today. Throughout the Protero-zoic, organisms that produced oxygen as a waste product (as plants do today) increased the amount of oxygen in Earth's atmosphere. Although life remained soft-bodied, some organisms evolved into sophisticated creatures with cells containing nuclei. The increase in atmospheric oxygen to nearly present-day concentrations toward the end of the Proterozoic may have encouraged single-celled organisms to evolve into multicellular algae and animals, which are preserved in the late Proterozoic fossil record.

Phanerozoic Eon The start of the Phanerozoic eon is marked by the first appearance of shelly fossils at the beginning of the Cambrian period, now dated at 542 million years ago. The name of this eon—from the Greek *phaneros* and *zoi* ("visible life")—certainly fits, because it comprises all three eras recognized in the fossil record: the Paleozoic (542 million to 251 million years ago), the Mesozoic (251 million to 65 million years ago), and the Cenozoic (65 million years ago to the present).

> *Precambrian time is divided into three eons: the Hadean, the Archean, and the Proterozoic. The Phanerozoic, a fourth eon, comprises eras since the beginning of the Cambrian period.*

Natural Environment: Clocking the Climate System

The Pliocene and Pleistocene epochs were times of rapid and dramatic global climate change. We can chart these climate changes from the isotopes contained in shelly fossils buried in deep-sea sediments. Deep-sea drilling vessels such as the *JOIDES Resolution* (see Figure 3.13) have taken cores from sedimentary beds around the world's oceans. Geologists can use the carbon-14 dating method to estimate when the shells recovered from these sediment cores were formed, and they can measure the stable isotopes of oxygen to estimate temperature of the seawater in which the shell-producing organisms lived.

The careful tabulation of both temperature and age estimates for many sedimentary layers has provided us with a precise record of global climate during the last 5 million years (**Figure 8.17**). The record shows a general

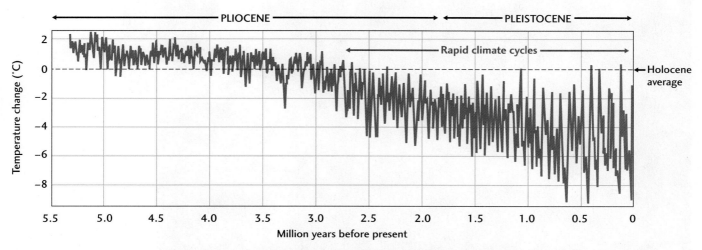

FIGURE 8.17 Changes in Earth's average surface temperature (jagged blue line) during the Pliocene and Pleistocene epochs, measured from temperature indicators in well-dated oceanic sediments. Zero change (dashed black line) corresponds to the average temperature during the Holocene epoch of the last 11,000 years. Note the rapid climate cycles since about 2.7 million years ago. The low temperatures during these cycles correspond to "ice ages." [*Courtesy of L. E. Lisiecki and M. E. Raymo.*]

cooling trend beginning about 3.5 million years ago and the subsequent development of rapid climate cycles that became especially large during the Pleistocene epoch. The low temperatures during these cycles, which were as much as 8°C below the average present-day temperature of Earth's surface, correspond to the Pleistocene "ice ages," when glaciers covered large areas of North America, Europe, and Asia.

Repeated cycles of glaciation have occurred with periods ranging from 40,000 to 100,000 years, and shorter-term cycles lasting a few thousand years or less are also evident. The effects of these climate cycles, such as rises and drops in sea level, can have profound effects on Earth's surface. We will explore these cycles and their causes in more detail in Chapter 10.

Precise dating of deep-sea sediments has allowed geologists to chart changes in the global climate system.

Perspectives on Geologic Time

In the dusty sheep country of far western Australia stands a small promontory of ancient red rocks called the Jack Hills (**Figure 8.18**). Geologists have pulverized truckloads of these rocks to isolate a few sand-sized crystals of zircon. Using the uranium-lead isotopic dating technique, they have found one small crystal fragment with an age of 4.4 billion years—the oldest mineral grain yet discovered in Earth's crust. How can we relate to such a mind-boggling span of time?

Imagine compressing the 4.56 billion years of Earth history into a single year, starting with the formation of Earth on January 1 and ending at midnight on December 31. Within the first week, Earth was organized into core, mantle, and crust. The oldest zircon grain from the Jack Hills crystallized on January 13. The first primitive organisms appeared in mid-March. By mid-June, stable continents had developed, and throughout the summer and early fall, the biological activity of evolving life increased the concentration of oxygen in the atmosphere. On November 18, at the beginning of the Cambrian period, complex organisms, including those with shells, appeared. On December 11, reptiles evolved, and late on Christmas Day, the dinosaurs became extinct. Modern humans, *Homo sapiens*, did not appear on the scene until 11:42 P.M. on New Year's Eve, and the last ice age did not end until 11:58 P.M. Three and a half seconds before midnight, Columbus landed on a West Indies island, and a couple of tenths of a second ago, you were born.

20 µm

FIGURE 8.18 The outcrop in the Jack Hills, in Western Australia, in which geologists have dated zircon grains as old as 4.4 billion years. [*Bruce Watson, Rensselaer Polytechnic Institute.*] *Inset:* Zircon crystals (ZrSiO$_4$) from the Hadean eon extracted from the Jack Hills. [*Chd GFDL.*]

GOOGLE EARTH PROJECT

In the Grand Canyon, in northern Arizona, the Colorado River cuts through sediments deposited over hundreds of millions of years of Earth history. This unimaginably long interval of time is marked by changes in the life-forms preserved in sedimentary rocks. Because plant and animal remains are deposited in and preserved in sediments, the age of a fossil is basically the same as that of the sedimentary bed in which it lies. This relationship, along with basic stratigraphic principles, allows us to determine the relative ages of sedimentary beds. Let's use this spectacular setting as a natural laboratory for understanding geologic time.

LOCATION   Bright Angel Trail, Grand Canyon Visitor Center, Arizona, United States

GOAL   Visualize the Grand Canyon sedimentary sequence

LINKED   Figure 8.12

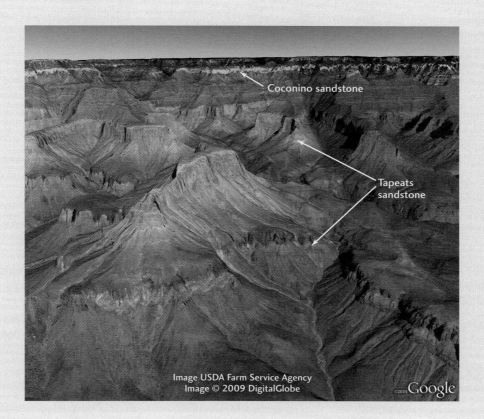

1. Type "Bright Angel Trail, Grand Canyon, Arizona" into the GE search window. Once you arrive there, zoom out to an eye altitude of 10 km. Use the cursor to measure the difference in elevation between the Bright Angel trailhead near the Grand Canyon Visitor Center and the Colorado River directly to the north. Which value best approximates this elevation difference?

   a. 30 m   c. 1300 m
   b. 100 m   d. 2500 m

2. Navigate down the canyon along the Bright Angel Trail from an eye altitude of about 2 km. Tilt your frame of view to the north so that you can gain an oblique view of the north wall of the canyon along the Colorado River. Trace the elevations of rock layers with distinctive

(Continued.)

colors across the landscape. What is the general orientation of the sedimentary rock layers near the surface?

a. Very nearly horizontal

b. Very nearly vertical

c. Tilted at about 45 degrees from the horizontal to the east

d. Tilted at about 45 degrees from the horizontal to the south

3. Viewing the north canyon wall from an eye altitude of about 2 km, locate the thin, white rock layer below the rim of the canyon and the thicker, tan rock layer just above the lowest exposures of red rock visible in the canyon. These formations are the Permian-aged Coconino sandstone and the Cambrian-aged Tapeats sandstone, respectively. Measure the vertical distance between these two formations and refer to Figure 8.12. What is your estimate of the sediment thickness between the two formations, and which geologic periods are missing?

a. 800 m of sediment with both the Ordovician and Silurian periods missing

b. 400 m of sediment with no geologic periods missing

c. 800 m of sediment with the Permian, Cambrian, and Devonian periods missing

d. 200 m of sediment with the Carboniferous period missing

4. Based on the relationship of the layered rock exposed within the canyon walls and the canyon itself, which of the following must have formed first?

a. The layer of rock nearest the bottom of the canyon

b. The layer of rock at the rim of the canyon

c. The Grand Canyon itself

d. The smaller side canyon that the Bright Angel Trail follows

Optional Challenge Question

5. Navigate to the following latitude and longitude along the canyon: 36°10′56″N; 113°06′52″W. View it from an eye altitude of 30 km and zoom in as necessary. Below is a volcanic feature that has produced basaltic lava flows that interact with the Colorado River at the base of the canyon. Based on the principle of superposition and the visible cross-cutting relationships, which of the following sequences of events seems most likely? (It may be helpful to tilt the frame of view to the north along the river to gain a better perspective of the sequence of events.)

a. A volcanic eruption produced lava flows, then the Colorado River cut through the lava flows, and finally layers of sedimentary rocks were deposited on either side of the river channel.

b. Sedimentary rocks were deposited, then a volcanic eruption produced lava flows that covered the sediments, and finally the Colorado River cut through the entire sequence to create a large canyon.

c. Sedimentary rocks were deposited, then the Colorado River cut through them to create a large canyon, and finally a volcanic eruption produced lava flows that flowed into the river.

d. A volcanic eruption produced lava flows on which sedimentary rocks were deposited, then the Colorado River cut through the entire sequence to create a large canyon.

◾ SUMMARY

How do we know whether one rock is older than another? We can determine the relative ages of rocks by studying the stratigraphy, fossils, and cross-cutting relationships of rock formations observed at outcrops. According to Steno's principles, an undeformed sequence of sedimentary beds is horizontal, with each bed younger than the layers beneath it and older than the ones above it. In addition, the fossils found in each bed reflect the organisms that were present when that bed was deposited. Knowing the faunal succession makes it easier to spot unconformities, which indicate gaps in the stratigraphic record where no rock was deposited or where existing rock was eroded away before the next strata were laid down.

How was a global geologic time scale created? By using faunal succession to match up rocks in outcrops around the world, geologists compiled a composite stratigraphic succession from which they developed a relative time scale. The use of isotopic dating allowed them to assign absolute dates to the eons, eras, periods, and epochs that constitute the geologic time scale. Isotopic dating is based on the decay of radioactive isotopes, in which unstable parent atoms are transformed into stable daughter atoms at a constant rate. By measuring the amounts of parent and daughter atoms, geologists can calculate the absolute ages of rocks. The isotopic clock starts ticking when radioactive isotopes are locked into minerals as igneous rocks crystallize or metamorphic rocks recrystallize.

What are the principal divisions of the geologic time scale? The geologic time scale is divided into four eons: the Hadean (4.56 billion to 3.9 billion years ago), Archean (3.9 billion to 2.5 billion years ago), Proterozoic (2.5 billion to 542 million years ago), and Phanerozoic (542 million years ago to the present). The Phanerozoic eon is divided into three eras, the Paleozoic, Mesozoic, and Cenozoic, each of which is divided into shorter periods. The boundaries of the eras and periods are marked by abrupt changes in the fossil record; many correspond to mass extinctions.

Why is the geologic time scale important to geology? The geologic time scale enables us to reconstruct the chronology of events that have shaped the planet. The time scale has been instrumental in studying plate tectonics and in estimating the rates of geologic processes too slow to be monitored directly. The creation of the relative time scale paralleled the development of paleontology and the theory of evolution. The assignment of absolute ages to the time scale revealed that Earth is 4.56 billion years old.

KEY TERMS AND CONCEPTS

absolute age (p. 249)
eon (p. 262)
epoch (p. 255)
era (p. 255)
geologic time scale (p. 255)
half-life (p. 260)
isotopic dating (p. 260)
mass extinction (p. 255)
period (p. 255)
principle of faunal
  succession (p. 251)

principle of original
  horizontality (p. 250)
principle of superposition
  (p. 250)
relative age (p. 249)
stratigraphic succession
  (p. 250)
stratigraphy (p. 250)
unconformity (p. 253)

EXERCISES

1. Many fine-grained muds are deposited at a rate of about 1 centimeter per 1000 years. At this rate, how long would it take to accumulate a sedimentary sequence half a kilometer thick?

2. How many formations can you count in the geologic cross section of the Grand Canyon in Figure 8.12? How many are the same formations observed in Zion Canyon? Are any of the formations observed in both the Grand Canyon and Bryce Canyon cross sections?

3. By comparing the sequence of formations in Figure 8.12 with the relative time scale in Figure 8.11, identify a major disconformity in the Grand Canyon sedimentary sequence. Which periods of geologic time are missing? What is the minimum amount of geologic time, measured in millions of years, that is missing? (*Hint:* Consult Figure 8.12.)

4. What type of unconformity would probably be produced on a continental margin that was broadly uplifted above sea level and then subsided below sea level? What type of unconformity might separate young flat-lying sediments from older metamorphosed sediments?

5. In studying an area of tectonic compression, a geologist discovers a sequence of older, more deformed sedimentary rocks on top of a younger, less deformed sequence, separated by an angular unconformity. What tectonic process might have created the angular unconformity?

6. Mass extinctions have been dated at 444 million years, 416 million years, and 359 million years. How are these events expressed in the geologic time scale of Figure 8.16?

7. A geologist discovers a distinctive set of fish fossils that dates from the Devonian period within a low-grade metamorphic rock. The rubidium-strontium isotopic age of the rock is determined to be only 70 million years. Give a possible explanation for the discrepancy.

8. Explain why the last eon of geologic history is named the Phanerozoic.

9. At the present rate of seafloor spreading, the entire seafloor is recycled every 200 million years. Assuming that the past rate of seafloor generation has been this fast or faster, calculate the minimum number of times the seafloor has been recycled since the end of the Archean eon.

10. Is carbon-14 a suitable isotope for dating geologic events in the Pliocene epoch?

Visual Literacy Task

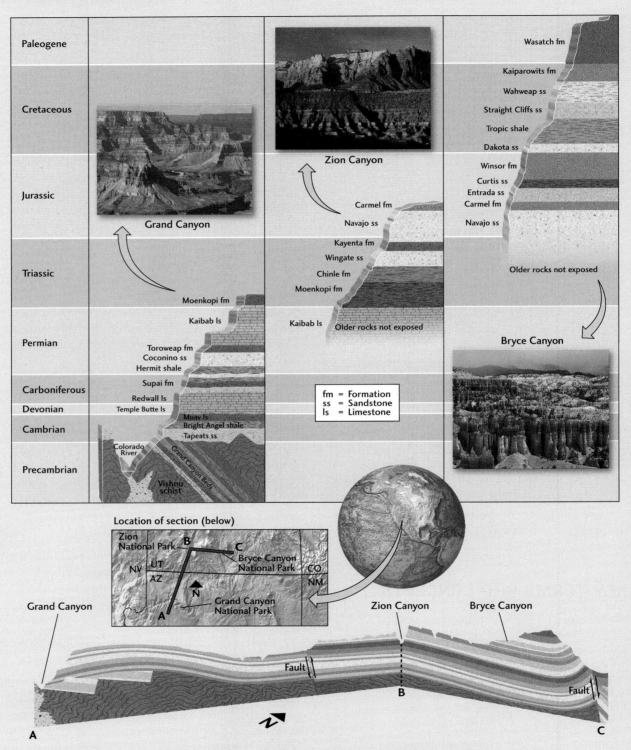

FIGURE 8.12 Stratigraphic sequence of the Colorado Plateau, reconstructed from strata exposed in Grand Canyon, Zion Canyon, and Bryce Canyon National Parks. [*Grand Canyon: John Wang/Photo Disc/Getty Images. Zion Canyon: David Muench/Corbis. Bryce Canyon: Tim Davis/Photo Researchers.*]

1. The photo of the Grand Canyon relates to what letter on the map insert and cross section?

a. A
b. B
c. C

2. When looking at the Bryce Canyon stratigraphic section, how can you easily figure out what rocks are beneath the Navajo sandstone?

a. Dig a hole with a shovel
b. Look at the Grand Canyon stratigraphic section
c. Look at the Zion Canyon stratigraphic section

3. What rock layers can you find at the surface in both the Grand Canyon and Bryce Canyon?

a. Moenkopi formation and Carmel formation
b. Kaibab limestone and Tapeats sandstone
c. Moenkopi formation and Kaibab limestone
d. No rock layers can be found at the surface in both areas

4. What is the order of canyons from highest to lowest stratigraphic elevation?

a. Bryce, Zion, Grand
b. Grand, Bryce, Zion
c. Grand, Zion, Bryce
d. Bryce, Grand, Zion

5. Based on the cross section and photo, how did Zion Canyon form?

a. Rocks were built up in layers along the sides of the canyon.
b. Rocks existed in layers and the canyon formed by erosion by a river.
c. Plate tectonics stretched the area out and the rocks faulted, forming a large valley

Thought Question: Explain how you can tell that the tilting of the Precambrian sedimentary rocks shown in the Grand Canyon section occurred *before* the deposition of the Navajo sandstone, whereas the tilting of the sedimentary formations shown in the cross section at the bottom of the figure occurred *after* the deposition of the Navajo sandstone.

9

History of Earth

Geologists on the Moon

HARRISON "JACK" SCHMITT is the only geologist to have done fieldwork on another celestial body. He was the lunar module pilot aboard the *Apollo 17* spacecraft that landed in the narrow Taurus-Littrow Valley, situated in the ancient highlands on the southeastern rim of Mare Serenitatis—the Moon's lava-filled Sea of Serenity. The mission was near-perfect from its spectacular midnight launch by a 393-foot-tall Saturn V rocket on December 7, 1972, until its splashdown in the South Pacific on December 19.

During their 75 hours in Taurus-Littrow, Schmitt and the mission commander, Gene Cernan, traversed 30 km of the valley floor in a "Moon buggy," investigating the volcanic rocks that partially fill the valley, the boulders that had rolled into the valley from the surrounding mountains, and the impact-generated soils that cover the valley floor. They set up an intricate series of geophysical experiments that measured the properties of the soils,

> . . . lunar highlands were formed very early in the history of the solar system— more than 4.4 billion years ago—and have a common ancestry with the silicate materials that constitute our own planet.

the crust, and the deep lunar interior, and they collected 110 kg of lunar rocks, more than any other Apollo mission. Isotopic dating of the rock samples they and other astronauts returned to Earth demonstrated that the lunar highlands were formed very early in the history of the solar system—more than 4.4 billion years ago—and have a common ancestry with the silicate materials that constitute our own planet. As we will see in this chapter, the Apollo samples provided the key to unlock the early geologic history not only of the Earth-Moon system but also of Mercury, Venus, and Mars.

Schmitt was trained at Caltech and at Harvard, where he received his Ph.D. in geology in 1964. He was recruited into NASA's first group of scientist-astronauts in 1965 while working at the U.S. Geological Survey's Astrogeology Center at Flagstaff, Arizona. He and astrogeologist Eugene Shoemaker, the first director of the Flagstaff center, played key roles in training Apollo crews to be competent geological observers when they were in lunar orbit and on the lunar surface. Throughout the Apollo program, Schmitt participated in the examination and evaluation of the returned lunar samples and helped the crews with their scientific reports.

Jack Schmitt was the last person to leave footprints in the lunar soil. Gene Shoemaker died in a car accident in July 1997 while prospecting for impact craters in central Australia. Shoemaker's ashes were carried aboard the *Lunar Prospector* space probe, an orbiter that mapped the Moon's surface in great detail for over a year. On July 31, 1999, the spacecraft was deliberately crashed into Shoemaker crater near the lunar south pole in a search for ancient ice deposits. None were detected. To date, Shoemaker, the founder of astrogeology, is the only person to have been buried on another celestial body. ◆

Geologist-astronaut Harrison "Jack" Schmitt, *Apollo 17* lunar module pilot, uses an adjustable scoop to retrieve samples of lunar rock on the rim of Camelot Crater in the Taurus-Littrow Valley. [*NASA*.]

TODAY, THE SOLAR SYSTEM is a well-ordered place, with planets moving in stately orbits around the Sun. But the crater-marked surface of the Moon and the occasional meteorite that crashes through Earth's atmosphere remind us of a more disorganized, chaotic time when the solar system was young and Earth's environment was much less hospitable.

Some of the most fascinating scientific questions concern the birth of our planet and its early evolution. How did its rocky mass come together and differentiate into a core, mantle, and crust? Why does Earth's surface, with its blue oceans and wandering continents, look so different from those of its planetary neighbors? And how did that surface evolve and become populated with the myriad species that now constitute our biosphere?

Geologists can draw from many lines of evidence to answer these questions. The rocks of continents preserve a record of geologic processes more than 4 billion years old, and even more ancient materials have been collected from meteorites and by space exploration. Astronauts trained in geology have explored the lunar surface, mapping outcrops and collecting dust and rock samples for analysis back on Earth.

In this chapter, we will explore the 4.5-billion-year history of Earth from the beginning of the solar system to the present day. We will see how Earth and

FIGURE 9.1 Space exploration has progressed from its modest beginnings to address fundamental questions such as the origin of the solar system. (*left*) Robert H. Goddard, one of the fathers of rocketry, fired this liquid oxygen-gasoline rocket on March 16, 1926, at Auburn, Massachusetts. (*right*) Seventy years later, on November 2, 1995, the Hubble Space Telescope (in orbit around Earth) took this stunning photograph of the Eagle Nebula. The dark, pillarlike structures are columns of cool hydrogen gas and dust that give birth to new stars. [(*left*) NASA. (*right*) NASA/ESA/STSci.]

the other planets formed around the Sun and how they differentiated into layered bodies. Then we'll see how tectonic processes began, and how they led to the formation of continents. Finally, we'll chart the evolution of life by examining the geologic record preserved in the ancient rocks of those continents.

ORIGIN OF THE SOLAR SYSTEM

Our search for the origins of the universe—and of our own small part of it—goes back to the earliest recorded mythologies. Today, the generally accepted scientific explanation is the Big Bang theory, which holds that our universe began about 13.7 billion years ago with a cosmic "explosion." Before that moment, all matter and energy were compacted into a single, inconceivably dense point. Although we know little of what happened in the first fraction of a second after time began, astronomers have a general understanding of the billions of years that followed. In a process that still continues, the universe has expanded and thinned out to form galaxies and stars. Geology explores the latter third of that time: the past 4.5 billion years, during which our *solar system*—the star that we call the Sun and the planets that orbit it—formed and evolved. In particular, geologists look to the formation of the solar system to understand Earthlike planets.

> Evidence from astronomy tells us that the universe began with the Big Bang: a cosmic explosion of matter and energy that expanded from a single densely packed point.

The Nebular Hypothesis

In 1755, the German philosopher Immanuel Kant suggested that the origin of the solar system could be traced to a rotating cloud of gas and fine dust, an idea called the **nebular hypothesis.** We now know that outer space beyond our solar system is not as empty as we once thought. Astronomers have recorded many clouds of the type that Kant surmised, and they have named them *nebulae* (plural of the Latin word for "fog" or "cloud") (**Figure 9.1**). They have also identified the materials that form these clouds. The gases are mostly hydrogen and helium, the two elements that make up all but a small fraction of our Sun. The dust-sized particles are chemically similar to materials found on Earth.

How could our solar system take shape from such a cloud? This diffuse, slowly rotating mass contracted under the force of gravity (**Figure 9.2**). Contraction, in turn,

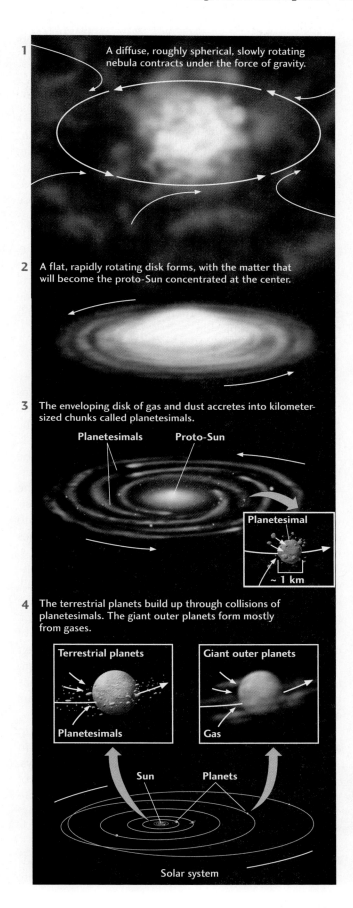

1 A diffuse, roughly spherical, slowly rotating nebula contracts under the force of gravity.

2 A flat, rapidly rotating disk forms, with the matter that will become the proto-Sun concentrated at the center.

3 The enveloping disk of gas and dust accretes into kilometer-sized chunks called planetesimals.

Planetesimals   Proto-Sun

Planetesimal

~ 1 km

4 The terrestrial planets build up through collisions of planetesimals. The giant outer planets form mostly from gases.

Terrestrial planets   Giant outer planets

Planetesimals   Gas

Sun   Planets

Solar system

FIGURE 9.2 The nebular hypothesis explains the formation of the solar system.

accelerated the rotation of the particles (just as ice skaters spin more rapidly when they pull in their arms), and the faster rotation flattened the cloud into a disk.

The Sun Forms

Under the pull of gravity, matter began to drift toward the center of the cloud, accumulating into a protostar, the precursor of our present Sun. Compressed under its own weight, the material in the proto-Sun became dense and hot. The internal temperature of the proto-Sun rose to millions of degrees, at which point nuclear fusion began. The Sun's nuclear fusion, which continues today, is the same nuclear reaction that occurs in a hydrogen bomb. In both cases, hydrogen atoms, under intense pressure and at high temperature, combine (fuse) to form helium. Some mass is converted into energy in the process. The Sun releases some of that energy as sunshine; an H-bomb releases it as an explosion.

The Planets Form

Although most of the matter in the original nebula was concentrated in the proto-Sun, a disk of gas and dust, called the **solar nebula,** remained to envelop it. The temperature of the solar nebula rose as it flattened into a disk. It became hotter in the inner region, where more of the matter accumulated, than in the less dense outer regions. Once formed, the disk began to cool, and many of the gases condensed; that is, they were transformed to their liquid or solid state, just as water vapor condenses into droplets on the outside of a cold glass and water solidifies into ice when it cools below the freezing point.

Gravitational attraction caused the dust and condensing material to clump together (accrete) into small, kilometer-sized chunks, or *planetesimals.* In turn, the planetesimals collided and stuck together, forming larger, Moon-sized bodies (see Figure 9.2). In a final stage of cataclysmic impacts, a few of the larger bodies—with their larger gravitational attraction—swept up the others to form the planets in their present orbits. Planetary formation happened rapidly, probably within 10 million years after the condensation of the nebula.

As the planets formed, those in orbits close to the Sun and those in orbits farther from the Sun developed in markedly different ways. Thus, the composition of the inner planets is quite different from that of the outer planets.

The planets, including Earth, formed from the gas and dust of a nebula that enveloped the proto-Sun.

Inner Planets The four inner planets, in order of closeness to the Sun, are Mercury, Venus, Earth, and Mars (Figure 9.3). They are also known as the Earthlike, or *terrestrial*, planets. In contrast to the outer planets, the four inner planets are small and are made up of rocks and metals. They formed close to the Sun, where conditions were so hot that most of the *volatile* materials (materials that most easily become gases) boiled away. Radiation and matter streaming from the Sun—the solar wind—blew away most of the hydrogen, helium, water, and other light gases and liquids on these planets. Thus, the inner planets were formed mostly from the dense matter that was left behind, which included the rock-forming silicates as well as metals such as iron and nickel. From isotopic dating of the meteorites that occasionally strike Earth and are believed to be remnants of this preplanetary process, we know that the inner planets began to accrete about 4.56 billion years ago (see Chapter 8). Computer simulations indicate that they would have grown to planetary size in a remarkably short time— perhaps as quickly as 10 million years or less.

The four inner planets formed primarily from dense materials such as silicate minerals, iron, and nickel.

Giant Outer Planets Most of the volatile materials swept from the region of the terrestrial planets were carried to the cold outer reaches of the solar system to form the giant outer planets—Jupiter, Saturn, Uranus, and Neptune—and their satellites. The giant planets were big enough and their gravitational attraction strong enough to enable them to hold onto the lighter nebular material. Thus, although they have rocky and metal-rich cores, they, like the Sun, are composed mostly of hydrogen and helium and the other light materials of the original nebula.

Small Bodies of the Solar System

Not all the material from the solar nebula ended up in planets. Some planetesimals collected between the orbits of Mars and Jupiter to form the *asteroid belt* (see Figure 9.3). This region now contains more than 10,000 **asteroids** with diameters larger than 10 km and about 300 larger than 100 km. The biggest is Ceres, which has a diameter of 930 km. Most **meteorites**—chunks of material from outer space that strike Earth—are tiny pieces of asteroids ejected from the asteroid belt during collisions with one another. Astronomers originally thought the asteroids were the remains of a large planet that had broken apart early in the history of the solar system, but it now appears they are pieces that never coalesced into a planet, probably due to the gravitational influence of Jupiter.

Another important group of small, solid bodies are the *comets*, aggregations of dust and ice that condensed in the cooler outer reaches of the solar nebula. Most of the

The inner planets are small and rocky.

The four giant outer planets and their moons are gaseous, with rocky cores.

Pluto is a snowball of methane, water, and rock.

FIGURE 9.3 The solar system. This diagram shows the relative sizes of the planets as well as the asteroid belt separating the inner and outer planets. Though considered one of the nine planets since its discovery in 1930, Pluto was demoted from that status by the International Astronomical Union in 2006. With this revision, there are only eight true planets, not nine.

comets—there are probably many millions of them with diameters larger than 10 km—orbit the Sun far beyond the outer planets, forming concentric "halos" around the solar system. Occasionally, collisions or near misses throw one of them into an orbit that penetrates the inner solar system. We can then observe it as a bright object with a tail of gases blown away from the Sun by the solar wind. Comets are intriguing to geologists because they provide clues about the more volatile components of the solar nebula, including water and carbon-rich compounds, which they contain in abundance.

EARLY EARTH: FORMATION OF A LAYERED PLANET

We know that Earth is a layered planet with a core, mantle, and crust surrounded by a fluid ocean and a gaseous atmosphere (see Chapter 2). How did Earth evolve from a hot, rocky mass into a living planet with continents, oceans, and a pleasant climate? The answer lies in **gravitational differentiation:** the transformation of random chunks of primordial matter into a body whose interior is divided into concentric layers that differ from one another both physically and chemically. Differentiation occurred early in Earth's history, when the planet got hot enough to melt.

Earth Heats Up and Melts

Although Earth probably started out as an accretion of planetesimals and other remnants of the solar nebula, it did not retain this form for long. To understand Earth's present layered structure, we must return to the time when Earth was still subject to violent impacts by planetesimals and larger bodies. As these objects crashed into the primitive planet, most of their energy of motion was converted into heat—another form of energy—and the heat caused melting. A planetesimal colliding with Earth at a typical velocity of 15 to 20 km/s would deliver as much energy as 100 times its weight in TNT. The impact energy of a body the size of Mars colliding with Earth would be equivalent to exploding several trillion 1-megaton nuclear bombs (a single one of which would destroy a large city), enough to eject a vast amount of debris into space and to generate enough heat to melt most of what remained of Earth.

Collision with a Mars-sized body ejected debris that formed the Moon and melted the outer part of Earth.

Many scientists now think that such a cataclysm did occur during the middle to late stages of Earth's accretion. A giant impact by a Mars-sized body created a shower of debris from both Earth and the impacting body and propelled it into space. The Moon aggregated from the debris (**Figure 9.4**). According to this theory, Earth re-formed as a body with an outer molten layer hundreds of kilometers thick—a *magma ocean*. The huge impact sped up Earth's rotation and changed its axis, knocking it from vertical with respect to Earth's orbital plane to its present 23° inclination. All this occurred about 4.51 billion years ago, between the beginning of Earth's accretion (4.56 billion years ago) and the formation of the

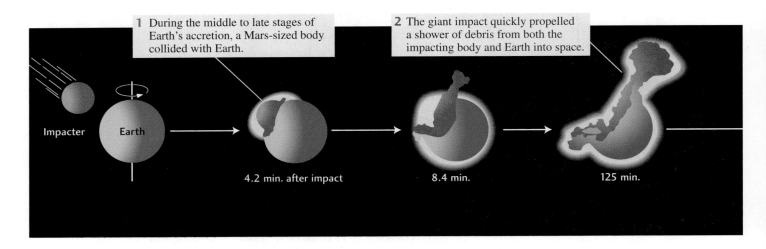

1 During the middle to late stages of Earth's accretion, a Mars-sized body collided with Earth.

2 The giant impact quickly propelled a shower of debris from both the impacting body and Earth into space.

Impacter   Earth

4.2 min. after impact

8.4 min.

125 min.

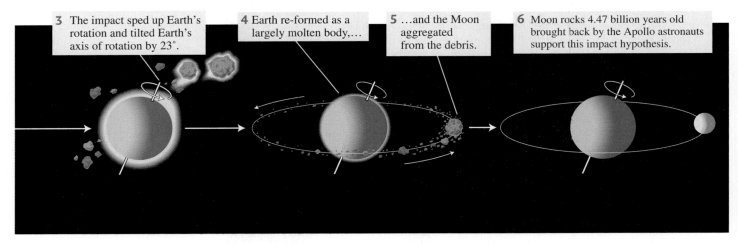

3 The impact sped up Earth's rotation and tilted Earth's axis of rotation by 23°.

4 Earth re-formed as a largely molten body,...

5 ...and the Moon aggregated from the debris.

6 Moon rocks 4.47 billion years old brought back by the Apollo astronauts support this impact hypothesis.

FIGURE 9.4 Computer simulation of the impact of a Mars-sized body on Earth. [*Solid-Earth Sciences and Society. Washington, D.C.: National Research Council, 1993.*]

oldest Moon rocks brought back by the Apollo astronauts (4.47 billion years old).

Another source of heat that contributed to melting early in Earth's history was radioactivity. When radioactive elements decay, they emit heat. Although present in only small amounts, radioactive isotopes of uranium, thorium, and potassium have continued to keep Earth's interior hot.

Differentiation of Earth's Core, Mantle, and Crust

As a result of the tremendous impact energy absorbed during Earth's formation, its entire interior was heated to a "soft" state in which its components could move around. Heavy material sank to become the core, releasing gravitational energy and causing more melting, and lighter material floated to the surface and formed the crust. The rising lighter matter brought interior heat to the surface, where it could radiate into space. In this way, Earth differentiated into a zoned planet with three main layers: a central core, a mantle, and an outer crust (**Figure 9.5**).

Earth's Core Iron, which is denser than most of the other elements, accounted for about a third of the primitive planet's material. The iron and other heavy elements, such as nickel, sank to form a central core, which begins at a depth of about 2890 km. By probing the core with seismic waves, scientists have found that it is molten on the outside but solid in a region called the inner core, which extends from a depth

In the young, hot Earth, iron and nickel sank to form the core, while the less dense materials floated to the surface, forming a primitive crust.

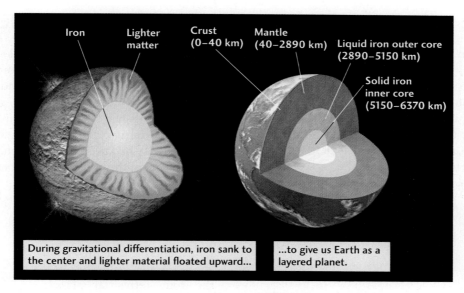

Iron   Lighter matter   Crust (0–40 km)   Mantle (40–2890 km)   Liquid iron outer core (2890–5150 km)   Solid iron inner core (5150–6370 km)

During gravitational differentiation, iron sank to the center and lighter material floated upward...

...to give us Earth as a layered planet.

FIGURE 9.5 Gravitational differentiation of early Earth resulted in a zoned planet with three main layers.

of about 5150 km to Earth's center at about 6370 km. Today the inner core is solid because the pressures at the center are too high for iron to melt.

Earth's Crust Other molten materials were less dense than iron and nickel and floated toward the surface of the magma ocean. There they cooled to form Earth's solid crust, which today ranges in thickness from about 7 km in the oceans to about 40 km on the continents. We know that oceanic crust is constantly generated by seafloor spreading and recycled back into the mantle by subduction. In contrast, continental crust began to accumulate early in Earth's history from silicates of relatively low density with a felsic composition and low melting temperatures. This contrast between dense oceanic crust and less dense continental crust is what helps drive oceanic crust into subduction zones, while continental crust resists subduction.

The 4.4-billion-year-old zircon grains recently found in Western Australia (see Chapter 8) are the oldest terrestrial material yet discovered. Chemical analysis indicates that they formed near the surface under relatively cool conditions and in the presence of water. This finding suggests that Earth had cooled enough for a crust to exist only 100 million years after the planet re-formed following the giant impact that formed the Moon.

Earth's Mantle Between the core and the crust lies the mantle, the layer that forms the bulk of the solid Earth. The mantle is made up of the material left in the middle zone after most of the denser material sank and the less dense material rose toward the surface. It is about 2900 km thick and consists of ultramafic silicate rocks containing more magnesium and iron than crustal silicates do. Convection in the mantle removes heat from Earth's interior (see Chapter 2).

Because the mantle was hotter early in Earth's history, it was probably convecting more vigorously than it does today. Some form of plate tectonics may have been operating even then, although the "plates" were probably much smaller and thinner, and the tectonic features were probably very different from the linear mountain belts and long mid-ocean ridges we now see on Earth's surface. Some scientists think that Venus today provides an analog for these long-vanished processes on Earth. We will compare tectonic processes on Earth and Venus shortly.

Earth's Oceans and Atmosphere Form

The oceans and atmosphere can be traced back to the "wet birth" of Earth itself. The planetesimals that aggregated into our planet contained ice, water, and other volatiles, such as nitrogen and carbon. Originally, these volatiles were locked up in minerals carried by the aggregating planetesimals. As Earth differentiated, water vapor and other gases were freed from these minerals, and released through volcanic activity.

The enormous volumes of gases spewed from volcanoes 4 billion years ago probably consisted of the same substances that are expelled from present-day volcanoes (though not necessarily in the same relative abundances): primarily hydrogen, carbon dioxide, nitrogen, water vapor, and a few others (**Figure 9.6**). Almost all of the hydrogen escaped to outer space, while the heavier gases enveloped the planet. Some of the air and water may also have come from volatile-rich bodies, such as comets, that struck the planet after it had formed. The comets, for example, are composed largely of frozen water (ice) plus frozen carbon dioxide and other gases. Countless comets may have bombarded Earth early in its history, contributing water and gases to the early oceans and atmosphere. The early atmosphere lacked the oxygen that makes up 21 percent of the atmosphere today. Oxygen did not enter the atmosphere until oxygen-producing organisms evolved, as described later in this chapter.

> *Earth's oceans and atmosphere developed when water and gases were released from its interior during differentiation. Water and gases may also have arrived with comets that bombarded Earth after it had already formed.*

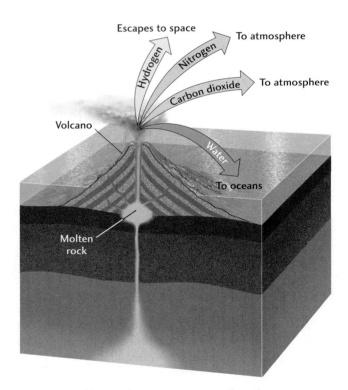

FIGURE 9.6 Early volcanic activity contributed enormous amounts of water vapor, carbon dioxide, and other gases to the atmosphere and oceans. Later, photosynthetic microbes removed carbon dioxide and added oxygen to the primitive atmosphere. Hydrogen, because it is lighter, escaped into space.

DIVERSITY OF THE PLANETS

By about 4.4 billion years ago, in less than 200 million years since its origin, Earth had become a fully differentiated planet. The core was still hot and mostly molten, but the mantle was fairly well solidified, and a primitive crust and continents had begun to develop. Oceans and atmosphere had formed, and the geologic processes that we observe today had been set in motion. But what about the early history of the other terrestrial planets?

Planetary Differentiation and Layering

Information transmitted from space probes indicates that the four terrestrial planets have all undergone differentiation into layered structures with iron-nickel cores, a silicate mantle, and an outer crust.

> *The four terrestrial planets have all undergone differentiation into layered structures with iron-nickel cores, silicate mantles, and an outer crust.*

Mercury's average density is nearly as great as Earth's, even though it is a much smaller planet. Accounting for differences in interior pressure (remember, higher pressures increase density), scientists have surmised that Mercury's iron-nickel core must make up about 70 percent of its mass, a record for solar system planets (Earth's core is only one-third of its mass). Perhaps Mercury lost part of its silicate mantle in a giant impact. Alternatively, the Sun could have vaporized part of its mantle during an early phase of intense radiation. Scientists are still debating these hypotheses.

Venus is close to Earth in mass and size (**Table 9.1**), and its core seems to be about the same size as Earth's. Though Mars is considerably smaller than Earth—only about one-tenth of Earth's mass—the relative size of its core appears to be similar, about half its surface radius.

As stated earlier, the Moon appears to have formed from remnants of Earth's collision with a Mars-sized body (see Figure 9.4). In bulk, the Moon's materials are lighter than Earth's, probably because the heavier matter of the giant impacting body remained embedded in Earth. The lunar core is therefore small, comprising only about 20 percent of the lunar mass.

The Man in the Moon: A Planetary Time Scale

If you look at the face of the Moon through binoculars on a clear night, you will see two distinct types of terrain: rough

TABLE 9.1	Characteristics of the Terrestrial Planets and Earth's Moon				
	Mercury	**Venus**	**Earth**	**Mars**	**Earth's Moon**
Radius (km)	2440	6052	6371	3390	1737
Mass (Earth = 1)	0.06 (3.3×10^{23} kg)	0.81 (4.9×10^{24} kg)	1.00 (6.0×10^{24} kg)	0.11 (6.4×10^{23} kg)	0.01 (7.3×10^{22} kg)
Mean density (g/cm³)	5.43	5.24	5.52	3.93	3.34
Orbit period (Earth days)	88	225	365	687	27
Distance from Sun (millions of km)	58	108	150	228	
Moons	0	0	1	2	0

areas that appear light-colored with lots of big craters and smooth, dark areas, usually circular in shape, where craters are small or nearly absent (**Figure 9.7**). The light-colored regions are the mountainous *lunar highlands*, which cover about 80 percent of the surface. The dark regions are low-lying plains called *lunar maria*, which is Latin for "seas," because they looked like seas to early Earth-bound observers.

The heavily cratered lunar surface we see today is that of a very old, geologically dead body, dating back to a period early in the history of the solar system known as the **Heavy Bombardment,** when crater-forming impacts were very frequent (**Figure 9.8**). In this way, the Moon

FIGURE 9.7 The Moon has two types of terrain: the lunar highlands, with many craters, and the lunar lowlands, or maria, with few craters. The maria appear darker due to the presence of widespread basalts that flowed across their surface over 3 billion years ago. The highlands are lighter because the abundant craters reflect sunlight better. [*NASA/JPL.*]

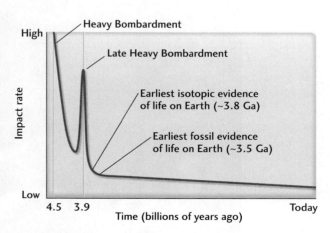

FIGURE 9.8 The number of planetary impacts has varied over the history of the solar system. After the planets formed, they continued to collide with the residual matter that still cluttered the solar system. These collisions tapered off over the first 500 million years of planetary development. However, there was another period of frequent impacts, known as the Late Heavy Bombardment, which peaked around 3.9 billion years ago. (Ga: billion years ago.)

> *The Moon's heavily cratered surface indicates that it is geologically dead; no active geologic processes are changing its surface features.*

contrasts with Earth, where most surfaces older than about 500 million years have been obliterated through the combined activities of the plate tectonic and climate systems.

In preparation for the Apollo missions to the Moon, geologists such as Gene Shoemaker (**Figure 9.9**) developed a relative time scale for the formation of lunar surfaces based on the following simple principles:

■ Craters are absent on a new geologic surface; older surfaces have more craters than younger surfaces.

■ Impacts by small bodies are more frequent than impacts by large bodies; thus, older surfaces have larger craters.

■ More recent impact craters cross-cut or cover older craters.

By applying these principles and by mapping the numbers and sizes of craters—a procedure known as *crater counting*—geologists showed that the lunar highlands are older than the maria. They interpreted the maria to be basins formed by the impacts of asteroids or comets that were subsequently flooded with basalts, which "repaved"

the basins. They were able to assign different parts of the Moon to geologic intervals analogous to those in the relative time scale worked out by nineteenth-century geologists for Earth.

Applying the isotopic dating methods described in Chapter 8 to rock samples brought back by the Apollo astronauts, geologists were able to calibrate an absolute time scale for the Moon's geologic intervals. Sure enough, the highlands turned out to be very ancient (4.4 billion to 4.0 billion years old) and the maria younger (4.0 billion to 3.2 billion years old). **Figure 9.10** plots these ages on the geologic time scale.

The relatively young ages of the maria turned out to be a puzzle, however. The best computer simulations of the Heavy Bombardment indicate that it should have been over rather quickly, perhaps in a few hundred million years or even less. Why, then, did some of the biggest impacts observed on the Moon—those that formed the maria—occur so late in lunar history?

The simulations missed an important event. The rate at which large objects struck the Moon did decrease quickly, as the simulations predicted, but then it spiked up again in a period known as the *Late* Heavy Bombardment, which occurred between about 4.0 billion and 3.8 billion years ago (see Figure 9.8). The explanation of this event is still controversial, but it looks as though small changes in

FIGURE 9.9 Astrogeologist Eugene Shoemaker (at left, pointing with hammer) leads an astronaut training trip on the rim of Meteor Crater, Arizona, in May 1967. (An aerial view of Meteor Crater is shown in Figure 2.7.) [*U.S. Geological Survey.*]

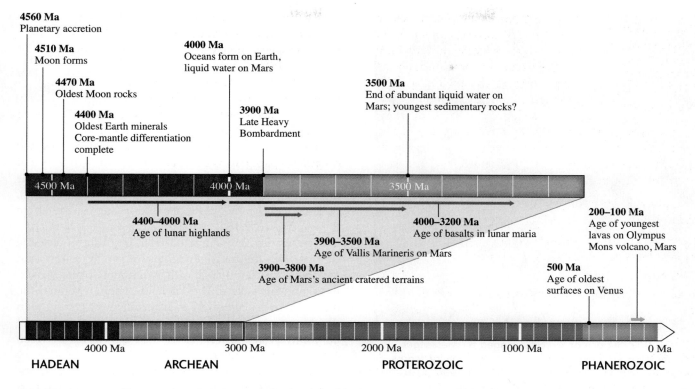

4560 Ma
Planetary accretion

4510 Ma
Moon forms

4470 Ma
Oldest Moon rocks

4400 Ma
Oldest Earth minerals
Core-mantle differentiation
complete

4000 Ma
Oceans form on Earth,
liquid water on Mars

3900 Ma
Late Heavy
Bombardment

3500 Ma
End of abundant liquid water on
Mars; youngest sedimentary rocks?

4500 Ma   4000 Ma   3500 Ma

4400–4000 Ma
Age of lunar highlands

3900–3500 Ma
Age of Vallis Marineris on Mars

3900–3800 Ma
Age of Mars's ancient cratered terrains

4000–3200 Ma
Age of basalts in lunar maria

200–100 Ma
Age of youngest
lavas on Olympus
Mons volcano, Mars

500 Ma
Age of oldest
surfaces on Venus

4000 Ma   3000 Ma   2000 Ma   1000 Ma   0 Ma

HADEAN   ARCHEAN   PROTEROZOIC   PHANEROZOIC

FIGURE 9.10 By calibrating the relative time scale developed by crater counting with the absolute ages of lunar rocks, geologists have constructed a geologic time scale for the terrestrial planets. (Ma: million years ago.)

the orbits of Jupiter and Saturn about 4 billion years ago (caused by their mutual gravitational interactions as they settled into their present orbits) perturbed the orbits of the asteroids. Some of the asteroids were thrown into the inner solar system, where they collided with the Moon and the terrestrial planets, including Earth. The Late Heavy Bombardment explains why so few rocks on Earth have ages greater than 3.9 billion years. It is this event that marks the end of the Hadean eon and the beginning of the Archean eon (see Figure 9.10).

Mercury: The Ancient Planet

The topography of Mercury is poorly understood. *Mariner 10* was the first and only spacecraft to visit Mercury when it flew by the planet in March 1974. It mapped less than half the planet, and we have little idea of what is on the other side.

The *Mariner 10* mission confirmed that Mercury has a geologically dormant, heavily cratered surface. It is the oldest surface of all the terrestrial planets (**Figure 9.11**). Between the large old craters lie younger plains, which are probably volcanic, like the lunar maria. Unlike Earth and Venus, Mercury shows very few features that are clearly due to tectonic forces having reshaped its surface.

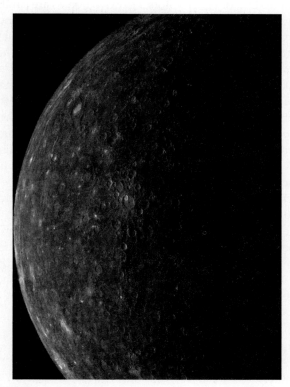

FIGURE 9.11 Mercury has a heavily cratered surface similar to that of Earth's Moon. [*NASA/JPL/Northwestern University.*]

FIGURE 9.12 Surface of Mercury imaged by *Messenger* showing Moon-like impact craters, as well as linear scarps suggestive of faulting. [*NASA/Johns Hopkins University, Applied Physics Laboratory/Carnegie Institution of Washington.*]

In many respects, the face of Mercury seems very similar to that of Earth's Moon. The two bodies are similar in size and mass, and most of their tectonic activity took place within the first billion years of their histories. There is one interesting difference, however. Mercury's face has several scars marked by scarps nearly 2 km high and up to 500 km long (**Figure 9.12**). Such features are common on Mercury, but they are rare on Mars and absent on the Moon. These cliffs appear to have resulted from horizontal compression of Mercury's brittle crust, which formed enormous thrust faults (see Chapter 7). Some scientists think they formed during the cooling of the planet's crust immediately after its formation.

On August 3, 2004, the first new mission to Mercury in 30 years was launched successfully. After 7 years in space, *Messenger* arrived on March 18, 2011, and entered a mapping orbit. *Messenger* will provide information about Mercury's surface composition, its geologic history, and its core and mantle, and it will search for evidence of ice and other frozen gases, such as carbon dioxide, at the poles.

Venus: The Volcanic Planet

Venus is our closest planetary neighbor, often brightly visible just before sunset. Yet in the early decades of space exploration, Venus frustrated scientists. The entire planet is shrouded in a dense fog of carbon dioxide, water vapor, and sulfuric acid, which prevents scientists from studying its surface with ordinary telescopes and spacecraft cameras. It was not until August 10, 1990, after traveling 1.3 billion kilometers, that the *Magellan* spacecraft arrived at Venus and took the first high-resolution pictures of its surface. *Magellan* did this using radar (shorthand for *r*adio *d*etection *a*nd *r*anging), which bounces radio waves off surfaces and can penetrate the fog (**Figure 9.13**).

> *Venus is a geologically active planet with a convecting mantle like Earth's. Its surface shows evidence of tectonic activity, especially volcanism.*

The images *Magellan* returned to Earth showed clearly that beneath the fog Venus is a surprisingly diverse and tectonically active planet with mountains, plains, volcanoes, and rift valleys. The lowland plains of Venus—the blue regions in Figure 9.13—have far fewer craters than the Moon's youngest maria, indicating that they must be younger still. Estimates of their age range between 1600 million and 300 million years.

The young plains are dotted with hundreds of thousands of small volcanic domes 2 to 3 km across and perhaps 100 m or so high, which formed over places where Venus's crust got very hot. There are larger, isolated volcanoes as well, up to 3 km high and 500 km across, similar to the shield volcanoes of the Hawaiian Islands (**Figure 9.14**a). *Magellan* also observed unusual circular features called *coronae* that appear to result from blobs of hot lava that rose, created a large bulge or dome in the surface, and then sank, collapsing the dome and leaving a wide ring that looks like a fallen soufflé (Figure 9.14b).

Because Venus has so much evidence of widespread volcanism, it has been called the Volcanic Planet. Venus has a convecting mantle like Earth's, in which hot material rises and cold material sinks (see Figure 2.16), but unlike Earth, it does not appear to have thick plates of rigid lithosphere. Instead, only a thin crust of frozen lava overlies the convecting mantle. As the convection currents push and stretch the surface, the crust breaks up into flakes or crumples like a rug, and blobs of hot lava bubble up to form large landmasses and volcanic deposits. Scientists have called this type of geology *flake tectonics*. When Earth was younger and hotter, it is possible that flakes, rather than plates, were the main expression of its tectonic activity.

Mars: The Red Planet

Of all the planets, Mars has a surface most similar to Earth's. Mars has features suggesting that liquid water

FIGURE 9.13 This topographic map of Venus is based on more than a decade of mapping, culminating in the 1990–1994 *Magellan* mission. Regional variations in elevations are illustrated by the highlands (tan colors) and the lowlands (blue and green colors). Vast lava plains are found in the lowlands. [*NASA/USGS.*]

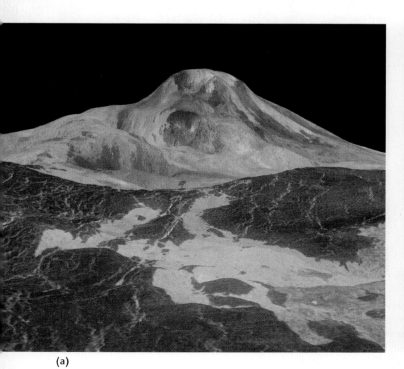

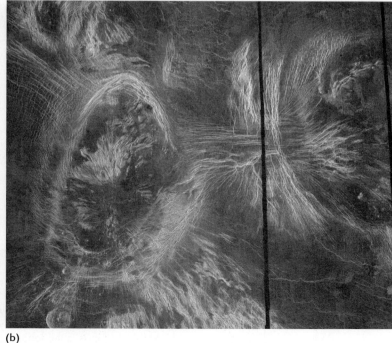

(a)   (b)

FIGURE 9.14 Venus is a tectonically active planet with many surface features. (a) Maat Mons, a volcanic mountain that may be up to 3 km high and 500 km across. (b) Volcanic features called coronae are not observed on any other planet except Venus. The visible lines that define the coronae are fractures, faults, and folds produced when a large blob of hot lava collapsed like a fallen soufflé. Each corona is a few hundred kilometers across. [*Images from NASA/USGS.*]

once flowed across its surface, and liquid water may still exist in its deep subsurface. And where there is water, there may be living organisms. No other planet in our solar system has as much chance of harboring extraterrestrial life as Mars.

The abundance of iron oxide minerals on the surface of Mars gives the Red Planet its name. Iron oxide minerals are common on Earth and tend to form where weathering of iron-bearing silicates occurs. We now know that many other minerals common on Earth, such as olivine and pyroxene, which form in basalt, are also present on Mars. But there are other relatively unusual minerals, such as sulfates, that record an earlier, wetter phase of the Martian climate, when liquid water may have been stable.

The topography of Mars shows a greater range of elevation than that of Earth or Venus (see Figure 9.13). Olympus Mons, at 25 km high, is a giant, recently active volcano—the tallest mountain in the solar system (**Figure 9.15**a). The Vallis Marineris canyon, 4000 km long and averaging 8 km deep, stretches the distance from New York to Los Angeles and is five times deeper than the Grand Canyon (Figure 9.15b).

Recently, geologists have discovered evidence of past glacial processes, when ice sheets similar to the ones that covered North America during the most recent ice age, flowed across the surface of Mars.

Finally, like the Moon, Mercury, and Venus, Mars has ancient cratered terrain that preserves a record of the Late Heavy Bombardment and is therefore likely to be older than 3.9 billion years (see Figure 9.10). But younger surfaces are also widespread on Mars, occupying much of its northern hemisphere. Some are paved with volcanic lavas, as on Venus, whereas others are covered with sediments and landslide deposits. The largest volcanoes, including Olympus Mons, appear to have erupted during the last 100 million years and may still be active. Our understanding of the surface processes

Parts of the Martian surface consist of ancient cratered terrain, but other parts are covered with lavas and sediments.

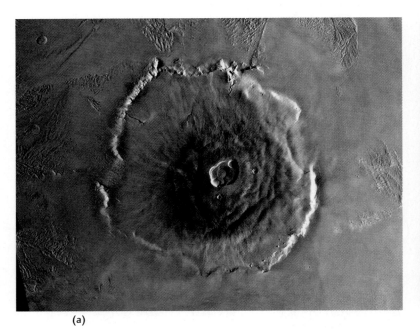

(a)

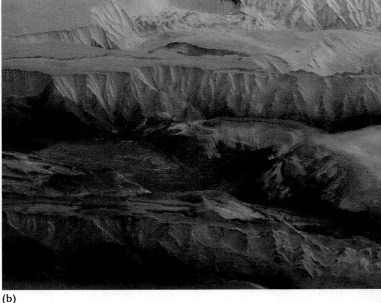

(b)

FIGURE 9.15 The topography of Mars shows a large range of elevations. (a) Olympus Mons is the tallest volcano in the solar system, with a summit almost 25 km above the surrounding plains. Encircling the volcano is an outward-facing scarp 550 km in diameter and several kilometers high. Beyond the scarp is a moat filled with lava, most likely derived from Olympus Mons. (b) Vallis Marineris is the longest (4000 km) and deepest (up to 10 km) canyon in the solar system. It is five times deeper than the Grand Canyon. In this image, the canyon is exposed as a series of fault basins whose sides have partially collapsed (such as at upper left), leaving piles of rock debris. The walls of the canyon are 6 km high here! The layering of the canyon walls suggests deposition of sedimentary or volcanic rocks prior to faulting. [(a) NASA/USGS. (b) ESA/DLR/FU Berlin.]

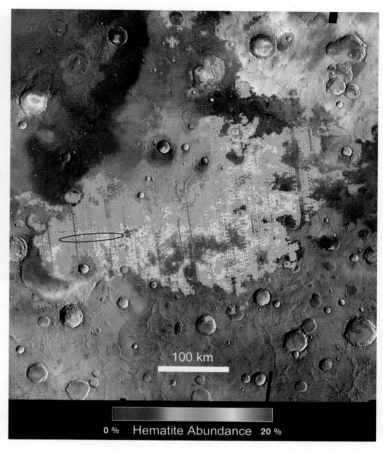

0 %   Hematite Abundance   20 %

FIGURE 9.16 Mars Exploration Rover landing sites. (*left*) *Spirit* has explored Gusev Crater, about 160 km in diameter, which is thought to have been filled with water, forming an ancient lake. A channel that might have supplied water to the crater is visible at the lower right. (*right*) *Opportunity* was sent to an area of Meridiani Planum where hematite—a mineral that often forms in water on Earth—is abundant. The image shows concentrations of hematite. The ellipse outlines the permissible landing area. [(*left*) NASA/JPL/ASU/MSSS. (*right*) NASA/ASU.]

on Mars dramatically improved when two golf-cart-sized robots landed on Mars in January 2004 (**Figure 9.16**).

The Mars Exploration Rovers, named *Spirit* and *Opportunity*, began their 300-million-kilometer journey from Cape Canaveral, Florida, to the Red Planet in June 2003, accompanied by the *Mars Express*, an orbiter equipped with geologic remote sensing tools. These missions succeeded beyond anyone's expectations, making 2004 and 2005 two of the greatest years in the history of space exploration.

Spirit failed to survive its fourth winter on Mars, and after repeated attempts to regain communication for almost a year, it was officially declared deceased in May 2011. At that time, both rovers had traveled more than 30 kilometers. As of this writing, *Opportunity* was just arriving at Endeavour Crater, where it was to explore the contact between very ancient rocks of Mars and the younger sedimentary sequence it had been traveling over for the past seven years.

One of the most exciting findings has been the discovery of a vertical succession of sedimentary rocks by the rover *Opportunity* at a site called Endurance crater (**Figure 9.17**), the first of its kind on another planet. These sediments accumulated in a shallow lake or sea, and the sulfate minerals they contain precipitated as water evaporated. This site has provided information similar to stratigraphic successions observed on Earth (compare, for example, Figure 6.18).

Another new orbiter, *Mars Reconnaissance Orbiter*, which started its mission in 2006, has collected a vast set of observations that show evidence of aqueous processes over broad regions of the planet. Its camera is taking stunning pictures of the surface of Mars at unprecedented resolution (25 cm/pixel). The *Phoenix* lander conducted operations in the polar region of Mars from June to November of 2008 and confirmed the presence of water ice just a few centimeters below the dusty surface.

The data collected by all these Mars missions provide compelling evidence for liquid

Evidence that liquid water once flowed on the Martian surface includes sulfide mineral deposits, deep canyons, and sedimentary sequences that accumulated in a shallow lake or sea.

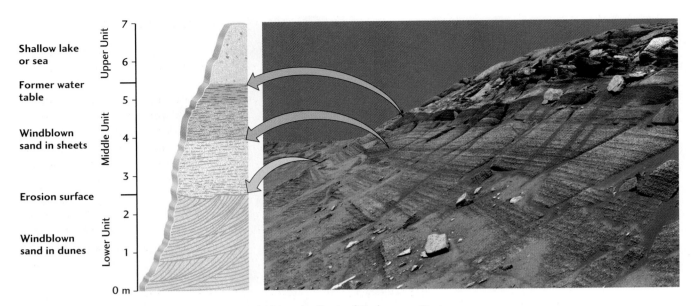

7 — Upper Unit
Shallow lake or sea
6 —
Former water table
5 —
Windblown sand in sheets — Middle Unit
4 —
3 —
Erosion surface
2 —
Windblown sand in dunes — Lower Unit
1 —
0 m —

FIGURE 9.17 Sedimentary sequence exposed along the flank of Endurance Crater, photographed by the rover *Opportunity.* (*left*) An interpretive drawing showing each stage in the history of the outcrop. (*right*) The vertical succession of layers in the outcrop preserves an excellent record of early Martian environments. [*NASA/JPL/Cornell.*]

water on the Martian surface at some point in the planet's history. But many mysteries remain. How much water was there? How long did it last? Did it ever rain, or was all the evidence produced by groundwater leaking to the surface? Did the water last long enough, and have the right composition, to allow life to get started? Only one thing is certain at this point: more missions will be required to answer these questions. One of those missions is preparing to launch in November, 2011, and is the largest rover ever built. Named *Curiosity*, this rover is now undergoing

final construction at the Kennedy Space Flight Center in Florida, prior to its planned launch on November 25, 2011. If all goes well, it will be on the surface of Mars at a site called Gale Crater in August 2012. It has all the tools required to address several of these questions, including the search for environments that might have been habitable for microorganisms. *Curiosity* is as large as a car and weighs almost one ton (**Figure 9.18**). (One of the authors of this textbook, John Grotzinger, is the Chief Scientist for the Mars Science Laboratory mission.)

FIGURE 9.18 The rover *Curiosity* is scheduled to land on Mars in August 2012. *Curiosity* weighs almost a ton, is as large as a car, and carries ten instruments that will make scientific measurements. With 17 cameras on board, it is the most capable rover ever sent to Mars. [*Jet Propulsion Laboratory.*]

EVOLUTION OF THE CONTINENTS

In contrast to the cratered face of the Moon, the exterior of our home planet preserves no vestiges of its fiery beginning. Evidence of the Heavy Bombardment has been obliterated by the resurfacing processes of the plate tectonic and climate systems. As we saw in Chapter 3, nearly two-thirds of Earth's surface—its entire oceanic crust—was created by seafloor spreading over the past 200 million years, an interval spanning a mere 4 percent of Earth's history. The story of all earlier events is entirely contained in the continental crust.

Although the oldest crustal minerals are older than the Late Heavy Bombardment (see Chapter 8), indicating that some type of continental crust existed in the Hadean eon, the main episodes of continent building occurred after this violent period. In this section, we describe the structure of the continents and the processes that formed them—and are still forming them today.

The North American Continent

Continents, like people, show a great variety of features that reflect their parentage and experience over time. Yet, also like people, continents share many similarities in their basic structure and growth patterns. Let's begin by outlining the major features of one particular continent: North America. The long-term tectonic evolution of the continent is reflected in its *tectonic provinces*—large-scale regions formed by a particular set of tectonic processes (**Figure 9.19**).

Canadian Shield The oldest parts of North America's crust, built during the most ancient episodes of deformation, tend to be found in the northern interior of the continent. This huge region was named the *Canadian Shield* by nineteenth-century geologists because the crystalline rocks that form the upper crust ("basement" rocks) emerge from the surrounding sediments like a shield

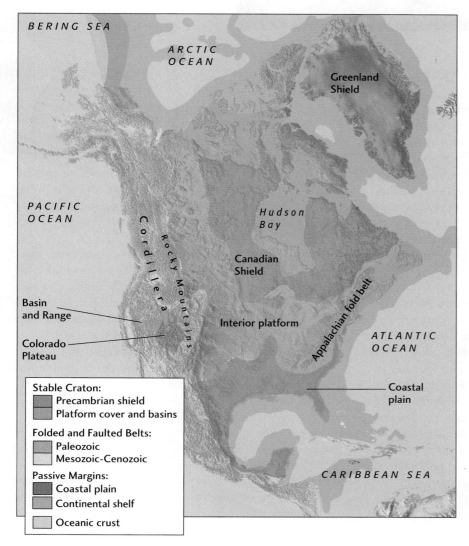

FIGURE 9.19 The major tectonic provinces of North America reflect the processes that formed the continent.

FIGURE 9.20 Aerial view of ancient, eroded metamorphic rocks in Nunavut, Canada, exposed on the surface of the Canadian Shield. [*Roy Tanami/Ursus Photography.*]

partially buried in the dirt of a battlefield. The Canadian Shield, which includes most of Canada and the closely connected landmass of Greenland, is *tectonically stable.* In other words, it has remained largely undisturbed by rifting, continental drift, or continent-continent collision throughout the Phanerozoic, but it has been eroded nearly flat (**Figure 9.20**). It consists primarily of Precambrian granitic and metamorphic rocks, such as gneisses, together with highly deformed and metamorphosed sedimentary and volcanic rocks, and it contains major deposits of iron, gold, copper, diamond, and nickel. Large portions of the shield were formed during the Archean eon, representing one of the oldest records of Earth's history.

Natural Resources: Archean Greenstone Belts The Canadian Shield is famous for its very abundant mineral resources. One important type of mineral deposit occurs in distinctive geologic units called "greenstone belts," where Archean-age island arcs were stitched together to form vast volcanic terrains composed dominantly of metamorphosed mafic rocks. The greenstones are composed of minerals with a green hue, such as chlorite, actinolite, and other green amphiboles. These mineral deposits formed where hydrothermal fluids circulated and precipitated a broad range of economically important minerals such as gold, silver, copper, zinc, lead, and molybdenum.

The Abitibi greenstone belt—one of the world's largest—contains an enormous volcanic caldera complex about 2.7 billion years old. The past 80 years of exploration have shown that the Abitibi belt contains over 83 massive sulfide deposits that collectively contain over 730 million tons of in-ground metal. **Figure 9.21** shows the distribution of rocks in the Abitibi greenstone belt, including significant gold deposits, localized along major faults.

> The Canadian Shield is a vast tectonic province of very old Precambrian granitic and metamorphic rocks that have been eroded nearly flat.

Interior Platform Extensive flat-lying (*platform*) sediments have been deposited on stable continental crust around the periphery of the Canadian Shield and also near its center, beneath Hudson Bay (see Figure 9.19). The vast, low-lying, sediment-covered region south and west of the Canadian Shield, which includes the Great Plains of Canada and the United States, is called the *interior platform.* The Precambrian basement rocks of the interior platform are a continuation of the Canadian Shield, although here they lie under nearly flat layers of Paleozoic sedimentary rocks, typically less than 2 km thick.

Within the interior platform are broad sedimentary basins, roughly circular or oval depressions where the sediments are thicker than in the surrounding areas (**Figure 9.22**). Most of these features are thermal subsidence basins, that is, regions that subsided when heated portions of the lithosphere cooled and contracted (see Chapter 6).

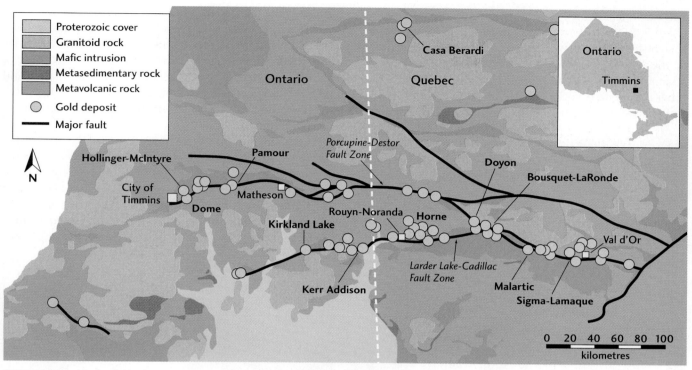

FIGURE 9.21 Map of Abitibi Greenstone Belt, showing preferred location of gold mineralization along major faults that cut metamorphosed volcanic and sedimentary rocks. Hot fluids circulated along these faults, concentrating precious metals from these rocks.

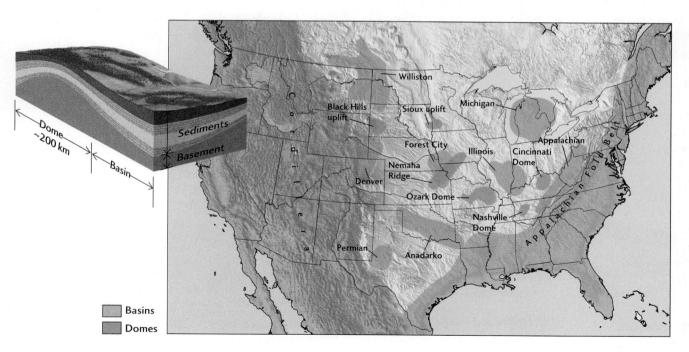

FIGURE 9.22 Map of the interior platform of North America, showing its basin and dome structure. The basins are nearly circular regions of thick sediments, whereas the domes are regions where the sediments are anomalously thin. Basement rocks are exposed on the tops of some domes, such as the Black Hills and the Ozark Dome.

An example is the Michigan Basin, a circular area of about 200,000 km^2 that covers most of the Lower Peninsula of Michigan. This basin subsided throughout much of the Paleozoic era and received sediments more than 5 km thick in its central, deepest part. The sandstones and other sedimentary rocks of these basins, laid down under tectonically quiet conditions, have remained unmetamorphosed and only slightly deformed to this day. The interior platform basins contain important deposits of uranium, coal, oil, and natural gas.

> In the interior platform, the rocks of the Canadian Shield are overlain by Paleozoic sediments.

Appalachian Fold Belt On the edges of these older tectonic provinces are younger metamorphic belts where most of the present-day mountain chains are found. These mountain chains form elongated topographic features near the margins of the continent. The two main examples are the *Cordillera*, which runs down the western edge of North America and includes the Rocky Mountains, and the *Appalachian fold belt*, which runs southwest to northeast on the continent's eastern margin from Alabama to Newfoundland.

The rock assemblages and structures of the old, eroded Appalachian Mountains resulted from the continent-continent collisions that formed the supercontinent Pangaea 470 million to 270 million years ago. The western side of the Appalachians is bounded by the *Allegheny Plateau*, a region of slightly uplifted, mildly deformed sediments that is rich in coal and oil. Moving east-

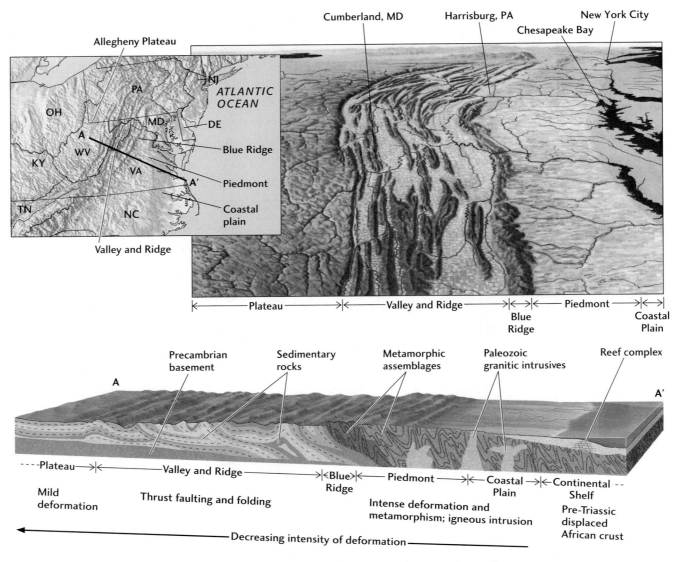

FIGURE 9.23 Appalachian Mountain fold belt province, shown in an aerial view to the northeast and an idealized cross section. The major intensity of deformation increases from west to east.
[*After S. M. Stanley, Earth System History. New York: W. H. Freeman, 2008. Aerial view from NASA.*]

ward, we encounter regions of increasing deformation (**Figure 9.23**):

■ The folded and faulted Paleozoic sedimentary rocks of the *Valley and Ridge province*

■ The eroded mountains of the *Blue Ridge province*, composed of highly metamorphosed Precambrian and Cambrian rocks, which were thrust as sheets over the sedimentary rocks of the Valley and Ridge province

■ The hilly *Piedmont*, containing Precambrian and Paleozoic metamorphosed sedimentary and volcanic rocks, intruded by granite and thrust over Blue Ridge rocks

■ The modern *coastal plain*, made up of relatively undisturbed sediments of Jurassic age and younger, underlain by rocks similar to those of the Piedmont

> The Appalachian fold belt was formed by continent-continent collisions during the assembly of Pangaea.

North American Cordillera The stable interior platform of North America is bounded on the west by a younger complex of mountain ranges and deformation belts. This region is part of the North American Cordillera, a mountain belt that extends the length of North America from Alaska to Guatemala. Across its middle section, between San Francisco and Denver, the Cordilleran system is about 1600 km wide and includes several different tectonic provinces: the Coast Ranges along the Pacific Ocean; the lofty Sierra Nevada; the Basin and Range province; the high tableland of the Colorado Plateau; and the rugged Rocky Mountains, which end abruptly at the edge of the Great Plains on the stable interior platform (**Figure 9.24**).

The history of the Cordillera is a complicated one that involves interactions among the Pacific, Farallon, and North American plates over the past 200 million years. Before the breakup of Pangaea, the Farallon Plate occupied most of the eastern Pacific Ocean. As North America moved westward, most of this plate's oceanic lithosphere was subducted eastward under the continent. The westward margin of the continent swept up island arcs and continental fragments, and the subduction zone eventually swallowed portions of the Pacific-Farallon spreading center, which converted the convergent boundary into the modern San Andreas transform-fault system

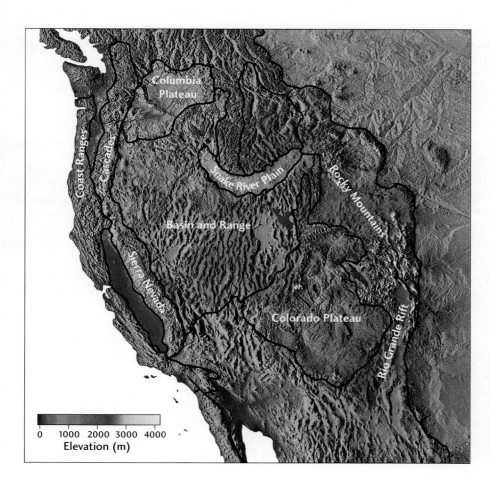

FIGURE 9.24 Topography of the North American Cordillera in the western United States. Computer manipulation of digitized elevation data produced this color shaded-relief map. The major tectonic provinces of the area are clearly visible, as if illuminated by a light source low in the west.

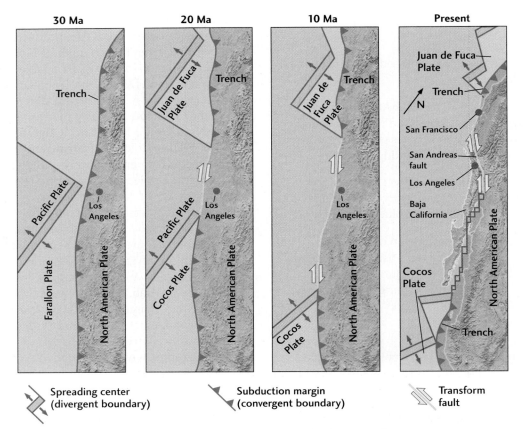

FIGURE 9.25 Interaction of the western coast of North America with the shrinking Farallon Plate as it was progressively subducted beneath the North American Plate, leaving the present-day Juan de Fuca and Cocos plates as small remnants. Large solid arrows show the present-day direction of relative movement between the Pacific and North American plates. (Ma: million years ago.) [*After W. J. Kious and R. I. Trilling, This Dynamic Earth: The Story of Plate Tectonics. Washington, D.C.: U.S. Geological Survey, 1996.*]

(**Figure 9.25**). Today, all that is left of the Farallon Plate are small remnants, including the Juan de Fuca and Cocos plates, which are still subducting beneath North America.

The *Basin and Range* province developed through the uplift and stretching of the crust in a northwest-southeast direction (see Figure 9.24). This extension began with the heating of the lithosphere by upwelling convection currents about 15 million years ago and continues to the present day (see Chapter 7). It has resulted in a wide zone of normal faulting extending from southern Oregon to Mexico and from eastern California to western Texas. The Basin and Range province is volcanically active and contains extensive hydrothermal deposits of gold, silver, copper, and other valuable metals. The Wasatch Range of Utah and the Teton Range of Wyoming (**Figure 9.26**) are being uplifted on the eastern edge of the Basin and Range province, while the Sierra Nevada of California is being uplifted and tilted on the province's western edge.

The *Colorado Plateau* seems to be an island of stability that has experienced no major extension or compression

since Precambrian time (see Figure 9.24). The broad uplift of the plateau has allowed the Colorado River to cut through flat-lying rock formations, creating the Grand Canyon. Geologists believe that this uplift was caused by the same type of lithospheric heating that led to stretching of the crust in the Basin and Range province.

> *The North American Cordillera was formed during the subduction of the Farallon Plate beneath the North American continent.*

Coastal Plain The Atlantic coastal plain and the offshore continental shelf (see Figure 9.19) are sediment-covered regions underlain by crust that was extended by normal faulting during the Triassic-Jurassic splitting of North America from Africa and the subsequent opening of the modern North Atlantic Ocean about 180 million years ago. The opening rift valleys formed sedimentary basins that trapped a thick series of nonmarine sediments. In

FIGURE 9.26 Image synthesized from the satellite data of the Teton Range, Wyoming. The sharp eastern face of the mountain range, which has a vertical relief of more than 2000 m, is the result of normal faulting along the northeastern edge of the Basin and Range province. The view is from the northeast looking to the southwest. Grand Teton, the mountain near the center of the image, rises to an altitude of 4200 meters. [NANA/Goddard Space Flight Center, Landsat 7 Team.]

the early Cretaceous period, the deeply eroded, sloping surface of the Atlantic coastal plain and continental shelf began to subside, accumulating sediments as much as 5 km thick.

The coastal plain and continental shelf of the Gulf of Mexico are similar to their Atlantic equivalents, separated from them by the Florida Peninsula, a large carbonate platform. The Mississippi, Rio Grande, and other rivers that drain the interior of the North American continent have delivered enough sediments to fill a basin some 10 to 15 km deep running parallel to the Gulf Coast. The Gulf coastal plane and shelf are rich reservoirs of oil and natural gas.

> *The Atlantic coastal plain and continental shelf were areas of rifting during the breakup of Pangaea and have since subsided.*

Tectonic Provinces Around the World

We will now expand our view from North America to Earth's other continents. Each continent has its own distinctive features, but a general pattern becomes evident when continental geology is viewed on a global scale (**Figure 9.27**). Continental shields and platforms make up the most stable parts of the continental lithosphere, called **cratons,** and contain the eroded remnants of ancient deformed rocks. The North American craton, for instance, comprises the Canadian Shield and the interior platform (see Figure 9.19).

Around these cratons are elongated mountain belts that were formed by later episodes of compressive deformation. The youngest orogenic (mountain-building) systems, such as the North American Cordillera, are found along the *active margins* of continents, where tectonic activity caused by relative plate movement continues to deform the continental crust.

The *passive margins* of continents—those that are attached to oceanic crust as part of the same plate and thus are not near plate boundaries—are zones of extended crust, stretched during the rifting that broke older continents apart and initiated seafloor spreading. This rifting often occurred along older mountain belts, such as the Appalachian fold belt.

The current distribution of continental tectonic provinces and their ages is like a giant puzzle in which the original pieces have been rearranged and

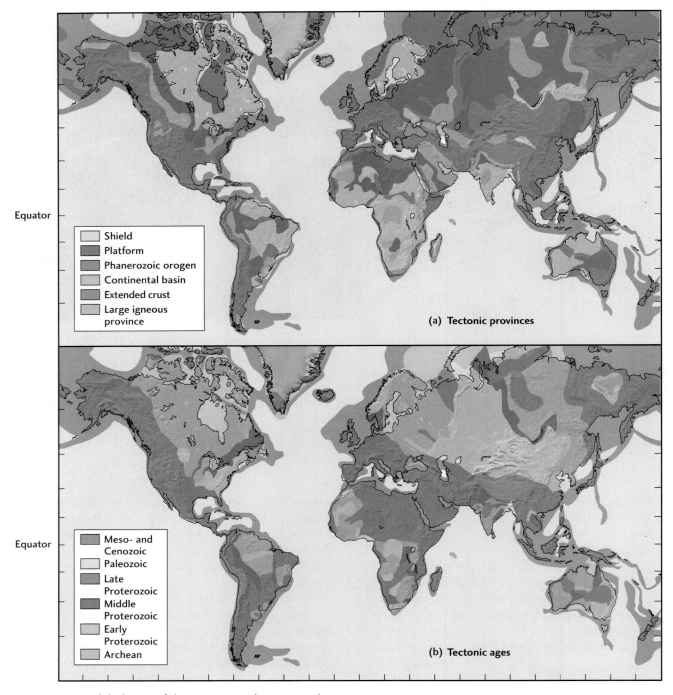

Shield
Platform
Phanerozoic orogen
Continental basin
Extended crust
Large igneous province

(a) Tectonic provinces

Meso- and Cenozoic
Paleozoic
Late Proterozoic
Middle Proterozoic
Early Proterozoic
Archean

(b) Tectonic ages

Equator

FIGURE 9.27 Global view of the continents, showing (a) their major tectonic provinces and (b) their tectonic ages. [*W. Mooney/USGS.*]

reshaped by continental rifting, continental drift, and continent-continent collisions over billions of years. Only the past 200 million years of plate movements can be reliably determined from existing oceanic crust (see Chapter 3). Earlier plate movements must be inferred from the indirect evidence found in continental rocks. From this evidence, we can gain insights into several key questions about continental evolution: What geologic processes built the continents we see today? How do these processes fit into the theory of plate tectonics? Can plate tectonics explain the formation of the cratons?

We will illustrate the answers to these questions using the evolution of North America as our prime example,

Continents are built by orogenic activity around stable parts of the continental lithosphere called cratons.

starting with its youngest provinces along its west coast and working backward in time to the Canadian Shield.

How Continents Grow

Over its 4-billion-year history, new crust has been added to the western coast of North America at an average rate of about 2 km³ per year. In the modern plate tectonic system, two basic processes work together to form new continental crust: magmatic addition and accretion.

Magmatic Addition The process of magmatic differentiation of low-density, silica-rich rock in Earth's mantle and *vertical transport* of this buoyant, felsic material from the mantle to the crust is called **magmatic addition.** Most new continental crust is born in subduction zones from magmas formed by fluid-induced melting of the subducting lithospheric slab and the mantle wedge above the slab (see Chapter 5). These magmas, which are of basaltic to andesitic composition, migrate toward the surface, pooling in magma chambers near the base of the crust. Here they incorporate crustal materials and differentiate further to form the felsic magmas that migrate into the upper crust, forming dioritic and granodioritic plutons capped by andesitic volcanoes.

Magmatic addition can emplace new crustal material directly at active continental margins. Subduction of the Farallon Plate beneath North America during the Cretaceous period, for example, created the batholiths along the western edge of the continent, including the rocks now exposed in Baja California and the Sierra Nevada. Subduction of the remnant Juan de Fuca Plate continues to add new material to the crust in the volcanically active Cascade Range of the Pacific Northwest, just as subduction of the Nazca Plate is building up the crust in the Andes of South America.

> Magmatic addition occurs primarily in subduction zones as mantle material undergoes fluid-induced melting and silica-rich magmas rise into the crust.

Accretion The integration of crustal material previously differentiated from mantle material into existing continental masses by *horizontal transport* during plate movements is called **accretion.** Geologic evidence for accretion can be found on the active margins of North America. In the Pacific Northwest and Alaska, the crust consists of a mix of odd pieces—island arcs, seamounts (extinct underwater volcanoes) and remnants of thickened oceanic plateaus, old mountain ranges and other slivers of continental crust—that were plastered onto the leading edge of the continent as it moved across Earth's surface (**Figure 9.28**). These pieces are sometimes referred to as *accreted terrains*.

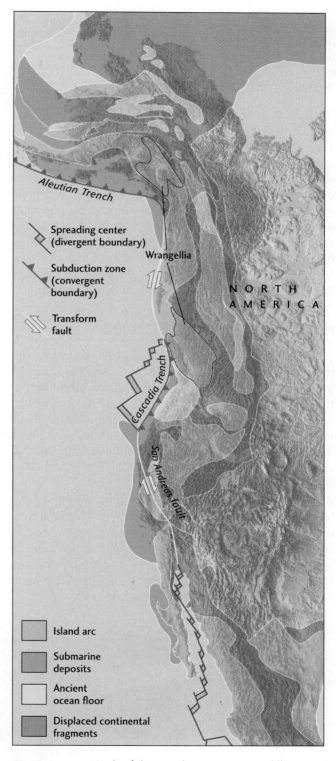

FIGURE 9.28 Much of the North American Cordillera has been formed by terrain accretion over the past 200 million years. Wrangellia, for example, is a former basalt plateau that was transported to its present location from 5000 km away in the Southern Hemisphere.
[After D. R. Hutchison, "Continental Margins." Oceanus 35 (Winter 1992–1993): 34–44. Modified from work of D. G. Howell, G. W. Moore, and T. J. Wiley.]

> *Accretion occurs when plate movement attaches a fragment of buoyant crust to the edge of a continent.*

The geologic arrangement of accreted terrains can be chaotic. Adjacent blocks of crust can contrast sharply in their rock types, the nature of their folding and faulting, and their history of magmatism and metamorphism. Geologists often find fossils indicating that these blocks originated in different environments and at different times from those of the surrounding area. For example, an accreted terrain comprising ophiolite suites (pieces of seafloor) that contain deep-water fossils might be surrounded by remnants of island arcs and continental fragments containing shallow-water fossils of a completely different age. The boundaries between terrains are almost always major faults that have undergone substantial slipping, although the nature of the faulting is often difficult to discern.

How Continents Are Modified

The Cordillera of western Canada includes many terrains accreted during the drift of North America since the breakup of Pangaea. The geology of this youthful part of the continent looks nothing like that of the ancient Canadian Shield, which lies directly east of the Cordillera. In particular, the accreted terrains do not show the high degree of melting or the high-grade metamorphism that characterize the Precambrian crust of the shield. Why such a difference?

The answer lies in the tectonic processes that have repeatedly modified the older parts of the continent throughout its long history. Continental crust is profoundly altered by orogeny—the mountain-building processes of folding, faulting, magmatic addition, and metamorphism. Orogenic processes have repeatedly modified the edges of cratons throughout their long history. Most periods of mountain building (orogenies) result from plate convergence. Orogenies can result when a continent rides forcefully over subducting oceanic crust, as in the Andean orogeny now under way in South America, but the most intense orogenies are caused by the collision of two or more continents.

As we saw in Chapter 7, when continents collide, thrust faulting caused by the convergence can stack the upper part of the crust into multiple thrust fault sheets tens of kilometers thick, deforming and metamorphosing the rocks they contain (**Figure 9.29**) and creating fold and thrust belts (see Chapter 7). Horizontal compression throughout the crust can double its thickness, causing the rocks in the lower crust to melt. This melting can generate huge amounts of granitic magma, which rises to form extensive batholiths in the upper crust.

> *Orogenies modify continents by metamorphosing and thickening continental crust and causing the lower crust to melt.*

The Alpine-Himalayan Orogeny To see orogeny in action today, we look to the great chains of high mountains that stretch from Europe through the Middle East and across Asia, known collectively as the *Alpine-Himalayan belt*. The breakup of Pangaea sent Africa, Arabia, and India northward, causing the Tethys Ocean to close as its lithosphere was subducted beneath Eurasia (see Figure 3.16).

The Himalaya, the world's highest mountains, are the most spectacular result of this modern episode of continent-continent collision (**Figure 9.30**). They were formed from overthrust slices of the old northern portion of India stacked one atop the other, as in Figure 9.29. The horizontal compression and formation of fold and thrust belts also thickened the crust north of India, causing the uplift of the huge Tibetan Plateau, which now has a

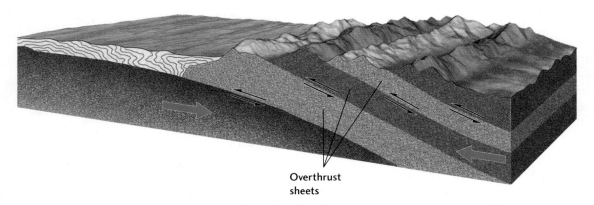

Overthrust
sheets

FIGURE 9.29 When continents collide, the continental crust can break into overthrust sheets stacked one above the other.

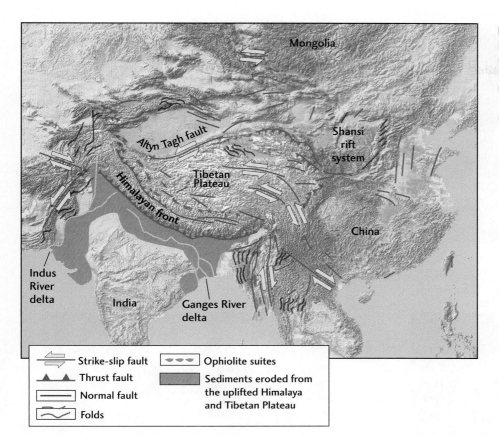

FIGURE 9.30 The collision between India and Eurasia has produced many spectacular tectonic features, including large-scale faulting and uplift. [*After P. Molnar and P. Tapponier, "The Collision Between India and Eurasia." Scientific American (April 1977): 30.*]

Strike-slip fault
Thrust fault
Normal fault
Folds
Ophiolite suites
Sediments eroded from the uplifted Himalaya and Tibetan Plateau

crustal thickness of 60 to 70 km (almost twice the thickness of normal continental crust) and stands nearly 5 km above sea level.

Further compression has pushed China and Mongolia eastward, out of India's way, like toothpaste squeezed from a tube. Most of this sideways movement has taken place along the major strike-slip faults shown in Figure 9.30. The mountains, plateaus, faults, and great earthquakes of Asia, extending thousands of kilometers from the Indian-Eurasian suture, are all results of the Alpine-Himalayan orogeny, which continues today as India plows into Asia at the rate of 40 to 50 mm/year.

Paleozoic Orogenies During the Assembly of Pangaea

If we go farther back in geologic time, we find abundant evidence of older orogenies caused by the earlier episodes of plate convergence that produced the supercontinent Pangaea. For example, at least three distinct orogenies were responsible for the Paleozoic deformation now exposed in the eroded Appalachian fold belt of the eastern United States (**Figure 9.31**).

The supercontinent Rodinia began to break up toward the end of the Proterozoic eon, spawning several paleocontinents (see Figure 3.16). One was the large continent of Gondwana. Two of the others were *Laurentia*, which

included the North American craton and Greenland, and *Baltica*, comprising the lands around the Baltic Sea. In the Cambrian period, Laurentia was rotated almost 90 degrees from its present orientation and straddled the equator; its southern (today, eastern) side was a passive continental margin. To its immediate south was the proto-Atlantic, or *Iapetus Ocean* (in Greek mythology, Iapetus was the father of Atlantis), Baltica lay off to the southeast, and Gondwana was thousands of kilometers to the south.

The island arc built up by the southward-directed subduction of Iapetus oceanic lithosphere collided with Laurentia in the middle to late Ordovician (470 million to 440 million years ago), causing the first episode of mountain building: the *Taconic orogeny*. (You can see some of the rocks accreted and deformed during this period if you drive the Taconic State Parkway, which runs east of the Hudson River for about 160 km north of New York City.) The second orogeny began when Baltica and a connected set of island arcs began to collide with Laurentia in the Devonian. The collision deformed southeastern Greenland, northwestern Norway, and Scotland in what European geologists refer to as the *Caledonian orogeny*. The deformation continued into present-day North America as the *Acadian orogeny*, as island arcs that would become

Middle Cambrian (510 Ma)
After the breakup of Rodinia, Laurentia straddled the equator. Its southern side was a passive continental margin, bounded on the south by the Iapetus Ocean.

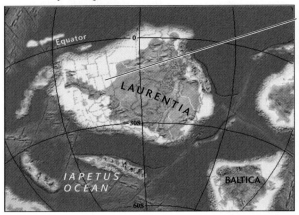

Late Ordovician (450 Ma)
The island arc built up by the southward-directed subduction of Iapetus lithosphere collided with Laurentia in the middle to late Ordovician, causing the Taconic orogeny.

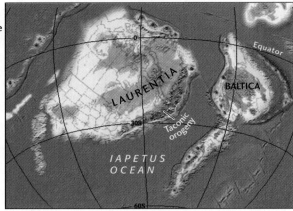

Early Devonian (400 Ma)
The collision of Laurentia with Baltica caused the Caledonian orogeny and formed Laurussia. The southward continuation of the convergence caused the Acadian orogeny.

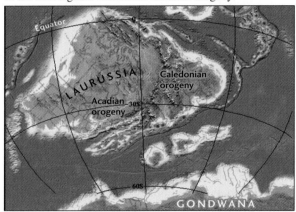

Early Carboniferous (340 Ma)
The collision of Gondwana with Laurussia began with the Variscan orogeny in what is now central Europe…

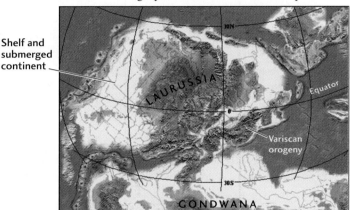

Late Carboniferous (300 Ma)
… and continued along the margin of the North American craton with the Appalachian orogeny. At the same time, Siberia converged with Laurussia in the Ural orogeny to form Laurasia, while the Hercynian orogeny created new mountain belts across Europe and northern Africa.

Early Permian (270 Ma)
The end product of these episodes of continental convergence was the supercontinent of Pangaea.

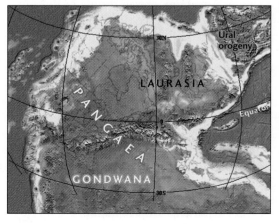

FIGURE 9.31 Paleogeographic reconstructions of the present North Atlantic region, showing the sequence of orogenic episodes that resulted from the assembly of Pangaea. (Ma: million years ago.) [Ronald C. Blakey, Northern Arizona University, Flagstaff.]

the terrains of maritime Canada and New England accreted to Laurentia in the middle to late Devonian 380 to 360 million years ago.

The grand finale in the assembly of Pangaea was the collision of the behemoth landmass of Gondwana with Laurasia and Baltica, by then joined into a continent named *Laurussia* (see Figure 9.31). The collision began around 340 million years ago with the *Variscan orogeny* in what is now central Europe and continued along the margin of the North American craton with the *Appalachian orogeny* (320 million to 270 million years ago). This phase of assembly pushed Gondwanan crust over Laurentia, lifting the Blue Ridge into a mountain chain that may have been as high as the modern Himalaya and causing much of the deformation now seen in the Appalachian fold belt.

The Wilson Cycle

From our brief look at the history of eastern North America, we can infer that the edges of many cratons have experienced multiple episodes of deformation in a general plate tectonic cycle that comprises four main phases (**Figure 9.32**):

1. Rifting during the breakup of a supercontinent

2. Passive margin cooling and sediment accumulation during seafloor spreading and ocean opening

3. Active margin volcanism and terrain accretion during subduction and ocean closure

4. Orogeny during the continent-continent collision that forms the next supercontinent

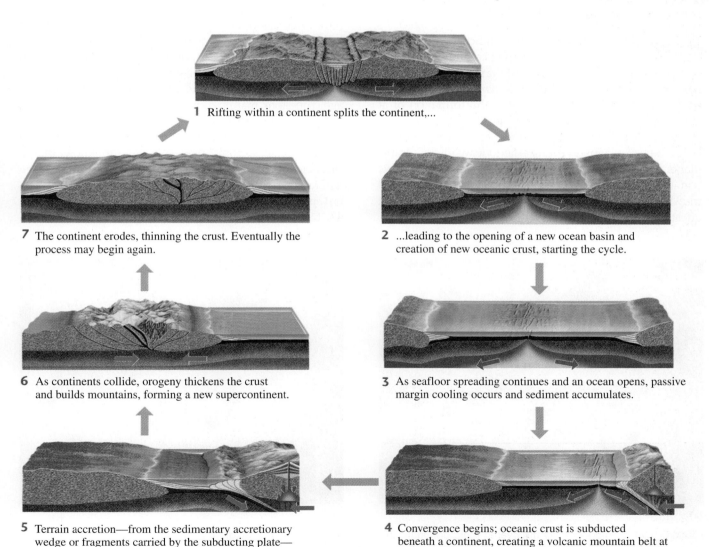

1 Rifting within a continent splits the continent,...

2 ...leading to the opening of a new ocean basin and creation of new oceanic crust, starting the cycle.

3 As seafloor spreading continues and an ocean opens, passive margin cooling occurs and sediment accumulates.

4 Convergence begins; oceanic crust is subducted beneath a continent, creating a volcanic mountain belt at the active margin.

5 Terrain accretion—from the sedimentary accretionary wedge or fragments carried by the subducting plate—welds material to the continent.

6 As continents collide, orogeny thickens the crust and builds mountains, forming a new supercontinent.

7 The continent erodes, thinning the crust. Eventually the process may begin again.

FIGURE 9.32 The Wilson cycle comprises the plate tectonic processes responsible for the formation and breakup of supercontinents and the opening and closing of ocean basins.

Earth
accretion

Oldest   Oldest
Earth   continental
minerals   rocks   Earliest cratons form

Assembly of early
supercontinent?

Assembly of Columbia
supercontinent

Assembly of Rodinia
supercontinent

Assembly of Pangaea
supercontinent

Alpine-
Himalayan
orogeny

4000 Ma   3000 Ma   2000 Ma   1000 Ma   0 Ma

HADEAN EON   **ARCHEAN EON**   **PROTEROZOIC EON**   **PHANEROZOIC EON**

FIGURE 9.33 Geologic time scale, showing some important events in the history of the continents. (Ma: million years ago).

> *The Wilson cycle is the sequence of continental tectonic events that occur during the assembly and breakup of supercontinents and the opening and closing of ocean basins.*

This idealized sequence of events was named the **Wilson cycle** after the Canadian pioneer of plate tectonics J. Tuzo Wilson, who first recognized its importance in the evolution of continents.

The geologic record suggests that the Wilson cycle has operated throughout the Proterozoic and Phanerozoic eons (**Figure 9.33**), resulting in the formation of at least two supercontinents prior to Rodinia. One of these supercontinents (*Columbia*) formed about 1.9 billion to 1.7 billion years ago. An even earlier one, whose assembly marks the transition from the Archean to the Proterozoic eon, formed about 2.7 billion to 2.5 billion years ago. Did the Wilson cycle also operate in the Archean eon? That question is still being debated by geologists.

Formation of Cratons

Every continental craton contains regions of ancient lithosphere that have been stable (i.e., undeformed) since the Archean eon (4.0 billion to 2.5 billion years ago). As we have seen, deformation has occurred at the edges of these stable landmasses and new crust has subsequently accreted around them. But how were these central parts of the cratons created in the first place?

We know that Earth was a hotter planet 4 billion years ago. Therefore, mantle convection during the Archean was more vigorous, although we don't know to what extent the movements of rocks on Earth's surface resembled modern plate tectonic processes. We do know that a silica-rich continental crust existed at this early stage in Earth's history. The Acasta gneiss, in the northwestern part of the Canadian Shield, has been dated to 4.0 billion years ago, and its rocks look very similar to

(a)

(b)

FIGURE 9.34 Geologists have discovered rock formations that demonstrate that continental crust existed on Earth's surface at the beginning of the Archean eon. (a) The Acasta gneiss from the Slave craton has been dated at 4.0 billion years old. (b) Amphibole-bearing rocks from the Nuvvuagittuq greenstone belt, northern Quebec, Canada, have been dated at 4.28 billion years old, making them the oldest rock formation yet discovered. [(a) Courtesy of Sam Bowring, Massachusetts Institute of Technology. (b) Jonathan O'Neil.]

modern gneisses (**Figure 9.34**a). Formations as old as 3.8 billion years have been found on many continents; most are metamorphic rocks evidently derived from even older continental crust. Geologists recently discovered an even older rock formation, nearly 4.3 billion years old, in northern Quebec (Figure 9.34b).

> Cratons, the most stable regions of today's continents, were formed in the Archean by the accretion of smaller rafts of lithosphere.

In the early part of the Archean eon, the continental crust that had differentiated from the mantle was very mobile. It may have been organized in small rafts that were rapidly pushed together and torn apart by intense tectonic activity—a version of the flake tectonics that appears to be happening on Venus today. The first continental crust with long-term stability began to form around 3.3 billion to 3.0 billion years ago. In North America, the oldest surviving example is the central Slave province in northwestern Canada where the Acasta gneiss is found, which stabilized around 3 billion years ago. Geologists have been able to show that this stabilization process involved not only the continental crust, but also chemical changes in the mantle portion of the continental lithosphere.

By the end of the Archean, 2.5 billion years ago, enough continental lithosphere had been stabilized in cratons to allow the formation of larger and larger continents by magmatic addition and accretion. The plate tectonic system was probably operating much as it does today. It is at about this time that we see the first evidence of major continent-continent collisions and the assembly of supercontinents. From this point onward in Earth's history, the history of the continents was governed by the plate tectonic processes of the Wilson cycle.

BIOLOGICAL EVENTS IN EARTH'S HISTORY

As the continents began to take shape during the Archean eon, so did early forms of life. We don't understand how life got started on Earth, but several lines of evidence—fossils as well as chemical traces of life—indicate that single-celled microorganisms, or **microbes,** were flourishing by 3.5 billion years ago. According to one evolutionary hypothesis, these primitive organisms, and all that came after, descended from a single "universal ancestor."

Not much is known about the earliest forms of life, but they must have been composed of carbon-rich compounds, like all subsequent organic substances, and they must have contained *genes*—molecules with instructions for growth and reproduction. Biologists have learned how to use the genetic information contained in living organisms to understand which forms of life are most closely related to one another. This knowledge has allowed them to organize the hierarchy of ancestors and descendants into a *universal tree of life*, with the hypothesized universal ancestor at its root. The tree has three main branches, called *domains:* Bacteria, Archaea, and Eukaryota (**Figure 9.35**).

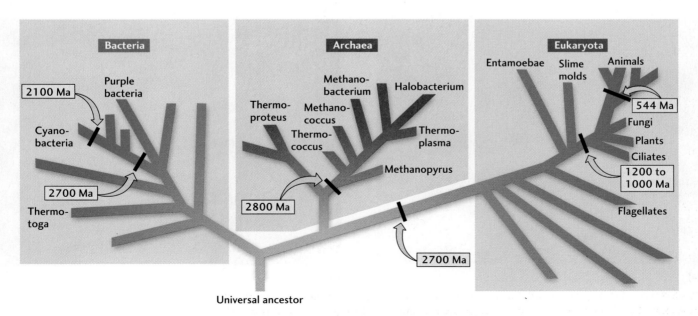

FIGURE 9.35 The universal tree of life shows how all organisms are related to one another. Organisms are subdivided into three great domains: the Bacteria, Archaea, and Eukaryota. These domains are all descended from a universal common ancestor. All three domains are dominated by microorganisms. Note that animals appear at the tip of the eukaryote branch. (Ma: million years ago.)

Early Life

Sometime early in Earth history, probably during the Archean eon, the microbes split into these three branches. The Bacteria and Archaea appear to have evolved first; all of their descendants have remained single-celled microorganisms. The Eukaryota, thought to be the youngest branch, have a more complex cellular structure, which includes a cell nucleus that contains the genetic material. Eukaryotes evolved from small, single-celled organisms into larger, multicellular organisms—an essential step in the evolution of animals and plants. We have identified fossils of multicellular organisms only in rocks younger than 1 billion years. It therefore appears that microbes were the only organisms on Earth for at least 2.5 billion years!

> Microfossils and chemical traces of life indicate that microbes were alive on Earth's surface as early as 3.5 billion years ago.

Precambrian microbes, like those living today, were tiny, only a few microns in size (1 micron = 10^{-6} m). The traces of individual microbes preserved in rocks are therefore called **microfossils.** Needless to say, such features were much harder to find than the macroscopic fossils of shells, bones, and twigs used by geologists to study the evolution of animals and plants during the Phanerozoic eon (recall that *Phanerozoic* means "visible life"). In fact, the first microfossils were not discovered until 1954, in the 2.1-billion-year-old Gunflint formation of southern Canada (**Figure 9.36**a).

Well-preserved microfossils have also been found in the 3.2-billion-year-old Fig Tree formation of South Africa. Tiny threads similar in size and appearance to modern microorganisms occur in formations in Western Australia as old as 3.5 billion years, though the interpretation of these older features as microfossils remains controversial.

Some of the best evidence for life in the early Archean comes not from microfossils, but from macroscopic structures called *stromatolites:* rocks with distinctive thin layers formed by microbial communities as matted sheets and domes in lakes and tidal flats (**Figure 9.36**b). Microbial communities can be observed building stromatolites today in intertidal environments such as Shark Bay, Western Australia (**Figure 9.37**). Stromatolites are common in Precambrian continental cratons and have been identified in sedimentary rocks of early Archean age (Figure 9.37b). Other types of indirect evidence include chemical traces of early life, such as the ratios of carbon isotopes found in some early Archean rocks, which show values that can be produced only by biological processes.

Origin of Earth's Oxygenated Atmosphere

As we learned earlier in this chapter, Earth's early atmosphere was dominated by carbon dioxide released from Earth's interior by volcanism. Our current oxygen-rich atmosphere was produced by early life through the basic metabolic process of **photosynthesis.** Through this

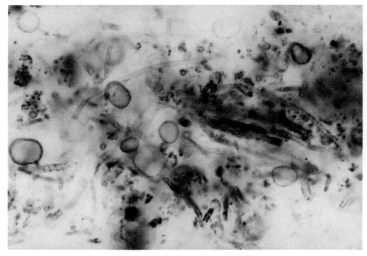

(a)

(b)

FIGURE 9.36 (a) Abundant microfossils are well preserved in the 2.1-billion-year-old Gunflint formation of southern Ontario, Canada. (b) Early Archean (3.4 billion years old) stromatolites in the Warrawoona formation, Western Australia. The conical shapes suggest that the microbial mats that formed these rocks may have grown toward the sunlight. [*Images courtesy of H. J. Hofmann.*]

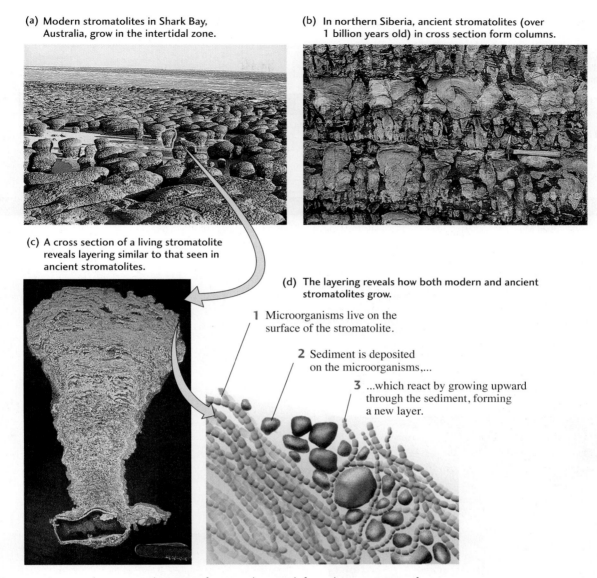

(a) Modern stromatolites in Shark Bay, Australia, grow in the intertidal zone.

(b) In northern Siberia, ancient stromatolites (over 1 billion years old) in cross section form columns.

(c) A cross section of a living stromatolite reveals layering similar to that seen in ancient stromatolites.

(d) The layering reveals how both modern and ancient stromatolites grow.

1 Microorganisms live on the surface of the stromatolite.

2 Sediment is deposited on the microorganisms,...

3 ...which react by growing upward through the sediment, forming a new layer.

FIGURE 9.37 Stromatolites are sedimentary features that result from the interaction of microorganisms with their environment. [*Images from John Grotzinger.*]

process, organisms such as green plants, algae, and certain types of bacteria use energy from sunlight to convert water and carbon dioxide to carbohydrates, such as sugar, and oxygen. The oxygen is released to the atmosphere, and the sugar is stored as an energy source for future use by the organism.

The oxygenation of Earth's atmosphere probably occurred in two main steps, separated by more than a billion years (**Figure 9.38**). The first major increase began with the evolution of the first group of microorganisms capable of photosynthesis, the *cyanobacteria*, perhaps as early as 2.7 billion years ago. The oxygen they produced reacted with iron dissolved in seawater, causing iron

oxide minerals, such as magnetite and hematite, and silica-rich minerals, such as chert and iron silicates, to precipitate and sink to the seafloor. These minerals accumulated in thin, alternating layers of sediments, called *banded iron formations* (**Figure 9.39**a). This process would have continued until most of the dissolved iron was used up, allowing oxygen to accumulate in the atmosphere.

Atmospheric oxygen concentrations reached an initial plateau around 2.1 billion to 1.8 billion years ago, when the first eukaryote fossils, of a type of algae, entered the

> *Earth's atmosphere was oxygenated by early life in two main steps separated by more than a billion years.*

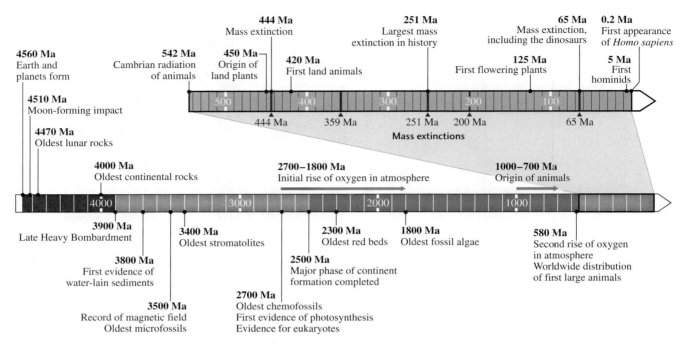

4560 Ma
Earth and
planets form

4510 Ma
Moon-forming impact

4470 Ma
Oldest lunar rocks

542 Ma
Cambrian radiation
of animals

450 Ma
Origin of
land plants

420 Ma
First land animals

444 Ma
Mass extinction

251 Ma
Largest mass
extinction in history

65 Ma
Mass extinction,
including the dinosaurs

0.2 Ma
First appearance
of *Homo sapiens*

125 Ma
First flowering plants

5 Ma
First
hominids

444 Ma   359 Ma   251 Ma   200 Ma   65 Ma

Mass extinctions

4000 Ma
Oldest continental rocks

2700–1800 Ma
Initial rise of oxygen in atmosphere

1000–700 Ma
Origin of animals

3900 Ma
Late Heavy Bombardment

3400 Ma
Oldest stromatolites

2300 Ma
Oldest red beds

1800 Ma
Oldest fossil algae

580 Ma
Second rise of oxygen
in atmosphere
Worldwide distribution
of first large animals

3800 Ma
First evidence of
water-lain sediments

2500 Ma
Major phase of continent
formation completed

3500 Ma
Record of magnetic field
Oldest microfossils

2700 Ma
Oldest chemofossils
First evidence of photosynthesis
Evidence for eukaryotes

FIGURE 9.38 Geologic time scale, showing major events in the history of life.
(Ma: million years ago.)

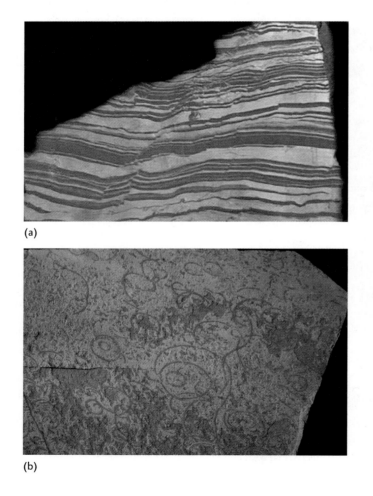

(a)

(b)

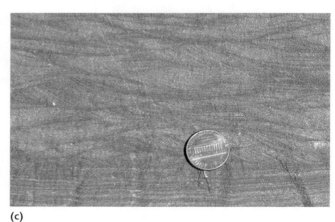

(c)

FIGURE 9.39 Unusual sedimentary rocks and new, larger
organisms mark the rise of oxygen in the atmosphere
between 2.7 billion and 2.1 billion years ago. (a) A
banded iron formation. (b) These fossils of *Grypania*, a
type of eukaryotic algae, are visible with the unaided eye.
(c) Red beds are mostly sandstones and shales cemented
together by iron oxide minerals. [(a) Pan Terra. (b) Courtesy of
H. J. Hofmann. (c) John Grotzinger.]

geologic record (Figure 9.39b). The large size of these organisms—at least 10 times larger than anything that came before them—is thought to be a consequence of the oxygen increase. This time also marks the first appearance of *red beds*, unusual river deposits of sandstones and shales bound together by iron oxide cement, which gives them their red color (Figure 9.39c).

After eukaryotic algae came on the scene, not much happened for over a billion years. Then, about 600 million years ago, atmospheric oxygen rose dramatically, almost to its modern level (see Figure 9.38). The reason for this second increase is still not understood, though it may be related to the enrichment of the deep sea with oxygen, perhaps by an increase in the burial of organic carbon by sedimentation. In any case, the consequences were dramatic: the first really large animals suddenly appeared, and all the modern groups of animals evolved shortly thereafter, ushering in the Phanerozoic eon with its wonderfully complex multicellular organisms.

Radiation of Life: The Cambrian Explosion of Animals

Perhaps the most remarkable geobiological event in Earth's history, aside from the origin of life itself, was the sudden appearance of large animals with shells and skeletons at the end of Precambrian time (**Figure 9.40**). This rapid development of new types of organisms—what biologists call an *evolutionary radiation*—had such an extraordinary effect on the fossil record that its culmination 542 million years ago is used to mark the most profound boundary of the geologic time scale: the beginning of the Phanerozoic eon, the Paleozoic era, and the Cambrian period (see Chapter 8 and Figure 9.33).

The radiation of animals during the early Cambrian was so fast that it is often called the **Cambrian explosion.** Every major

> *All major groups of animals on Earth today originated in the early Cambrian period during a rapid evolutionary radiation called the Cambrian explosion.*

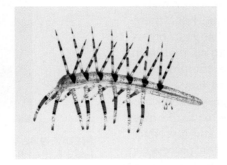

Namacalathus *Hallucigenia* **Trilobites**

FIGURE 9.40 Fossils that record the Cambrian explosion. Precambrian organisms such as *Namacalathus* (left) were the first organisms to use calcite in making shells. These organisms became extinct at the Precambrian-Cambrian boundary, paving the way for a strange new group of organisms, including *Hallucigenia* (center) and the more familiar trilobites (right). In each example, the fossils are shown on top and the reconstructed organism is shown on the bottom. [left, top: John Grotzinger; left, bottom: W. A. Watters; center, top: National Museum of Natural History/Smithsonian Institution; center, bottom: Chase Studio/ Photo Researchers; right, top: Courtesy of Musée cantonal de géologie, Lausanne. Photo by Stéphane Ansermat; right, bottom: Chase Studio/Photo Researchers.]

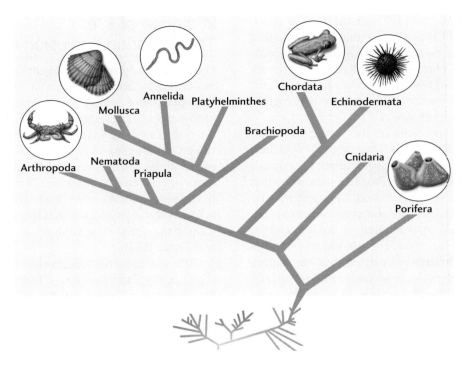

FIGURE 9.41 Every major group of animals alive today originated during a great evolutionary radiation in the early Cambrian, known as the Cambrian explosion.

animal group that exists on Earth today, as well as a few more that have since become extinct, originated during the Cambrian explosion. **Figure 9.41** shows the tree of life for the major groups of animals; all the branches on this tree (*phyla*) originated during the Cambrian explosion.

Geologists have raised a major question about the Cambrian explosion: What allowed these early animals to develop such complex body shapes so rapidly and therefore to become so diverse? Systematic change in organisms over many generations is referred to as **evolution.** Evolution is driven by **natural selection,** the process by which populations of organisms adapt to new environments. The theory of *evolution by natural selection* states that, over many generations, individuals with the most favorable traits are most likely to survive and reproduce, passing those traits on to their offspring. This process can lead eventually to the emergence of new species.

One hypothesis for the cause of the Cambrian explosion is that the genes of the early animals changed in some way that made it possible for them to exceed some sort of evolutionary barrier. The stage was set by the development of multicellularity in late Precambrian time (**Figure 9.42**), which opened up new evolutionary possibilities. For example, the development of skeletons may have been an important trigger: once one group of

FIGURE 9.42 A fossilized animal embryo from the latest part of Precambrian time. Such fossils show that multicellular animals had evolved before the Precambrian-Cambrian boundary and are the ancestors of the animals that evolved during the Cambrian explosion. [*Courtesy Shuhai Xiao, Virginia Tech.*]

animals had evolved hard parts, the others had to as well, or they would have been eliminated through competition.

The right genes were probably not enough to account for such a major radiation, however. The dramatic environmental transformations that occurred near the end of Precambrian time, including major changes in climate and ocean chemistry as well as the increase in atmospheric oxygen, may have set the evolutionary stage for the Cambrian explosion.

Extinction of Life: The Demise of the Dinosaurs

Another dramatic event in Earth's history was the extinction of the dinosaurs at the end of the Cretaceous period (about 65 million years ago). This event defines the *Cretaceous-Tertiary boundary* of the geologic time scale and also marks the end of the Mesozoic era (see Figure 8.13).

When groups of organisms are no longer able to adapt to changing environmental conditions or compete with more successful groups of organisms, they become extinct. An interval when many groups of organisms become extinct at the same time is called a *mass extinction* (see Chapter 8). The Cretaceous-Tertiary boundary represents one of the greatest mass extinctions in Earth's history (**Figure 9.43**). Entire global ecosystems were obliterated, and about 75 percent of all species on Earth,

both on land and in the oceans, were extinguished forever. The dinosaurs are only one of several groups that became extinct at the end of the Cretaceous period, but they are certainly the most prominent. Other groups, such as ammonites, marine reptiles, certain types of clams, and many types of plants and plankton, also perished.

We are now virtually certain that the cause of the Cretaceous-Tertiary mass extinction was a gigantic asteroid impact (see Figure 1.23). In 1980, geologists discovered a very thin layer of dust containing the distinctive element *iridium*, which is typically found in extraterrestrial bolides, in sediments deposited at the end of the Cretaceous (**Figure 9.44**). This extraterrestrial dust was subsequently found at many other locations around the world, but always exactly at the Cretaceous-Tertiary boundary. The geologists argued that the accumulation of this much iridium-bearing dust in such a thin layer would require an asteroid about 10 km in diameter to hit Earth, explode, and send its cosmic detritus across the globe.

> The impact of a 10-km asteroid 65 billion years ago killed off 75 percent of all species, including all the dinosaurs. This mass extinction marks the Cretaceous-Tertiary boundary in the geologic time scale.

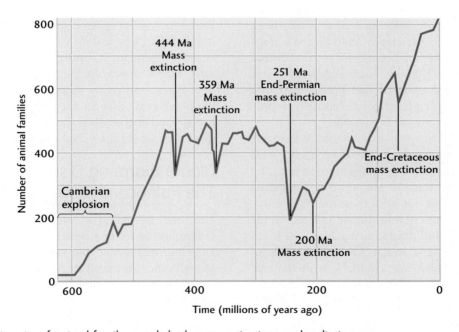

FIGURE 9.43 The diversity of animal fossils reveals both mass extinctions and radiations. This graph shows the number of "shelly" animal families found in the fossil record during the last 600 million years; each family comprises many species. During a radiation, such as the Cambrian explosion, the number of new animal families increases. During a mass extinction, such as the one at the end of the Cretaceous period, the number of families decreases. (Ma: million years ago.)

FIGURE 9.44 The pocketknife marks a light-colored layer of clay containing both extraterrestrial materials and materials from local rocks at the Chicxulub impact site that accumulated in the Raton Basin of the southwestern United States following the asteroid impact that killed the dinosaurs. Such deposits have been found worldwide. [From David Kring and Daniel Dura, "The Day the World Burned." *Scientific American* (December 2003): 104. © December 2003 by *Scientific American*. All rights reserved.]

In the early 1990s, geologists found the smoking gun: a huge crater, almost 200 km in diameter and 1.5 km deep, buried under sediments near a town on Mexico's Yucatán Peninsula, called Chicxulub. The material that had been blown out of the Chicxulub crater, including the fragments of the asteroid itself, formed a blanket that covered all of North America and possibly South America as well. Additional, finer-grained material had been lofted into the atmosphere, enveloping Earth and eventually depositing a thin layer of dust around the world.

The immediate aftermath of this impact is the stuff that disaster movies are made of. The blast itself would have been hundreds of millions of times more devastating than the most powerful nuclear weapon ever detonated and 6 million times more powerful than the 1980 eruption of Mount St. Helens. It would have created winds of unimaginable fury and tsunamis as high as 1 km (100 times higher than the Indian Ocean tsunamis of 2004). The sky would have turned black with massive amounts of dust and vapor. A global firestorm may have resulted as the flaming fragments from the blast fell back to Earth (**Figure 9.45**).

The direct effects of the impact itself would have been devastating for many organisms. Most organisms near the impact zone would have been instantly wiped out. But worse yet would have been the aftermath over months and years to come, which scientists think led to the actual mass extinction. The high concentration of debris in the atmosphere would have blocked out the Sun, vastly reducing the light energy available for photosynthesis. In addition to the solid debris, poisonous sulfur- and nitrogen-bearing gases would have been injected into the atmosphere, where they would have reacted with water vapor to form toxic sulfuric and nitric acids that would have rained down on Earth. The combination of these two effects and many others would have killed most of the plants and other photosynthetic organisms at the base of the food chain. Once the food sources were gone, the dinosaurs would have died off as well.

Global Warming Disaster: The Paleocene–Eocene Mass Extinction

The mass extinction at the Paleocene-Eocene boundary (55 million years ago; see Figure 8.13) was not one of the largest of such events but it is significant in the evolution of life because it paved the way for the mammals, including the primates, to radiate into an important group. Unlike the mass extinction that wiped out the dinosaurs, it had no extraterrestrial cause. Instead, it was the result of abrupt global warming. Earth scientists are very interested in the details of what happened because global warming, this time produced by human activities, may threaten life in the coming decades (see Chapter 14).

We now believe that the global warming at the end of the Paleocene epoch occurred when the ocean suddenly

FIGURE 9.45 Artist's rendition of the Cretaceous-Tertiary scene after the asteroid impact.
[*Alfred Kamajian.*]

belched an enormous amount of methane—a very potent greenhouse gas—into the atmosphere. The resulting global warming was the primary cause of death that led to the mass extinction. But where did all that methane come from?

When marine organisms die, they settle to the seafloor, where they accumulate as organic debris. Some of this carbon-rich debris is buried in sediments under oxygen-poor conditions, where it is consumed by microbes that produce methane (CH_4) as a waste product. The methane accumulates in the pores of seafloor sediments. If the seafloor is as cold as it is in our present climate (about 3°C), the methane combines with water to form a frozen solid (a methane-water ice), which is stored within the sediments. Geologists searching for oil have found layers with abundant methane-water ices in the upper 1500 m of sediments along many continental margins. If temperatures rise by even a few degrees, however, the methane-water ice melts, and the methane is quickly transformed into a gas.

At the end of the Paleocene epoch, average temperatures in the deep sea increased enough to start this process. Once some of the methane-water ice thawed, methane bubbled up through the oceans and entered the atmosphere and reinforced the global warming trend. This, in turn, raised temperatures on the seafloor even further,

which accelerated the rate of thawing. This cycle resulted in a sudden—and catastrophic—release of methane that caused average global temperatures to rise dramatically. As much as 2 *trillion* tons of carbon, in the form of methane, may have escaped to the atmosphere in this methane over a period as short as 10,000 years or less!

Because methane reacts easily with oxygen, the release of methane also caused oxygen levels in the oceans to plummet. Marine organisms were essentially suffocated when oxygen concentrations dropped below a critical level. The oxygen decrease and temperature rise were devastating to organisms in the deep sea, and up to 80 percent of bottom feeders, such as clams, became extinct.

Today, the global inventory of all methane deposits is estimated at 10 to 20 trillion tons of carbon present as methane, far more than what was released to cause the Paleocene-Eocene mass extinction. We are adding carbon dioxide to the atmosphere through the burning of fossil fuels, causing the climate to warm significantly. If this trend continues and the oceans warm up, it is possible that the current methane deposits could thaw. We would

> *A massive release of methane from the seafloor around 55 million years ago caused abrupt global warming and led to the mass extinction that marks the Paleocene-Eocene boundary.*

be wise to pay attention to the lessons of our geologic history.

The Mother of All Mass Extinctions: Whodunit?

The Cretaceous-Tertiary and Paleocene-Eocene extinctions are clear-cut examples of dramatic changes in Earth's environment that caused the catastrophic collapse of ecosystems and led to mass extinction. They were big, but not the biggest.

The extinction at the end of the Permian period was even worse (see Figure 9.43)—95 percent of all species became extinct! Here, there is no smoking gun, and it seems unlikely that something as simple as an asteroid impact could explain how almost every species on Earth was killed. Not surprisingly, the absence of clear-cut evidence has resulted in a long list of hypotheses. Some scientists point to extraterrestrial events, such as a comet impact or an increase in the solar wind. Others argue for events that come from Earth itself, such as an increase in volcanism, depletion of oxygen in the oceans, or a sudden release of carbon dioxide from the oceans. As in the Paleocene-Eocene extinction, a sudden release of methane from the oceans has also been proposed.

Recently, it has been shown that the mass extinction at the end of the Permian, which occurred 251 million years ago, precisely matches the age of an enormous deposit of flood basalts in Siberia. Flood basalts, as we learned in Chapter 5, are extrusive igneous rocks formed from huge volumes of lava that pour out across the surface of Earth in a relatively short time. In Siberia, volcanic fissures spewed out some 3 million cubic kilometers of basaltic lava, covering an area of 4 million square kilometers, almost twice the size of Alaska. Isotopic dating of the basalt shows that all this lava poured out within 1 million years or less. It is hard to escape the conclusion that the Permian mass extinction was somehow related to this catastrophic eruption. These tremendous volcanic emissions would have injected enormous amounts of carbon dioxide and sulfur dioxide gases into the atmosphere. Carbon dioxide contributes to global warming, and sulfur dioxide is the principal source of acid rain (see Chapter 1). Both are harmful to life if atmospheric concentrations get too high.

The cause of the greatest mass extinction of all time, which wiped out 95 percent of all species at the end of the Permian period, is still being debated.

Whatever the cause of the Permian mass extinction, one point is clear: just as in the Cretaceous-Tertiary and Paleocene-Eocene mass extinctions, the ultimate cause was the collapse of ecosystems. We know that this collapse occurred, but we don't know exactly how. The message that we should take away from this history lesson is that the past may repeat itself. As we will see in Chapter 14, the decisions that humans make today will inevitably affect our environment—we just don't know exactly how, at least not yet.

GOOGLE EARTH PROJECT

With current technology, we are able to send "robotic geologists" to the surface of Mars to study its current and ancient environments until humans can go there one day. These rovers have many of the tools that human geologists would use, including cameras, microscopes, spectrometers, and tools to sample the rocks. The two Mars Exploration Rovers, *Spirit* and *Opportunity*, were launched in 2003 and landed on Mars in early 2004. In this exercise, we will use Google Mars to follow in *Opportunity*'s tracks as we explore Meridiani Planum, the region where it landed.

Open Google Earth and click on the planet icon at the top (that looks like Saturn). Select "Mars." In the "Fly To" search window on the left, type in "*Opportunity.*" Google Mars will navigate to the area where *Opportunity* landed, and which it has been traversing since 2004.

As your cursor hovers over the Google Mars screen, a hand will appear. You can use the hand to click and drag the region of interest. Click and drag so that the screen is centered on *Opportunity*'s landing site (marked by an American flag). Zoom in to see the traverse route the rover has taken (marked by a red line).

LOCATION   Meridiani Planum, the region of Mars where the rover *Opportunity* landed

GOAL   Use Google Mars tools to measure traverse distances and decipher crater morphology

LINKED   Figure 9.16

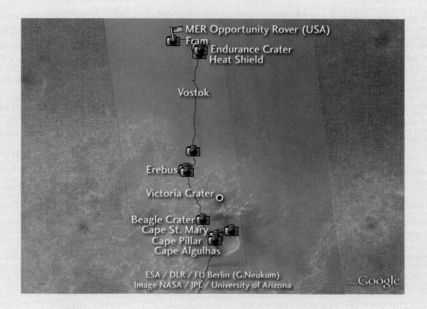

1. Where did *Opportunity* land? (You may have to zoom in on the landing site to see place-names.)

   a. Eagle Crater

   b. Endurance Crater

   c. Victoria Crater

   d. Gusev Crater

2. What is the latitude and longitude of *Opportunity*'s landing site?

   a. 2°03'05" S, 5°29'44" W

   b. 1°56'51" S, 5°30'30" W

   c. 1°56'42" S, 5°31'16" W

   d. 2°25'01" S, 5°30'10" W

3. Which is the largest crater *Opportunity* has explored?

   a. Erebus Crater

   b. Victoria Crater

   c. Vostok Crater

   d. Endurance Crater

4. In which direction overall has the rover driven?

   a. East   c. North

   b. West   d. South

5. What might you expect to see on the floor and in the walls of Victoria Crater?

   a. Dunes on the floor, rock outcrops in the walls

   b. Dunes on the floor, piles of dust in the walls

   c. Alien footprints on the floor, ancient riverbeds in the walls

   d. The spacecraft heat shield on the floor, plant roots in the walls

Optional Challenge Question

6. Using the GE ruler tool, select "path," and then select "meters" as the unit of measure. Draw a pathway following the traverse *Opportunity* has taken from Eagle Crater to the western rim of Victoria Crater (red line). (Ignore the rover paths within and around craters.) What is the approximate distance the rover traversed?

   a. 10,000 m   c. 4500 m

   b. 7300 m   d. 2700 m

■ SUMMARY

How did our solar system originate? According to the nebular hypothesis, the Sun and its planets formed when a cloud of gas and dust condensed about 4.5 billion years ago. The inner planets, including Earth, differ from the giant outer planets in their composition.

How did Earth form and evolve over time? Earth probably grew by the accretion of colliding planetesimals. Soon after it formed, it was struck by a large body about the size of Mars. Matter ejected into space from both Earth and the impacting body reassembled to form the Moon. The impact generated enough heat to melt much of Earth's outer layers. Radioactivity also contributed to early heating and melting. Heavy matter, rich in iron, sank toward Earth's center, and lighter matter floated upward to form the outer layers that became the crust and the continents. Still lighter gases formed Earth's atmosphere and oceans. In this way, Earth was transformed into a differentiated planet with distinct layers.

What are some major events in the early history of the solar system? The age of the solar system, as determined from isotopic dating of meteorites, is about 4.56 billion years. Earth and the other terrestrial planets had formed within about 10 million years. The impact that formed the Moon occurred about 4.5 billion years ago. Minerals as old as 4.4 billion years have survived in Earth's crust. The lunar maria were formed by impacts during the Late Heavy Bombardment, which peaked around 3.9 billion years ago and marks the end of the Hadean eon on Earth.

How can planetary surfaces be dated? Rocks returned from the surface of the Moon by the Apollo missions have been dated using isotopic methods. The lunar highlands show ages from 4.47 billion to about 4.0 billion years. The lunar maria show ages from 4.0 billion to 3.2 billion years. These isotopic ages allowed geologists to calibrate the relative time scale they had developed by mapping and counting craters.

Do other planets have plate tectonics? Venus is the only planet other than Earth that has features indicating tectonic activity resulting from mantle convection. But Venus does not appear to have thick plates. Instead, it has a thin crust of frozen lava that breaks up into flakes or crumples like a rug. A similar process, which geologists refer to as flake tectonics, may have occurred on Earth when it was younger and hotter.

What are the major tectonic provinces of North America? The continent's most ancient crust is exposed in the Canadian Shield. South of the Canadian Shield is the interior platform, where Precambrian basement rocks are covered by layers of Paleozoic sedimentary rocks. These older regions form the continental craton. Around the edges of the craton are elongated mountain chains. The most important of these orogenic belts are the Cordillera, which runs down the western edge of North America, and the Appalachian fold belt, which trends southwest to northeast on its eastern margin. The coastal plain and continental shelf of the Atlantic Ocean and Gulf of Mexico are parts of a passive continental margin that subsided after rifting during the breakup of Pangaea.

How do continents grow? Two plate tectonic processes, magmatic addition and accretion, add crust to continents. Buoyant silica-rich rocks are produced by magmatic differentiation, primarily in subduction zones and added to the continental crust by vertical transport. Accretion occurs when preexisting crustal material is attached to existing continental masses by horizontal plate movement.

How do orogenies modify continents? Orogenies caused by plate convergence can deform continental crust. Thrust faulting can stack the upper part of the crust into multiple thrust fault sheets tens of kilometers thick, deforming and metamorphosing the rocks they contain. Compression throughout the crust can double its thickness, causing the rocks in the lower crust to melt and form extensive batholiths.

What is the Wilson cycle? The Wilson cycle is a sequence of tectonic events that occur during the assembly and breakup of supercontinents and the opening and closing of ocean basins. It has four main phases: rifting during the breakup of a supercontinent; passive margin cooling and sediment accumulation during seafloor spreading and ocean opening; active magmatic addition and accretion during subduction and ocean closure; and orogeny during continent-continent collision.

What evidence do we have of life on early Earth? The oldest possible fossils on Earth are 3.5 billion years old and appear to be the remnants of microbes. Photosynthetic bacteria had probably evolved by 2.7 billion years ago. Banded iron formations, red beds, and the appearance of eukaryotic algae testify to an initial rise in atmospheric oxygen concentrations by 2.1 billion years ago. A second, more dramatic rise in oxygen concentrations occurred near the end of Precambrian time and might have triggered the evolution of animals.

What is the difference between radiation and extinction? Evolutionary radiation is the rapid development of new types of organisms. When groups of organisms are no longer able to adapt to changing environmental conditions or to compete with more successful groups of organisms, they become extinct. In a mass extinction, many groups of organisms become extinct at the same time. The greatest radiation of animals in Earth's history occurred during the early Cambrian, when all the animal phyla living today evolved. Several mass extinctions have occurred throughout the Phanerozoic eon. A major mass extinction occurred at the end of the Cretaceous period, when an asteroid hit Earth and 75 percent of all species were wiped out. Global warming by release of methane caused a mass extinction at the Paleocene-Eocene boundary. The cause of the greatest mass extinction of all time, which wiped out 95 percent of all species at the end of the Permian period, is still being debated.

■ KEY TERMS AND CONCEPTS

accretion (p. 295)

asteroid (p. 274)

Cambrian explosion (p. 305)

craton (p. 293)

gravitational differentiation
  (p. 275)

evolution (p. 306)

Heavy Bombardment
  (p. 279)

magmatic addition (p. 295)

meteorite (p. 274)

microbe (p. 301)

microfossil (p. 302)

natural selection (p. 306)

nebular hypothesis (p. 273)

photosynthesis (p. 302)

solar nebula (p. 274)

Wilson cycle (p. 300)

■ EXERCISES

1. How and why do the inner planets differ from the giant outer planets?

2. What caused Earth to differentiate into a layered planet, and what was the result?

3. Mercury's average density is less than Earth's, but the relative size of its core is larger than that of Earth's core. How can you explain this?

4. What aspects of the geology of Earth and the Moon are consistent with high impact rates during the Late Heavy Bombardment?

5. Why is the topography of the North American Cordillera higher than that of the Appalachian Mountains? How long ago were the Appalachians at their highest elevation?

6. Describe the tectonic province in which you live.

7. Are the interiors of continents usually younger or older than their margins? Explain your answer using the concept of the Wilson cycle.

8. Two continents collide, thickening the crust from 35 km to 70 km and forming a high plateau. After hundreds of millions of years, the plateau is eroded down to sea level. (a) What kinds of rocks might be exposed at the surface by this erosion? (b) Estimate the crustal thickness after the erosion has occurred. (c) Where in North America has this sequence of events been recorded in surface geology?

9. How many times have the continents been joined in a supercontinent since the end of the Archean eon? Use this number to estimate the typical duration of a Wilson cycle and how fast plate tectonic processes move continents.

10. During an evolutionary radiation, organisms evolve rapidly. What would the geologic record look like if a slow evolutionary radiation occurred during the time of an unconformity? How would you distinguish between a rapid radiation and the effects of an unconformity?

11. Compare and contrast the events that led to the Permian, Cretaceous-Tertiary, and Paleocene-Eocene mass extinctions. Are there particular processes that contributed to these mass extinctions that we, as members of a global society, should be aware of? Which processes may be induced by human activities?

Visual Literacy Task

1 Rifting within a continent splits the continent,...

7 The continent erodes, thinning the crust. Eventually the process may begin again.

2 ...leading to the opening of a new ocean basin and creation of new oceanic crust, starting the cycle.

6 As continents collide, orogeny thickens the crust and builds mountains, forming a new supercontinent.

3 As seafloor spreading continues and an ocean opens, passive margin cooling occurs and sediment accumulates.

5 Terrain accretion—from the sedimentary accretionary wedge or fragments carried by the subducting plate—welds material to the continent.

4 Convergence begins; oceanic crust is subducted beneath a continent, creating a volcanic mountain belt at the active margin.

FIGURE 9.32 The Wilson cycle comprises the plate tectonic processes responsible for the formation and breakup of supercontinents and the opening and closing of ocean basins.

1. What do the red and blue arrows indicate?

a. Red shows plate motion at divergent boundaries, and blue shows plate motion at convergent boundaries.

b. Red shows plate motion at convergent boundaries, and blue shows plate motion at divergent boundaries.

c. Red and blue show plate motion at both divergent and convergent boundaries, just at different points in time.

2. What must happen in order for an ocean basin to begin getting smaller?

a. A divergent boundary must form.

b. A convergent boundary must form.

c. Mountains must form

3. **What happens to the divergent plate boundary between Stage 4 and Stage 5?**

a. It gets subducted.
b. It stops diverging and turns into permanent seafloor.
c. It becomes a sedimentary accretionary wedge.

4. **If the supercontinent Pangaea was at Stage 7, when there was one major continent and one large ocean, what was present before Pangaea?**

a. Several smaller continents with no oceans
b. Only a large ocean
c. Earth formed with Pangaea already existing
d. Both smaller continents and large oceans, depending on how far in the past

5. **How do tectonic plates change over time?**

a. They can divide into separate plates.
b. They can combine with other plates.
c. They can both divide and combine.
d. They neither divide nor combine.

Thought Question: How would you answer a fellow student who asked, "Once a plate boundary forms, is it a permanent feature?"

The Climate System and Glaciation

The Ice Hunter

LONNIE THOMPSON HAS SPENT more time at elevations above 18,000 feet than any other person on Earth. Trekking to the Himalaya and the Andes and beyond, he has risked blood clots and temporary blindness in the name of a single pursuit: preserving tens of thousands of years of weather history coded deep in the planet's fast-vanishing glaciers. "No scientist has taken bigger risks to track ancient weather patterns and help us understand the anomaly of current climate trends," says Al Gore.

Thompson stores his prehistoric glacial samples at Ohio State University in vaults kept at subarctic temperatures, and he studies the dust particles and air trapped within the ice. From this atmospheric evidence, he has reconstructed a meticulous calendar of temperatures dating back 750,000 years. The upshot: "It proves that the warming trends of today are vastly more dramatic than what we've seen over the last 5000 years," says Thompson.

> "You could have knocked me over with a feather the day I discovered, firsthand, that glaciers contain a frozen history of the Earth."

Growing up on a small farm in West Virginia, Thompson studied geology so that he could work in the coal industry. But he got sidetracked in grad school when he examined the first ice core ever extracted by American scientists. "You could have knocked me over with a feather the day I discovered, firsthand, that glaciers contain a frozen history of the Earth," he recalls. Now, in his work at the Byrd Polar Research Center at Ohio State, Thompson, defying frostbite and hurricane-force winds, possesses a will to survive on a par with Lance Armstrong's. Photographs he has taken provide disturbing views of the world's melting glaciers—including the ice cap on Mount Kilimanjaro, which is expected to disappear entirely by 2015.

Thompson dismisses skeptics who contend that the current warming trend is due to a natural cycle. "Name one who has ever really studied climate or collected data," he says. "I bet you can't." Glaciers, he adds, "have no political agenda. They don't care if you're a Democrat or a Republican. Science is about what is, not what we believe or hope. And it shows that global warming is wiping out invaluable geological archives right before our eyes." ◆

—From *Rolling Stone*, November 3, 2005

Glaciologist Lonnie Thompson at an altitude of 5300 m (17,390 ft) on Tibet's Dasuopu Glacier. Ice coring on this glacier provides evidence of abnormal global warming during the twentieth century. [*Lonnie Thompson/ Byrd Polar Research Center/Ohio State University.*]

CHAPTER OUTLINE

N O ASPECT OF EARTH SCIENCE is more important to our continued well-being than the study of the climate system. Throughout geologic time, evolutionary radiations and extinctions of organisms have been closely connected to changes in climate. Even the short history of our own species is deeply imprinted by climate change: agricultural societies began to flourish only about 11,500 years ago, when the harsh climate of the most recent ice age rapidly transformed into the mild and steady climate of the Holocene epoch. Now, a globalized human society based on a petroleum-fueled economy is injecting greenhouse gases into the atmosphere at an ever-increasing rate, with potentially dire consequences: global warming, sea level rise, and unfavorable changes in weather patterns. The climate system is a huge, incredibly complex machine, and, like or not, our hands are on the controls. We're in the driver's seat, with pedal to the metal, so we had better understand how the machinery works!

In this chapter, we will examine the main components of the climate system and the ways in which these components interact to produce the climate we live in today. We will investigate the geologic record of climate change, particularly the ice ages that spread vast glaciers across the northern continents, and we will see how those glaciations have modified the landscape.

An understanding of the climate system will equip us to study the wide range of geologic processes that shape the face of our planet—weathering, erosion, sediment transport, and the interaction of the plate tectonic and climate systems—which will be the topics of the next two chapters. The material presented here will also prepare us for the final topic of this textbook: a geological perspective on the resource needs and environmental impacts of human society.

 ## NATURAL ENVIRONMENT: Components of the Climate System

At any point on Earth's surface, the amount of energy received from the Sun changes on daily, yearly, and longer-term cycles associated with Earth's movement through the solar system. This cyclical variation in the input of solar energy, known as *solar forcing*, cause changes in the surface environment: temperatures rise during the day and fall at night, and they rise in summer and fall in winter. The term **climate** refers to the average conditions at a point on Earth's surface and their variation during these cycles of solar forcing.

> *Climate is described by the average weather conditions at a point on Earth's surface during daily, yearly, and longer-term cycles of solar forcing.*

Climate is described by daily and seasonal statistics of the atmospheric temperature near Earth's surface (the *surface temperature*), as well as humidity, cloud cover, rate of rainfall, wind speed, and other weather conditions. **Table 10.1** gives an example of temperature statistics for New York City, which include measures of temperature

TABLE 10.1	Daily and Seasonal Temperatures (°F) in Central Park, New York City			
Data type[a]	January 1	April 1	July 1	October 1
Record high	62	83	100	88
Mean high	40	54	84	69
Mean low	28	39	66	54
Record low	−5	12	52	36

[a]*Mean temperatures are for 1970–2008; record temperatures are for 1869–2008.*

variability (record highs and lows) as well as average values. In addition to these common weather statistics, a full scientific description of climate includes the non-atmospheric components of the surface environment, such as soil moisture and streamflow on land, as well as sea surface temperature and the velocity of currents in the ocean.

The **climate system** includes all the components of the Earth system and all the interactions among the components that determine how climate varies in space and time

(**Figure 10.1**). The main components of the climate system are the atmosphere, hydrosphere, cryosphere, lithosphere, and biosphere. Each component plays a different role in the climate system, and the role depends on the component's ability to store and transport mass and energy.

The Atmosphere

Earth's atmosphere is the most mobile and rapidly changing part of the climate system. Like Earth's interior, the

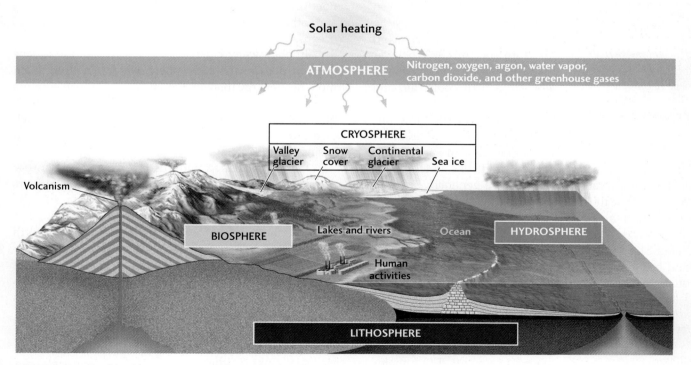

FIGURE 10.1 Earth's climate system involves complex interactions among many components.

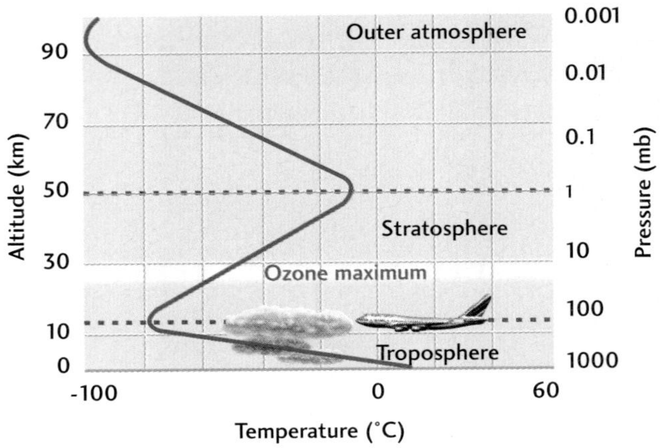

FIGURE 10.2 Layers of the atmosphere, showing the variation of temperature (indicated by the blue line) and pressure with altitude.

atmosphere is layered (**Figure 10.2**). About three-fourths of its mass is concentrated in the layer closest to Earth's surface, the **troposphere,** which has an average thickness of 11 km. Above the troposphere is the **stratosphere,** a dryer layer that extends to an altitude of about 50 km. The outer atmosphere, above the stratosphere, has no abrupt cutoff; it slowly becomes thinner and fades away into outer space.

The Sun's heat causes convection in the troposphere, which, combined with Earth's rotation, sets up prevailing wind belts that transport heat from the warmer equatorial regions to the cooler polar regions.

The troposphere convects vigorously due to the uneven heating of Earth's surface by the Sun (*tropos* is the Greek word for "turn" or "mix"). When air is warmed, it expands, becomes less dense than cooler air, and tends to rise. Conversely, cool air tends to sink. The resulting convection patterns in the troposphere (which we'll examine more closely in Chapter 12), combined with Earth's rotation, set up a series of prevailing wind belts. In temperate regions, the prevailing winds have a generally eastward flow, such that they transport a typical parcel of air eastward around the globe in about a month (which is why it takes a few days for storms to blow across the continental United States). The spiral-like circulation of air in these wind belts also transports heat energy from the warmer equatorial regions to the cooler polar regions.

The atmosphere is composed mostly of nitrogen gas (78 percent) and oxygen gas (21 percent).

The atmosphere is a mixture of gases, mainly nitrogen (78 percent by volume in dry air) and oxygen (21 percent by volume in dry air). The remaining 1 percent consists of argon (0.93 percent), carbon dioxide (0.035 percent), and other minor gases (0.035 percent). Water vapor is concentrated in the troposphere in highly variable amounts (up to 3 percent, but typically about 1 percent). Water vapor and carbon dioxide are the principal greenhouse gases in the atmosphere.

Another minor constituent of the atmosphere is ozone (O_3^+), a highly reactive greenhouse gas produced primarily by the effect of solar radiation on atmospheric oxygen. Most atmospheric ozone is found in the stratosphere, where its concentration reaches a maximum at an altitude of 25 to 30 km (see Figure 10.2). As we saw in Chapter 1, this stratospheric ozone layer filters out incoming ultraviolet radiation, protecting the biosphere at Earth's surface from its potentially damaging effects.

The Hydrosphere

The *hydrosphere* comprises all the liquid water on, over, and under Earth's surface, including oceans, lakes, streams, and groundwater. Almost all liquid water is in the oceans (1350 million cubic kilometers); lakes, streams, and groundwater constitute a mere 1 percent (15 million cubic kilometers). However small, these continental components of the hydrosphere play a vital role in the climate system. They are reservoirs for moisture on land and provide the transport system for returning precipitation and transporting salt and other minerals to the oceans.

The hydrosphere comprises all the liquid water on Earth, most of which is in the oceans.

Although water circulates more slowly in the oceans than air does in the atmosphere, water can store much more heat energy than air. For that reason, ocean currents transport heat energy very effectively. Prevailing winds blowing across the ocean generate surface currents, which give rise to large-scale circulation patterns within ocean basins (**Figure 10.3**a).

Oceanic circulation patterns involve vertical convection as well as horizontal movement. The Gulf Stream, for example, flows from the Gulf of Mexico and the Caribbean Sea along the western Atlantic margin, carrying warm water that warms the climate of the North Atlantic and Europe. In the North Atlantic, that water cools and becomes more saline (because less fresh water enters the sea from rivers at high latitudes than evaporates from the ocean surface). Cool water is denser than warm water, and salty water is denser than less salty water; therefore, cooler, saltier water sinks. In this way, a subsurface cold current is created that flows southward as part of a global pattern of

Ocean currents, like prevailing wind belts, transport heat energy from warmer regions near the equator to cooler regions near the poles.

(a)

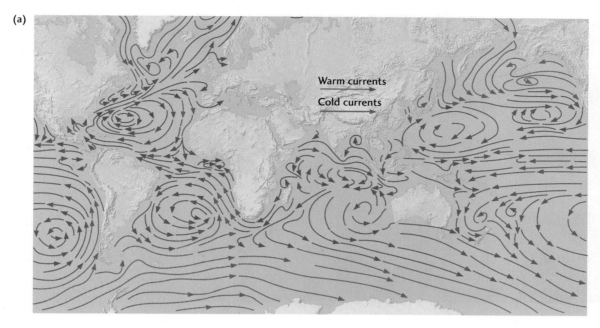

Warm currents

Cold currents

(b)

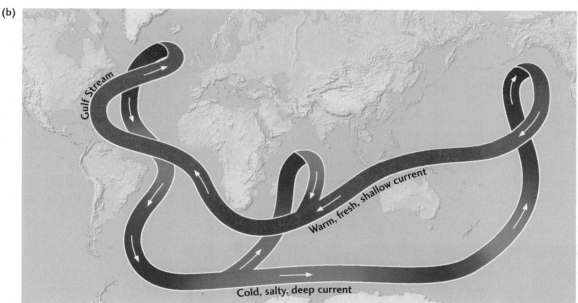

Gulf Stream

Warm, fresh, shallow current

Cold, salty, deep current

FIGURE 10.3 Two major circulation systems in the oceans. (a) Currents at the surface of the oceans are generated by winds. [*U.S. Naval Oceanographic Office.*] (b) A schematic representation of thermohaline circulation, which acts like a conveyor belt to transport heat from warm equatorial regions to cool polar regions.

thermohaline circulation—so called because it is driven by differences in temperature and salinity.

On a global scale, thermohaline circulation acts like an enormous conveyor belt running through the oceans that moves heat from the equatorial regions toward the poles (Figure 10.3b). Changes in this circulation pattern can strongly influence global climate.

The Cryosphere

The ice component of the climate system is called the *cryosphere.* It comprises 33 million cubic kilometers of ice, primarily the ice caps of the polar regions. Today, continental glaciers cover about 10 percent of the land surface (15 million square kilometers), storing about 75 percent

FIGURE 10.4 Several valley glaciers flow together in the St. Elias Mountains, Kluane National Park, Yukon, Canada. [*Stephen J. Krasemann/DRK Photo.*]

of the world's fresh water. Floating ice includes sea ice in the open oceans as well as frozen lake and river water. The role of the cryosphere in the climate system differs from that of the liquid hydrosphere because ice is relatively immobile and because it reflects almost all of the solar energy that falls on it.

Thought of as a mineral, ice has some unusual properties. Its melting temperature is very low (0°C), hundreds of degrees below the melting temperatures of silicate minerals. Most minerals are denser in their solid form than in their liquid form, which is why magma rises buoyantly through the lithosphere. Ice, on the other hand,

is less dense than its liquid form (water), which is why icebergs float on the ocean. And although ice may seem hard, it is much softer than most minerals.

Because ice is so easily deformed, it flows readily downhill like a viscous fluid. *Glaciers* are large masses of ice on land that move under the force of gravity. **Valley glaciers** are rivers of ice that form in the cold heights of mountain ranges and flow down valleys (**Figure 10.4**). A valley glacier usually occupies the complete width of a valley and may bury its floor under hundreds of meters of ice (see Figure 10.4). In warmer, low-latitude climates, valley glaciers are found only at the heads of valleys on

FIGURE 10.5 Iceberg calving at Wrangell-St. Elias National Park, Alaska. Calving occurs when huge blocks of ice break off at the edge of a glacier that has moved to a shoreline. [*Tom Bean.*]

FIGURE 10.6 Sentinel Range, Antarctica. These mountains rise more than 4000 m, protruding through the thick ice of the Antarctic continental glacier.

the highest mountain peaks. Valley glaciers that flow down coastal mountain ranges at high latitudes may terminate at the ocean's edge, where masses of ice break off and form icebergs—a process called *iceberg calving* (**Figure 10.5**).

A **continental glacier** is a thick, slow-moving sheet of ice (sometimes called an *ice sheet*) that covers a large part of a continent or other large landmass (**Figure 10.6**). Today, the world's largest continental glaciers overlie much of Greenland and Antarctica, covering about 10 percent of Earth's land surface and storing about 75 percent of the world's fresh water. In Greenland, 2.6 million cubic kilometers of ice occupy 80 percent of the island's total area (**Figure 10.7**). The upper surface of the ice sheet resembles an extremely wide convex lens. At its highest point, in the middle of the island, the ice is more than 3200 m thick. From this central area, the ice surface slopes to the sea on all sides. At the mountain-rimmed coast, the ice sheet breaks up into narrow tongues resembling valley glaciers that wind through the mountains to reach the sea, where icebergs form by calving.

Though very large, the Greenland Glacier is dwarfed by the Antarctic ice sheet, which covers an area of about 13.6 million square kilometers and reaches thicknesses of 4000 m (**Figure 10.8**). The total volume of Antarctic ice, about 30 million cubic kilometers, constitutes over 90 percent of the cryosphere. As in Greenland, the ice forms a dome in the center and slopes down to the margins of the continent.

The cryosphere comprises all the ice on Earth, most of which is found in the Antarctic continental glacier.

MAP OF GREENLAND

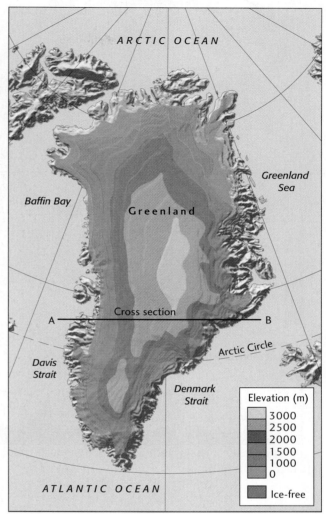

(a) Cross section of Greenland

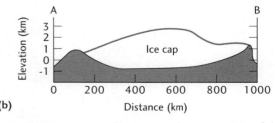

(b)

FIGURE 10.7 Topographic map and cross section of the Greenland continental glacier. (a) Extent and elevation of the Greenland ice sheet. [*After R. F. Flint, Glacial and Quaternary Geology. New York: Wiley, 1971.*] (b) Generalized cross section of south-central Greenland shows the lenslike shape of the glacier. Ice shelves are shown in white.

Parts of Antarctica are rimmed by thinner sheets of ice—*ice shelves*—floating on the ocean and attached to the main glacier on land. The best known is the Ross Ice Shelf, a thick layer of ice about the size of Texas (see Figure 10.8).

MAP OF ANTARCTICA

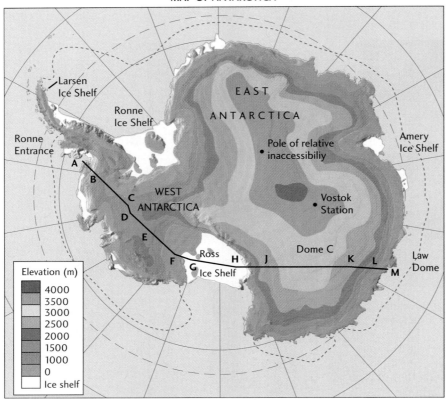

FIGURE 10.8 Topographic map and cross section of the Antarctic continental glacier. (a) Extent and elevation of the Antarctic ice sheet. Ice shelves are shown in white. (b) Generalized cross section of the ice sheet and the land beneath it. [*After U. Radok, "The Antarctic Ice." Scientific American (August 1985): 100; based on data from the International Antarctic Glaciological Project.*]

(a) Cross section of Antarctica

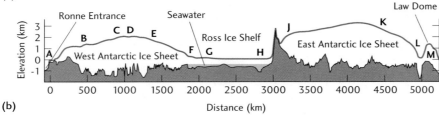

(b)

Floating ice also includes *sea ice* in the open ocean. During the winter, sea ice typically covers 14 million to 16 million square kilometers of the Arctic Ocean (**Figure 10.9**) and 17 million to 20 million square kilometers of the Antarctic Ocean, shrinking to about one-third of that area in the summer.

The seasonal exchange of water between the cryosphere and the hydrosphere is an important process of the climate system. About one-third of the land surface is covered by seasonal snows, almost entirely (all but 2 percent) in the Northern Hemisphere. Melting snow is the source of much of the fresh water in the hydrosphere. In the U.S. Sierra Nevada and Rocky Mountains, for

> *About one-third of Earth's land surface is covered by seasonal snow, an important source of fresh water when it melts.*

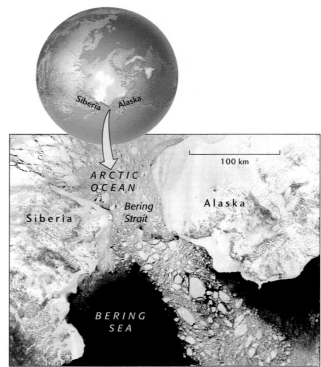

FIGURE 10.9 The volume of sea ice varies seasonally. This satellite image shows Arctic sea ice flowing through the Bering Strait in May 2002. [*NASA MODIS satellite.*]

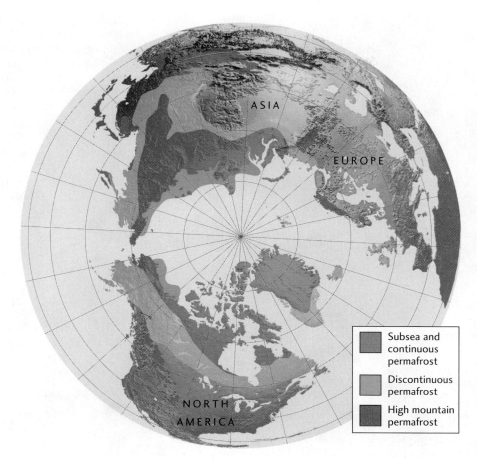

FIGURE 10.10 Map of the Northern Hemisphere with the North Pole at its center, showing the distribution of permafrost on the northern continents. The large area of high mountain permafrost at the top of the map is on the Tibetan Plateau. [*T. L. Pewe, Arizona State University.*]

example, 60 to 70 percent of annual precipitation is snowfall, which is later released as water during spring snowmelt and river runoff.

In very cold regions, the ground stays frozen throughout the year. Perennially frozen soil, or *permafrost*, today covers as much as 25 percent of Earth's total land area and constitutes an important part of the cryosphere (**Figure 10.10**). Permafrost is defined solely by temperature, not by soil moisture content, overlying snow cover, or location: any rock or soil remaining at or below 0°C for 2 or more years is permafrost. Permafrost covers about 82 percent of Alaska and 50 percent of Canada, as well as great parts of Siberia. Outside the polar regions, it is present in high, mountainous areas such as the Tibetan Plateau.

The Lithosphere

The part of the lithosphere that is most important to the climate system is the land surface, which makes up about 30 percent of Earth's total surface area (**Figure 10.11**). The composition of the land surface affects the way it absorbs solar energy or releases it to the atmosphere. As the temperature of the land surface rises, more heat energy is radiated back into the atmosphere, and more water evaporates from the land surface and enters the atmosphere. Because evaporation uses considerable energy, it causes the surface to cool. Consequently, soil moisture and other factors that influence evaporation—such as vegetation cover and the subsurface flow of water—are very important in controlling atmospheric temperatures.

Topography has a direct effect on climate through its influence on atmospheric circulation. Air masses that flow over mountain ranges dump rain on the windward side, creating a rain shadow on the leeward side of the mountains (see Figure 11.4).

> *The land surface, the part of the lithosphere that is most important to the climate system, absorbs and releases solar energy.*

FIGURE 10.11 Snapshot of the climate system taken by sensors on several spacecraft, showing cloud cover (in white), variations in sea surface temperature (from the warmest in red to the coolest in dark blue), and land surface properties, including density of vegetation (from the lowest density in brown to the highest in green). [*R. B. Husar/NASA Visible Earth.*]

At much longer time scales, geologists have documented many changes in the climate system that result from plate tectonic processes. The overall asymmetry of the continents—a direct consequence of plate movements—induces hemispheric asymmetries in the global climate system. Changes in the shape of the seafloor due to seafloor spreading cause changes in sea level, and the drift of continents over the poles leads to the growth of continental glaciers. The movements of continents can also block ocean currents or open gateways through which they can flow. This can inhibit or facilitate the global transfer of heat. For example, if future tectonic activity were to close the narrow channel between the Bahamas and Florida through which the Gulf Stream flows, temperatures in western Europe might drop drastically.

Volcanism in the lithosphere affects climate by changing the composition and properties of the atmosphere. Volcanic eruptions inject tiny liquid drops and solid particles, called *aerosols*, into the stratosphere, blocking solar radiation and temporarily lowering atmospheric temperatures on a global scale. After the massive April 1815 eruption of Mount Tambora in Indonesia, New England suffered through a "year without a summer" in 1816. According to a diarist in Vermont, "no month passed without a frost, nor one without a snow," and crop failures were common. Careful studies have shown that recent large volcanic eruptions—including those of Krakatau (1883), El Chichón (1982), and Mount Pinatubo (1991)—each produced an average dip of 0.3°C in global surface temperatures about 14 months after the eruption (local temperature variations can be much larger, of course). Temperatures returned to normal in about 4 years.

> Plate movements can cause long-term climate changes by rearranging continents and ocean basins.

The Biosphere

The *biosphere* comprises all the organisms living on and beneath Earth's surface, in its atmosphere, and in its waters. Life is ubiquitous at Earth's surface, but the amount of life at any location depends on local climate conditions, as we can see from the satellite images of plant and algal biomass in Figure 1.38.

The amount of energy contained and transported by living organisms is relatively small—less than 0.1 percent of incoming solar energy is used by plants in photosynthesis and thus enters the biosphere. The biosphere, however, is strongly coupled to the climate system. For example, terrestrial vegetation can affect both atmospheric temperature, because plants absorb solar radiation and release it as heat, and atmospheric moisture, because they soak up groundwater and release it as water vapor. Organisms also regulate the composition of the atmosphere by taking up or releasing greenhouse gases such as carbon dioxide and methane. Through photosynthesis, plants and algae transfer carbon dioxide from the atmosphere to the biosphere. Some of it is precipitated in the oceans as shells made of calcium carbonate or buried as organic matter in sediments. The biosphere thus plays a central role in the carbon cycle (discussed in Chapter 14).

Humans, of course, are part of the biosphere, though hardly an ordinary part. Our influence over the biosphere is growing rapidly, and we have become the most active agents of environmental change. As an organized society, we behave in fundamentally different ways from other species. For example, we can study climate change scientifically and modify our actions according to what we have learned.

THE GREENHOUSE EFFECT

The Sun is a yellow star that puts out about half its radiant energy as visible light. The other half is split between infrared waves, which have lower energy intensities than visible light (and which we perceive as heat), and ultraviolet waves, which have higher energy intensities. The average amount of solar radiation Earth's surface receives throughout the year is 342 watts per square meter of surface area (342 W/m^2, 1 watt = 1 joule per second; a joule is a unit of energy or heat). In comparison, the average amount of heat flowing out of Earth's deep interior by mantle convection is minuscule, only 0.06 W/m^2. Essentially all the energy driving the climate system ultimately comes from the Sun (**Figure 10.12**).

We know that the global surface temperature, averaged over daily and seasonal cycles, remains constant. Therefore, Earth's surface must be radiating energy back into space at a rate of precisely 342 W/m^2. Any less would cause the surface to heat up; any more would cause it to cool down. In other words, Earth maintains a *radiation balance:* an equilibrium between incoming and outgoing radiant energy. How is this equilibrium achieved?

> *A balance between incoming solar energy and outgoing radiation maintains the average global temperature of Earth's surface.*

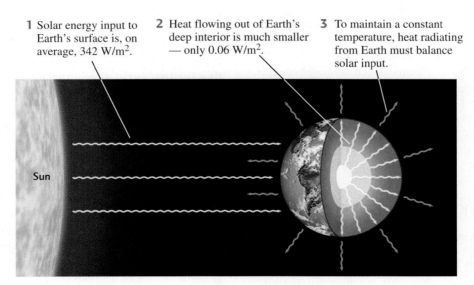

1 Solar energy input to Earth's surface is, on average, 342 W/m^2.

2 Heat flowing out of Earth's deep interior is much smaller — only 0.06 W/m^2.

3 To maintain a constant temperature, heat radiating from Earth must balance solar input.

Sun

FIGURE 10.12 Earth's energy balance is achieved by the radiation of incoming solar energy back into space. Heat gain from Earth's interior is negligible in comparison to that from solar energy.

A Planet Without Greenhouse Gases

Suppose Earth were a rocky sphere like the Moon, with no atmosphere at all. Some of the sunlight falling on the surface would be reflected back into space, and some would be absorbed by the rocks, depending on the color of the surface. A perfectly white planet would reflect all the solar energy falling on it, whereas a perfectly black planet would absorb it all. The fraction of the solar energy reflected by a surface is called its **albedo** (from the Latin word *albus*, meaning "white"). Although the full Moon looks bright to us, the rocks on its surface are mainly dark basalts, so its albedo is only about 7 percent. In other words, the Moon is dark gray—very nearly black.

The energy radiated by a black body increases rapidly as its temperature increases. A cold bar of iron is black and gives off little heat. If you heat the bar to 100°C, it gives off warmth in the form of infrared waves (like a steam radiator). If you heat the bar to 1000°C, it becomes bright orange, radiating heat at visible wavelengths (like the burner on an electric stove).

A black body exposed to the Sun heats up until its temperature is just the right value for it to radiate the incoming solar energy back into space. The same principle applies to a "gray body" like the Moon, except that the reflected energy must be excluded from the radiation balance. And, in the case of rotating bodies like the Moon and Earth, day-and-night cycles must be taken into account. The Moon's daytime temperatures rise to 130°C, and its nighttime temperatures drop to –170°C. Not a pleasant environment!

Earth rotates much faster than the Moon (once per day rather than once per month), which evens out the day-and-night extremes of temperature. Earth's albedo, at about 31 percent, is much higher than the Moon's because Earth's blue oceans, white clouds, and ice caps are more reflective than dark lunar basalts. If our atmosphere did not contain greenhouse gases, the average surface temperature required to balance the absorbed solar radiation would be about –19°C (–2°F), cold enough to freeze all the water on the planet. Instead, Earth's average surface temperature remains a balmy 14°C (57°F). The difference of 33°C is a result of the greenhouse effect.

Earth's Greenhouse Atmosphere

Greenhouse gases such as water vapor, carbon dioxide, methane, and ozone absorb energy—that coming directly from the Sun as well as that radiated by Earth's surface—and reradiate it as infrared energy in all directions, including downward to Earth's surface. In this way, the gases act like the glass in a greenhouse, allowing light energy to pass through but trapping heat in the atmosphere. This trapping of heat, which increases the temperature at the surface relative to the temperature higher in the atmosphere, is known as the **greenhouse effect.**

Greenhouse gases in Earth's atmosphere, primarily water vapor, carbon dioxide, methane, and ozone, trap solar energy, thereby helping to maintain Earth's equable temperatures.

How Earth's atmosphere balances incoming and outgoing radiation is illustrated in **Figure 10.13**. Incoming solar radiation that is not directly reflected is absorbed by Earth's atmosphere and surface. To achieve radiation balance, Earth radiates this same amount of energy back into space as infrared energy. Because of the heat trapped by the greenhouse gases, the amount of energy transported away from Earth's surface, both by radiation and by the flow of warm air and moisture from the surface, is significantly larger than the amount Earth receives as direct solar radiation. The excess is exactly the energy received as Earthward infrared radiation from the greenhouse gases. It is this "back radiation" that causes Earth's surface to be 33°C warmer than it would be if the atmosphere contained no greenhouse gases.

Balancing the System Through Feedbacks

How does the climate system actually achieve the radiation balance illustrated in Figure 10.13? Why does the greenhouse effect yield an overall warming of 33°C and not some larger or smaller amount? The answers to these questions are not simple because they depend on interactions among the many components of the climate system. The most important of those interactions involve feedbacks.

Feedbacks come in two basic types: **positive feedbacks,** in which a change in one component is *enhanced* by the changes it induces in other components, and **negative feedbacks,** in which a change in one component is *reduced* by the changes it induces in other components. Positive feedbacks tend to amplify changes in a system, whereas negative feedbacks tend to stabilize the system against change.

Here are some of the feedbacks within the climate system that can significantly affect the temperatures achieved by radiation balance:

■ *Water vapor feedback.* A rise in temperature increases the amount of water vapor that moves from Earth's surface into the atmosphere through evaporation. Water vapor is a greenhouse gas, so this increase enhances the greenhouse effect, and the temperature rises further—a positive feedback.

■ *Albedo feedback.* A rise in temperature reduces the accumulation of ice and snow in the cryosphere, which

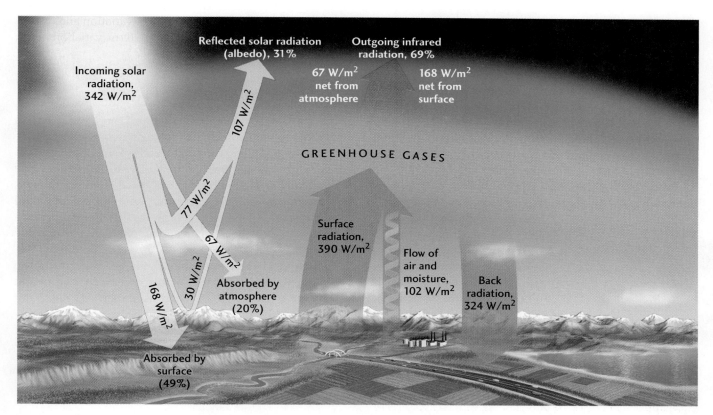

FIGURE 10.13 To maintain radiation balance, Earth radiates as much energy into outer space, on average, as it receives from the Sun (342 W/m²). Radiation and flows of warm air and moisture transport more energy away from Earth's surface (492 W/m²) than it receives (168 W/m²). The greenhouse gases in the atmosphere reflect most of this energy back to Earth's surface as infrared radiation (324 W/m²). [*IPCC, Climate Change 2001: The Scientific Basis.*]

decreases Earth's albedo and increases the energy its surface absorbs. This increased warming of the atmosphere enhances the temperature rise—another positive feedback.

■ *Radiation feedback.* A rise in atmospheric temperature greatly increases the amount of infrared energy radiated back into space, which moderates the temperature rise—a negative feedback. This "radiative damping" stabilizes Earth's climate against major changes, keeping the oceans from freezing up or boiling off and thus maintaining an equable habitat for water-loving life.

■ *Plant growth feedback.* Increasing atmospheric CO_2 concentrations stimulates plant growth. Growing plants remove CO_2 from the atmosphere by converting it to carbon-rich organic matter, thus reducing the greenhouse effect—another negative feedback.

Feedbacks can involve much more complex interactions among components of the climate system. For example, an increase in atmospheric water vapor produces more clouds. Because clouds reflect solar energy, they increase the planetary albedo, which sets up a negative feedback between atmospheric water vapor and temperature. On the other hand, clouds absorb infrared radiation efficiently, so increasing the cloud cover enhances the greenhouse effect, thus providing a positive feedback between atmospheric water vapor and temperature. Does the net effect of clouds produce a positive or negative feedback?

Scientists have found it surprisingly difficult to answer this question. The components of our climate system are joined through an amazingly complex web of interactions on a scale far beyond experimental control. Consequently, it is often impossible to gather data that isolate one type of feedback from all the others. Scientists must therefore turn to computer models to understand the inner workings of the climate system. The role of computer models in making climate predictions will be discussed in Chapter 14.

> *Changes in the climate system can be amplified by positive feedbacks; negative feedbacks can stabilize the climate system against change.*

CLIMATE VARIATION

Earth's climate varies considerably from place to place—its poles are frigid and arid, its tropics sweltering and humid. Comparable variations in climate can also occur over time. The geologic record shows us that periods of global warmth have alternated with periods of glacial cold many times in the past. This climate variation is erratic; dramatic changes can happen in just a few decades or evolve over time scales of many millions of years.

Some climate variation can be attributed to factors outside the climate system, such as solar forcing and changes in the distribution of land and sea surfaces caused by continental drift. Others result from variations within the climate system itself, such as the growth of continental glaciers that increase Earth's albedo. In this section, we

> *Climate variations can be caused by factors outside the climate system, such as solar forcing, or effects within the climate system, such as changes in albedo.*

will examine several types of climate variation and discuss their causes, beginning with short-term variations on a regional scale.

Regional Variations in Climate

Local and regional climates are much more variable than the average global climate: averaging over large surface areas, like averaging over time, tends to smooth out small-scale fluctuations. Over periods of years to decades, the predominant regional variations result from interactions between atmospheric circulation and the sea and land surfaces. They generally occur in distinct geographic patterns, although their timing and amplitudes can be highly irregular.

One of the best-known examples is a warming of the eastern Pacific Ocean that occurs every 3 to 7 years and lasts for a year or so. Peruvian fishermen call this event **El Niño** ("the boy child" in Spanish) because the warming typically reaches the surface waters off the coast of

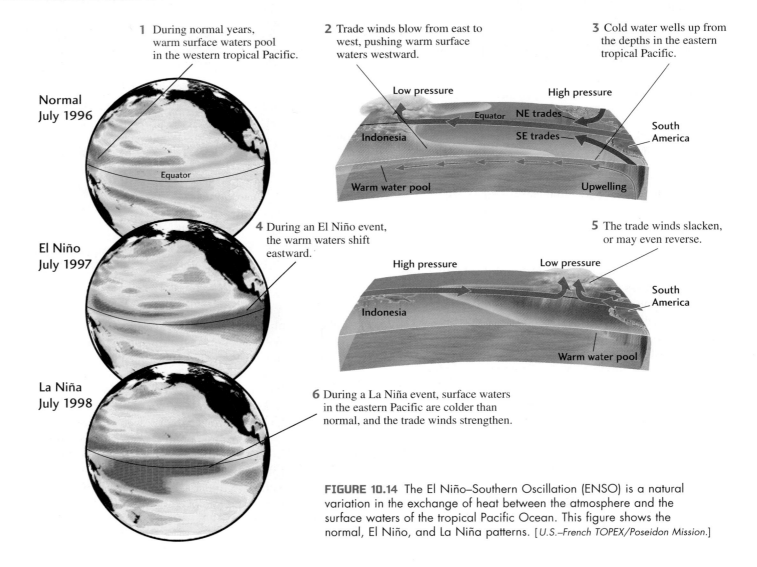

1 During normal years, warm surface waters pool in the western tropical Pacific.

2 Trade winds blow from east to west, pushing warm surface waters westward.

3 Cold water wells up from the depths in the eastern tropical Pacific.

Normal July 1996

Low pressure High pressure

Equator NE trades

Indonesia SE trades South America

Equator

Warm water pool Upwelling

El Niño July 1997

4 During an El Niño event, the warm waters shift eastward.

5 The trade winds slacken, or may even reverse.

High pressure Low pressure

Indonesia South America

Warm water pool

La Niña July 1998

6 During a La Niña event, surface waters in the eastern Pacific are colder than normal, and the trade winds strengthen.

FIGURE 10.14 The El Niño–Southern Oscillation (ENSO) is a natural variation in the exchange of heat between the atmosphere and the surface waters of the tropical Pacific Ocean. This figure shows the normal, El Niño, and La Niña patterns. [*U.S.–French TOPEX/Poseidon Mission.*]

South America around Christmastime. El Niño events can decimate fish populations, which depend on the upwelling of cold water for their nutrient supply, and can thus be disastrous for coastal human populations that depend on fishing.

Scientists have shown that El Niño and a complementary cooling event called *La Niña* ("the girl child") are part of a natural variation in the exchange of heat between the atmosphere and the tropical Pacific Ocean. This variation is known as the El Niño–Southern Oscillation, or ENSO (**Figure 10.14**). Normally, atmospheric pressure gradients cause prevailing winds, called trade winds, to blow from east to west along the equator, pushing the warm tropical waters westward. This movement of water causes colder water to well up from the ocean depths off Peru. Sporadically, the trade winds weaken or occasionally even reverse direction, cutting off the upwelling and equalizing water temperatures across the tropical Pacific (an El Niño event). At other times, the trade winds strengthen, enhancing the temperature difference between the eastern and western Pacific (a La Niña event).

Climate scientists have identified similar patterns of weather and climate variation in other regions. One example is the North Atlantic Oscillation, a highly irregular fluctuation in the atmospheric pressure between Iceland and the Azores that has a strong influence on the movement of storms across the North Atlantic and thus affects weather conditions throughout Europe and parts of Asia. A better understanding of these patterns is improving long-range weather forecasting and may provide important information about the regional effects of human-induced climate change.

Natural Hazards: The Dangers of El Niño

In addition to disrupting the eastern Pacific fishery, El Niño has been implicated in triggering changes in wind and rain patterns over much of the globe. The 1997–1998 El Niño was the strongest on record and it contributed to droughts in Australia and Indonesia, severe hurricanes and typhoons in the Pacific, heavy rains in Peru, Ecuador, and Kenya, and storms in California that caused landslides and floods (**Figure 10.15**). According to one estimate, the global disruption in weather patterns and ecosystems may have cost 23,000 lives and caused $33 billion in damage.

ENSO enhances the global variability of climate. Without El Niño and La Niña events, we would have a more benign climate. The study of the climate system is improving our ability to forecast these events, which allows people to anticipate their hazardous effects. However, there is a growing concern among climate scientists

FIGURE 10.15 Storm waves associated with the 1997–1998 El Niño attacking homes along the Pacific coastline in Del Mar, California. [*Reuters.*]

that the increase in greenhouse gases caused by fossil fuel burning and the associated global warming will amplify climate variability in the future, leading to stronger El Niño and La Niña events.

Twentieth-Century Warming: Human Fingerprints of Global Change

How do we know that Earth's climate is changing, or that the changes are the result of our own activities? Humans have been tracking global temperatures for some time. The most basic device for measuring climate, the thermometer, was invented in the early seventeenth century, and Daniel Fahrenheit set up the first standard temperature scale in 1724. By 1880, temperatures around the world were being reported by enough meteorological stations on land and on ships at sea to allow accurate estimation of Earth's average annual surface temperature.

Although the average annual surface temperature fluctuates substantially from year to year and decade to decade, the overall trend has been upward (**Figure 10.16**). Between the end of the nineteenth century and the beginning of the twenty-first, the average annual surface temperature rose by about 0.6°C. This increase is referred to as the **twentieth-century warming** (Figure 10.16a). Fossil-fuel burning and deforestation since the industrial revolution have substantially increased the concentration of CO_2 in the atmosphere during the same interval.

We know that human activities are responsible for the increasing concentrations of CO_2 in the atmosphere because the carbon isotopes of fossil fuels have a distinctive signature that precisely matches the changing isotopic composition of atmospheric carbon. But how certain can we be that the twentieth-century warming was a direct consequence of the CO_2 increase—that is, a result of an *enhanced greenhouse effect*—and not some other kind of change associated with natural climate variation?

To answer this question, we have to look at a longer history of climate. Although direct temperature measurements are not available

> *Global temperatures fluctuate considerably from year to year, but the increase in temperatures during the twentieth century was abnormal compared with changes in the preceding millennium.*

before the nineteenth century, climate indicators such as tree rings and ice cores have allowed scientists to reconstruct a temperature record for the Northern Hemisphere over the last millennium (Figure 10.16b). This record shows an irregular but steady global cooling of about 0.2°C in the nine centuries between 1000 and 1900. The fluctuations in mean temperature during any one of these centuries were less than a few tenths of a degree. Against this background, the twentieth-century warming appears to be abnormal.

Based on this observation and a variety of other data, almost all experts on Earth's climate are now convinced that the twentieth-century warming was in large part anthropogenic and that the warming will continue into the twenty-first century as human activities continue to raise the concentrations of greenhouse gases in the atmosphere. We will return to this issue in Chapter 14.

> *The twentieth-century warming was due primarily to the anthropogenic increase in atmospheric greenhouse gases.*

Reconstructing Past Climates

In 1830, the geologist Charles Lyell famously stated that "the present is the key to the past." Modern geologists are

(a) 1850–2000 The twentieth-century warming is correlated with the increase in atmospheric CO_2 concentrations since the industrial revolution.

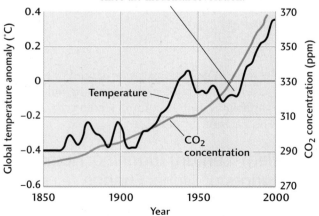

(b) 1000–2000 The twentieth-century warming is clearly anomalous when compared with climate variation over the last millennium.

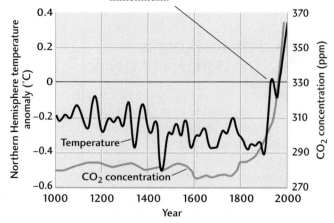

FIGURE 10.16 A comparison of average annual surface temperature anomaly (black lines) with atmospheric CO_2 concentrations (blue lines) shows a recent warming trend that is correlated with increases in atmospheric CO_2 concentrations. (a) Average global annual surface temperature anomalies, calculated from thermometer measurements, between 1850 and 2000. (b) Average annual surface temperature anomalies for the Northern Hemisphere, estimated from tree rings, ice cores, other climate indicators, and atmospheric CO_2 concentrations, for the last millennium. In both these figures, the temperature anomaly is defined as the difference between the observed temperature and the temperature average for the period 1961–1990. [IPCC, Climate Change 2001: The Scientific Basis.]

FIGURE 10.17 Russian scientists at the Vostok Station carefully remove an ice core from a drill. The layers produced by annual cycles of ice formation are visible in the core. Vostok Station, at an altitude of 3500 m near the center of Antarctica (see Figure 10.8a), is an especially grueling place to do research. Its average annual temperature is only –55°C, and the lowest reliably measured temperature on Earth's surface, –89.2°C, was recorded there in 1983. [*R. J. Delmas, Laboratoire de glaciologie et géophysique de l'environnement, Centre National de la Recherche Scientifique.*]

fond of turning his wording around to say that the past is the key to the present, and Figure 10.16b illustrates why: knowing how the climate system has changed in the past can help us understand how it might change in the future. For this reason, *paleoclimatology* is one of the hottest fields in geology.

Some of the best data on past climates come from cores of ice recovered by drilling into the thick continental glaciers of Antarctica and Greenland (**Figure 10.17**). The cores show layers produced by annual cycles of ice formation from snow. The age of the ice can be determined by carefully counting the layers, working from the top down, just as the age of wood in a tree can be determined by counting tree rings. The composition of the atmosphere, including the concentrations of greenhouse gases such as carbon dioxide and methane, can be measured in tiny

bubbles of air trapped at the time the ice formed.

The atmospheric temperature at which each layer in the ice core formed can be estimated by measuring the proportions of two oxygen isotopes found in the ice: oxygen-16, the most common oxygen isotope, and oxygen-18, a heavier and less common form. Water containing the heavier isotope evaporates less readily and condenses more rapidly as the air gets colder. Therefore, the ratio of oxygen-18 to oxygen-16 in the ice is a good indicator of how cold the air was when the snow fell; the lower the proportion, the colder the air.

One of the best ice core records, stretching back more than 400,000 years, has come from drilling of the East Antarctic Ice Sheet at Russia's Vostok Station (**Figure 10.18**). This record shows a great deal of short-term variation in temperature, but the largest swings correspond to

> *Ice cores taken from continental glaciers provide records of atmospheric temperature and composition over the last several hundred thousand years.*

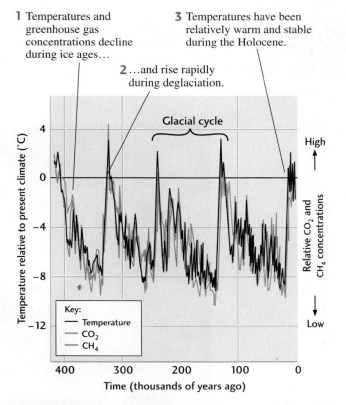

FIGURE 10.18 Three types of data were recovered from the Vostok Station ice cores, which were drilled to a depth of 3600 m in the East Antarctic Ice Sheet. Temperatures were estimated from oxygen isotope ratios. Carbon dioxide and methane concentrations came from measurements of air samples trapped as tiny bubbles within the Antarctic ice. [*IPCC, Climate Change 2001: The Scientific Basis.*]

glacial cycles, which form a sawtooth pattern of peaks and troughs with an average spacing of about 100,000 years. As you can also see from Figure 10.18, the concentrations of carbon dioxide and methane are higher in ice layers formed at higher temperatures, suggesting that past climate variations were strongly influenced by the greenhouse effect.

Glacial Cycles

The most dramatic features observed in the climate record of the last 400,000 years are glacial cycles. In Figure 10.18, we can see that each glacial cycle is marked by a gradual decline of about 6° to 8°C from a warm **interglacial period** to a cold *glacial period*, or **ice age.** As the climate cools, water is transferred from the hydrosphere to the cryosphere. The amount of sea ice increases, and more snow falls on the continents in winter than melts in summer, increasing the volume and area of the polar ice caps and decreasing the volume of the oceans. As the polar ice caps expand into lower latitudes, they reflect more solar energy back into space, and Earth's surface temperatures fall further—an example of albedo feedback. The ice age ends abruptly with a rapid rise in temperature during a

> *Earth's climate has alternated between cold glacial periods and warmer interglacial periods.*

short interval of deglaciation. Water is transferred from the cryosphere to the hydrosphere as the ice caps melt and sea level rises.

The most recent ice age is known as the *Wisconsin glaciation.* Temperatures began to drop about 120,000 years ago, but they reached their lowest values only about 21,000 to 18,000 years ago (the Wisconsin *glacial maximum*). Temperatures then rebounded to warm interglacial levels 11,500 years ago, marking the end of the Pleistocene and the beginning of the Holocene. Three other ice ages, which reached their minimum temperatures about 140,000, 260,000, and 350,000 years ago, are visible in the Vostok record (see Figure 10.18).

During the Wisconsin glaciation, the volume of the cryosphere exceeded 100 million cubic kilometers, more than three

> *The Wisconsin glaciation peaked about 18,000 years ago, when ice sheets up to 3 km thick covered the continents of the Northern Hemisphere.*

times the amount of ice on Earth today (**Figure 10.19**). Ice sheets with thicknesses of 2 to 3 km built up over North America, Europe, and Asia. Continental shelves that today are under water were exposed by a drop in sea level of about 130 m, increasing the land area. The Wisconsin glaciation is only the latest in a series of ice ages that we will study later in this chapter.

FIGURE 10.19 The extent of continental glaciers and sea ice in the Northern Hemisphere at the Wisconsin glacial maximum, around 20,000 years ago. The continental shelves were exposed by the lowering of the sea level, exemplified here by the expanded coastline of Florida. [*Wm. Robert Johnston.*]

Short-Term Variations Within Glacial Cycles

Within glacial cycles, temperatures do not vary smoothly over time (see Figure 10.18). Superimposed on the 100,000-year glacial cycles are climate fluctuations of shorter duration, some nearly as large as the changes from glacial to interglacial periods. Geologists have combined the information from cores in continental and valley glaciers, lake sediments, and deep-sea sediments to reconstruct a decade-by-decade and in some cases a year-by-year history of short-term climate variations during the most recent glacial cycle. Here we summarize some of the basic features of this remarkable chronicle.

- During the Wisconsin glaciation, Earth's climate was highly variable, with shorter (1000-year) temperature oscillations occurring within longer (10,000-year) cycles. The most extreme variations appear to have been in the North Atlantic region, where average local temperatures rose and fell by as much as 15°C. Massive discharges of icebergs and fresh water resulting from the sudden warmings altered thermohaline circulation in the oceans and dumped large amounts of glacial material into deep-sea sediments.

Within glacial cycles, there have been many short-term variations in climate, some marked by abrupt global warming.

- The main phase of warming after the Wisconsin glaciation occurred between 14,500 and 10,000 years ago. It was not a smooth transition; rather it occurred in two main stages, with a pause in deglaciation and a return to cold conditions between 13,000 and 11,500 years ago. The extremely abrupt increases in temperature at 14,500 and 11,500 years ago are perhaps the most astonishing aspect of this jerky transition. Broad regions of Earth experienced almost simultaneous changes from ice age to interglacial temperatures during intervals as short as 30 to 50 years. Evidently, atmospheric circulation can reorganize very rapidly, flipping the entire climate system from one state (glacial cold) to another (interglacial warmth) in less than a human lifetime! This observation raises the possibility that human-induced global climate change could involve abrupt shifts to a new (and unknown) climate state, rather than just a gradual warming.

- The current interglacial period has been unusually long and stable when compared with the previous interglacial periods of the Pleistocene; in fact, the Holocene appears to be the longest and most stable warm period over the last 400,000 years. The warmest temperatures occurred at the beginning of this epoch.

Geologists have documented regional variations of about 5°C on time scales of 1000 years or so, but the global changes during this period are much smaller, with a total range of only 2°C. Holocene conditions were no doubt favorable for the rapid rise of agriculture and civilization that followed the end of the Wisconsin glaciation. Scientists have speculated that the production of greenhouse gases caused by deforestation, agriculture, and other preindustrial human activities have kept the gases at high enough atmospheric concentrations to prevent the climate from plunging into another ice age.

The Holocene interglacial period has been the longest and most stable warm period during the last 400,000 years.

GLACIAL LANDSCAPES

Glaciation is the most important geologic process of the climate system. Many spectacular features of the landscapes we see today were created by recent episodes of glaciation. In mountainous regions, glaciers have gouged out steep-walled valleys, scraped their surfaces down to bedrock, and plucked huge blocks from their rocky floors. During ice ages, glaciers have pushed across entire continents, carving far more topography than do rivers and wind.

The Glacial Budget

For a glacier to form, temperatures must be low enough to keep snow on the ground year-round. These conditions occur at high latitudes, because the Sun's rays strike Earth at low angles there (see Figure 12.4), and at high altitudes, because the atmosphere becomes steadily cooler up to altitudes of about 10 km (see Figure 10.2). Therefore, the height of the *snow line*—the altitude above which snow does not completely melt in summer—generally decreases toward the poles, where snow and ice cover the ground year-round even at sea level. Near the equator, glaciers form only on mountains that are higher than about 5000 m.

The precipitation of snow and the formation of glaciers require moisture as well as cold. Moisture-laden winds tend to drop most of their snow on the windward side of a high mountain range, so the leeward side is likely to be dry and unglaciated. Parts of the high Andes of South America, for instance, lie in a belt of prevailing easterly winds. Glaciers form on the moist eastern slopes, but the dry western side has little snow and ice. In arid climates, glaciers are unlikely to form unless the temperature is so frigid throughout the year that very little snow melts, as in Antarctica.

A fresh snowfall is a fluffy mass of loosely packed snowflakes. As these small, delicate ice crystals age on the ground, they shrink and become grains, and the mass of snowflakes compacts to form a dense, granular snow. As new snow falls and buries the older snow, the older granular snow compacts, eventually producing solid glacial ice as the grains recrystallize and are cemented together. The amount of ice added to the glacier annually is its **accumulation.**

> Glaciers form where temperatures are cold enough that snow covers the ground year-round. Such conditions are found at high latitudes and at high altitudes.

When ice accumulates to a mass sufficient for gravity to act on it, a glacier is born. Ice, like water, flows downhill under the pull of gravity. The ice moves down a mountain valley or downward from the dome of ice at the center of a continental glacier. In either case, the glacier flows to lower altitudes where temperatures are warmer and it loses ice, a process called **ablation.**

> When glaciers accumulate enough ice and become sufficiently massive, they begin to flow downhill, eventually losing ice through ablation.

The two mechanisms by which glaciers lose the most ice are melting and iceberg calving (see Figure 10.5), although wind erosion and, in cold climates, transformation of ice directly into water vapor can also contribute. Most of the ablation takes place at the glacier's leading edge. Therefore, even when a glacier is advancing downward or outward from its center, the ice front may be retreating.

The difference between accumulation and ablation, called the *glacial budget,* determines the growth or shrinkage of a glacier (**Figure 10.20**). When accumulation equals ablation over a long period, the glacier remains a constant size, even as it continues to flow downslope from the area where it forms. Such a glacier accumulates snow and ice in its upper reaches as an equal amount is ablated in its lower reaches. If accumulation exceeds ablation, the glacier grows; if ablation exceeds accumulation, the glacier shrinks.

> The growth or shrinkage of a glacier is determined by its glacial budget: the difference between accumulation and ablation.

Glacial budgets vary from year to year. Over the past several thousand years, many glaciers have maintained a constant average size, though some show evidence of growth or shrinkage in response to short-term regional climate variations. In the last century, however, mountain glaciers in many low-latitude regions have been shrinking in response to global warming, as shown in **Figure 10.21** and as described in the opening story of this chapter.

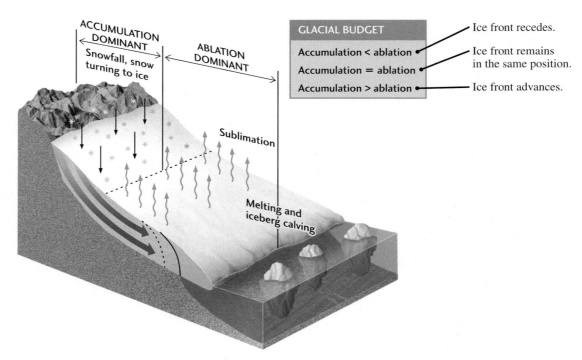

FIGURE 10.20 Accumulation takes place mainly by snowfall over a glacier's colder upper regions. Ablation takes place mainly by melting and iceberg calving in the warmer lower regions. The relationship between accumulation and ablation is the glacial budget.

(a)

(b)

FIGURE 10.21 Photographs of Qori Kalis Glacier in Peru from the same vantage point (a) in July 1978 and (b) in July 2004. Between 1998 and 2001, the ice front retreated an average of 155 m/year, an alarming 32 times faster than the average annual retreat from 1963 to 1978. [*Lonnie G. Thompson, Byrd Polar Research Center, Ohio State University.*]

Glacial Erosional Landforms

Ice is a far more efficient agent of erosion than water or wind. A valley glacier only a few hundred meters wide can tear up and crush millions of tons of bedrock in a single year. The ice carries this heavy load of sediment to the ice front, where it is dropped as the ice melts. The total amount of sediment deposited in the world's oceans per year has been several times larger during recent ice ages than during interglacial periods.

As a glacier drags rocks along its base, the rocks scratch or groove the bedrock beneath it. The abrasions are termed *striations*. The orientation of striations shows us the direction of ice movement—an especially important factor in the study of continental glaciers, which lack obvious valleys. By mapping striations over wide areas formerly covered by a continental glacier, we can reconstruct the glacier's flow patterns (**Figure 10.22**).

A flowing valley glacier carves a series of erosional forms (**Figure 10.23**). At the head of a glacial valley, the plucking and tearing action of the ice tends to carve out an amphitheater-like hollow called a *cirque*, usually shaped like half an inverted cone. With continued erosion, cirques at the heads of adjacent valleys gradually meet at the mountaintops, producing sharp, jagged crests called *arêtes* along the divide. As the glacier flows down from its cirque, it excavates a new valley or deepens an existing stream valley, creating a characteristic **U-shaped valley** (see Figure 10.23). Glacial valley floors are wide and flat and have steep walls, in contrast to the V-shaped valleys of many mountain streams (see Chapter 12).

FIGURE 10.22 Glacial striations on bedrock in Glacier Bay National Park, Alaska. Striations are evidence of the direction of ice movement and are especially important clues for reconstructing the movement of continental glaciers. [*Carr Clifton.*]

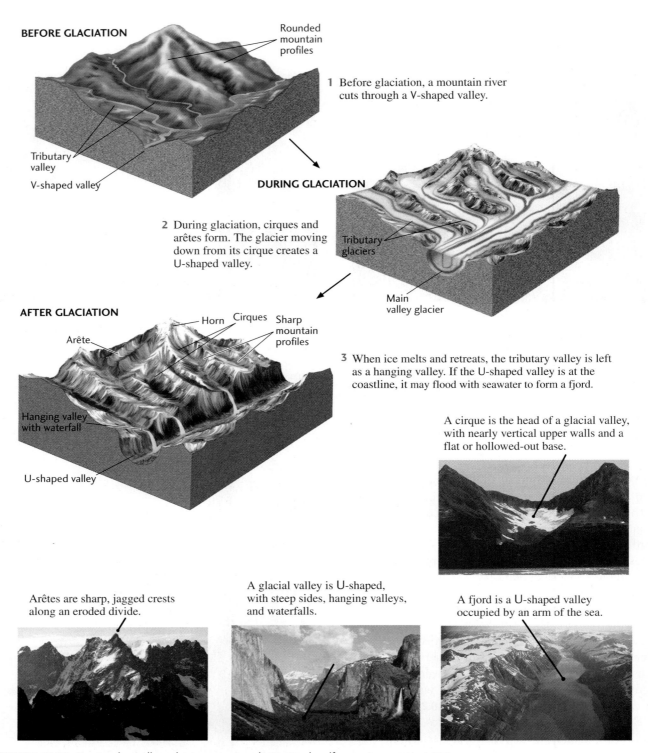

BEFORE GLACIATION

Rounded mountain profiles

1 Before glaciation, a mountain river cuts through a V-shaped valley.

Tributary valley

V-shaped valley

DURING GLACIATION

2 During glaciation, cirques and arêtes form. The glacier moving down from its cirque creates a U-shaped valley.

Tributary glaciers

Main valley glacier

AFTER GLACIATION

Horn   Cirques   Sharp mountain profiles

Arête

Hanging valley with waterfall

U-shaped valley

3 When ice melts and retreats, the tributary valley is left as a hanging valley. If the U-shaped valley is at the coastline, it may flood with seawater to form a fjord.

A cirque is the head of a glacial valley, with nearly vertical upper walls and a flat or hollowed-out base.

Arêtes are sharp, jagged crests along an eroded divide.

A glacial valley is U-shaped, with steep sides, hanging valleys, and waterfalls.

A fjord is a U-shaped valley occupied by an arm of the sea.

FIGURE 10.23 Erosion by valley glaciers creates distinctive landforms. [*cirque: Marli Miller; arêtes: Julien Beausseron; glacial valley: Tom Bean/Corbis; fjord: Steve McCutcheon/Visuals Unlimited.*]

As valley glaciers flow, they pick up and transport heavy loads of sediment, carving U-shaped valleys and other characteristic erosional landforms.

Glaciers and streams also differ in how their tributaries form junctions. Although the ice surface is level where a tributary glacier joins a main valley glacier, the floor of the main valley may be carved much more deeply than that of the tributary valley. When the ice melts, the tributary valley is left as a *hanging valley*—one whose floor lies high above the main valley floor (see Figure 10.23). After the ice is gone and streams occupy the valleys, the junction is marked by a waterfall as the stream in the hanging valley plunges over the steep cliff separating it from the main valley below.

Valley glaciers at coastlines may erode their valley floors far below sea level. When the ice retreats, these steep-walled valleys—which still maintain a U-shaped profile—are flooded with seawater (see Figure 10.23). These arms of the sea carved out by glaciers, called **fjords,** create the spectacular rugged scenery for which the coasts of Alaska, British Columbia, Norway, and New Zealand are renowned.

Glacial Sedimentary Landforms

Glaciers transport eroded rock materials of all kinds and sizes downstream, eventually depositing them where the ice melts. Ice is a very effective transport agent because the material it picks up does not settle out like the load of sediment carried by a stream. Ice can carry huge blocks many meters in diameter that no other transport agent could budge. Some glacial ice is so full of rock material that it is dark and looks like sediment cemented with ice.

When glacial ice melts, it deposits a poorly sorted, heterogeneous load of boulders, pebbles, sand, and clay. This heterogeneous material puzzled geologists who were not aware of its glacial origins. They called it *drift* because it seemed to have drifted in somehow from other areas. A wide range of particle sizes is the characteristic that differentiates glacial sediment from the much better sorted material deposited by streams and winds.

Ice-Laid Deposits Some drift is deposited directly by melting ice. This unstratified and poorly sorted sediment is known as **till,** and it may contain all sizes of rock fragments from clay to boulders (**Figure 10.24**). The large boulders often contained in till are called *erratics* because

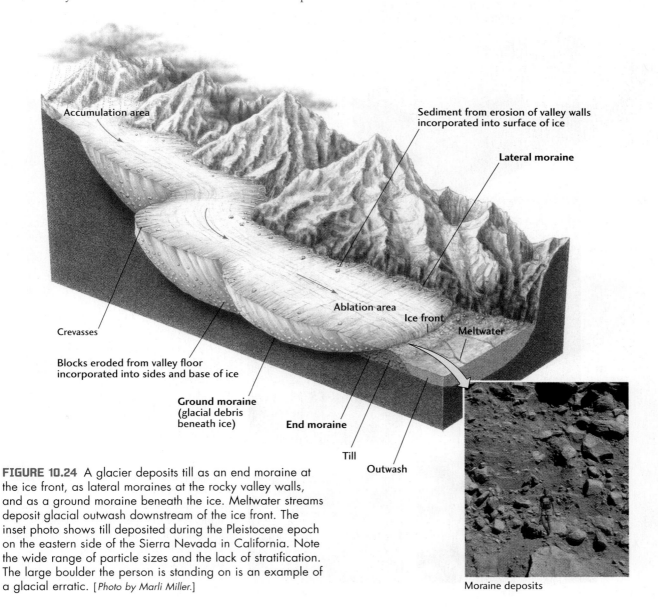

FIGURE 10.24 A glacier deposits till as an end moraine at the ice front, as lateral moraines at the rocky valley walls, and as a ground moraine beneath the ice. Meltwater streams deposit glacial outwash downstream of the ice front. The inset photo shows till deposited during the Pleistocene epoch on the eastern side of the Sierra Nevada in California. Note the wide range of particle sizes and the lack of stratification. The large boulder the person is standing on is an example of a glacial erratic. [*Photo by Marli Miller.*]

Accumulation area

Sediment from erosion of valley walls incorporated into surface of ice

Lateral moraine

Crevasses

Blocks eroded from valley floor incorporated into sides and base of ice

Ablation area

Ice front

Meltwater

Ground moraine (glacial debris beneath ice)

End moraine

Till

Outwash

Moraine deposits

of their seemingly random composition, often very different from that of local rocks.

A **moraine** is an accumulation of rocky, sandy, and clayey material carried by glacial ice or deposited as till. There are many types of moraines, each named for its position with respect to the glacier that formed it (see Figure 10.24). One of the most prominent in size and appearance is an *end moraine*, formed at the ice front. As the ice flows steadily downstream, it brings more and more sediment to its melting edge. The unsorted material accumulates there as a hilly ridge of till.

Some continental glacial terrains display prominent landforms called *drumlins*—large, streamlined hills of till and bedrock that parallel the direction of ice movement (**Figure 10.25**). Drumlins form when the sediment-rich layer at the base of a glacier encounters a knob of bedrock or other obstacle and the excess pressure squeezes out water and drops the sediment.

> When a glacier melts at the ice front, it deposits a poorly sorted sediment called till, forming an end moraine.

Water-Laid and Windblown Deposits Other deposits of drift are laid down as the ice melts and releases water and sediment. Meltwater flowing in tunnels within and beneath the ice and in streams at the ice front may pick up,

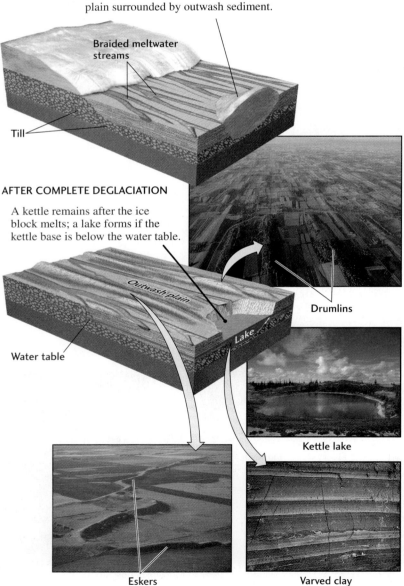

DURING ICE MELTING

A large block of melting ice is isolated from the main ice mass on an outwash plain surrounded by outwash sediment.

Braided meltwater streams

Till

AFTER COMPLETE DEGLACIATION

A kettle remains after the ice block melts; a lake forms if the kettle base is below the water table.

Outwash plain

Lake

Water table

Drumlins

Kettle lake

Eskers

Varved clay

FIGURE 10.25 Water-laid glacial deposits. *Drumlins,* Rochester, New York. [*John Shelton.*] *Kettle lake,* near headwaters of the Thelon River, Canada. [*Galen Rowell/ Corbis.*] Pleistocene *varved clay,* from an excavation in Stockholm, Sweden. The light layers are the coarse sediments deposited in a lake during warm seasons. The dark layers are the fine clays deposited when the lake was frozen in winter. [*John Shelton.*] *Eskers,* near Dahlen, North Dakota. [*Tom Bean.*]

transport, and deposit some of the drift. Like any other waterborne sediment, this material is stratified and well sorted and may be cross-bedded. Drift that has been picked up and distributed by meltwater is called *outwash*, and it often forms broad sedimentary plains downstream

of melting glaciers, known as outwash plains. Strong winds can blow fine-grained material from outwash plains and transport it over long distance and deposit it as *loess*.

Deposits of outwash by glacial meltwater take a variety of forms. *Eskers* are long, narrow, winding ridges of sand and gravel deposited by meltwater flowing in tunnels along the bottom of a melting glacier (see Figure 10.25). *Kames* are small hills of sand and gravel created when drift fills a hole in a glacier and is left behind when the glacier recedes.

Glacial terrains are dotted with **kettles,** hollows or undrained depressions that often have steep sides and may be occupied by ponds or lakes. Kettles are formed when retreating glaciers leave behind isolated blocks of

ice in outwash plains. A block of ice a kilometer in diameter may take 30 years or more to melt. By the time the block has melted completely, the ice front has

retreated so far that little outwash reaches the area to fill the hole. The sand and gravel that were deposited as outwash around the block of ice now surround a depression (see Figure 10.25).

Glacial meltwater may build up behind deposits of till to form lakes. Some lakes formed by continental glaciers were huge, many thousands of square kilometers in extent. The till dams that created these lakes were sometimes breached and carried away at a later time, causing the lakes to drain rapidly and creating huge floods. In eastern Washington State, an area called the Channeled Scablands (**Figure 10.26**) is covered by broad, dry stream channels, relics of torrential floodwaters draining from Lake Missoula, a large glacial lake, now completely emptied. From the giant ripples, sandbars, and coarse gravels found there, geologists have estimated that this flood discharged

FIGURE 10.26 The Channeled Scablands in eastern Washington State contain unique erosional features formed by catastrophic flooding that resulted from the draining of Lake Missoula, a huge glacial lake. This aerial photograph shows Dry Falls, a 350-foot-high, 3-mile-wide group of scalloped cliffs created by the flood. [*Dave Rahm/ Easterbrook Photo and Image Center.*]

21 million cubic meters of water per second, flowing as fast as 30 m/s. For comparison, the discharge rate of the Mississippi River in full flood is less than 50,000 m³/s.

Valley glaciers may deposit silts and clays on the bottom of a lake formed at the edge of the ice. These deposits are characterized by a series of alternating coarse and fine layers called *varves*, formed by seasonal freezing of the lake surface (see Figure 10.25).

PLEISTOCENE GLACIATIONS

Louis Agassiz, a Swiss geologist, was the first to propose (in 1837) that the glaciers he observed in the Alps must have been much larger and thicker in the geologically recent past. He suggested that during a past ice age, Switzerland was covered by an extensive continental glacier almost as thick as its mountains were tall, similar to the one in Greenland today. Among the evidence he cited was the obvious glacial sculpting of the high Alpine peaks, such as the mighty Matterhorn (**Figure 10.27**). Agassiz's hypothesis was controversial and was not immediately accepted.

Agassiz emigrated to the United States in 1846 and became a professor at Harvard University, where he continued his studies in geology and other sciences. His research took him to many places in the northern parts of Europe and North America, from the mountains of Scandinavia and New England to the rolling hills of the American Midwest. In all these diverse regions, Agassiz saw signs of glacial erosion and sedimentation (**Figure 10.28**). In the flat country of the American Great Plains, he observed deposits of glacial drift that reminded him of the end moraines of Swiss valley glaciers. The heterogeneous material of the drift, including erratic boulders, convinced him of its glacial origin, and the freshness of the soft sediments indicated that they were deposited in the recent past. The areas covered by this drift were so vast that the ice that deposited them must have been a continental glacier much larger than the ones that now cover Greenland and Antarctica.

Agassiz expanded his ice age hypothesis, proposing that a great continental glaciation had extended the polar ice caps far into regions that now enjoy more temperate climates. For the first time, people began to talk about ice ages.

FIGURE 10.27 The high mountains of the Alps, such as the famous Matterhorn, shown here, were sculpted by a continental glacier nearly as thick as the peaks are tall. These obviously sculpted peaks provided compelling evidence for an ice age in the recent geologic past. [*Hubert Stadler/Corbis.*]

FIGURE 10.28 Irregular hills alternate with lakes in a terrain of glacial till in Coteau des Prairies, South Dakota. Such landscapes provide evidence for the great continental glaciations of the Pleistocene ice ages. [*John S. Shelton.*]

The Geologic Record of Glaciations

Soon after Agassiz's hypothesis became widely accepted in the mid-nineteenth century, geologists discovered that there had been multiple ice ages during the Pleistocene epoch (1.8 million to 11,000 years ago), with warmer interglacial periods between them. As they mapped glacial deposits in more detail, they became aware of several distinct layers of drift, the lower ones corresponding to earlier ice ages. Between these older layers of glacial material were soils containing fossils of warm-climate plants. These fossils provided evidence that the glaciers had retreated as the climate warmed. By the early part of the twentieth century, scientists were convinced that at least four major glaciations had affected North America and Europe during the Pleistocene epoch. In North America, these ice ages, from youngest to oldest, are named after the U.S. states where the evidence of glacial advance is best preserved: Wisconsin, Illinois, Kansas, and Nebraska.

There were multiple glacial cycles during the Pleistocene epoch.

This simple scheme was modified in the late twentieth century, when geologists examining oceanic sediments found fossil evidence of warming and cooling of the oceans. Sediments that had accumulated continuously in undisturbed ocean basins contained a much more complete geologic record of climate variations than did continental glacial deposits, and they could be more precisely dated. By analyzing oxygen isotopes ratios in marine sediments from around the world, geologists were able to construct a record of climate history millions of years into the past. This record showed a more much complicated history of glacial advance and retreat.

Milankovitch Cycles

The major ups and downs in the marine sediment record (**Figure 10.29**) during the Pleistocene epoch match the glacial cycles in the ice core record (see Figure 10.18). Why does the climate fluctuate in such a pattern? Changes in the amount of solar energy received by Earth's surface are an obvious possibility. We know that it gets cold in winter because the amount of sunlight that falls at a particular latitude decreases due to the tilt of Earth's axis. Could periods of glacial cold be explained by decreases in the solar input over much longer time scales?

The answer appears to be yes. There are indeed small periodic variations in the amount of radiation Earth receives from the Sun. These variations are caused by **Milankovitch cycles,** periodic variations in Earth's movement around the Sun. They are named after the Serbian geophysicist who first calculated them in the early twentieth century. Three kinds of Milankovitch cycles can be correlated with climate variation (**Figure 10.30**).

First, the shape of Earth's orbit around the Sun changes periodically, becoming more circular at some times and more elliptical at others. The degree of ellipticity of Earth's orbit around the Sun is known as *eccentricity*. A nearly circular orbit has low eccentricity, and a more elliptical orbit has high eccentricity (Figure 10.30a). The amount of solar radiation Earth

Milankovitch cycles are small periodic variations in the orientation of Earth's axis and in the shape of its orbit around the Sun.

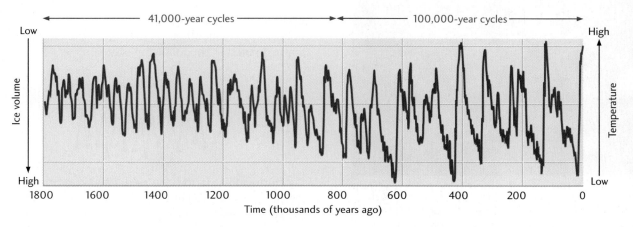

FIGURE 10.29 Changes in global climate over the last 1.8 million years, as inferred from oxygen isotope ratios in marine sediments. The peaks indicate interglacial periods (high temperatures, low ice volumes, high sea level), and the valleys indicate ice ages (low temperatures, high ice volumes, low sea level). [*After L. E. Lisiecki and M. E. Raymo, Paleoceanography 20 (2005): 1003.*]

(a) Eccentricity (100,000 years)

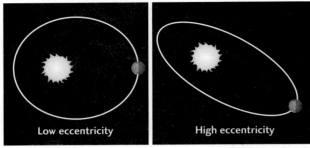

Low eccentricity High eccentricity

(b) Tilt (41,000 years)

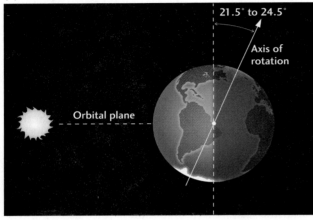

21.5° to 24.5°

Axis of rotation

Orbital plane

(c) Precession (23,000 years)

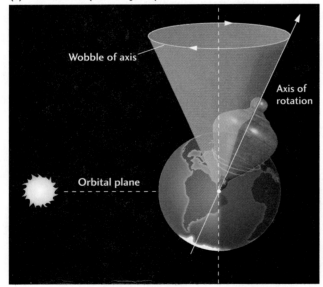

Wobble of axis

Axis of rotation

Orbital plane

FIGURE 10.30 Three kinds of Milankovitch cycles (much exaggerated in these diagrams) affect the amount of solar radiation Earth receives. (a) Eccentricity is the degree of ellipticity of Earth's orbit. (b) Tilt is the angle between Earth's axis of rotation and the angle perpendicular to the orbital plane. (c) Precession is the wobble of the axis of rotation. One can imagine this motion by thinking of the wobble of a spinning top.

receives varies slightly with eccentricity. The length of one cycle of variation in eccentricity is about 100,000 years.

Second, the angle or *tilt* of Earth's axis of rotation changes periodically. Today this angle is 23.5°, but it cycles between 21.5° and 24.5° with a period of about 41,000 years. These variations also slightly change the amount of radiation Earth receives from the Sun (Figure 10.30b).

Third, Earth's axis of rotation wobbles like a top, giving rise to a pattern of variation called *precession* with a time period of about 23,000 years (Figure 10.30c). Precession, too, modifies the amount of radiation Earth receives from the Sun, though by less than variations in eccentricity and tilt.

You can see lots of small ups and downs in the climate record of Figure 10.29, but in the last half-million years, the record reveals a sawtooth pattern of major glacial cycles that looks roughly like this:

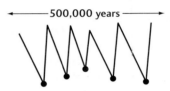

◄——— 500,000 years ———►

In particular, you can count five glacial maxima, in which ice volumes are high and temperatures are low (shown in the sketch above as black dots). This pattern, which is also visible in Figure 10.29, reveals an average time interval between glacial maxima of 100,000 years. The 100,000-year spacing of ice ages closely matches the times of high eccentricity, when Earth receives less radiation, on average, from the Sun—a Milankovitch cycle.

Now let's move backward in time to examine the first half-million years of the record shown in Figure 10.29, from 1.8 to 1.3 million years ago. Again we see many small fluctuations, but major glacial maxima and minima occur more frequently than they do in the later record, as approximated in the following sketch:

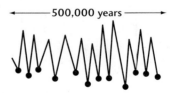

◄——— 500,000 years ———►

During this period, we find 12 glacial maxima, with an average spacing between them of about 41,000 years (500,000 years/12 cycles = 41,667 years per cycle). This shorter interval is very close to the 41,000-year pattern of variation in the tilt angle of Earth's axis—another Milankovitch cycle! Like the variation in eccentricity, the variation in tilt is very small, only about 3° (see Figure 10.30b), but it's evidently enough to trigger ice ages.

The small changes in solar radiation caused by Milankovitch cycles, however, cannot completely explain the magnitude of the drops in Earth's surface temperature from interglacial periods to ice ages. Some type of positive feedback must be operating within the climate system to amplify the solar forcing. The data in Figure 10.16 strongly suggest that this feedback involves greenhouse gases. Atmospheric concentrations of carbon dioxide and methane precisely track the temperature variations throughout the glacial cycles—warm interglacial periods are marked by high concentrations, cold glacial periods by low concentrations. Exactly how this feedback works has not yet been fully explained, but it demonstrates the importance of the greenhouse effect in long-term climate variations.

Many other aspects of this story are not yet understood. For example, you will notice in Figure 10.29 that the 41,000-year periodicity dominates the climate record up to about 1 million years ago. Then the highs and lows become more variable, eventually shifting to the 100,000-year periodicity about 700,000 years ago. What caused this transition? Climate scientists are still scratching their heads.

In fact, we don't really know what triggered the Pleistocene ice ages in the first place. The climate record shows that the 41,000-year glacial cycles were not confined to the Pleistocene but extended back at least into the Pliocene epoch (5.3 million to 1.8 million years ago), when Antarctica became covered in ice (see Figure 8.17). The global cooling of Earth's climate that preceded these glaciations began during the Miocene epoch (23 million to 5.3 million years ago). Its cause continues to be debated, although most geologists believe it is somehow related to continental drift. According to one hypothesis, the collision of the Indian subcontinent with Eurasia and the resulting Himalayan orogeny led to an increase in the weathering of silicate rocks and the chemical reactions of weathering decreased the amount of carbon dioxide in the atmosphere. Other hypotheses are based on changes in oceanic circulation associated with the opening of the Drake Passage between South America and Antarctica (25 million to 20 million years ago) or the closing of the Isthmus of Panama between North and South America (about 5 million years ago). Perhaps the cooling resulted from a combination of these events.

THE RECORD OF ANCIENT GLACIATIONS

The Pleistocene glacial cycles were not unique in Earth's history. Since the early part of the twentieth century, we have known from glacial striations and lithified ancient tills, called **tillites,** that glaciers covered parts of the

continents several times in the distant geologic past, long before the Pleistocene. Tillites record major continental glaciations during the Permian, Carboniferous, and Ordovician periods and at least twice during Precambrian time (**Figure 10.31**).

The Permian-Carboniferous glaciation covered much of southern Gondwana about 300 million years ago, leaving deposits that have been preserved as tillites across much of the Southern Hemisphere. The Ordovician glaciation was more limited in its distribution and is best preserved in northern Africa.

The oldest confirmed glaciation occurred during the Proterozoic eon, about 2.4 billion years ago. Its glacial deposits are preserved in Wyoming, along the Canadian portion of the Great Lakes, in northern Europe, and in South Africa. Some geologists argue for an even older glaciation in the Archean eon about 3 billion years ago, but this interpretation is disputed.

The youngest Proterozoic glaciation, which spanned a period between 750 million and about 600 million years ago, involved several ice ages separated by warm interglacial periods. Glacial deposits of this age have been found on every continent (Figure 10.31a). Curiously, the reconstruction of paleocontinents indicated that the ice sheets extended to much lower latitudes than during the Pleistocene glaciations, perhaps all the way to the equator! This evidence has provoked some geologists to speculate that Earth may have been completely covered by ice, from pole to pole—a bold hypothesis called *Snowball Earth* (see Figure 10.31d).

According to the Snowball Earth hypothesis, there was ice everywhere—even the oceans were frozen. The average global temperatures would have been about –40°C, like those of the Antarctic today. Except for a few warm spots near volcanoes, very little life would have survived. How could such an apocalyptic event have occurred? And how could it have ended, returning us to the climate we know today? The answers may lie in the feedbacks that occur within the climate system.

According to one scenario, as Earth initially cooled, ice sheets at the poles spread outward, their white surfaces reflecting more and more sunlight away from Earth. This increase in Earth's albedo cooled the planet, which further expanded the ice sheets. This self-reinforcing process continued until it reached the tropics, encasing the planet in a layer of ice as much as 1 km thick. This scenario is an example of albedo feedback gone wild.

Earth remained buried in ice for millions of years, but the few volcanoes that poked above the surface slowly pumped carbon dioxide into the atmosphere. When the concentration of carbon dioxide reached a critical level,

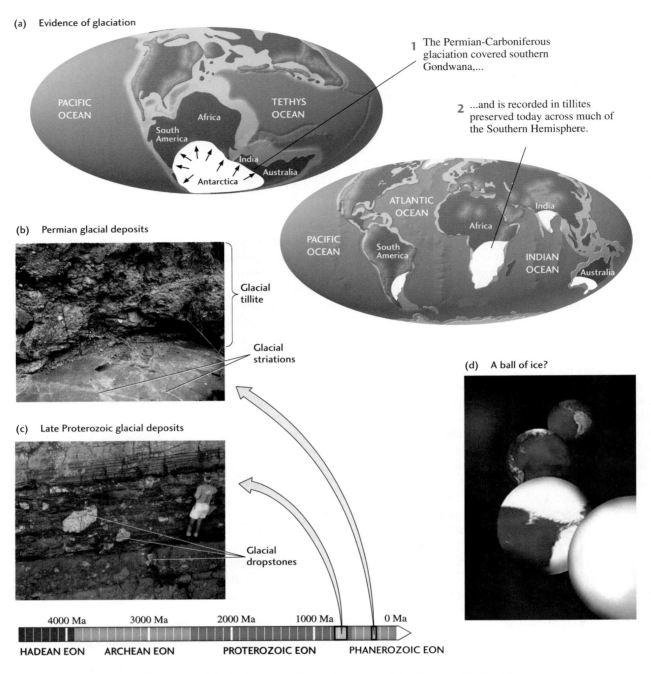

(a) Evidence of glaciation

PACIFIC OCEAN

TETHYS OCEAN

Africa

South America

India

Australia

Antarctica

1 The Permian-Carboniferous glaciation covered southern Gondwana,...

2 ...and is recorded in tillites preserved today across much of the Southern Hemisphere.

ATLANTIC OCEAN

PACIFIC OCEAN

South America

Africa

India

INDIAN OCEAN

Australia

(b) Permian glacial deposits

Glacial tillite

Glacial striations

(c) Late Proterozoic glacial deposits

Glacial dropstones

(d) A ball of ice?

| 4000 Ma | 3000 Ma | 2000 Ma | 1000 Ma | 0 Ma |

HADEAN EON ARCHEAN EON PROTEROZOIC EON PHANEROZOIC EON

FIGURE 10.31 Ancient glaciations. (a) The first map shows the extent of the Permian-Carboniferous glaciation, which occurred more than 350 million years ago. At that time, the southern continents were assembled into the giant continent Gondwana, and the ice cap was situated in the Southern Hemisphere, centered over Antarctica, which is home to a continental glacier today. The second map shows the distribution of Permian-Carboniferous glacial deposits today. (b) Permian glacial deposits from South Africa. (c) Late Proterozoic glacial deposits. (d) The development of a late Proterozoic Snowball Earth. Geologists debate the extent to which ice covered the globe, but some think even the oceans became frozen. [*Photos by John Grotzinger.*]

the temperature rose, the ice melted, and Earth again became a greenhouse.

The Snowball Earth hypothesis is very controversial, and some geologists disagree with the idea that the oceans were completely frozen. Nevertheless, the evidence for glaciation at low latitudes is strong, and the hypothesis serves as an example of the potential of feedbacks in Earth's climate system to produce extreme change. Geologists have their work cut out for them in trying to understand the extremes of Earth's climate system.

GOOGLE EARTH PROJECT

Glaciers are the most visible features of Earth's cryosphere. Their movements erode the rocks beneath them and deposit huge amounts of sediments. Glaciers have created some spectacular features on Earth's surface in the recent geologic past that can be easily seen using Google Earth.

In this Google Earth Project, you will explore glaciers and glacial landscapes at a number of locations around the world. For this project, you will need to turn on the Terrain Layer and, in the 3D View window of Options, choose "Decimal degrees" in the Show Lat/Long box and "Meters, Kilometers" in the Show Elevation box. You can navigate to the initial geographic position for each exercise by typing the listed coordinates into the "Fly To" search window and clicking on the Search button. You can then navigate by using the Zoom slider to zoom the eye altitude in or out and the Look joystick to rotate the compass azimuth of the view or to tilt the view toward the horizontal. (For these exercises, it's better to turn off "Automatically tilt while zooming" in the Navigation window of Options.) Use the Move joystick to translate your position while maintaining the same Look angles.

LOCATION	Glaciers and glacial landscapes around the world
GOAL	Learn how to identify the types of glaciers and glacial features
LINKED	Figures 10.23 and 10.24

1. Navigate to 61.385° N, 148.500° W in south central Alaska, zoom to an eye altitude of 4.0 km, rotate the view to look east, and tilt the view to see an expanse of ice: the Knik glacier. Using your cursor, explore the elevation of the ice surface to observe the direction of its slope. Based on this information, which of the following is the best description of the ice mass?

   a. Continental glacier flowing outward from its highest point near the center of the glacier

   b. Continental glacier flowing westward from its highest point on the east side of the glacier

   c. Valley glacier flowing westward

   d. Valley glacier flowing eastward

2. Navigate to 64.400° N, 16.800° W on the south side of Iceland, zoom to an eye altitude of 150 km, and examine the large mass of ice below you, which the Icelanders call the Vatnajökull Glacier. Using the ruler tool, measure the size of the glacier, and using the cursor,

find the region of the glacier with the highest elevation. Explore the glacier looking for evidence of flow. Based on this information, which of the following is the best description of the ice mass?

   a. Continental glacier flowing outward from its highest point near the center of the glacier

   b. Continental glacier flowing westward from its highest point on the east side of the glacier

   c. Valley glacier flowing westward

   d. Valley glacier flowing eastward

3. Navigate to 37.730° N, 119.580° W in Yosemite National Park, California, zoom to an eye altitude of 3 km, rotate the view to look northeast, and tilt the view to see Yosemite Valley. Observe the shape of the valley perpendicular to its axis and, using your cursor, explore the elevation of the valley floor to

(Continued.)

observe the direction of its slope. Which of the following is the best description of Yosemite Valley?

a. V-shaped valley cut by a stream flowing to the southwest

b. U-shaped valley cut by a glacier flowing to the southwest

c. V-shaped valley cut by a stream flowing to the northeast

d. U-shaped valley cut by a glacier flowing to the northeast

4. Navigate to 45.100° S, 167.020° E on the west coast of South Island, New Zealand. Zoom to an eye altitude of 1 km, rotate the view to look southeast, and tilt the view to see a water-filled valley in the mountainous terrain. Explore the extent of this water-filled valley. Which of the following terms best describes it?

a. Glacial lake   **c.** Kettle lake

b. Hanging valley   **d.** Fjord

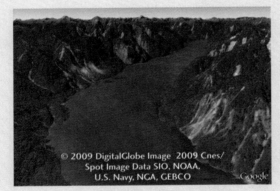

5. Navigate to 43.765° N, 110.730° W in Grand Teton National Park, zoom to an eye altitude of 7 km, and examine Jenny Lake, which lies at the eastern mouth of a large valley. Use your cursor to profile the elevation. You will observe that the lake is rimmed on its eastern side by a narrow ridge, green with trees, that rises up to 30 m above the lake

surface. Zoom in to 3.5 km, rotate the view to look west, and tilt the view to look up the valley; using the Move joystick, move eastward so that you can view the position of the lake relative to the Teton mountain front. Which of the following terms best describes the ridge that encircles Jenny Lake?

a. Esker   **c.** End moraine

b. Drumlin   **d.** Lateral moraine

Optional Challenge Question

6. Navigate to 46.014° N, 7.616° E in the Swiss Alps, zoom to an eye altitude of 3.5 km, and observe the cracks in the glacial ice. Using your cursor, explore the elevation of the ice surface to observe how the slope changes. Which of the following is the best explanation of the cracks?

a. Crevasses along the side of a valley glacier caused primarily by a bend in flow direction

b. Crevasses across a valley glacier caused primarily by a slope increase in the direction of flow

c. Crevasses across a valley glacier caused primarily by a blockage of flow by an end moraine

d. Crevasses along the side of a valley glacier caused primarily by a constriction of the flow by the valley wall

SUMMARY

What is the climate system? The climate system includes all the components of the Earth system and all the interactions among the components that determine how climate behaves in space and time. The main components of the climate system are the atmosphere, hydrosphere, cryosphere, lithosphere, and biosphere. Each component plays a role in the climate system that depends on its ability to store and transport mass and energy.

What is the greenhouse effect? When Earth's surface is warmed by the Sun, it radiates heat back into the atmosphere. Carbon dioxide and other greenhouse gases absorb some of this infrared radiation and reradiate it in all directions, including downward to Earth's surface. This radiation maintains the atmosphere at a warmer temperature than it would be if there were no greenhouse gases, similar to the warmer air temperature maintained in a greenhouse.

Was the twentieth-century warming caused by human activities? The observed increase of about 0.6°C in Earth's average annual surface temperature during the twentieth century is correlated with a significant rise in the atmospheric concentrations of CO_2 and other greenhouse gases. Much of this rise is due to fossil-fuel burning, deforestation, and other human activities. Most experts on Earth's climate are now convinced that the twentieth-century warming was human-induced and that the warming will continue into the twenty-first century as atmospheric concentrations of greenhouse gases continue to rise.

How has Earth's climate changed over time? Natural variations in climate occur on a wide range of scales in both time and space. Some variations result from factors outside the climate system, such as solar forcing and changes in the distribution of land and sea surfaces caused by continental drift. Others result from variations within the climate system itself. Short-term regional climate variations include the El Niño-Southern Oscillation. Long-term global climate variations are exemplified by the Pleistocene glacial cycles, during which surface temperatures changed by as much as 6° to 8°C. The Holocene has been the longest and most stable warm period of the last 400,000 years.

How do glaciers form? Glaciers form where climates are cold enough that snow, instead of melting completely in summer, is transformed into ice. As snow accumulates, the glacier thickens, until it becomes so massive that gravity starts to pull it downhill. The glacial budget is the difference between accumulation and ablation. If accumulation equals ablation, the size of the glacier remains constant. If the amount of ice ablated by melting, iceberg calving, and transformation into water vapor is greater than the amount of ice accumulated in the glacier's upper reaches, the glacier shrinks. Conversely, if accumulation exceeds ablation, the glacier grows.

How do glaciers shape the landscape? Glaciers scrape, pluck, and grind rocks into sizes ranging from boulders to fine dust. Valley glaciers erode cirques and arêtes at their heads, excavate U-shaped and hanging valleys, and create fjords by eroding their valleys below sea level at the coast. Glaciers transport huge quantities of sediment to the ice front, where melting releases them. The sediments may be deposited directly by the melting ice as till or picked up by meltwater streams and laid down as outwash

What are ice ages, and what causes them? Studies of the geologic ages of glacial deposits on land and in marine sediments show that continental ice sheets advanced and retreated many times during the Pleistocene epoch. Each glacial period corresponded to a global lowering of sea level that exposed large areas of continental shelves. During interglacial periods, sea level rose and submerged the shelves. The favored explanation for these glacial cycles is the effect of Milankovitch cycles. These very small periodic variations in the eccentricity of Earth's orbit and the angle of its axis alter the amount of solar radiation received at Earth's surface. These changes have been amplified by positive feedbacks involving concentrations of greenhouse gases.

KEY TERMS AND CONCEPTS

ablation (p. 336)	Milankovitch cycle (p. 343)
accumulation (p. 336)	moraine (p. 340)
albedo (p. 328)	negative feedback (p. 328)
climate (p. 318)	positive feedback (p. 328)
climate system (p. 319)	stratosphere (p. 320)
continental glacier (p. 323)	thermohaline circulation
El Niño (p. 330)	(p. 321)
fjord (p. 339)	till (p. 339)
glacial cycle (p. 334)	tillite (p. 345)
greenhouse effect (p. 328)	troposphere (p. 320)
greenhouse gas (p. 328)	twentieth-century warming
ice age (p. 334)	(p. 332)
interglacial period (p. 334)	U-shaped valley (p. 337)
kettle (p. 341)	valley glacier (p. 322)

EXERCISES

1. What is a greenhouse gas, and how does it affect Earth's climate?

2. What is the role of continental glaciers in climate variation?

3. Give an example not discussed in this chapter of a positive feedback and a negative feedback in the climate system.

4. Why do scientists think the twentieth-century warming is associated with human activities?

5. What information about glacial cycles has been obtained by studying ice cores?

6. What does glacial shrinkage tell us about the balance between ablation and accumulation?

7. What type of sedimentary deposit marks the farthest advance of a glacier?

8. Why are kettles described as water-laid sedimentary deposits rather than ice-laid deposits?

9. Why does sea level drop during ice ages?

10. What variation in Earth's orbit correlates most strongly with glacial cycles during the past 500,000 years?

11. Do Milankovitch cycles fully explain the warming and cooling of the global climate during Pleistocene glacial cycles?

Visual Literacy Task

MAP OF GREENLAND

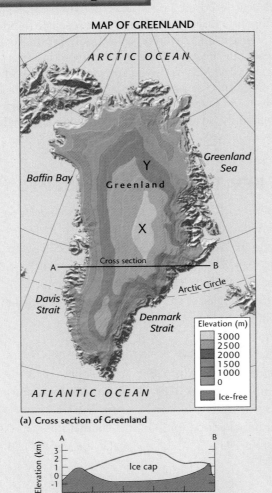

(a) Cross section of Greenland

FIGURE 10.7 Topographic map and cross section of the Greenland continental glacier. (a) Extent and elevation of the Greenland ice sheet. [*After R. F. Flint, Glacial and Quaternary Geology. New York: Wiley, 1971.*] (b) Generalized cross section of south-central Greenland shows the lenslike shape of the glacier. Ice shelves are shown in white.

MAP OF ANTARCTICA

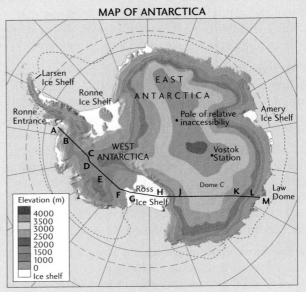

(a) Cross section of Antarctica

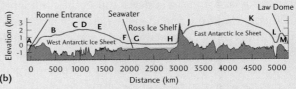

(b)

FIGURE 10.8 Topographic map and cross section of the Antarctic continental glacier. (a) Extent and elevation of the Antarctic ice sheet. Ice shelves are shown in white. (b) Generalized cross section of the ice sheet and the land beneath it. [*After U. Radok, "The Antarctic Ice." Scientific American (August 1985): 100; based on data from the International Antarctic Glaciological Project.*]

1. **Compare the vertical and horizontal scales for the Greenland cross section. How does the width of the Greenland glacier compare to its thickness?**

a. The width is about the same as the thickness.
b. The width is about 3 times the thickness.
c. The width is about 200 times the thickness.

2. **What does the color green represent?**

a. Forests
b. Ice 0 to 1000 m above sea level
c. Swamps
d. Land exactly at sea level (0 m)

3. **What on the cross section shows the same information as the colors on the map?**

a. The height of the blue line
b. The thickness between the blue line and gray/brown at the bottom of the cross section
c. The top of the gray/brown at the bottom of the cross section
d. The thickness of the gray/brown at the bottom of the cross section

4. **If you were traveling from X to Y on the map of Greenland, in what direction are you traveling?**

a. Uphill and north
b. Uphill and south
c. Downhill and north
d. Downhill and south

5. **In general, what can you say about the thickness of ice and elevation of land?**

a. The elevation of land is lowest when the thickness of ice is greatest.
b. The elevation of land is highest when the thickness of ice is greatest.
c. The elevation of land and thickness of ice are not related.

Thought Question: Explain how you would respond to a fellow student who said, "The colors on the map represent the thickness of the ice."

The Hydrologic Cycle and Groundwater

Water Management in California

CALIFORNIA IS IN THE MIDST of intense public policy debates about how to manage its water supply: where it comes from, how much is available, and who should get it. Several years of dry weather have depleted reservoirs and groundwater supplies. Although the exceptional winter rain and snow of 2010–2011 have provided much-needed relief, the residents of California need to plan carefully for water usage during times of drought.

California makes an excellent case study of significant and diverse water management issues. Imagine the following virtual tour of the state. In the northern part of the state, water allocation policies threaten the population of salmon on the Klamath River, creating

> How [do we] manage the water supply: where it comes from, how much is available, and who should get it [?]

recurring conflict. To the south, some residents of the Imperial Valley are still upset about a decision to divert Colorado River water, to be used for irrigation of their land, to the municipality of San Diego. In the east-central part of the state, the success of a hard-fought battle to restore salmon to the San Joaquin River depends on continued cooperation among various stakeholder groups and continued downstream improvements in river quality. In west-central parts of the state, cities and farms have been ordered to reduce their water usage in order to restore flows to the Russian River where steelhead trout spawn. And across the state, flood-prone communities have petitioned the federal government to relax laws that impose strict regulations on floodplain designations so that property can be developed and residents can reduce costly flood insurance.

A tabulation of key environmental statistics shows why California has become a major management concern in recent decades: 22 percent of the state's remaining native fish species are listed as threatened or endangered, and another 45 percent are imperiled or qualified for listing. More than 90 percent of California's lakes, rivers, and streams are listed as "impaired," meaning they cannot be used for one or more of their intended uses (drinking, irrigation, fishing, swimming, and so on). Looking ahead, the challenges and conflicts of water management are likely to intensify as population growth increases pressure on California's water resources. ◆

Aerial view of a water-carrying aqueduct in Outer Los Angeles, California. [*Ron Chapple Stock/Photolibrary.*]

MOST OF US HAVE HEARD the lines from Samuel Taylor Coleridge's "Rime of the Ancient Mariner": "Water, water, everywhere/Nor any drop to drink." About 70 percent of Earth's surface is covered in water, but only a fraction of it is available for human consumption. Sometimes it takes a prolonged drought, like the one in the southwestern United States, to remind us how important water is to us.

The United States, one of the heaviest users of water in the world, has been steadily increasing its water consumption for over 100 years. In the 30 years between 1950 and 1980 alone, water use increased over 10 times, from 34 billion gallons a day to about 370 billion gallons a day. Some, but not all, of this increase is a result of population growth. The good news is that because of increased awareness, water consumption per person in the United States actually fell by about 20 percent between 1980 and 1985 and has remained relatively constant since then despite continued population growth (**Figure 11.1**). Many developed countries have started to emphasize more efficient use of their finite water resources.

Water is also essential to a wide variety of geologic processes. It is important in weathering and erosion, both as a solvent of minerals in rock and soil and as a transport

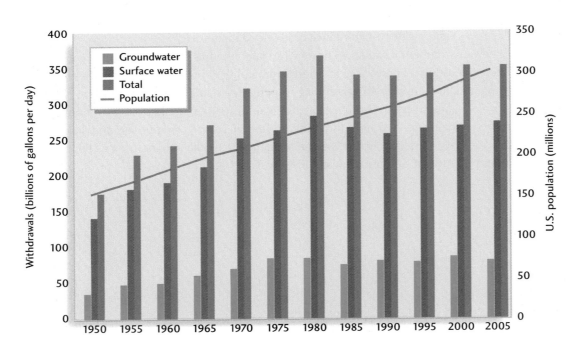

FIGURE 11.1 Estimated use of water, by source, in the United States from 1950 to 2005.
[*Data from U.S. Geological Survey.*]

agent that carries away dissolved and weathered materials. Rivers and glacial ice help to shape the landscape of the continents. Water that sinks into Earth's crust forms large reservoirs of groundwater.

We saw in Chapter 10 that water exchanged between the oceans and atmosphere forms a critical link in Earth's climate system. Climate scientists now recognize that understanding the cycling of water on, over, and under Earth's surface is one of the most important steps in climate prediction.

> Hydrology is the study of the movements and characteristics of water on and under Earth's surface

In this chapter, we will study the movements and characteristics of water on, over, and under Earth's surface. This field of study, known as hydrology, is becoming more important to all of us as the demand on limited water supplies increases and our need to understand Earth's climate patterns is heightened. To address these concerns, we must understand not only where to find water, but also how water cycles through the Earth system.

FLOWS AND RESERVOIRS

We can see water in Earth's lakes, oceans, and polar ice caps, and we can see water moving over Earth's surface in streams and glaciers. It is harder to see the massive amounts of water stored in the atmosphere and underground, or the flows of water into and out of the storage places. As water evaporates, it moves into the atmosphere as water vapor. As it falls from the sky as rain and sinks into the ground, it becomes **groundwater**—the mass of water that flows beneath Earth's surface.

Each place that stores water is referred to as a **reservoir.** Earth's largest natural reservoirs, in the order of their size, are oceans, glaciers and polar ice, groundwater, lakes and rivers, the atmosphere, and the biosphere. **Figure 11.2** shows the distribution of water among the reservoirs.

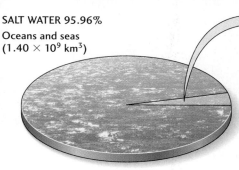

SALT WATER 95.96%
Oceans and seas
(1.40×10^9 km^3)

FRESH WATER 4.04%
Glaciers and polar ice 2.97%
(4.34×10^7 km^3)

Atmosphere 0.001%
(1.5×10^4 km^3)

Lakes and rivers 0.009%
(1.27×10^5 km^3)

Underground waters 1.05%
(1.54×10^7 km^3)

Biosphere 0.0001%
(2×10^3 km^3)

FIGURE 11.2 Distribution of water on Earth. [Data from J. P. Peixoto and M. Ali Kettani, "The Control of the Water Cycle." Scientific American (April 1973): 46; E. K. Berner and R. A. Berner, Global Environment. Upper Saddle River, N.J.: Prentice Hall, 1996, pp. 2–4.]

Although the total amount of water in rivers and lakes is very small compared with the amount in the oceans, these reservoirs are important to human populations because they contain fresh water. The amount of groundwater is more than 100 times the amount in rivers and lakes, but much of it is unusable because it contains large quantities of dissolved material.

> **Reservoirs are places where water is stored.**

Reservoirs gain water from inflows, such as rain and streams running in, and lose water from outflows, such as evaporation and streams running out. If inflow equals outflow, the size of the reservoir stays the same, even though water is constantly entering and leaving it. These flows mean that any given quantity of water spends a certain average time, called the *residence time,* in a reservoir.

How Much Water Is There?

Earth's total water supply is enormous—about 1.4 billion cubic kilometers distributed among the various reservoirs. If all that water covered the land area of the United States, it would submerge the 50 states under a layer of water about 145 km deep. This total is constant, even though the flows from one reservoir to another may vary from day to day, year to year, and century to century. Over these geologically short time intervals, there is neither a net gain nor a net loss of water to or from Earth's interior; nor is there any significant loss from the atmosphere to outer space.

The Hydrologic Cycle

All water on Earth cycles among the various reservoirs in the oceans, in the atmosphere, and on and under the land surface. The cyclical movement of water—from the ocean to the atmosphere by evaporation, to Earth's surface by precipitation, to streams through runoff and groundwater, and back to the ocean—is the **hydrologic cycle (Figure 11.3)**. Because organisms use water, small amounts are also stored in the biosphere.

The Sun's heat drives the hydrologic cycle, mainly by evaporating water from the oceans and transporting it as water vapor in the atmosphere. Under the right temperature and humidity conditions, water vapor condenses into the tiny droplets of water that form clouds and eventually fall as rain or snow—together known as **precipitation.** Some of the precipitation that falls on land soaks into the ground by **infiltration,** a process by which water enters rock or soil through cracks or small pores between particles. Part of this groundwater

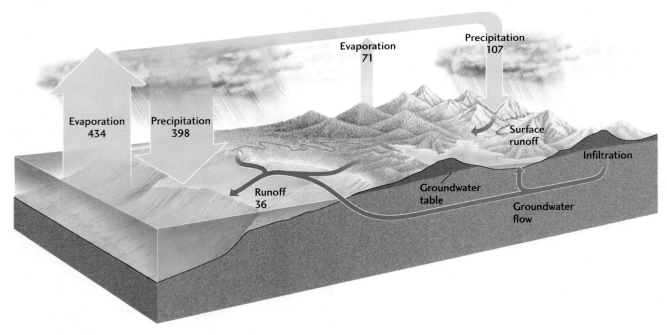

FIGURE 11.3 The hydrologic cycle is the movement of water through Earth's crust, atmosphere, oceans, lakes, and streams. The numbers indicate the amount of water (in thousands of cubic kilometers per year) that flows among these reservoirs annually.

evaporates through the soil surface. Another part moves through the biosphere as it is absorbed by plant roots, carried up to the leaves, and returned to the atmosphere as water vapor. Most groundwater, however, flows slowly underground. The residence time of water in groundwater reservoirs is long, but it eventually returns to the surface in springs that empty into rivers and lakes and thus returns to the oceans.

The precipitation that does not infiltrate the ground runs off the land surface, gradually collecting in streams and rivers as **runoff.** Some runoff may later seep into the ground or evaporate from rivers and lakes, but most of it eventually flows into the oceans.

Snowfall may be converted into ice in glaciers, which return water to the oceans by melting and runoff. A much smaller amount is returned to the atmosphere by transformation from a solid (ice) directly into a gas (water vapor).

Most of the water that evaporates from the oceans returns to them as precipitation. The remainder falls over land and either evaporates or returns to the ocean as runoff.

Figure 11.3 shows how the total flows among reservoirs balance one another in the hydrologic cycle. The land surface, for example, gains water from precipitation and loses the same amount of water by evaporation and runoff. The ocean gains water from runoff and precipitation and loses the same amount by evaporation. As you can see from Figure 11.3, more water evaporates from the oceans than falls on them as rain. This loss is balanced by the water returned as runoff from the continents. Thus, on a global scale, the size of each reservoir stays constant. Variations in climate, however, produce local variations in the balance among evaporation, precipitation, runoff, and infiltration.

> The hydrologic cycle, which is driven by the Sun's heat, represents the cyclical flow of water among the oceans, the atmosphere, and the land.

How Much Water Can We Use?

Almost all the water used by human society is *fresh water*—water that is not salty. Artificial desalination of seawater (the removal of salt) is producing small but steadily growing amounts of fresh water in areas such as the arid Middle East. In the natural world, however, fresh water is supplied only by rain, rivers, lakes, groundwater, and water melted from snow or ice on land. All these waters are ultimately supplied by precipitation. Therefore, the practical limit

> Our supply of water can be replenished only by precipitation.

to the amount of natural fresh water that we can ever envision using is the amount steadily supplied to the continents by precipitation.

HYDROLOGY AND CLIMATE

For most practical purposes, geologists focus on local hydrology—the amount of water in reservoirs of a region and how it flows from one reservoir to another—rather than global hydrology. The strongest influence on local hydrology is the local climate, especially temperature and precipitation levels. In warm areas where rain falls frequently throughout the year, water supplies—both at the surface and underground—are abundant. In warm arid or semiarid regions, it rarely rains, and water is a precious resource. People who live in icy climates rely on meltwaters from snow and ice. In some parts of the world, seasons of heavy rain, called *monsoons*, alternate with long dry seasons during which water supplies shrink, the ground dries out, and vegetation shrivels.

Wherever we live, climate and the geology of the landscape strongly influence the amounts of water that move from one reservoir to another. Geologists are especially interested in how changes in precipitation and evaporation affect water supplies by altering the amounts of infiltration and runoff, which determine groundwater levels. Over the longer term, climate may affect water supplies in another way: if the sea level rises as a result of global warming, groundwater in low-lying coastal regions may become salty as seawater invades formerly fresh groundwater reservoirs.

Humidity, Rainfall, and Landscapes

Many geographic variations in climate are related to the average temperature of the air and the average amount of water vapor it contains, both of which affect levels of precipitation. The **relative humidity** is the amount of water vapor in the air, expressed as a percentage of the total amount of water the air could hold at the same temperature if it were saturated. When the relative humidity is 50 percent and the temperature is 15°C, for example, the amount of moisture in the air is one-half the maximum amount the air could hold at 15°C.

Warm air can hold much more water vapor than cold air. When unsaturated warm air cools enough, it becomes supersaturated, and some of its water vapor condenses into water droplets, which form clouds. We can see clouds because they are made up of visible water droplets rather than invisible water vapor. When enough moisture has condensed into clouds and the droplets have grown too heavy to stay suspended by air currents, they fall as rain.

Most of the world's rain falls in warm, humid regions near the equator, where both the air and the surface waters of the ocean are warmed by the Sun. Under these conditions, a great deal of ocean water evaporates, resulting in high relative humidity. When the humid air over these tropical oceans rises and blows over nearby continents, it cools, condenses, and becomes supersaturated. The result is heavy rainfall over the land, even at great distances from the coast.

Unlike tropical climates, polar climates tend to be very dry. The polar oceans and the air above them are cold, so little ocean water evaporates, and the air can hold little moisture. Between the tropical and polar extremes are the temperate climates, where rainfall and temperatures are moderate.

Certain landscape features can alter precipitation patterns. Mountain ranges, for example, form **rain shadows,** areas of low rainfall on their leeward (downwind) slopes. Humid winds rising over high mountains cool and precipitate on the windward slopes, losing much of their moisture by the time they reach the leeward slopes (**Figure 11.4**). The air warms again as it drops to lower elevations on the other side of the mountain range. Because the warmer air can hold more moisture, relative

humidity declines, decreasing the likelihood of precipitation even more. There is a rain shadow on the eastern side of the Cascade Range of Oregon, where the prevailing winds blow inland from the Pacific Ocean. These moist winds release heavy rainfall on the mountains' western slopes. The eastern slopes, on the other side of the range, are dry and barren.

Natural Hazards: Droughts

Droughts—periods of months or years when precipitation is much lower than normal—can occur in all climates, but arid regions are especially vulnerable to their effects. Lacking replenishment from precipitation, streams may shrink and dry up, artificial reservoirs may evaporate, and the soil may dry and crack while vegetation dies. As human populations grow, demands on reservoirs increase, so a drought can deplete already inadequate water supplies.

The severest drought of the past few decades has affected lands along the southern border of the Sahara Desert (**Figure 11.5**), where tens of thousands of lives have been lost to famine. This long drought has expanded the desert and has effectively destroyed farming and grazing in the area, as we'll see in Chapter 12.

Another prolonged but less severe drought affected most of California from 1987 until February 1993, when torrential rains arrived. During the drought, groundwater and surface reservoirs dropped to their lowest levels in

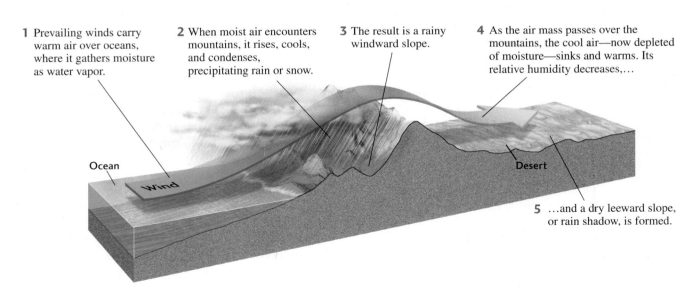

1 Prevailing winds carry warm air over oceans, where it gathers moisture as water vapor.

2 When moist air encounters mountains, it rises, cools, and condenses, precipitating rain or snow.

3 The result is a rainy windward slope.

4 As the air mass passes over the mountains, the cool air—now depleted of moisture—sinks and warms. Its relative humidity decreases,…

Ocean

Wind

Desert

5 …and a dry leeward slope, or rain shadow, is formed.

FIGURE 11.4 Rain shadows are areas of low rainfall on the leeward (downwind) slopes of a mountain range.

FIGURE 11.5 A small village in Niger can barely subsist on the crops planted here because of the ongoing drought. For decades, Niger has faced recurring food crises. [*Tomas van Houtryve/VII Network/Corbis.*]

15 years. Some restrictions on water use were instituted, but a move to reduce the extensive use of water supplies for irrigation encountered strong political resistance from farmers and the agricultural industry.

Ten years later, the most recent drought to affect the western United States began. If you live in the state of Arizona, California, Colorado, New Mexico, Nevada, or Utah, chances are you would have heard a lot about reducing water use between 2000 and 2010. Some cities imposed a variety of restrictions, such as allowing watering of lawns only on alternate days, banning planting of new grass, shutting off water to decorative fountains, installing low-volume flush toilets and no-water urinals, and increasing fees for excessive water use. In Phoenix, city departments were required to permanently reduce their water use by at least 5 percent. In 2007, Denver increased its water rates by 7 percent, while at the same time offering businesses a rebate on their water bills for improving the efficiency of processes that use water. In Las Vegas, which continues to grow at a rate that is one of the highest in the country, the Southern Nevada Water Authority estimated that the city was short 64 million gallons of water a day in 2010.

Lake Powell, one of the West's greatest artificial reservoirs, served as a very visible gauge for this drought (see Figure 1.19). Lake Powell was full during the summer of 1999, with reservoir storage at 97 percent of its capacity. In 2000, however, when rainfall totals for the year added up to only 30 percent of the average, it became clear that the stage was set for the first year of a possible drought and receding lake levels. By 2005, Lake Powell had dropped 145 feet from its 1999 high, and water storage was at only 33 percent of reservoir capacity. The last time Lake Powell had been this low was in 1969. Through the spring and early summer of 2005, the water surface elevation increased. By 2007, the lake level had risen by 50 feet, though it was still well below its normal levels.

> *Droughts are periods when precipitation is much lower than normal.*

 ## Natural Resources: Who Should Get Water?

As threats of drought and water shortages loom, the use of water enters the arena of public policy debate. Until recently, most people in the United States have taken their water supply for granted. In the near future, however, many areas of the country will experience water shortages more and more frequently. These shortages will create conflict among several sectors of society—residential, industrial, agricultural, and recreational—over who has

Public supply, 11 percent

[Richard R. Marella, USGS.]

Public supply water intake, Bay County, Fla.

Irrigation, 34 percent

[Jeff Vanuga, USDA NRCS.]

Gated-pipe flood irrigation, Fremont County, Wyo.

Aquaculture, less than 1 percent

[AP Photo/Kevin J. Kilmer]

Striped Bass Farm, New Haven, Illinois

Mining, less than 1 percent

[Nancy L. Barber, USGS.]

Pegmatite mine, Kings Mountain, N.C.

Domestic, less than 1 percent

[Alan M. Cressler, USGS.]

Domestic well, Early County, Ga.

Livestock, less than 1 percent

[Jeff Vanuga, USDA NRCS.]

Livestock watering, Rio Arriba County, N. Mex.

Industrial, 5 percent

[Joe Sohm/Digital Vision/Getty Images.]

Paper mill, Savannah, Calif.

Thermoelectric power, 48 percent

[Alan M. Cressler, USGS.]

Cooling towers, Burke County, Ga.

FIGURE 11.6 Total water withdrawals by category in 2005. [*Data from U.S. Geological Survey, USGS Circular 1344.*]

FIGURE 11.7 Center pivot irrigation on wheat growing in Yuma County, Colorado. [*Gene Alexander/USDA NRCS.*]

the greatest rights to the water supply. **Figure 11.6** shows U.S. water use by category in 2005.

In recent years, widely publicized droughts and restrictions on water use in California, Florida, Colorado, and many other places have warned the public that the nation faces major water shortages. Public concern waxes and wanes, however, as periods of drought and abundant rainfall come and go, and governments are not pursuing long-term solutions with the urgency that they deserve. Here are some facts to ponder:

■ A human can survive with about 2 liters of water per day. In the United States, the per capita use for individuals is about 250 liters per day. If industrial, agricultural, and energy production are considered, then the per capita usage rises to about 6000 liters per day.

■ Agriculture uses about 34 percent of the water withdrawn from U.S. reservoirs, and thermoelectric power generation uses about 49 percent (**Figure 11.7**).

■ Although the western United States receives one-fourth of the country's rainfall, their per capita water use (mostly for irrigation) is 10 times greater than that of the eastern states, and at much lower prices. In California, for example, which imports most of its water, 85 percent of the water is used for irrigation, 10 percent for municipalities and personal consumption, and 5 percent for industry. A 15 percent reduction in irrigation use would almost double the amount of water available for use by cities and industries.

> *Droughts and the resulting water shortages generate conflict among sectors of society over who should get water.*

■ Global climate change may lead to reduced rainfall in western states, worsening the problems there and making long-term solutions even more urgent.

The Hydrology of Runoff

A dramatic example of how precipitation affects local stream and river runoff can be seen when flash flooding occurs after torrential rains. When levels of precipitation and runoff are measured over a large area (such as all the states drained by a major river) and over a long

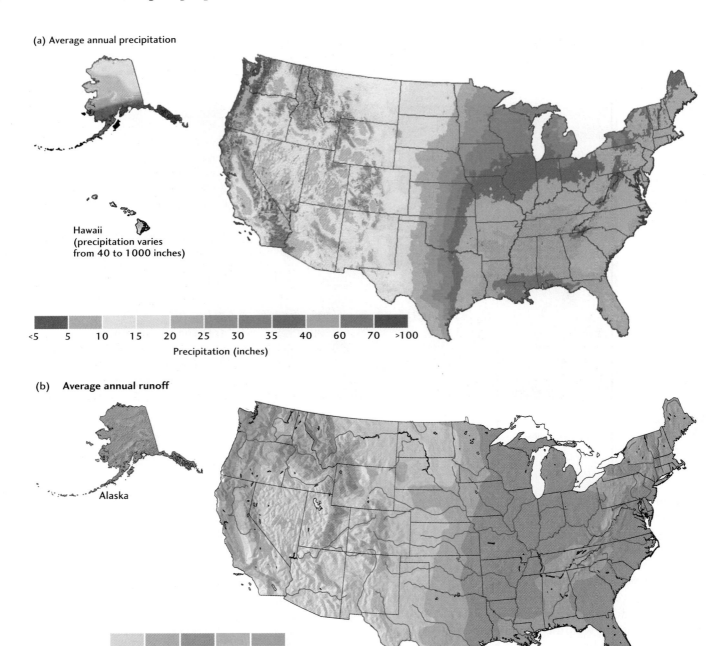

(a) Average annual precipitation

Hawaii
(precipitation varies
from 40 to 1000 inches)

<5   5   10   15   20   25   30   35   40   60   70   >100

Precipitation (inches)

(b)   **Average annual runoff**

Alaska

0   2.5   5   50   100   >100

Runoff (inches)

FIGURE 11.8 (a) Average annual precipitation in the United States. [*Data from U.S. Department of Commerce, Climatic Atlas of the United States, 1968.*] (b) Average annual runoff in the United States. [*Data from USGS Professional Paper 1240-A, 1979.*]

period (such as a year), the relationship between them is less direct, but still strong. The maps in **Figure 11.8** illustrate this relationship. When we compare them, we see that in areas of low precipitation—such as Southern California, Arizona, and New Mexico—only a small fraction of precipitation ends up as runoff. In dry regions, much of the

precipitation is lost by evaporation and infiltration. In humid areas, such as the southeastern United States, a much higher proportion of the precipitation runs off in rivers. A large river may carry large amounts of water from an area with high rainfall to an area with low rainfall. The Colorado River, for example, begins in an area of moderate rainfall in

TABLE 11.1	Water Flows of Some Major Rivers
River	**Water Flow (m³/s)**
Amazon, South America	175,000
La Plata, South America	79,300
Congo, Africa	39,600
Yangtze, Asia	21,800
Brahmaputra, Asia	19,800
Ganges, Asia	18,700
Mississippi, North America	17,500

Colorado and then carries its water through arid western Arizona and Southern California.

Rivers and streams carry most of the world's runoff. The millions of small and medium-sized streams carry about half the world's runoff; about 70 major rivers carry the other half. And the Amazon River of South America carries almost half of that. The Amazon carries about 10 times more water than the Mississippi, the largest river of North America (**Table 11.1**). These major rivers transport great volumes of water because they collect it from large networks of streams and rivers that cover very large areas. The Mississippi, for example, collects its water from a network of streams that covers about two-thirds of the United States (**Figure 11.9**).

Runoff collects and is stored in natural lakes as well as in artificial reservoirs created by the damming of streams. Wetlands, such as swamps and marshes, also act as reservoirs for runoff (**Figure 11.10**). If these reservoirs are large enough, they can absorb short-term inflows from major rainfall events, holding some of the water that would otherwise spill over riverbanks. During dry seasons or droughts, these reservoirs release water to streams or to water systems built for human use. Thus, they help to control flooding by smoothing out seasonal or yearly variations in runoff and releasing steady flows of water downstream. For this reason, many governments have laws that regulate the artificial draining of wetlands for real estate development. Destruction of wetlands also threatens biological diversity because

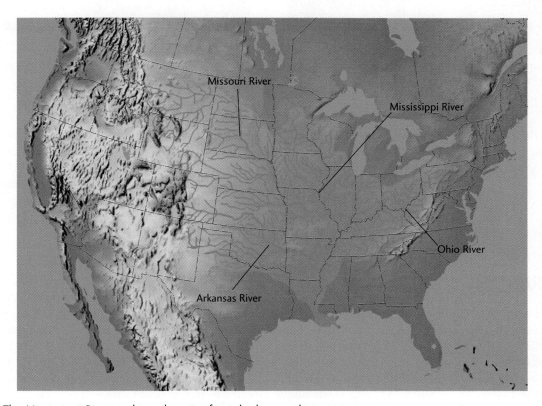

FIGURE 11.9 The Mississippi River and its tributaries form the largest drainage network in the United States.

DRY PERIOD: LOW RUNOFF

In dry periods, streams bring in small amounts of water…

…and carry away small amounts.

WET PERIOD: HIGH RUNOFF

In wet periods, streams bring in large amounts of water,…

…which is stored…

…and slowly released during dry periods.

FIGURE 11.10 Like a natural lake or an artificial reservoir behind a dam, a wetland (such as a swamp or marsh) stores water during times of rapid runoff and slowly releases it during periods of little runoff.

wetlands are breeding grounds for a great many types of plants and animals. In the United States, more than half the wetlands that existed before European settlement are now gone. California and Ohio have kept only 10 percent of their original wetlands.

Runoff is carried in streams and rivers and collects in natural lakes, artificial reservoirs, and wetlands.

GROUNDWATER

Groundwater forms as raindrops and melting snow infiltrate soil and other unconsolidated surface materials and even sink into the cracks and crevices of bedrock. The enormous reservoir of groundwater stored beneath Earth's surface equals about 29 percent of all the fresh water stored in lakes and rivers, glaciers and polar ice, and the

Groundwater is formed as rain and melting snow infiltrate porous rocks and soils.

atmosphere. For thousands of years, people have drawn on this resource, either by digging shallow wells or by storing water that flows out onto the surface at natural springs. Springs are direct evidence of water moving below the surface (**Figure 11.11**).

How Water Flows Through Soil and Rock

When water moves into and through the ground, what determines where and how fast it flows? With the exception of caves, there are no large open spaces for pools or rivers of water underground. The only spaces available for water are pores and cracks in soil and bedrock. Some pores, however small and few, are found in every kind of rock and soil, but large amounts of pore space are most often found in sandstones and limestones.

Recall from Chapter 5 that the amount of pore space in rock, soil, or sediment determines its *porosity*—the percentage of its total volume that is taken up by pores. This pore space consists mainly of the spaces between grains and in cracks (**Figure 11.12**). It can vary from a small percentage of the total volume to as much as 50 percent where rock has been dissolved. In sedimentary rocks, porosity is typically 5 to 15 percent. Most metamorphic and igneous rocks have little pore space, except where fracturing has occurred.

Although a rock's porosity tells us how much water it can hold if all its pores are filled, it gives us no information about how rapidly water can flow through those pores. Water travels through a porous material by winding between grains and through cracks. The smaller the pore spaces and the more complex the path, the more slowly the water travels. The capacity of a solid to allow fluids to pass through it is its **permeability.** Generally, permeability increases as porosity increases.

> *The porosity of a material tells us how much water it can hold. Its permeability tells us how rapidly water can flow through it.*

Both porosity and permeability are important factors to geologists searching for groundwater supplies. In general, a good groundwater reservoir is a body of rock, sediment, or soil with both high porosity (so that it can hold large amounts of water) and high permeability (so that the water can be pumped from it easily). Well drillers in temperate climates, for example, know that

FIGURE 11.11 Groundwater flows from a cliff in Vasey's Paradise, Marble Canyon, Grand Canyon National Park, Arizona, where hilly topography allows it to flow out onto the surface in a natural spring. [*Larry Ulrich.*]

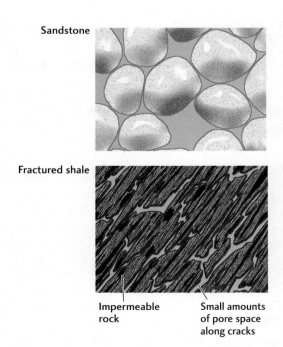

Sandstone

Fractured shale

Impermeable rock

Small amounts of pore space along cracks

FIGURE 11.12 Pores in rocks are normally filled partly or entirely with water. (Pores in oil- or gas-bearing sandstones and limestones are filled with oil or gas.)

TABLE 11.2	Porosity and Permeability of Various Rock Types	
Rock Type	Porosity (Pore Space That May Hold Fluid)	Permeability (Ability to Allow Fluids to Pass Through)
Gravel	Very high	Very high
Coarse- to medium-grained sand	High	High
Fine-grained sand and silt	Moderate	Moderate to low
Sandstone, moderately cemented	Moderate to low	Low
Fractured shale or metamorphic rocks	Low	Very low
Unfractured shale	Very low	Very low

they are most likely to find a good supply of water if they drill into porous sand or sandstone beds not far below the surface. A rock with high porosity but low permeability may contain a great deal of water, but because the water flows so slowly, it is hard to pump it out of the rock. **Table 11.2** summarizes the porosity and permeability of various rock types.

The Groundwater Table

As well drillers bore deeper into soil or rock, the samples they bring up become wetter. At shallow depths, the material is unsaturated—the pores contain some air and are not completely filled with water. This level is called the **unsaturated zone** (often termed the *vadose zone*). Below it is the **saturated zone** (often termed the *phreatic zone*), in which the pores are completely filled with water. The saturated and unsaturated zones can be in unconsolidated material or in bedrock. The boundary between the two zones is the **groundwater table,** usually shortened to *water table* (**Figure 11.13**). When a hole is drilled below the water table, water from the saturated zone flows into the hole and fills it to the level of the water table.

> The groundwater table is the boundary between the saturated zone, where water fills all available pore space, and the unsaturated zone, where pores are partly filled with air.

Groundwater moves under the force of gravity, so some of the water in the unsaturated zone may be on its way down to the water table. A fraction of that water, however, remains in the unsaturated zone, held in small pore spaces by surface tension—the attraction between the water molecules and the surfaces of the grains.

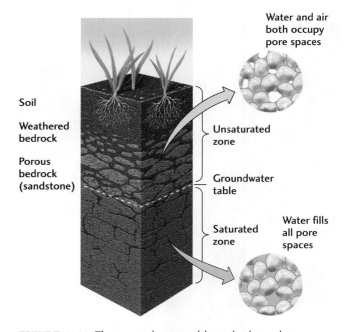

FIGURE 11.13 The groundwater table is the boundary between the unsaturated zone and the saturated zone. The saturated and unsaturated zones can be in unconsolidated material or in bedrock.

Surface tension is what keeps the sand on a beach moist, even though gravity acts to move the water to spaces below.

If we were to drill wells at several sites and measure the elevations of the water levels in those wells, we could construct a map of the water table. A cross section of the landscape might look like the one shown in **Figure 11.14.** The water table follows the general shape of the

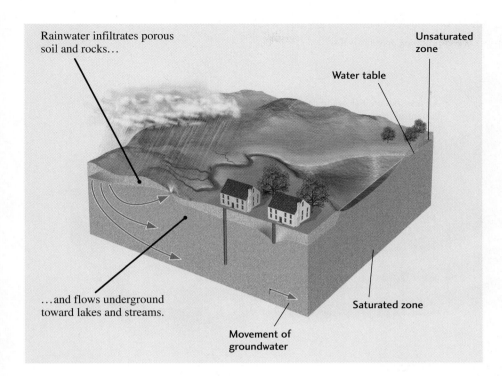

Rainwater infiltrates porous soil and rocks...

Unsaturated zone

Water table

...and flows underground toward lakes and streams.

Movement of groundwater

Saturated zone

FIGURE 11.14 Cross section of a landscape illustrating the dynamics of a water table. Wells of different depths may tap different parts of the aquifer, where groundwater may be absent at shallower levels.

surface topography, but its slopes are gentler. The water table is exposed at the surface in riverbeds and lake beds and at springs. Under the influence of gravity, groundwater moves downhill from places where the water table elevation is high—under a hill, for example—to places where the water table elevation is low—such as a spring where it flows out onto the surface.

> *Groundwater flows downhill from areas where the water table elevation is high to areas where it is lower.*

Water enters and leaves the saturated zone through recharge and discharge. **Recharge** is the infiltration of water into any subsurface formation. Rain and melting snow are the most common sources of recharge. **Discharge,** the movement of groundwater to the surface, is the opposite of recharge. When a stream channel lies at an elevation below the water table, water discharges from the groundwater into the stream. Groundwater is also discharged through springs and by pumping from artificial wells.

Aquifers

Rock formations through which groundwater flows in sufficient quantity to supply wells are called **aquifers.** Groundwater may flow in unconfined or confined aquifers. In *unconfined aquifers*, the water travels through formations of more or less uniform permeability that extend

to the surface. The level of the groundwater reservoir in an unconfined aquifer is the same as the height of the water table (as in Figure 11.14).

Many permeable formations, however—typically sandstones—are bounded above and below by low-permeability beds, such as shales. These relatively impermeable formations are called **aquicludes.** Groundwater either cannot flow through them or flows through them very slowly. When aquicludes lie both over and under an aquifer, they form a *confined aquifer.*

The aquicludes above a confined aquifer prevent rainwater from infiltrating the aquifer directly. Instead, a confined aquifer is recharged by precipitation over a *recharge area*, often a topographically higher upland characterized by outcropping rocks. Here there is no aquiclude preventing infiltration, so the rainwater travels down to and through the aquifer underground (**Figure 11.15**).

Water moving through a confined aquifer—known as an **artesian flow**—is under pressure. At any point in the aquifer, the pressure is equivalent to the

> *Unconfined aquifers have relatively uniform permeability that extends to the surface. Confined aquifers are bounded by impermeable formations known as aquicludes.*

weight of all the water in the aquifer above that point. If we drill a well into a confined aquifer at a point where the elevation of the ground surface is lower than that of the water table in the recharge area, the water will flow

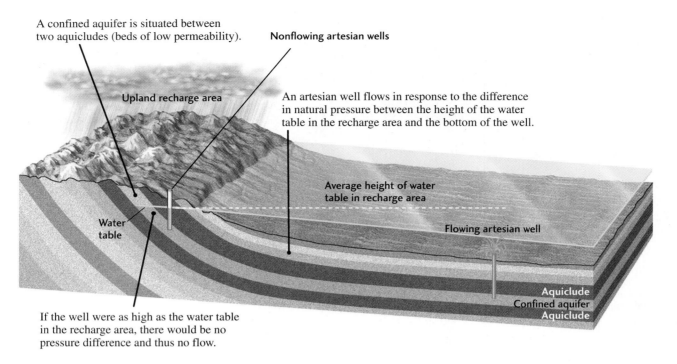

A confined aquifer is situated between two aquicludes (beds of low permeability).

Nonflowing artesian wells

Upland recharge area

An artesian well flows in response to the difference in natural pressure between the height of the water table in the recharge area and the bottom of the well.

Average height of water table in recharge area

Water table

Flowing artesian well

If the well were as high as the water table in the recharge area, there would be no pressure difference and thus no flow.

Aquiclude
Confined aquifer
Aquiclude

FIGURE 11.15 A permeable formation situated between two aquicludes forms a confined aquifer, through which water flows under pressure.

FIGURE 11.16 Small artesian well at St. Andrews picnic area, Jeykll Island, Georgia. Water flows up through the ground because of regional groundwater pressure. [*Alan Cressler.*]

out of the well spontaneously (**Figure 11.16**). Such wells are called *artesian wells*, and they are extremely desirable because no energy is required to pump the water to the surface. The water is brought up by its own pressure.

In more complex geological environments, water tables may be more complicated. For example, if a relatively impermeable mudstone layer forms an aquiclude within an otherwise permeable sandstone formation, the aquiclude may lie below the water table of a shallow aquifer and above the water table of a deeper aquifer (**Figure 11.17**). The water table in the shallow aquifer is called a *perched water table* because it is "perched" above the main water table in the deeper aquifer. Many perched water tables are small, only a few meters thick and restricted in area, but some extend for hundreds of square kilometers.

> Artesian flows occur in confined aquifers and may be under sufficient pressure to drive water in wells to the surface without pumping.

Balancing Recharge and Discharge

When recharge and discharge are balanced, the groundwater reservoir in an aquifer and the elevation of the water table remain constant, even though water is continually flowing through the aquifer. For recharge to balance discharge, rainfall must be frequent enough to compensate

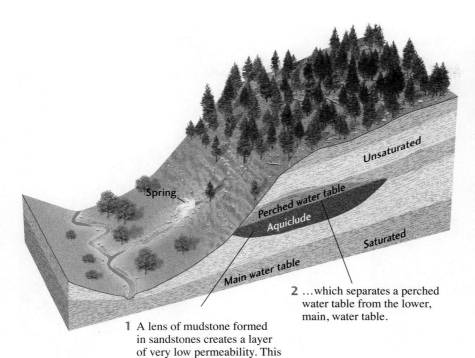

FIGURE 11.17 A perched water table forms in geologically complex situations—in this case, where a mudstone aquiclude is located above the main water table in a sandstone aquifer. The dynamics of the perched water table's recharge and discharge may be different from those of the main water table.

Unsaturated

Spring

Perched water table

Aquiclude

Saturated

Main water table

2 ...which separates a perched water table from the lower, main, water table.

1 A lens of mudstone formed in sandstones creates a layer of very low permeability. This layer forms a shallow aquiclude,...

for runoff in streams and the outflow from springs and wells.

But recharge and discharge are usually not equal because recharge varies with rainfall from season to season. Typically, the water table drops in dry seasons and rises in wet seasons. A longer period of low recharge,

such as during a prolonged drought, will be followed by a longer-term imbalance and a greater lowering of the water table (**Figure 11.18**).

An increase in discharge, usually due to increased pumping from wells, can also produce a long-term imbalance and a lowering of the water table. Shallow wells

During the wet season, abundant rainfall increases infiltration, which recharges the groundwater and raises the water table.

Abundant rainfall recharges groundwater

During the dry season, evaporation outpaces infiltration, and the water table drops.

Evaporation discharges groundwater

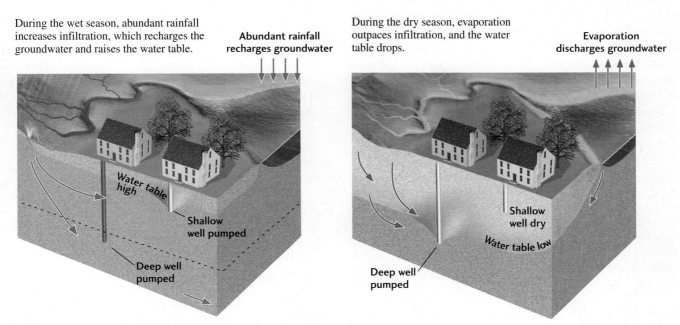

Water table high

Shallow well pumped

Deep well pumped

Shallow well dry

Water table low

Deep well pumped

FIGURE 11.18 The water table fluctuates seasonally in response to the balance between recharge and discharge.

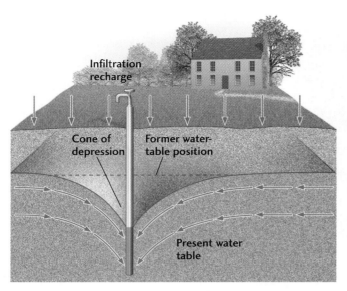

FIGURE 11.19 When discharge from a well exceeds recharge, the well draws down the water table into a cone-shaped depression. The water level in the well is lowered to the depressed level of the water table.

FIGURE 11.20 Excessive pumping of groundwater in Antelope Valley, California, has led to fissures and sinklike depressions on Rogers Lakebed at Edwards Air Force Base. This fissure, formed in January 1991, is about 625 m long. [*James W. Borcher/USGS.*]

may end up in the unsaturated zone and go dry. When a well pumps water from an aquifer faster than recharge can replenish it, the water table in the aquifer is lowered in a cone-shaped area around the well, called a *cone of depression* (**Figure 11.19**). The water level in the well is lowered to the depressed level of the water table. If the cone of depression extends below the bottom of the well, that well goes dry. If the bottom of the well is above the base of the aquifer, extending the well deeper into the aquifer may allow more water to be withdrawn, even at continued high pumping rates. If the rate of pumping is maintained and the well is deepened so much that the entire aquifer is tapped, however, the cone of depression can reach the bottom of the aquifer and deplete it. The aquifer will recover only if the pumping rate is reduced enough to give it time to recharge.

Excessive withdrawals of water may not only deplete the aquifer; they may also cause another undesirable environmental effect. As water pressure in the pore spaces falls, the ground surface overlying the aquifer may subside, creating sinklike depressions (**Figure 11.20**).

> *Groundwater supplies remain constant only when recharge is equal to discharge.*

The Speed of Groundwater Flows

The speed at which water moves underground strongly affects the balance between discharge and recharge.

Most groundwaters flow slowly, a fact of nature that is responsible for our groundwater supplies. If groundwater flowed as rapidly as streams, aquifers would run dry after a period without rain, just as many small streams do. But the slow flow of groundwater also makes rapid recharge impossible if groundwater levels are lowered by excessive pumping.

> *The speed at which groundwater flows through an aquifer depends on the slope of the water table, the distance between the recharge area and the well where groundwater is being withdrawn, and the permeability of the aquifer.*

Although all groundwaters flow through aquifers slowly, some flow more slowly than others. The rate of flow depends on several factors (**Figure 11.21**): it increases as the slope of the water table increases; it increases as the distance decreases between where the groundwater is recharged and the well where it is withdrawn; and it increases as the permeability of the aquifer increases.

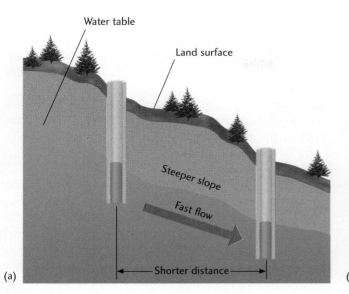

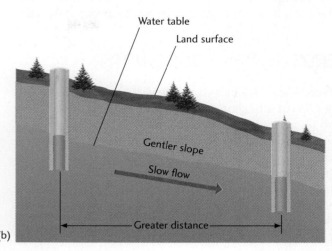

FIGURE 11.21 The rate of groundwater flow depends on several factors, including the slope of the water table and the distance between the recharge zone and the well.

HUMAN EFFECTS ON GROUNDWATER RESOURCES

Large parts of North America rely solely on groundwater for their water needs (**Figure 11.22**). The demand for groundwater resources has grown as populations have increased and uses such as irrigation have expanded (see Figure 11.1). Many areas of the Great Plains and other parts of the Midwest rest on sandstone formations, most of which are confined aquifers that function like the one shown in Figure 11.15. Thousands of wells have been drilled into these aquifers, most of which transport water over hundreds of kilometers and thus constitute a major water resource. The aquifers are recharged from outcrops in the western high plains, some very close to the foothills of the Rocky Mountains. From there, the water runs downhill in an easterly direction over hundreds of kilometers. Extensive pumping withdraws water from such aquifers faster than the slow flow from distant recharge areas can fill them, so the reservoirs of groundwater they contain are being depleted. The depletion of the Ogallala aquifer, described in Chapter 1, is an example of the problems this region is facing.

A variety of innovative approaches are being used to enhance the sustainability of groundwater resources. In some areas, efforts to reduce excessive discharge have been supplemented by attempts to increase the recharge of aquifers. On Long Island, New York, for example, the water authority drilled a large system of recharge wells to pump treated wastewater into

> *Aquifers are depleted when the rate of pumping exceeds the rate of recharge.*

the ground. The water authority also constructed large, shallow basins over natural recharge areas to increase infiltration by catching and diverting runoff, including stormwater and industrial waste drainage. The officials in charge of the program knew that urban development can decrease recharge by interfering with infiltration. As urbanization progresses, the impermeable materials used to pave large areas for streets, sidewalks, and parking lots increase runoff and prevent water from infiltrating the ground. Such decreases in natural infiltration may deprive aquifers of much of their recharge. One remedy is to catch and use stormwater runoff in a systematic program of

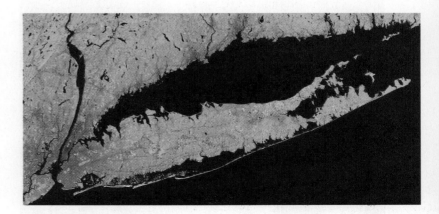

FIGURE 11.22 Long Island, New York, rests on bedrock and is surrounded by salt water on its south shore (Atlantic Ocean) and brackish water on the North Shore (Long Island Sound). The island's 7.5 million residents rely on groundwater for all their water needs. [*NASA/Goddard Space Flight Center, Scientific Visualization Studio.*]

artificial recharge, as the Long Island water authority did. The multiple efforts of the water authority helped rebuild the Long Island aquifer, though not to its original levels.

EROSION BY GROUNDWATER

Every year, thousands of people visit caves, either on tours of popular attractions such as Mammoth Cave, Kentucky, or in adventurous explorations of little-known caves. Caves are produced by the dissolution of limestone—or, rarely, of other soluble rocks such as evaporites—by groundwater. Huge amounts of limestone have been dissolved to make some caves. Mammoth Cave, for example, has tens of kilometers of large and small interconnected chambers. The Big Room at Carlsbad Caverns, New Mexico, is more than 1200 m long, 200 m wide, and 100 m high (**Figure 11.23**).

Limestone is widespread in the upper parts of Earth's crust, but caves form only where these relatively soluble rocks are at or near the surface and enough carbon dioxide–rich or sulfur dioxide–rich water infiltrates the surface to dissolve extensive areas of this relatively soluble rock. The atmospheric carbon dioxide contained in rainwater creates a weak acid that enhances the dissolution of limestone. Water that infiltrates soil may pick up even more carbon dioxide from plant roots, microbes, and other soil-dwelling organisms that give off this gas. As this carbon dioxide–rich water moves down to the water table, through the unsaturated zone to the saturated zone, it creates openings as it dissolves carbonate minerals. These openings are enlarged as the limestone dissolves along joints and fractures, forming a network of rooms and passages. Such networks form extensively in the saturated zone, where—because the caves are filled with water—dissolution takes place over all surfaces, including floors, walls, and ceilings.

We can explore caves that were once below the water table but are now in the unsaturated zone because the water table has dropped. In

> *Caves are produced when acidic groundwater dissolves limestone or other highly soluble rocks.*

these caves, now air-filled, water saturated with calcium carbonate may seep through the ceiling. As each drop of water drips from the cave's ceiling, some of its dissolved carbon dioxide evaporates, escaping to the cave's atmosphere. Its evaporation makes the calcium carbonate in the groundwater solution less soluble, so each water droplet precipitates a small amount of calcium carbonate on the ceiling. These deposits accumulate, just as an icicle grows, in a long, narrow spike of carbonate called a *stalactite* suspended from the ceiling. When the drop of water falls to the cave floor, more carbon dioxide escapes, and another small amount of calcium carbonate is precipitated on the floor below the stalactite. These deposits also accumulate, forming a *stalagmite*. Eventually, a stalactite and a stalagmite may grow together to form a column (see Figure 11.23).

In some places, dissolution may thin the roof of a limestone cave so much that it collapses suddenly, producing a **sinkhole**—a small, steep depression in the land surface above the cave (**Figure 11.24**). Sinkholes are characteristic of a distinctive type of topography known as *karst*, named for a region in the northern part of Slovenia.

FIGURE 11.23 Chinese Theater, in the Big Room of Carlsbad Caverns, New Mexico. Stalactites from the ceiling and stalagmites from the floor have joined to form a column. [David Muench.]

FIGURE 11.24 Large sinkhole formed by the collapse of a shallow underground cavern in Winter Park, Florida. Such collapses can occur so suddenly that moving cars are buried. [Leif Skoogfors/Woodfin Camp.]

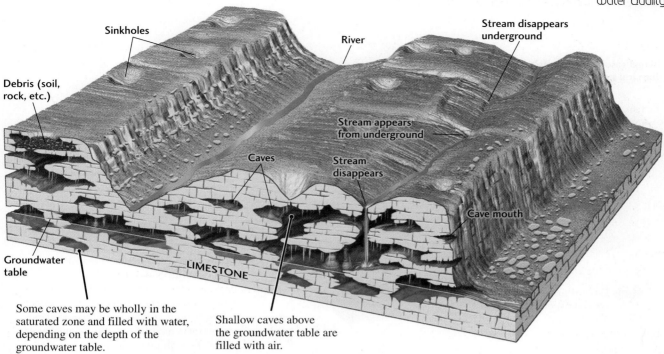

Sinkholes

River

Stream disappears underground

Debris (soil, rock, etc.)

Stream appears from underground

Caves

Stream disappears

Cave mouth

Groundwater table

LIMESTONE

Some caves may be wholly in the saturated zone and filled with water, depending on the depth of the groundwater table.

Shallow caves above the groundwater table are filled with air.

FIGURE 11.25 Some major features of karst topography are caves, sinkholes, and disappearing streams.

FIGURE 11.26 The tower karst of southeastern China is a spectacular terrain that features isolated hills with nearly vertical slopes. [*Dennis Cox/Alamy.*]

Karst topography is an irregular, hilly type of terrain characterized by sinkholes, caves, and a lack of surface streams (**Figure 11.25**). Underground drainage channels replace the normal surface drainage system of small and large streams. The short, scarce streams often end in sinkholes, detouring underground and sometimes reappearing miles away.

Karst terrains often have environmental problems, including the potential for catastrophic cave-ins and surface subsidence. Karst topography is found in limestone terrains of Indiana, Kentucky, and Florida and on the Yucatán Peninsula of Mexico. It is also well developed on uplifted coral limestone terrains formed from tropical volcanic island arcs in the late Cenozoic era. The spectacular tower karst of China forms when cave networks collapse to form sinkholes, which then expand and merge to leave behind "towers" (**Figure 11.26**).

Karst topography is an irregular, hilly type of terrain where dissolution of rocks by groundwater has occurred.

WATER QUALITY

Unlike in many other parts of the world, almost all public water supplies in North America are free of bacterial contamination, and the vast majority are free enough of chemicals to drink safely. Yet as more rivers become polluted and more aquifers are contaminated by toxic wastes from surface dumps, North Americans are likely to see changes in water quality.

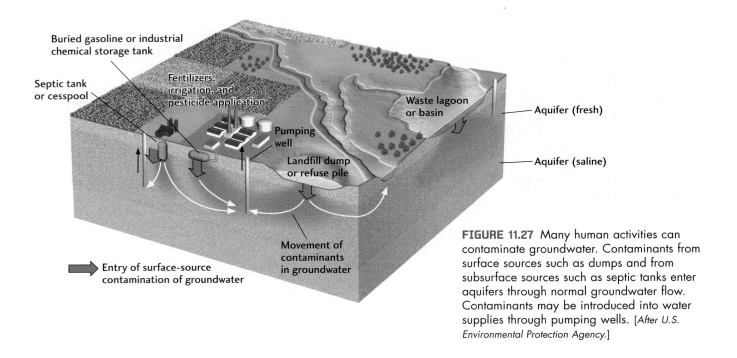

Buried gasoline or industrial chemical storage tank

Septic tank or cesspool

Fertilizers, irrigation, and pesticide application

Pumping well

Landfill dump or refuse pile

Waste lagoon or basin

Aquifer (fresh)

Aquifer (saline)

Movement of contaminants in groundwater

Entry of surface-source contamination of groundwater

FIGURE 11.27 Many human activities can contaminate groundwater. Contaminants from surface sources such as dumps and from subsurface sources such as septic tanks enter aquifers through normal groundwater flow. Contaminants may be introduced into water supplies through pumping wells. [*After U.S. Environmental Protection Agency.*]

Natural Environment: Contamination of Our Water Supply

The quality of groundwater is often threatened by a variety of contaminants. Generally, these contaminants are chemicals including radioactive waste, though microorganisms can also have negative effects on human health under certain conditions.

Contaminants A number of human activities produce chemicals that can contaminate groundwater (**Figure 11.27**). The disposal of chlorinated solvents—such as trichloroethylene (TCE), widely used as a cleaner in industrial processes—poses a formidable problem. These solvents persist in the environment because they are difficult to remove from contaminated waters. The burning of coal and the incineration of municipal and medical waste also emit mercury into the atmosphere, which then contaminates water supplies. Buried gasoline storage tanks can leak, and road salt inevitably drains into the soil and ultimately into aquifers. Rain can wash agricultural pesticides, herbicides, and fertilizers into the soil, from which they percolate downward into aquifers. In some agricultural areas where nitrate fertilizers are heavily used, groundwaters contain high concentrations of nitrate. In one recent study, 21 percent of the shallow wells sampled exceeded the maximum amounts of nitrate (10 ppm) allowed in drinking water in the United States. Such

high nitrate levels pose a danger of "blue baby" syndrome (an inability to maintain healthy oxygen levels) to infants 6 months old and younger.

A highly publicized type of chemical contamination includes radioactive waste. Because radioactive substances exist for so long—thousands to millions of years in many cases—there is no easy solution to the problem of groundwater contamination by radioactive waste. When radioactive wastes are buried underground, they may be leached by groundwater and find their way into aquifers. Storage tanks and burial sites at the atomic weaponry plants in Oak Ridge, Tennessee, and Hanford, Washington, have already leaked radioactive wastes into shallow groundwaters.

Finally, a surprisingly common type of groundwater contamination is related to the presence of microorganisms that are introduced into aquifers. The most widespread causes of groundwater contamination by microbes are leaky residential septic tanks and cesspools. These containers, widely used in neighborhoods that lack full sewer networks, are settling tanks buried at shallow depths in which bacteria decompose the solid wastes from household sewage. To prevent contamination of drinking water, cesspools should be replaced by septic tanks, which must be installed at sufficient distance from water wells in shallow aquifers.

Groundwater can be contaminated by chemicals such as chlorinated solvents, gasoline, road salt, and fertilizers. Radioactive wastes and microorganisms may pose high local risks.

Reversing Contamination

Can we reverse the contamination of groundwater supplies? Yes, but the process is costly and very slow. The faster an aquifer recharges, the easier it is to decontaminate. If the recharge rate is rapid, fresh water moves into the aquifer as soon as we close off the sources of contamination, and in a relatively short time, the water quality is restored. Even a fast recovery, however, can take a few years.

The contamination of slowly recharging groundwater reservoirs is more difficult to reverse. The rate of groundwater movement may be so slow that contamination from a distant source takes a long time to appear. By the time it does, it is too late for rapid remediation. Even after the recharge area has been cleaned up, some contaminated deep aquifers extending hundreds of kilometers from the recharge area may not be free of contaminants for many decades.

> Once contaminated, groundwater is extremely difficult and expensive to decontaminate. Because groundwater flows very slowly, remediation processes take a long time.

When public water supplies are polluted, we can pump the water and then treat it chemically to make it safe, but this is an expensive procedure. Alternatively, we can try to treat the water while it remains underground. In one moderately successful experimental procedure, contaminated water was funneled into a buried bunker full of iron filings that detoxified the water by reacting with the contaminants. The reactions produced new, nontoxic compounds that attached themselves to the iron filings.

Is the Water Drinkable?

Much of the water in groundwater reserves is unusable not because it has been contaminated by human activities, but because it naturally contains large quantities of dissolved materials. Water that tastes agreeable and is not dangerous to human health is called **potable** water. The amounts of dissolved materials in potable waters are very small, usually measured by weight in parts per million (ppm). Potable groundwaters of good quality typically contain about 150 ppm total dissolved materials, because even the purest natural waters contain some dissolved substances derived from weathering. Only distilled water contains less than 1 ppm dissolved materials.

The many cases of groundwater contamination have led to the establishment of water quality standards based on medical studies. These studies have concentrated on the effects of ingesting average amounts of water containing various quantities of contaminants. For example, the U.S. Environmental Protection Agency has set the maximum allowable concentration of arsenic, a well-known poison, at 0.01 ppm (**Figure 11.28**). Natural contamination of groundwater by arsenic is particularly acute in Bangladesh where groundwater provides 97 percent of the drinking water supply.

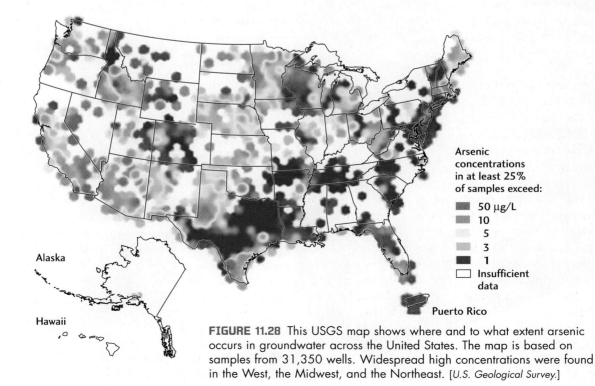

Arsenic concentrations in at least 25% of samples exceed:

- 50 µg/L
- 10
- 5
- 3
- 1
- Insufficient data

FIGURE 11.28 This USGS map shows where and to what extent arsenic occurs in groundwater across the United States. The map is based on samples from 31,350 wells. Widespread high concentrations were found in the West, the Midwest, and the Northeast. [*U.S. Geological Survey.*]

Groundwater is almost always free of solid particles when it seeps into a well from a sand or sandstone aquifer. The complex passageways of the pore networks in the rock or sand act as a fine filter, removing small particles of clay and other solids and even straining out microbes and some large viruses. Limestone aquifers may have larger pores and so may filter water less efficiently.

Some groundwaters, although perfectly safe to drink, simply taste bad. Some have a disagreeable taste of "iron" or are slightly sour. Groundwaters passing through limestone dissolve carbonate minerals and carry away calcium, magnesium, and bicarbonate ions, making the water "hard." Hard water may taste fine, but soap used with hard water does not lather readily. Water passing through waterlogged forests or swampy soils may contain dissolved organic compounds and hydrogen sulfide, which give the water a disagreeable smell similar to rotten eggs.

How do these differences in taste and quality arise in safe drinking waters? Some of the highest-quality, best-tasting public water supplies come from lakes and artificial surface reservoirs, many of which are simply collecting places for rainwater. Groundwaters that pass through rocks that weather only slightly taste just as good. Sandstones made up largely of quartz, for example, contribute little in dissolved substances, and thus waters passing through them have a pleasant taste.

As we have seen, the contamination of groundwater in relatively shallow aquifers is a serious problem, and remediation is difficult. But are there deeper groundwaters that we can use?

> *Groundwater usually lacks solid particles, which are filtered out by pore networks, but it may contain dissolved substances that impart a disagreeable odor or taste.*

WATER DEEP IN THE CRUST

Most crustal rocks below the groundwater table are saturated with water. Even in the deepest wells drilled for oil, some 8 or 9 km deep, geologists find water in permeable formations. At these depths, groundwaters move so

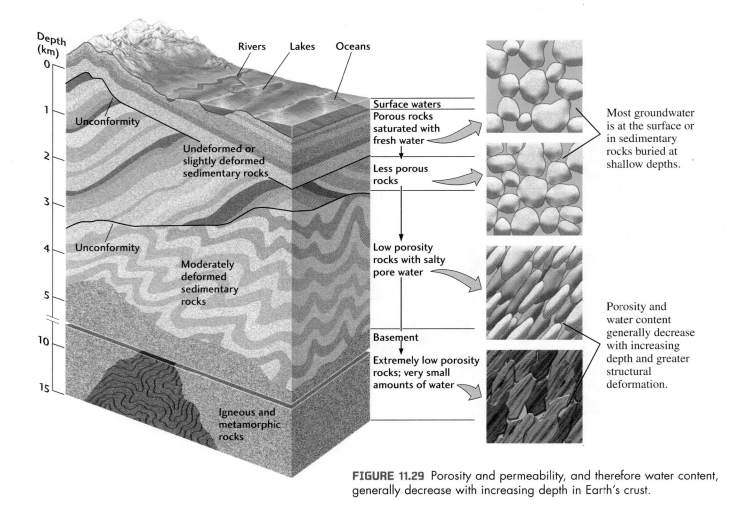

Depth (km)

Rivers   Lakes   Oceans

Unconformity

Undeformed or slightly deformed sedimentary rocks

Unconformity

Moderately deformed sedimentary rocks

Igneous and metamorphic rocks

Surface waters
Porous rocks saturated with fresh water
Less porous rocks
Low porosity rocks with salty pore water
Basement
Extremely low porosity rocks; very small amounts of water

Most groundwater is at the surface or in sedimentary rocks buried at shallow depths.

Porosity and water content generally decrease with increasing depth and greater structural deformation.

FIGURE 11.29 Porosity and permeability, and therefore water content, generally decrease with increasing depth in Earth's crust.

slowly—probably less than a centimeter per year—that they have plenty of time to dissolve minerals from the rocks through which they pass. Thus, dissolved materials become more concentrated in these waters than in near-surface waters, making them unpotable. For example, groundwaters that pass through salt beds, which dissolve quickly, tend to contain large concentrations of sodium chloride.

At depths greater than 12 to 15 km, deep in the basement igneous and metamorphic rocks that underlie the sedimentary formations of the upper crust, porosities and permeabilities are very low due to the tremendous weight of the overlying rocks. Although these rocks contain very little water, they are saturated (**Figure 11.29**). Even some mantle rocks are presumed to contain water, although in very minute quantities.

> *Most rocks deep in Earth's crust contain very little water because their porosities and permeabilities are very low.*

Hydrothermal Waters

Natural hot springs are found in Yellowstone National Park; in Hot Springs, Arkansas; in Reykjavík, Iceland; and in many other places. Hot springs exist where **hydrothermal waters**—hot waters deep in the crust—migrate rapidly upward without losing much heat and emerge at the surface, sometimes at boiling temperatures.

Hydrothermal waters are loaded with chemical substances dissolved from rocks at high temperatures. As long as the water remains hot, the dissolved material stays in solution. However, as hydrothermal waters reach the surface, where they cool quickly, they may precipitate various minerals, such as opal (a form of silica) and calcite or aragonite (forms of calcium carbonate). Crusts of calcium carbonate produced at some hot springs build up to form the rock travertine, which can form impressive deposits such as those seen at Mammoth Hot Springs in Yellowstone National Park (**Figure 11.30**). While still below the surface, hydrothermal waters deposit some of the world's richest metallic ores as they cool, as we saw in Chapter 4.

Most hydrothermal waters of the continents come from surface waters that percolate downward to deeper regions of the crust (**Figure 11.31**). These surface waters originate primarily as **meteoric waters**—rain, snow, or other forms of water derived from the atmosphere (from the Greek *meteoron*, "phenomenon in the sky," which also gives us the word *meteorology*). These waters may be very old. It has been determined, for example,

FIGURE 11.30 Travertine deposits at Mammoth Hot Springs, Yellowstone National Park, form large lobelike masses made of aragonite and calcite. [*John Grotzinger.*]

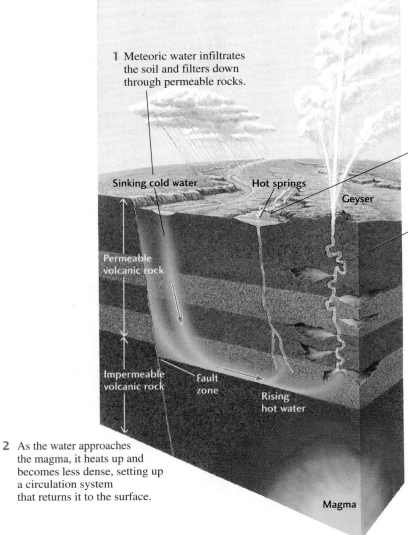

1 Meteoric water infiltrates the soil and filters down through permeable rocks.

Sinking cold water

Hot springs

Geyser

3 Hot springs occur where heated groundwater is discharged at the surface.

4 Geysers occur where an irregular network of pores and cracks slows the flow of water. Steam and boiling water are released to the surface under pressure, resulting in intermittent eruptions.

Permeable volcanic rock

Impermeable volcanic rock

Fault zone

Rising hot water

2 As the water approaches the magma, it heats up and becomes less dense, setting up a circulation system that returns it to the surface.

Magma

FIGURE 11.31 Circulation of water over a magma body produces geysers and hot springs.

Hydrothermal waters form when groundwater is heated deep in the crust, where it accumulates high concentrations of dissolved minerals. When these waters are discharged at Earth's surface, they cool rapidly and precipitate those minerals.

that the water at Hot Springs, Arkansas, derives from rain and snow that fell more than 4000 years ago and slowly infiltrated the ground.

Water that escapes from magma can also contribute to hydrothermal waters. In areas of igneous activity, sinking meteoric waters are heated as they encounter hot masses of rock. These hot waters then mix with water released from the nearby magma.

Hot Springs and Geysers

Hydrothermal waters may return to the surface as hot springs or geysers. Hot springs flow steadily; geysers erupt hot water and steam intermittently.

Geysers are probably connected to the surface by a system of very irregular and crooked fractures, recesses, and openings in rock, in contrast to the more regular and direct plumbing of hot springs. This irregular plumbing sequesters some water in recesses,

Geysers form when hydrothermal waters boil in a restricted space, where pressure builds and triggers an eruption.

thus helping to prevent the deeper waters from mixing with shallower waters and cooling. The bottom waters are heated by contact with hot rock. When they reach the boiling point, steam starts to ascend and heats the shallower waters, increasing the pressure and triggering an eruption. After the pressure is released, the geyser becomes quiet as the fractures slowly refill with water.

In 1997, geologists reported the results of a novel technique to study geysers. They lowered a miniature video camera to about 7 m below the surface of a geyser. They found that the geyser shaft was constricted at that point. Farther down, the shaft widened into a large chamber containing a wildly boiling mixture of steam, water, and what appeared to be carbon dioxide bubbles. These direct observations confirmed our ideas about how geysers work.

Geologists have turned to hydrothermal waters in the search for new and clean sources of energy. Northern California, Iceland, Italy, and New Zealand have already harnessed the steam produced by hot springs and geysers to drive electricity-generating turbines. Hydrothermal waters may soon be put to wider use for producing power, as we will see in Chapter 14.

Although hydrothermal waters provide human society with economically valuable ore deposits as well as electric power, these waters do not contribute to our drinking water supplies, primarily because they contain so much dissolved material.

SUMMARY

How does water move through the Earth system in the hydrologic cycle? The water movements of the hydrologic cycle maintain a balance among the major reservoirs of water on Earth. Evaporation from the oceans and the land transfers water to the atmosphere. Precipitation returns water from the atmosphere to the oceans and the land surface. Runoff returns part of the precipitation that falls on land to the ocean. The remainder infiltrates the ground and forms groundwater.

How does water move below the ground? Groundwater forms as precipitation infiltrates the ground and travels through porous and permeable formations. The groundwater table is the boundary between the unsaturated and saturated zones of an aquifer. Groundwater moves downhill under the influence of gravity, eventually emerging at springs where the water table intersects the ground surface. Groundwater may flow through unconfined aquifers in formations of uniform permeability or in confined aquifers, which are bounded by aquicludes. Confined aquifers produce artesian flows and spontaneously flowing artesian wells. The rate of groundwater flow depends on several factors, including the slope of the water table, the permeability of the aquifer, and the distance of flow.

What factors govern human use of groundwater resources? As the human population grows, the demand for groundwater increases greatly, particularly where irrigation is widespread. As discharge exceeds recharge, many aquifers, such as those of the Great Plains of North America, are being depleted, and there is no prospect of renewal for many years. Artificial recharge may help to renew some aquifers. The contamination of groundwater by industrial wastes, radioactive wastes, and sewage reduces the potability of some groundwater and limits our water resources.

What geologic processes are affected by groundwater? Erosion by groundwater in limestone terrains produces karst topography, with caves, sinkholes, and disappearing streams. The heating of meteoric waters deep in the crust forms hydrothermal waters, which may return to the surface as geysers and hot springs. At great depths in the crust, rocks contain extremely small quantities of water because their porosities are very low.

■ KEY TERMS AND CONCEPTS

aquiclude (p. 367)	meteoric waters (p. 377)
aquifer (p. 367)	permeability (p. 365)
artesian flow (p. 367)	potable (p. 375)
discharge (p. 367)	precipitation (p. 356)
drought (p. 358)	rain shadow (p. 358)
groundwater (p. 355)	recharge (p. 367)
groundwater table (p. 366)	relative humidity (p. 357)
hydrologic cycle (p. 356)	reservoir (p. 355)
hydrology (p. 355)	runoff (p. 357)
hydrothermal waters (p. 377)	saturated zone (p. 366)
infiltration (p. 356)	sinkhole (p. 372)
karst topography (p. 373)	unsaturated zone (p. 366)

■ EXERCISES

1. What are the main reservoirs of water at or near Earth's surface?

2. How do mountains form rain shadows?

3. What is an aquifer?

4. Why does water from an artesian well flow to Earth's surface without pumping?

5. What is the difference between the saturated and unsaturated zones of an aquifer?

6. How do aquicludes form a confined aquifer?

7. How do the slope of the water table and the permeability of an aquifer affect flow rate?

8. How does groundwater create karst topography?

9. What are the sources of water in hot springs?

10. What are some common contaminants in groundwater?

Visual Literacy Task

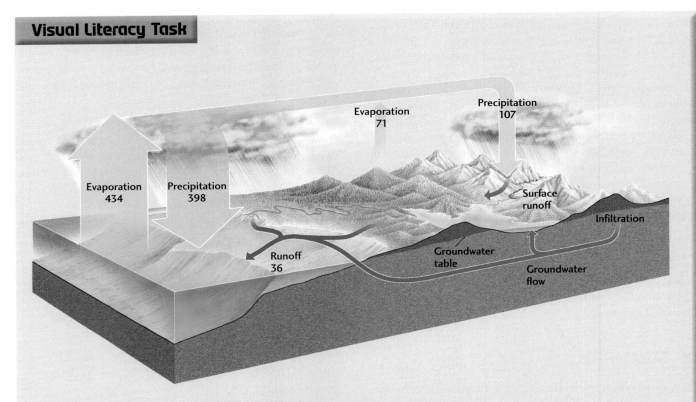

FIGURE 11.3 The hydrologic cycle is the movement of water through Earth's crust, atmosphere, oceans, lakes, and streams. The numbers indicate the amount of water (in thousands of cubic kilometers per year) that flows among these reservoirs annually.

1. What do the dark blue arrows indicate?

a. The movement of water into and out of the atmosphere only
b. The movement of water on and through land only
c. The movement of water in rivers and streams only

2. Once water evaporates from the ocean, what does it do next?

a. Precipitates into the ocean
b. Precipitates onto the land
c. Flows across the land
d. Either precipitates into the ocean or onto the land
e. Either precipitates onto the land or flows across the land

3. Why is there runoff in the diagram?

a. There is more precipitation than evaporation on land.

b. There is more precipitation in the ocean than precipitation on land.

c. There is more evaporation in the ocean than evaporation on land.

d. There is more precipitation in the ocean than evaporation plus precipitation on land.

4. How much does the total amount of water in the oceans change each year?

a. It increases by 36 km^3.

b. It increases by 398 km^3.

c. It decreases by 434 km^3.

d. It stays roughly the same.

5. What are the total evaporation and total precipitation around Earth?

a. Both are 505,000 km^3/yr.

b. One is 832,000 km^3/yr, and one is 178,000 km^3/yr.

c. One is 434,000 km^3/yr, and one is 398,000 km^3/yr.

d. One is 541,000 km^3/yr ,and one is 469,000 km^3/yr.

Thought Question: Explain how you would respond to a fellow student who asserted, "There are streams because there is more precipitation than evaporation around the globe."

Shaping Earth's Surface: Streams, Coastlines, and Wind

Beaches Made of Diamonds: Consequences of Coastal Geology

ALONG THE FOG-SHROUDED COASTLINE OF NAMIBIA lies the world's most fabulous concentration of diamonds. This deposit is known to contain at least 50 million carats of 95 percent gem-quality diamonds. It was formed entirely by surface processes involving currents that concentrated the diamonds in river and beach gravels.

The Orange River, which divides Namibia from South Africa along Africa's western coastline, reaches far inland to the core of South Africa. There, magmatic intrusions that once tapped the Earth's mantle are being slowly eroded, along with their diamond-bearing nodules. Ultimately all natural diamonds derive from the Earth's mantle. But once exposed at the Earth's surface, they become concentrated in rivers that then join together to form a single trunk river that then disperses its sediments—including the diamonds—into the Atlantic Ocean. The Orange River itself concentrates the diamonds, which, because of their high density, become sorted along with gravel deposits formed in channels. Finer sediments move by in the current, leaving a "lag deposit" of gravel and diamonds.

> In the Atlantic Ocean, along the coastline of Namibia, we find the world's greatest concentration of diamonds, carried and deposited there by river currents.

When these gravels debouch into the Atlantic Ocean, they become subject to unusually strong ocean currents originating in Antarctica that push the diamond-laden sediments northward along the coastline. This process, again involving a current pushing against the sediment, causes further concentration of the diamonds. At this terminal stage in the transport process the diamonds form beaches. Mining techniques are simple: large vacuums are used to suck the diamonds out of the beach deposit and they are picked and graded by hand. ◆

Detrital diamonds are deposited along with coarse sand and gravel along the coastline of Namibia. They are mined by a simple system of vacuums, which suck them up to be passed through a processor that excludes the sand and concentrates the diamonds. On the right side of the waste pile of sand is the bare bedrock that was vacuumed clean in the search for precious stones. [*George Steinmetz/Corbis.*]

CHAPTER OUTLINE

THE MOST FABLED SHIPWRECKS have left people stranded along the desert coastlines of the world, particularly in Africa. Both the Sahara (in northern Africa) and Namib (in southern Africa) deserts have claimed the lives of many helpless men, women, and children marooned on desert beaches, often adjacent to rivers that only rarely flow with fresh water. The same currents that concentrate diamonds in the southernmost part of the continent generate treacherous sand shoals along the entire coastline of southwestern Africa. There is no more treacherous coastline than that of Namibia, between Angola and South Africa. Mariners call it "the Skeleton Coast" and dread it. Its white sands are strewn with the skeletons of ships and travelers. Maps mark it merely as the Kaokoveld, which, freely translated from the local indigenous dialect, means "Coast of Loneliness."

As recently as 50 years ago, the Skeleton Coast had not been completely charted, and even today charts are unreliable. The shoreline is in constant motion, moving westward, farther and farther out to sea. There is visible evidence of this movement in the wrecks of ships that are today high and dry in the sand, far from the water's edge. One vessel, stranded in 1909, is today well over half a mile inland.

In one epic drama of survival, the *Dunedin Star* ran aground on the Skeleton Coast on November 29, 1942, while carrying passengers and cargo from England to South Africa. In a first attempt at rescue, a second ship was wrecked even before reaching the *Dunedin Star*, stranding its own crew farther down the coast. Then an airplane was dispatched to deliver water, food, and medical supplies to the survivors, but when it landed, it became stuck in the loose sand, and its crew also became stranded. A second airplane was then sent to drop water and supplies to the survivors of the two shipwrecks and the airplane, but this second airplane crashed into the ocean, also stranding its crew, who survived and were able to swim to the coast. Finally, a convoy of trucks was sent to rescue all four groups of survivors, who just barely managed to make it back to civilization, almost a month after the wreck of the *Dunedin Star.*

Earth's surface is affected by a wide range of processes that regulate where water forms oceans, streams, and lakes, how the atmosphere circulates to form trade winds and deserts, and how our familiar landscapes are created. Why do deserts form? How do streams form, and why do they flow to the edges of continents? Why does the coastline have such a distinctive shape? This diverse list of apparently different topics has a common thread: moving currents of air and water. As these currents interact with Earth's surface, they sculpt it, they control where it rains, they provide navigational corridors into the deep interiors of continents, they transport vast quantities of gravel, sand, and silt, forming beaches and great fields of dunes both on land and under water, and they create navigational hazards at sea.

In this chapter, we'll discuss how water and air flow in currents and how these currents erode rock and carry sediments. Streams form as a result of interactions between the plate tectonic and climate systems, and we will see how they carve valleys and

develop vast networks of channels. We will examine the processes that affect coastlines, including the effects of waves, tides, and damaging storms. Finally, we will describe desert environments, where wind plays its greatest role in shaping landscapes.

EARTH'S SURFACE IS SHAPED BY CURRENTS

Currents of water and air flow over the entire surface of Earth. As they move, they help shape the land we live on, erode mountains, create the paths of rivers, form beaches, and control the locations of deserts.

Rivers and Streams

We use the word **stream** for any body of water, large or small, that flows over the land surface, and the word **river** for any major branch of a large stream system. Most streams run through well-defined troughs called **channels,** which allow water to flow over long distances. Streams are the lifelines of the continents. Their appearance directly records the interaction of climate and plate tectonic processes. Tectonic processes lift up the land, producing the steep topography and slopes of mountainous regions. Climate determines where rain and snow will fall. Rainwater runs downhill, eroding the rocks and soils of the mountains, forming channels and carving out valleys as it gathers into streams. Streams carry back to the sea the bulk of the precipitation that falls on land and much of the sediment produced by erosion of the land surface. Worldwide, streams carry about 25 billion tons of siliciclastic sediment and an additional 2 billion to 4 billion tons of dissolved matter each year.

Streams have always been important to human civilization. A stream flows through almost every town and city in most parts of the world. These streams have served as commercial waterways for barges and steamers and as water resources for resident populations and industries (**Figure 12.1**). Before airplanes, people traveled across the oceans to coastal ports where they could trade goods and resupply their ships with food and water. Commonly these coastal ports were located on deltas, where great rivers met the ocean, allowing further access to the interiors of continents. The Nile River, for example, was vital to the agricultural economy of ancient Egypt and remains important to Egypt today. Living near a river also entails risks, however. When rivers flood, they destroy lives and property, sometimes on a huge scale.

> *All flowing bodies of water are called streams, and rivers are the major branches of a stream system. Most streams run in channels that allow water to flow over long distances.*

Coastlines and Ocean Currents

Rivers meet oceans along Earth's **coastlines,** the broad zones where land meets the sea. Coastlines can give rise to a great diversity of landscapes, depending on what kinds of currents are active in the ocean and how tectonically active the coastal region is (**Figure 12.2**). Along the coast of North Carolina, for example, long stretches of sandy beaches extend for miles (Figure 12.2a). Here, tectonic activity is limited, and it is the currents produced by breaking waves that mold the coastline. The Oregon coastline, on the other hand, is dominated by rocky cliffs, showing that even though the effect of waves is considerable, it is tectonic uplift along faults that shapes that landscape (Figure 12.2b).

FIGURE 12.1 "Three rivers junction" in Pittsburgh, Pennsylvania, where the Allegheny and Monongahela rivers join to form the Ohio River. [*Jupiter Images/ Comstock Premium/Alamy.*]

(a)

(b)

FIGURE 12.2 (a) Waves break along the shoreline and transport sand to form a beach at Cape Hatteras National Seashore, North Carolina. [*Stephen Saks/Lonely Planet.*] (b) Waves pound a rocky shoreline at Cape Arago State Park, Coos County, Oregon. [*Steve Terrill/Corbis.*]

The major surface processes operating along coastlines are waves and tides. Together, they erode even the most resistant rocky shores. Waves and tides create currents, which transport sediments produced by erosion of the land and deposit them on beaches and in shallow waters along the shore. Coastlines are where hurricanes and other tropical storms affect the continents most strongly, causing devastation not just because of their strong winds but also because of coastal flooding.

> The landscapes of coastlines are shaped by tectonic processes and by currents driven by tides and waves.

Deserts: The Role of Wind

In deserts, rainfall is infrequent, water is scarce, and strong winds can howl for days on end. Some coastlines, like that of Namibia, are flanked by broad deserts. Other deserts cover broad regions of the interiors of continents. The locations of deserts are determined by the global flow of air, and wind plays a significant role in shaping their landscapes. Wind is much like water in its ability to erode, transport, and deposit sediments, and it is capable of moving enormous quantities of sand, silt, and dust over large regions of continents and oceans. The ancient Greeks called the god of winds Aeolus, and geologists today use the term **eolian** for the geologic processes powered by wind.

> The locations of deserts are determined by global wind patterns.

Recently, concern over the expansion of Earth's deserts has increased. The population of southern Spain, for example, increasingly wonders whether the Sahara has jumped the Mediterranean Sea and is now encroaching on southern Europe. The process of desertification, in which land is degraded by decreases in rainfall resulting from various factors such as climate variations and human activities, has become a major focus of scientists trying to understand Earth's climate system.

THE FLOW OF WATER AND WIND

All currents, both water and air, share some basic characteristics of flowing fluids. We can illustrate two kinds of fluid flow by using lines of motion called *streamlines* (**Figure 12.3**). In **laminar flow,** the simplest kind of fluid movement, straight or gently curved streamlines run parallel to one another without mixing or crossing between layers. The slow movement of thick syrup over a pancake, with strands of unmixed melted butter flowing in parallel but separate paths, is a laminar flow. **Turbulent flow** has a more complex pattern of movement, in which streamlines mix, cross, and form swirls and eddies. Fast-moving river waters typically show this kind of motion. Turbulence—the degree to which there are irregularities and eddies in the flow—may be low or high.

> In laminar flow, streamlines run parallel to each other. In turbulent flow, streamlines mix, cross, and form swirls and eddies.

Whether a flow is laminar or turbulent depends primarily on three factors:

1. Its velocity (rate of movement)

2. Its depth

3. Its viscosity

Laminar flow

Turbulent flow

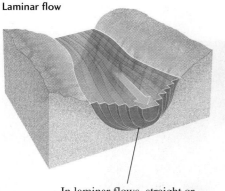
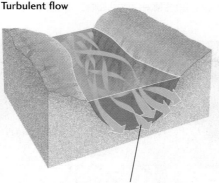

FIGURE 12.3 The two basic patterns of fluid flow: laminar flow and turbulent flow. The photograph shows the transition from laminar to turbulent flow in water along a flat plate, revealed by the injection of a dye. Flow is from left to right. [ONERA.]

In laminar flows, straight or gently curved streamlines run parallel without mixing or crossing.

In turbulent flows, streamlines mix, cross, and form swirls and eddies.

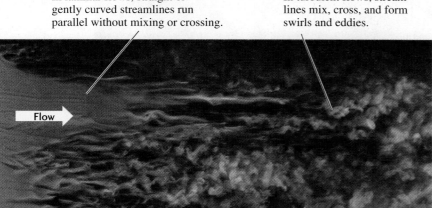

Flow

Viscosity is a measure of a fluid's resistance to flow. The more viscous (the thicker) a fluid is, the more it resists flow. The higher the viscosity, the greater the tendency for laminar flow. The viscosity of most fluids, including water, decreases as the temperature increases. Given enough heat, a fluid's viscosity may decrease sufficiently to change a laminar flow into a turbulent one.

Water has low viscosity in the range of temperatures at Earth's surface. For this reason alone, most streams in nature tend toward turbulent flow. In addition, the rapid movement of water in most streams makes them turbulent. In nature, we are likely to see laminar flows of water only in thin sheets of rain runoff flowing slowly down nearly level slopes. In cities, we may see small laminar flows in street gutters. Because most streams and rivers are broad and deep and flow quickly, their flows are almost always turbulent.

Whether a flow is laminar or turbulent depends on its velocity, its depth, and its viscosity.

A stream may show turbulent flow over much of its width and laminar flow along its edges, where the water is shallower and is moving more slowly. The flow velocity is highest near the center of a stream. We commonly refer to a rapid flow of water as a strong current.

Like flows of water in streams, air flows are nearly always turbulent. The extremely low density and viscosity of air make it turbulent even at the velocity of a light breeze.

Wind is a natural flow of air that is parallel to the surface of the rotating planet. Winds obey all the laws of fluid flow that apply to water in streams. Unlike flows of water in stream channels, however, winds are generally unconfined by solid boundaries, except for the ground surface and narrow valleys. Air flows are free to spread out in all directions, including upward into the atmosphere. Furthermore, the much lower density of wind makes it less powerful than water currents, even though wind speeds are often much greater than those in currents of water.

Winds vary in speed and direction from day to day, but over the long term they tend to come mainly from one direction because Earth's atmosphere flows in global prevailing wind belts (**Figure 12.4**). In the temperate latitudes, which include most of North America, the prevailing winds come from the west and so are referred to as the *westerlies*. Temperate climates are located at latitudes between 30° and 60° N and 30° and 60° S. In the tropics, which are between 30° S and 30° N of the equator, the *trade winds* (named for an archaic use of the word *trade* to mean "track" or "course") blow from the east.

These prevailing wind belts arise because the Sun warms a given amount of land surface most intensely at the equator, where the Sun's rays are almost perpendicular to Earth's surface. The Sun heats the land less intensely at high latitudes and at the poles because there its rays strike Earth's surface at an angle (see

Global prevailing wind belts result from variable heating of Earth's surface by the Sun.

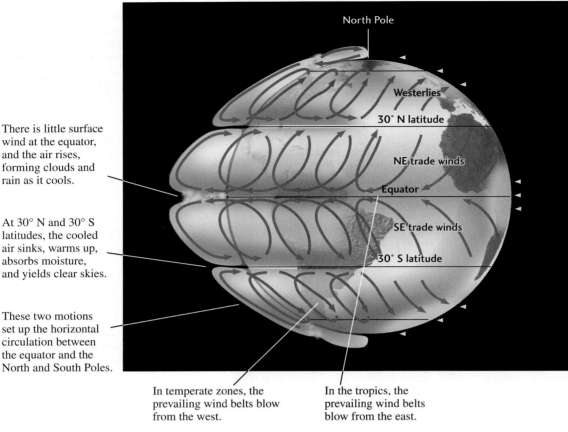

There is little surface wind at the equator, and the air rises, forming clouds and rain as it cools.

At 30° N and 30° S latitudes, the cooled air sinks, warms up, absorbs moisture, and yields clear skies.

These two motions set up the horizontal circulation between the equator and the North and South Poles.

North Pole

Westerlies

30° N latitude

NE trade winds

Equator

SE trade winds

30° S latitude

At the poles, the Sun's rays strike the surface at an angle and so are spread out over greater areas, yielding colder temperatures.

At the equator, the Sun's rays are almost perpendicular to the surface, concentrating heat in this region.

In temperate zones, the prevailing wind belts blow from the west.

In the tropics, the prevailing wind belts blow from the east.

FIGURE 12.4 Earth's atmosphere circulates in prevailing wind belts created by variable solar heating and by Earth's rotation.

Figure 12.4). Hot air, which is less dense than cold air, rises at the equator and flows toward the poles, gradually sinking and releasing water vapor as it cools. The sinking air reaches ground level at about 30° S and 30° N, then flows back along Earth's surface toward the equator to form the trade winds. This cool, dry air warms and absorbs moisture as it sinks, producing clear skies and arid climates. It overlies many of the world's deserts, such as the Sahara.

SEDIMENT TRANSPORT AND DEPOSITION

Currents of water and air vary in their ability to erode and carry sand grains and other sediments. Laminar flows of water can lift and carry only the smallest, lightest clay-sized particles. Turbulent flows of water, depending on their speed, can move particles ranging in size from clay to pebbles and cobbles. Winds cannot normally carry particles larger than sand grains. As turbulence lifts particles into a flow, the flow carries them downcurrent. Turbulence also rolls and slides larger particles. A current's **suspended load** includes all the material temporarily or permanently suspended in the flow. Its **bed load** is the material the current carries along Earth's surface by sliding and rolling (**Figure 12.5**). The *bed* is the layer of loose sediment that interacts with the current.

Turbulent flows can generally transport larger sediment particles than can laminar flows. Particles suspended in a current are its suspended load, whereas particles that slide and roll across its bed are its bed load.

Suspension, Saltation, and Settling

A current's ability to carry sediment depends on a balance between turbulence, which lifts particles, and the competing downward pull of gravity, which makes them settle out of the current and become part of the bed. Small grains of silt and clay are easily lifted into the current and settle slowly, so they tend to stay in suspension. Larger particles, such as medium- and coarse-grained sand, settle more quickly. Most larger particles therefore stay suspended in the current only a short time before they settle.

Sand grains in a current typically move by **saltation**—an intermittent jumping motion along the bed. The grains are sucked up into the flow by turbulent eddies, move with the current for a short distance, and then fall back to the bed (see Figure 12.5). If you were to

1 Current flowing over a bed of gravel, sand, silt, and clay carries a **suspended load** of finer particles…

3 As current velocity increases, the suspended load grows,…

5 Particles move by saltation, jumping along a bed. At a given current velocity, smaller particles jump higher and travel farther than larger particles.

2 …and a **bed load** of material sliding and rolling along the bottom.

4 …and the increased force of the flow generates an increase in the bed load.

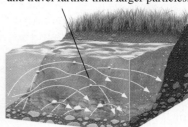

FIGURE 12.5 A current flowing over a bed of unconsolidated material can transport particles in three ways: sliding and rolling, saltation, and suspension.

stand in a rapidly flowing sandy stream of water, you might see a cloud of saltating sand grains moving around your ankles. The bigger the grain, the longer it will tend to remain on the bed before it is picked up. Once a large grain is in the current, it will settle quickly. The smaller the grain, the more frequently it will be picked up, the higher it will "jump," and the longer it will take to settle.

> *During saltation, particles are lifted into a flowing current, move downcurrent a short distance, and then land on the bed.*

Thus, turbulent flows transport sediment by suspension (clays), saltation (sands), and rolling and sliding along the bed (sand and gravel). To study how a particular river carries sediments, geologists and engineers measure the relationship between particle size and the force the flow exerts on the particles in the suspended and bed loads. This relationship shows them how much sediment a particular flow can move and how rapidly it can move it. This information allows them to design dams and bridges or to estimate how quickly artificial reservoirs behind dams will fill with sediments.

As we saw in Chapter 6, the greater a current's velocity, the larger the sediment particles it can transport. Thus, when the current slows, the largest particles settle out first. Geologists can infer the velocities of ancient currents from the sizes of particles in sedimentary rocks.

Windblown Dust

Most North Americans are familiar with rainstorms or snowstorms: high winds accompanied by heavy precipitation. We may have less experience of dry storms, during which high winds blowing for days on end carry enormous amounts of sand, silt, and dust.

Air has a staggering capacity to hold dust in suspension. Large dust storms may carry more than 100 million tons of dust and deposit it in layers several meters thick. Fine-grained particles from the Sahara have been found as far away as England and have been traced across the Atlantic Ocean to Florida. Scientists on oceanographic research vessels have measured airborne dust far out to sea, and today it can be observed directly from space (**Figure 12.6**).

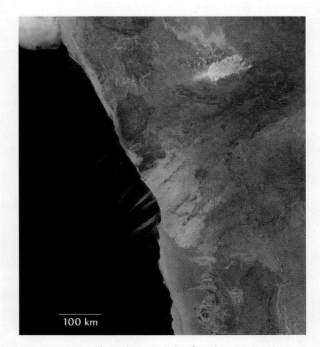

100 km

FIGURE 12.6 Satellite photograph of a dust storm originating in the Namib Desert in September 2002. Dust and sand are being transported from right (east) to left (west) by strong winds blowing out to sea. These sediments can be transported for hundreds to thousands of kilometers across the ocean. [*NASA.*]

> *Flows of air can hold large amounts of dust in suspension. Large dust storms may deposit suspended particles in layers several meters thick.*

Dust includes microscopic rock and mineral fragments of all kinds, especially silicates, as might be expected from their abundance as rock-forming minerals. Two of the most important sources of silicate minerals in dust are clays from soils on dry plains and volcanic ash from eruptions.

Windblown dust has complex effects on climate. Mineral dust in the atmosphere scatters the incoming visible (light) radiation from the Sun and absorbs the outgoing infrared (heat) radiation from Earth.

Sediment Bed Forms: Dunes and Ripples

When a current transports sand grains by saltation, the grains tend to form dunes and ripples (see Chapter 6). **Dunes** are mounds or ridges of sand up to many meters high that form in flows of both wind and water over a sandy bed (**Figure 12.7**). As a dune grows, the whole mound starts to migrate downcurrent by the combined movements of a host of individual grains as sand is eroded from the windward side and deposited on the leeward (downcurrent) side. Sand grains constantly saltate to the top of the low-angled windward slope and then fall over into the wind shadow on the leeward slope, as shown in **Figure 12.8**. The grains gradually build up a steep, unstable accumulation on the upper part of the leeward slope. Periodically, this accumulation gives way and spontaneously slips or cascades down this *slip face*, as it is called, to a new slope at a lower angle.

Ripples are very small dunes—with heights ranging from less than a centimeter to several centimeters—whose

FIGURE 12.7 Sand dunes in the Namib Desert, in southwestern Africa, are among the tallest in the world. [*John Grotzinger.*]

1 A ripple or dune advances by the movements of individual grains of sand. The whole form moves forward slowly as sand erodes from the windward slope and is deposited on the leeward slope.

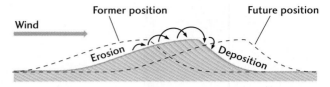

2 Particles of sand arriving on the windward slope of the dune move by saltation over the crest,...

3 ...where the wind velocity decreases and the sand deposited slips down the leeward slope.

4 This process acts like a conveyor belt that moves the dune forward.

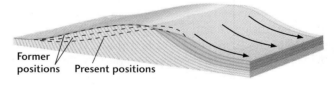

5 The dune stops growing vertically when it reaches a height at which the wind is so fast that it blows the sand grains off the dune as quickly as they are brought up.

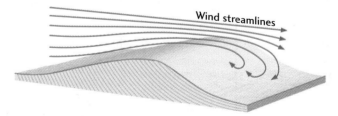

FIGURE 12.8 Sand dunes grow and move as wind transports sand particles by saltation.

FIGURE 12.9 Wind ripples in sand at Stovepipe Wells, Death Valley, California. Although complex in form, ripples are always transverse (at right angles) to the wind direction. [*Tom Bean.*]

long dimension is formed at right angles to the current (**Figure 12.9**). Although underwater ripples and dunes are harder to observe than are those produced on land by air currents, they form in the same way and are just as common. The steady downcurrent transfer of grains across their ridges causes ripples and dunes to migrate downcurrent. The speed of this migration is much slower than the movement of individual grains and very much slower than the current.

The shapes of ripples and dunes and their migration speeds change as the velocity of the current increases. At the lowest current velocities, few grains are saltating,

and the sediment bed is flat. At slightly higher velocities, the number of saltating grains increases. A rippled bed forms, and the ripples migrate downstream (**Figure 12.10**). As the velocity increases further, the ripples grow larger and migrate faster until, at a certain point, dunes replace the ripples. Both ripples and dunes have a cross-bedded structure, and as the current flows over their tops, it can actually reverse and flow backward along their leeward side. As the dunes grow larger, small ripples form on them. The ripples tend to climb over the backs of the dunes because they migrate more quickly than the dunes.

> *Transport of individual sand particles by saltation forms dunes and ripples. Ripples usually form first and grow into dunes as current velocity increases.*

STREAMS: TRANSPORT FROM MOUNTAINS TO OCEANS

In this section we'll look at how moving currents of water shape the continents and transport sediments to the oceans. These currents modify the surfaces of continents to form mountain valleys and major rivers. The sediments produced by weathering and erosion and transported in streams are ultimately delivered to the edges of continents, where deltas are formed in low-lying regions.

Streamflow

The flow of a stream may appear steady when you view it from a bridge for a few minutes or canoe along it for a few hours, but its volume and velocity at a single place may change appreciably from month to month and season to season. Streams are constantly changing, shifting from low waters to floods over a few hours or days and reshaping their valleys over longer periods (**Figure 12.11**).

(*Continued on page 394.*)

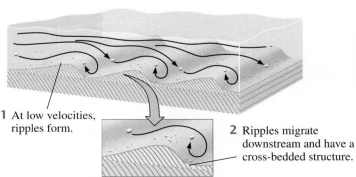

1 At low velocities, ripples form.

2 Ripples migrate downstream and have a cross-bedded structure.

3 Higher velocities produce a larger bed form—a dune. Then, smaller ripples may piggyback up the dune.

4 Ripples migrate faster, so they tend to climb over the backs of dunes.

FIGURE 12.10 The form of a sediment bed changes with increasing current velocity. [*After D. A. Simmons and E. V. Richardson, "Forms of Bed Roughness in Alluvial Channels." American Society of Civil Engineers Proceedings 87 (1961): 87–105.*]

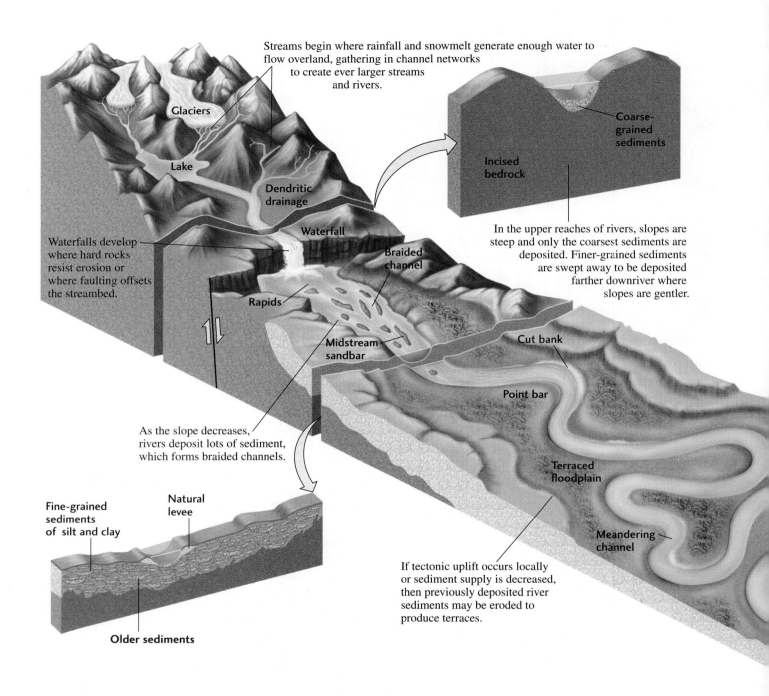

FIGURE 12.11 Stream networks transport water and sediments from their headwaters to the ocean.

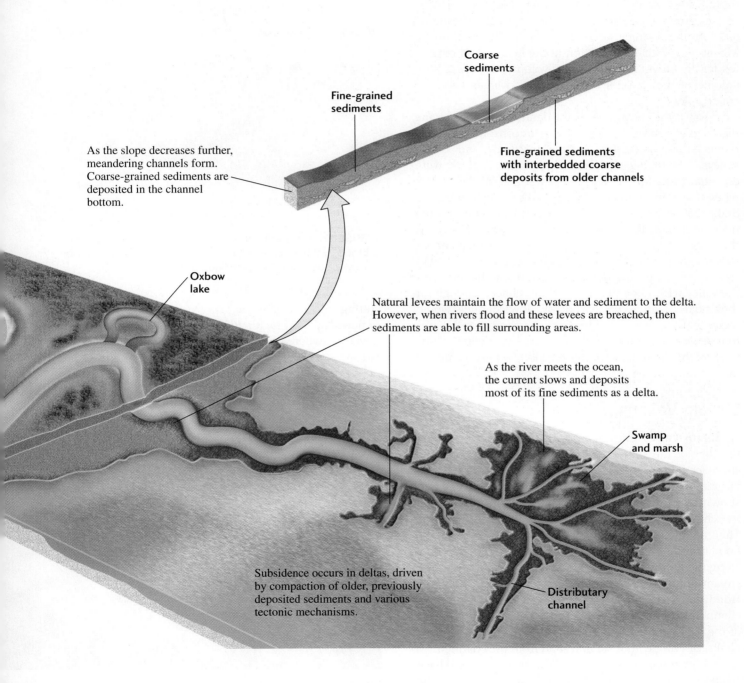

Coarse
sediments

Fine-grained
sediments

Fine-grained sediments
with interbedded coarse
deposits from older channels

As the slope decreases further,
meandering channels form.
Coarse-grained sediments are
deposited in the channel
bottom.

Oxbow
lake

Natural levees maintain the flow of water and sediment to the delta.
However, when rivers flood and these levees are breached, then
sediments are able to fill surrounding areas.

As the river meets the ocean,
the current slows and deposits
most of its fine sediments as a delta.

Swamp
and marsh

Subsidence occurs in deltas, driven
by compaction of older, previously
deposited sediments and various
tectonic mechanisms.

Distributary
channel

The flow and dimensions of a stream also change as it moves downslope, from narrow valleys in its upland headwaters to broader floodplains in its middle and lower courses. Most of these longer-term changes are adjustments in the normal (nonflood) volume and velocity of flow as well as the depth and width of the channel.

From their headwaters to their termination points in lakes and oceans, all streams react to changes in climate (such as changes in precipitation) and tectonic processes (uplift or subsidence of Earth's crust). Near their headwaters, streams gather into ever larger streams, eventually forming a single large strand, as in the case of the Mississippi River. Precipitation in the headwaters may be felt far downriver, where a river's volume may exceed the volume of the channel and then spill over its banks to create a flood.

As streams move across the land surface—in some places over bedrock, in others over unconsolidated sediments—they erode the surface materials and create valleys. Identifying and mapping stream valleys were essential tasks for Meriwether Lewis and William Clark when they first explored and mapped the western United States 200 years ago. As they traveled upstream and the river branched, they had to choose which branch was the larger of the two. They used two observations about streams to help them make this choice: the width of the stream valley and the depth of the stream channel. Was the valley wide enough and the channel deep enough for their boats? Narrow valleys and shallow channels would mean that the branch led into a much shorter and therefore less desirable route; wider valleys and deeper channels, on the other hand, promised a longer passage up the main branch of the river.

Stream Valleys

A **stream valley** encompasses the entire area between the tops of the slopes on both sides of the stream. The cross-sectional profile of many stream valleys is V-shaped, but many other valleys have a broad, low profile like that shown in **Figure 12.12**. At the bottom of the valley is the channel, the trough through which the water runs. The channel carries all the water during normal, nonflood times. At low water levels, the stream may run only along the bottom of the channel. At high water levels, the stream occupies most of the channel. In broad valleys, a **floodplain**—a flat area about level with the top of the channel—lies on either side of the channel. It is the part of the valley that is flooded when the river spills over its banks, carrying with it silt and sand from the channel.

In high mountains, stream valleys are narrow and have steep walls, and the channel may occupy most or all of the valley bottom. A small floodplain may be visible only at low water levels. In such valleys, the stream is actively

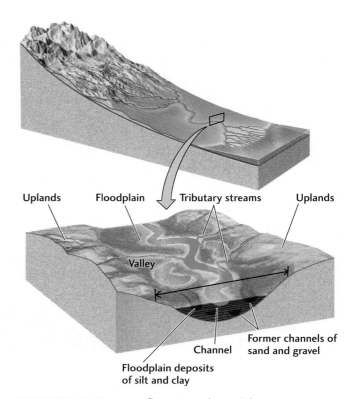

FIGURE 12.12 A stream flows in a channel that moves over a broad, flat floodplain in a wide valley. Floodplains may be narrow or absent in steep-walled valleys.

cutting into the bedrock, a process that is characteristic of newly uplifted highlands in tectonically active areas. In lowlands, where tectonic uplift has long since ceased, the stream shapes its valley by eroding sediment particles and

FIGURE 12.13 Aerial view of the meandering Adelaide River, in Australia. Its appearance is typical of meandering streams in lowland environments. [*Peter Bowater/Photo Researchers.*]

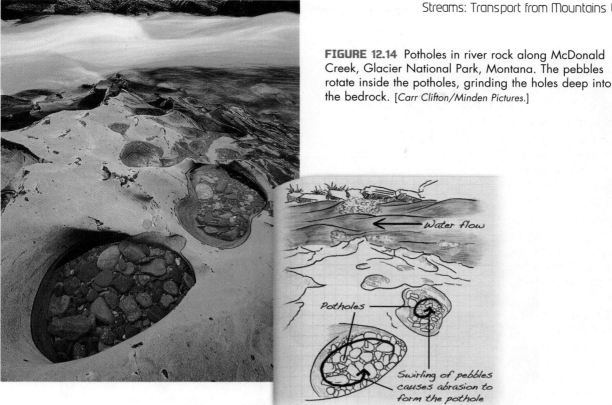

FIGURE 12.14 Potholes in river rock along McDonald Creek, Glacier National Park, Montana. The pebbles rotate inside the potholes, grinding the holes deep into the bedrock. [*Carr Clifton/Minden Pictures.*]

transporting them downstream. With a long time to operate, these processes produce gentle slopes and floodplains many kilometers wide (**Figure 12.13**).

Stream valleys are formed when streams break apart and erode rock by abrasion. The sand and pebbles the stream carries create a sandblasting action that wears away even the hardest rock. In some streambeds, pebbles and cobbles rotating inside swirling eddies grind deep *potholes* into the bedrock (**Figure 12.14**). At low water, pebbles and sand can be seen lying quietly at the bottom of exposed potholes.

> A stream valley contains a channel through which the water runs. Normally, the channel carries all the water, but during floods the water spills out of the channel onto the adjacent floodplain.

Stream channels begin where rainwater, draining off the surface of the land, flows so fast that it abrades the soil and bedrock and carves into it (**Figure 12.15**). Once a small gully forms, it captures more of the flow of water over the land surface, and thus the tendency of the stream to cut downward increases. As the gully progressively deepens, the rate of downcutting increases as more water is captured.

Rock erosion is particularly strong at rapids and waterfalls. *Rapids* are places in a stream where the flow is extremely fast because the slope of the streambed suddenly steepens, typically at rocky ledges. The speed and turbulence of the water quickly break rocks into smaller pieces that are carried away by the strong current.

FIGURE 12.15 Streams create gullies when the action of water flowing across Earth's surface causes erosion of soil or bedrock. The smallest gullies converge to form larger stream channels, and farther downslope they become river channels. These gullies were formed in the desert of Oman by occasional rainstorms that inundate the surface with rapidly flowing water that erodes the bedrock. [*Petroleum Development Oman.*]

FIGURE 12.16 This waterfall on the Iguaçú River, Brazil, is retreating upstream as falling water and sediment pound the cliff's base and undercut it. Looking downstream, from the center to the upper left of the photo, one can see steep walls, the remnants of the waterfall's retreat upstream. [*Donald Nausbaum.*]

> *Streams carrying coarse sediments are capable of eroding solid rock. The erosion is greatest at rapids and waterfalls, where flows are fastest and most turbulent.*

The tremendous impact of huge volumes of plunging water and tumbling boulders quickly erodes streambeds below waterfalls. Waterfalls also erode the underlying rock of the cliffs that form the falls. As erosion undercuts these cliffs, the upper streambed collapses, and the falls recede upstream (**Figure 12.16**).

Stream Channel Patterns

As a stream channel makes its way along the bottom of a valley, it may run straight in some stretches and take a snaking, irregular path in others, sometimes splitting into multiple channels. The channel may flow along the center of the floodplain or hug one edge of the valley.

Meanders On a great many floodplains, stream channels follow curves and bends called **meanders** (see Figure 12.13). Meanders are the normal pattern for streams flowing through plains or lowlands, where their channels typically cut through unconsolidated sediments—fine sand, silt, or mud—or easily eroded bedrock. Meanders are less pronounced but still common in streams flowing down steeper slopes over harder bedrock. In such terrain, meandering stretches may alternate with long, relatively straight ones.

A stream that has cut deeply into the curves and bends of its channel may produce incised meanders (**Figure 12.17**). Other streams may meander on somewhat wider floodplains bounded by steep, rocky valley walls. Meanders on a floodplain migrate over periods of many years as the stream erodes the outside banks of bends, where the current is strongest (**Figure 12.18**). As the outside banks are eroded, sediments are deposited to form curved sandbars called **point bars** along the inside banks, where the current is slower. In this way, meanders slowly shift position from side to side, as well as downstream, in a snaking motion something like that of a long rope being snapped.

As meanders migrate, sometimes unevenly, the bends may grow closer and closer together until finally the river

Incised meanders   Point bar

FIGURE 12.17 This section of the San Juan River, Utah, is a good example of an incised meander belt, a deeply eroded, meandering, V-shaped valley with virtually no floodplain. *[Tom Bean.]*

bypasses one of them, often during a major flood. The river takes a new, shorter course like that shown in Figure 12.18. In its abandoned path, it leaves behind an *oxbow lake*—a crescent-shaped, water-filled loop.

> *Meanders migrate over time as a stream cuts into the outsides of bends and deposits sediments on the insides of bends as point bars.*

Braided Streams Some streams have many channels instead of a single one. A **braided stream** is one whose channel divides into an interlacing network of channels, which then rejoin in a pattern resembling braids of hair

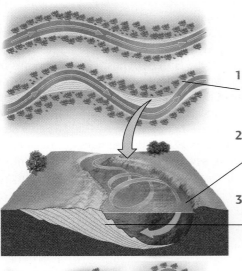

1 Meanders shift from side to side in a snaking motion.

2 The current is faster at outside banks, which are eroded,…

3 …and sediments are deposited at inside banks where the current is slower, forming point bars.

Point bars

4 As the erosion and deposition process continues, the bends grow closer and the point bars bigger.

5 During a major flood, when velocity and water volume increase, the river takes a new, shorter course, cutting across the loop.

6 The abandoned loop remains as an oxbow lake.

Point bar

Oxbow lake

FIGURE 12.18 Meanders migrate over a period of many years. (*top*) Meanders in an Alaskan River. (*bottom*) Oxbow lake in Blackfoot River valley, Montana. [(*top*) Peter Kresan. (*bottom*) James Steinberg/Photo Researchers.]

Channels in braided streams

FIGURE 12.19 This stretch of the Chitina River, Alaska, is a braided stream. [*Tom Bean.*]

(**Figure 12.19**). Braided streams are found in many settings, from broad lowland valleys to wide, sediment-filled rift valleys adjacent to mountain ranges. Braids tend to form in streams with large variations in volume of flow combined with a high sediment load and banks that are easily eroded. They are well developed, for example, in the sediment-choked streams formed at the edges of melting glaciers.

Braided streams have interlacing networks of channels.

Alluvial Fans Sometimes streams leave narrow mountain valleys and enter broad, relatively flat valleys at lower elevations. Along sharply defined mountain fronts, typically at steep fault scarps, streams drop large amounts of sediment in cone- or fan-shaped accumulations called **alluvial fans** (**Figure 12.20**). This deposition results from the sudden decrease in current velocity that occurs as the channel widens abruptly. Coarse materials, from boulders to sand, dominate on the steep upper slopes of the fan. Lower down, finer sands, silts, and muds are deposited.

Alluvial fans are deposits of sediment that form where current velocity decreases as streams emerge from narrow mountain valleys.

Natural Environment: Floodplains and the Development of Cities

A stream channel migrating over the floor of a valley creates a floodplain. As floodwaters spread out over the floodplain, the velocity of the water slows, and the current loses its ability to carry sediment. The floodwater velocity drops most quickly along the immediate borders of the channel. As a result, the current deposits large amounts of coarse sediment, typically sand and gravel, along a narrow strip at the edge of the channel. Successive floods build up **natural levees,** ridges of coarse material that confine the stream within its banks between floods, even when water levels are high (**Figure 12.21**). Where natural levees have reached a height of several meters and the stream almost fills the channel, the floodplain level is below the stream

FIGURE 12.20 An alluvial fan (Tucki Wash) in Death Valley, California. Alluvial fans are large cone- or fan-shaped accumulations of sediment deposited when stream velocity slows, as at a mountain front. [*Marli Miller.*]

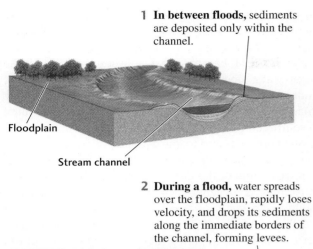

1 **In between floods,** sediments are deposited only within the channel.

Floodplain

Stream channel

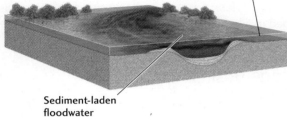

2 **During a flood,** water spreads over the floodplain, rapidly loses velocity, and drops its sediments along the immediate borders of the channel, forming levees.

Sediment-laden floodwater

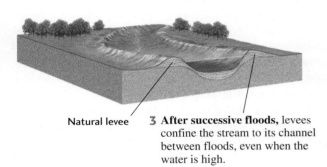

Natural levee

3 **After successive floods,** levees confine the stream to its channel between floods, even when the water is high.

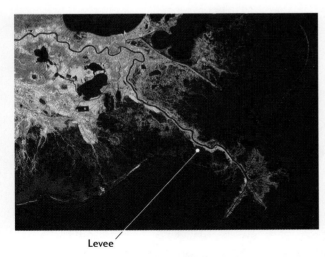

Levee

FIGURE 12.21 Floods form natural levees along the banks of a river. The false color LANDSAT image shows natural levees along the main channel of the Mississippi River near South Pass, Louisiana. [*USGS National Wetlands Research Center.*]

FIGURE 12.22 Floodwaters from the Red and Wild Rice rivers take over the Forst River area of Fargo, North Dakota, on April 18, 1997. [*Jim Mone/AP Photo.*]

level. You can walk the streets of an old river town built on a floodplain, such as Vicksburg, Mississippi, and look up at the levee, knowing that the river waters are rushing by above your head.

> *Floods form natural levees by depositing coarse sediments at the margins of stream channels.*

Floodplains are natural sites for urban settlements because they combine easy transportation along a river with access to fertile agricultural lands. Such sites, however, remain subject to the floods that formed the floodplains (**Figure 12.22**).

About 4000 years ago, cities began to dot floodplains in Egypt along the Nile, in the ancient land of Mesopotamia along the Tigris and Euphrates rivers, and in Asia along the Indus River of India and the Yangtze and Huang Ho of China. Later, many of the capital cities of Europe were built on floodplains: Rome on the Tiber, London on the Thames, Paris on the Seine. Floodplain cities in North America include St. Louis on the Mississippi, Cincinnati on the Ohio, and Montreal on the St. Lawrence.

> *Floodplains are natural sites for urban development because they provide river transportation as well as fertile soil for agriculture.*

 ## Natural Hazards: Floods

There are benefits to living on or near a floodplain, but cities built on floodplains are also prone to destructive floods. Floods occur when a stream's **discharge**—the volume of water that passes a given point at a given time—increases such that more water flows into the stream than can flow out. When this happens the excess water spills over the banks.

Streams flood regularly, some at infrequent intervals, others almost every year. Some floods are large, with very high water levels lasting for days. At the other extreme are minor floods that barely break out of the channel before they recede. Small floods are more frequent, occurring every 2 or 3 years on average. Large floods are generally less frequent, usually occurring every 10, 20, or 30 years.

Floods periodically destroyed sections of ancient and modern cities on the lower parts of the floodplains, but each time the inhabitants rebuilt them. Today, most large cities are protected by artificial levees that strengthen and heighten the river's natural levees. Extensive systems of dams can help control flooding that would affect these cities, but they cannot eliminate the risk entirely. In 1973, for example, the Mississippi went on a rampage with a flood that continued for 77 consecutive days at St. Louis. The river reached a record 4.03 m above flood stage (the height at which the river first overflows the channel banks). In 1993, the Mississippi and its tributaries broke loose again, shattering the old record in a disastrous flood that has been officially designated the second worst flood in U.S. history (behind the flooding of New Orleans by Hurricane Katrina in 2005).

The flooding of the Mississippi during April and May of 2011 rivaled the flood of 1993, one of the worst in the past century. The two storms that produced the 2011 floods conspired with springtime snowmelt to create the perfect situation for a 500-year flood. The second of the two storms also unleashed the deadliest outbreak of tornadoes since 1925: 250 tornadoes were reported (**Figure 12.23**). The tornado that touched down at Joplin, Missouri, on May 22, 2011, became the deadliest tornado to hit the United States since 1947, killing at least 123 and injuring hundreds more. The connection between storms, rainfall, tornados, and flooding is well known to the residents of the south-central United States.

What are cities and towns in this position to do? Some have urged a halt to all construction and development on the lowest parts of the floodplains. Some have called for the elimination of federally subsidized disaster funds for rebuilding in such areas. Harrisburg, Pennsylvania, hit hard by a flood in 1972, turned some of its devastated riverfront area into a park. Still other towns have moved off the floodplain to a completely new location, as described in Chapter 1. Yet the benefits of living on floodplains continue to attract people to those sites.

Drainage Basins

Every topographic rise between two streams, whether it measures a few meters or a thousand, forms a **divide**—a ridge of high ground along which all rainfall runs off down one side or the other. A **drainage basin** is an area of land, bounded by divides, that funnels all its water into the network of streams draining the area (**Figure 12.24**). Drainage basins occur at many scales, from a ravine surrounding a small stream to a great region drained by a major river and its tributaries (**Figure 12.25**).

A continent has several major drainage basins separated by major divides. In North America, the continental

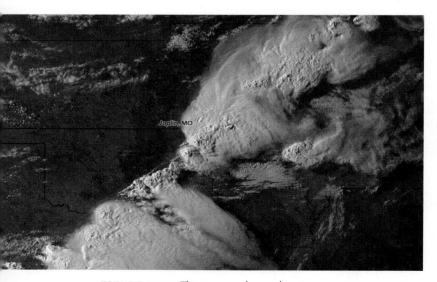

FIGURE 12.23 This image shows the storm system moments before spawning the tornado that struck Joplin, Missouri, on May 22, 2011. [*National Oceanic and Atmospheric Administration.*]

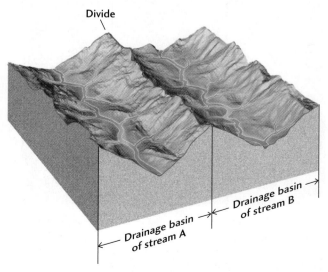

FIGURE 12.24 Drainage basins are separated by divides.

> A drainage basin is an area bounded by divides that contains a network of streams draining the area.

divide formed by the Rocky Mountains separates all waters flowing into the Pacific Ocean from all those entering the Atlantic. Lewis and Clark followed the Missouri River upstream to its headwaters at the continental divide in western Montana. When they crossed over the divide, they descended from the headwaters of the Columbia River, which they followed to the Pacific Ocean.

A map showing the courses of all the large and small streams in a drainage basin reveals a pattern of connections called a *drainage network*. If you followed a stream from its mouth to its headwaters, you would see that it steadily divides into smaller and smaller tributaries, forming a drainage network that shows a characteristic branching pattern (**Figure 12.26**). Branching is a general property of many kinds of networks in which material is collected and distributed. The network of the human circulatory system, for example, distributes blood to the body through a branching system of arteries and collects it through a corresponding system of veins.

> The streams in a drainage basin are joined in a drainage network with a characteristic branching pattern.

Dendritic drainage is characterized by branches similar to the limbs of a tree.

Rectangular drainage, developed on a strongly jointed rocky terrain, tends to follow the joint pattern.

Trellis drainage develops in valley and ridge terrain, where rocks of varying resistance to erosion are folded into anticlines and synclines.

Radial drainage patterns develop on a single large peak, such as a large dormant volcano.

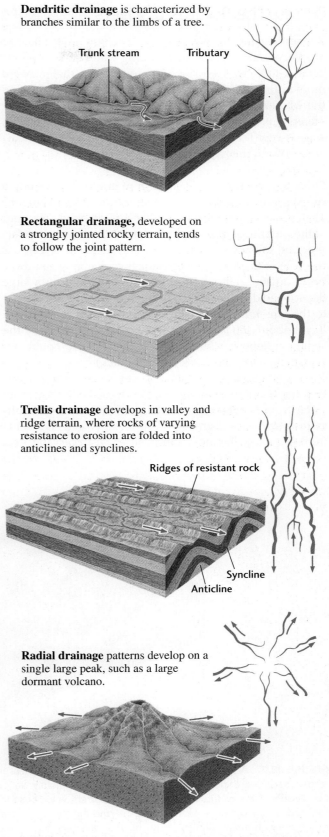

FIGURE 12.26 Some typical drainage network patterns.

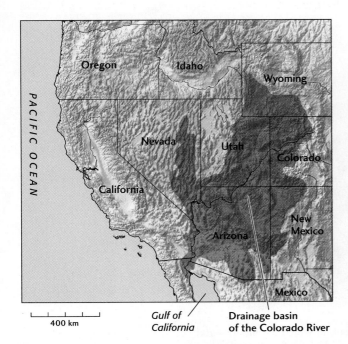

FIGURE 12.25 The drainage basin of the Colorado River covers about 630,000 km², constituting a large part of the southwestern United States. The basin is surrounded by divides that separate it from the neighboring drainage basins. [*After U.S. Geological Survey.*]

Deltas: The Mouths of Rivers

Sooner or later, all rivers end as they flow into a lake or an ocean, mix with the surrounding water, and—no longer able to travel downslope—gradually lose their forward momentum. The largest rivers, such as the Amazon and the Mississippi, can maintain some current many kilometers out to sea. Where smaller rivers enter a turbulent, wave-swept sea, the current disappears almost immediately beyond the river's mouth.

Delta Sedimentation As its current gradually dies out, a river progressively loses its power to transport sediments. The coarsest material, typically sand, is dropped first, right at the mouths of most rivers. Finer sands are dropped farther out, followed by silt and, still farther out, by clay. As the floor of the lake or ocean slopes to deeper water away from the shore, the deposited materials build up a large flat-topped deposit called a **delta.** (We owe the name *delta* to the Greek historian Herodotus, who traveled through Egypt about 450 B.C. The roughly triangular shape of the sediment deposit at the mouth of the Nile prompted him to name it after the Greek letter Δ, delta.)

As a river approaches its delta, it reverses its upstream-branching drainage pattern. Instead of collecting more water from tributaries, it discharges water into **distributaries**—smaller streams that receive water and sediments from the main channel, branch off *downstream*, and thus distribute the water and sediments into many channels. Materials deposited on top of the delta, typically sand, make up horizontal *topset beds*. Downstream, on the outer front of the delta, fine-grained sand and silt are deposited to form gently inclined *foreset beds*, which resemble large-scale cross-beds. Spread out on the seafloor seaward of the foreset beds are thin, horizontal *bottomset beds* of mud, which are eventually buried as the delta continues to grow. **Figure 12.27** shows how these structures form in a typical large marine delta.

> *Where a river enters a lake or ocean, a delta is formed as the current slows and drops its sediments.*

The Growth of Deltas As a delta builds outward into the ocean, the mouth of its river advances seaward, leaving new land in its wake. Much of this land is just a few meters above sea level. The delta of the Mississippi River, like many other major river deltas, has been growing for millions of years. About 150 million years ago, it started out around what is now the junction of the Ohio and the Mississippi rivers, at the southern tip of Illinois. It has advanced about 1600 km since then, creating almost the entire states of Louisiana and Mississippi as well as major parts of adjacent states. **Figure 12.28** shows the growth of the Mississippi delta over the past 6000 years as well as the direction its growth is likely to take in the future.

> *Deltas can build new areas of land very quickly.*

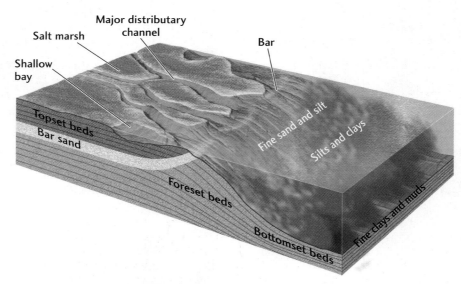

FIGURE 12.27 A typical large marine delta, many kilometers in extent, in which the fine-grained foreset beds are deposited at a very low angle, typically only 4° to 5° or less. Sandbars form at the mouths of the distributaries, where the current velocity suddenly decreases. The delta builds forward by the advance of the sandbars and the topset, foreset, and bottomset beds. Between distributary channels, shallow bays fill with fine-grained sediment and become salt marshes. This general structure is found on the Mississippi delta.

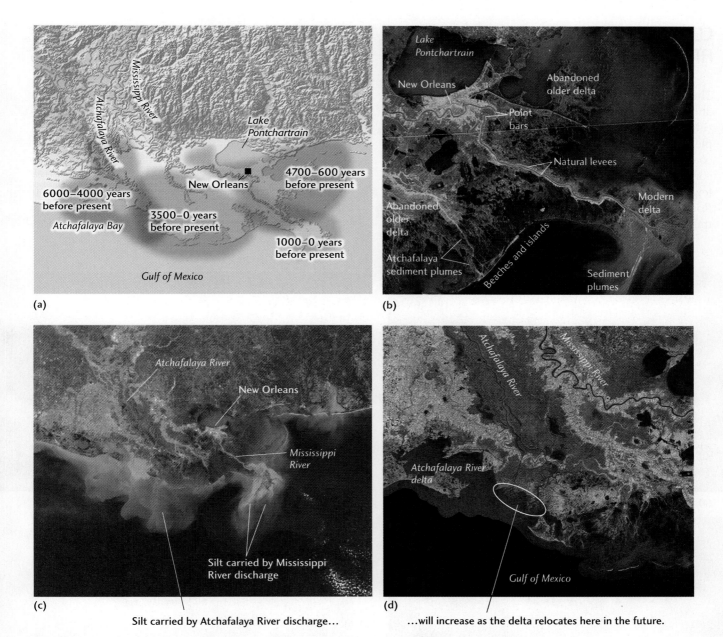

(a)

(b)

(c)

(d)

Silt carried by Atchafalaya River discharge...

...will increase as the delta relocates here in the future.

FIGURE 12.28 Over the past 6000 years, the Mississippi River has built its delta first in one direction and then in another as water flow has shifted from one major distributary to another. (a) The modern delta was preceded by deltas deposited to the east and west. (b) The infrared-sensitive film used to record this satellite image of the Mississippi delta causes vegetation to appear red, relatively clear water to appear dark blue, and water with suspended sediment to appear light blue. At the upper left are New Orleans and Lake Pontchartrain. Well-defined natural levees and point bars can be seen at the center. At the lower left are beaches and islands that formed as waves and currents transported river-deposited sand from the delta. (c) Satellite photograph of the Mississippi delta. (d) This image shows the discharge of sediment into the Gulf of Mexico from the Mississippi River delta and the Atchafalaya River. A major flood could divert the main flow of the Mississippi into the Atchafalaya, causing a new delta to form. Construction of artificial levees by the Army Corps of Engineers has prevented this occurrence so far. [(b) From G. T. Moore, "Mississippi River Delta from Landsat2," Bulletin of the American Association of Petroleum Geologists (1979). (c) NASA. (d) USGS National Wetlands Research Center.]

COASTAL PROCESSES: WAVES, TIDES, AND HURRICANES

The landscapes of coastlines, the broad regions where land and streams meet the oceans, present striking contrasts (**Figure 12.29**). On the coast of North Carolina, for example, long, straight, sandy beaches stretch for miles along low coastal plains (Figure 12.29a). In most of New England, by contrast, rocky cliffs bound elevated shores, and many of the few beaches that exist are made of gravel (Figure 12.29b). Rocky coastlines also occur along the south coast of Australia, where wave erosion generates spectacular landforms (Figure 12.29c). Many of the seaward edges of islands in the tropics, such as those in the Caribbean Sea, are coral reefs, the delight of divers (Figure 12.29d).

The major geologic forces operating at the **shoreline**—the line where the water surface intersects the land—are ocean currents created by waves and tides. Together, they erode even the most resistant rocky shores. Waves and tides transport sediments produced by erosion of the land and deposit them on beaches and in shallow waters along the shore.

Coastal processes that affect shorelines result from interactions within the climate system and the solar system. Tides are caused by gravitational interactions

(a)

(b)

(c)

(d)

FIGURE 12.29 Coastlines exhibit a variety of geologic forms. (a) Long, straight, sandy beach coastline, Pea Island, North Carolina. (b) Rocky coastline, Mount Desert Island, Maine. This formerly glaciated coastline has been uplifted since the end of the last ice age, about 10,000 years ago. (c) Twelve Apostles, Port Campbell, Australia, a group of stacks that developed from cliffs of sedimentary rock. These remnants of shore erosion are left as the shoreline retreats under the action of waves. (d) Coral reef along the Florida coastline. [(a) Courtesy of Bill Birkemeier/U.S. Army Corps of Engineers. (b) Neil Rabinowitz/Corbis. (c) Kevin Schafer. (d) Hays Cummins, Miami University.]

between Earth and the Sun and Moon, and coastal surf and storms—such as hurricanes—result from interactions between the atmosphere and the hydrosphere.

How can we monitor these processes and gauge their effects? Ordinary waves and tides are responsible for the slow, steady change of the shape of coastlines and the movement of sediments up and down the coast. These patterns may be modified or accelerated when storms arrive at the coast. What is present today may be gone tomorrow, and vice versa.

As we learned earlier in this chapter, currents are the key to understanding geologic processes at Earth's surface, and coastal processes are no exception. Let's look at the various types of currents that shape our shorelines.

Currents Created by Waves

Centuries of observation have taught us that waves are constantly changing. In quiet weather, waves with calm troughs between them roll regularly into the shore. In the high winds of a storm, however, waves move in a confusion of shapes and sizes. To understand the dynamics of shorelines and to make sensible decisions about coastal development, we need to understand how waves work.

The wind blowing over the surface of the ocean creates waves by transferring its energy of motion from air to water. As a gentle breeze of 5 to 20 km/hour starts to blow over a calm sea surface, ripples—little waves less than a centimeter high—take shape. As the speed of the wind

increases to about 30 km/hour, the ripples grow to full-sized waves. Stronger winds create larger waves and blow off their tops to make whitecaps. The height of waves depends on three factors:

- Wind speed
- Length of time over which the wind blows
- Distance the wind travels over water

Storms blow up large, irregular waves that radiate outward from the storm center, like the ripples moving outward from a pebble dropped into a pond. As the waves travel out from the storm center in ever-widening circles, they become more regular, changing into low, broad, rounded waves called *swell*, which can travel hundreds of kilometers. Several storms at different distances from a shoreline, each producing its own pattern of swell, account for the often irregular intervals between waves approaching the shore.

The height of waves depends on wind speed, how long the wind blows, and the distance the wind travels over water.

The Surf Zone Swell becomes higher as it approaches the shore, where it assumes the familiar sharp-crested wave shape (**Figure 12.30**). These waves are called *breakers* because, as they come closer to shore, they break and form surf—a foamy, bubbly surface. The *surf zone* is the belt along which breaking waves collapse as they

FIGURE 12.30 Wave motion is influenced by water depth and shape of the shoreline. [*Galen Rowell/Corbis.*]

approach the shore. Breaking waves pound the shore, eroding and carrying away sand, weathering and breaking up solid rock, and destroying structures built close to the shoreline.

The transformation from swell to breakers starts where water depth decreases to less than one-half the distance between waves. At that point, wave motion just above the bottom becomes restricted. The restricted motion of the water molecules slows the wave, which causes it to steepen (**Figure 12.31**a). As the wave rolls toward the shore, it becomes so steep that the water can no longer support itself, and the wave breaks with a crash in the surf zone. Gently sloping bottoms cause waves to break farther from shore, while steeply sloping bottoms make waves break closer to shore. Where rocky shores are bordered by deep water, the waves break directly on the rocks with a force amounting to tons per square meter, throwing water high into the air (see Figure 12.2b). It is not surprising that concrete seawalls built to protect structures along the shore quickly start to crack and must be repaired constantly.

After breaking at the surf zone, the waves, now reduced in height, continue to move in, breaking again right at the shoreline. They run up onto the sloping front of the beach, forming an uprush of water called *swash*. The water then runs back down again as *backwash*. Swash can carry sand and even large pebbles and cobbles if the waves are high enough. The backwash carries the particles seaward again.

> *Surf is created when waves arrive at the shoreline, where water depth decreases, and break. Water from breaking waves flows onto the beach as swash, then flows back toward the sea as backwash.*

This back-and-forth motion of the water near the shore is strong enough to carry sand grains and even gravel. Wave action in water as deep as about 20 m can move fine sand. Large waves caused by intense storms can scour the bottom at much greater depths, down to 50 m or more. At shallower water depths, these big storms transport sediments in an offshore direction, often depleting beaches of their fine sand.

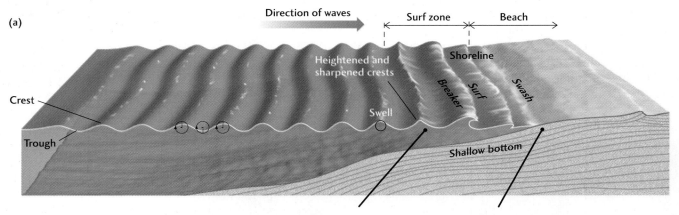

(a)

Direction of waves

Surf zone   Beach

Heightened and sharpened crests

Shoreline

Crest

Breaker   Surf   Swash

Swell

Trough

Shallow bottom

When the bottom shallows to about one-half the wavelength, the wave slows.

As waves approach the shore, they become too steep to support themselves and break in the surf zone, running up the beach in a swash.

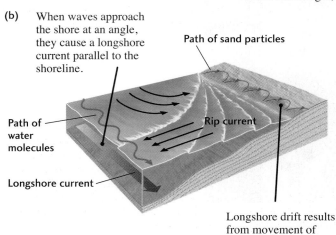

(b) When waves approach the shore at an angle, they cause a longshore current parallel to the shoreline.

Path of sand particles

Path of water molecules

Rip current

Longshore current

Longshore drift results from movement of sand particles by swash and backwash.

FIGURE 12.31 Breaking waves create longshore drift and longshore currents. (a) Swell becomes breaking waves as it approaches the shoreline. (b) When breaking waves approach the shoreline at an angle, their swash and backwash move sand and water in a zigzag pattern.

Wave Motion and Longshore Currents Far from shore, the lines of swell are parallel to one another, but as the waves approach a beach over a shallowing bottom, they gradually bend to a direction more parallel to the shore. When a wave encounters the shallowing bottom of the shoreline, the front of the wave slows. Then the next part of the wave meets the bottom and it, too, slows. Thus, in a continuous transition along the wave crest, the wave line bends toward the shore as it slows. As a result, most waves still approach the shore at a shallow angle.

> Because most waves approach the beach at a shallow angle, the swash and backwash of breaking waves transport sand down the beach in a zigzag motion known as longshore drift.

As the waves break on the shore, the swash moves up the beach slope at a slight angle, and the backwash runs down the slope in the opposite direction. The combination of these two motions moves the water a short way down the beach (Figure 12.31b). Sand grains carried by swash and backwash are thus moved along the beach in a zigzag motion known as *longshore drift*.

Waves approaching the shoreline at an angle can also cause a **longshore current,** a shallow-water current that flows parallel to the shore. The movement of swash and backwash in and out from the shore at an angle creates a zigzag path of water molecules that transports sediments along the shallow bottom in the same direction as the longshore drift. Much of the transport of sand along many beaches results from longshore currents. Longshore currents are prime determiners of the shape and extent of sandbars and other depositional shoreline features. At the same time, because of their ability to erode loose sand, longshore currents may remove large amounts of sand from a beach. Longshore drift and longshore currents working together are potent processes in the transport of sand on beaches and in very shallow waters. In slightly deeper waters (less

> Waves breaking at an angle to the shore can produce longshore currents in shallow water that flows parallel to the shore.

than 50 m), longshore currents—especially those running during large storms—strongly affect the bottom.

Currents Created by Tides

Tides are caused by the gravitational attraction between Earth and the Moon. The strength of this attraction varies across Earth's surface. On the side of Earth closest to the Moon, the ocean water experiences a greater gravitational attraction than the average for the whole of the solid Earth. This pull produces a bulge in the water. On the side of Earth farthest from the Moon, the solid Earth, being closer to the Moon than the water, is pulled toward the Moon more than the water is, and the water therefore appears to be pulled away from Earth as another bulge. Thus, two bulges of water occur on Earth's oceans: one on the side nearest the Moon, and the other on the side farthest from the Moon (**Figure 12.32**). As Earth rotates, these bulges stay approximately aligned: one always faces the Moon; the other is always directly opposite the Moon. These bulges of water passing over the rotating Earth are the high tides.

> The gravitational attraction between Earth and the Moon produces bulges of ocean water that create high tides as Earth rotates.

Although tides occur regularly everywhere, the difference between high and low tides varies in different parts of the ocean. As Earth rotates, the tidal bulges of water move along the surface of the ocean, encountering obstacles, such as continents and islands, that hinder the flow of water. In the middle of the Pacific Ocean—in Hawaii, for example, where there is little to obstruct the flow of the tides—the difference between low and high tides is only 0.5 m. Near Seattle, where the shape of the shoreline along Puget Sound is very irregular and the tidal flow must move through narrow passageways, the difference between the two tides is about 3 m. Extraordinary tides occur in a few places, such as the Bay of Fundy in eastern Canada, where the tidal range can be more than 12 m. Many people living along the coast need to know

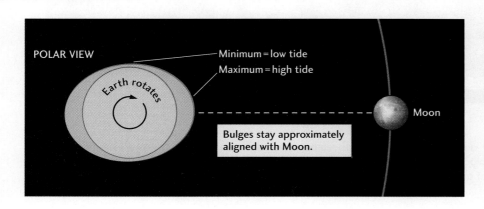

FIGURE 12.32 Tides are caused by the gravitational attraction of Earth and the Moon. The Moon's gravitational pull causes two bulges of water on Earth's oceans, one on the side of Earth nearest the Moon and the other on the side farthest from the Moon. As Earth rotates, these bulges remain aligned with the Moon and pass over Earth's surface, creating the high tides.

when tides will occur, so governments publish tide tables showing predicted tide heights and times.

Tides moving near shorelines cause currents that can reach speeds of a few kilometers per hour. As the tide rises, water flows in toward the shore as a *flood tide*, moving through narrow passages into inlets and bays, into shallow coastal marshes, and up small streams. As the tide passes the high stage and starts to fall, water flows out as an *ebb tide*, and low-lying coastal areas are exposed. Tidal currents meander across *tidal flats*, the muddy or sandy areas that lie above low tide but are flooded at high tide (**Figure 12.33**). Where obstacles restrict tidal flow and increase tidal range, such as in the Bay of Fundy, current velocities become very high. Large sand ridges many meters high may be formed in these tidal channels.

> *The difference between high and low tides depends on the presence of obstacles that hinder tidal flow.*

Natural Hazards: Hurricanes and Coastal Storm Surges

Few things in nature can compare with the destructive force of a hurricane. Regarded as the greatest storms on Earth, hurricanes can pound shorelines and coastal

FIGURE 12.33 Tidal flats, such as this one at Mont-Saint-Michel, France, may be extensive areas covering many square kilometers, but most often are narrow strips seaward of the beach. When a very high tide advances on a broad tidal flat, it may move so rapidly that areas are flooded faster than a person can run. The beachcomber is well advised to learn the local tides before wandering. [*Thierry Prat/Corbis Sygma.*]

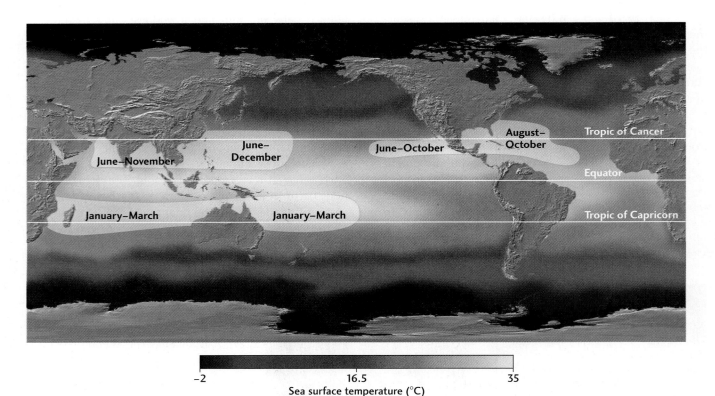

-2   16.5   35
Sea surface temperature (°C)

FIGURE 12.34 Hurricanes arise in summer and early fall when ocean temperatures are warmest. The light-shaded areas indicate places where hurricanes are most common. The times of year when they are most abundant are also shown. [*NASA/GSFC.*]

regions with sustained winds, lasting many hours, of 260 km/hour (160 miles/hour) or more, combined with intense rainfall and a storm surge caused by the storm's extremely low atmospheric pressure. Earth scientists have estimated that during its life cycle, a hurricane can expend as much energy as 10,000 nuclear bombs!

Hurricanes form over tropical parts of Earth's oceans, between 8° and 20° latitude, in areas of high humidity, light winds, and warm sea surface temperatures (typically 26°C or greater). These conditions usually occur in the summer and early fall in the tropical North Atlantic and North Pacific oceans. For this reason, hurricane "season" in the Northern Hemisphere runs from June through November (**Figure 12.34**).

Most hurricanes that affect the Atlantic Ocean and Gulf of Mexico originate just off the coast of West Africa and intensify as they move westward across the open ocean. Once sustained wind speeds reach 37 km/hour (23 miles/hour), the storm system is called a *tropical depression*. In the Northern Hemisphere, the increasing winds begin to circulate in a counterclockwise pattern around the storm's area of lowest pressure, which ultimately becomes the "eye" of the hurricane (**Figure 12.35**). When the wind speeds reach 119 km/hour (74 miles/hour), the storm is classified as a hurricane.

> Hurricanes are storms that form over tropical parts of the oceans and have sustained wind speeds of at least 119 km/hour.

As a hurricane intensifies, a dome of seawater— known as a **storm surge**—rises above the level of the surrounding ocean surface. Large swells, high surf, and wind-driven waves ride atop this dome as it floods coastal land areas, causing extensive damage to structures and the shoreline environment (**Figure 12.36**). Any landmass in the path of a storm surge is affected to a greater or lesser extent, depending on a number of factors. The stronger the storm and the shallower the offshore waters, the higher the storm surge.

The storm surge is the most deadly of a hurricane's associated hazards, as underscored by Hurricane Katrina

> A storm surge is a dome of water formed during a hurricane that can flood coastal regions. A storm surge can be the most damaging force in a hurricane.

FIGURE 12.35 Hurricane Katrina on August 28, 2005, a few hours before it struck New Orleans. In the Northern Hemisphere, winds circulate in a counterclockwise direction around the "eye" of the hurricane, which is the location of lowest surface pressure. Katrina was one of the most powerful hurricanes on record, with maximum sustained winds of up to 280 km/hour (175 miles/hour) and gusts up to 360 km/hour (225 miles/hour). [*NASA/Jeff Schmaltz, MODIS Land Rapid Response Team.*]

FIGURE 12.36 Hurricane Katrina caused storm surges along the coastline, resulting in the complete destruction of areas of the city. Here, water spills over a levee along the Inner Harbor Navigational Canal and floods the inner city of New Orleans. [*Vincent Laforet-Pool/Getty Images.*]

in 2005. The magnitude of a hurricane is usually described in terms of its wind speed, but coastal flooding causes many more deaths than high wind. Boats ripped from their moorings, utility poles, and other debris floating atop a storm surge often demolish buildings not destroyed by extreme winds. Even without the weight of floating debris, a storm surge can severely erode beaches and highways and undermine bridges. Because much of the densely populated Atlantic and Gulf Coast shorelines of the United States lie less than 3 m above mean sea level, the danger from storm surges is tremendous.

THE SHAPING OF SHORELINES

At shorelines, we can observe the constant motion of ocean waters and their effects on the shore. Current environmental problems such as coastal erosion and pollution of shallow waters have made the geology of shorelines and shallow seas a critical area of research. Waves, longshore currents, and tidal currents interact with the geologic structures and plate tectonic processes of the coast to shape shorelines into a multitude of forms (see Figure 12.28). We can see these factors at work in the most popular of shorelines: beaches.

Beaches

A **beach** is a shoreline made up of sand and pebbles. Beaches may change shape from day to day, week to week, season to season, and year to year. Waves and tides sometimes broaden and extend a beach by depositing sand and sometimes narrow it by carrying sand away.

> *A beach is a shoreline made up of sand and pebbles. Its shape is often changed by the action of waves and tides.*

Many beaches are straight stretches of sand that range from 1 km to more than 100 km long; others are smaller crescents of sand between rocky headlands.

Belts of dunes border the landward edge of many beaches; bluffs or cliffs of sediment or rock border others. A beach may have a *tide terrace*—a flat, shallow area between the upper beach and an outer bar of sand—on its seaward side (**Figure 12.37**).

The Structure of a Beach The major parts of a beach are shown in **Figure 12.38**. These parts may not all be present at all times on any particular beach. Farthest out is the *offshore*, bounded by the surf zone, where the bottom begins to become shallow enough for waves to break. The *foreshore* includes the surf zone; the tide terrace; and, right at the shore, the swash zone, a slope dominated by the swash and backwash of the waves. The *backshore* extends from the swash zone up to the highest level of the beach.

> *A beach can be subdivided into three parts: offshore, foreshore, and backshore.*

The Sand Budget of a Beach A beach is a scene of constant movement. Each wave moves sand back and forth with swash and backwash. Both longshore drift and longshore currents move sand down the beach. At the end of a beach and to some extent along it, sand is removed and deposited in deep water. In the backshore or along sea cliffs, sand and pebbles are freed by erosion and replenish the beach. The wind that blows over the beach transports sand, sometimes offshore into the water and sometimes onshore onto the land.

All these processes together maintain a balance between the addition and removal of sand, resulting in a beach that may appear to be stable but is actually exchanging its material with the environments on all sides. **Figure 12.39** illustrates the sand budget of a beach—the inputs and outputs caused by erosion, sedimentation, and transport. At any point along a beach, it gains sand from the inputs: material along the backshore is eroded, longshore drift and longshore currents bring sand to the beach, and rivers that enter the ocean along the shore bring in

FIGURE 12.37 Tide terrace exposed at low tide. This shallow depression between an outer ridge (a sandbar at high tide) and the upper beach is rippled by the tidal flow in many places. [*James Valentine.*]

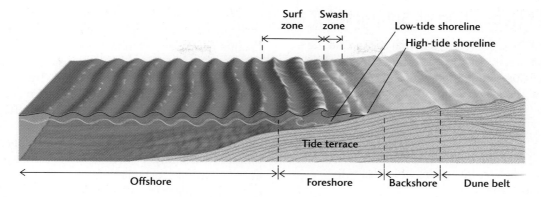

FIGURE 12.38 Profile of a beach, showing its major parts.

sediment. The beach loses sand from the outputs: winds carry sand to backshore dunes, longshore drift and currents carry it downcurrent, and deep-water currents and waves transport it during storms.

If the total input balances the total output, the beach will keep the same general form. If input and output are not balanced, the beach grows or shrinks. Temporary imbalances are natural over weeks, months, or years. A series of large storms, for example, might move large amounts of sand from the beach to deeper waters on the far side of the surf zone, narrowing the beach. Then weeks of mild weather and low waves might move the sand onto the shore and rebuild a wide beach. Without this constant shifting of sands, beaches might be unable to recover from the effects of trash, litter, and other kinds of pollution. Within a year or two, even oil from spills is transported or buried out of sight, although the tarry residue may later be uncovered in spots.

> *The form of a beach depends on a balance between inputs and outputs of sand. If the balance is upset, the beach may grow or shrink.*

SAND BUDGET	
INPUTS	**OUTPUTS**
Sediments eroded from backshore cliffs by waves	Sediments transported to backshore dunes by offshore winds
Sediments eroded from upcurrent beach by longshore drift and current	Sediments transported downcurrent by longshore drift and current
Sediments brought in by rivers	Sediments transported to deep water by tidal currents and waves

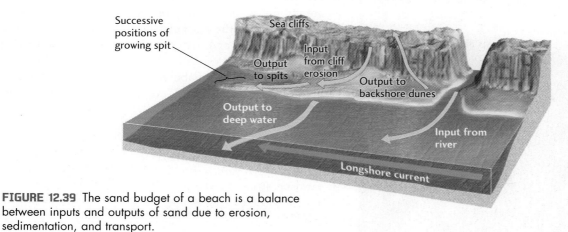

FIGURE 12.39 The sand budget of a beach is a balance between inputs and outputs of sand due to erosion, sedimentation, and transport.

Erosion and Deposition at Shorelines

The topography of the shoreline, like that of the land interior, is a product of tectonic forces elevating or depressing Earth's crust, erosion wearing it down, and sedimentation filling in the low spots. Thus, several factors are directly at work:

■ Uplift of the coastal region, which leads to erosion at the coast

■ Subsidence of the coastal region, which leads to sediment buildup at the coast

■ Nature of the rocks or sediments at the shoreline

■ Changes in sea level, which affect the drowning or emergence of a shoreline

■ Average and storm wave heights

■ Heights of the tides, which affect both erosion and sedimentation

Erosional Coastal Forms Erosion is active along tectonically uplifted rocky coasts. Along these coasts, prominent cliffs and headlands jut into the sea, alternating with narrow inlets and irregular bays with small beaches. Waves crash against the rocky shorelines, undercutting cliffs and causing huge blocks of rock to fall into the water, where they are gradually eroded away. As the sea cliffs retreat, isolated remnants called *stacks* may be left standing in the sea, far from the shore (see Figure 12.29c). Erosion by waves also planes the rocky surface beneath the surf zone and creates a **wave-cut terrace,** which is sometimes visible at low tide (**Figure 12.40**). Wave erosion over long periods may straighten shorelines as headlands retreat faster than recesses and bays.

Where relatively soft sediments or sedimentary rocks make up the coastal region, the slopes are gentler and the heights of shoreline bluffs are lower. Waves erode these softer materials efficiently, and erosion of bluffs on such shores may be extraordinarily rapid. The high sea cliffs of soft glacial materials along the Cape Cod National Seashore in Massachusetts, for instance, are retreating about a meter each year. Since Henry David Thoreau walked the entire length of the beach below those cliffs in the mid-nineteenth century and wrote of his travels in *Cape Cod,* about 6 km² of coastal land have been eaten away by the ocean, equivalent to about 150 m of beach retreat.

In recent decades, more than 70 percent of the total length of the world's sand beaches has retreated at a rate of at least 10 cm/year, and 20 percent of the total length has retreated at a rate of more than 1 m/year. Much of this loss can be traced to the damming of rivers, which decreases the sediment supply to the shoreline.

> *Wave action is strong enough to erode rocky coastlines. Softer sediments and sedimentary rocks can be more easily eroded.*

Depositional Coastal Forms Sediment builds up in areas where subsidence depresses Earth's crust along a coastline. Such coastlines are characterized by long, wide beaches and wide, low-lying coastal plains of sedimentary rock. Shoreline forms along these coastlines include sandbars, low-lying sandy islands, and extensive tidal flats. Long beaches grow longer as longshore currents carry sand to the downcurrent end of the beach. There it builds up, first forming a submerged sandbar, then rising above the surface and extending the beach by a narrow addition called a **spit.**

Offshore, long sandbars may build up into **barrier islands** that form a barricade between open ocean waves and the main shoreline. Barrier islands are common, especially along low-lying coasts composed of sediments that are easily eroded and transported or of poorly cemented sedimentary rocks where longshore currents are strong. Some of the most prominent barrier islands are found along the coast of New Jersey; at Cape Hatteras, North Carolina; and along the Texas coast of the Gulf of Mexico, where one—Padre Island—is 130 km long. As the sandbars build up above the waves, vegetation takes hold, stabilizing the islands and helping them resist wave

FIGURE 12.40 Multiple wave-cut terraces on the California coastline. Each terrace records a distinctly different sea level elevation. [*Photo by Dan Muhs/USGS. Daniel R. Muhs, Kathleen R. Simmons, George L. Kennedy, and Thomas K. Rockwell, "The Last Interglacial Period on the Pacific Coast of North America: Timing and Paleoclimate." Geological Society of America Bulletin (May 2002): 569–592.*]

erosion during storms. Barrier islands are separated from the coast by tidal flats or shallow lagoons. Like beaches on the main shore, barrier islands are affected by natural changes in climate or in wave and current patterns, as well as by real estate development. Disruption or devegetation can lead to increased erosion, and barrier islands may even disappear beneath the sea surface. Barrier islands may also grow larger and more stable if sedimentation increases.

> Spits, sandbars, and barrier islands may form along coastlines where sediments are easily eroded and transported.

Over hundreds of years, these shorelines may undergo significant changes. Hurricanes and other intense storms, such as the "storm of the century" that hit the Eastern Seaboard of the United States in March 1993, may form new inlets, elongate spits, or breach existing spits and barrier islands. Such changes have been documented by aerial photographs taken at various time intervals. The shoreline of Chatham, Massachusetts, at the elbow of Cape Cod, changed enough in the past 160 years or so that a lighthouse had to be moved. **Figure 12.41** illustrates the many changes that have taken place in the configuration of the barrier islands to the north and to the

(a) Beach near Chatham Light

The 1987 breach in the barrier spit, shown at the right below, closed again before this photo was taken.

(b)

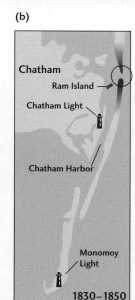
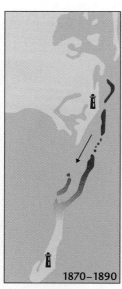

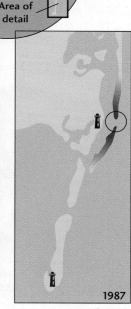

1830–1850	1870–1890	1910–1930	1950–1970	1987
The circle shows the approximate location of the 1846 breach in barrier island. Ram Island later disappears.	The beach south of the inlet breaks up and migrates southwest toward the mainland and Monomoy.	The southern beach has disappeared, and its remnants soon will connect Monomoy to the mainland.	The northern beach steadily grows with cliff sediment; Monomoy breaks from the mainland.	The 140-year cycle begins again with the Jan. 2 breach in the barrier spit across from the Chatham Light (circle).

FIGURE 12.41 Migrating barrier islands at Chatham, Massachusetts, at the southern tip of Cape Cod. (a) Aerial view of Monomoy Point. This spit has advanced into deep water to the south (*foreground*) from barrier islands along the main body of the Cape to the north (*background*). (b) Changes in the shoreline at Chatham over the past 160 years. [(a) *Steve Dunwell/The Image Bank.* (b) *After Cindy Daniels, Boston Globe (February 23, 1987).*]

long spit of Monomoy Point, as well as several breaches of the barrier islands. Many homes in Chatham are now at risk, but there is little that the residents or the state can do to prevent these beach processes from taking their natural course.

OFFSHORE: THE CONTINENTAL SHELF AND BEYOND

Offshore, beyond the shoreline, lies the continental shelf. At its edge is a continental slope that descends more or less steeply into the depths of the ocean. At the foot of the slope is a continental rise, a gently sloping apron of sediment extending to the abyssal plain at the bottom of the ocean basin (**Figure 12.42**).

Continental shelves consist of essentially flat-lying sediments several kilometers thick. These sediments accumulate mostly as a result of transport by waves and tidal currents. The continental shelf is one of the most economically valuable parts of the ocean. Georges Bank off New England and the Grand Banks of Newfoundland, for example, are among the world's most productive fishing grounds. Recently, oil-drilling platforms have been used to extract huge quantities of

> The continental shelf is formed by thick accumulations of sediments transported offshore by waves and tides.

oil and gas from the continental shelf, especially off the Gulf Coast of Louisiana and Texas.

Seaward of the continental shelf, the seafloor drops off abruptly. Here, the waters are too deep for the seafloor to be affected by waves and tidal currents. As a consequence, sediments that have been carried across the shallow continental shelf by waves and tides come to rest on the continental slope. Deposits of sand, silt, and mud on both the continental slope and the continental rise indicate active transport of sediments into deeper waters. These sediments are transported by **turbidity currents**—turbulent flows of muddy water down the slope. Because of its suspended load of mud, the turbid water is denser than the overlying clear water and flows beneath it.

Turbidity currents start when the sediment draped over the edge of the continental shelf slumps onto the continental slope (see Figure 12.42). Such a sudden submarine landslide, which can occur spontaneously or be triggered by an earthquake, throws mud into suspension, creating a dense, turbid layer of water near the bottom. This turbid layer starts to flow, accelerating down the slope. As the turbidity current reaches the foot of the slope, it slows. Some of the coarser sandy sediment starts to settle, often forming a *submarine fan*—a deposit something like an alluvial fan on land. Stronger turbidity currents may continue across the continental rise, cutting channels into submarine fans. Where these currents reach

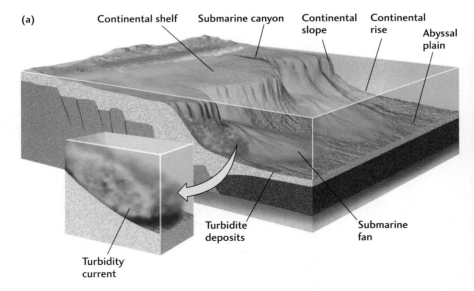

(a)

Continental shelf · Submarine canyon · Continental slope · Continental rise · Abyssal plain

Turbidite deposits · Submarine fan · Turbidity current

(b) Slump

FIGURE 12.42 Turbidity currents transport sediments from the continental shelf into deeper waters. (a) Slumps on the continental slope generate turbidity currents, which flow down the continental slope and continental rise to the abyssal plain. These currents can erode and transport large quantities of sand down the continental slope. (b) A slump at the edge of the continental shelf. [*U.S. Navy.*]

the abyssal plain, they spread out and come to rest in beds of sand, silt, and mud called *turbidites*.

Some of the world's largest oil and gas fields are located in very deep water where turbidites have accumulated, forming an ideal environment for trapping these fossil fuels. Their exploitation is a tricky and expensive business, as deep-sea wells may cost hundreds of millions of dollars to drill.

> *Turbidity currents transport sediments from the continental shelf down the continental slope to the abyssal plain.*

So far we have considered environments and processes dominated by flows of water. Let's turn our attention to deserts, where water is scarce and wind has a major role in sculpting the landscape.

THE DESERT ENVIRONMENT

The deserts of the world are among the most hostile environments for humans. Yet many of us are fascinated by these hot, dry, apparently lifeless zones, full of bare rocks and sand dunes. Desert environments amount to one-fifth of Earth's land area, about 27.5 million square kilometers.

Of all Earth's environments, the desert is where wind is best able to do its work of erosion, transportation, and sedimentation. Today's deserts may have been wet regions in the past but may have dried out as a result of long-term climate change. The dry climate of deserts creates harsh yet fragile conditions, where human impacts last for decades.

Where Deserts Are Found

Rainfall is the major factor determining the locations of the world's great deserts (**Figure 12.43**). The Sahara and Kalahari deserts of Africa and the Great Australian Desert get extremely low amounts of rainfall, normally less than 25 mm/year and in some places less than 5 mm/year. These deserts are found between 30° N and 30° S latitudes, where prevailing wind patterns cause dry air to sink to ground level and atmospheric pressures are high (see Figure 12.4). Because the relative humidity is extremely low, clouds are rare, and the chance for precipitation is very small. The Sun beats down week after week.

Deserts also exist at midlatitudes—between 30° N and 50° N and between 30° S and 50° S—in regions where rainfall is low because moisture-laden winds either are

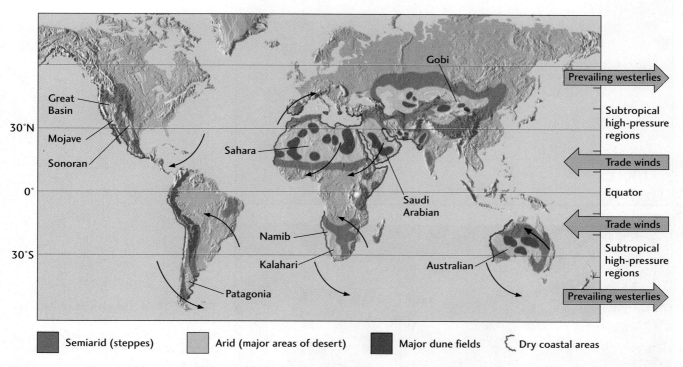

FIGURE 12.43 Major desert areas of the world (exclusive of polar deserts). Notice the relationships of their locations to prevailing wind belts and major mountain ranges. Notice, too, that sand dunes make up only a small proportion of the total desert area. [*After K. W. Glennie, Desert Sedimentary Environments. New York: Elsevier, 1970.*]

blocked by mountain ranges or must travel great distances from the ocean, their source of moisture. The Great Basin and Mojave deserts of the western United States, for example, lie in rain shadows created by the western coastal mountains (see Figure 11.9). The Gobi and other deserts of central Asia are so deep in the interior of the continent that the winds reaching them have precipitated all their ocean-derived moisture long before they arrive there.

> *Deserts develop at latitudes where global wind patterns limit rainfall. They also form where mountains create rain shadows, in the interior of continents, and in polar regions.*

Another kind of desert is found in polar regions. There is little precipitation in these cold, dry areas because the frigid air can hold only small amounts of moisture. The dry valley region of southern Victoria Land in Antarctica is so dry and cold that its environment resembles that of Mars.

The Role of Climate Change Changes in a region's climate may transform semiarid lands into deserts, a process called **desertification.** Climate changes that we do not fully understand may decrease precipitation for decades or even centuries. After such a dry period, a region may return to a milder, wetter climate. Over the past 10,000 years, climates of the Sahara appear to have oscillated between drier and wetter conditions. We have evidence from orbiting satellites above the Sahara that an extensive system of river channels existed a few thousand years ago (**Figure 12.44**). Now dry and buried by more recent sand deposits, these ancient drainage systems carried abundant running water across the northern Sahara during wetter periods.

The Sahara Desert may now be expanding northward. The Desert Watch project, led by the European Space Agency, reports that over 300,000 km^2 of Europe's Mediterranean coast (an area almost as large as the state of New York), with a population of 16 million, has been enduring the longest drought in recorded history. During 2005, fires raged along the southern Spanish coast, and temperatures set new record highs for weeks on end. Was this merely a long, hot summer, or are these the initial symptoms of desertification, made worse by overpopulation and overdevelopment within the fragile ecosystems of dry landscapes? Evidence to support the latter scenario is building.

> *Long-term climate change can create deserts where milder, wetter climates once were present. This process is called desertification.*

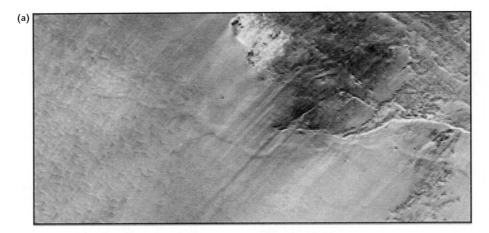

(a)

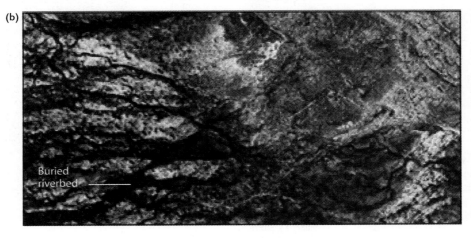

(b)

Buried riverbed

FIGURE 12.44 The climate of the Sahara was not always as arid as it is today. (a) Remote sensing techniques that look only at Earth's surface see nothing but sand in the Sahara. (b) Remote sensing techniques that penetrate a few meters below the surface see a dense network of buried streambeds. [NASA/JPL Imaging Radar Team.]

Natural Environment: The Role of Humans in Desertification

Climate change in the Sahara and other deserts occurs naturally, but human activities are responsible for some desertification. The growth of human populations and increased agriculture and animal grazing may result in the expansion of deserts. When population growth and periods of drought coincide, the results in semiarid regions can be disastrous. In Spain, the greatest urban and agricultural expansion is taking place in the driest region. A report published in 2007 argues that 36% of Spain is now suffering from desertification and that Spain is dealing with the worst drought of the past 70 years. In May of 2011 Spain became the second country (after France) to appeal to the European Union for drought relief aid citing "exceptional climate conditions" and "the serious economic crisis affecting the agricultural sector" of the past year, as well as difficulties for farmers in acquiring bank loans to cover production losses. Former farmlands have been stripped of vegetation due to overfarming (up to four crops per year), which depletes water and strips soils. A tourism boom and its accompanying development are literally paving over the dry lands and desiccating the countryside that is left. In 2004, more than 350,000 new homes were built on Spain's Mediterranean coast, many with backyard swimming pools and nearby golf courses requiring large amounts of water. In isolation, any one of these human activities might not have a negative effect. Together, however, they add up to desertification.

Human activities, such as depletion of water sources, overfarming, and urbanization, can result in desertification.

Desert Weathering

Weathering and transportation work in the same way in deserts as they do everywhere, but with a different balance. In deserts, physical weathering predominates over chemical weathering. Chemical weathering of feldspars and other silicates to clay minerals proceeds slowly because the water required for the reactions is lacking. The little clay that does form is usually blown away by strong winds before it can accumulate. Slow chemical weathering and rapid wind transport combine to prevent the buildup of any significant thickness of soil, even where sparse vegetation binds some of the particles. Thus, desert soils are thin and patchy. Sand, gravel, rock fragments of many sizes, and bare bedrock are characteristic of much of the desert surface.

Physical weathering predominates over chemical weathering in deserts because of the lack of water.

The Colors of the Desert The rusty, orange-brown colors of many weathered surfaces in the desert come from the ferric iron oxide minerals hematite and limonite. These minerals are produced by the slow chemical weathering of iron silicate minerals such as pyroxene. The iron oxides, even when present only in small amounts, stain the surfaces of sands, gravels, and clays.

Desert varnish is a distinctive dark brown, sometimes shiny coating found on many rock surfaces in the desert. It is a mixture of clay minerals with smaller amounts of manganese and iron oxides. Desert varnish probably forms when silicate minerals weather in the presence of dew to form iron and manganese oxides that adhere to exposed rock surfaces. The process is so slow that Native American inscriptions scratched in desert varnish hundreds of years ago still appear fresh, with a stark contrast between the dark varnish and the light unweathered rock beneath (**Figure 12.45**). Desert varnish requires thousands of years to form, and some particularly ancient varnishes in North America are of Miocene age. However, recognizing varnish as such on ancient sandstones is difficult.

Desert varnish is a shiny, dark coating on rock surfaces that forms primarily by accumulation of iron and manganese oxides.

Desert Streams Wind plays a larger role in erosion in the desert than it does elsewhere, but it cannot compete with the erosive power of streams. Even though it rains so seldom that most streams flow only intermittently, streams do most of the erosional work in the desert when they do flow.

Even the driest desert gets occasional rain. In sandy and gravelly areas of deserts, rainfall infiltrates the soil and permeable bedrock. There, some of it evaporates.

FIGURE 12.45 Petroglyphs scratched in desert varnish by Native Americans at Newspaper Rock, Canyonlands, Utah. The scratches are several hundred years old but appear fresh on the varnish, which has accumulated over thousands of years. [*Peter Kresan.*]

A smaller amount eventually reaches the groundwater table far below—in some places, as much as hundreds of meters below the surface. Desert oases form where the groundwater table comes close enough to the surface that roots of palms and other plants can reach it.

When rain occurs in heavy cloudbursts, so much water falls in such a short time that infiltration cannot keep pace, and the bulk of the water runs off into streams. Unhindered by vegetation, the runoff is rapid and may cause flash floods along valley floors that have been dry for years. Thus, a large proportion of streamflows in the desert consists of floods. When floods occur, they have great erosive power because most of the loose sediment is not held in place by vegetation. Streams may become so choked with sediment that they look more like fast-moving mudflows than rivers. The abrasiveness of this sediment load moving rapidly at flood velocities makes such streams efficient eroders of bedrock valleys.

> *Most desert streams flow only intermittently, but a large proportion of streamflows in deserts are floods.*

Desert Sediments and Sedimentation

Deserts are composed of a diverse set of depositional environments. The environments may change dramatically when rain suddenly forms raging rivers and widespread lakes. Prolonged dry periods intervene, during which sediments are blown into sand dunes.

Alluvial Deposits As sediment-laden flash floods dry up, they leave distinctive deposits on the floors of desert valleys. In many cases, a flat fill of coarse sediment covers the entire valley floor, and the ordinary differentiation of the stream into channel, natural levees, and floodplain is absent (**Figure 12.46**). The sediments of many other desert valleys clearly show the intermixing of stream-deposited channel and floodplain sediments with eolian sediments. This pattern creates deposits containing extensive layers of eolian sandstone separated by stream flood sediments and ancient river sandstones.

Large alluvial fans are prominent features at mountain fronts in deserts because desert streams deposit much of their sediment load on the fans (see Figure 12.20). The rapid infiltration of stream water into the permeable sediments that make up the fan deprives the stream of the water required to carry the sediment load any farther downstream.

> *Alluvial deposits in deserts are generally spread across the valley floor.*

Eolian Sediments By far the most dramatic sedimentary accumulations in deserts are sand dunes. Dune fields range in size from a few square kilometers to the "seas of sand" found in Namibia (see Figure 12.7). These sand seas—or *ergs*—may cover as much as 500,000 km^2, twice the area of the state of Nevada. Although film and television portrayals might lead one to think that deserts are mostly sand, actually only one-fifth of the world's desert area is covered by sand (see Figure 12.43). The other four-fifths are rocky or covered with desert pavement—a coarse, gravelly surface that forms as finer-grained particles are eroded away. Sand covers only a little more than one-tenth of the Sahara Desert, and sand dunes are far less common in the deserts of the southwestern United States.

> *Eolian sediments, typified by vast dune fields, are common in deserts.*

(a)

(b)

FIGURE 12.46 (a) A desert valley during a summer thunderstorm at Saguaro National Park, Arizona. (b) The same valley a day after the flooding. The coarse sediment deposited by such sudden desert floods may cover the entire valley floor. [*Peter Kresan.*]

FIGURE 12.47 A desert playa lake in Death Valley, California. [*David Muench.*]

Evaporite Sediments Playa lakes are permanent or temporary lakes that occur in arid mountain valleys or basins where water pools after rainstorms (**Figure 12.47**). As the lake water evaporates, dissolved minerals are concentrated and gradually precipitated. Playa lakes are sources of evaporite minerals such as sodium carbonate, borax (sodium borate), and other unusual salts. The water in playa lakes may be deadly to drink because of the high concentrations of dissolved minerals. Desert streams carry large amounts of dissolved salts, and these salts accumulate in playa lakes when the streams redissolve evaporite minerals deposited by evaporation from earlier runoff. If evaporation is complete, the lakes become *playas*, flat beds of clay that are sometimes encrusted with precipitated salts.

> *Playa lakes form where water pools after rainstorms and dissolved minerals are concentrated and precipitated as the water evaporates.*

■ SUMMARY

How do water and air behave as fluids, and how do they transport and deposit sediments? A fluid can move in either laminar or turbulent flow, depending on its velocity, viscosity, and depth. The turbulence that characterizes most water and wind currents is responsible for transporting sediments by suspension (clays), saltation (sands), and rolling and sliding along the bed (sand and gravel). The tendency for particles to be carried in suspension is countered by the gravitational force that pulls them to settle to the bottom.

Where do winds form and how do they flow? Earth is encircled by belts of prevailing winds that develop because the Sun warms Earth most intensely at the equator, which causes the air to rise there and flow toward the poles. As the air moves toward the poles, it gradually cools and begins to sink. This cool, dense air then flows back along Earth's surface to the equator.

How do winds deposit sand dunes and dust? Dunes migrate downwind as sand grains saltate up their gentler windward slopes and fall over onto their steeper downwind slip faces. As the velocity of dust-laden winds decreases, the dust settles in thick layers.

How do stream valleys and their channels and floodplains evolve? As a stream flows, it carves a valley with steep to gently sloping walls and a floodplain on either side of its channel. The channel may be straight, meandering, or braided. During normal, nonflood periods, the channel carries the flow of water and sediments within its banks. When stream discharge increases to flood stage, the sediment-laden water overflows the banks of the channel and inundates the floodplain. The velocity of the floodwater decreases as it spreads over the floodplain. The water drops sediments, which build up natural levees and floodplain deposits.

How do drainage networks work as collection systems and deltas as distribution systems for water and sediment? Rivers and their tributaries constitute an upstream-branching drainage network that collects water and sediments from a drainage basin. Each drainage basin is separated from other drainage basins by a divide. Where it meets the ocean, the river drops its sediments to form a delta, branching downstream into distributaries that spread sediments across the delta.

What processes shape shorelines? Waves and tides, interacting with plate tectonic processes, shape the landscapes of coastlines, which vary from beaches and tidal flats to uplifted rocky coasts. Winds blowing over the ocean generate swell; as those waves approach the shore, they are transformed into breakers in the surf zone. Waves approaching the shoreline at an angle result in longshore drift and longshore currents, which transport sand along beaches. Tides, which are generated by the gravitational pull of the Moon on ocean water, can generate currents that transport sediments to tidal flats and form bars in tidal channels.

How do hurricanes affect coastal areas? Hurricanes are intense storms with extremely high winds and very low atmospheric pressures. The low pressure results in the formation of a dome of seawater, known as a storm surge. As

GOOGLE EARTH PROJECT

Water, one of the most prolific weathering and transport agents on Earth, is constantly moving material from one location to another. Google Earth is an ideal tool for interpreting and appreciating this uniquely surficial process. Large rivers such as the Mississippi illustrate how efficiently river systems can gather sediment from mountainous regions of a continent (a source area) and transport it to the ocean, where deltas form (a sink area). What kinds of drainage and channel patterns do you find in the Mississippi drainage basin? How does the slope of the river channel change as one moves downstream? These questions and many more can be explored though the GE interface.

LOCATION   Missouri-Mississippi drainage basin, United States

GOAL   Understand source-to-sink transportation of sediment by river systems; observe meandering rivers with point bars, eroded outside banks, and oxbow lakes

LINKED   Figure 12.11

This image shows the continental scale of the Mississippi River from near its point of origin (Ft. Benton) to where it enters the Gulf of Mexico near New Orleans

1. Type "Ft. Benton, Montana, United States" into the GE search window. Once you arrive there, zoom out to an eye altitude of 35 km. You will be looking down on the Missouri River, the longest tributary of the Mississippi River. Examine the stretch of river that flows from the southwest to the northeast through town and describe the channel pattern you see.

 a. Distributary

 b. Braided

 c. Meandering

 d. Artificially straightened

2. Using the cursor, determine the change in the elevation of the Missouri River channel over the 525 km between Ft. Benton, Montana, and Williston, North Dakota. Now compare that value with the change in the elevation of the Mississippi River channel over the same distance between Memphis, Tennessee, and Baton Rouge, Louisiana, to the south. Which relationship is most accurate?

 a. The slope of the Missouri River is steeper than that of the Mississippi River.

 b. The slope of the Mississippi River is steeper than that of the Missouri River.

 c. The slopes of the two rivers are nearly equal.

 d. The slopes of the rivers cannot be compared from the information given.

3. From an eye altitude of about 500 km, follow the Mississippi River south from its inception in Lake Itasca, Minnesota. The Mississippi River was used to denote the boundaries along portions of the states of Wisconsin, Iowa, Illinois, Missouri, Kentucky, Arkansas, Tennessee, and Mississippi. Notice that with the "Borders and Labels" layer activated, you can compare the locations of state lines (determined by the original surveyed location of the river channel) with the location of the modern river. How has the river changed over time? (*Hint:* Refer to Figure 18.3. One can also view changing channel patterns with the GE time function.)

   a. The river channel has straightened its course in all locations.

   b. The river has shortened its path by widening its channel.

   c. The river channel has become more sinuous over its entire length.

   d. The river has cut off meanders in some places and lengthened them in others.

4. Based on your inspection of the river's characteristics at each of the following cities, which city seems most vulnerable to seasonal flooding by the Mississippi River? (*Hint:* Look for evidence of levees and at the proximity of the channel to each city.)

   a. Cairo, Illinois

   b. Biloxi, Mississippi

   c. St. Louis, Missouri

   d. Memphis, Tennessee

Optional Challenge Question

5. Using the GE search window, navigate to New Orleans, Louisiana. Zoom out to an eye altitude of 310 km to appreciate the close relationship between the city and the Mississippi delta. Now zoom in on the delta itself to observe the deposition of sediment from the Mississippi River channel in the Gulf of Mexico. Why is sediment deposited in this location in particular, and why in such large quantities?

   a. Hurricanes originating in the Gulf of Mexico drive sediment from Florida over to Louisiana.

   b. The Mississippi River meets the ocean here, so the river current slows down, and this decrease in current velocity causes the sediments in the current to drop out and be deposited.

   c. The Army Corps of Engineers dumped all the sediment here when it dredged the Mississippi River.

   d. Sediments were shed from the Appalachian Mountains when they were uplifted in the Cretaceous period.

the storm surge moves onshore, it floods low-lying areas, often causing more extensive damage than the storm's high winds.

What forces shape offshore environments? Waves and tides affect the continental shelf, but the continental slope is shaped primarily by turbidity currents. These deep-water currents carry large loads of sediment from the continental shelf down the continental slope to the continental rise and abyssal plain, where they form deposits known as turbidites.

How do wind and water combine to shape the desert environment and its landscape? Deserts occur in subtropical regions of constant high atmospheric pressure, in the rain shadows of mountain ranges, and in the interiors of continents. Physical weathering predominates over chemical weathering because of the lack of water. Most desert soils are thin, and bare rock surfaces are common. Wind plays a larger role in shaping the landscape in deserts than it does elsewhere, but streams are responsible for most erosion and sedimentation in deserts even though they flow only intermittently.

KEY TERMS AND CONCEPTS

alluvial fan (p. 398)
barrier island (p. 412)
beach (p. 410)
bed load (p. 388)
braided stream (p. 397)
channel (p. 385)
coastline (p. 385)
delta (p. 402)
desertification (p. 416)
discharge (p. 399)
distributary (p. 402)
divide (p. 400)
drainage basin (p. 400)
dune (p. 390)
eolian (p. 386)
floodplain (p. 394)
laminar flow (p. 386)
longshore current (p. 407)

meander (p. 396)
natural levee (p. 398)
playa lake (p. 419)
point bar (p. 396)
ripple (p. 390)
river (p. 385)
saltation (p. 388)
shoreline (p. 404)
spit (p. 412)
storm surge (p. 409)
stream (p. 385)
stream valley (p. 394)
suspended load (p. 388)
tides (p. 407)
turbidity current (p. 414)
turbulent flow (p. 386)
wave-cut terrace (p. 412)

■ EXERCISES

1. What is the difference between the way wind transports dust and the way it transports sand?

2. What is a natural levee, and how is it formed?

3. What is a delta distributary?

4. How are ocean waves formed?

5. How does the Moon interact with Earth to create tidal currents?

6. What is a storm surge?

7. Where do turbidity currents form?

8. What is desertification?

9. What are the geologic processes that form playa lakes?

10. Why might the flow of a very small, shallow stream be laminar in winter and turbulent in summer?

11. In some places, engineers have artificially straightened a meandering stream. If such a straightened stream were then left free to adjust its course naturally, what changes would you expect?

12. In a 100-year period, the southern tip of a long, narrow, north-south beach has become extended about 200 m to the south by natural processes. What shoreline processes could have caused this extension?

13. There are large areas of sand dunes on Mars. From this fact alone, what can you infer about conditions on the Martian surface?

14. How does desert weathering differ from or resemble weathering in more humid climates?

Visual Literacy Task

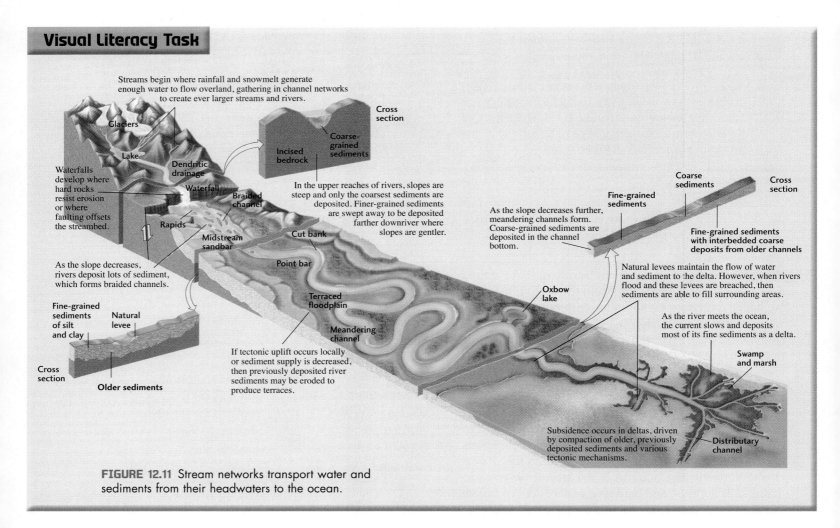

FIGURE 12.11 Stream networks transport water and sediments from their headwaters to the ocean.

1. **How does the slope change along the length of a stream?**

a. Gentle at the beginning, steep at the end
b. About the same from beginning to end
c. Steep at the beginning, gentle at the end

2. **How does sediment size and the amount of sediment change over the length of the stream?**

a. Sediment size gets smaller and sediment amount decreases.
b. Sediment size gets smaller and sediment amount increases.
c. Sediment size gets larger and sediment amount decreases.
d. Sediment size gets larger and sediment amount increases.

3. **As shown in the cross sections of Figure 12.11, what makes up the stream valley and nearby surrounding area?**

a. Sediments in the steep parts and bedrock in the more gentle parts
b. Bedrock in the steep parts and sediments in the more gentle parts
c. Bedrock the length of the stream

4. **How does the slope of the stream play a role in whether or not the stream erodes or incises the bedrock?**

a. Only streams in steep areas erode or incise the bedrock.
b. Only streams in flatter areas erode or incise the bedrock.
c. The slope does not affect whether or not the stream can erode or incise the bedrock.

5. **How does the form of the channel change with the slope of the stream?**

a. Steepest slope = erode bedrock; intermediate slope = braided channel; gentlest slope = meandering channel.
b. Steepest slope = braided channel; intermediate slope = meandering channel; gentlest slope = erode bedrock.
c. Steepest slope = meandering channel; intermediate slope = erode bedrock; gentlest slope = braided channel.
d. The form of the channel can vary independent of the steepness of the slope.

Thought Question: How would you respond to a fellow student who claimed, "Streams have the same characteristics from where they start to where they end."

Earthquakes and Earth Structure

13

The Tohoku Megaquake and Tsunami

ON A HILL ALONG THE TOHOKU COASTLINE of northeastern Honshu, in the fishing hamlet of Aneyoshi, sits a stone monument of uncertain age, inscribed with Japanese characters that read, "High dwellings are the peace and harmony of our descendants. Remember the calamity of the great tsunamis. Do not build any homes below this point." Aneyoshi, now part of the city of Miyako, was once more conveniently located down by the sea where the fishermen tied their boats, but only four of its residents survived the tsunami of 1896 and only two survived the tsunami of 1933. The stone reminds people why they now live on higher ground.

> "High dwellings are the peace and harmony of our descendants. Remember the calamity of the great tsunamis."

History became prophecy at 2:46 P.M. on March 11, 2011, when the offshore thrust fault that separates Japan from the Pacific Plate began to slip. The rupture started on a small patch of the fault surface 30 kilometers beneath the ocean, about 100 km southeast of Aneyoshi, and accelerated outward on a shallow-dipping fault, reaching speeds of nearly 3 km/s (more than 6000 miles/hour). By the time it stopped several minutes later, the Pacific Plate had moved under Japan by as much as 40 m along a fault surface the size of South Carolina. Seismic waves from the Tohoku megaquake, which measured magnitude 9, propagated over Earth's surface and through its deep interior, causing the planet to ring like a bell for many days.

The thrusting of Honshu eastward and upward over the Pacific Plate raised the seafloor as much as 10 m almost instantly, displacing several hundred billion tons of water, which flowed away from the fault in a huge tsunami. In less than an hour, the water waves, slower than the seismic waves but much more deadly, passed into the bays and inlets of the Japanese coastline like an undulating monster, gaining height as they approached the shore. Funneled into harbors, the waves created immense walls of water—*tsunami* is Japanese for "harbor wave"—which inundated the nearshore communities, sweeping up boats, cars, and buildings, in some places traveling several kilometers inland.

The fast-moving swath of horrific devastation was captured on videos from helicopters overhead and by survivors who made it to high ground and the tops of buildings. The tsunami overran the seawalls designed to protect the city center of Miyako, destroying all but 30 of the 1000 boats in its famous fishing fleet and killing many hundreds who could not or did not escape in time. Though the exact number remains uncertain, the death toll along the Tohoku coastline exceeded 20,000. One of the highest levels reached by the enormous wave—the greatest in recent Japanese history—was 39 m (128 ft) above the shoreline, just below the Aneyoshi tsunami stone. The residents in their houses above the stone were safe. ◆

The tsunami from the 2011 Tohoku earthquake crashes over a seawall designed to protect the city of Miyako from destructive sea waves.
[AP Photo/Mainichi Shimbun, Tomohiko Kano.]

The tsunami stone of Aneyoshi.
[Ko Sasaki/The New York Times.]

CHAPTER OUTLINE

EARTHQUAKES RIVAL ALL OTHER NATURAL DISASTERS in the threat they pose to human life and property. Our fragile "built environment" is necessarily anchored in Earth's active crust, making it vulnerable to seismic movements and their secondary effects, such as landslides and tsunamis.

To cope with the all-too-frequent death and destruction caused by earthquakes, we have long sought to improve our ability to predict where and when earthquakes might occur and our understanding of what happens when they do. Science has shown that seismic activity can be understood in terms of the basic machinery of plate tectonics. As a result, attempts to reduce earthquake risk have become increasingly tied to the quest for a more fundamental understanding of our geologically active Earth.

This chapter will examine what happens during an earthquake, how scientists use seismic waves to locate and measure earthquakes, and what can be done to reduce earthquake risk. We will also see how seismic waves have been used to image Earth's layered structure—its crust, mantle, and core—and how they are now being used to image the motions of mantle convection.

We have already discussed why earthquakes are concentrated on the faults that form the boundaries between shifting tectonic plates (see Chapter 3), as well as the basic types of faulting produced in brittle crust by the forces acting at those boundaries (see Chapter 7). Here we will see that the pattern of faulting in many earthquake-prone areas, such as California, can be much more complex than the idealized boundaries of simple plate tectonics.

These complexities contribute to the difficulty of earthquake prediction, which remains an unsolved scientific problem. But we will examine several important ways in which geological knowledge can be used to reduce earthquake risk.

WHAT IS AN EARTHQUAKE?

The movement of lithospheric plates generates enormous forces at the boundaries between plates. These forces act to deform brittle crustal rocks in ways that can be described by the concepts of stress, strain, and strength. *Stress* is the local force per unit area that causes rocks to deform. *Strain* is the relative amount of deformation, expressed as the percentage of distortion (for example, compression of a rock by 1 percent of its length). Rocks *fail*—that is, they lose cohesion and break into two or more parts—when they are stressed beyond a critical value, called their *strength*.

An **earthquake** occurs when brittle rocks under stress suddenly fail along a geologic fault. Most large earthquakes are caused by ruptures of preexisting faults, where past earthquakes have already weakened the rocks on the fault surface. The two blocks of rock on either side of the fault slip suddenly, releasing energy in the form of *seismic waves*, which we feel as ground shaking. When the fault slips, the stress is reduced, dropping to a value below the rock strength. After the earthquake, the stress begins to increase again, eventually leading to another large earthquake (**Figure 13.1**).

The faults involved in this repeated earthquake cycle are called *active faults*, and they

> Earthquakes occur when rocks on a fault surface are stressed beyond their strength and the fault slips.

ROCKS DEFORM ELASTICALLY, THEN REBOUND DURING AN EARTHQUAKE RUPTURE

A A farmer builds a stone wall across a right-lateral strike-slip fault a few years after its last rupture.

B Over the next 150 years, the relative motion of the blocks on either side of the fault causes the ground and the stone wall to deform.

C Just before the next rupture, a new fence is built across the already deformed land.

D The fault slips, lowering the stress, and the elastic rebound restores the blocks to their prestressed state. Both the rock wall and the fence are shifted equal amounts along the fault.

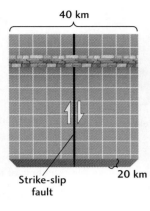

40 km

Strike-slip
fault

20 km

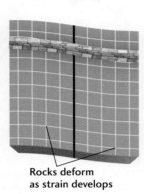

Rocks deform
as strain develops

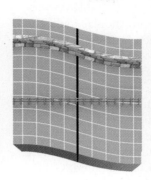

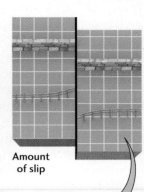

Amount
of slip

STRESS BUILDS UNTIL IT EXCEEDS ROCK STRENGTH

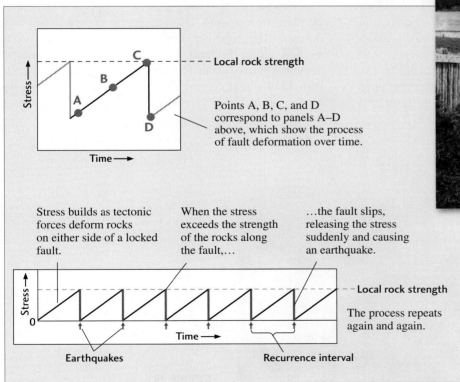

— Local rock strength

Points A, B, C, and D
correspond to panels A–D
above, which show the process
of fault deformation over time.

Stress builds as tectonic
forces deform rocks
on either side of a locked
fault.

When the stress
exceeds the strength
of the rocks along
the fault,…

…the fault slips,
releasing the stress
suddenly and causing
an earthquake.

— Local rock strength

The process repeats
again and again.

Earthquakes

Recurrence interval

A fence built across the San
Andreas fault near Bolinas,
California, is offset by nearly 4 m
after the great San Francisco
earthquake of 1906.

FIGURE 13.1 The elastic rebound theory explains the earthquake cycle. According to the theory, stress on rocks builds up over time as a result of plate movements. Earthquakes occur when the stress exceeds rock strength. Rocks under stress deform elastically, then rebound during an earthquake. Panels A–D show deformation at the points labeled A–D in the bottom panel. [*Photo by G. K. Gilbert/USGS.*]

are concentrated in the zones that form plate boundaries, where most of the stress and strain caused by plate movements is concentrated.

The Elastic Rebound Theory

On a fine April morning in 1906, the citizens of northern California were awakened by the roar and violent shaking caused by the breaking of the San Andreas fault. The rupture began just west of the Golden Gate and extended over 400 km, from the mission town of San Juan Bautista all the way to Cape Mendocino (**Figure 13.2**), creating the most destructive earthquake the United States has yet experienced. The ensuing fires destroyed the city of San Francisco; by the time the flames died away, nearly 3000 of its inhabitants were dead (**Figure 13.3**).

The 1906 San Francisco earthquake was studied more than any earthquake had been up to that time. Geologists were able to map the earthquake-producing rupture in considerable detail by observing the displacements of structures that crossed the San Andreas fault, such as the fence in Figure 13.1. In 1910, one of the scientists who studied the rupture, Henry Fielding Reid of Johns Hopkins University, used these observations to formulate the **elastic rebound theory,** which explains why earthquakes recur on active faults in Earth's crust.

Picture a strike-slip fault between two crustal blocks and imagine that surveyors had painted straight lines on the ground, running perpendicular to the fault and

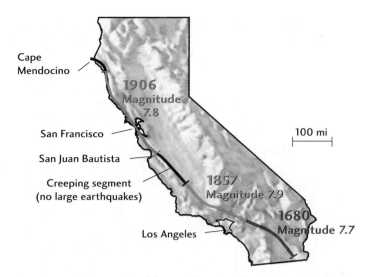

FIGURE 13.2 Map of California, showing the segments of the San Andreas fault that ruptured in 1680, 1857, and 1906. [*Southern California Earthquake Center.*]

extending from one block to the other, as in Figure 13.1a. The two blocks are being pushed in opposite directions by plate movements. The weight of the overlying rock presses them together, however, so friction locks them in place along the fault. They do not move, just as a car does not move when the emergency brake is engaged. Instead of slipping along the fault as stress builds up, the blocks are strained elastically near the fault, as shown

FIGURE 13.3 This photograph, taken from a balloon by George Lawrence 5 weeks after the great earthquake of April 18, 1906, shows the devastation of San Francisco caused by the quake and subsequent fire. The view is looking over Nob Hill toward the business district. [*Corbis.*]

by the bent lines in Figure 13.1b. By *elastically*, we mean that the blocks would spring back and return to their undeformed, stress-free shape if the fault were suddenly to unlock.

As the slow plate movements continue to push the blocks in opposite directions, the elastic strain in the rocks—evidenced by the bending of the survey lines—continues to build up over decades, centuries, or even millennia. At some point, the strength of the rocks is exceeded (Figure 13.1c). The frictional bond that locks the fault can no longer hold, and somewhere along the fault, it breaks. The blocks slip suddenly, and the rupture extends over a section of the fault.

Figure 13.1d shows how the two blocks have rebounded—sprung back to their undeformed state—after the earthquake. The bent survey lines have straightened, and the two blocks have been displaced. (Note that a fence built just before the rupture has been bent by the displacement.) The distance of the displacement is called the **fault slip.** After the fault has slipped, it locks up again. The steady movement of the blocks on either side of the fault causes the stresses to begin to rise again, and the cycle repeats.

The energy that is slowly built up by elastic strain as two blocks are pushed in opposite directions is like the elastic energy stored in a rubber band when it is slowly stretched. The sudden release of energy when a fault slips is like the violent backlash, or *rebound*, that occurs when the rubber band breaks. Some of this elastic energy is radiated as seismic waves,

> According to the elastic rebound theory, stressed rocks are strained elastically, so that when a fault breaks, the rocks on either side of the fault return to their undeformed, stress-free shape.

which can cause violent shaking many kilometers away from the fault.

The elastic rebound theory implies that there should be a periodic buildup and release of elastic energy at faults. However, most active faults, including the San Andreas, rarely exhibit this simple behavior. For instance, all the strain accumulated since the last earthquake may not be released in the next—that is, the rebound may be incomplete—or the stress on one fault may change because of earthquakes on nearby faults (**Figure 13.4**). This irregularity is one reason why earthquakes are so difficult to predict.

Fault Rupture During Earthquakes

The point at which fault slipping begins is the **focus** of the earthquake (**Figure 13.5**). The **epicenter** is the geographic point on Earth's surface directly above the focus. For example, you might hear in a news report: "The U.S. Geological Survey reports that the epicenter of last night's destructive earthquake in California was located 6 kilometers east of Los Angeles City Hall. The depth of the focus was 10 kilometers."

The focal depths of most earthquakes occurring in continental crust range from about 2 to 20 km. Continental earthquakes below 20 km are rare, because under the high temperatures and pressures found at those greater depths, continental crust behaves as a ductile rather than a brittle material (just as hot wax flows when stressed, whereas cold wax breaks; see Chapter 7). In subduction zones, however, where cold oceanic lithosphere plunges into the mantle, earthquakes can originate at depths greater than 600 km.

The fault rupture does not happen all at once. It begins at the focus and expands outward along the fault surface, typically at 2 to 3 km/s (see Figure 13.5). The rupture stops where the stresses become insufficient to continue breaking the fault (where the rocks are stronger) or where the rupture enters ductile material in which it can no longer propagate. As we will see later in this chapter, the magnitude of an earthquake is related to the total area of fault rupture. Most earthquakes are very small, with rupture dimensions much less than the depth of the focus, so that the rupture never breaks the ground surface. In large, destructive earthquakes, however, surface breaks are common. The great 1906 San Francisco earthquake, for example, caused surface displacements averaging about 4 m along the 400-km section of the San Andreas fault that ruptured in that event (see the inset in Figure 13.1 and Figure 13.2). Fault ruptures in the largest earthquakes can

> When a fault slips, stored elastic energy is released as seismic waves, which travel outward from the earthquake focus.

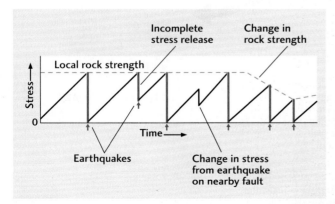

FIGURE 13.4 Irregularities in the earthquake cycle can be caused by incomplete stress release, changes in stress caused by earthquakes on nearby faults, and local variations in rock strength.

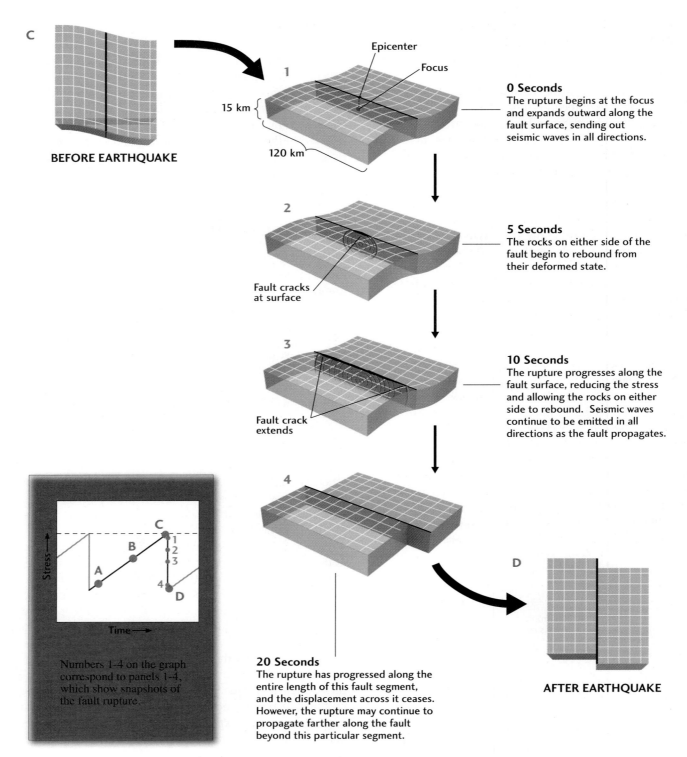

C

BEFORE EARTHQUAKE

Epicenter

Focus

15 km

120 km

1

0 Seconds
The rupture begins at the focus and expands outward along the fault surface, sending out seismic waves in all directions.

2

Fault cracks at surface

5 Seconds
The rocks on either side of the fault begin to rebound from their deformed state.

3

Fault crack extends

10 Seconds
The rupture progresses along the fault surface, reducing the stress and allowing the rocks on either side to rebound. Seismic waves continue to be emitted in all directions as the fault propagates.

4

D

AFTER EARTHQUAKE

20 Seconds
The rupture has progressed along the entire length of this fault segment, and the displacement across it ceases. However, the rupture may continue to propagate farther along the fault beyond this particular segment.

Stress

C

B

A

1
2
3

4

D

Time

Numbers 1-4 on the graph correspond to panels 1-4, which show snapshots of the fault rupture.

FIGURE 13.5 During an earthquake, fault slipping begins at the focus and spreads out along the fault surface. Panel C is just before the earthquake; panel D is just after, as in Figure 13.1. Panels 1–4 are snapshots of the fault rupture corresponding to the numbered points on the graph.

extend for more than 1000 km, and the fault slip can be tens of meters. Generally, the longer the fault rupture, the greater the fault slip.

As we have seen, the sudden slipping of the blocks at the time of the earthquake reduces the stress on the fault and releases much of the stored elastic energy. Most of this stored energy is converted to frictional heat in the fault zone or dissipated by rock fracturing, but part of it is released as seismic waves that travel outward from the rupture, much as waves ripple outward from the spot where a stone is dropped into a still pond. The focus of the earthquake generates the first seismic waves, but slipping parts of the fault continue to generate waves until the rupture stops. In a large event, the propagating fracture continues to produce waves for many tens of seconds. These waves radiate all along the fault rupture and can cause damage far from the epicenter. Towns along the San Andreas fault far north of San Francisco were badly damaged in the 1906 earthquake.

Foreshocks and Aftershocks

Almost all large earthquakes trigger smaller earthquakes called **aftershocks.** Aftershocks follow the triggering event, or *mainshock*, in sequences, and their foci are distributed in and around the rupture plane of the mainshock (**Figure 13.6**). Aftershock sequences exemplify the complexities of earthquakes that cannot be described by simple elastic rebound theory. Although fault slipping during the mainshock decreases the stress along most of the rupture surface, it can increase the stress on parts of the fault surface that did not slip or where the slip was incomplete, as well as in surrounding regions. Aftershocks happen where that stress exceeds the rock strength.

The number and sizes of aftershocks depend on the magnitude of the mainshock, and their frequencies decrease with time after the mainshock. The aftershocks of a magnitude 5 earthquake might last for only a few weeks, whereas those of a magnitude 7 earthquake can continue for several years. The size of the largest aftershock is usually about one magnitude unit smaller than the mainshock. According to this rough rule of thumb, a magnitude 7 earthquake might have an aftershock as large as magnitude 6 and sometimes even larger.

In populated regions, the shaking from large aftershocks can be very dangerous, compounding the damage caused by the mainshock. On September 4, 2010, a magnitude 7.1 earthquake west of Christchurch, the second largest city in New Zealand, caused extensive damage, but nobody was killed and only a few people were injured. However, a magnitude 6.3 aftershock struck right beneath the center of Christchurch on February 22, 2011, collapsing many buildings and killing more than 160 people. The economic losses

JUST BEFORE EARTHQUAKE

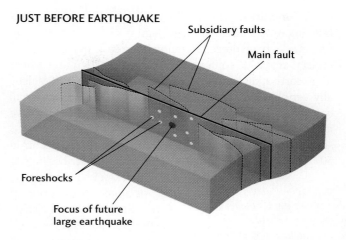

DURING EARTHQUAKE

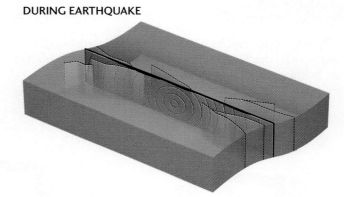

JUST AFTER EARTHQUAKE

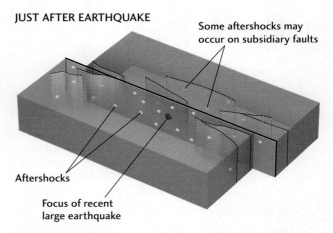

FIGURE 13.6 Aftershocks are smaller shocks that follow a large earthquake (the mainshock). Foreshocks occur near but before the mainshock.

from this aftershock, estimated at $16 billion, were several times greater than the losses caused by the mainshock five months before. Two other strong aftershocks (magnitudes 5.5 and 6.0) hit the city on June 13, 2011, injuring dozens more and causing $4 billion in additional damages. More aftershocks can be expected in the years to come.

A **foreshock** is a small earthquake that occurs near but before a mainshock (see Figure 13.6). One or more foreshocks have preceded many large earthquakes, so scientists have tried to use them to predict when and where large earthquakes might happen. Unfortunately, it is usually very hard to distinguish foreshocks from other small earthquakes that occur randomly and frequently on active faults, so this prediction method has only rarely proved successful. The magnitude 7.2 foreshock that occurred 50 hours before the Tohoku megaquake (described in the essay that opens this chapter) did not appear to be unusual at the time. In some sense, the megaquake can be classified as an anomalously big "aftershock" triggered by the first event.

> *Most large earthquakes are followed by smaller earthquakes called aftershocks. They are sometimes preceded by one or more foreshocks.*

As this example illustrates, foreshocks, mainshocks, and aftershocks can be classified in a definitive way only after an earthquake sequence has ended; during the sequence, we cannot be sure whether the mainshock—the biggest event in the sequence—is yet to come.

MEASURING EARTHQUAKES USING SEISMIC WAVES

The **seismograph,** an instrument that records the seismic waves generated by earthquakes, is to the Earth scientist what the telescope is to the astronomer: a tool for peering into inaccessible regions (**Figure 13.7**). The ideal seismograph would be a device affixed to a stationary frame not attached to Earth. When the ground shook, the seismograph would measure the changing distance between the frame, which did not move, and the vibrating ground, which did. As yet, we have no way to position a seismograph that is not attached to Earth—although modern space technology, such as the Global Positioning System, is beginning to remove this limitation. So we compromise. We attach a dense mass, such as a piece of

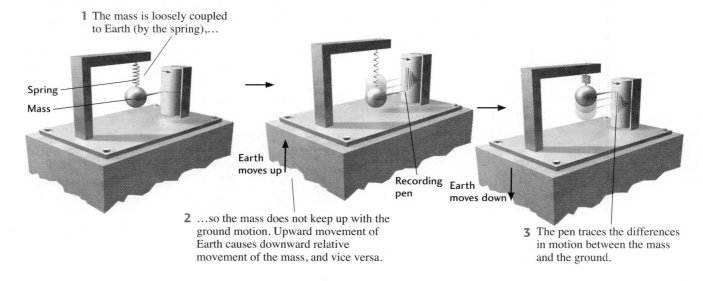

1 The mass is loosely coupled to Earth (by the spring),…

Spring
Mass

Earth moves up

2 …so the mass does not keep up with the ground motion. Upward movement of Earth causes downward relative movement of the mass, and vice versa.

Recording pen   Earth moves down

3 The pen traces the differences in motion between the mass and the ground.

(b)   Seismograph designed to detect horizontal movement

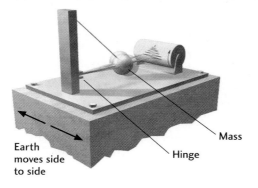

Earth moves side to side

Mass

Hinge

FIGURE 13.7 A seismograph consists of a dense mass (such as a steel ball) attached to a recording device. Because of its inertia and its loose coupling to Earth through (a) a spring or (b) a hinge, the mass does not keep up with the motion of the ground.

steel, to Earth so loosely that the ground can vibrate up and down or side to side without causing much movement of the mass.

This loose attachment can be achieved by suspending the mass from a spring (Figure 13.7a). When seismic waves move the ground up and down, the mass tends to remain stationary according to the principle of inertia (an object at rest tends to stay at rest), but the mass and the ground move relative to each other because the spring can compress or stretch. In this way, the vertical displacement of the ground caused by seismic waves can be recorded by a pen on chart paper or, almost always these days, digitally recorded on a computer. Such a record is called a *seismogram*.

Loose attachment of the mass can also be achieved with a hinge. A seismograph that has its mass suspended on hinges like a swinging gate (Figure 13.7b) can record the horizontal displacement of the ground. Modern seismographs can detect ground oscillations of less than a billionth of a meter—an astounding feat, considering that such small displacements are of atomic size!

Seismic Waves

Install a seismograph anywhere, and within a few hours it will record the passage of seismic waves generated by an earthquake somewhere on Earth. The waves travel from the earthquake focus through Earth and arrive at the seismograph in three distinct groups. The first waves to arrive are called primary waves, or **P waves.** The secondary waves, or **S waves,** follow. Both

> *P waves are compressional waves that can travel through solid, liquid, or gaseous materials. S waves are shear waves that can travel only through solid materials.*

P waves and S waves travel through Earth's interior. Afterwards come the slower **surface waves,** which travel around Earth's surface (**Figure 13.8**).

P waves in rock are similar to sound waves in air, except that P waves travel through the solid rock of Earth's crust at about 6 km/s, which is about 20 times faster than sound waves travel through air. Like sound waves, P waves are *compressional waves*, so called because they travel through solid, liquid, or gaseous materials as a succession of compressions and expansions. P waves can be thought of as push-pull waves: they push or pull particles of matter in the direction of their path of travel (**Figure 13.9**).

S waves travel through solid rock at a little more than half the velocity of P waves. They are *shear waves* that displace material at right angles to their path of travel. Shear waves cannot travel through liquids or gases.

The velocities at which P and S waves travel are higher when the resistance to their movement is greater. It takes more force to compress solids than to shear them, so P waves always travel faster than S waves through a solid, which is why the P waves from an earthquake arrive at a seismograph before the S waves. This physical principle also explains why S waves cannot travel through air, water, or Earth's liquid outer core: gases and liquids put up no resistance to shear.

Surface waves are confined to Earth's surface and outer layers, like waves on the ocean. Their velocity is slightly less than that of S waves. One type of surface wave sets up a rolling motion in the ground; another type shakes the ground sideways (see Figure 13.9).

People have felt seismic waves and witnessed their destructiveness throughout history, but not until the close of the nineteenth century were scientists able to devise seismographs to record them accurately. Seismic waves enable seismologists to locate earthquakes and

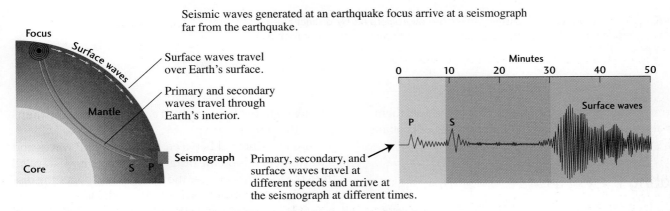

Seismic waves generated at an earthquake focus arrive at a seismograph far from the earthquake.

Surface waves travel over Earth's surface.

Primary and secondary waves travel through Earth's interior.

Primary, secondary, and surface waves travel at different speeds and arrive at the seismograph at different times.

FIGURE 13.8 The three types of seismic waves travel by different routes and at different speeds to a seismograph, which records them.

Seismic waves are characterized by distinct types of ground deformation.

P-wave motion	S-wave motion	Surface-wave motion

P-wave motion

P waves (primary waves) are compressional waves that travel quickly through rock.

Compressional-wave crest

P waves travel as a series of contractions and expansions, pushing and pulling particles in the direction of their path of travel.

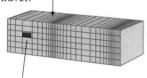

The red square charts the contraction and expansion of a section of rock.

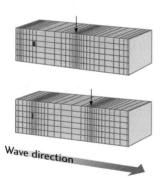

Wave direction

S-wave motion

S waves (secondary waves) travel at about half the speed of P waves.

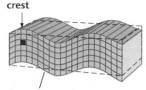

Shear-wave crest

S waves are shear waves that push material at right angles to their path of travel.

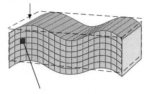

The red square shows how a section of rock shears from a square to a parallelogram as the S wave passes.

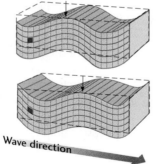

Wave direction

Surface-wave motion

Surface waves ripple across Earth's surface, where air above the surface allows free movement. There are two types of surface waves.

In one type, the ground surface moves in a rolling, elliptical motion that decreases with depth beneath the surface.

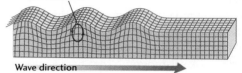

Wave direction

In the second type, the ground shakes sideways, with no vertical motion.

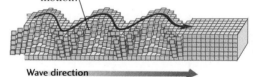

Wave direction

FIGURE 13.9 The three types of seismic waves are characterized by distinct types of ground deformation. The red squares show the distortion of a section of rock as a wave passes through it.

determine the nature of the faulting that produces them. They also provide our most important means of probing Earth's deep interior.

Locating the Focus

Locating a quake's focus is like deducing the distance of a lightning strike from the time interval between the flash of light and the sound of thunder—the greater the distance to the lightning bolt, the longer the time interval. Light travels faster than sound, so the lightning flash may

be likened to the P waves of an earthquake and the thunder to the slower S waves.

The time interval between the arrival of P waves and S waves depends on the distance the waves have traveled from the focus: the longer the interval, the longer the distance the waves have traveled. Seismologists have used networks of sensitive seismographs around the world and highly accurate clocks to time the arrival of seismic waves from earthquakes as well as from underground nuclear explosions at known locations. To estimate the distance to a new quake's focus, seismologists read from a seismogram the amount of time that elapsed between the arrival of the first P waves and the arrival of the S waves. Then they use a table or a graph like the one shown in **Figure 13.10** to determine the distance from the seismograph station to the focus. If they can determine the distances from three or more seismographic stations, they can locate the focus. They can also deduce the time of the quake at the focus because the arrival time of the P waves at each station is known, and it is possible to determine from a graph or table how long those waves took to reach the station. Today, this entire process is done automatically by computers, which use the data from a large network of seismographs to determine each earthquake's epicenter, depth of the focus, and origin time.

> *Seismologists can determine the location of an earthquake's focus by using the time interval between the arrival of P waves and S waves at three or more seismographic stations.*

1 Seismic waves from an earthquake move out concentrically from the focus and arrive at distant seismographic stations at different times.

2 Because P waves travel almost twice as fast as S waves, the interval between their arrival times increases with distance.

3 By matching the observed interval to known travel-time curves, a seismologist can determine the distance from the station to the quake epicenter.

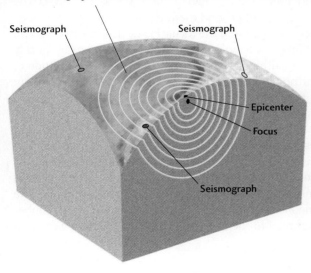

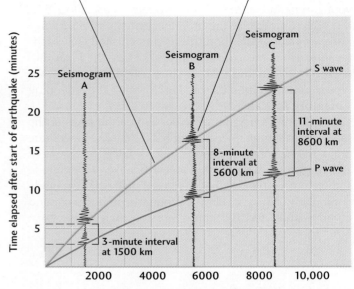

4 A seismologist then draws a circle with a radius calculated from the travel-time curves around each seismographic station.

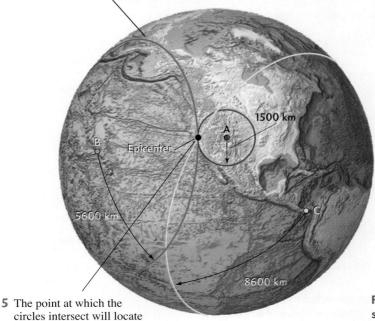

5 The point at which the circles intersect will locate the earthquake's epicenter.

FIGURE 13.10 Readings from three or more seismographic stations can be used to determine the location of an earthquake's focus.

Measuring the Size of an Earthquake

Locating earthquakes is only one step on the way to understanding them. We must also determine their sizes, or *magnitudes.* Other things being equal (such as distance from the focus and regional geology), an earthquake's magnitude is the main factor that determines the intensity of the seismic waves it produces and thus the earthquake's potential destructiveness.

Richter Magnitude In 1935, Charles Richter, a California seismologist, devised a simple procedure that assigned a numerical size to each earthquake, now called the *Richter magnitude* (**Figure 13.11**). Richter studied astronomy as a young man and learned how astronomers use a logarithmic scale to measure the brightness of stars over a huge range of values. Adapting this idea to earthquakes, Richter took as his measure of earthquake size the logarithm of the largest ground movement registered by a standard type of seismograph at a standard distance, thus defining a **magnitude scale.**

On Richter's magnitude scale, two earthquakes at the same distance from a seismograph differ by one magnitude unit if the size of their ground motions differs by a factor of 10. The ground movement of an earthquake of magnitude 6, therefore, is 10 times that of an earthquake of magnitude 5. Similarly, a magnitude 7 earthquake produces ground motions that are 100 times greater than those of a

magnitude 5 earthquake. The energy released as seismic waves increases even more strongly with earthquake magnitude, by a factor of about 32 for each Richter unit. A magnitude 7 earthquake releases 32 × 32, or about 1000, times the energy of a magnitude 5 earthquake. According to this energy scale, the Tohoku megaquake was a million times more powerful than a magnitude 5 event!

Seismic waves gradually weaken as they move away from the focus, so to make his procedure work for any seismograph, Richter had to find a way to correct the measurement of ground motion for the distance between the seismograph and the focus. He devised a simple graph that allowed seismologists at different locations to quickly come up with nearly the same value for the magnitude of an earthquake no matter how far their instruments were from the focus (see Figure 13.11). His procedure came to be used throughout the world.

Moment Magnitude Although "Richter scale" has become a household term, seismologists prefer a measure of earthquake size that is more directly related to the physical properties of the faulting that causes the earthquake. The *seismic moment* of an earthquake is defined as a number proportional to the product of the area of faulting and the

> *The Richter magnitude scale estimates the size of an earthquake using the logarithm of the largest ground motion registered by a seismograph.*

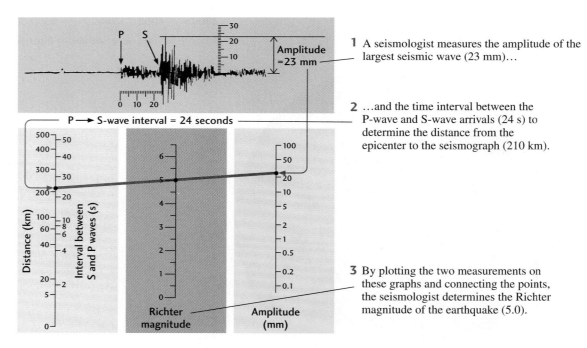

1 A seismologist measures the amplitude of the largest seismic wave (23 mm)…

2 …and the time interval between the P-wave and S-wave arrivals (24 s) to determine the distance from the epicenter to the seismograph (210 km).

3 By plotting the two measurements on these graphs and connecting the points, the seismologist determines the Richter magnitude of the earthquake (5.0).

FIGURE 13.11 The maximum amplitude of ground motion, corrected by the P-S wave interval, is used to assign a Richter magnitude to an earthquake. [*California Institute of Technology.*]

average fault slip. The corresponding *moment magnitude* increases by about one unit for every 10-fold increase in the area of faulting. Although Richter's method and the moment method produce roughly the same numerical values, the moment magnitude can be measured more accurately from seismograms, and it can also be determined directly from field measurements of the fault.

Large earthquakes occur much less often than small ones. This observation can be expressed by a very simple relationship between earthquake frequency and magnitude (**Figure 13.12**). Worldwide, approximately 1,000,000 earthquakes with magnitudes greater than 2 take place each year, and this number decreases by a factor of 10 for each magnitude unit. Hence, there are about 100,000 earthquakes with magnitudes greater than 3, about 1000 with magnitudes greater than 5, and about 10 with magnitudes greater than 7.

> *The moment magnitude estimates the size of an earthquake from the area of faulting and the fault slip.*

According to these statistics, there should be, on average, about 1 earthquake with a magnitude greater than 8 per year and 1 earthquake with a magnitude greater than 9 every 10 years. In fact, the very largest earthquakes, such as the ones that occurred on thrust faults in the subduction zones off Japan in 2011 (moment magnitude 9.0), Sumatra in 2004 (moment magnitude 9.2), Alaska in 1964 (moment magnitude 9.1), and Chile in 1960 (moment magnitude 9.5) and 2010 (moment magnitude 8.8), are almost this common when averaged over many decades. However, even the largest subduction zone faults are too small to create magnitude 10 earthquakes, so seismologists believe that events of such extreme size do not follow this rule; that is, they are much less frequent than once per century.

> *Large earthquakes occur less often than small ones. The number of quakes per year decreases by a factor of 10 with each magnitude unit.*

Shaking Intensity Earthquake magnitude by itself does not describe seismic hazard because the shaking that causes destruction generally weakens with distance from the fault rupture. A magnitude 8 earthquake in a remote area far from the nearest city might cause no human or economic losses, whereas a magnitude 6 quake immediately beneath a city is likely to cause serious damage. The destruction in Christchurch by the earthquakes of February 22 and June 13, 2011, illustrates this important point.

In the late nineteenth century, before Richter invented his magnitude scale, seismologists and earthquake engineers developed **intensity scales** for estimating the intensity of shaking directly from an earthquake's destructive effects. **Table 13.1** shows the intensity scale that remains in most common use today, called the *modified Mercalli intensity scale* after Giuseppe Mercalli, the Italian scientist who proposed it in 1902. This scale assigns a value, given as a Roman numeral from I to XII, to the intensity of the

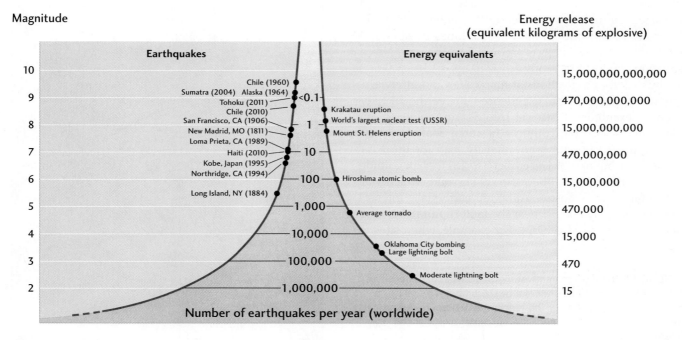

FIGURE 13.12 Relationship between moment magnitude, energy release, and number of earthquakes per year worldwide. Several other large sources of sudden energy release are included for comparison. [*Adapted from IRIS Consortium.*]

TABLE 13.1	Modified Mercalli Intensity Scale

Intensity Level	Description
I	Not felt.
II	Felt only by a few people at rest. Suspended objects may swing.
III	Felt noticeably indoors. Many people do not recognize it as an earthquake. Parked cars may rock slightly.
IV	Felt indoors by many, outdoors by few. Dishes, windows, doors rattle. Parked cars rock noticeably.
V	Felt by most; many awakened. Some dishes, windows broken. Unstable objects overturned.
VI	Felt by all. Some heavy furniture moves. Damage slight.
VII	Slight to moderate damage in well-built structures; considerable damage in poorly built structures; some chimneys broken.
VIII	Considerable damage in well-built structures. Damage great in poorly built structures. Fall of chimneys, factory stacks, columns, monuments, walls.
IX	Damage great in well-built structures, with partial collapse. Buildings shifted off foundations.
X	Some well-built wooden structures destroyed; most masonry and frame structures destroyed. Rails bent.
XI	Few if any masonry structures remain standing. Bridges destroyed. Rails bent greatly.
XII	Damage total. Lines of sight and level are distorted. Objects thrown into the air.

shaking at a particular location. For example, a location where an earthquake is barely felt by a few people is assigned an intensity of II, whereas one where it was felt by nearly everyone is given an intensity of V. Numbers at the upper end of the scale describe increasing amounts of damage. The narrative attached to the highest value, XII, is tersely apocalyptic: "Damage total. Lines of sight and level are distorted. Objects thrown into the air." Not a situation you want to be in!

By making observations at many sites and interviewing many people who experienced an earthquake, or even by examining historical records, seismologists can make maps showing contours of equal shaking intensity. **Figure 13.13** shows an intensity map for the New Madrid earthquake of December 16, 1811, a magnitude 7.7 event near the southern tip of Missouri, which was felt as far away as Boston. Although earthquake intensities are generally highest near the fault rupture, they also depend on the local geology. For example, when sites at

equal distances from the rupture are compared, the shaking tends to be more intense on soft sediments (especially water-saturated sediments near shorelines) than on hard

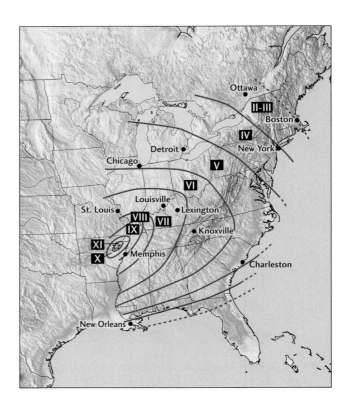

FIGURE 13.13 Measurements of modified Mercalli intensities for the New Madrid earthquake of December 16, 1811, a magnitude 7.7 event near the juncture of Missouri, Arkansas, and Tennessee. Regions near the epicenter experienced intensities greater than X, and intensities as high as VI were observed 200 km from the epicenter (see Table 13.1). [*Carl W. Stover and Jerry L. Cossman, USGS Professional Paper 1527, 1993.*]

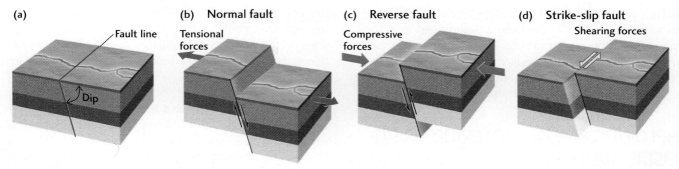

FIGURE 13.14 The three main types of fault mechanisms that initiate earthquakes and the stresses that cause them. (a) Fault before movement takes place. (b) Normal faulting due to tensional stress. (c) Reverse faulting due to compressive stress. (d) Strike-slip faulting due to shearing stress (in this case, left-lateral).

basement rocks. Intensity maps thus provide engineers with crucial data for designing structures to withstand seismic shaking.

Determining Fault Mechanisms

The pattern of ground shaking also depends on the orientation of the fault rupture and the direction of slipping, which together specify the **fault mechanism** of an earthquake. The fault mechanism tells us whether the rupture was on a *normal*, *reverse*, or *strike-slip* fault. If the rupture was on a strike-slip fault, the fault mechanism also tells us whether the motion was *right-lateral* or *left-lateral* (see Figure 7.7 for the definition of these terms). We can then use this information to infer the regional pattern of tectonic forces (**Figure 13.14**).

For shallow ruptures that break the surface, we can sometimes determine the fault mechanism from field observations of the fault scarp. As we have seen, however, most ruptures are too deep to break the surface, so we must deduce the fault mechanism from seismograms.

For large earthquakes at any depth, this task turns out to be easy because there are enough seismographic stations around the world to surround any earthquake's focus. In some directions from the focus, the very first movement of the ground recorded by a seismograph—a P wave—is a *push away* from the focus, causing upward motion on a vertical seismograph. In other directions, the initial ground motion is a *pull toward* the focus, causing downward motion on a vertical seismograph. In other words, the slipping on a fault looks like a push if you view it from one direction but like a pull if you view it from another. The locations of pushes and pulls can be plotted and divided into four sections based on the positions of the seismographic stations, as shown in **Figure 13.15**. One of the two boundaries between the sections will be

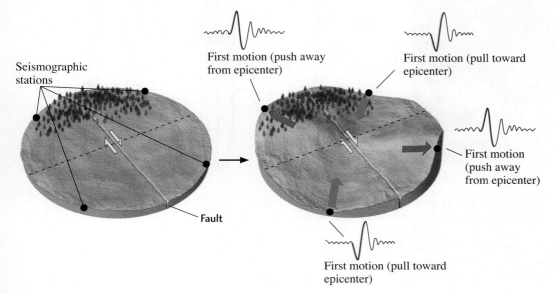

FIGURE 13.15 The motion marking the first P wave arriving from a fault rupture at each of several seismographic stations is used to determine the orientation of the fault and the direction of fault slipping. The case shown here is the rupture of a right-lateral strike-slip fault. Note that the alternating pattern of pushes and pulls would remain the same if a plane perpendicular to the fault ruptured with left-lateral displacement. Seismologists can usually choose between the two possibilities using additional information, such as field mapping of the fault scarp or the alignment of aftershocks along the fault.

the fault orientation; the other will be a plane perpendicular to the fault. The slip direction is determined from the arrangement of pushes and pulls. In this manner, without surface evidence, seismologists can deduce whether the horizontal crustal forces that triggered an earthquake were tensional, compressive, or shearing forces.

EXPLORING EARTH'S INTERIOR USING SEISMIC WAVES

The same seismic waves that cause terrible shaking during earthquakes can illuminate Earth's deepest regions, allowing us to construct models of its internal layering, make three-dimensional images of crustal geology and mantle convection, and understand the workings of its central core.

Paths of Seismic Waves Through Earth

If Earth were made of a single material with constant properties from the surface to the center, P and S waves would travel from the focus of an earthquake to a distant seismographic station along straight lines through Earth's interior (just as the Sun's rays travel in straight lines through the atmosphere). When the first global networks of seismographs were installed about a century ago, however, seismologists discovered that the structure of Earth's interior was much more complicated (see Chapter 2). Observations of long-distance seismic waves showed that

the paths of P and S waves are *refracted* (bent) upward through the mantle and can be *reflected* by sharp boundaries, such as Earth's solid surface and the boundary between its mantle and core. These types of wave paths are illustrated in **Figure 13.16**.

Waves Refracted by Earth's Internal Boundaries From the travel times and the amount of upward refraction of seismic waves, seismologists were able to conclude that P waves travel much faster through rock at great depths than they do through rock at Earth's surface. This was hardly surprising, because rocks subjected to great pressures in Earth's interior are squeezed into tighter crystal structures. The atoms in these tighter structures are more resistant to further compression, which causes P waves to travel through them more quickly.

> *The travel times and paths of P and S waves can be used to construct models of Earth's internal layers.*

Seismologists were very surprised, however, by what they found at progressively greater distances from an earthquake focus (**Figure 13.17**). After the P waves and S waves had traveled beyond about 11,600 km, they suddenly disappeared! Like airplane pilots and ship captains, seismologists prefer to measure distances traveled on Earth's surface in angular degrees, from 0° at the earthquake focus to 180° at a point on the opposite side of Earth. Each degree measures 111 km at the surface, so 11,600 km corresponds to an angular distance of 105°, as shown in

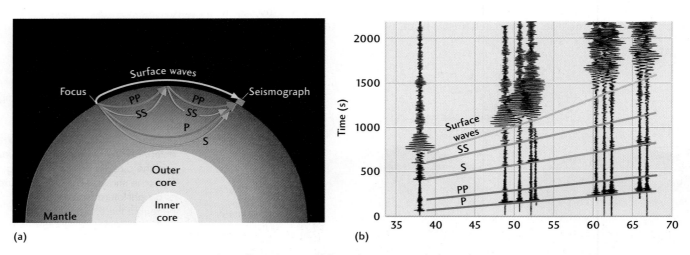

(a)

(b)

FIGURE 13.16 (a) P and S waves may be reflected upward from the core-mantle boundary and may also be reflected from Earth's surface. A seismic wave that has been reflected once from Earth's surface is labeled with a double letter (PP or SS). (b) Seismograms recorded at various distances from an earthquake focus in the Aleutian Islands, Alaska. The colored lines identify the arrival times of the P and S waves, the surface waves, and the PP and SS waves reflected from Earth's surface.

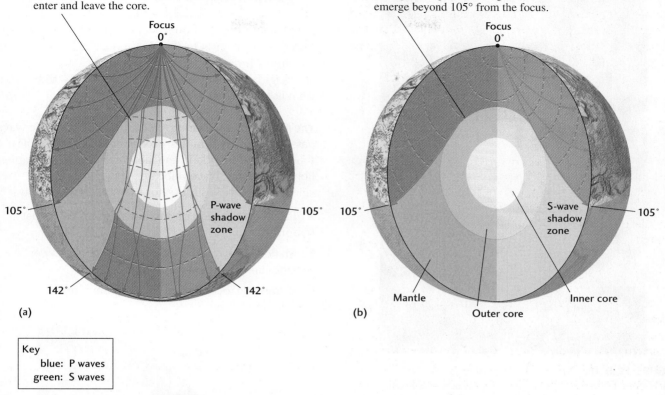

P waves cannot reach the surface within the shadow zone because of the way they are refracted when they enter and leave the core.

Although S waves reach the core, they cannot travel through its liquid outer region, and therefore never emerge beyond 105° from the focus.

FIGURE 13.17 Earth's core creates P-wave and S-wave shadow zones. Seismic wave paths from an earthquake focus through Earth's interior are shown by solid lines. The dashed lines show the progress of the waves at 2-minute intervals. Distances are measured in angular degrees from the earthquake focus. (a) The P-wave shadow zone extends from 105° to 142°. (b) The larger S-wave shadow zone extends from 105° to 180°.

Figure 13.17. When seismologists looked at seismograms recorded beyond 105° from the focus, they did not see the distinct P- and S-wave arrivals that were so clear on seismograms recorded at shorter distances. Beyond about 15,800 km from the focus (142°), the P waves suddenly reappeared, although they were much delayed beyond their expected travel times. The S waves never reappeared.

In 1906, the British seismologist R. D. Oldham put these observations together to provide the first evidence that Earth has a liquid outer core. S waves cannot travel through the liquid outer core, he argued, because it is liquid, and liquids have no resistance to shearing. Thus, there is an S-wave **shadow zone** beyond 105° from the earthquake focus (see Figure 13.17b). The propagation of P waves is more complicated (see Figure 13.17a). At 105°, their paths just miss the core, whereas waves that would have traveled to greater distances encounter the core-mantle boundary. At that boundary, P-wave velocity drops by a factor of almost two. Therefore, the waves are refracted downward into the core and emerge at greater distances after the delay caused by their detour through the core. This refraction effect forms a P-wave shadow zone at angular distances between 105° and 142°.

Waves Reflected by Earth's Internal Boundaries When seismologists looked at records of seismic waves made at angular distances of less than 105° from an earthquake focus, they found waves that must have been reflected from the core-mantle boundary. In 1914, a German seismologist, Beno Gutenberg, used the travel times of these reflected waves to determine an accurate depth for the core-mantle boundary: just under 2900 km. Twenty-two years later, the Danish seismologist Inge Lehmann

The absence of shear waves on seismograms recorded at large distances from earthquake foci revealed the liquid state of Earth's core.

Seismologists have used seismic waves reflected by sharp boundaries between Earth's layers to estimate the depths of the boundaries.

FIGURE 13.18 The Danish seismologist Inge Lehmann discovered Earth's inner core in 1936. [*Courtesy of Beverley Bolt.*]

(**Figure 13.18**) discovered Earth's inner core, a solid sphere at Earth's center of only about one-third the core radius. Waves reflected from the inner core–outer core boundary showed that the depth of this boundary is about 5150 km.

Seismic Exploration of Near-Surface Layering

Seismic waves can also be used to probe the shallow parts of Earth's crust. This technique, called *seismic profiling*, has a number of practical applications. Seismic waves generated by artificial sources, such as dynamite explosions, are reflected by geologic structures at shallow depths in the crust. Recording these reflections has proved to be the most successful method for finding deeply buried oil and gas reservoirs (see Figure 1.11). This type of seismic exploration is now a multibillion-dollar industry. Reflected seismic waves are also used to measure the depth of water tables and the thickness of glaciers. At sea, compressional waves can be generated by mechanical devices similar to loudspeakers, and oceanographic ships routinely use the underwater sound they produce to measure the depth of the ocean and the thickness of sediments on the seafloor.

Layering and Composition of Earth's Interior

Seismograms of earthquakes around the world have allowed seismologists to devise a model of Earth's interior layering (**Figure 13.19**), which we will explore by taking an imaginary downward journey through Earth's interior from its outer crust to its inner core.

The Crust By measuring the velocities of seismic waves passing through samples of various materials typical of the crust and mantle in the laboratory, we can compile a library of seismic wave velocities. Rough values for P-wave velocities in igneous rocks, for example, are as follows:

- Felsic rocks typical of the upper continental crust (granite): 6 km/s

- Mafic rocks typical of oceanic crust or the lower continental crust (gabbro): 7 km/s

- Ultramafic rocks typical of the upper mantle (peridotite): 8 km/s

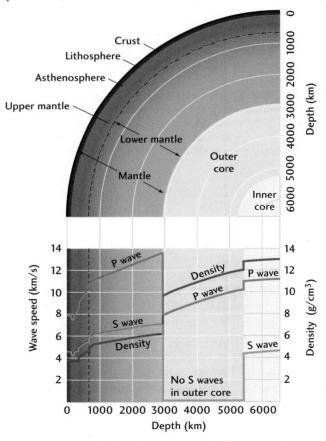

FIGURE 13.19 Earth's layering as revealed by seismology. The lower diagram shows changes in P-wave and S-wave velocities and rock density with depth. The upper diagram is a cross section through Earth on the same depth scale, showing how these changes are related to the major layers.

These velocities vary because they depend on a rock's density and its resistance to compression and shear, all of which vary with chemical composition and crystal structure. In general, higher densities correspond to higher P-wave velocities; typical densities for granite, gabbro, and peridotite are 2.6 g/cm^3, 2.9 g/cm^3, and 3.3 g/cm^3, respectively.

We know from measurements of P-wave velocities that the upper part of the continental crust is made up mostly of low-density granitic rocks. The measurements also show that no granite exists on the deep seafloor; the crust there consists entirely of basalt and gabbro overlain by sediments. The velocity of P waves increases abruptly to 8 km/s at the Mohorovičić discontinuity, or Moho, which marks the base of the crust (see Chapter 2). The seismic data show that Earth's crust is thin (about 7 km) under oceans, thicker (about 33 km) under the stable, flat-lying continents, and thickest (as much as 70 km) under high mountains.

The Mantle The P-wave velocity of 8 km/s indicates that the mantle below the Moho is made mostly of peridotite, a dense, ultramafic rock composed primarily of olivine and pyroxene. These minerals have less silica and more magnesium and iron than those in typical crustal rocks (see Chapter 5).

S-wave velocities have been used to explore the layering of the mantle. The layering of the upper mantle (**Figure 13.20**) is primarily caused by the effects of increasing pressure and temperature on peridotite. The pressure in Earth's interior increases steadily with depth. The increase of temperature with depth, which defines a curve known as the **geotherm,** is more complicated. In **Figure 13.21,** the geotherm (in yellow) is compared with the temperatures at which Earth materials begin to melt (in red).

The mantle just below the Moho is relatively cold. Like the crust, it is part of the lithosphere, the rigid layer that forms the tectonic plates (see Chapter 2). On average, the thickness of the lithosphere is about 100 km, but it is highly variable geographically, ranging from almost no thickness near spreading centers, where new oceanic lithosphere is forming from hot, rising mantle material, to over 200 km beneath the cold, stable continental cratons.

The temperature rises rapidly with depth in the lithosphere, and near 100 km it gets hot enough that some minerals in the peridotite begin to melt. Although the amount of melting is small (in most places, less than 1 percent), it is sufficient to cause an abrupt drop in the rigidity of the rock, which decreases the S-wave velocity, forming a **low-velocity zone.** Because partial melting allows peridotite to flow more easily, geologists identify the low-velocity zone with the top part of the *asthenosphere*—the weak, ductile layer on which the lithospheric

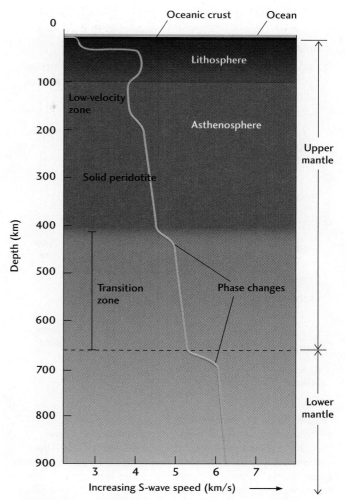

FIGURE 13.20 The structure of the mantle beneath old oceanic lithosphere, showing S-wave velocities to a depth of 900 km. Changes in S-wave velocity mark the strong, brittle lithosphere, the weak, ductile asthenosphere, and a transition zone, in which increasing pressure forces rearrangements of atoms into denser and more compact crystal structures (phase changes).

plates slide. This idea fits nicely with evidence that the asthenosphere is the source of most basaltic magma (see Chapter 5).

In the asthenosphere, the melting temperature of rock rises because of the increasing pressure, whereas mantle convection causes the geotherm to flatten (for the same reason that stirring your bathwater evens out the temperatures). Therefore, the geotherm drops below the melting curve as we go deeper into the mantle, and the S-wave velocities begin to increase again.

> *Below the lithosphere, seismic velocities decrease because of the partial melting of peridotite, which contributes to the ductile properties of the asthenosphere.*

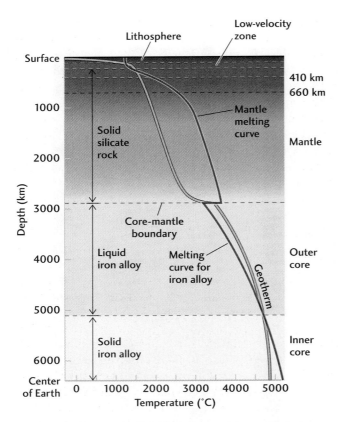

FIGURE 13.21 An estimate of Earth's geotherm, which describes the increase in temperature with depth (yellow line). The geotherm first rises above the melting curve—the temperature at which peridotite begins to melt (red line)—in the upper mantle, forming the partially molten low-velocity zone. It does so again in the outer core, where the iron-nickel alloy is in a liquid state. The geotherm falls below the melting curve throughout most of the mantle and in the solid inner core.

About 400 km below the surface, S-wave velocity increases by about 10 percent within a narrow zone less than 20 km thick. This jump in S-wave velocity can be explained by a **phase change** in olivine, the major mineral constituent of the upper mantle, whose ordinary crystal structure is transformed into a denser, more closely packed structure at high pressures. When olivine is subjected to high pressures in the laboratory, the atoms that form its crystal shift into a more compact arrangement at the temperatures and pressures corresponding to depths of about 410 km. Moreover, the jumps in the P- and S-wave velocities measured in the laboratory match the increase observed for seismic waves at this depth.

In the region from 410 to 660 km below the surface, mantle properties change slowly as depth increases. Near 660 km, however, the S-wave velocity abruptly increases again, indicating a

At two depths in the mantle, 410 km and 660 km below the surface, seismic velocities jump, indicating phase changes in the crystal structures of mantle minerals.

second major phase change in olivine to an even more closely packed crystal structure. Laboratory experiments have confirmed the existence of another major mineralogical phase change at pressures and temperatures found at this depth.

Because it contains two major phase changes (and several minor ones), the layer between 400 and 700 km in depth is called the *transition zone.* Below the transition zone, the seismic wave velocities increase gradually and do not show any more unusual features until close to the core-mantle boundary (see Figure 13.19). This relatively homogeneous region, more than 2000 km thick, is called the **lower mantle.**

The reflection of seismic waves at a depth of 2890 km reveals a sharp core-mantle boundary, where density increases dramatically and seismic wave velocities drop.

The Core-Mantle Boundary At the **core-mantle boundary,** about 2890 km below the surface, we encounter the most extreme change in properties found anywhere in Earth's interior. From the way seismic waves reflect from this boundary, seismologists can tell that it is a very sharp interface. Here the material changes abruptly from a solid silicate rock to a liquid iron alloy. Because of the complete loss of rigidity, the S-wave velocity drops from about 7.5 km/s to zero and the P-wave velocity drops from more than 13 km/s to about 8 km/s, causing the core shadow zones. Density, on the other hand, increases by about 4.5 g/cm³ (see Figure 13.19). This large density difference, which is even greater than the increase in density from atmosphere to lithosphere at Earth's solid surface, keeps the core-mantle boundary very flat (you could probably skateboard on it!) and prevents any large-scale mixing of the mantle and core.

The core-mantle boundary appears to be a very active place. Heat conducted out of the core increases the temperatures at the base of the mantle by as much as 1000°C (see Figure 13.21). Indeed, the paths of seismic waves that pass near the base of the mantle show peculiar complications, suggesting a region of exceptional geologic activity. Some geologists believe this hot region to be the source of mantle plumes that rise all the way to Earth's surface, creating volcanic hot spots such as Hawaii and Yellowstone.

The lowest boundary layer of the mantle, a region about 300 km thick, may be the ultimate graveyard of some subducted lithospheric material, such as the dense, iron-rich parts of the oceanic crust. It is possible that this region experiences an upside-down version of the tectonics we see at Earth's surface. For example, accumulations of heavy, iron-rich material might form chemically distinct "anticontinents" that are constantly pushed to and fro across the core-mantle boundary by convection currents. Seismologists are teaming with other geologists who study mantle and core convection to learn more about whatever geologic processes might be active in this strange place.

The Core Many lines of evidence support the hypothesis that Earth's core is made up of iron and nickel. These metals are abundant in the universe (see Chapter 2); in addition, they are dense enough to explain the mass of the core (about one-third of Earth's total mass) and to be consistent with the theory that the core formed by gravitational differentiation (see Chapter 9).

Laboratory measurements at appropriately high pressures and temperatures have led to a slight revision of this hypothesis. A pure iron-nickel alloy turns out to be about 10 percent too dense to match the data for the outer core. Therefore, it has been proposed that the core includes minor amounts of some lighter element. Oxygen and sulfur are leading candidates, although the precise composition remains the subject of research and debate.

Seismology tells us that the core below the mantle is liquid, but the core is not liquid to the very center of Earth. As Lehmann first discovered, P waves that penetrate to depths of 5150 km suddenly speed up, indicating the presence of an *inner core*, a metallic sphere two-thirds the size of the Moon. Seismologists have shown that the inner core transmits shear waves, confirming early speculations that it is solid. In fact, some calculations suggest that the inner core spins at a slightly faster rate than the mantle, acting like a "planet within a planet."

The very center of the planet is not a place you would want to be. The pressures are immense, over 4 million times the atmospheric pressure at Earth's surface. And it's very hot. In Figure 13.20, you can see that the geotherm rises above the melting curve in the outer core but falls below it in the inner core, which explains why the former is liquid and the latter is solid.

Visualizing Earth's Three-Dimensional Structure

So far, we have investigated how the properties of Earth's materials vary with depth. Such a one-dimensional description would suffice if our planet were perfectly symmetrical, but of course it is not. At the surface, we can see *lateral variations* (geographic differences) in Earth's structure associated with oceans and continents and with the basic features of plate tectonics: mid-ocean ridges at spreading centers, deep-sea trenches at subduction zones, and mountain belts lifted up by continent-continent collisions.

Below the crust, we can expect that convection will cause changes in temperature from one part of the mantle to another. Downwelling currents, such as those associated with subducted lithospheric plates, will be relatively cold, whereas upwelling currents, such as those associated with mantle plumes, will be relatively hot. Computer models tell us that lateral variations in temperature due to mantle convection should be on the order of several hundred degrees. From laboratory experiments on rocks, we know that such temperature differences should cause small variations in seismic wave velocities from place to place.

Seismic tomography is an adaptation of a medical technique commonly used to map the human body, called computerized axial tomography (CAT). CAT scanners construct three-dimensional images of organs by measuring small differences in X rays that sweep the body in many directions. Similarly, we can use the velocities of seismic waves from earthquakes, as recorded on thousands of seismographs all over the world, to sweep Earth's interior in many different directions and construct a three-dimensional image of what's inside.

Seismic tomography has revealed features in the mantle that are clearly associated with mantle convection. In the 1990s, researchers at Harvard University constructed the tomographic model of the mantle. Their model is displayed in **Figure 13.22** as a cross section of Earth and as a series of global maps at depths ranging from just below the crust down to the core-mantle boundary. Near Earth's surface (Figure 13.22b), you can clearly see the structure of plate tectonics. The upwelling of hot mantle material along the mid-ocean ridges is visible in warm colors; the cold lithosphere in old ocean basins and beneath the continental cratons is visible in cool colors.

At greater depths, the features become more variable and less coherent with surface tectonic features, reflecting what is probably a complex pattern of mantle convection. Some large-scale features stand out particularly well. You will notice that, just above the core-mantle boundary (Figure 13.22e), there is a red region of relatively low S-wave velocities beneath the central Pacific Ocean, surrounded by a broad blue ring of higher S-wave velocities. Seismologists have speculated that the high velocities represent a "graveyard" of oceanic lithosphere subducted beneath the Pacific's volcanic arcs—the Ring of Fire—during the last 100 million years or so.

> Seismic tomography is a technique that maps small variations in the velocities of seismic waves to create three-dimensional images of Earth's interior.

The cross section through the mantle (Figure 13.22a) clearly reveals material from the once-large Farallon Plate, which has been almost completely subducted under North America (see Chapter 9). The obliquely sinking slab material (in blue) appears to have penetrated the entire mantle. The image also indicates sinking colder rock beneath Indonesia. In addition, a large yellow blob

(a) Tomographic cross section

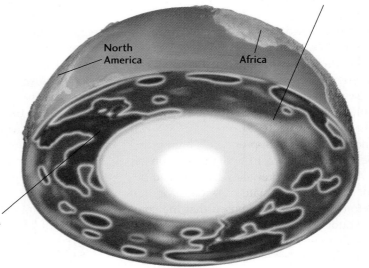

A tomographic cross section through Earth reveals hot regions, such as a superplume rising from Earth's core beneath South Africa,…

North America

Africa

…and cooler regions, such as the descending remnants of the Farallon Plate under the North American Plate.

(b–e) Global maps at four different depths

(b)

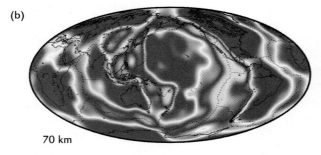

70 km

Near Earth's surface, hot rocks in the asthenosphere lie beneath oceanic spreading centers.

(c)

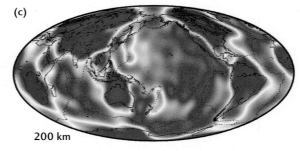

200 km

Moving deeper, we see the cold lithosphere of stable continental cratons and the warmer asthenosphere beneath ocean basins.

(d)

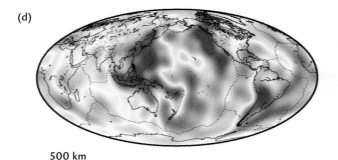

500 km

Deeper in the mantle, the features no longer match the continental positions.

(e)

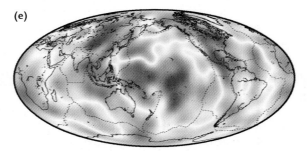

2800 km (near core-mantle boundary)

Near the core-mantle boundary, the colder regions around the Pacific may be the "graveyards" of sinking lithospheric slabs.

FIGURE 13.22 Three-dimensional model of Earth's mantle created by seismic tomography. Regions with faster S-wave velocities (blue and purple) indicate colder, denser rock; regions with slower S-wave velocities (red and yellow) indicate hotter, less dense rock. (a) Cross section of Earth. (b–e) Global maps at four different depths. [*S-wave velocities by G. Ekström and A. Dziewonski, Harvard University; cross section (a) from M. Gurnis, Scientific American (March 2001): 40; maps (b–e) by L. Chen and T. Jordan, University of Southern California.*]

of hotter rock, thought to be a "superplume," can be seen rising at an angle from the core-mantle boundary to a position beneath southern Africa. This hot, buoyant mass pushing up the cooler material above it may explain the uplifted, mile-high plateaus of South Africa. The other visible blobs of hotter and cooler material may be evidence of material exchanges among the lithosphere, the mantle, and the layer of hotter material at the core-mantle boundary.

Seismologists thus use the waves produced by earthquakes to understand the forces in Earth's interior that are responsible for the plate tectonic processes observed at Earth's surface. Of course, it is those processes that generate earthquakes in the first place. By mapping patterns of earthquake faulting, we can learn more about the forces acting within and between the lithospheric plates.

EARTHQUAKES AND PATTERNS OF FAULTING

As we have seen, seismologists are using networks of sensitive seismographs to locate earthquakes around the world, measure their magnitudes, and deduce their fault mechanisms. The **seismicity map** in **Figure 13.23** shows the epicenters of earthquakes recorded around the world over several decades.

The Big Picture: Earthquakes and Plate Tectonics

The most obvious features of the global seismicity map, known to geologists for many decades, are the belts of seismic activity that mark the major plate boundaries (**Figure 13.24**). The fault mechanisms observed for earthquakes in these belts are consistent with the types of faulting along different types of plate boundaries that we discussed in Chapter 7.

Divergent Boundaries The narrow belts of shallow earthquakes that run through ocean basins coincide with mid-ocean ridge crests and their offsets on transform faults. The first P waves recorded from these ridge-crest quakes indicate that they are caused by normal faulting. The faults strike parallel to the ridge and dip toward the mid-ocean rift valley. Normal faulting implies that tensional forces are at work as the plates are pulled apart during seafloor spreading, which explains why rift valleys develop at the

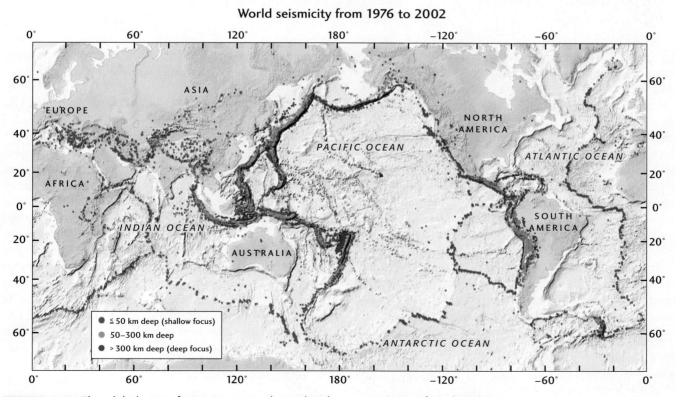

FIGURE 13.23 This global map of seismic activity shows that the concentration of earthquakes is along the boundaries between major lithospheric plates. [*The map is based on data from the Harvard CMT catalog; the plot is by M. Boettcher and T. Jordan.*]

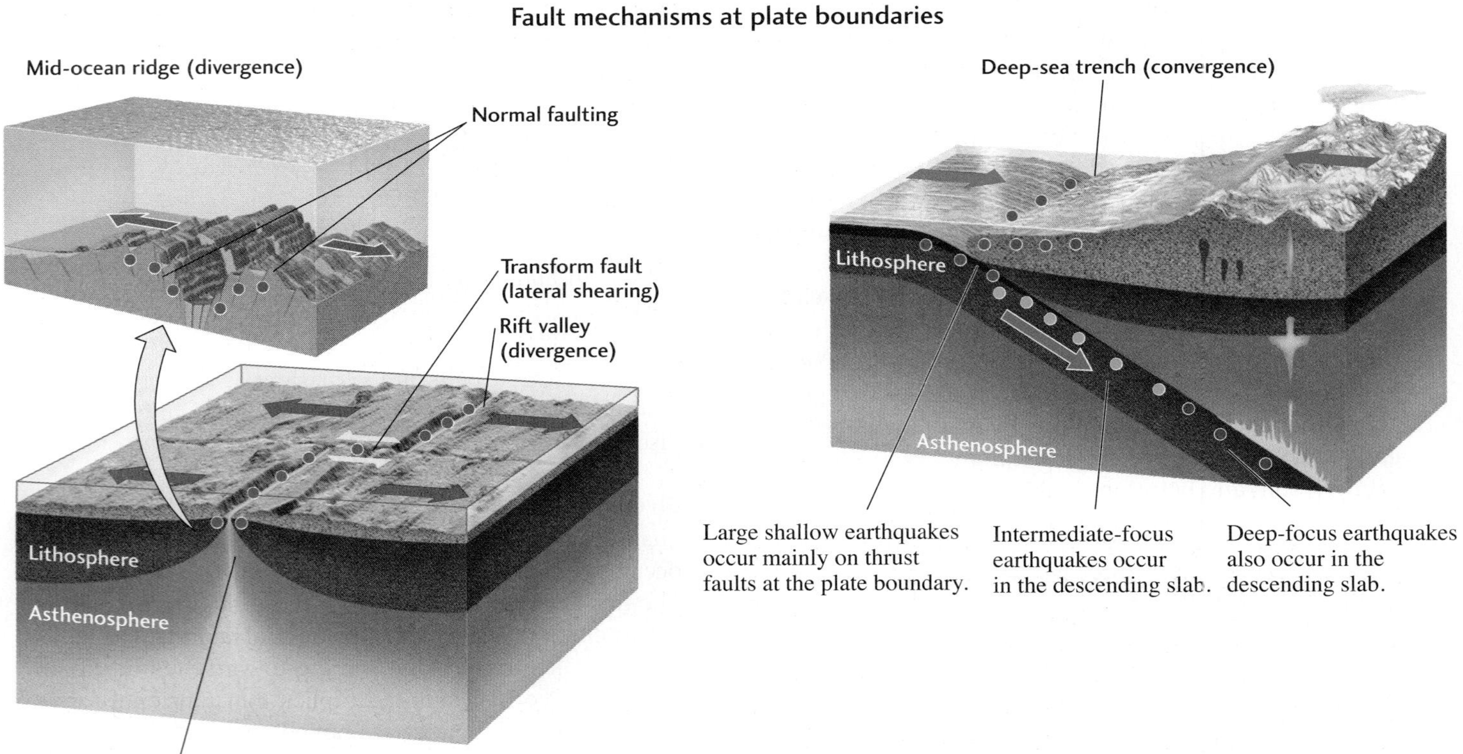

Fault mechanisms at plate boundaries

Mid-ocean ridge (divergence)

Normal faulting

Transform fault
(lateral shearing)

Rift valley
(divergence)

Lithosphere

Asthenosphere

Deep-sea trench (convergence)

Lithosphere

Asthenosphere

Large shallow earthquakes
occur mainly on thrust
faults at the plate boundary.

Intermediate-focus
earthquakes occur
in the descending slab.

Deep-focus earthquakes
also occur in the
descending slab.

Shallow earthquakes coincide with normal
faulting at divergent boundaries and with
strike-slip faulting at transform-fault boundaries.

FIGURE 13.24 Most earthquakes occur at plate boundaries. The fault
mechanisms observed at different types of plate boundaries conform to the
predictions of plate tectonic theory.

ridge crests. Earthquakes also have normal fault mechanisms in zones where continental crust is being pulled apart, such as in the East African rift valleys and in the Basin and Range province of western North America.

Transform-Fault Boundaries Earthquake activity is even greater along the transform-fault boundaries that offset mid-ocean ridge segments. These earthquakes have strike-slip fault mechanisms, just as one would expect where plates slide past each other in opposite directions. Moreover, for earthquakes along these transform faults, the slip direction indicated by the fault mechanisms is left-lateral where the ridge crest steps right and right-lateral where it steps left. These directions are the opposite of what would be needed to create the offsets of the ridge crest but are consistent with the direction of slip predicted by sea-floor spreading. In the mid-1960s, seismologists used this property of transform faults to support the hypothesis of seafloor spreading (see Figure 13.24). Slip directions on transform faults that run through continental crust, such as California's San

Tensional forces at divergent plate boundaries cause normal faulting. Strike-slip faulting occurs along transform faults, including those that offset mid-ocean ridge segments.

Andreas fault and New Zealand's Alpine fault (both right-lateral), also agree with the predictions of plate tectonics.

Convergent Boundaries The world's largest earthquakes, such as the 2011 Tohoku earthquake (magnitude 9.0), described in the story that opens this chapter, and the 2004 Sumatra earthquake (magnitude 9.2), occur at convergent plate boundaries. The fault mechanisms of these great earthquakes show that they were caused by horizontal compression along *megathrusts*, the huge thrust faults that form the boundaries where one plate is subducted beneath another (see Figure 13.24). Both of these recent earthquakes displaced the seafloor and the water above it, generating deadly tsunamis. The tsunami from the Tohoku earthquake killed more than 20,000 people living along the Japanese coast, and it propagated across the Pacific Ocean, causing damage in Hawaii and California. The Sumatra earthquake sent a tsunami across the Indian Ocean that drowned more than 200,000 people living on coastlines from Indonesia and Thailand to Africa.

Earth's deepest earthquakes also occur at convergent boundaries. Almost all earthquakes originating below 100 km rupture the descending plate in a subduction zone. The fault mechanisms of these deep earthquakes

> *Earth's largest earthquakes occur along megathrusts under compressive stress at convergent boundaries. Ruptures of descending plates in subduction zones cause Earth's deepest earthquakes.*

show a variety of orientations, but they are consistent with the deformation expected within the descending plate as gravity pulls it back into the convecting mantle. The deepest earthquakes take place in the oldest—and therefore coldest—descending plates, such as those beneath South America, Japan, and the island arcs of the western Pacific Ocean.

Intraplate Earthquakes Although most earthquakes occur at plate boundaries, a small percentage of global seismic activity originates within plate interiors (see Figure 13.23). The foci of these *intraplate earthquakes* are relatively shallow, and most occur on continents. Among these earthquakes are some of the most famous in American history: a sequence of three large events near New Madrid, Missouri, in 1811–1812; the Charleston, South Carolina, earthquake of 1886; and the Cape Ann earthquake, near Boston, Massachusetts, in 1755. Many intraplate earthquakes occur on old faults that were once parts of ancient plate boundaries. These faults no longer form

plate boundaries but remain zones of crustal weakness that concentrate and release intraplate stresses.

One of the deadliest intraplate earthquakes (magnitude 7.6) occurred near Bhuj in the state of Gujarat, in western India, in 2001. It is estimated that some 20,000 lives were lost. The epicenter was 1000 km south of the boundary between the Indian Plate and the Eurasian Plate. Indian geologists believe that the compressive stresses responsible for the Bhuj earthquake originated in the continuing northward collision of India with Eurasia. Apparently, strong crustal forces can still develop and cause faulting within a lithospheric plate far from modern plate boundaries—in this case, triggering a previously unknown thrust fault at a depth of about 20 km.

Regional Fault Systems

Although the fault mechanisms of most major earthquakes conform to the predictions of plate tectonic theory, a plate boundary can rarely be described as a single fault, particularly when the boundary involves continental crust. Rather, the zone of deformation between two moving plates usually is made up of a network of interacting faults—a *fault system*. The fault system in Southern California provides an interesting example (**Figure 13.25**).

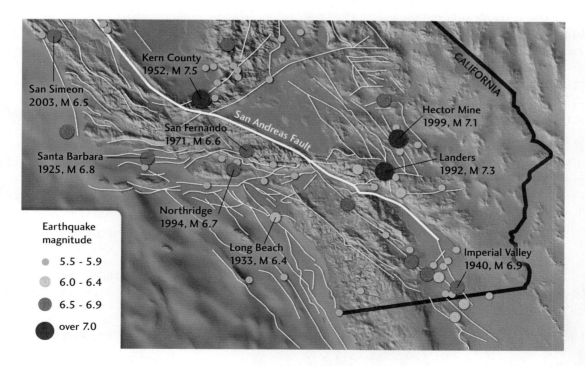

FIGURE 13.25 A map of the fault system of Southern California, showing the surface traces of the San Andreas fault (thick white line) and its subsidiary faults (thin white lines). Colored circles show the epicenters of earthquakes with magnitudes greater than 5.5 during the last 100 years. Significant earthquakes are labeled with their names, years, and magnitudes. [*Southern California Earthquake Center.*]

The "master fault" of this system is our old nemesis, the San Andreas—a right-lateral strike-slip fault that runs northwestward through California from the Salton Sea near the Mexican border until it goes offshore in the northern part of the state (see Figure 13.2). There are a number of subsidiary faults on either side of the San Andreas that generate large earthquakes. In fact, most of the damaging earthquakes in Southern California during the last century have occurred on the subsidiary faults.

Why is the San Andreas fault system so complex? Part of the explanation has to do with the geometry of the San Andreas fault itself. A bend in the fault creates compressive forces that cause thrust faulting in the area north of Los Angeles (as described in Chapter 7). Thrust faulting in the "Big Bend" was responsible for two recent deadly earthquakes, the San Fernando earthquake of 1971 (magnitude 6.6, 65 people killed) and the Northridge earthquake of 1994 (magnitude 6.9, 58 people killed). Over the past several million years, thrust faulting associated with the Big Bend has raised the San Gabriel Mountains to elevations of 1800 to 3000 m.

Another complication is the plate extension taking place east of California in the Basin and Range province, which spans the state of Nevada and much of Utah and Arizona (see Chapters 7 and 9). This broad zone of extensional deformation is connected with the San Andreas fault system through a series of faults that run along the eastern side of the Sierra Nevada and through the Mojave Desert. Faults of this system were responsible for the 1992 Landers earthquake (magnitude 7.3) and the 1999 Hector Mine earthquake (magnitude 7.1), as well as the 1872 Owens Valley earthquake (magnitude 7.6).

NATURAL HAZARDS: Earthquake Destructiveness

In just the last decade, earthquakes worldwide have killed more than 700,000 people and disrupted the economies of entire regions. The United States has been relatively lucky, although two earthquakes on the San Andreas fault — the 1989 Loma Prieta earthquake (magnitude 7.1), which occurred 80 km south of San Francisco, and the 1994 Northridge earthquake in Los Angeles—were among the costliest natural disasters in the nation's history. Damage amounted to more than $10 billion for Loma Prieta and $40 billion for Northridge owing to the proximity of the fault ruptures to urban centers. About 60 people died in each event, but the death toll would have been many times higher if stringent building codes had not been in place (**Figure 13.26**).

Destructive earthquakes are even more frequent in Japan than in California. The recorded history of earthquake disasters, going back 2000 years, has left an indelible impression on the Japanese people. Not surprisingly, Japan is the best prepared nation in the world to deal with earthquakes, with impressive public education campaigns, building codes, and warning systems. Yet, despite this preparedness, more than 5600 people were killed in a devastating earthquake (moment magnitude

FIGURE 13.26 Sixteen people died in the Northridge Meadows apartment building in Los Angeles during the 1994 Northridge earthquake. The victims lived on the first floor and were crushed when the upper levels collapsed. Many more buildings would have collapsed if the newer buildings in the area had not been constructed according to stringent building codes. [*Nick Ut, Files/AP Photo.*]

FIGURE 13.27 This elevated expressway, in Kobe, Japan, was overturned during an earthquake in 1995. [*Tom Wagner/Corbis/SABA.*]

6.8) that struck Kobe on January 16, 1995 (**Figure 13.27**). The large numbers of casualties and structural failures (50,000 buildings destroyed) resulted partly from the less stringent building codes that were in effect before 1980, when much of the city was built, and from the location of the earthquake rupture so close to the city. The tsunami of the 2011 Tohoku earthquake caused an even greater loss of life (more than 20,000). The disaster was compounded by meltdowns and explosions at the Fukushima-Daiichi power plant, one of the world's largest nuclear facilities.

How Earthquakes Cause Damage

Earthquakes proceed as chain reactions in which the primary effects of earthquakes—faulting and ground shaking—trigger secondary effects, including landslides and tsunamis as well as destructive processes within the built environment, such as collapsing structures and fires.

Faulting and Shaking The *primary hazards* of earthquakes are the ruptures in the ground surface that occur when faults break the surface, the permanent subsidence and uplift of the ground surface caused by faulting, and the ground shaking caused by seismic waves radiated during the quake. Seismic waves can shake structures so hard that they collapse. The ground accelerations near the epicenter of a large earthquake can approach and even exceed the acceleration of gravity, so an object lying on the ground surface can literally be thrown into the air. Very few structures built by human hands can survive such severe shaking, and those that do are severely damaged. The collapse of buildings and other structures is the leading cause of casualties and economic damage during earthquakes. Death tolls can be especially high in densely populated areas of developing countries, where buildings are often constructed from brick and mortar without steel reinforcement. A magnitude 7 earthquake on January 12, 2010, destroyed 250,000 residences and 30,000 commercial buildings in Haiti's capital city, Port-au-Prince, killing more than 200,000 people (see Figure 1.29). Improving construction practices so that buildings are able to withstand seismic shaking is the key to avoiding such tragedies.

> *The primary effects of earthquakes, faulting and ground shaking, induce secondary effects, such as landslides and tsunamis, as well as destructive processes within the built environment, such as fires.*

Landslides and Other Types of Ground Failure The primary hazards of faulting and ground shaking generate a number of *secondary hazards.* Among secondary hazards are landslides and other forms of ground failure that give rise to mass movements of Earth materials. When seismic waves shake water-saturated soils, the soils can behave like a liquid and become unstable. The ground simply flows away, taking buildings, bridges, and everything else with it. Such soil *liquefaction* destroyed the residential area of Turnagain Heights near Anchorage, Alaska, in the 1964 earthquake (**Figure 13.28**); the Nimitz Freeway near San Francisco in the 1989 Loma Prieta earthquake; and areas of Kobe in the 1995 earthquake.

In some instances, landslides cause more damage than the ground shaking itself. A 1970 earthquake in Peru triggered an immense avalanche of rock and snow (up to 50 million cubic meters) that destroyed the mountain towns of Yungay and Ranrahirca (see Figure 1.35). Of the more than 66,000 people killed in the earthquake, about 18,000 died in the avalanche.

Tsunamis A large earthquake that occurs beneath the ocean can generate a destructive sea wave, or **tsunami** (Japanese for "harbor wave"). Tsunamis are by far the deadliest and most damaging hazards associated with the world's largest earthquakes: the megathrust

> *The largest tsunamis are caused by displacement of the seafloor during ruptures of subduction megathrusts.*

events that occur in subduction zones. When a megathrust ruptures, it can push the seafloor landward of the deep-sea trench upward by as much as 10 m, displacing a large mass of the overlying ocean water. This disturbance flows outward in waves that travel across the ocean at speeds of up to 800 km/hour (about as fast as a commercial jetliner). In the deep sea, a tsunami is hardly noticeable, but when it approaches shallow coastal waters, the waves slow down and pile up, inundating the shoreline in walls of water that can reach heights of tens of meters (**Figure 13.29**).

The destructive power of a great tsunami was brought home by the terrifying video images captured on March 11, 2011, as the Tohoku tsunami swept over the shoreline of northeastern Japan. In the coastal city of Miyako, the height of the water mass reached an astounding 38 m (123 ft!) above normal sea level, destroying nearly everything in its path (see the opening essay of this chapter). In low-lying regions near the port city of Sendai, the tsunami traveled up to 10 km inland, transporting huge floating debris fields of buildings, boats, cars, and trucks (**Figure 13.30**). The waves propagated across the entire Pacific Ocean, attaining heights of more than 2 m along the coast of Chile, 16,000 km away.

Tsunami warning systems in Japan and the circum-Pacific region worked according to design. The warning times along the Japanese coast nearest the earthquake were too short for complete evacuation (less than an hour), and more than 20,000 people were killed. Nevertheless, the system is credited with saving many lives.

(a)

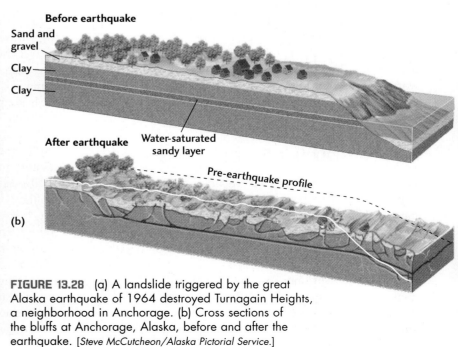

(b)

FIGURE 13.28 (a) A landslide triggered by the great Alaska earthquake of 1964 destroyed Turnagain Heights, a neighborhood in Anchorage. (b) Cross sections of the bluffs at Anchorage, Alaska, before and after the earthquake. [*Steve McCutcheon/Alaska Pictorial Service.*]

TSUNAMI GENERATION

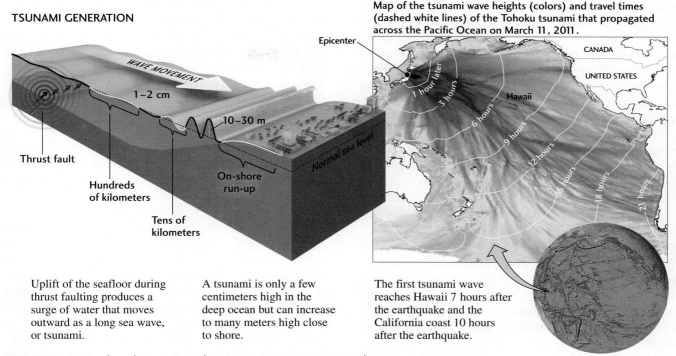

WAVE MOVEMENT

1–2 cm

10–30 m

Normal sea level

Thrust fault

Hundreds of kilometers

Tens of kilometers

On-shore run-up

Epicenter

Map of the tsunami wave heights (colors) and travel times (dashed white lines) of the Tohoku tsunami that propagated across the Pacific Ocean on March 11, 2011.

CANADA

UNITED STATES

Hawaii

1 hour later
3 hours
6 hours
9 hours
12 hours
15 hours
18 hours
21 hours

Uplift of the seafloor during thrust faulting produces a surge of water that moves outward as a long sea wave, or tsunami.

A tsunami is only a few centimeters high in the deep ocean but can increase to many meters high close to shore.

The first tsunami wave reaches Hawaii 7 hours after the earthquake and the California coast 10 hours after the earthquake.

FIGURE 13.29 Earthquakes on megathrusts may generate tsunamis that can propagate across ocean basins. [*NOAA, Pacific Marine Environmental Laboratory.*]

No tsunami warning system was in place in the Indian Ocean when the magnitude 9.2 Sumatra earthquake of December 26, 2004, unleashed an ocean-wide tsunami that swept over low-lying coastal areas from Indonesia and Thailand to Sri Lanka, India, and the east coast of Africa (**Figure 13.31**). Within 15 minutes, the first wave ran up the Sumatran coastline. Few eyewitnesses survived there, but geologic investigations after the tsunami

FIGURE 13.30 Video image taken from a helicopter, showing the tsunami surge carrying debris across farmland near Sendai, Japan, following the Tohoku earthquake of March 11, 2011. [*AP Photo/NHK TV.*]

FIGURE 13.31 The tsunami caused by the 2004 Sumatra earthquake struck without warning on a beach in Phuket, Thailand. [*Courtesy of David Rydevik.*]

indicated that the maximum wave height on the beaches of the west-facing coast was about 15 m, and the run-up attained heights of 25 to 35 m, reaching inland up to 2 km and wiping out most built structures, vegetation, and human life in its path (**Figure 13.32**). It is believed that more than 150,000 people perished along the Sumatran coastline, though no one will ever be sure because many bodies were washed out to sea.

The tsunami moved faster in the deep sea to the west and more slowly in the shallow waters to the east, so that the waves arrived at the heavily populated coastlines of Sri Lanka and Thailand at nearly the same time, some 2 hours after the earthquake. In Sri Lanka and Thailand, more than 40,000 and 15,000 people were killed, respectively. The tsunami struck India at 3 hours (15,000 deaths), the Maldive Islands at 4 hours (108 deaths), and Somalia on the African coast at 9 hours (298 deaths) after the quake. This tragedy motivated an international project to set up a tsunami warning system in the Indian Ocean so that, next time, coastal residents can be forewarned.

Disturbances of the seafloor caused by landslides or volcanic eruptions can also produce tsunamis. The 1883 explosion of Krakatau, a volcano in Indonesia, generated a tsunami that reached 40 m in height and drowned 36,000 people on nearby coasts.

Fires The secondary hazards of earthquakes also include destructive processes that stem from the nature of the built environment itself, such as the fires ignited by ruptured gas lines or downed electrical power lines. Damage to water mains in an earthquake can make fire fighting all but impossible—a circumstance that contributed to the burning of San Francisco after the 1906 earthquake (see Figure 13.3). Most of the 140,000 fatalities in the 1923 Kanto earthquake, one of Japan's greatest disasters, resulted from fires in the cities of Tokyo and Yokohama.

Reducing Earthquake Risk

In Chapter 1, we discussed why it is important to distinguish between hazard and risk when considering how to prepare for natural disasters. Here we illustrate those issues with earthquakes.

Seismic hazard describes the frequency and intensity of earthquake shaking and ground disruption that can be expected over the long term at some specified location. Seismic hazard, which depends on the proximity of the site to active faults that might generate earthquakes, can be expressed in the form of a seismic hazard map. **Figure 13.33** displays the national seismic hazard map produced by the U.S. Geological Survey.

In contrast, **seismic risk** describes the *damage* that can be expected over the long term in a specified region, such as a county or state, usually measured in terms of casualties and dollar losses per year. A region's risk depends not only on its seismic hazard, but also on its exposure to seismic damage (its population and density of buildings and other infrastructure) and its fragility (the vulnerability of its built structures to seismic shaking). Because so many geologic and economic variables must be considered,

FIGURE 13.32 This small headland near Banda Aceh on the west coast of Sumatra was previously covered by dense jungle to the waterline, but it was stripped clean to a height of about 15 m by the 2004 tsunami. [*José Borrero, University of Southern California/Tsunami Research Group.*]

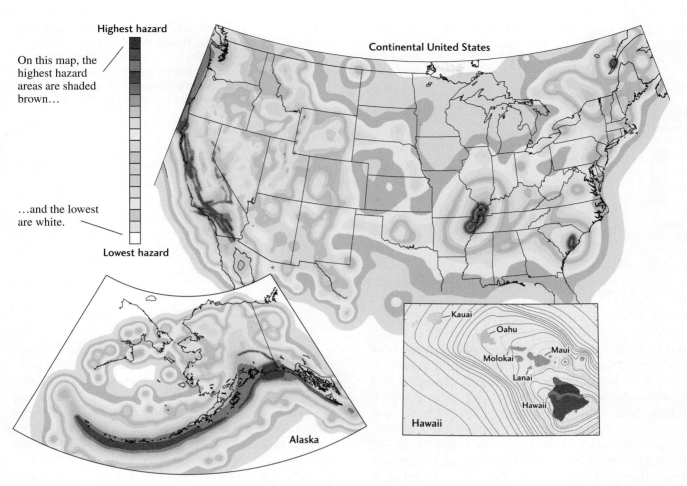

FIGURE 13.33 Seismic hazard map for the United States. The region of highest hazard lies along the San Andreas fault system in California, with a branch extending into eastern California and western Nevada. High hazard levels are also found along the coast of the Pacific Northwest. In the central and eastern United States, the areas of highest hazard are near New Madrid, Missouri, and Charleston, South Carolina; in eastern Tennessee; and in portions of the Northeast. [*U.S. Geological Survey, http://geohazards.cr.usgs.gov/eq/.*]

estimating seismic risk is a complex job. The results of the first comprehensive study of seismic risk in the United States, published by the Federal Emergency Management Agency in 2001, are presented in **Figure 13.34**.

The differences between seismic hazard and seismic risk can be appreciated by comparing the two types of national maps. For instance, although the seismic hazard levels in Alaska and California are both high (see Figure 13.33), California's exposure to seismic damage is much greater, yielding a much larger total risk (see Figure 13.34). California leads the nation in

Seismic hazard describes the intensity of seismic shaking and ground disruption expected at a specified location. Seismic risk describes the damage that can be expected at the location.

seismic risk, with about 75 percent of the national total; in fact, a single county, Los Angeles, accounts for 25 percent. Nonetheless, the problem is truly national: 46 million people in metropolitan areas outside California face substantial earthquake risks. Those areas include Hilo, Honolulu, Anchorage, Seattle, Tacoma, Portland, Salt Lake City, Reno, Las Vegas, Albuquerque, Charleston, Memphis, Atlanta, St. Louis, New York, Boston, and Philadelphia.

Not much can be done about seismic hazard because we have no way to prevent or control earthquakes. However, there are many important steps that society can take to reduce seismic risk if the hazard is properly characterized.

Hazard Characterization The first step is to follow the advice of the old proverb: "Know thy enemy." We still

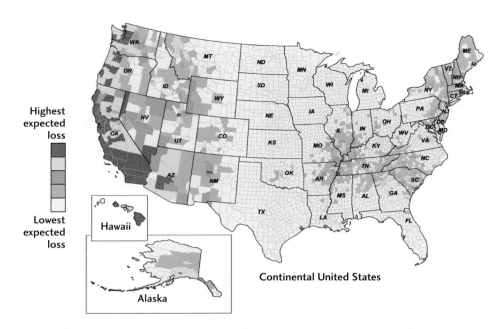

Highest
expected
loss

Lowest
expected
loss

Hawaii

Alaska

Continental United States

FIGURE 13.34 Seismic risk map for the United States. The map shows current annualized earthquake losses (AEL) on a county-by-county basis. [*Federal Emergency Management Agency, Report 366, Washington, D.C., 2001.*]

have much to learn about the sizes and frequencies of ruptures on active faults. For example, it is only in the past couple of decades that we have come to appreciate that an earthquake in the Cascadia subduction zone, which stretches from northern California through Oregon and Washington to British Columbia, could produce a tsunami as large as the one that devastated the Indian Ocean region in 2004 and Japan in 2011. These dangers became apparent when geologists found evidence of a magnitude 9 earthquake that occurred in 1700, before any written historical accounts of the area existed. This monstrous rupture caused major ground subsidence along the Cascadia coastline and left a record of flooded, dead coastal forests.

A tsunami at least 5 m high hit Japan, where historical records pin down its exact date (January 26, 1700). Geologists know that the Juan de Fuca Plate is being subducted under the North American Plate at a rate of about 40 mm/year. They have debated whether this movement occurs seismically or by slow creep, but expert opinion pegs the average time between magnitude 9 earthquakes in the Cascadia subduction zone at 500 to 600 years.

Although we have a good understanding of seismic hazards in some parts of the world—the United States and Japan, in particular—we know much less about others. During the 1990s, the United Nations sponsored an effort to map seismic hazards worldwide as part of the

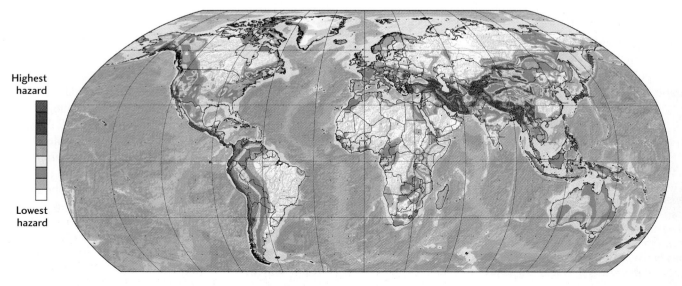

Highest
hazard

Lowest
hazard

FIGURE 13.35 Global seismic hazard map. [*K. M. Shedlock et al., Seismological Research Letters 71(2000): 679–686.*]

International Decade of Natural Disaster Reduction. This effort resulted in the first global seismic hazard map, shown in **Figure 13.35**. The map is based primarily on historical earthquakes, so it may underestimate the hazard in some regions where the historical record is short. Much more needs to be done to characterize seismic hazards on a global scale.

Land-Use Policies The exposure of buildings and other structures to earthquake hazard can be reduced by policies that restrict land uses in high-hazard areas. This approach works well where the hazard is localized, as in the case of faults that are known to be active or areas where a shallow water table makes soil liquefaction during earthquakes very likely. Erecting buildings on active faults, as was done in the case of the residential developments pictured in **Figure 13.36** is clearly unwise, because few buildings can withstand the deformation to which they might be subjected during a quake. In the 1971 San Fernando earthquake, a fault ruptured under a densely populated area of Los Angeles, destroying almost 100 structures. The state of California responded in 1972 with a law that restricts the construction of new buildings across an active fault. If an existing residence is on or very near a fault, real estate agents are required to disclose the information to potential buyers. A notable omission is that the act does not cover publicly owned or industrial facilities.

Siting of nuclear power plants and other critical industrial facilities to avoid seismic and tsunami hazards would seem to be an obvious priority, but the experience in Japan shows how additional considerations, such as the need for water to cool the reactors, can lead to unwise compromises. Two nuclear facilities along the Japanese coastline have been severely damaged by earthquakes in the last few years, the Kashiwazaki-Kariwa facility in 2007 (see Figure 1.9) and the Fukushima-Daiichi facility in 2011 (**Figure 13.37**). The heightened public concern about the earthquake safety of nuclear plants has led the Japanese government to shut down the Hamaoka nuclear plant near Tokyo, increasing the power shortages that the region has experienced since the Tohoku tsunami.

Earthquake Engineering Although land-use policies help to reduce the risk from localized hazards such as ground ruptures, soil liquefaction, and tsunamis, they are less effective where seismic shaking is distributed across large regions. The risks from seismic shaking can best be reduced by good engineering and construction. Standards for the design and construction of new buildings are regulated by building codes enacted by state and local governments. A **building code** specifies the forces a structure must be able to withstand, based on the maximum intensity of shaking expected. In the aftermath of an earthquake, engineers study buildings that were damaged and recommend modifications to building codes that could reduce future damage from similar earthquakes.

U.S. building codes have been largely successful in preventing loss of life during earthquakes. In the 20-year period from 1981 to 2011, for example, 146 people died in 11 severe earthquakes in the western United States, whereas more than 1 million were killed by earthquakes worldwide. Nevertheless, more can be done. Damage from inevitable earthquakes could be reduced by retrofitting older, more vulnerable structures to be seismically safe, as well as by using specialized construction materials and advanced engineering methods in new construction,

FIGURE 13.36 Housing tracts constructed within the San Andreas fault zone, on the San Francisco Peninsula, before California passed legislation restricting this practice. The red line indicates the approximate fault trace, along which the ground ruptured and slipped about 2 m during the earthquake of 1906. [*Michael Rymer.*]

FIGURE 13.37 Areal photo taken by a small, unmanned drone on March 24, 2011, showing the reactor containment buildings of the Fukushima Daiichi nuclear power plant that were damaged by explosions after the Tohoku tsunami crippled the plant. [*AP Photo/Air Photo Service.*]

such as putting entire buildings on movable supports to isolate them from the shaking.

Emergency Response Public authorities must plan ahead and be prepared with emergency supplies, rescue teams, evacuation procedures, fire-fighting plans, and other steps to minimize the consequences of a severe earthquake. Once an earthquake happens, networks of seismographs can transmit signals automatically to central processing facilities. In a fraction of a minute, computers can pinpoint the earthquake focus, measure its magnitude, and determine its fault mechanism. If equipped with strong-motion sensors that accurately record the most violent shaking, these automated systems can also deliver accurate maps in nearly real time showing where the ground shaking was strong enough to cause significant damage. Such information can help emergency managers and other officials deploy equipment and personnel as quickly as possible to save people trapped in rubble and to reduce further property losses from fires and other secondary hazards. Bulletins about the magnitude and area of the shaking can also be channeled through the mass media to reduce public confusion during disasters and allay the fears aroused by minor tremors.

> *The best defense against earthquakes is to design the built environment to withstand seismic shaking and potentially damaging secondary effects.*

Earthquake Early Warning With the technology just described, it is possible to detect earthquakes in the early stages of fault rupture, rapidly predict the intensity of the future ground motions, and warn people before they experience the intense shaking that might be damaging. The worst shaking is usually caused by seismic shear waves and surface waves, which travel at only half the speed of the compressional waves (see Figure 13.10), much slower than an electronic warning message. Earthquake early warning systems detect strong shaking near an earthquake's epicenter and transmit alerts ahead of the destructive seismic waves.

Potential warning times depend primarily on the distance between the user and the earthquake epicenter. There is a "blind zone" near an earthquake epicenter where early warning is not feasible, but at more distant sites, warnings can be issued from a few seconds up to about one minute prior to the strong ground shaking. Such warnings can be used to reduce the harm to people and infrastructure during earthquakes. Potential applications include alerting people to "drop, cover, and hold on," move to safer locations, or otherwise prepare for shaking (e.g., surgeons in operating rooms), as well as many types of automated actions: stopping elevators at the nearest floor, opening firehouse doors, slowing rapid-transit vehicles and high-speed trains to avoid accidents, shutting down pipelines and gas lines to minimize fire hazards, shutting down manufacturing operations to decrease potential damage to equipment, saving vital computer information to avoid losses of data, and controlling structures by active and semiactive systems to reduce building damage.

Earthquake early warning systems have already been deployed in at least five countries—Japan, Mexico,

FIGURE 13.38 A woman walks with her bicycle on a seawall of Taro, Japan, a town that was destroyed by the March 11, 2011, earthquake and tsunami. The tsunami breached this seawall, which was one of the highest in Japan. [*Reuters/Carlos Barria.*]

Romania, Taiwan, and Turkey—and a prototype system is being developed in the United States. Japan is the only country with a nationwide system that provides public alerts. A national seismic network of nearly 1000 seismological stations is used to detect earthquakes and issue warnings, which are transmitted via the Internet, satellite, and wireless networks to cell phone users, desktop computers, and automated control systems that stop trains, place sensitive equipment in a safe mode, and isolate hazards while the public takes cover. The Japanese system has proven to be effective in several recent events, including the great Tohoku earthquake of March 11, 2011.

Tsunami Warning The great tsunamis generated by the Sumatra earthquake of 2004 and the Tohoku earthquake of 2011 illustrate the issues associated with tsunami warning. Because tsunami waves travel 10 times more slowly than seismic waves, there is enough time after a large suboceanic quake occurs, sometimes many hours, to warn people on distant shorelines of an impending disaster. Warnings broadcast by the Pacific Tsunami Warning Center after the Tohoku earthquake allowed islands such as Hawaii and the western coastlines of the Americas to be evacuated prior to the tsunami arrival, minimizing the loss of life. Unfortunately, no such system had been installed in the Indian Ocean, so the tsunami from the great 2004 Sumatra earthquake struck with essentially no warning, killing tens of thousands.

The most difficult situations arise in areas located close to active offshore faults, where tsunamis arrive so quickly that there is little time for warning. The Tohoku tsunami hit the coastline nearest the epicenter less than an hour after the earthquake. Despite the immediate activation of the Japanese tsunami warning system, many lives were lost. Some of the Japanese coastal communities were sheltered behind tall seawalls designed to withstand a large tsunami, but most of these were overwhelmed by the height of the waves, which in some places exceeded 30 m (**Figure 13.38**). Near the shoreline, the best warning system is a very simple one: if you feel a strong earthquake, move quickly away from the coastal lowlands to higher ground!

Seven Steps to Earthquake Safety

As the exclamation at the end of the last paragraph emphasizes, individuals living in seismically active areas need to prepare for earthquakes and know how to respond when one strikes. Here are seven steps to earthquake safety recommended by the Southern California Earthquake Center, which you can use to protect yourself and your family (**Figure 13.39**).

Before an earthquake:

1. *Identify potential hazards in your home and begin to fix them.* As buildings are becoming better designed to withstand seismic shaking, more of the damage and injuries that occur result from falling objects. You should secure items in your home that are heavy enough to cause damage or injury if they fall or valuable enough to be a significant loss if they break.

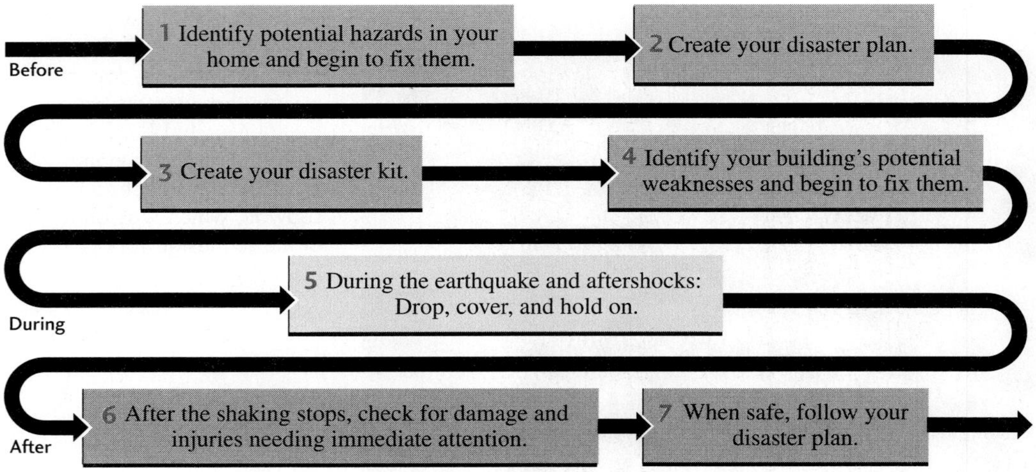

Before

| 1 Identify potential hazards in your home and begin to fix them. | 2 Create your disaster plan. |

| 3 Create your disaster kit. | 4 Identify your building's potential weaknesses and begin to fix them. |

During

5 During the earthquake and aftershocks: Drop, cover, and hold on.

After

| 6 After the shaking stops, check for damage and injuries needing immediate attention. | 7 When safe, follow your disaster plan. |

FIGURE 13.39 Seven steps to earthquake safety. For further information, see *Putting Down Roots in Earthquake Country,* Southern California Earthquake Center, 32 pp., 2004. This pamphlet is available online at http://www.scec.org.

2. *Create your disaster plan.* Plan now what you will do before, during, and after an earthquake. The plan should include safe spots you can go to during the shaking, such as under sturdy desks and tables; a safe spot outside your home where you can go after the shaking stops; and contact phone numbers, including someone outside the area who can be called to relay information in case local communications are disrupted.

3. *Create your disaster kit.* Stock your disaster kit with essential items. Your personal kit should include medications, a first-aid kit, a whistle, sturdy shoes, high-energy snacks, a flashlight with extra batteries, and personal hygiene supplies. Your home kit should include a fire extinguisher, wrenches to turn off gas and water mains, a portable radio, drinking water, food supplies, and extra clothing.

4. *Identify your building's potential weaknesses and begin to fix them.* Consult a building inspector or contractor to identify potential safety problems. Common problems include inadequate foundations, unbraced cripple walls, weak first stories, unreinforced masonry, and vulnerable pipes.

During an earthquake:

5. *Drop, cover, and hold on.* During an earthquake or severe aftershock, drop to the floor, take cover under a sturdy desk or table, and hold on to it so that it doesn't move away from you. Wait there until the shaking stops. Stay away from danger zones, such as those near the exterior walls of buildings, near windows, and under architectural façades.

Individuals living in seismically active areas need to prepare for earthquakes and know how to respond when one strikes.

After an earthquake:

6. *After the shaking stops, check for damage and injuries needing immediate attention.* Take care of your own situation first; get to a safe location and remember your disaster plan. If you are trapped, protect your mouth, nose, and eyes from dust; signal for help using a cell phone or whistle or by knocking loudly on a solid part of the building three times every few minutes (rescuers will be listening for such knocks). Check for injuries and treat people needing assistance. Check for fires, gas leaks, damaged electrical systems, and spills. Stay away from damaged structures.

7. *When safe, follow your disaster plan.* Be in communication by turning on your portable radio and listening for advisories. Call your out-of-area contact to report your status if possible, and then stay off the phone except for emergencies (chances are cell phone systems will be overloaded). Check your food and water supplies and check on your neighbors.

NATURAL HAZARDS: Can Earthquakes Be Predicted?

If we could predict earthquakes reliably, communities could be prepared, people could be evacuated from dangerous locations, and many aspects of the impending disaster might be averted. How well can we predict earthquakes?

Predicting an earthquake means specifying its time, location, and size. By combining plate tectonic theory with detailed geologic mapping of regional fault systems, geologists can reliably predict which faults are likely to produce earthquakes over the long term. However, specifying precisely *when* a particular fault will rupture in a large earthquake has turned out to be very difficult.

Long-Term Forecasting

Ask a seismologist to predict the time of the next large earthquake at a particular location and the response is likely to be "The longer the time since the last big quake, the sooner the next one will be." As we have seen, the **recurrence interval**—the time required to accumulate the strain that will be released by fault slipping in a future earthquake—can be calculated from the rate of relative plate movement and the expected fault slip, as estimated from the displacements observed in past earthquakes.

> *The earthquake recurrence interval on a fault can be estimated by dividing the expected fault slip by the fault slip rate.*

Geologists can also estimate the intervals between large earthquakes up to several thousand years in the past by finding and dating soil layers that were offset by fault displacements (**Figure 13.40**). Although these two methods usually give similar results, the uncertainty of the predictions turns out to be large—as much as 50 percent of the average recurrence interval. In Southern California, for example, the recurrence interval for

the San Andreas fault is estimated to be 110 to 180 years, but the observed intervals between individual earthquakes can be appreciably shorter or longer than this average value. One part of this fault experienced a great earthquake in 1857, whereas another part (the southernmost) appears to have remained locked since a large earthquake that occurred around 1680 (see Figure 13.2). Therefore, an earthquake can be expected at any time—tomorrow, or decades from now.

Because of the large uncertainties (decades to centuries) of this method of earthquake prediction, it is called *long-term forecasting* to distinguish it from what most people would really want—a *short-term prediction* of a large rupture on a specific fault accurate to within days or even hours of the actual event.

Short-Term Prediction

There have been a few successful short-term earthquake predictions. In 1975, an earthquake was predicted only hours before it occurred near Haicheng, in northeastern China. Chinese seismologists used what they considered to be *precursors* to make their predictions: swarms of tiny earthquakes and a rapid deformation of the ground several hours before the mainshock. Almost a million people, prepared in advance by a public education campaign, evacuated their homes and factories in the hours before the quake. Although many towns and villages were destroyed and a few hundred people were killed, many lives were saved. The very next year, however, an unpredicted earthquake struck the Chinese city of Tangshan, killing more than 240,000. Obvious precursors such as those seen in Haicheng have not occurred in subsequent large events.

Although many schemes have been proposed, we have not yet found a reliable method for consistently predicting large earthquakes from minutes to weeks ahead of time. Although we cannot say that short-term earthquake prediction is impossible, seismologists are pessimistic about the feasibility of short-term prediction in the near future.

Some useful statements can be made about how the earthquake probabilities change over short periods of time, however. We know that earthquakes have aftershocks, so the chances of having a potentially damaging earthquake tend to go up during periods of increased seismic activity. Short-term forecasts based on how earthquakes cluster in time are currently being used to help Californians assess seismic risks.

Medium-Term Forecasting

Uncertainties in earthquake forecasting can be reduced by studying the behavior of regional fault systems. One

FIGURE 13.40 Geologist Gordon Seitz examines layers of rock and peat that have been distorted by prehistoric earthquakes in a trench crossing the San Jacinto fault, a major strand of the San Andreas fault system in Southern California. By dating the peat using the carbon-14 method, geologists can reconstruct the history of large earthquakes on this fault. Such information helps scientists to forecast future events. [*Courtesy of Tom Rockwell, San Diego State University.*]

strategy is to generalize the elastic rebound theory. The simple version of the theory depicted in Figure 13.1 describes how the tectonic stress that builds steadily on an isolated fault segment is released in a periodic sequence of fault ruptures. However, as we have seen in the case of Southern California (see Figure 13.25), faults are rarely isolated. Instead, they are connected together in complex networks. Thus, a rupture on one fault segment changes the stresses throughout the surrounding region. Depending on the geometry of the fault system, these changes can either increase or decrease the likelihood of earthquakes on nearby fault segments. In other words, when and where earthquakes happen in one part of a fault system influences when and where they happen elsewhere in the system.

If Earth scientists can understand how variations in stress raise or lower earthquake probabilities, they might be able to forecast potentially damaging earthquakes over intervals as short as a few years or maybe even a few months, although still with substantial uncertainties. Monitoring of such events on networks of seismographs could then provide a regional "stress gauge." Someday you might

> *Earthquake forecasting can be improved through a better understanding of how variations in stress change the frequency of seismic events in a regional fault system.*

hear a news report that says, "The National Earthquake Prediction Evaluation Council estimates that, during the next year, there is a 50 percent probability of a magnitude 7 or larger earthquake on the southern segment of the San Andreas fault."

The ability to issue *medium-term forecasts* would raise some difficult questions, however. How should society respond to a threat that is neither imminent nor long term? A medium-term forecast would give the probability of an earthquake only on time scales of months to years—not precise enough to evacuate areas that might be damaged. What effect would such predictions have on property values and other investments in the threatened region? False alarms would be common. How should communities deal with this uncertainty? These are questions more suited to be addressed by policy makers than by scientists.

■ SUMMARY

What is an earthquake? An earthquake is a shaking of the ground that occurs when brittle rocks being stressed by tectonic forces break suddenly. When they break, the elastic energy built up over years of slow deformation is released rapidly, and some of it is radiated as seismic waves.

Where do earthquakes occur? Most earthquakes originate at plate boundaries, where stresses are concentrated and straining of the crust is intense. The foci of most continental earthquakes are above 20 km; below that depth, the crust behaves as a ductile material. In subduction zones, however, where cold oceanic lithosphere plunges back into the mantle, earthquakes can occur at depths as great as 690 km.

What are the three types of seismic waves? Two types of waves travel through Earth's interior: P (primary) waves, which are transmitted by all forms of matter and move fastest, and S (secondary) waves, which are transmitted only by solids and move at a little more than half the velocity of P waves. P waves are compressional waves that travel as a succession of compressions and expansions. S waves are shear waves that displace material at right angles to their path of travel. Surface waves are confined to Earth's surface and outer layers. They travel more slowly than S waves.

What is earthquake magnitude, and how is it measured? Earthquake magnitude is a measure of the size of an earthquake. Richter magnitude is proportional to the logarithm of the amplitude of the largest ground movement recorded by seismographs. Seismologists prefer to use the moment magnitude because it is derived from the physical properties of the faulting that causes the earthquake: the area of faulting and the average fault slip.

How frequently do earthquakes occur? About 1,000,000 earthquakes with magnitudes greater than 2 take place each year. This number decreases by a factor of 10 for each magnitude unit. Hence, there are about 100,000 earthquakes with magnitudes greater than 3, about 1000 with magnitudes greater than 5, and about 10 with magnitudes greater than 7. The largest earthquakes, with magnitudes of 9 to 9.5, are rare and are confined to thrust faults in subduction zones.

What do seismic waves reveal about the layering of Earth's crust and mantle? Correlations of seismic wave velocities with rock types have revealed that the continental crust is made up mostly of low-density granitic rocks and that the deep seafloor is composed of basalt and gabbro. The crust and outer part of the mantle make up the rigid lithosphere. Beneath the lithosphere lies the asthenosphere, the weak, ductile layer of the mantle on which the lithospheric plates ride. At the top of the asthenosphere, the temperature is high enough to partially melt peridotite, forming an S-wave low-velocity zone. Below 200 to 250 km, S-wave velocities again increase with depth. At 410 km and 660 km below the surface, S-wave velocities show jumps caused by phase changes in mantle minerals. Below 660 km lies the lower mantle, a layer more than 2000 km thick in which seismic wave velocities increase steadily.

What do seismic waves tell us about the layering of Earth's core? Seismic waves reflected from the core-mantle boundary locate this sharp boundary at a depth of 2890 km. The failure of S waves to penetrate below the core-mantle boundary indicates that the outer core is fluid. A jump in P-wave velocity marks the boundary between the liquid outer core and the solid inner core at a depth of 5150 km. Several lines of evidence show that the core is composed mostly of iron and nickel, with minor amounts of some lighter element, such as oxygen or sulfur.

What has seismic tomography revealed about structures in the mantle? Seismologists can use seismic tomography to create three-dimensional images of Earth's interior. Regions where seismic wave velocities increase indicate relatively cool, dense rock; regions where they decrease indicate relatively hot, less dense rock. Tomographic images reveal the structures of plate tectonics close to Earth's surface, from the upwelling of hot mantle material under mid-ocean ridges to the cold lithosphere that extends deep beneath continental cratons. They also reveal many features of mantle convection, such as the sinking of lithospheric slabs into the lower mantle and the rising of plumes from deep within the mantle.

What governs the type of faulting that occurs in an earthquake? The fault mechanism of an earthquake is determined by the type of plate boundary at which it occurs. Normal faulting, caused by tensional forces, occurs at divergent boundaries. Strike-slip faulting, caused by shearing forces, occurs along transform-fault boundaries. The largest earthquakes, caused by compressive forces, occur on megathrusts at convergent boundaries. A small number of earthquakes occur far from plate boundaries, mostly within the continents.

What causes the hazards of earthquakes? Faulting and ground shaking during an earthquake can damage or destroy buildings and other infrastructure. They can also trigger secondary hazards, such as landslides and fires. Earthquakes on the seafloor can trigger tsunamis, which may cause widespread destruction when they reach shallow coastal waters.

What can be done to reduce the risks of earthquakes? Land-use regulation can restrict new building near active fault zones. Construction in high-hazard areas can be regulated by building codes so that structures will be strong enough to withstand the expected intensity of seismic shaking. Systems using networks of seismographs and other sensors are being developed to provide early warnings of earthquakes and tsunamis. Public authorities can plan ahead, be prepared, and put early warning systems in place. People living in earthquake-prone areas can be informed about how to prepare and what to do when an earthquake occurs.

Can scientists predict earthquakes? Scientists can characterize the level of seismic hazard in a region, but they cannot consistently predict earthquakes with the accuracy that would be needed to alert a population hours to weeks in advance. The best hope of making such predictions in the future may lie in a better understanding of how variations in stress raise or lower the frequency of seismic events in a regional fault system.

■ KEY TERMS AND CONCEPTS

aftershock (p. 431)
building code (p. 457)
core-mantle boundary
  (p. 444)
earthquake (p. 426)
elastic rebound theory
  (p. 428)
epicenter (p. 429)
fault mechanism (p. 439)
fault slip (p. 429)
focus (p. 429)
foreshock (p. 432)
geotherm (p. 443)
intensity scale (p. 437)
lower mantle (p. 444)

low-velocity zone (p. 443)
magnitude scale (p. 436)
P wave (p. 433)
phase change (p. 444)
recurrence interval (p. 461)
S wave (p. 433)
seismic hazard (p. 454)
seismic risk (p. 454)
seismic tomography
  (p. 445)
seismicity map (p. 447)
seismograph (p. 432)
shadow zone (p. 441)
surface wave (p. 433)
tsunami (p. 452)

■ EXERCISES

1. Seismographic stations report the following S-wave–P-wave time differences for an earthquake: Dallas, S-P = 3 minutes; Los Angeles, S-P = 2 minutes; San Francisco, S-P = 2 minutes. Use a map of the United States (or better yet, a globe) and the travel time curves in Figure 13.10 to obtain a rough location for the epicenter of the earthquake.

2. Describe two scales for measuring the size of an earthquake. Which is the more appropriate scale for measuring the amount of faulting that caused the earthquake? Which is more appropriate for measuring the amount of shaking experienced by a particular observer?

3. How much more energy is released by a magnitude 7.5 earthquake than by a magnitude 6.5 earthquake?

4. In Southern California, a magnitude 5 earthquake occurs about once per year. Approximately how many magnitude 4 earthquakes would you expect each year? How many magnitude 2 earthquakes?

5. What evidence suggests that the asthenosphere is partially molten?

6. What evidence indicates that Earth's outer core is molten and composed mostly of iron and nickel?

7. How can features of mantle convection, such as rising and descending convection currents, be imaged by seismic tomography?

8. How does the distribution of earthquake foci correlate with plate boundaries?

9. What are the fault mechanisms of earthquakes at the three types of plate boundaries?

10. In Figure 13.1, the right-lateral fault offsets the fence line to the right. In Figure 13.24, the mid-ocean ridge crest is also offset to the right. Why, then, is the transform fault in Figure 13.24 left-lateral?

11. The belts of shallow-focus earthquakes at divergent plate boundaries, shown by the blue dots in Figure 13.23, are wider and more diffuse in the continents than in the oceans. Why? (In answering this question, you might want to review Chapter 7.)

12. Why are earthquakes with focal depths greater than 20 km infrequent in continental lithosphere?

13. Why do the largest earthquakes occur on megathrusts at subduction zones and not, say, on continental strike-slip faults?

14. Destructive earthquakes occasionally occur within plates, far from plate boundaries. Why?

15. At a location along the boundary fault between the Nazca Plate and the South American Plate, the relative plate movement is 80 mm/year. The last great earthquake, in 1880, showed a fault slip of 12 m. When should local residents begin to worry about another great earthquake?

16. Taking into account the possibility of false alarms, mass hysteria, economic depression, and other possible negative consequences of earthquake prediction, do you think the objective of predicting earthquakes should have a high priority?

Visual Literacy Task

1. In what layer of Earth do earthquakes occur?

a. Lithosphere
b. Asthenosphere
c. Both the lithosphere and asthenosphere

2. Earthquakes on the subducting ocean plate are measured at what range of depths?

a. From the surface down to 50 km below the surface of Earth
b. From the surface down to more than 300 km below the surface of Earth
c. Between 50 km and 300 km below the surface of Earth

3. Can the size of earthquakes be determined based on the information in the diagram?

a. Yes, the color of the dots show size.
b. Yes, the depth of the dots show size.
c. No, the size is not shown.

4. Based on the depth of earthquakes, what type of plate boundary exists in Area A?

a. Convergent
b. Divergent
c. Not enough information on the diagram to determine the type of plate boundary

5. What is happening in Area B?

a. The eastern ocean plate is subducting down under the western ocean plate.
b. The western ocean plate is subducting down under the eastern ocean plate.
c. The two ocean plates are pulling apart.
d. The two ocean plates are sliding past each other.

Thought Question: How would you respond to a fellow student who asserted, "The depth of earthquakes tells you nothing about the plate boundary."

World seismicity from 1976 to 2002

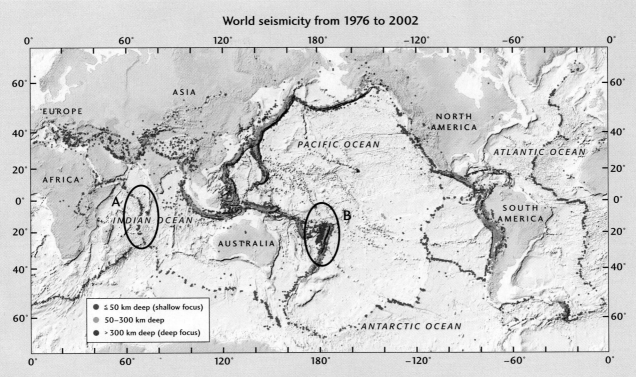

FIGURE 13.23 This global map of seismic activity shows that the concentration of earthquakes is along the boundaries between major lithospheric plates. [*The map is based on data from the Harvard CMT catalog; the plot is by M. Boettcher and T. Jordan.*]

Fault mechanisms at plate boundaries

Mid-ocean ridge (divergence)

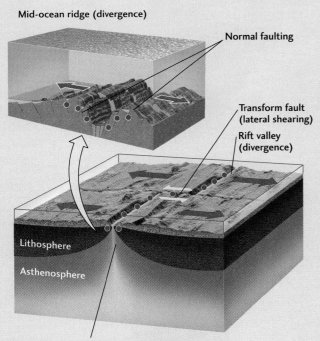

Normal faulting

Transform fault (lateral shearing)

Rift valley (divergence)

Lithosphere

Asthenosphere

Shallow earthquakes coincide with normal faulting at divergent boundaries and with strike-slip faulting at transform-fault boundaries.

Deep-sea trench (convergence)

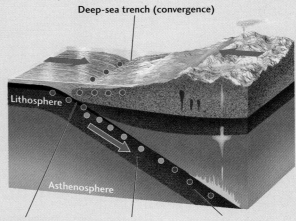

Lithosphere

Asthenosphere

Large shallow earthquakes occur mainly on thrust faults at the plate boundary.

Intermediate-focus earthquakes occur in the descending slab.

Deep-focus earthquakes also occur in the descending slab.

FIGURE 13.24 Most earthquakes occur at plate boundaries. The fault mechanisms observed at different types of plate boundaries conform to the predictions of plate tectonic theory.

Human Impact on Earth's Environment

14

The Carbon Crisis

ON A HOT AFTERNOON IN LATE-SUMMER LOS ANGELES, the carbon crisis is easy to observe. The freeways are crowded with returning vacationers. Smog gives the atmosphere a gray-brown tinge that turns to burnt gold in the sunset. The California power grid is stretched thin by increasing energy demands. Rains fell earlier in the year, but the hills are now tinderboxes, and disputes over the state's dwindling water supplies are growing more contentious. The Great Recession has ended, and the price of oil on the spot market again exceeds $100 per barrel. The price of gasoline has jumped by more than 30 percent in the last year. The radio reports that the volume of Arctic sea ice has reached an all-time low and may melt entirely in the late summers 50 years from now. The magnificent glaciers of Kilimanjaro, the Andes, Switzerland, and Tibet are not expected to last even that long. Ecosystems are becoming more isolated, and some are collapsing, banishing their species to shrinking terrains, where many will face extinction.

> The real voyage of discovery consists not in seeking new lands, but in seeing with new eyes.
>
> —Marcel Proust

These problems all stem from interactions between human civilization and its environment, and all of them are connected by the cycling of carbon through the Earth system. The crux of the problem can be stated simply enough: our economy depends on the burning of fossil fuels, a nonrenewable resource, which produces carbon dioxide, a powerful greenhouse gas.

How shall we confront the carbon crisis? From the perspective of Earth science, an old environmental adage—"Think globally, act locally"—makes sense. The carbon crisis must be understood and dealt with on a global scale. A plausible goal for the next half-century might be to stabilize our total carbon emissions at current rates, rather than allowing them to grow. As we will see, even this stopgap measure would require the implementation of new technologies on a Herculean scale. Our leaders are not likely to stomach such actions without a strong consensus among the electorate to move beyond "business as usual." To sustain the environment as we know it, cooperation will have to trump competition. Any consensus will have to be forged within our local communities by people who understand the crisis.

At the outset of our journey together, we offered to give you new eyes to view our planet, its riches, and its problems. Now we urge you to keep your eyes wide open as you chart the future course of Spaceship Earth. ◆

—T.H.J. and J.P.G., Los Angeles, September 2011

This single stack is part of the Mosenergo Thermal Power Plant, a natural gas–fired power plant, in Moscow, Russia. The plant supplies heat to more than 1.6 million people in and around Moscow. [iStockphoto/Thinkstock.]

CHAPTER OUTLINE

THE HUMAN HABITAT IS A THIN INTERFACE where Earth meets sky, where the global geosystems—the climate system, the plate tectonic system, and the geodynamo—interact to provide a life-sustaining environment. We have increased our standard of living by discovering clever ways to exploit this environment: to grow food, extract minerals, build structures, transport materials, and manufacture goods of all kinds. One result has been an explosion in the human population.

In the early Holocene, about 10,000 years ago, when the climate had warmed from the last ice age and agriculture began to flourish, roughly 100 million people were living on the planet (**Figure 14.1**). The world population grew slowly, taking about 5000 years to double. The first doubling, to 200 million, was achieved early in the Bronze Age, when humans learned how to mine ores and refine them into metals such as copper and tin (of which bronze is an alloy). The second doubling, to 400 million, was not achieved until the late Middle Ages. But once industrialization began, the global population took off, climbing to 1 billion in 1804, 2 billion in 1927, and 4 billion in 1974. By the mid-twentieth century, the doubling time for the human population had dropped to only 47 years—less than a human lifetime. The world population exceeded 7 billion in early 2012, and our numbers are expected to reach 8 billion by 2025.

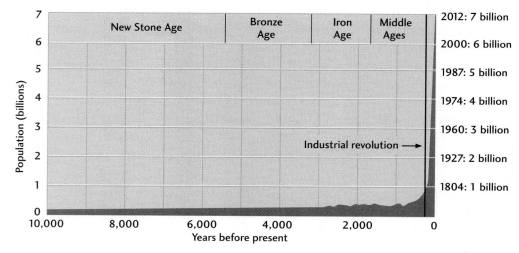

FIGURE 14.1 Human population growth over the last 10,000 years. The global population is expected to reach 8 billion by 2025.

As our population has exploded, our appetites for energy and other natural resources have become voracious. The demand for natural resources is skyrocketing as civilization expands and people around the world strive to improve the quality of their lives. Our energy usage, for example, has increased by 1000 percent over the last 70 years and is now going up twice as fast as the human population. The view of Earth from space shows a glowing lattice of highly energized urbanization spreading rapidly across the planet's surface (see Figure 1.2). Because the natural resources of Spaceship Earth are necessarily limited, the progress of civilization cannot be taken for granted. Environmental conditions and overall prosperity are not improving in some parts of the world, and the prospects for detrimental changes to the global environment loom large. Balancing our use of resources against detrimental effects on our environment and thus on the ability of our environment to sustain a prosperous society raises new challenges for Earth scientists and for all of human society.

In this final chapter, we will survey the energy resources that power our civilization and examine how energy production and other human activities are affecting the global environment. We will focus on two pressing questions: What are our future energy choices? How might future energy production cause unwanted global change? In particular, we will explore how these two issues are closely connected through the flow of carbon from one component of the Earth system to another.

FIGURE 14.2 Coal-fired power plants such as this one in San Juan County, New Mexico, are converting fossil carbon previously buried in Earth's crust into atmospheric carbon dioxide. [*Larry Lee Photography/Corbis.*]

THE CARBON ECONOMY

Fossil fuels—coal, oil, and natural gas—are mixtures of combustible compounds rich in hydrogen and carbon. We mine these *hydrocarbons* from reservoirs in the lithosphere, where they have accumulated over long periods of geologic time, and we burn them in massive quantities to produce energy for our homes, industries, and transportation (**Figure 14.2**).

Figure 14.3 depicts how energy was produced and used in the United States in 2007. This industrialized nation, with 5 percent of the world's population, consumed about four and a half times more energy per person than the global average. Production of energy from all sources totaled 101.5 quadrillion quads. Nonrenewable fossil fuels—crude oil, natural gas, and coal—provided 85 percent of that total, and renewable biomass accounted for another 3.6 percent. You will notice that the flow of energy through the system was not particularly efficient: about 39 percent of the energy performed useful work,

while 60 percent was wasted. The United States is not alone its dependence on fossil fuels. Modern civilization runs primarily on fossil fuels—oil, natural gas, and coal—and we can call this energy system a **carbon economy.**

You can also see that the system released about 1.8 gigatons (Gt) of carbon into the atmosphere, primarily as carbon dioxide (1 Gt = 1 billion tons = 10^{12} kg). Worldwide the total amount of carbon jetted by humans into the atmosphere was about 7 Gt. That's a lot of carbon!

Carbon plays a major role in the Earth system. Carbon compounds are continually moving between the atmosphere, biosphere, hydrosphere, cryosphere, and lithosphere in a natural *carbon cycle*. In the prehuman world, the exchange of carbon between the lithosphere and the other components of the Earth system was regulated by the slow rates at which geologic processes buried and unearthed organic matter.

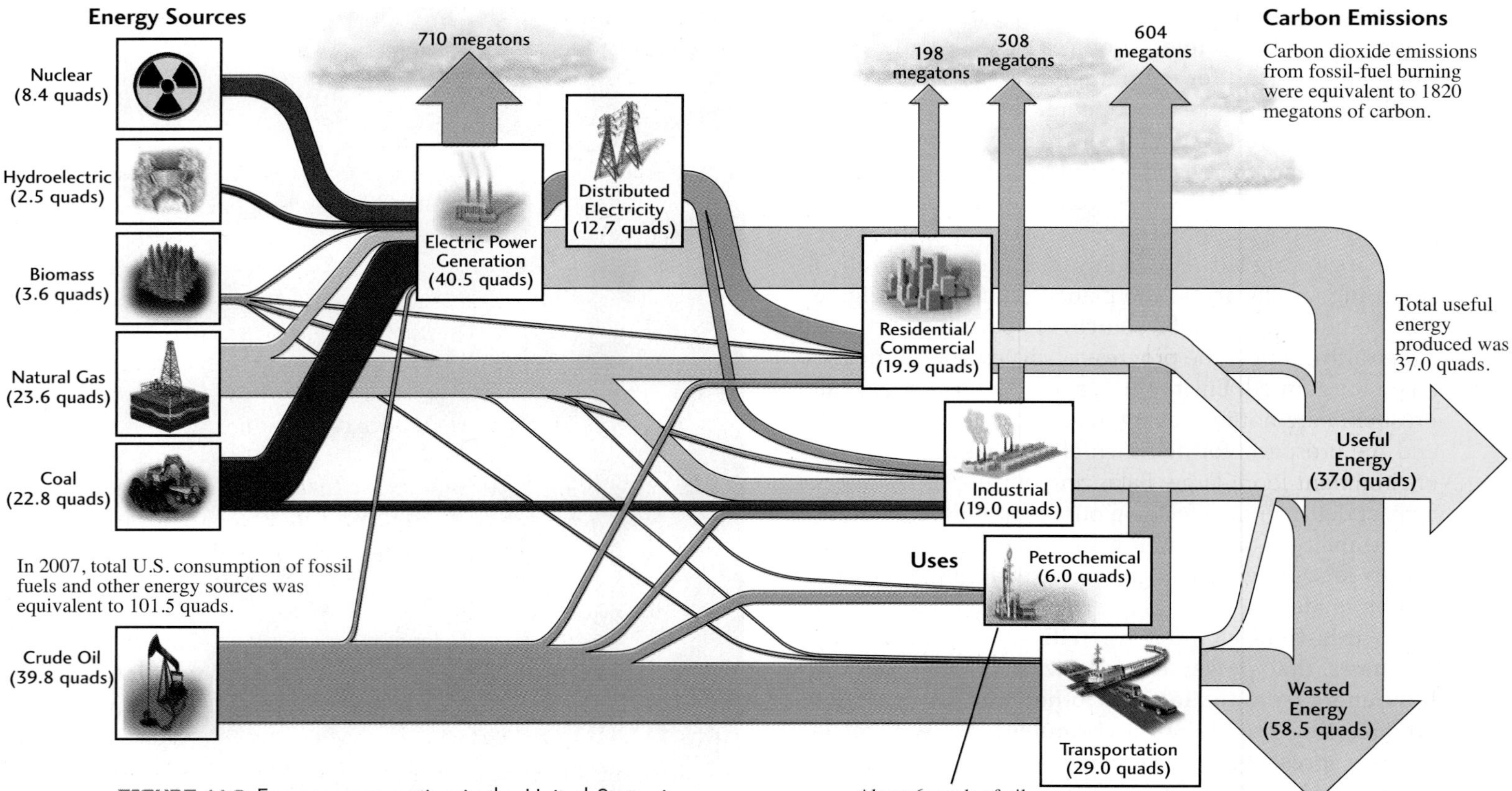

FIGURE 14.3 Energy consumption in the United States in 2007 (in quads). Energy from primary fuel sources (boxes on left side) is delivered to the residential, commercial, industrial, and transportation sectors (boxes in middle to right side). Not represented are small contributions to electric power generation from solar (0.1 quad), wind (0.3 quad), and geothermal (0.3 quad) energy. [*Adapted from a special study by Lawrence Livermore National Laboratory based on data from the Energy Information Administration.*]

Our dependence on carbon-based fuels has disrupted the natural carbon cycle by pumping huge amounts of carbon from the lithosphere directly into the atmosphere.

The natural carbon cycle has been disrupted by the rise of the carbon economy, which is now pumping huge amounts of carbon from the lithosphere directly into the atmosphere.

The massive inflow of carbon has increased atmospheric concentrations of carbon dioxide from preindustrial levels of about 280 parts per million (ppm) to more than 390 ppm today (see Figure 10.13). If the global economy continues to grow and the burning of fossil fuels continues unabated, the amount of carbon dioxide in the atmosphere will double from its preindustrial levels by mid-century. Because CO_2 is a greenhouse gas, the increase is likely to lead to enhancement of the greenhouse effect and global climate warming.

Our current reliance on the carbon economy poses some difficult questions: How long will our fossil-fuel resources last? To what extent will the increase in atmospheric carbon dioxide concentrations caused by fossil-fuel combustion adversely affect the global climate? How quickly will we need to replace fossil fuels with alternative energy sources? How can we use our resources to achieve sustainable development—to meet the needs of the present without compromising the ability of future generations to meet their own needs? As we rewrite the operating manual for Spaceship Earth to address these problems, we will be constrained by the availability of fossil fuels and alternative sources of energy, which we will now consider in more detail.

NATURAL RESOURCES: Fossil Fuels

In Chapter 1, we provided an overview of renewable and nonrenewable natural resources. We distinguished *reserves*—the confirmed supplies of materials that are exploitable under current economic conditions—from *resources*—the total amounts of those materials, which are not precisely known but "guesstimated" based on past experience or theoretical models.

Figure 14.4 gives a rough estimate of the world's remaining nonrenewable energy resources of all types, which total about 360,000 quads. The largest resource is uranium (240,000 quads), which can be used to produce nuclear energy, followed by coal (67,500 quads).

Calculations based on these numbers can be deceptive, however. For example, simply dividing total resources by current annual consumption, which is about 500 quads (see Figure 1.10), might lead one to conclude (mistakenly) that many hundreds of years' worth of resources remain before we have to worry about depletion of energy supplies. The economics of this issue are much more complicated, however, because some energy sources will give out before others, the various sources of energy are not readily interchangeable, and the environmental costs of converting some of them into useful forms of energy may be too great.

We can better understand what energy resources will be available in the future by considering the geologic

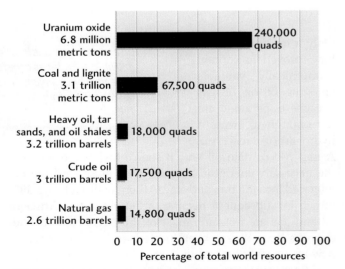

FIGURE 14.4 A rough estimate of total remaining nonrenewable world energy resources amounts to about 360,000 quads. Amounts are given in conventional units of weight (metric tons) or volume (barrels) and by energy content (quads). [*World Energy Council.*]

circumstances in which these resources are found and the problems related to their recovery and use.

How Do Oil and Gas Form?

Economically valuable deposits of our most important energy resources, crude oil and natural gas, develop under special environmental and geologic conditions. Both come from the organic debris of former life: plants and microorganisms (such as bacteria and algae) that have been buried, transformed, and preserved in sediments.

Oil and gas begin to form in sedimentary basins where the production of organic matter is high and the supply of oxygen in the sediments is inadequate to decompose all the organic matter they contain. Many offshore thermal subsidence basins on continental margins satisfy both these conditions. In such environments, and to a lesser degree in some river deltas and inland seas, the rate of sedimentation is high, and organic matter is buried and protected from decomposition.

During millions of years of burial, chemical reactions triggered by the elevated temperatures and pressures found deep in the sediments slowly transform some of the organic material in these *source beds* into combustible hydrocarbons. The simplest hydrocarbon is methane gas (CH_4), the compound we call **natural gas.** Raw petroleum, or **crude oil,** includes a diverse class of liquids composed of more complex hydrocarbons.

Crude oil forms at a limited range of pressures and temperatures, known as the **oil window,** usually located at depths between about 2 and 5 km. Above the oil window, temperatures are too low (generally below 50°C) for the maturation of organic material into hydrocarbons. Below the oil window, temperatures are so high (greater than 150°C) that the hydrocarbons that form are broken down into methane, producing only natural gas.

As burial progresses, compaction of the source beds forces crude oil and natural gas into adjacent beds of permeable rock (such as sandstones or porous limestones), which act as *hydrocarbon reservoirs*. The relatively low densities of oil and gas cause them to rise so that they float atop the water that almost always occupies the pores of permeable rock formations.

> Crude oil and natural gas form in sediments rich in organic material and poor in oxygen. High temperatures and pressures transform the organic material into combustible hydrocarbons.

Where Do We Find Oil and Gas?

The conditions that favor large-scale accumulation of oil and natural gas are combinations of geologic structures

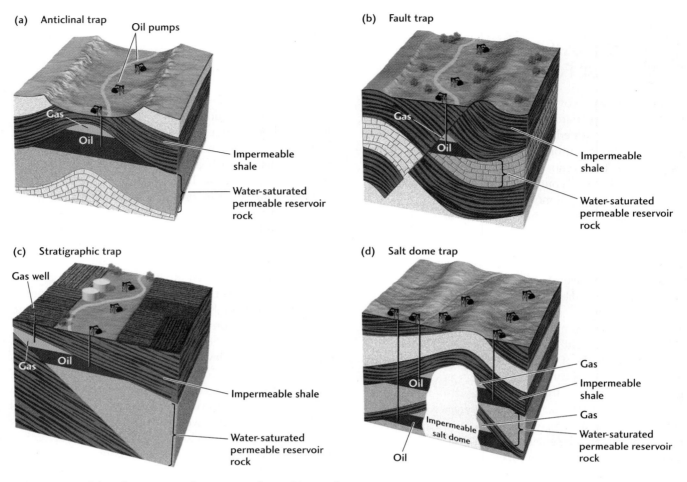

(a) Anticlinal trap

Oil pumps

Gas

Oil

Impermeable shale

Water-saturated permeable reservoir rock

(b) Fault trap

Gas

Oil

Impermeable shale

Water-saturated permeable reservoir rock

(c) Stratigraphic trap

Gas well

Gas

Oil

Gas

Impermeable shale

Water-saturated permeable reservoir rock

(d) Salt dome trap

Gas

Impermeable shale

Gas

Oil

Impermeable salt dome

Water-saturated permeable reservoir rock

Oil

FIGURE 14.5 Oil and gas accumulate in traps formed by geologic structures. Four types of traps are illustrated here.

and rock types that create an impermeable barrier to upward migration—an **oil trap** (**Figure 14.5**). Some oil traps, called *structural traps*, are created by structural deformation. One type of structural trap is formed by an anticline in which an impermeable layer of shale overlies a permeable sandstone formation (Figure 14.5a). The oil and gas accumulate at the crest of the anticline—the gas highest, the oil next—both floating on the groundwater that saturates the sandstone. Similarly, an angular unconformity or displacement at a fault may place a dipping permeable limestone formation opposite an impermeable shale, creating another type of structural trap (Figure 14.5b). Other types of oil traps are created by the original pattern of sedimentation, as when a dipping permeable sandstone formation thins out against an impermeable shale (Figure 14.5c). These structures are called *stratigraphic traps.* Oil can also be trapped against an impermeable mass of salt in a *salt dome trap* (Figure 14.5d).

In their search for petroleum resources, geologists have mapped thousands of oil traps throughout the world.

Only a fraction of them have proved to contain economically valuable amounts of oil or gas because traps alone are not enough to create a hydrocarbon reservoir.

Oil and gas accumulate where geologic structures create impermeable barriers to their upward migration, forming oil traps.

A trap will contain oil only if source beds were present, the necessary chemical reactions took place, and the oil migrated into the trap and stayed there without being disturbed by subsequent heating or deformation. Although oil and gas are not rare, most of the large, easy-to-find deposits have already been located, and the discovery of new resources is becoming more difficult.

Efforts are always under way to find more efficient ways to extract oil and natural gas from deep rock formations. Drilling holes deep into Earth's crust has become a very sophisticated and expensive business (see Figure 1.5). Petroleum engineers use three-dimensional models to steer drill bits on swooping paths into the richest parts

Proved reserves at end 2009
(thousand million barrels)

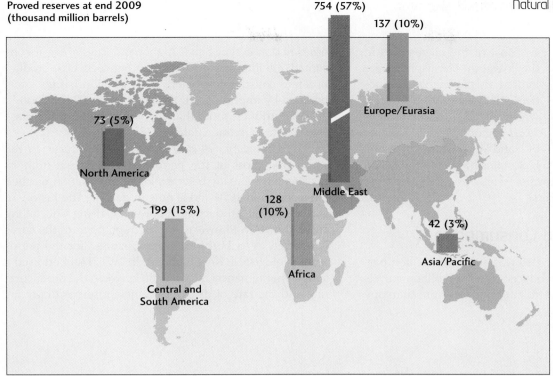

FIGURE 14.6 Estimated world oil reserves at the end of 2009 by region. [*British Petroleum Statistical Review of World Energy 2010.*]

of a reservoir. To coax oil out of stubborn formations, they inject water and carbon dioxide down strategically positioned drill holes to fracture the rock and push the oil into areas where it can be more efficiently pumped through other drill holes. These methods have increased the fraction of oil that can be extracted from known oil fields, increasing oil reserves.

Distribution of Oil Reserves

The worldwide reserves of oil are estimated to be about 1.3 trillion barrels (1 barrel = 42 gallons); these reserves are broken down by region in **Figure 14.6**. The oil fields of the Middle East—including Iran, Kuwait, Saudi Arabia, Iraq, and the Baku region of Azerbaijan—contain almost 60% of the world's total. Here sediments rich in organic material have been folded and faulted by the closure of the ancient Tethys Ocean, forming a nearly ideal environment for oil accumulation. The extensive reservoirs discovered in this vast convergence zone include the enormous Ghawar field in Saudi Arabia, the world's largest. Ghawar began production in 1948 and is expected to produce at least 88 billion barrels of oil over its lifetime (**Figure 14.7**).

FIGURE 14.7 The Ghawar oil field in Saudi Arabia is the world's largest, with a total reserve of at least 88 billion barrels. [*Robert Azzi/Woodfin Camp Associates.*]

Worldwide reserves of oil are clustered in the Middle East, where closure of the Tethys Ocean created a nearly ideal environment for oil accumulation.

Most of the oil reserves in the Western Hemisphere are located in the highly productive Gulf Coast–Caribbean area, which includes the Louisiana-Texas region, Mexico, Colombia, and Venezuela. Thirty-one U.S. states have commercial oil reserves, and small, noncommercial resources can be found in most of the others.

Oil Production and Consumption

Global oil production in 2010 was around 31 billion barrels. The United States produced about 2.7 billion barrels of crude oil and other petroleum liquids, more than any

other nation except Saudi Arabia and Russia, but it consumed 7 billion barrels. This gap between U.S. production and consumption, more than 4 billion barrels, must be filled by importing oil. This imbalance, $252 billion in 2010, contributes more than any other factor to the massive U.S. foreign trade deficit.

The United States is a "mature" oil producer, in the sense that most of the petroleum resources within its borders have already been exploited; production reached a maximum in 1970 and is now in decline. The history of U.S. crude oil production follows the rising and falling curve shown in **Figure 14.8**a. The high point of the curve is referred to as **Hubbert's peak,** named for the petroleum geologist M. King Hubbert. In 1956, Hubbert used a simple mathematical relationship between the production rate and the rate of discovery of new reserves to predict

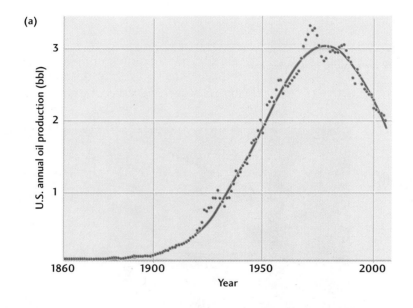

(a)

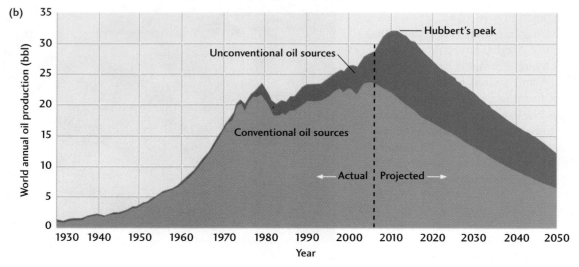

(b)

FIGURE 14.8 (a) U.S. annual oil production from 1860 to 2000. Production peaked in 1970 and subsequently declined. [*Modified from K. Deffeyes, Hubbert's Peak. Princeton, N.J.: Princeton University Press, 2001.*] (b) A pessimistic projection of world oil production, made by Colin Campbell in 2006, which shows Hubbert's peak occurring around 2011. "Unconventional oil sources" include oil wells drilled in deep water. [*Colin J. Campbell, "The General Depletion Picture," Association for the Study of Oil and Gas Newsletter, 2007.*]

that U.S. oil production, which was growing rapidly at the time, would actually begin to decline sometime in the early 1970s. His arguments were roundly dismissed as overly pessimistic, but history has proved him right.

When Will We Run Out of Oil?

At the current production rate, the world will consume all of the known oil reserves in just 40 years. Does that mean we will run out of oil by mid-century? No, because oil resources are much greater than oil reserves.

In fact, we will never really "run out" of oil. As the resources diminish, prices will eventually rise to such high levels that buyers will not be able to afford to waste oil by burning it as a fuel. Its main use will then be as a raw material to produce plastics, fertilizers, and a host of other *petrochemical* products. The petrochemical industry is already a very big business, consuming 7 percent of global oil production. As oil geologist Ken Deffeyes has noted, future generations will probably look back on the Petroleum Age with a certain amount of disbelief: "They burned it? All those lovely organic molecules, and they just burned it?"

The key issue is not when oil will run out, but when oil production will stop rising and begin to decline. This milestone—Hubbert's peak for world oil production—is the real tipping point; once it is reached, the gap between supply and demand will grow rapidly, driving oil prices sky-high.

So how close are we to Hubbert's peak? The answer to this question is the subject of considerable debate. Oil optimists believe there is enough undiscovered oil to satisfy world demand for several decades into the future. Oil pessimists, on the other hand, believe that we are fast approaching Hubbert's peak and predict a scenario like that shown in Figure14.8b. Their views are supported by the same type of analysis Hubbert used to predict the 1970 peak in U.S. oil production. In particular, the pessimists note that the rate of discovery of new sources, which determines the growth of reserves, is declining too rapidly to be consistent with the USGS's optimistic scenario. Some assert that the world is already past its peak oil production, although several more years of data will be needed to confirm this claim.

> *Oil optimists and oil pessimists disagree on how soon world oil production will reach Hubbert's peak.*

Oil Production and the Environment

Extracting fossil fuels can have a number of detrimental effects on the environment. On April 20, 2010, an explosion aboard the drilling platform *Deepwater Horizon* killed 11 men and injured 17 others. This blowout resulted in the largest marine oil spill in history, releasing 5 million barrels of crude into the Gulf of Mexico during the next three months (**Figure 14.9**). The oil spill caused significant environmental damage to ecosystems along the

FIGURE 14.9 Oil slick in the Gulf of Mexico imaged on May 24, 2010, by NASA's Terra satellite, 34 days after the explosion of the *Deepwater Horizon*. [*NASA Earth Observatory.*]

FIGURE 14.10 The oil spilled by the *Deepwater Horizon* blowout harmed wildlife along the Gulf Coast. [*AP Photo/ Bill Haber.*]

Gulf Coast (**Figure 14.10**), and its long-term effects are not yet known.

This accident, like earlier spills off the Yucatán coast in 1979 and Santa Barbara, California, in 1969, renewed the long-term debate about the regulation of drilling in fragile

habitats, such as Alaska's Arctic National Wildlife Refuge (ANWR) (**Figure 14.11**). The total petroleum resource in ANWR has not been fully evaluated, but it could be as much as 40 billion barrels of oil. The USGS estimates that if oil prices were high enough, 6 billion to 16 billion barrels of this oil could be produced economically using current technologies. There is no doubt that these resources would contribute to the national economy. But oil and gas production would require the building of roads, pipelines, and housing in a very delicate ecological environment that is a particularly important breeding area for caribou, musk-oxen, snow geese, and other wildlife. Policy makers must weigh economic benefits against possible long-term environmental losses in making this decision.

> *The drilling and transportation of oil can pose serious threats to local and regional environments.*

Natural Gas

Natural gas is a premium fuel for several reasons. In combustion, methane combines with atmospheric oxygen, releasing energy in the form of heat and producing only carbon dioxide and water. Natural gas therefore burns much more cleanly than coal or oil, which also produce ash, sulfur dioxide (the major cause of acid rain, see Chapter 1), and other pollutants. It emits 30 percent less carbon dioxide per unit of energy than oil and more than 40 percent less than coal.

Natural gas is easily transported across continents through pipelines. It accounts for about 24 percent of all fossil-fuel consumption in the United States each year. More than half of U.S. homes and a great majority of commercial and industrial buildings are connected to a network of underground pipelines that draw gas from fields in the United States, Canada, and Mexico. Getting the gas from source to market across oceans has been more difficult. The construction of tankers and ports that can handle liquefied natural gas (LNG) is beginning to solve this problem, although the potential dangers (e.g., the risks of large explosions) have made LNG facilities controversial in the communities where they would be located.

The world's resources of natural gas are comparable to its crude oil resources (see Figure 14.4) and may exceed them in the decades ahead. Estimates of natural gas resources have been rising in recent years, because exploration has increased and geologic traps have been

FIGURE 14.11 Herd of caribou in the Arctic National Wildlife Refuge. An intense environmental controversy surrounds proposals to drill for oil and natural gas in this pristine region. [*Prisma Bildagentur AG/Alamy.*]

identified in new settings, such as very deep formations, overthrust belts, coal beds, tight (less permeable) sandstones, and shales.

As in the case of oil, new technologies have increased the efficiency of gas production and the types of rock formations from which gas can be extracted. In a method called *hydraulic fracturing* ("fracking" in the lingo of the oil industry), large amounts of water are injected through the drilling pipe into a hard rock formation, such as a shale, to create tiny fissures in the rock, which allows the gas to flow more readily into pipe.

This technology has created a boom in the extraction of natural gas from deep shale formations, such as the Marcellus shale that underlies the northern Appalachian Mountains and the Allegheny Plateau of the eastern United States. The production of "shale gas" has increased tenfold in the last decade and now accounts for almost one-quarter of U.S. natural gas production. The environmental costs associated with shale gas can be steep,

> *Natural gas, a resource as abundant as oil, is a premium fuel because it burns more cleanly and produces less carbon dioxide than coal or oil.*

however. Fracking uses huge amounts of water, and wastes from shale gas production can foul the local water supply.

Coal

The abundant plant fossils found in **coal beds** show that coal is a biological sediment formed from large accumulations of plant material in wetlands. As the luxuriant plant growth of a wetland dies, leaves, twigs, and branches fall to the waterlogged soil. Rapid burial and immersion in water protect the dead twigs, branches, and leaves from complete decay because the bacteria that decompose vegetative matter are cut off from the oxygen they need. The plant material accumulates and gradually turns into *peat*, a porous brown mass of organic matter in which twigs, roots, and other plant parts can still be recognized (**Figure 14.12**). The accumulation of peat in oxygen-poor environments can be seen in modern swamps and peat bogs. When dried, peat burns readily because it is 50 percent carbon.

Over time, with continued burial, the peat is compressed and heated. Chemical transformations increase the peat's already high carbon content, and it becomes *lignite*, a very soft, brownish black, coal-like material

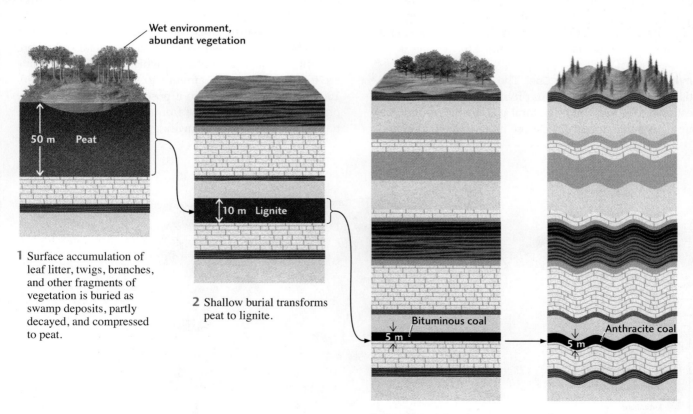

Wet environment, abundant vegetation

50 m   Peat

10 m   Lignite

Bituminous coal

5 m

Anthracite coal

5 m

1 Surface accumulation of leaf litter, twigs, branches, and other fragments of vegetation is buried as swamp deposits, partly decayed, and compressed to peat.

2 Shallow burial transforms peat to lignite.

3 Further burial under hundreds to thousands of meters of sediment transforms lignite to soft (bituminous) coal.

4 Continued burial and structural deformation, plus heat, metamorphose soft coal to hard (anthracite) coal.

FIGURE 14.12 The formation of coal beds begins with the deposition of vegetation in oxygen-poor environments.

containing about 70 percent carbon. The higher temperatures and structural deformation that accompany greater depths of burial may metamorphose the lignite into *subbituminous* and *bituminous coal*, or soft coal, and ultimately into *anthracite*, or hard coal. The higher the grade of metamorphism, the harder and more vitreous the coal, and the higher its carbon content, which increases its energy content. Anthracite is more than 90 percent carbon.

> Coal is formed when wetland vegetation accumulates in an oxygen-poor environment and undergoes burial, compression, and metamorphism.

Coal Resources There are huge resources of coal in sedimentary rocks. Only about 2.5 percent of the world's coal reserves have been used. According to some estimates, the amount of coal remaining is about 3.1 trillion metric tons, which is capable of producing 67,500 quads of energy (see Figure 14.4). About 85 percent of the world's coal resources is concentrated in the former Soviet Union, China, and the United States; these countries are also the largest coal producers. Domestic coal resources in the United States (**Figure 14.13**) would last for a few hundred years at the nation's current rates of use—about a billion tons a year. Coal has supplied an increasing proportion of U.S. energy needs since 1975, when the price of oil began to rise; it currently accounts for about 22 percent of energy consumption in the nation (see Figure 1.6).

Environmental Costs of Coal The extraction and combustion of coal present serious problems that make it a less desirable fuel than oil or natural gas. Underground coal mining is a dangerous profession; more than 4000 miners are killed each year in China alone. Many more coal miners suffer from black lung, a debilitating inflammation of the lungs caused by the inhalation of coal particles. Surface or "strip" mining, the removal of soil and surface sediments to expose coal beds, is safer for miners but can ravage the countryside if the land is not restored. An especially destructive type of surface mining common in the Appalachian Mountains of the eastern United States is "mountaintop removal," in which up to 300 vertical meters of a mountain crest are blasted away to expose underlying coal beds (**Figure 14.14**).

Coal is a notoriously dirty fuel. When burned, it produces 25 percent more CO_2 per unit of energy than oil, on average, and 70 percent more than natural gas. Most coal also contains appreciable amounts of pyrite, which is released into the atmosphere as noxious sulfur-containing gases when the coal is burned. Acid rain, which forms when these gases combine with rainwater, has become a severe problem in Canada, Scandinavia, the northeastern United States, and eastern Europe (see Chapter 1).

> Coal produces more carbon dioxide than other fossil fuels, and its combustion releases sulfur oxides into the atmosphere, which cause acid rain.

An inorganic residue, called coal ash, remains after coal is burned. Coal ash contains all the metals that were present in the coal, some of which, such as mercury, are toxic. Coal ash can amount to several tons for every 100 tons of coal burned, so it poses a significant disposal problem. It can also escape from smokestacks, creating a health risk to people downwind.

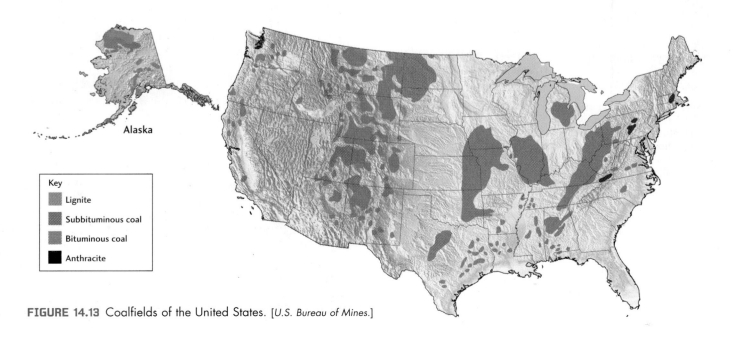

FIGURE 14.13 Coalfields of the United States. [*U.S. Bureau of Mines.*]

Key
- Lignite
- Subbituminous coal
- Bituminous coal
- Anthracite

Alaska

FIGURE 14.14 Mountaintop removal mining in the Appalachian Mountains of West Virginia. [*Rob Perks, NRDC.*]

United States government regulations require industries that burn coal to adopt technologies for "clean" coal combustion, which have reduced emissions of sulfur and toxic chemicals such as mercury. Federal laws also mandate the restoration of land disrupted by strip mining and the reduction of danger to miners. These measures are expensive and add to the cost of coal but it is still an inexpensive fuel compared with oil and natural gas.

> *Although the extraction and burning of coal present a number of risks to human life and the environment, its use is likely to increase because it is so abundant relative to other fossil fuels.*

Unconventional Hydrocarbon Resources

Extensive deposits of hydrocarbons occur in two other forms: source beds that are rich in organic material but never reached the oil window and formations that once contained oil but have since "dried out" to form *heavy oil* or a tarlike substance called *natural bitumen* (not to be confused with bituminous coal).

A hydrocarbon of the first type is *oil shale*, a fine-grained, clay-rich sedimentary rock containing relatively large amounts of organic matter. In the 1970s, oil companies began trying to commercialize the extensive oil shales of western Colorado and eastern Utah, but those efforts were largely abandoned by the 1980s as oil prices fell, concerns over environmental damage increased, and technical problems persisted. The efficiency of producing energy from oil shale is very low, and the environmental costs per unit of energy are very high. For example, shale oil and combustible gas can be extracted from these rocks, but the process requires huge amounts of water, a scarce resource in the western United States. Nevertheless, renewed interest in energy production from domestic oil shale has been sparked by higher oil prices and is being encouraged by national energy policies.

One deposit of the second type, the *tar sands* of Alberta, Canada, is estimated to contain a hydrocarbon reserve equivalent to 180 billion barrels of oil and a total resource perhaps 10 times that amount. More than 400 million barrels of oil are now extracted from the Alberta tar sands each year, and Canadian production is projected to increase fivefold by 2030, providing as much as 5 percent of world demand.

Development of the tar sands, like oil shales, raises important environmental concerns, however. It takes 2 tons of mined sand to produce 1 barrel of oil, leaving lots of waste sand, which is polluting. Moreover, production of oil from tar sands is an energy-intensive process that sucks up about two-thirds of the energy they ultimately render—a very low energy efficiency—and emits six times more CO_2 than conventional oil production.

> *The production of fossil fuels from unconventional sources, such as tar sands and oil shales, is limited by high energy costs and negative environmental effects.*

ALTERNATIVE ENERGY RESOURCES

As we continue to deplete our fossil-fuel resources, alternative energy resources will have to take up more and more of the demand. How quickly will this transition to a post-petroleum economy occur? Which alternative sources of energy have the greatest potential to replace oil?

Nuclear Energy

The first large-scale use of the radioactive isotope uranium-235 to produce energy was in the atomic bomb in 1944, but the nuclear physicists who observed the vast energy released when its nucleus split spontaneously (a phenomenon called *fission*) foresaw the possibility of peaceful applications of this new energy source. After World War II, countries around the world built nuclear reactors to produce **nuclear energy.** In these reactors, the fission of uranium-235 releases heat that is used to make steam, which then drives turbines to create electricity. A typical commercial reactor produces about 1000 megawatts of electric power (1 megawatt = 1 million watts). Nuclear power supplies a substantial fraction of the electric energy used by some countries, such as France (76 percent) and Sweden (52 percent), but this proportion is much smaller in the United States (21 percent). Overall, the nation's 110 nuclear reactors account for about 8 percent of total U.S. energy demand. The early expectation that nuclear fuels would provide a large, low-cost, environmentally safe source of energy has not been realized, primarily because of problems with

> *Nuclear energy has the potential to meet the world's electric energy needs for hundreds of years, but its promise has been compromised by safety and security concerns.*

reactor safety, disposal of radioactive wastes, and nuclear security.

Uranium Reserves Uranium is found as a trace element in some granite at an average concentration of only 0.00016 percent of the rock. Moreover, only a small portion of this element is uranium-235; other more abundant isotopes of uranium (for example, uranium-238) are not radioactive enough to be used as fuel. Uranium is nevertheless the world's largest minable energy resource by far, with a potential energy-generating capacity of at least 240,000 quads (see Figure 14.4). Minable concentrations are typically found as small quantities of uraninite, a uranium oxide mineral (also called pitchblende), in veins in granites and other felsic igneous rocks. When groundwater is present, uranium in igneous rocks near Earth's surface may oxidize and dissolve, be transported in groundwater, and later be reprecipitated as uraninite in sedimentary rocks.

> *Uranium is the world's largest minable energy resource.*

Hazards of Nuclear Energy Accidents caused by design flaws and human errors at nuclear power plants have raised questions about the safety of nuclear energy. Damage to a nuclear reactor at Three Mile Island, Pennsylvania, in 1979 released radioactive debris. Though very little radioactivity escaped the containment building and no one was harmed, it was a close call. Much more serious was the complete meltdown of a nuclear reactor in the Ukrainian town of Chernobyl in 1986, which contaminated hundreds of square miles of land surrounding the reactor. Deaths from cancer caused by exposure to the fallout may be in the thousands.

FIGURE 14.15 Aerial view of the north entrance to the Yucca Mountain Nuclear Waste Repository being developed at the Nevada Test Site, north of Las Vegas. Yucca Mountain is the high ridge to the right of the entrance. [*U.S. Department of Energy.*]

The most serious disaster occurred when the tsunami from the great Tohoku earthquake of March 11, 2011, inundated the Fukushima Daiichi nuclear power plant on the northeastern coast of Honshu, Japan (see Figure 13.37). The reactors shut down as designed, but the tsunami destroyed the backup diesel generators, cutting power to the water pumps that were supposed to cool the still-hot reactors. Three of the six reactors suffered complete or partial meltdowns, and explosions of hydrogen gas generated during the meltdowns destroyed the reactor containment buildings, releasing radioactive debris into the atmosphere. Water sprayed to cool the damaged reactors carried radioactive material into the ocean. At the time this textbook was written, efforts to stabilize the reactors were ongoing and the full extent of the disaster was not yet clear.

The uranium consumed in nuclear reactors leaves behind dangerous radioactive wastes. A system of safe long-term waste disposal is not yet available, and reactor wastes are being held in temporary storage at reactor sites. (Spent fuel rods stored on site contributed to the radioactive debris released at Fukushima.) Many scientists believe that geological containment—the burial of nuclear wastes in deep, stable, impermeable rock formations—would provide safe storage of the most dangerous wastes for the hundreds of thousands of years required before they decay. France and Sweden have built such underground nuclear waste repositories. A similar facility was being developed at Yucca Mountain, Nevada (**Figure 14.15**), but local opposition caused the federal government to terminate funding

Concerns about the safety and security of nuclear reactors and the unresolved problem of nuclear waste disposal may limit the use of nuclear energy.

for the site in 2010. At present, the United States has no long-term plan for nuclear waste disposal.

Solar Energy

The energy Earth receives from sunlight is about 18,000 times more than we currently use to power all human activities. **Solar energy** is the prime example of a resource that cannot be depleted by usage—the Sun will continue to shine for at least the next several billion years.

Although using solar energy to heat water for homes, industries, and agriculture is economically profitable with existing technology, the methods for the large-scale conversion of sunlight into electricity are still inefficient and expensive. Nevertheless, the solar generation of electricity is increasing rapidly as larger power plants are being built in response to voter mandates and government subsidies. Upon its completion in 2011, the Mojave Solar Park, being built by Pacific Gas and Electric company in the California desert (**Figure 14.16**), will become the world's largest solar electric power plant in both generating capacity (up to 550 megawatts) and land size (23 km²).

Efforts to increase the use of solar energy are focused on improving the efficiency of converting sunlight to electricity.

In the United States, solar energy conversion rose from 0.065 quad in 2004 to 0.10 quad in 2009, a 50 percent increase in just five years. Yet this amounts to only 0.1 percent of total U.S. energy consumption. Optimistic projections indicate that, worldwide, solar conversion could increase to as much as 12 quads per year in a decade or so, which would be about 2 percent of total

FIGURE 14.16 The Kramer Junction solar facility in the Mojave Desert, near Boron, California. Nearby is the Mojave Solar Park, which is the largest solar electric power plant in the world. [*Sandia National Laboratories/DOE.*]

energy production. A more realistic figure is probably less than 6 quads per year.

Hydroelectric Energy

Hydroelectric energy is derived from water moving under the force of gravity that is made to drive a turbine that generates electricity. Waterfalls or artificial reservoirs behind dams usually provide the water. Hydroelectric energy depends on the Sun, whose energy drives the climate system and produces rainfall; thus, like solar energy it is renewable. It is also relatively clean, risk-free, and cheap.

The Three Gorges Dam, on the Yangtze River in China (**Figure 14.17**), is the world's largest hydroelectric facility. It is capable of generating 22,500 megawatts, close to 5 percent of China's total electricity demand. The project has been controversial, however, because the damming of the Yangtze has caused flooding that has displaced over a million people.

In the United States, hydroelectric dams deliver about 3 quads annually, or about 3 percent of the nation's annual energy consumption. The U.S. Department of Energy has identified more than 5000 sites where new hydroelectric dams could be built and operated economically. Such expansion would be resisted, however, because the dams would drown farmlands and wilderness areas under artificial reservoirs but add only a few percentage points to U.S. energy production. This is one energy issue on which

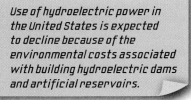

Use of hydroelectric power in the United States is expected to decline because of the environmental costs associated with building hydroelectric dams and artificial reservoirs.

most people agree, that the environmental costs would exceed the economic gains. For this reason, most energy experts expect that the proportion of the nation's energy produced by hydroelectric power will actually decline in the future.

Wind Energy

Humans have harnessed wind power since ancient times, using sails to propel ships, windmills for mechanical power to mill grain, and wind pumps to draw water and drain fields. Today, the generation of electricity by high-efficiency wind turbines is a fast-growing source of renewable energy. Winds farms containing hundreds of turbines can produce as much electric power as a mid-sized nuclear reactor (**Figure 14.18**).

Worldwide, the amount of wind-generated electric power grew tenfold between 2000 and 2010. Denmark now produces 21 percent of its electric power by wind, Portugal 18 percent. In the United States, electricity from wind sources increased threefold from 2005 to 2010, and wind now accounts for about 11 percent of U.S. renewable energy production.

The U.S. Department of Energy estimates that winds sufficient for power generation blow across 6 percent of the land area of the continental United States and that they have the potential to supply more than one and a half times the nation's current electricity consumption. However, harvesting wind energy at this scale would require placing millions of windmills, each over 100 m tall, across hundreds of thousands of square kilometers of land and sea. Changes to the landscape required for

FIGURE 14.17 The Three Gorges Dam on China's Yangtze River is about 2335 m (7660 feet) long and 185 m (616 feet) high. Its 32 generators are capable of producing 22,500 megawatts of hydroelectric power. [AP Photo/Xinhua Photo, Xia Lin.]

FIGURE 14.18 The Roscoe Wind Farm in Texas. Texas is the largest producer of wind-powered electric power in the United States. [*Joel Sartore/National Geographic Stock.*]

industrial wind farming have made the siting of new facilities a controversial environmental issue in some regions.

Geothermal Energy

Earth's internal heat can be tapped to drive electric generators and heat homes. **Geothermal energy** depends on the heating of water as it passes through a region of hot rock (a *heat reservoir*) that may be hundreds or thousands of meters beneath Earth's surface. Hot water or steam is brought to the surface through boreholes drilled for the purpose. Usually, the water is naturally occurring groundwater that seeps downward along fractures in rock. Less typically, the water is artificially introduced by pumping from the surface.

At least 46 countries now use some form of geothermal energy. By far the most abundant source is naturally occurring groundwater that has been heated to the relatively low temperatures of 80° to 180°C and is used for residential, commercial, and industrial heating. Warm underground water drawn from a reservoir in the Paris sedimentary basin now heats more than 120,000 dwellings in France. Iceland, as we saw in Chapter 5, sits on the Mid-Atlantic Ridge, where hot mantle material wells up as the North American and Eurasian plates separate. Reykjavík, the capital of Iceland, is almost entirely heated by geothermal energy derived from this magma source.

Geothermal energy reservoirs with temperatures above 180°C are useful for generating electricity. They are present primarily in regions of recent volcanism as hot, dry rock, natural hot water, or natural steam. The latter two sources are limited to the few areas where surface water seeps down through underground faults and fractures to reach deep rocks heated by recent magmatic activity. Naturally occurring water heated above the boiling point and naturally occurring steam are highly prized resources. The world's largest facility for producing electricity from natural steam, located at The Geysers, 120 km north of San Francisco, generates more than 600 megawatts of electricity (**Figure 14.19**). Some 70 geothermal electricity-generating plants operate in California, Utah, Nevada,

> *In regions where geothermal gradients are steep, Earth's internal heat can be tapped as a source of energy.*

FIGURE 14.19 The Geysers, one of the world's largest supplies of natural steam. Its geothermal energy is converted into electricity for San Francisco, 120 km to the south. [*Pacific Gas and Electric.*]

and Hawaii, producing 2800 megawatts of power—enough to supply about a million people.

In 2010, five major geothermal power plants in Iceland produced more than a quarter of the nation's electric energy. According to one Icelandic estimate, as much as 40 quads of electricity could be generated each year from accessible geothermal energy sources, but so far only a tiny fraction of that amount, about 0.15 quad per year, is actually being generated. Another 0.12 quad of geothermal energy is used for direct heating.

Like most of the other energy sources we have examined, geothermal energy presents some environmental problems. Regional ground subsidence can occur if hot groundwater is withdrawn without being replaced. In addition, geothermally heated waters can contain salts and toxic materials dissolved from the hot rock.

Biofuels

Before the coal-fired industrial revolution of the mid-nineteenth century, the burning of wood and other biomass derived from plants and animals (e.g., whale oil, dried buffalo dung) satisfied most of society's energy needs (see Figure 1.6). Even today, the energy derived from biomass exceeds the total derived from all other renewable resources.

Biomass is an attractive alternative to fossil fuels because, at least in principle, it is *carbon-neutral*; that is, the CO_2 produced by the combustion of biomass is eventually removed from the atmosphere by plant photosynthesis and used to produce new biomass. In particular, liquid **biofuels** derived from biomass, such as *ethanol* (ethyl alcohol: C_2H_6O), could replace gasoline as our main automobile fuel.

The use of biofuels in transportation is hardly new. The first four-stroke internal combustion engine, invented by Nikolaus Otto in 1876, ran on ethanol, and the diesel engine, patented by Rudolf Diesel in 1898, ran on vegetable oil. Henry Ford's Model T car, first produced in 1903, was designed to operate completely on ethanol. But soon thereafter, petroleum from the new reserves discovered in Pennsylvania and Texas became widely available, and cars and trucks were converted almost entirely to petroleum-based gasoline and diesel fuel (see Chapter 1).

Ethanol can be mixed with gasoline to run most car engines built today. It is produced mainly from corn in the United States and from sugarcane in Brazil. For the last 30 years, the Brazilian government has been pushing to replace imported oil with domestic ethanol; today, more than 30 percent of Brazil's

> The biofuel ethanol, which is produced from plant biomass, is an alternative to gasoline that can reduce the net anthropogenic emission of carbon dioxide.

FIGURE 14.20 Switchgrass, a perennial plant native to the Great Plains, is an efficient source of ethanol, the most popular biofuel. Here, geneticist Michael Casler harvests switchgrass seed as part of a breeding program to increase the plant's ethanol yield. [*Wolfgang Hoffmann*.]

automobile fuels come from sugarcane, saving the country about $50 billion in oil imports.

The United States has initiated a crash program to develop more efficient methods for producing biofuels, with the goal of reducing oil imports by at least 75 percent before 2025. A promising biomass crop is switchgrass, a perennial plant native to the Great Plains (**Figure 14.20**). Switchgrass has the potential to produce up to 1000 gallons of ethanol per acre per year, compared with 665 gallons for sugarcane and 400 gallons for corn, and it can be cultivated on grasslands of marginal use for other types of agriculture.

What about the environmental benefits of biofuels? Can they really be carbon-neutral? If the energy used to fertilize the plants, transform them into biofuels, and deliver the biofuels to market comes primarily from fossil fuels, the answer is no. The widespread use of biofuels for transportation would no doubt reduce the pumping of carbon from the lithosphere to the atmosphere, but experts are still arguing about the magnitude of the reduction.

This question raises a more general one: What will be the net effect of our various energy-producing strategies on carbon dioxide concentrations in the atmosphere? To answer this question, we need to understand how carbon moves through the Earth system—the carbon cycle, our next topic.

NATURAL ENVIRONMENT: The Carbon Cycle

In the past 200 years, the atmospheric CO_2 concentrations have risen from 280 ppm to 390 ppm, an increase of

40 percent. Earth's atmosphere has not contained this much CO_2 for at least the last 400,000 years and probably for the last 20 million years. Atmospheric CO_2 concentrations are now increasing at the unprecedented rate of 0.4 percent per year (see Figure 10.13), faster than at any time in recent geologic history.

Yet the situation could be worse. Over the decade from 1990 to 1999, human activities emitted about 8.0 gigatons (Gt) of carbon into the atmosphere. (Note that emissions are calculated in gigatons of carbon, not carbon dioxide; see Exercise 10 at the end of this chapter.) Fossil-fuel burning and other industrial activities emitted about 6.4 Gt of carbon each year, and the burning of forests and other changes in land use emitted an additional 1.6 Gt. If all of that carbon had stayed in the air, the atmospheric CO_2 increase would have been closer to 0.9 percent per year, more than twice the observed rate of 0.4 percent. Instead, 4.8 Gt of carbon were removed from the atmosphere each year by natural processes. Where did all that carbon go?

We will answer this question by examining the **carbon cycle**—the continual movement of carbon between different components of the Earth system. What are the interactions between the carbon cycle and the climate system? Is the net feedback between these systems positive or negative—that is, will changes in the carbon cycle enhance or reduce global warming? Finally, how might humans intervene in the carbon cycle to stabilize the climate system and prevent global warming?

> *The carbon cycle describes the movement of carbon among its principal geochemical reservoirs: the atmosphere, the lithosphere, the hydrosphere, and the biosphere.*

Geochemical Cycles and How They Work

In Chapter 11, we saw that the hydrologic cycle involves the transfer of a chemical substance (water) between reservoirs. Similarly, when discussing the carbon cycle, we view the major components of the Earth system—atmosphere, hydrosphere, cryosphere, lithosphere, and biosphere—as **geochemical reservoirs** for the storage of carbon and other terrestrial chemicals, and we consider the processes that transport those chemicals among reservoirs.

Geochemical Fluxes Geochemical cycles are patterns of the flow, or *flux*, of chemicals from one reservoir to another. These cycles include the evaporation and precipitation of water, the airborne transfer of dust from the ground surface to the oceans, volcanic discharges, and gas exchange between the sea surface and the atmosphere (**Figure 14.21**). On time scales relevant to the climate system, the major fluxes into and out of a reservoir are usually in approximate balance. Sedimentation in the oceans, for example, balances the influx of sediments from rivers. As seafloor sediments are buried, they become part of the oceanic crust. There they stay until they move into the mantle through subduction or become part

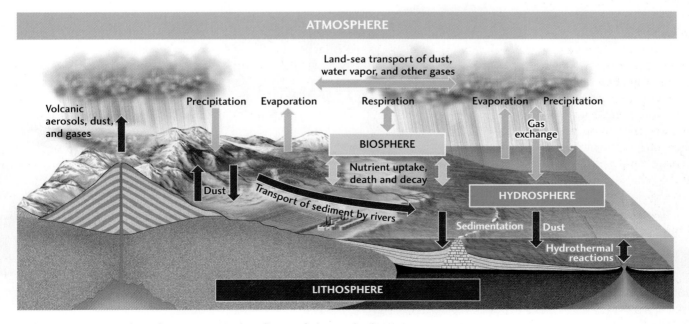

FIGURE 14.21 A number of processes result in fluxes of chemical substances between components of the climate system.

of the continental crust through accretion. Over the long term, tectonic uplift exposes crustal rocks to weathering and erosion, maintaining the balance of fluxes among the reservoirs.

The biosphere is a unique reservoir because each individual organism is constantly interacting with its environment. The most important fluxes into and out of the biosphere are the inflow and outflow of atmospheric gases by respiration, the inflow of nutrients from the lithosphere and hydrosphere, and the outflow of chemicals through the death and decay of organisms. When the biosphere plays a key role in the flux of a chemical, geologists refer to the cycle that describes that flux as a *biogeochemical cycle*. The carbon cycle, which depends critically on the pumping of carbon into and out of the atmosphere by living organisms, is clearly a biogeochemical cycle.

> Geochemical cycles describe the flux of a chemical among its geochemical reservoirs. Biogeochemical cycles are cycles in which the biosphere plays a key role.

Residence Times Reservoirs gain chemicals from inflows and lose chemicals from outflows. If inflow equals outflow, the amount of the chemical in the reservoir stays the same, even though the chemical is constantly entering and leaving it. On average, a molecule of the chemical spends a certain amount of time, called the **residence time,** in a reservoir.

We can visualize a chemical's residence time in the ocean, for example, as the average time that elapses between the entry of a molecule into the ocean and its removal through sedimentation or some other process. For example, the residence time of sodium in the ocean is extremely long—about 48 million years—because sodium is highly soluble in seawater (that is, the capacity of the reservoir to store sodium is high) and because rivers contain relatively small amounts of sodium (its inflow into the reservoir is low). In contrast, iron has a residence time in the ocean of only about 100 years because its solubility in seawater is very low and the inflow from rivers is relatively high.

The residence times of chemicals in the atmosphere are usually shorter than those in the ocean because the atmosphere is a smaller reservoir than the ocean and because fluxes into and out of the atmosphere can be relatively high. Sulfur dioxide, for example, has a residence time in the atmosphere of hours to weeks, and oxygen, which makes up about 21 percent of the atmosphere, has a residence time of 6000 years. An exception is the most common atmospheric gas, nitrogen, which constitutes

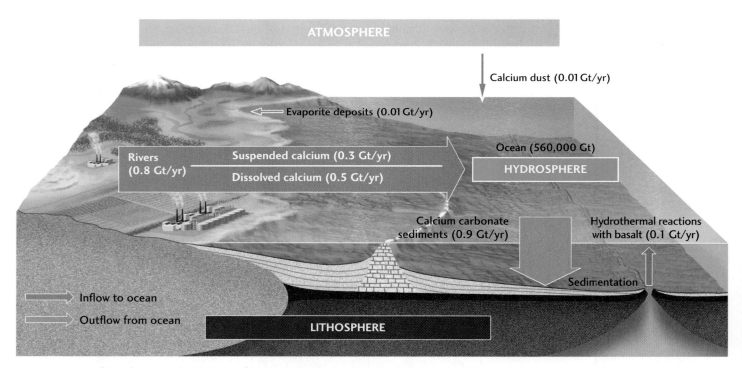

FIGURE 14.22 The calcium cycle, showing fluxes into and out of the ocean. Fluxes are given in gigatons (Gt; 10^{12} kg) per year. The inflow of calcium to the ocean approximately balances the outflow.

about 78 percent of the atmosphere; because it is abundant and stable, its residence time in the atmosphere is almost 400 million years. A molecule of nitrogen that entered the atmosphere in the late Paleozoic era, about 300 million years ago, is still likely to be there!

An Example: The Calcium Cycle Before we examine the carbon cycle in more detail, let's take a look at the calcium cycle, which provides a simple illustration of the concepts involved in geochemical cycles (**Figure 14.22**). The ocean contains about 560,000 Gt of calcium, dissolved in a total ocean mass of about 1.4×10^9 Gt. Calcium steadily enters this reservoir in large quantities through the rivers of the world, which transport dissolved and suspended calcium derived from the weathering of carbonate rocks and other minerals such as gypsum and calcium-rich plagioclase feldspars. A much smaller amount enters the ocean via transport by windblown dust.

If the ocean received this continuous inflow of calcium liberated by weathering without there being any way to remove it, the ocean would quickly become supersaturated with calcium ions (Ca^{2+}). Calcium ions are removed from seawater when they react with carbonate ions (CO_3^{2-}) to form calcium carbonate ($CaCO_3$), which is precipitated out of the seawater by marine organisms as shells and coral. A smaller amount of calcium is precipitated as gypsum in evaporites.

The amount of calcium the ocean can hold is much larger than the inflow and outflow of calcium, so calcium has a fairly long residence time in the ocean. By dividing the total annual influx (0.9 Gt/year) by the ocean's calcium capacity (560,000 Gt), we obtain a residence time of about 600,000 years.

As fossil-fuel burning drives up the CO_2 concentration in the atmosphere, more CO_2 dissolves into seawater, where it reacts to form carbonic acid (H_2CO_3), causing the oceans to become more acidic. This process, called **ocean acidification,** decreases the carbonate concentration in seawater, which raises the solubility of calcium, making it more difficult for carbonate-producing organisms to survive.

The Cycling of Carbon

Carbon cycles among four main reservoirs: the atmosphere; the oceans, including marine organisms; the land surface, including terrestrial plants and soils; and the deeper lithosphere (**Figure 14.23**).

The greatest carbon flux, 120 Gt per year, is the cycling of CO_2 between the biosphere and the atmosphere by

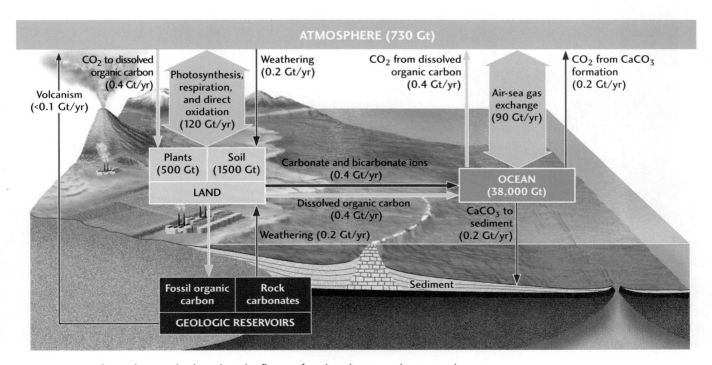

FIGURE 14.23 The carbon cycle describes the fluxes of carbon between the atmosphere and its other principal reservoirs. Fluxes are given in units of Gt per year. [*IPCC, Climate Change 2001: The Scientific Basis. Cambridge: Cambridge University Press, 2001.*]

land plants and animals through photosynthesis, respiration, and decay. Plants take in this entire amount during photosynthesis and respire about half of it back into the atmosphere. The other half is incorporated into plant tissues—leaves, wood, and roots—as organic carbon. Animals eat the plants, and microorganisms decompose them; both processes result in the breakdown of plant tissues and the respiration of CO_2. Much of the organic carbon, about three times the total plant mass, is stored in soils. A significant fraction (about 4 gigatons/year) reenters the atmosphere through direct oxidation by forest fires and other combustion of plant material. A smaller fraction (0.4 Gt/year) is dissolved in surface waters and transported by rivers to the ocean, where it is respired back into the atmosphere by marine organisms.

> *The greatest flux in the carbon cycle is between the biosphere and the atmosphere.*

The exchange of CO_2 directly across the interface between the oceans and the atmosphere amounts to a carbon flux of about 90 Gt per year. This flux depends on many factors, including air and sea temperatures and the composition of the seawater, but it is particularly sensitive to wind velocity, which increases the transfer of CO_2 and other gases by stirring up the surface water. Carbon dioxide dissolved in seawater escapes from solution and enters the atmosphere by evaporating from sea spray, while atmospheric CO_2 enters the ocean by dissolving in sea spray and rain or directly across the sea surface.

The weathering of carbonate rocks removes about 0.2 Gt of carbon per year from the lithosphere and an equal amount from the atmosphere. The CO_2 is dissolved in surface waters, primarily as bicarbonate ions, and is transported by rivers to the oceans. Here shell-forming marine organisms reverse the weathering reaction, precipitating calcium carbonate and releasing an equal amount of carbon into the atmosphere as CO_2.

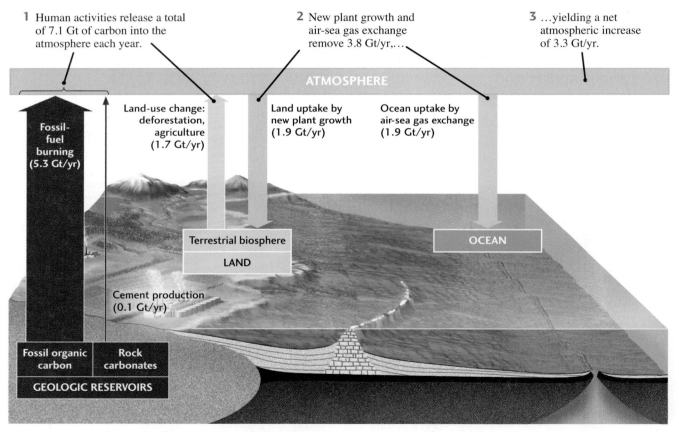

1 Human activities release a total of 7.1 Gt of carbon into the atmosphere each year.

2 New plant growth and air-sea gas exchange remove 3.8 Gt/yr,…

3 …yielding a net atmospheric increase of 3.3 Gt/yr.

ATMOSPHERE

Fossil-fuel burning (5.3 Gt/yr)

Land-use change: deforestation, agriculture (1.7 Gt/yr)

Land uptake by new plant growth (1.9 Gt/yr)

Ocean uptake by air-sea gas exchange (1.9 Gt/yr)

Terrestrial biosphere
LAND

OCEAN

Cement production (0.1 Gt/yr)

Fossil organic carbon — Rock carbonates

GEOLOGIC RESERVOIRS

FIGURE 14.24 Humans add CO_2 to the atmosphere by burning fossil fuels; by releasing carbon through deforestation, agriculture, and other land-use changes; and by producing cement. Much of the CO_2 emitted into the atmosphere by human activities is absorbed by the oceans and by plant growth on land. The remainder stays in the atmosphere, increasing the concentration of CO_2. The fluxes shown in this figure (given in gigatons per year) are for 1990–1999. [*IPCC, Climate Change 2007: The Physical Science Basis. Cambridge: Cambridge University Press, 2007.*]

Geologic processes regulate the carbon cycle over the long term.

Other geologic processes are also involved in the carbon cycle. Volcanism releases minor amounts of CO_2 into the atmosphere, and the weathering of silicate rocks removes CO_2 from the atmosphere (through the reaction of silicates with carbon dioxide to produce calcium carbonate and silica). The net flux of carbon by these processes is relatively small (less than 0.1 Gt/year), so they are usually neglected in considerations of short-term effects on climate. Over the long term, however, their effects can be substantial. According to one controversial hypothesis, mentioned in Chapter 10, the uplifting of the Himalaya and the Tibetan Plateau, which began about 40 million years ago, increased weathering rates enough to reduce the concentration of CO_2 in the atmosphere, and this reduction may have contributed to the subsequent climate cooling that led to the Pleistocene glaciations.

Human Perturbations of the Carbon Cycle

With this background, let's return to the fate of anthropogenic carbon emissions. We have good estimates of the extent to which human activities have perturbed the carbon cycle during the late twentieth century.

Figure 14.24 shows what happened to the 8 gigatons of carbon that was added to the atmosphere by human activities in the 1990s, a decade that has been exceptionally well studied. Only 40 percent of the total (3.2 Gt/year) remained in the atmosphere as CO_2. The rest was absorbed in nearly equal amounts by the oceans and the land surface. On land, the uptake was primarily by the growth of temperate and tropical forests, caused in part by the rapid growth of new forests on land previously used for agriculture and in part by the "fertilization effect" of increasing CO_2. Plants love the stuff, and they grow more as atmospheric CO_2 concentrations go up.

Between the 1980s and 1990s, the total emissions rose from an average of 6.8 Gt per year to 8.0 Gt per year (**Figure 14.25**). Remarkably, however, the net amount added to the atmosphere each year as CO_2 remained the same (3.2 Gt per year), because all of the increase was absorbed by the oceans (from 1.8 Gt/yr to 2.2 Gt/yr) and land surface (from 1.7 Gt/yr to 2.6 Gt/yr). The land and oceans are doing more than their fair share of absorbing our extra carbon dioxide!

The oceans absorb much of the excess carbon dioxide emitted by human activities, but continued increases in those emissions may decrease their capacity to do so.

Emissions due to deforestation and other land-use changes increased by 0.2 Gt/year from the 1980s to the 1990s,...

... and fossil-fuel emissions increased by 1.0 Gt/year.

Atmospheric carbon accumulation actually *decreased* by about 0.1 Gt/year from the 1980s to the 1990s,...

... and oceanic accumulation of carbon increased by only 0.4 Gt/year.

Therefore, the balance between accumulation and emissions required a 0.9 Gt/year increase in the carbon absorbed by the land surface.

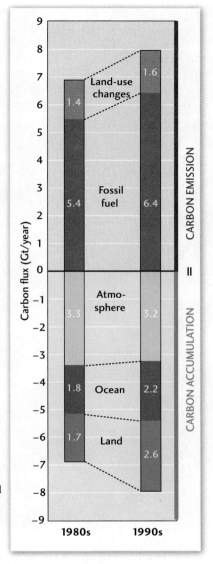

FIGURE 14.25 Comparison of anthropogenic carbon emissions and accumulation in the 1980s (left bar) and 1990s (right bar). Total emissions rose, but the net amount added to the atmosphere remained unchanged because all of the increase was absorbed by the oceans and land surface, the latter primarily by forest growth. [*Data from IPCC, Climate Change 2007: The Physical Science Basis. Cambridge: Cambridge University Press, 2007.*]

How will these numbers change as atmospheric CO_2 concentrations continue to rise? Scientists believe that the percentage of human carbon emissions taken up by the oceans will decrease as more CO_2 is dissolved in seawater, causing it to become more acidic. As noted previously, ocean acidification increases the solubility of calcium in seawater, making it more difficult for key marine organisms—such as corals and some plankton—to form their

FIGURE 14.26 Shell-forming marine organisms, such as these corals in the Great Barrier Reef of Australia, are threatened by ocean acidification. [© 2004 Richard Ling, www.rling.com.]

Now that we have described the carbon cycle, we are better prepared to examine the problem of global climate change in more detail. We introduced the problem in Chapter 1 and examined the evidence for recent anthropogenic warming in Chapter 10. Here we will focus on two key questions: How will the modifications of the carbon cycle brought about by our carbon economy change the climate system? What might be the effects of those changes on our environment?

GLOBAL CLIMATE CHANGE

We have seen in earlier chapters that the history of the Earth system is punctuated by extreme events—short periods of rapid global change. Geologists have ranked these changes according to their effects on the fossil record, and they have used major transitions in life-forms to delineate the eras, periods, and epochs of the geologic time scale (see Figure 8.13).

The time scale's most recent division was set at about 11,000 years ago, when the Pleistocene epoch, a period characterized by a cold climate with sporadic ice ages, gave way to a warm, stable period—the Holocene epoch (see Chapter 10). Average global temperatures jumped by 7°C over an interval that may have been as short as a few decades, sea level rose, ecosystems migrated poleward with the receding ice, and many large mammals went extinct.

At about the same time, our own species, *Homo sapiens*, began to alter the environment. In particular, humans developed new social structures and technologies that allowed them to grow crops, raise livestock, and build cities—the novel landscaping processes of civilization.

Holocene Climate Stability

Some scientists think that if human civilization had not come along, Earth's climate might by now be plunging into another ice age, driven by decreasing amounts of solar energy (due to the Milankovitch cycles described in Chapter 10) and accompanied by decreasing atmospheric concentrations of greenhouse gases. According to paleoclimatologist William Ruddiman, the expansion of civilization began to release significant amounts of carbon dioxide and methane from the biosphere to the atmosphere as early as 8000 years ago, primarily through deforestation and the rise of

external calcium carbonate skeletons (**Figure 14.26**). Recent studies have shown that coral reefs are already in trouble, and if the present trends continue, ocean acidification could begin to kill off certain types of marine organisms within the next few decades. In addition, the enhanced greenhouse effect resulting from rising atmospheric CO_2 concentrations will warm the oceans, which will further decrease their capacity to absorb carbon from the atmosphere—an example of positive feedback in the climate system (see Chapter 10). The overall magnitude of these effects is not yet certain, but it has become clear that the changes in the carbon cycle caused by human activities are likely to amplify greenhouse warming.

agriculture. Ruddiman hypothesizes that this extra source of greenhouse gases has been responsible for extending the warm interglacial period beyond its natural limit.

Whatever the reasons, the Holocene warm period has been unusually long and stable when compared with the previous interglacial periods of the Pleistocene epoch (see Chapter 10). Measurements from ice cores indicate that, from the end of the Pleistocene until about 200 years ago, the average global temperature fluctuated by only a few degrees, and the atmospheric concentrations of the major greenhouse gases stayed within a rather narrow range. The average CO_2 concentration, for example, fluctuated only between 260 and 280 ppm—less than a 10 percent variation over this entire period.

The Holocene epoch has been an unusually long and stable interglacial period. Anthropogenic emissions of greenhouse gases may have prevented the onset of another ice age.

Industrialization and Global Warming

Since the beginning of the industrial era, fossil-fuel burning, deforestation, land-use changes, and other human activities have caused a significant rise in the concentrations of greenhouse gases in the atmosphere. **Figure 14.27** shows the atmospheric concentrations of three greenhouse gases—carbon dioxide, methane, and nitrous oxide—over the past 10,000 years. In all three cases, there is a remarkable correspondence with the history of the human population: the concentrations remained relatively constant throughout the Holocene but shot upward after the industrial revolution (compare Figure 14.27 with Figure 14.1).

The global atmospheric concentration of methane has increased by almost 150 percent from its preindustrial value, and that of carbon dioxide has increased by 40 percent. In both cases, the observed increases can be explained by human activities, predominantly agriculture and fossil-fuel use. Methane's greenhouse effect is weaker than that of carbon dioxide, so even though its relative concentration has gone up more, its contribution to greenhouse warming is only about 30 percent as large. The postindustrial increase in nitrous oxide, primarily from agriculture, has been 18 percent; its contribution to the greenhouse effect is a small fraction of carbon dioxide's.

Atmospheric concentrations of greenhouse gases began to increase rapidly with the industrialization of society and the explosion of the human population.

The increases in greenhouse gas concentrations have been accompanied by a rise in average temperatures at Earth's surface (see Figure 10.13), with most of the warming occurring

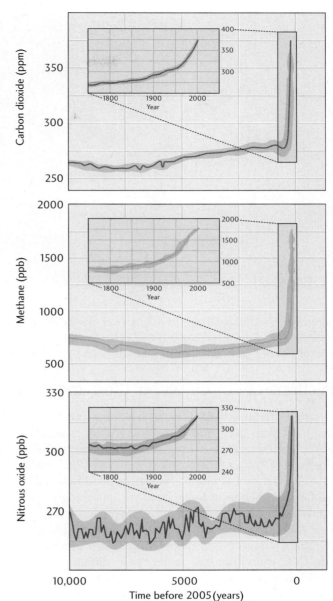

FIGURE 14.27 Atmospheric concentrations of carbon dioxide, methane, and nitrous oxide over the last 10,000 years (large panels) and since 1750 (inset panels). These measurements, compiled by the Intergovernmental Panel on Climate Change, were derived from ice cores and atmospheric samples. Shaded bands show the uncertainties in the measurements. [*IPCC, Climate Change 2007: The Physical Science Basis. Cambridge: Cambridge University Press, 2007.*]

in the latter half of the twentieth century. **Figure 14.28** is a map of the average temperature differences between two time periods, 1995–2004 and 1940–1980. The average global increase was 0.42°C, but the changes have not been geographically uniform: they have been greater than

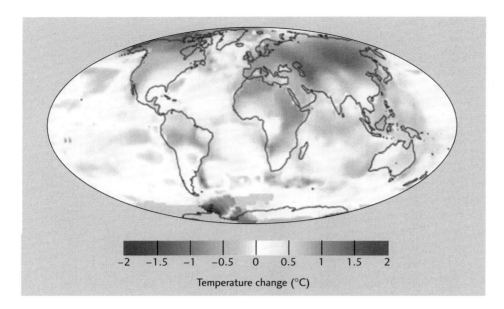

FIGURE 14.28 Differences between average surface temperatures measured during the period 1995–2004 and those measured at the same location during the period 1940–1980 (the baseline period). [*NASA/Goddard Institute for Space Studies.*]

average over the land surface and less over the oceans. In particular, you can see that the rise observed in large regions of the northern continents exceeds 1°C.

Predictions of Future Global Warming

The warming trend observed in the twentieth century is continuing. Up to 2010, the 10 warmest years recorded since accurate temperature measurements began in 1880 all occurred after 1997, and the warmest of all were 2005 and 2010. How much hotter will the planet get, and how will this global warming affect local climates? Answering such questions requires a sequence of uncertain predictions.

The Intergovernmental Panel on Climate Change (IPCC; see Chapter 1) has modeled increases in atmospheric CO_2 concentrations under three different scenarios (**Figure 14.29**):

A. Continued reliance on fossil fuels as our major energy source, sometimes called the business-as-usual scenario

B. Greater use of cleaner alternative energy sources, including nuclear energy and renewable resources

C. An even more rapid conversion from fossil fuels to cleaner alternatives

> *Predictions of future climate change depend on a series of uncertain assumptions, but even the most favorable scenarios involve significant global warming.*

In the business-as-usual scenario, the atmospheric CO_2 concentration is predicted to exceed 900 ppm by 2100—more than three times the preindustrial value. Even the rapid-conversion scenario yields twice the preindustrial value, which climate scientists consider to be a dangerous level.

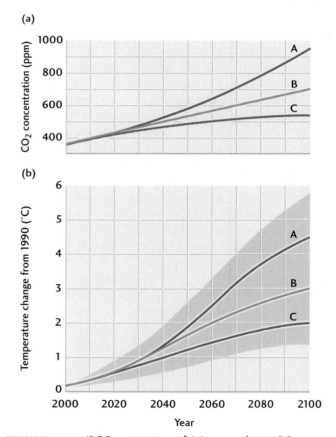

FIGURE 14.29 IPCC projections of (a) atmospheric CO_2 concentrations and (b) average surface temperatures over the twenty-first century based on three economic scenarios: (A) business as usual (continued reliance on fossil fuels), (B) a more balanced use of fossil and nonfossil fuels, and (C) rapid conversion to cleaner and more resource-efficient energy technologies. Shaded band on panel (b) shows the prediction uncertainties due to incomplete knowledge of the climate system. [*IPCC, Climate Change 2001: The Scientific Basis. Cambridge: Cambridge University Press, 2001.*]

To calculate how changes in greenhouse gas concentrations and other factors, such as land-use changes, will change temperatures at Earth's surface, we need a **climate model**—a sophisticated computer code that is capable of predicting the circulation patterns of air and water as they are driven by solar energy, including small disturbances such as storms in the atmosphere and eddies in the oceans. Such models represent the properties of the atmosphere and oceans on three-dimensional grids comprising millions of geographic points (**Figure 14.30**).

Figure 14.31 maps the temperature increases predicted by a climate model that was developed by the United Kingdom's Hadley Centre for Climate Change. These predictions were based on the current rates of greenhouse gas emissions (the business-as-usual scenario). The predicted geographic pattern of temperature changes displays some similarities with the observed pattern of twentieth-century warming in Figure 14.28. In particular, the warming is greater over land than over the oceans, and the temperate and polar regions of the Northern Hemisphere show the most warming.

The IPCC has used its scenarios for greenhouse gas increases and a number of alternative climate models to predict average global surface temperatures (Figure 14.29b). It found that the range of likely global temperature increases during the twenty-first century is

> According to current climate models, global temperatures are likely to increase by 1° to 6°C during the twenty-first century.

FIGURE 14.30 Numerical climate models are used to predict future global climate change. This climate model, developed with support from the U.S. Department of Energy, portrays interactions among the atmosphere, hydrosphere, cryosphere, and land surface. [*Warren Washington and Gary Strand/National Center for Atmospheric Research.*]

1° to 6°C. The lower values in this range can be achieved only through rapid reductions in fossil-fuel burning and the introduction of clean and resource-efficient energy technologies.

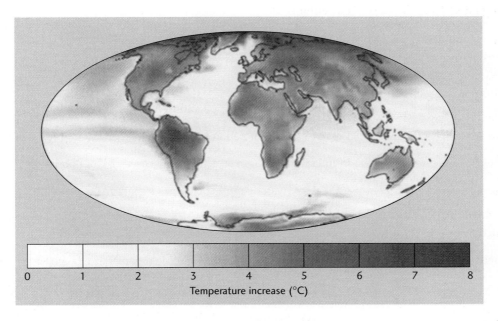

Temperature increase (°C)

FIGURE 14.31 Average surface temperatures predicted for 2070–2100, expressed as differences from average surface temperatures measured at the same location during the period 1960–1990 (the baseline period). These predictions were made using business-as-usual projections of CO_2 and other greenhouse gas emissions (see Figure 14.29a). [*Data from Hadley Centre for Climate Change; map courtesy of Robert A. Rohde.*]

NATURAL HAZARDS: Effects of Climate Change

It seems clear that human emissions of greenhouse gases will cause further global warming and may result in major changes in the climate system. These changes have the potential to affect civilization in both positive and negative ways. Some regional climates may improve—now may be the time to buy land in Canada—while others may deteriorate. The authors' home city of Los Angeles, for example, is likely to become hotter and drier. Some potential effects of climate change are listed in **Table 14.1**.

Changes in Regional Weather Patterns

The IPCC has documented a number of current trends in regional weather patterns that are likely to continue:

■ The frequency of heavy precipitation events has increased over many land areas in a manner consistent with observed temperature increases and the resulting increases in atmospheric water vapor concentrations. Increased precipitation has been observed in eastern parts of North and South America, northern Europe, and northern and central Asia.

■ Drying has been observed in the Sahel (**Figure 14.32**), the Mediterranean, southern Africa, and parts of southern Asia. More intense and longer droughts have been observed over wider areas since the 1970s, particularly in the tropics and subtropics.

■ Widespread changes in temperature extremes have been observed over the last 50 years. Cold days, cold nights, and frost have become less frequent, while hot days, hot nights, and heat waves have become more frequent.

■ Intense hurricane activity has increased in the North Atlantic. This increase is consistent with increases in tropical sea surface temperatures. Although there is no clear trend in the annual number of hurricanes, the number of very strong hurricanes (category 4 and 5 storms) has almost doubled over the past three decades.

Global warming is expected to produce heavy precipitation in some regions and drought in others, more frequent heat waves, and more intense hurricane activity.

TABLE 14.1	Potential Effects of Climate Change on Various Systems
Systems	**Potential Effects**
Forests and other terrestrial ecosystems	Migration of vegetation; reduction in ecosystem ranges; altered ecosystem composition
Species diversity	Loss of diversity; migration of species; invasion of new species
Coastal wetlands	Inundation of wetlands; migration of wetland vegetation
Aquatic ecosystems	Loss of habitat; migration to new habitats; invasion of new species
Coastal resources	Inundation of coastal structures; increased risk of flooding
Water resources	Changes in water supplies; changes in patterns of drought and flooding; changes in water quality
Agriculture	Changes in crop yields; shifts in relative productivity among regions
Human health	Shifts in ranges of infectious disease organisms; changes in patterns of heat-stress and cold-weather afflictions
Energy	Increase in cooling demand; decrease in heating demand; changes in hydroelectric energy resources.

Source: Office of Technology Assessment, U.S. Congress.

FIGURE 14.32 Members of the Mali Gao tribe digging for edible roots during the Sahel drought of 1984 and 1985. Global warming is expected to increase droughts in this and other subtropical regions. [*Frans Lemmers/Alamy.*]

Changes in the Cryosphere

Nowhere are the effects of global warming more evident than in the polar regions. The amount of sea ice in the Arctic Ocean is decreasing, and the downward trend seems to be accelerating. The sea ice cover in September 2007 was the lowest for that month since the keeping of satellite records began in 1978: 4.1 million square kilometers, down by 45 percent from the 1978–1988 average of 7.4 million square kilometers. The minimum was nearly that low in 2010 (4.6 million square kilometers), and 2011 may break the 2007 record. According to climate models, the north polar ice cap will continue to shrink rapidly, and much of the Arctic Ocean will become ice-free within a few decades (**Figure 14.33**). The shrinkage of sea ice is already severely disrupting Arctic ecosystems.

Temperatures at the top of the permafrost layer in the Arctic have gone up by 3°C since the 1980s, and the melting of permafrost is destabilizing structures such as the Trans-Alaska oil pipeline (**Figure 14.34**). The maximum area covered by seasonally frozen ground has decreased by about 7 percent in the Northern Hemisphere since 1900, with a decrease in spring of up to 15 percent. Valley glaciers at lower latitudes retreated during the

> *Climate warming is most evident in the cryosphere: reduction of Arctic sea ice and permafrost, retreat of mountain glaciers, and breakoffs from Antarctic ice shelves.*

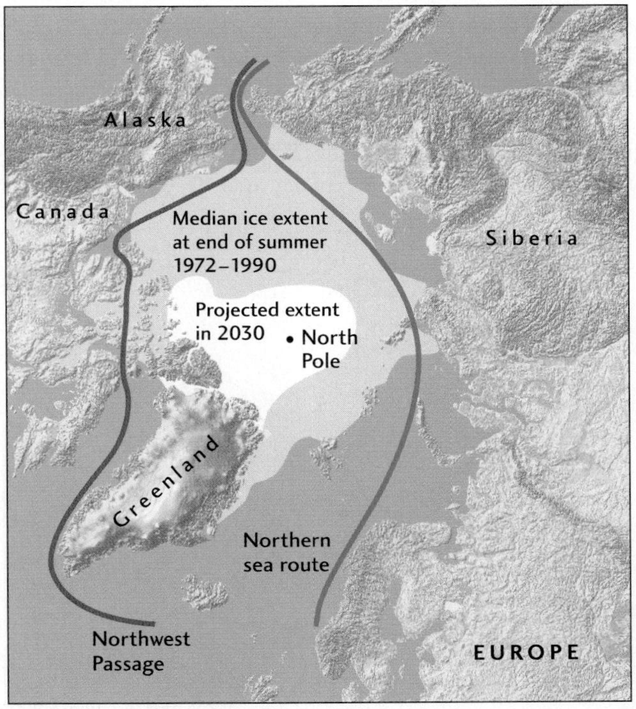

FIGURE 14.33 Global warming is melting the Arctic ice cap. This map of the Arctic compares the average extent of the polar ice cap at the end of the summer during the period 1972–1990 with its projected extent in 2030. One benefit for human society could be the opening of the Northwest Passage and other shorter sea routes between the Atlantic and Pacific oceans within the twenty-first century. [*U.S. Navy.*]

FIGURE 14.34 Permafrost melting could destabilize structures at high latitudes, such as the Trans-Alaska oil pipeline. Its 1300-km (800-mile) route from Prudhoe Bay to Valdez includes 675 km of permafrost. Where this pipeline crosses permafrost, it is perched on specially designed vertical supports. Because thawing of the permafrost would cause the supports to become unstable, they are outfitted with heat pumps designed to keep the ground around them frozen. [Galen Rowell/Corbis.]

twentieth-century warming (**Figure 14.35**). Fieldwork by Lonnie Thompson (see the story at the beginning of Chapter 10) and other geologists has demonstrated that rates of glacial retreat and snow cover loss are increasing in both hemispheres. According to a study by the USGS, Glacier National Park in northern Montana will lose the last of its glaciers by 2030.

Using high-resolution radar satellite mapping, geologists have observed that several Antarctic glaciers have retreated more than 30 km in just 3 years and that enormous pieces of ice have snapped off the Antarctic ice shelves. In March 2000, an iceberg larger than Delaware (just under 10,000 km^2) calved from the Ross Ice Shelf (see Figure 10.6b). In February and March of 2002, a portion of the Larsen Ice Shelf larger than Rhode Island (about 3250 km^2) shattered and separated from the eastern side of the Antarctic Peninsula (**Figure 14.36**). The fracturing of this piece of the ice sheet produced thousands of icebergs. The breakup of the Wilkins Ice Shelf (14,000 km^2) began in 2008 and is continuing.

FIGURE 14.35 Photographs of Qori Kalis Glacier in Peru from the same vantage point (*left*) in July 1978 and (*right*) in July 2004. Between 1998 and 2001, the ice front retreated an average of 155 meters per year, an alarming 32 times faster than the average annual retreat from 1963 to 1978. [Lonnie G. Thompson, Byrd Polar Research Center, Ohio State University.]

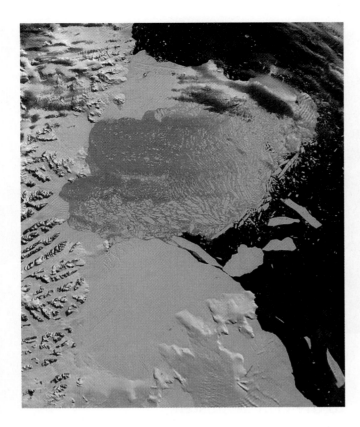

FIGURE 14.36 Collapse of the Larsen Ice Shelf. This satellite image was taken on March 7, 2002, toward the end of a 2-month period during which a huge piece of the ice shelf separated from land and splintered into thousands of icebergs. The darkest colors on the right-hand side of the image represent open seawater. The white parts are icebergs, the remaining parts of the ice shelf, and glaciers on land. The bright blue area is a mixture of seawater and highly fractured ice. The area of the image is about 150 km by 185 km. [NASA/GSFC/LaRC/JPL, MISR Team.]

ice shelves float because ice is less dense than seawater, so they displace a mass of seawater equal to their own weight (**Figure 14.37**). When they melt, the conversion of ice to water does not change the total mass of the oceans, and there is no change in sea level. (Do an experiment: the level of the drink in your glass doesn't change when the ice cubes melt, for the same reason.)

Melting ice can cause sea level to rise only if it is on land, not floating in water. The melting of glaciers sheds water and new icebergs into the sea, which is why we should be concerned by the observed rapid increase in glacial melting. The destruction of ice shelves contributes to sea level rise only if it desta-bilizes the continental glaciers that feed the ice shelves, causing

> Although the melting of floating ice will not raise sea level, the melting of glaciers on land and the warming of ocean water have the potential to do so.

Sea Level Rise

If the ice shelves around Antarctica continue to collapse, will sea level rise? It turns out that even if all of Earth's ice shelves were to break off into the ocean over the next few years, sea level wouldn't change much at all. Like icebergs,

If an iceberg in the ocean melts, sea level does not change.

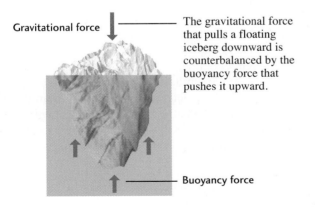

If ice on land melts or ruptures and slides into the sea, sea level rises.

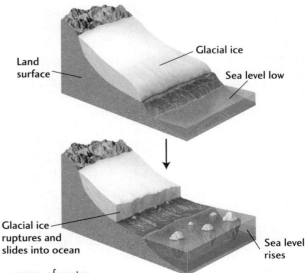

FIGURE 14.37 Ice shelves and icebergs float and thus displace a mass of water equal to their own mass; therefore, their melting by global warming will not change sea level. The melting of ice sheets on land, however, injects new water into the oceans, causing sea level to rise.

FIGURE 14.38 Aerial view of a coastal region of Bangladesh flooded by a storm surge in May 2009. This low-lying country would be subject to disastrous flooding if sea level rises due to global warming. [Jayanta Shaw/Reuters.]

the glaciers to flow more rapidly into the oceans. Such an acceleration of glacial flow was observed after the Larsen Ice Shelf collapse.

Sea level also rises as the oceans warm, because seawater expands with temperature. The IPCC estimates that sea level rose about 170 mm because of the twentieth-century warming and is currently rising at about 3 mm per year. So far, most of the rise, about two-thirds, has been due to the thermal expansion of seawater by ocean warming.

As continental glaciers melt at a faster rate, sea level rise is expected to accelerate. Climate models indicate that sea level could rise by a meter or more during the twenty-first century, creating serious problems for low-lying countries such as Bangladesh (**Figure 14.38**), as well as the Eastern Seaboard and Gulf Coast of the United States, where flooding during coastal storm surges could become much worse.

> *Climate models indicate that sea level could rise by a meter or more during the twenty-first century.*

Species and Ecosystem Migration

As local and regional climates change, ecosystems will change with them. Many plant and animal species will have difficulty adjusting to rapid climate change or migrating to more suitable climates. Those that cannot cope with the rapid warming could become extinct. Global warming is already being blamed for a variety of adverse ecological effects, such as the disruption of Arctic ecosystems as sea ice begins to melt, and the spread of tropical diseases like malaria as more of the world experiences a tropical climate.

The Potential for Catastrophic Changes to the Climate System

The current atmospheric concentrations of carbon dioxide and methane far exceed anything seen in the last 650,000 years. Our climate system is therefore entering unknown territory. Some observers think that the credibility of climate change projections suffers from the "Chicken Little" problem: too many people are running around yelling, "The sky is falling!" Yet most scientists think that those projections may be too conservative because they do not properly take into account some of the positive feedbacks within the climate system that could greatly enhance climate change. Ocean acidification, which we discussed earlier in this chapter, is a good example. Here are a few more:

■ *Destabilization of continental glaciers.* The surface melting of the Greenland glacier in 2010 was the largest on record, and there are indications that glacial streams within the ice sheet are accelerating much faster than expected. If the Greenland and Antarctic glaciers begin to shed ice faster than snowfall can generate new glacial ice, sea level could begin to rise much faster than current projections.

■ *Shutoff of thermohaline circulation.* Changes in precipitation and evaporation patterns are decreasing the salinity of seawater at mid- and high latitudes. Some scientists have speculated that this change could substantially reduce global thermohaline circulation (see Figure 10.3), which is driven by differences in temperature and salinity. Major changes in the Gulf Stream and other aspects of the climate system could result.

■ *Methane release from seafloor sediments and permafrost.* Recall from Chapter 9 that a massive release of methane from shallow seafloor sediments about 55 million years ago may have caused abrupt global warming and led to the mass extinction at the Paleocene-Eocene boundary. Today, there is far more methane stored in shallow seafloor sediments and in permafrost than was released at the end of the Paleocene. If global warming begins to thaw these methane deposits, another runaway cycle of extreme warming could begin.

> *Positive feedbacks within the climate system could greatly enhance anthropogenic climate change, with potentially catastrophic effects.*

Welcome to the Anthropocene: A New Geologic Epoch

In 2003, atmospheric chemist and Nobel laureate Paul Crutzen proposed the recognition of a new geologic epoch—the **Anthropocene,** or Age of Humans—beginning about 1780, when James Watt's coal-powered steam engine launched the industrial revolution. The global changes that mark the Holocene-Anthropocene boundary are just now getting under way, so a future scientist with a full record of the next few thousand years may place the geologic boundary at a somewhat later date. Crutzen's main point, however, is that global changes are proceeding so rapidly that such quibbles are likely to be minor. As with many previous geologic boundaries, the main marker will be a mass extinction.

> *Changes to Earth's environment caused by rapid industrialization, including a major drop in biodiversity, may mark a new geologic epoch, the Anthropocene.*

Between 1850 and 1880, as much as 15 percent of the land surface was deforested, and rates of deforestation have continued to rise. According to the United Nations, over 150,000 km^2 of tropical rain forests—about 1 percent of the total resource—are being converted each year to other land uses, mostly agricultural ones. In 1950, forests covered approximately 25 percent of Haiti (a Caribbean island country the size of Maryland); its forested area now stands at less than 2 percent (**Figure 14.39**). Other developing nations face a similar problem.

FIGURE 14.39 The Caribbean island country of Haiti is now 98 percent deforested.
[*Daniel Morel/AP World Wide.*]

Given these rates of habitat loss, it is not surprising that the number of extant species, the most important measure of *biodiversity*, is declining. Biologists estimate that there are over 10 million different species alive on the planet today, although only 1.5 million have been officially classified. Extinction rates are difficult to quantify, but most knowledgeable scientists believe that up to one-fifth of all species will disappear during the next 30 years, and as many as one-half may go extinct during the twenty-first century. One respected biologist, Peter Raven, has put the problem bluntly:

We are confronting an episode of species extinction greater than anything the world has experienced for the past 65 million years. Of all the global problems that confront us, this is the one that is moving the most rapidly and the one that will have the most serious consequences. And, unlike other global ecological problems, it is completely irreversible.

Some observers, such as the sociobiologist E. O. Wilson, have gone so far as to call the current rapid worldwide decline in biodiversity the "Sixth Extinction," placing it in the same rank as the "Big Five" mass extinctions of the Phanerozoic eon (see Figure 2.19 and Chapter 9). Others consider this claim to be premature, however, because even the rapid losses in biodiversity we see today will not necessarily affect the fossil record as profoundly as, say, the impact-related mass extinction at the end of the Cretaceous period, or even the less severe mass extinction associated with global warming at the Paleocene-Eocene boundary.

EARTH SYSTEM ENGINEERING AND MANAGEMENT

By any measure, the problems we face in confronting global change are daunting. If the human population and its per capita energy use continue to grow at their current rates, fossil-fuel burning will cause the rate of carbon emissions to nearly double in the next 50 years, from 8 Gt per year to at least 15 Gt per year. The CO_2 concentration in the atmosphere would reach 650 ppm and continue to increase thereafter, with potentially disastrous consequences. Controlling our carbon emissions—one of civilization's most important tasks—will require an extraordinary collaboration of Earth scientists, policy makers, and the public.

Energy Policy

There is little question that we will need to make changes in our energy sources. One set of questions policy makers must tackle is how much money we should spend to curb anthropogenic carbon emissions and whether the benefits will justify the costs. Too much spending could depress the economy and cause job losses; yet preventing the most drastic effects of climate change might be less costly than coping with disasters after they happen.

A partial solution—and certainly the most economical one—is to improve energy use efficiency and reduce waste. In a real sense, using energy more efficiently is like discovering a new source of fuel. Some experts believe that the United States could reduce its emissions of greenhouse gases by as much as 50 percent from 1990 levels by implementing efficiency measures that cost relatively little—for example, insulating buildings, replacing incandescent light bulbs with fluorescent bulbs, increasing the fuel efficiency of motor vehicles, and making greater use of natural gas. The savings could amount to hundreds of billions of dollars a year, much of which is now being spent on expensive imported oil. These modest steps would offer substantial fringe benefits as well, including lowered manufacturing costs and improved air quality.

The most economical way to reduce carbon emissions is to improve the efficiency of our energy system.

Many observers would say that fossil fuels are simply too cheap in the United States. Carbon emissions are not taxed as they are in many other developed nations; consequently, there is little incentive for energy conservation or conversion to new energy sources. The full economic costs of fossil fuels include the costs of cleaning up atmospheric pollution, oil spills, and other environmental damage; the costs of trade deficits; and the military costs of defending oil supplies, as well as the costs of global warming. If these costs were included in energy pricing, alternative energy sources would become much more competitive with fossil fuels. Such *full-cost accounting* has not been politically popular in the United States, however.

We also face the issue of fairness in international politics. The United States, Canada, the European Union, and Japan—with only 25 percent of the world's population—are responsible for about 75 percent of the global increase in atmospheric greenhouse gases. These rich industrial nations are better able to pay the costs of reducing their greenhouse gas emissions more easily than the developing countries. China, for example, depends on its huge coal deposits for its economic growth, and its turbocharged economy became the world's leader in greenhouse gas emissions in 2007 (**Figure 14.40**). Developing nations argue that they will need financial and technological support from the developed countries to help them cope with the demand to reduce emissions. Policy makers have come to agree that our problems cannot be solved on a national scale and will have to be addressed through international cooperation.

FIGURE 14.40 A coal-fired power plant near Zibo, a city in Shandong Province, China. In 2007, China superseded the United States as the nation with the highest rate of greenhouse gas emissions. The carbon economies of China, India, and other developing countries will have a huge influence on future climates. [© David Lyons/Alamy.]

Alternative Energy Resources

As we have seen, no single alternative energy source will be able to quickly replace fossil fuels. However, some renewable energy resources, such as solar power, wind power, and biofuels, are becoming more important contributors to our energy system. If these technologies were aggressively implemented during the next 50 years, together they could reduce carbon emissions by gigatons per year.

Another step that could be taken is to increase the use of nuclear energy. The capacity of nuclear power plants, which today is approximately 350 gigawatts, could easily be tripled in the next 50 years, but this option is unattractive to many people, for reasons of safety and security. The potential exists for cleaner nuclear technologies, such as *fusion power*—the use of small, controlled thermonuclear explosions to generate energy. But scientific progress toward this goal has been slow, and conceptual breakthroughs will be required.

> *Increased use of alternative energy resources during the next 50 years could reduce carbon emissions by gigatons per year.*

Engineering the Carbon Cycle

What about the possibility of engineering the carbon cycle to reduce the accumulation of greenhouse gases in the atmosphere? Several promising technologies aim to reduce greenhouse gas emissions by pumping the carbon dioxide generated by fossil-fuel combustion into reservoirs other than Earth's atmosphere—a procedure known as **carbon sequestration.**

One obvious reservoir for carbon is the biosphere. Earlier in this chapter, we saw that forests withdraw CO_2 from the atmosphere in surprisingly large amounts. Land-use policies that would not only slow the current high rates of deforestation but also encourage reforestation and other biomass production might help to reduce the accumulation of atmospheric carbon.

Biotechnology might provide ways of increasing the capacity of the biosphere to sequester carbon. One possibility is the engineering of genetically modified bacteria that would be capable of metabolizing methane, sequestering the carbon it contains, and giving off hydrogen. Hydrogen is the ultimate clean fuel; burning it produces only water.

One straightforward technology for carbon sequestration—underground storage of CO_2—offers considerable promise. Carbon dioxide captured from oil and gas wells is already being pumped back into the ground as a means of moving oil toward the wells. If capture and underground storage of the carbon dioxide from coal-fired power plants were economically feasible, the world's abundant coal resources would become much more attractive as a replacement for petroleum.

Stabilizing Carbon Emission

The strategies and technologies we have just discussed may seem promising, but will they be enough? Under the business-as-usual scenario, as we have seen, carbon emissions are expected to increase by at least 7 Gt per year during the next half-century. How can this increase be stopped? In other words, what would it take to *stabilize carbon emissions at current levels?*

Two scientists from Princeton University, Stephen Pacala and Robert Socolow, have provided a simple quantitative framework to answer this question. They begin by admitting that there is no single solution to the problem—no "silver bullet." Instead, they break the problem into what they call **stabilization wedges,** each of which offsets the projected growth of carbon emissions by 1 gigaton per year in the next 50 years (**Figure 14.41**). Therefore, one wedge roughly corresponds to one-seventh of the problem.

Implementing each stabilization wedge will be a monumental task. To achieve wedge 1, for example, the average gasoline mileage of the world's entire fleet of passenger vehicles, which will grow to 2 billion by mid-century, will have to be increased from 30 miles per gallon (mpg) to 60 mpg. This calculation assumes that a car is driven 10,000 miles per year, the current annual average. An alternative, not shown in Figure 14.41, would be to maintain gas mileage at 30 mpg but reduce the average

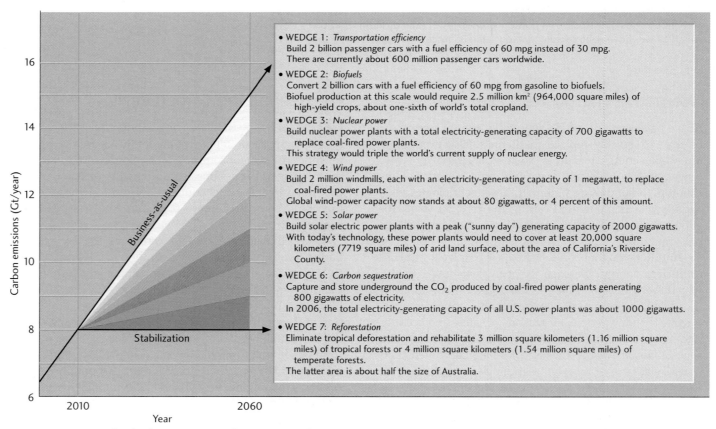

FIGURE 14.41 Under the business-as-usual scenario, carbon emissions are expected to increase by at least 7 gigatons per year during the next 50 years. The problem of stabilizing carbon emissions at their 2010 value can be broken into seven stabilization wedges, each representing a reduction in emissions of 1 Gt per year by 2056. [After S. Pacala and R. Socolow, *Science* 305: 968-972 (2004).]

amount of driving by half, to 5000 miles per year. Yet another alternative (wedge 2) would be to convert all cars to biofuels. Growing this much biofuel would take up one-sixth of the world's total cropland, so this strategy would adversely affect agricultural productivity and food supplies.

Some of the stabilization wedges involve controversial or expensive technologies, such as expanding nuclear power by a factor of three (wedge 3), increasing the number of large windmills into the millions (wedge 4), or covering large desert areas with solar panels (wedge 5). At least one of the proposed wedges, the capture and storage of carbon emitted from coal power plants (wedge 6), is at the margin of current technological feasibility. The last option, elimination of tropical deforestation and the reforestation of huge

> *Implementing at least seven stabilization wedges will be necessary to keep carbon emissions at their current level.*

additional land areas (wedge 7), is favored by many people in principle, but it would be difficult to achieve without imposing severe restrictions on developing countries such as Brazil.

The stabilization of carbon emissions at current emission rates would reduce but not eliminate the threat of global climate change. The stabilization scenario (scenario C in Figure 14.29) would still allow the atmospheric concentration of CO_2 to grow to 500 ppm, almost twice the preindustrial value. Further reductions in carbon emissions during the second half of the twenty-first century would be necessary to maintain the concentration below this value. Climate models indicate that such a scenario would still increase the average global temperature by about 2°C, more than three times the total twentieth-century warming (see Figure 14.29b).

Nevertheless, the continued rise of atmospheric CO_2 concentrations is not inevitable. The available inventory of stabilization wedges provides a technological framework for policy makers. Taking on the stabilization problem involves other difficulties such as developing broad public consensus and creating international agreements. Yet, as the Pacala-Socolow analysis demonstrates, there is still time for actions that can substantially reduce anthropogenic global change. Whether we can grasp this opportunity will depend on our understanding of the problem, its potential solutions, and the consequences of inaction.

◼ SUMMARY

In what sense is human civilization a global geosystem? Human society's use and production of energy now competes with the plate tectonic and climate systems in modifying Earth's environment. Most of the energy used by human civilization comes from carbon-based fuels. The rise of this carbon economy has altered the natural carbon cycle by creating a huge new flow of carbon from the lithosphere to the atmosphere. If that flow continues unabated, CO_2 concentrations in the atmosphere will double by the mid-twenty-first century.

What is the origin of oil and natural gas? Oil and natural gas form from organic matter deposited in oxygen-poor sedimentary basins, typically on continental margins. Under elevated temperatures and pressures, the buried organic matter is transformed into liquid and gaseous hydrocarbons. Oil and gas accumulate where geologic structures called oil traps create impermeable barriers to their upward migration.

Why is there concern about the world's oil supply? Oil is a nonrenewable resource: at current rates of use, it will be depleted far faster than geologic processes can replenish it. Therefore, as oil is withdrawn from the reservoirs of the world, its availability will diminish and its price will rise. The key issue is not when oil will run out but when global oil production will reach Hubbert's peak—when it will stop rising and begin to decline. Oil optimists argue that oil resources will meet demand for decades to come; oil pessimists think we are within a few years of Hubbert's peak.

What is the origin of coal, and what are the consequences of burning it? Coal is formed by the burial, compression, and metamorphism of wetland vegetation. There are huge resources of coal in sedimentary rocks. Coal combustion is a major source of atmospheric CO_2 as well as sulfur oxides that cause acid rain. Coal mining and toxic substances produced by coal burning present risks to human life and to the environment. Because of its abundance and low cost, however, the use of coal is likely to increase.

What are the prospects for alternative energy sources? Alternative energy sources include nuclear power, biofuels, and solar, hydroelectric, wind, and geothermal energy. Taken together, these energy sources currently supply only a small percentage of world energy demand. Nuclear energy produced by the fission of uranium, the world's most abundant minable resource, could be a major energy source, but only if the public can be assured of its safety and security. With advances in technology and reductions in cost, renewable sources such as solar energy, wind energy, and biomass could become major contributors in the twenty-first century.

What is the carbon cycle? The carbon cycle describes the flux of carbon among its four principal reservoirs: the atmosphere, biosphere, oceans, and terrestrial biosphere. Major fluxes of carbon between these reservoirs include gas exchange between the atmosphere and the ocean surface; the movement of CO_2 between the biosphere and the atmosphere through photosynthesis, respiration, and decay; the transport of dissolved organic carbon in surface waters to the ocean; and the weathering and precipitation of calcium carbonate.

How much global warming will there be in the twenty-first century, and what will be its consequences? Atmospheric concentrations of greenhouse gases will continue to rise throughout the twenty-first century, primarily because of fossil-fuel combustion and other human activities. The magnitude of the increase will depend on whether human society takes active steps to limit its greenhouse gas emissions. Projections of climate warming during the twenty-first century are highly uncertain, but the range of likely warming is 1° to 6°C. This warming will disrupt ecosystems and increase the rate of species extinctions. The oceans will warm and expand, raising sea level as much as a meter, and they will become more acidic. The north polar ice cap will continue to shrink rapidly, and much of the Arctic Ocean is expected to become ice-free.

How might we stabilize carbon emissions at their current levels? If human civilization continues to rely on fossil fuels, anthropogenic carbon emissions will increase by at least 7 Gt per year during the next 50 years. This problem could be addressed by implementing a series of stabilization wedges, each offsetting the projected growth of carbon emissions by a substantial fraction of the increase.

■ KEY TERMS AND CONCEPTS

Anthropocene (p. 499)	geothermal energy (p. 483)
biofuel (p. 484)	Hubbert's peak (p. 474)
carbon cycle (p. 485)	natural gas (p. 471)
carbon economy (p. 469)	nuclear energy (p. 480)
carbon sequestration (p. 501)	ocean acidification (p. 487)
	oil trap (p. 472)
climate model (p. 493)	oil window (p. 471)
coal bed (p. 477)	residence time (p. 486)
crude oil (p. 471)	solar energy (p. 481)
geochemical reservoir (p. 485)	stabilization wedge (p. 502)

■ EXERCISES

1. Describe some of the ways in which human civilization is fundamentally different from the natural geosystems we have studied in this textbook.

2. Which fossil fuel produces the least amount of CO_2 per unit of energy: oil, natural gas, or coal?

3. What are the prerequisites required for oil traps to contain oil?

4. Explain which of the following factors are important in estimating the future supply of oil and natural gas: (a) the rate of oil and gas accumulation, (b) the rate of depletion of known reserves, (c) the rate of discovery of new reserves, (d) the total amount of oil and gas now present on Earth.

5. Are you an oil optimist or an oil pessimist? Explain why.

6. An aggressive drilling program in the Arctic National Wildlife Refuge could produce as much as 16 billion barrels of oil. At current consumption rates, for how many years would this resource supply U.S. oil demand?

7. Which three countries have the largest coal reserves?

8. What issues related to the use of nuclear energy can be addressed by geologists?

9. Contrast the risks and benefits of nuclear fission and coal combustion as energy sources.

10. In the 1990s, emissions of carbon into the atmosphere from fossil-fuel burning and other industrial activities averaged about 6.4 Gt per year, almost all of it in the form of carbon dioxide. What was the mass of carbon dioxide emitted?

11. From the information given in Figure 14.23, estimate the residence time of CO_2 (a) in the ocean and (b) in the atmosphere.

12. If human activities keep pumping CO_2 into the atmosphere at a steadily increasing rate and Earth's climate warms significantly in the next 100 years, how might the global carbon cycle be affected?

13. Do you think we should act now to reduce carbon emissions or delay until the functioning of climate change is better understood?

14. What do you think will be the major sources of the world's energy in the year 2030? In the year 2100?

15. Is it justified to insist that developing countries such as China and India agree to limit their future carbon emissions, even though they have historically used much less fossil fuel than developed countries such as the United States?

16. Do you think that future scientists and engineers will be able to modify the natural carbon cycle to prevent catastrophic changes in the Earth system?

17. Do you think a geologist several thousand years in the future will consider the industrial revolution to be the beginning of a new geologic epoch?

Visual Literacy Task

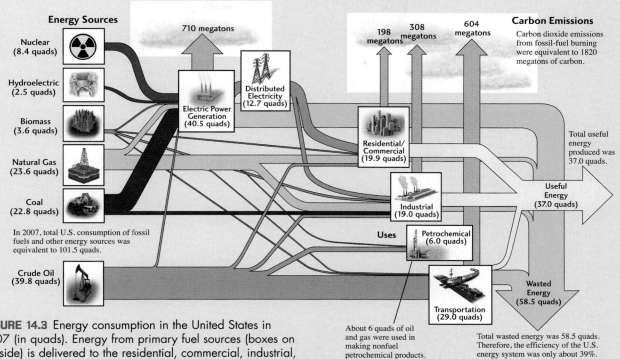

FIGURE 14.3 Energy consumption in the United States in 2007 (in quads). Energy from primary fuel sources (boxes on left side) is delivered to the residential, commercial, industrial, and transportation sectors (boxes in middle to right side). Not represented are small contributions to electric power generation from solar (0.1 quad), wind (0.3 quad), and geothermal (0.3 quad) energy. [*Adapted from a special study by Lawrence Livermore National Laboratory based on data from the Energy Information Administration.*]

1. What is most crude oil used for?

a. Electric power generation
b. Residential/commercial consumption
c. Industrial consumption
d. Making petrochemicals
e. Transportation

2. What do the pictures in the middle of the diagram represent?

a. Where energy comes from
b. Amount of carbon emitted
c. How energy is used
d. Waste products of energy

3. What does the width of the lines connecting pictures represent?

a. Type of energy source
b. Amount of energy
c. Amount of carbon emissions
d. How energy is used

4. Which energy use releases the most carbon into the atmosphere?

a. Electric power generation
b. Residential/commercial consumption
c. Industrial consumption
d. Transportation

5. What happens when electric power is generated?

a. It is almost all distributed to residential/commercial and industrial uses.
b. More of it is distributed to residential/commercial and industrial uses than is wasted.
c. More of it is wasted than is distributed to residential/commercial and industrial uses.

Thought Question: How would you respond to a fellow student who claimed, "Cars that run on electricity result in no carbon being emitted to the atmosphere."

Appendixes

Appendix 1

Conversion Factors

LENGTH

1 centimeter	0.3937 inch
1 inch	2.5400 centimeters
1 meter	3.2808 feet; 1.0936 yards
1 foot	0.3048 meter
1 yard	0.9144 meter
1 kilometer	0.6214 mile (statute); 3281 feet

LENGTH

1 mile (statute)	1.6093 kilometers
1 mile (nautical)	1.8531 kilometers
1 fathom	6 feet; 1.8288 meters
1 angstrom	10^{-8} centimeter
1 micrometer	0.0001 centimeter

VELOCITY

1 kilometer/hour	27.78 centimeters/second
1 mile/hour	17.60 inches/second

AREA

1 square centimeter	0.1550 square inch
1 square inch	6.452 square centimeters
1 square meter	10.764 square feet; 1.1960 square yards
1 square foot	0.0929 square meter
1 square kilometer	0.3861 square mile
1 square mile	2.590 square kilometers
1 acre (U.S.)	4840 square yards

VOLUME

1 cubic centimeter	0.0610 cubic inch
1 cubic inch	16.3872 cubic centimeters
1 cubic meter	35.314 cubic feet
1 cubic foot	0.02832 cubic meter
1 cubic meter	1.3079 cubic yards
1 cubic yard	0.7646 cubic meter
1 liter	1000 cubic centimeters; 1.0567 quarts (U.S. liquid)
1 gallon (U.S. liquid)	3.7853 liters

MASS

1 gram	0.03527 ounce
1 ounce	28.3495 grams
1 kilogram	2.20462 pounds
1 pound	0.45359 kilogram

PRESSURE

1 kilogram/square centimeter	0.96784 atmosphere; 0.98067 bar; 14.2233 pounds/square inch
1 bar	0.98692 atmosphere; 10^5 pascals

ENERGY

1 joule	0.239 calorie; 9.479×10^{-4} Btu
1 British thermal unit (Btu)	251.9 calories; 1054 joules
1 quad	10^{15} Btu

POWER

1 watt	0.001341 horsepower (U.S.); 3.413 Btu/hour

Degrees	
F	C
210	100
200	
190	90
180	80
170	
160	70
150	
140	60
130	
120	50
110	
100	40
90	30
80	
70	20
60	
50	10
40	
30	0
20	
10	10
0	
10	20

Appendix 2

Numerical Data Pertaining to Earth

Equatorial radius	6378 kilometers
Polar radius	6357 kilometers
Radius of sphere with Earth's volume	6371 kilometers
Volume	1.083×10^{27} cubic centimeters
Surface area	5.1×10^{18} square centimeters
Percent surface area of oceans	71
Percent surface area of land	29
Average elevation of land	623 meters
Average depth of oceans	3.8 kilometers
Mass	5.976×10^{27} grams
Density	5.517 grams/cubic centimeters
Gravity at equator	978.032 centimeters/second/second
Mass of atmosphere	5.1×10^{21} grams
Mass of ice	$25–30 \times 10^{21}$ grams
Mass of oceans	1.4×10^{24} grams
Mass of crust	2.5×10^{25} grams
Mass of mantle	4.05×10^{27} grams
Mass of core	1.90×10^{27} grams
Mean distance to Sun	1.496×10^{8} kilometers
Mean distance to Moon	3.844×10^{5} kilometers
Ratio: Mass of Sun/mass of Earth	3.329×10^{5}
Ratio: Mass of Earth/mass of Moon	81.303
Total geothermal energy reaching Earth's surface each year	10^{21} joule; 2.39×10^{20} calories; 949 quads
Earth's daily receipt of solar energy	14,137 quads; 1.49×10^{22} joules
U.S. energy consumption, 2007	101.5 quads

Appendix 3

Chemical Reactions

Electron Shells and Ion Stability

Electrons surround the nucleus of an atom in a unique set of concentric spheres called electron shells. Each shell can hold a certain maximum number of electrons. In the chemical reactions of most elements, only the electrons in the outermost shells interact. In the reaction between sodium (Na) and chlorine (Cl) that forms sodium chloride (NaCl), the sodium atom loses an electron from its outer shell of electrons, and the chlorine atom gains an electron in its outer shell (see Figure 4.4, page 98).

Before reacting with chlorine, the sodium atom has one electron in its outer shell. When it loses that electron, its outer shell is eliminated and the next shell inward, which has eight electrons (the maximum that this shell can hold), becomes the outer shell. The original chlorine atom had seven electrons in its outer shell, with room for

a total of eight. By gaining an electron, its outer shell is filled. Many elements have a strong tendency to acquire a full outer electron shell, some by gaining electrons and some by losing them in the course of a chemical reaction. The stability of ions with fully occupied outer shells is related to the interactions of electrons in various orbitals around the nucleus.

Many chemical reactions entail gains and losses of several electrons as two or more elements combine. The element calcium (Ca), for example, becomes a doubly charged cation, Ca^{2+}, as it reacts with two chlorine atoms to form calcium chloride. (In the chemical formula for calcium chloride, $CaCl_2$, the presence of two chloride ions is symbolized by the subscript 2.) Chemical formulas thus show the relative proportion of atoms or ions in a compound. Common practice is to omit the subscript 1 next to single ions in a formula.

The periodic table.

The periodic table organizes the elements (from left to right along a row) in order of atomic number (the number of protons), which also means increasing the numbers of electrons in the outer shell. The third row from the top, for example, starts at the left with sodium (atomic number 11), which has one electron in its outer shell. The next is magnesium (atomic number 12), which has two electrons in its outer shell, followed by aluminum (atomic number 13), with three, and silicon (atomic number 14), with four. Then come phosphorus (atomic number 15), with five; sulfur (atomic number 16), with six; and chlorine (atomic number 17), with seven. The last element in this row is argon (atomic number 18), with eight electrons, the maximum possible, in its outer shell. Each column in the table forms a vertical grouping of elements with similar electron-shell patterns.

Elements That Tend to Lose Electrons

The elements in the leftmost column of the table all have a single electron in their outer shells and have a strong tendency to lose that electron in chemical reactions. Of this group, hydrogen (H), sodium (Na), and potassium (K) are found in major abundance at Earth's surface and in its crust.

The second column from the left includes two more elements of major abundance, magnesium (Mg) and calcium (Ca). Elements in this column have two electrons in their outer shells and a strong tendency to lose both of them in chemical reactions.

Elements That Tend to Gain Electrons

Toward the right side of the table, the two columns headed by oxygen (O), the most abundant element in the Earth, and fluorine (F), a highly reactive toxic gas, group the elements that tend to gain electrons in their outer shells. The elements in the column headed by oxygen have six of the possible eight electrons in their outer shells and tend to gain two electrons. Those in the column headed by fluorine have seven electrons in their outer shells and tend to gain one.

Other Elements

The columns between those farthest on the left and those farthest on the right have varying tendencies to gain, lose, or share electrons. The column toward the right side of the table headed by carbon (C) includes silicon (Si), of major abundance in the Earth. Both silicon and carbon tend to share electrons.

The elements in the last column on the right, headed by helium (He), have full outer shells and thus no tendency either to gain or to lose electrons. As a result, these elements, in contrast with those in other columns, do not react chemically with other elements, except under very special conditions.

Appendix 4

Properties of the Most Common Minerals of Earth's Crust

	Mineral or Group Name	Structure or Composition	Varieties and Chemical Composition	Form, Diagnostic Characteristics	Cleavage, Fracture	Color	Hardness
LIGHT-COLORED MINERALS, VERY ABUNDANT IN EARTH'S CRUST IN ALL MAJOR ROCK TYPES	FELDSPAR	FRAMEWORK SILICATES	*POTASSIUM FELDSPARS* $KAlSi_3O_8$ Sanidine Orthoclase Microcline	Cleavable coarsely crystalline or finely granular masses; isolated crystals or grains in rocks, most commonly not showing crystal faces	Two at right angles, one perfect and one good; pearly luster on perfect cleavage	White to gray, frequently pink or yellowish; some green	6
			PLAGIOCLASE FELDSPARS $NaAlSi_3O_8$ Albite $CaAl_2Si_2O_8$ Anorthite		Two at nearly right angles, one perfect and one good; fine parallel striations on perfect cleavage	White to gray, less commonly greenish or yellowish	
	QUARTZ		SiO_2	Single crystals or masses of 6-sided prismatic crystals; also formless crystals and grains or finely granular or massive	Very poor or nondetectable; conchoidal fracture	Colorless, usually transparent; also slightly colored smoky gray, pink, yellow	7
	MICA	SHEET SILICATES	*MUSCOVITE* $KAl_3Si_3O_{10}(OH)_2$	Thin, disc-shaped crystals, some with hexagonal outlines; dispersed or aggregates	One perfect; splittable into very thin, flexible, transparent sheets	Colorless; slight gray or green to brown in thick pieces	2–$2\frac{1}{2}$
DARK-COLORED MINERALS, ABUNDANT IN MANY KINDS OF IGNEOUS AND METAMORPHIC ROCKS			*BIOTITE* $K(Mg, Fe)_3AlSi_3O_{10}(OH)_2$	Irregular, foliated masses; scaly aggregates	One perfect; splittable into thin, flexible sheets	Black to dark brown; translucent to opaque	$2\frac{1}{2}$–3
			CHLORITE $(Mg,Fe)_5(Al,Fe)_2Si_3O_{10}(OH)_8$	Foliated masses or aggregates of small scales	One perfect; thin sheets flexible but not elastic	Various shades of green	2–$2\frac{1}{2}$
	AMPHIBOLE	DOUBLE CHAINS	*TREMOLITE-ACTINOLITE* $Ca_2(Mg,Fe)_5Si_8O_{22}(OH)_2$	Long, prismatic crystals, usually 6-sided; commonly in fibrous masses or irregular aggregates	Two perfect cleavage directions at 56° and 124° angles	Pale to deep green Pure tremolite white	5–6
			HORNBLENDE Complex Ca, Na, Mg, Fe, Al silicate				
	PYROXENE	SINGLE CHAINS	*ENSTATITE-HYPERSTHENE* $(Mg,Fe)_2Si_2O_6$	Prismatic crystals, either 4- or 8-sided; granular masses and scattered grains	Two good cleavage directions at about 90°	Green and brown to grayish or greenish white	5–6
			DIOPSIDE $(Ca,Mg)_2Si_2O_6$			Light to dark green	
			AUGITE Complex Ca, Na, Mg, Fe, Al silicate			Very dark green to black	

(Continued)

Mineral or Group Name	Structure or Composition	Varieties and Chemical Composition	Form, Diagnostic Characteristics	Cleavage, Fracture	Color	Hardness
LIGHT-COLORED MINERALS, TYPICALLY AS ABUNDANT CONSTITUENTS OF SEDIMENTS AND SEDIMENTARY ROCKS						
OLIVINE	ISOLATED TETRAHEDRA	$(Mg,Fe)_2SiO_4$	Granular masses and disseminated small grains	Conchoidal fracture	Olive to grayish green and brown	$6\frac{1}{2}$–7
GARNET		Ca, Mg, Fe, Al silicate	Isometric crystals, well formed or rounded; high specific gravity, 3.5–4.3	Conchoidal and irregular fracture	Red and brown, less commonly pale colors	$6\frac{1}{2}$–7
CALCITE	CARBONATES	$CaCO_3$	Coarsely to finely crystalline in beds, veins, and other aggregates; cleavage faces may show in coarser masses; calcite effervesces rapidly in acid	Three perfect cleavages, at oblique angles; splits to rhombohedral cleavage pieces	Colorless, transparent to translucent; variously colored by impurities	3
DOLOMITE		$CaMg(CO_3)_2$	but dolomite effervesces slowly and only if crushed into powder			$3\frac{1}{2}$–4
CLAY MINERALS	HYDROUS ALUMINO-SILICATES	*KAOLINITE* $Al_2Si_2O_5(OH)_4$ *ILLITE* Similar to muscovite $+Mg,Fe$ *SMECTITE* Complex Ca, Na, Mg, Fe, Al silicate $+ H_2O$	Earthy masses in soils; bedded; in association with other clays, iron oxides, or carbonates; plastic when wet; montmorillonite swells when wet	Earthy, irregular	White to light gray and buff; also gray to dark gray, greenish gray, and brownish depending on impurities and associated minerals	$1\frac{1}{2}$–$2\frac{1}{2}$
GYPSUM	SULFATES	$CaSO_4 \cdot 2H_2O$	Granular, earthy, or finely crystalline masses; tabular crystals	One perfect, splitting to fairly thin slabs or sheets; two other good cleavages	Colorless to white; transparent to translucent	2
ANHYDRITE		$CaSO_4$	Massive or crystalline aggregates in beds and veins	One perfect, one nearly perfect, one good; at right angles	Colorless, some tinged with blue	3–$3\frac{1}{2}$
HALITE	HALIDES	$NaCl$	Granular masses in beds; some cubic crystals; salty taste	Three perfect cleavages at right angles	Colorless, transparent to translucent	$2\frac{1}{2}$
OPAL-CHALCEDONY	SILICA	SiO_2 [Opal is an amorphous variety; chalcedony is a formless microcrystalline quartz.]	Beds in siliceous sediments and chert; in veins or banded aggregates	Conchoidal fracture	Colorless or white when pure, but tinged with various colors by impurities in bands, especially in agates	5–$6\frac{1}{2}$
DARK-COLORED MINERALS, COMMON IN MANY ROCK TYPES						
MAGNETITE	IRON OXIDES	Fe_3O_4	Magnetic; disseminated grains, granular masses; occasional octahedral isometric crystals; high specific gravity, 5.2	Conchoidal or irregular fracture	Black, metallic luster	6

Mineral or Group Name	Structure or Composition	Varieties and Chemical Composition	Form, Diagnostic Characteristics	Cleavage, Fracture	Color	Hardness
HEMATITE	IRON OXIDES	Fe_2O_3	Earthy to dense masses, some with rounded forms, some granular or foliated; high specific gravity, 4.9–5.3	None; uneven, sometimes splintery fracture	Reddish brown to black	5–6
"LIMONITE"		*GOETHITE* [the major mineral of the mixture called "limonite," a field term] $FeO(OH)$	Earthy masses, massive bodies or encrustations, irregular layers; high specific gravity, 3.3–4.3	One excellent in the rare crystals; usually an early fracture	Yellowish brown to dark brown and black	5–5½
KYANITE	ALUMINO-SILICATES	Al_2SiO_5	Long, bladed or tabular crystals or aggregates	One perfect and one poor, parallel to length of crystals	White to light-colored or pale blue	5 parallel to crystal length 7 across crystals
SILLIMANITE		Al_2SiO_5	Long, slender crystals or fibrous, felted masses	One perfect parallel to length, not usually seen	Colorless, gray to white	6–7
ANDALUSITE		Al_2SiO_5	Coarse, nearly square prismatic crystals, some with symmetrically arranged impurities	One distinct; irregular fracture	Red, reddish brown, olive-green	7½
FELDSPATHOIDS		*NEPHELINE* $(Na,K)AlSiO_4$	Compact masses or as embedded grains, rarely as small prismatic crystals	One distinct; irregular fracture	Colorless, white, light gray; gray-greenish in masses, with greasy luster	5½–6
		LEUCITE $KAlSi_2O_6$	Trapezohedral crystals embedded in volcanic rocks	One very imperfect	White to gray	5½–6
SERPENTINE		$Mg_6Si_4O_{10}(OH)_8$	Fibrous (asbestos) or platy masses	Splintery fracture	Green; some yellowish brownish or gray; waxy or greasy luster in massive habit; silky luster in fibrous habit	4–6
TALC		$Mg_3Si_4O_{10}(OH)_2$ masses or aggregates	Foliated or compact masses or aggregates	One perfect, making thin flakes or scales; soapy feel	White to pale green; pearly or greasy luster	1
CORUNDUM		Al_2O_3	Some rounded, barrel-shaped crystals; most often as disseminated grains or granular masses (emery)	Irregular fracture	Usually brown, pink, or blue; emery black Gemstone varieties: ruby, sapphire	9

LIGHT-COLORED MINERALS, MAINLY IN METAMORPHIC AND IGNEOUS ROCKS AS COMMON OR MINOR CONSTITUENTS

(Continued)

	Mineral or Group Name	Structure or Composition	Varieties and Chemical Composition	Form, Diagnostic Characteristics	Cleavage, Fracture	Color	Hardness
DARK-COLORED MINERALS, COMMON IN METAMORPHIC ROCKS	EPIDOTE	SILICATES	$Ca_2(Al,Fe)Al_2Si_3O_{12}(OH)$	Aggregates of long prismatic crystals, granular or compact masses, embedded grains	One good, one poor at greater than right angles; conchoidal and irregular fracture	Green, yellow-green, gray, some varieties dark brown to black	6–7
	STAUROLITE		$Fe_2Al_9Si_4O_{22}(O,OH)_2$	Short prismatic crystals, some cross-shaped, usually coarser than matrix of rock	One poor	Brown, reddish, or dark brown to black	$7–7\frac{1}{2}$
METALLIC LUSTER, COMMON IN MANY ROCK TYPES, ABUNDANT IN VEINS	PYRITE	SULFIDES	FeS_2	Granular masses or well-formed cubic crystals in veins and beds or disseminated; high specific gravity, 4.9–5.2	Uneven fracture	Pale brass-yellow	$6–6\frac{1}{2}$
	GALENA		PbS	Granular masses in veins and disseminated; some cubic crystals; very high specific gravity, 7.3–7.6	Three perfect cleavages at mutual right angles, giving cubic cleavage fragments	Silver-gray	$2\frac{1}{2}$
	SPHALERITE		ZnS	Granular masses or compact crystalline aggregates; high specific gravity, 3.9–4.1	Six perfect cleavages at 60° to one another	White to green, brown, and black; resinous to submetallic luster	$3\frac{1}{2}–4$
	CHALCOPYRITE		$CuFeS_2$	Granular or compact masses; disseminated crystals; high specific gravity, 4.1–4.3	Uneven fracture	Brassy to golden-yellow	$3\frac{1}{2}–4$
	CHALCOCITE		Cu_2S	Fine-grained masses; high specific gravity, 5.5–5.8	Conchoidal fracture	Lead-gray to black; may tarnish green or blue	$2\frac{1}{2}–3$
MINERALS FOUND IN MINOR AMOUNTS IN A VARIETY OF ROCK TYPES AND IN VEINS OR PLACERS	RUTILE	TITANIUM OXIDES	TiO_2	Slender to prismatic crystals; granular masses; high specific gravity, 4.25	One distinct, one less distinct; conchoidal fracture	Reddish brown, some yellowish, violet, or black	$6–6\frac{1}{2}$
	ILMENITE		$FeTiO_3$	Compact masses, embedded grains, detrital grains in sand; high specific gravity, 4.79	Conchoidal fracture	Iron-black; metallic to submetallic luster	5–6
	ZEOLITES	SILICATES	Complex hydrous silicates; many varieties of minerals, including analcime, natrolite, phillipsite, heulandite, and chabazite	Well-formed radiating crystals in cavities in volcanics, veins, and hot springs; also as fine-grained and earthy bedded deposits	One perfect for most	Colorless, white, some pinkish	4–5

Glossary

Specific minerals are defined and described in Appendix 4.

ablation The total amount of ice that a *glacier* loses each year. (Compare *accumulation*.)

absolute age The actual number of years elapsed from a geologic event until now. (Compare *relative age*.)

accretion A process of continental growth in which buoyant fragments of *crust* are attached (accreted) to existing continental masses by horizontal transport during plate movements. (Compare *magmatic addition*.)

accumulation The amount of snow added to a *glacier* annually. (Compare *ablation*.)

acid rain Acid *precipitation* caused by *anthropogenic* emissions of sulfur dioxide into the atmosphere, where it reacts with water vapor to form sulfuric acid.

aftershock An *earthquake* that occurs as a consequence of a previous earthquake of larger magnitude. (Compare *foreshock*.)

albedo The fraction of *solar energy* reflected by a surface. (From the Latin *albus*, meaning "white.")

alluvial fan A cone- or fan-shaped accumulation of *sediment* deposited where a *stream* widens abruptly as it leaves a mountain front and enters a broad, relatively flat *valley*.

andesite An *intermediate igneous rock* with a composition between that of *dacite* and that of *basalt*; the *extrusive* equivalent of *diorite*.

Anthropocene The "Age of Man," a geologic epoch beginning about 1780, when the coal-powered steam engine launched the industrial revolution; proposed by atmospheric chemist Paul Crutzen to recognize the speed and magnitude of the changes industrial society is causing in the *Earth system*.

anthropogenic Arising from human activity.

anticline An arch-like fold of layered *rocks* that contains older rock layers in the core of the fold. (Compare *syncline*.)

aquiclude A relatively impermeable *formation* that bounds an *aquifer* above or below and acts as a barrier to the flow of *groundwater*.

aquifer A porous *formation* that stores and transmits *groundwater* in sufficient quantity to supply wells.

artesian flow A spontaneous flow of *groundwater* through a confined *aquifer* to a point where the elevation of the ground surface is lower than that of the *groundwater table*.

asteroid One of the more than 10,000 small celestial bodies orbiting the Sun, most of them between the orbits of Mars and Jupiter.

asthenosphere The weak, *ductile* layer of *rock* that constitutes the lower part of the upper *mantle* (below the *lithosphere*) and over which the lithospheric plates slide. (From the Greek *asthenes*, meaning "weak.")

atom The smallest unit of an element that retains the physical and chemical properties of that element.

barrier island A long offshore sandbar that builds up to form a barricade between open ocean waves and the main *shoreline*.

basalt A dark, fine-grained, *mafic igneous rock* composed largely of plagioclase feldspar and pyroxene; the *extrusive* equivalent of *gabbro*.

batholith A great irregular mass of *intrusive igneous rock* that covers at least 100 km²; the largest type of *pluton*.

beach A *shoreline* environment made up of *sand* and *pebbles*.

bed load The material a *stream* carries along its bed by sliding and rolling. (Compare *suspended load*.)

bedding The formation of parallel layers, or beds, of *sediment* as particles are deposited.

bedding sequence A sequence of interbedded and vertically stacked layers of different *sedimentary rock* types.

bioclastic sediment A shallow-water *sediment* made up of fragments of shells or skeletons directly precipitated by marine organisms and consisting primarily of two calcium carbonate *minerals*—calcite and aragonite—in variable proportions.

biofuel A fuel, such as ethanol, derived from biomass.

biological sediment A *sediment* formed near its place of deposition as a result of direct or indirect mineral *precipitation* by organisms. (Compare *chemical sediment*.)

biosphere The component of the *Earth system* that contains all of its living organisms.

bioturbation The process by which organisms rework existing *sediments* by burrowing through them.

blueschist A *metamorphic rock* formed under high pressures and moderate temperatures, often containing glaucophane, a blue amphibole.

bolide A chunk of *rock* or other debris from interplanetary space.

bomb A *pyroclast* 2 mm or larger, usually consisting of a blob of *lava* that cools in flight and becomes rounded or a chunk torn loose from previously solidified volcanic rock. (Compare with *volcanic ash*.)

braided stream A *stream* whose *channel* divides into an interlacing network of channels, which then rejoin in a pattern resembling braids of hair.

breccia A volcanic *rock* formed by the *lithification* of large *pyroclasts*. (Compare *tuff*.)

brittle Pertaining to a material that undergoes little *deformation* under increasing stress until it breaks suddenly. (Compare *ductile*.)

building code A set of standards for the design and construction of new buildings that specifies the intensity of shaking a structure must be able to withstand during an *earthquake*.

caldera A large, steep-walled, basin-shaped depression formed by a violent *volcanic eruption* in which large volumes of *magma* are discharged rapidly from a large *magma chamber*, causing the overlying volcanic structure to collapse catastrophically through the roof of the emptied chamber.

Cambrian explosion The rapid *evolutionary radiation* of animals during the early Cambrian period, after almost 3 billion years of very slow evolution, in which all the major branches of the animal tree of life originated within about 10 million years.

carbon cycle The continual movement of carbon among different components of the *Earth system*.

carbon economy The economy of modern industrial civilization, so called because it runs primarily on *fossil fuels.*

carbon sequestration The pumping of CO_2 generated by fossil-fuel combustion into reservoirs other than the atmosphere.

carbonate rock A *sedimentary rock* formed from *carbonate sediment.*

carbonate sediment A *sediment* formed from the accumulation of *carbonate minerals* directly or indirectly *precipitated* by marine organisms.

cementation A diagenetic change in which *minerals* are *precipitated* in the pores between *sediment* particles and bind them together.

channel A well-defined trough through which the water in a *stream* flows.

chemical and biological sedimentary environment An area of sediment accumulation characterized by precipitates of chemical or biological origin, principally carbonates or sulfates, and most commonly formed in marine settings.

chemical bond An electrostatic attraction between negatively charged electrons and positively charged protons that forms in a *chemical reaction.*

chemical reaction An interaction between the *atoms* of two or more chemical elements in certain fixed proportions that produces a new chemical compound.

chemical sediment A *sediment* formed at or near its place of deposition from dissolved materials that *precipitate* from water. (Compare *biological sediment.*)

chemical weathering *Weathering* in which the *minerals* in a *rock* are chemically altered or dissolved. (Compare *physical weathering.*)

chert A *sedimentary rock* made up of chemically or biologically *precipitated* silica.

clastic sediment A *sediment* formed by the transportation and accumulation of particles produced by the *weathering* of preexisting *rock.* (Compare *biological sediment, chemical sediment.*)

clay A *siliciclastic sediment* in which most of the particles are less than 0.0039 mm in diameter and which consists largely of clay minerals; the most abundant component of fine-grained *sedimentary rocks.*

claystone A *sedimentary rock* made up exclusively of *clay*-sized particles.

cleavage (1) The tendency of a *crystal* to break along planar surfaces. (2) The geometric pattern produced by such breakage.

climate The average conditions of Earth's surface environment and their variation.

climate model Any representation of the *climate system* that can reproduce one or more aspects of its behavior.

climate system The global *geosystem* that includes all the components of the *Earth system,* and all the interactions among these components, needed to determine climate on a global scale and how it changes over time.

coal A *biological sedimentary rock* composed almost entirely of organic carbon formed by the *diagenesis* of wetland vegetation.

coal bed A layer or stratum of *coal.*

coastline The narrow zones where land meets the sea.

color A property of a *mineral* imparted by transmitted or reflected light.

compaction A diagenetic decrease in the volume and *porosity* of a *sediment* as its particles are squeezed closer together by the weight of overlying sediments.

compressive force A force that squeezes or shortens a body. (Compare *shearing force, tensional force.*)

conglomerate A *sedimentary rock* composed of pebbles, cobbles, and boulders; the lithified equivalent of *gravel.*

contact metamorphism Metamorphism resulting from heat and pressure in a small area, as in rocks in contact with and near an igneous intrusion.

continental drift The large-scale movements of continents across Earth's surface driven by the *plate tectonic system.*

continental glacier A thick, slow-moving sheet of ice that covers a large part of a continent or other large landmass. (Compare *valley glacier.*)

continental shelf A broad, flat, submerged platform, consisting of a thick layer of flat-lying shallow-water *sediment,* that extends from the *shoreline* to the edge of the *continental slope.*

convergent boundary A boundary between lithospheric plates where the plates move toward each other and one plate is recycled into the *mantle.* (Compare *divergent boundary, transform fault.*)

core The dense central part of Earth below the *core-mantle boundary,* composed principally of iron and nickel. (See also *inner core, outer core.*)

core-mantle boundary The boundary between Earth's *core* and its *mantle,* about 2890 km below Earth's surface.

country rock The *rock* surrounding an igneous intrusion.

crater (1) A bowl-shaped pit found at the summit of most *volcanoes,* centered on the vent. (2) A depression caused by the impact of a *meteorite.*

craton A stable region of ancient continental *crust,* often made up of continental *shields* and *platforms.*

cross-bedding A *sedimentary structure* consisting of beds deposited by currents of wind or water and inclined at angles as much as 35° from the horizontal.

crude oil An organic sediment formed by *diagenesis* from organic material in the pores of sedimentary rocks; a diverse class of liquids composed of complex hydrocarbons. (Also called *petroleum.*)

crust The thin outer layer of Earth, averaging from about 8 km thick under the oceans to about 40 km thick under the continents, consisting of relatively low-density silicates that melt at relatively low temperatures.

crystal An ordered three-dimensional array of atoms in which the basic arrangement is repeated in all directions.

crystal habit The shape in which a *mineral*'s individual *crystals* or aggregates of crystals grow.

crystallization The formation of a solid *mineral* from a gas or liquid whose constituent *atoms* come together in the proper chemical proportions and ordered three-dimensional arrangement.

dacite A light-colored, fine-grained *intermediate igneous rock* with a composition between that of *rhyolite* and that of *andesite*; the *extrusive* equivalent of *granodiorite.*

decompression melting The spontaneous melting of rising *mantle* material as it reaches a level where pressure decreases below a critical point, without the introduction of any additional heat. (Compare *fluid-induced melting.*)

delta A large, flat-topped deposit of *sediments* formed where a *river* enters an ocean or lake and its current slows.

density The mass per unit volume of a substance, commonly expressed in grams per cubic centimeter (g/cm³). (Compare *specific gravity.*)

desertification The transformation of semiarid lands into deserts.

diagenesis The physical and chemical changes, caused by pressure, heat, and chemical reactions, by which buried *sediments* are lithified to form *sedimentary rocks.*

differentiation The transformation of an accretion of matter into a body whose interior is divided into concentric layers that differ both physically and chemically.

dike A sheetlike *discordant igneous intrusion* that cuts across layers of bedded *country rock.* (Compare *sill.*)

diorite A coarse-grained *intermediate igneous rock* with a composition between that of *granodiorite* and that of *gabbro;* the *intrusive* equivalent of *andesite.*

dip The amount of tilting of a *rock* layer; the angle at which a rock layer inclines from the horizontal, measured at right angles to the *strike.*

dip-slip fault A *fault* on which the relative movement of opposing blocks of rock has been up or down the *dip* of the fault plane.

discharge (1) The volume of *groundwater* leaving an *aquifer* in a given time. (Compare *recharge.*) (2) The volume of water that passes a given point in a given time as it flows through a *channel* of a certain width and depth.

distributary A smaller *stream* that receives water and *sediment* from the main *channel* of a *river,* branches off downstream, and thus distributes the water and sediment into many channels; typically found on a *delta.*

divergent boundary A boundary between lithospheric plates where two plates move apart and new *lithosphere* is created. (Compare *convergent boundary, transform fault.*)

divide A ridge of high ground along which all rainfall runs off down one side or the other.

dolostone An abundant *carbonate rock* composed primarily of dolomite and formed by the *diagenesis* of *carbonate sediments* and *limestones.*

drainage basin An area of land, bounded by *divides,* that funnels all its water into the network of *streams* draining the area.

drought A period of months or years when precipitation is much lower than normal.

ductile Pertaining to a material that undergoes smooth and continuous *deformation* under increasing stress without fracturing and does not spring back to its original shape when the stress is released. (Compare *brittle.*)

dune An elongated mound or ridge of *sand* formed by a current of wind or water.

Earth system The collection of Earth's open, interacting, and often overlapping *geosystems.*

earthquake The violent motion of the ground that occurs when brittle *rock* under stress suddenly breaks along a *fault.*

El Niño An anomalous warming of the eastern tropical Pacific Ocean that occurs every 3 to 7 years and lasts for a year or so.

elastic rebound theory A theory of *faulting* and *earthquake* generation holding that, as the crustal blocks on either side of a *fault* are deformed by tectonic forces, they remain locked in place by friction, accumulating elastic strain energy, until they fracture and rebound to their undeformed state.

environment The complex of physical, chemical, and biological factors (such as *climate, soil,* and living organisms) that act on an organism and ultimately determine its form and survival.

eolian Pertaining to wind.

eon The largest division of geologic time, including multiple *eras.*

epicenter The geographic point on Earth's surface directly above the *focus* of an *earthquake.*

epoch A division of geologic time representing one subdivision of a *period.*

era A division of geologic time representing one subdivision of an *eon* and including multiple *periods.*

erosion The set of processes that loosen *soil* and *rock* and move them downhill or downstream.

evaporite rock A *sedimentary rock* formed from *evaporite sediment.*

evaporite sediment *Chemical sediment* that is *precipitated* from evaporating seawater or lake water.

evolution Systematic change in organisms over time, driven by the process of *natural selection.*

exhumation The transportation of subducted *metamorphic rocks* back to Earth's surface.

extrusive igneous rock A fine-grained or glassy *igneous rock* formed from *magma* that erupts at Earth's surface as *lava* and cools rapidly. (Compare *intrusive igneous rock.*)

fault mechanism The orientation of the fault rupture and the slip direction of a *fault* that caused an *earthquake.*

fault slip The distance of the displacement of the two blocks of *rock* on either side of a *fault* that occurs during an *earthquake.*

faulting The process by which tectonic forces cause *rock* to break and slip on both sides of a fracture.

felsic rock Light-colored *igneous rock* that is poor in iron and magnesium and rich in high-silica *minerals* such as quartz, orthoclase feldspar, and plagioclase feldspar. (Compare *mafic rock, ultramafic rock.*)

fjord A former glacial valley with steep walls and a U-shaped profile, now flooded with seawater.

flexural basin A type of *sedimentary basin* that develops at a *convergent boundary* where one lithospheric plate pushes up over the other and the weight of the overriding plate causes the underlying plate to bend or flex downward.

flood Inundation that occurs when increased *discharge,* resulting from a short-term imbalance between inflow and outflow, causes a *stream* to overflow its banks.

floodplain A flat area about level with the top of a *channel* that lies on either side of the channel; the part of a *valley* that is flooded when a *stream* overflows its banks.

fluid-induced melting Melting of *rock* induced by the presence of water, which lowers its melting point. (Compare *decompression melting.*)

focus The point along a *fault* at which slipping initiates an *earthquake.*

folding The process by which tectonic forces deform an area of *crust* so that layers of *rock* are pushed into folds.

foliated rock *Metamorphic rock* that displays *foliation.* Foliated rocks include *slate, phyllite, schist,* and *gneiss.* (Compare *granoblastic rock.*)

foliation A set of flat or wavy parallel *cleavage* planes produced by *deformation* under directed pressure; typical of regionally metamorphosed rock.

foreshock A small *earthquake* that occurs in the vicinity of, but before, a main shock. (Compare *aftershock*.)

formation A distinct set of *rock* layers that can be identified throughout a region by its physical properties and possibly by the assemblage of *fossils* it contains.

fossil A trace of an organism that has been preserved in the *geologic record*.

fossil fuel An energy *resource* formed by the burial and heating of dead organic matter, such as *coal, crude oil,* or *natural gas.*

fractional crystallization The process by which the *crystals* formed in cooling *magma* are segregated from the remaining liquid *rock,* usually by settling to the floor of the *magma chamber.*

fracture The tendency of a *crystal* to break along irregular surfaces other than *cleavage* planes.

gabbro A dark gray, coarse-grained *igneous rock* containing an abundance of mafic *minerals,* particularly pyroxene; the *intrusive* equivalent of *basalt.*

geochemical reservoir A component of the *Earth system* where a chemical is stored at some point in its *geochemical cycle.*

geodynamo The global *geosystem* that produces Earth's *magnetic field,* driven by *convection* in the *outer core.*

geologic cross section A diagram showing the geologic features that would be visible if vertical slices were made through part of the *crust.*

geologic map A two-dimensional map representing the *rock formations* exposed at Earth's surface.

geologic record Information about geologic events and processes that has been preserved in *rocks* as they have formed at various times throughout Earth's history.

geologic time scale A worldwide history of geologic events that divides Earth's history into intervals, many of which are marked by distinctive sets of *fossils* and bounded by times when those sets of fossils changed abruptly.

geology The branch of Earth science that studies all aspects of the planet: its history, its composition and internal structure, and its surface features.

geosystem A specialized subsystem of the *Earth system* that produces specific types of geologic activity.

geotherm The curve that describes how Earth's temperature increases with depth.

geothermal energy Energy produced when underground water is heated as it passes through a subsurface region of hot *rock.*

glacial cycle A climate cycle alternating between cold *glacial periods,* or *ice ages,* during which temperatures decline, water is transferred from the hydrosphere to the cryosphere, ice sheets expand into lower latitudes, and sea level falls, and warm *interglacial periods,* during which temperatures rise abruptly, water is transferred from the cryosphere to the hydrosphere, and sea level rises.

global change Change in the *climate system* that has worldwide effects on the *biosphere,* atmosphere, and other components of the *Earth system.*

gneiss A light-colored, poorly *foliated,* high-grade *metamorphic rock* with coarse bands of segregated light and dark *minerals* throughout.

graded bedding A series of beds that progresses from large sediment particles at the bottom to small particles at the top, usually indicative of a weakening of the current that deposited the particles.

granite A felsic, coarse-grained *igneous rock* composed of quartz, orthoclase feldspar, sodium-rich plagioclase feldspar, and micas; the *intrusive* equivalent of *rhyolite.*

granoblastic rock A nonfoliated *metamorphic rock* composed mainly of *crystals* that grow in equant shapes, such as cubes and spheres, rather than in platy or elongate shapes. Granoblastic rocks include *hornfels, quartzite, marble, greenstone, amphibolite,* and *granulite.* (Compare *foliated rock.*)

granodiorite A light-colored, coarse-grained *intermediate igneous rock* that is similar to *granite* in containing abundant quartz, but whose predominant feldspar is plagioclase, not orthoclase; the *intrusive* equivalent of *dacite.*

gravel The coarsest *siliciclastic sediment,* consisting of particles larger than 2 mm in diameter and including pebbles, cobbles, and boulders.

greenhouse effect A global warming effect that results when a planet with an atmosphere containing *greenhouse gases* radiates *solar energy* back into space less efficiently than it would without such an atmosphere.

greenhouse gas A gas that absorbs and reradiates energy when it is present in a planet's atmosphere. Greenhouse gases in Earth's atmosphere include water vapor, carbon dioxide, and methane.

greenschist (1) A low-grade *metamorphic rock* formed from mafic volcanic rock and containing abundant chlorite. (2) The metamorphic grade above the zeolite grade.

groundwater The mass of water that flows beneath Earth's surface.

groundwater table The boundary between the *unsaturated zone* and the *saturated zone.*

half-life The time required for one-half the original number of parent *atoms* in a radioactive *isotope* to decay.

hardness A measure of the ease with which the surface of a *mineral* can be scratched.

Heavy Bombardment A time early in the early history of the solar system when planets were subjected to very frequent crater-forming impacts.

hot spot A region of intense, localized volcanism found far from a plate boundary; hypothesized to be the surface expression of a *mantle plume.*

Hubbert's peak The high point of a bell-shaped curve representing the rate of oil production; the point at which oil production will peak and then begin to decline.

hurricane A great storm that forms over the warm surface waters of tropical oceans (between 8° and 20° latitude) in areas of high humidity and light winds, producing winds of at least 119 km/hour (74 miles/hour) and large amounts of rainfall.

hydrologic cycle The cyclical movement of water from the ocean to the atmosphere by evaporation, to the surface by *precipitation,* to *streams* through *runoff* and *groundwater,* and back to the ocean.

hydrology The science that studies the movements and characteristics of water on and under Earth's surface.

hydrothermal solution A hot water solution formed when circulating *groundwater* or seawater comes into contact with a hot magmatic intrusion, reacts with it, and carries off significant quantities of elements and *ions* released by the reaction, which may be deposited later as *ore minerals.*

hydrothermal waters Hot water deep in the *crust*.

hypothesis A tentative explanation of a phenomenon, based on observational data and experiments, that serves as a basis for further investigation.

ice age The cold period of a *glacial cycle*, during which Earth cools, water is transferred from the hydrosphere to the cryosphere, ice sheets expand, and sea level drops. (Also called a *glacial period*. Compare *interglacial period*.)

igneous rock A *rock* formed by the solidification of *magma*. (From the Latin *ignis*, meaning "fire.")

infiltration The movement of water into *rock* or *soil* through cracks or small pores between particles.

inner core The central part of Earth below a depth of 5150 km, consisting of a solid sphere, composed of iron and nickel, suspended within the liquid *outer core*.

intensity scale A scale for estimating the intensity of a destructive geologic event, such as an earthquake or a hurricane, directly from the event's destructive effects.

interglacial period The warm period of a *glacial cycle* during which ice sheets melt, water is transferred from the cryosphere to the hydrosphere, and sea level rises. (Compare *ice age*.)

intermediate igneous rock An *igneous rock* midway in composition between mafic and felsic, neither as rich in silica as *felsic rock* nor as poor in it as *mafic rock*.

intrusive igneous rock A coarse-grained *igneous rock* formed from *magma* that intrudes into *country rock* deep in Earth's *crust* and cools slowly. (Compare *extrusive igneous rock*.)

ion An *atom* or group of atoms that has an electrical charge, either positive or negative, because of the loss or gain of one or more electrons.

island arc A chain of volcanic islands formed on the overriding plate at a *convergent boundary* by *magma* that rises from the *mantle* as water released from the subducting lithospheric slab causes *fluid-induced melting*.

isochron A *contour* that connects rocks of equal age.

isotope One of two or more forms of *atoms* of the same element that have different numbers of neutrons, and therefore different *atomic masses*.

isotopic dating The use of naturally occurring radioactive elements to determine the ages of *rocks*.

joint A crack in a *rock* along which there has been no appreciable movement.

karst topography An irregular, hilly type of terrain characterized by *sinkholes*, caves, and a lack of surface *streams*; formed in regions with humid climates, abundant vegetation, extensively jointed *limestone formations*, and appreciable *hydraulic gradients*.

kettle A hollow or undrained depression that often has steep sides and may be occupied by a pond or lake; formed in glacial deposits when *outwash* is deposited around a residual block of ice that later melts.

lahar A torrential mudflow of wet volcanic debris.

laminar flow Fluid movement in which straight or gently curved streamlines run parallel to one another without mixing or crossing between layers. (Compare *turbulent flow*.)

landslide A downhill flow of *rock, mud,* or *soil* that occurs when the friction that keeps the material from slipping downslope is overcome by the downward force of gravity, often due to water saturation.

lava *Magma* that flows out onto Earth's surface.

limestone A *carbonate rock* composed mainly of calcium carbonate in the form of the *mineral* calcite.

lithification The conversion of *sediment* into solid *rock* by *compaction* and *cementation*.

lithosphere The strong, rigid outer shell of Earth that comprises the *crust* and the uppermost part of the *mantle* down to an average depth of about 100 km. (From the Greek *lithos*, meaning "stone.")

longshore current A shallow-water current that runs parallel to the shore.

lower mantle A relatively homogeneous region of the *mantle* about 2200 km thick, extending from the *phase change* at about 660 km in depth to the *core-mantle boundary*.

low-velocity zone A layer near the base of the *lithosphere*, beginning at a depth of about 100 km, where *S-wave* speed abruptly decreases, marking the top part of the *asthenosphere*.

luster The way the surface of a *mineral* reflects light. (See Table 4.3.)

mafic rock Dark-colored *igneous rock* containing *minerals* such as pyroxenes and olivines that are rich in iron and magnesium and relatively poor in silica. (Compare *felsic rock, ultramafic rock*.)

magma Hot, molten *rock*.

magma chamber A large pool of *magma* that forms in the *lithosphere* as rising magmas melt and push aside surrounding solid *rock*.

magmatic addition A process of continental growth in which low-density, silica-rich *rock* differentiates in the *mantle* and is transported vertically to the *crust*. (Compare *accretion*.)

magmatic differentiation A process by which *rocks* of varying composition arise from a uniform parent *magma* as various *minerals* are withdrawn from it by *fractional crystallization* as it cools, changing its composition.

magnetic anomaly One in a pattern of long, narrow bands of high or low magnetic intensity on the seafloor that are parallel to and almost perfectly symmetrical with respect to the crest of a mid-ocean ridge.

magnetic field The region of influence of a magnetized body or an electric current.

magnetic time scale The detailed history of Earth's *magnetic field* reversals as determined by measuring the *thermoremanent magnetization* of *rock* samples.

magnitude scale A scale for estimating the size of an *earthquake* using the logarithm of the largest ground motion registered by a *seismograph* (Richter magnitude) or the logarithm of the area of the *fault* rupture (moment magnitude).

mantle The region that forms the main bulk of Earth, between the *crust* and the *core*, containing *rocks* of intermediate *density*, mostly compounds of oxygen with magnesium, iron, and silicon.

mantle plume A narrow, cylindrical jet of hot, solid material rising from deep within the *mantle*, thought to be responsible for intraplate volcanism.

mass extinction A short interval during which a large proportion of the species living at the time disappear from the *geologic record*.

meander A curve or bend in a *stream* that develops as the stream erodes the outer bank of a bend and deposits *sediment* against the inner bank.

metamorphic facies Groupings of *metamorphic rocks* of various *mineral* compositions formed under different grades of metamorphism from different parent rocks.

metamorphic rock *Rock* formed by high temperatures and pressures that cause changes in the mineralogy, texture, or chemical composition of any kind of preexisting rock while maintaining its solid form. (From the Greek *meta*, meaning "change," and *morphe*, meaning "form.")

metasomatism Change in a *rock*'s composition by fluid transport of chemical substances into or out of the rock.

meteoric water Rain, snow, or other forms of water derived from the atmosphere.

meteorite A chunk of material from outer space that strikes Earth.

microbe A single-celled microorganism.

microfossil A trace of an individual *microorganism* preserved in the *geologic record*.

Milankovitch cycle A pattern of periodic variations in Earth's movement around the Sun that affects the amount of solar energy received at Earth's surface. Milankovitch cycles include the eccentricity of Earth's orbit; the tilt of Earth's axis of rotation; and precession, Earth's wobble about its axis of rotation.

mineral A naturally occurring, solid crystalline substance, generally inorganic, with a specific chemical composition.

mineralogy (1) The branch of geology that studies the composition, structure, appearance, stability, occurrence, and associations of *minerals*. (2) The relative proportions of a *rock*'s constituent minerals.

Mohs scale of hardness An ascending scale of mineral *hardness* based on the ability of one *mineral* to scratch another. (See Table 3.2.)

moraine An accumulation of rocky, sandy, and clayey material carried by glacial ice and deposited as *till*.

mud A fine-grained *siliciclastic sediment*, mixed with water, in which most of the particles are less than 0.062 mm in diameter.

mudstone A blocky, poorly bedded, fine-grained *sedimentary rock* produced by the *lithification* of *mud*.

natural gas Methane gas (CH_4), the simplest hydrocarbon.

natural hazard An event produced by natural processes that has the potential to kill people and damage buildings and other man-made structures.

natural levee A ridge of coarse material built up by successive *floods* that confines a *stream* within its banks between floods, even when water levels are high.

natural resource A supply of energy, water, or raw material used by human civilization that is available from the natural environment. (See also *resource*.)

natural selection The process by which populations of organisms adapt to new environments.

nebular hypothesis The idea that the solar system originated from a diffuse, slowly rotating cloud of gas and fine dust (a "nebula") that contracted under the force of gravity and eventually evolved into the Sun and planets.

negative feedback A process in which one action produces an effect (the feedback) that tends to counteract the original action and stabilize the system against change. (Compare *positive feedback*.)

nonrenewable resource A *natural resource* that is produced at a rate much slower than the rate at which human civilization is using it up; for example, *fossil fuels*. (Compare *renewable resource*.)

normal fault A *dip-slip fault* in which the *hanging wall* moves downward relative to the *foot wall*, extending the structure horizontally.

nuclear energy Energy produced by the fission of the radioactive *isotope* uranium-235, which can be used to make steam and drive turbines to create electricity.

obsidian A dense, glassy volcanic *rock*, usually of felsic composition.

ocean acidification A process in which carbon dioxide from the atmosphere dissolves into the ocean and reacts with seawater to form carbonic acid (H_2CO_3), increasing the acidity of the ocean.

oil trap An impermeable barrier that blocks the upward migration of *crude oil* or *natural gas*, allowing them to collect beneath the barrier.

oil window The limited range of pressures and temperatures, usually found at depths between about 2 and 5 km, at which *crude oil* forms.

ore A *mineral* deposit from which valuable metals can be recovered profitably.

orogeny Mountain building by tectonic forces, particularly through the *folding* and *faulting* of *rock* layers, often with accompanying volcanism. (From the Greek *oros*, meaning "mountain," and *gen*, meaning "be produced.")

outer core The layer of Earth extending from the *core-mantle boundary* to the *inner core*, at depths of 2890 to 5150 km, composed of molten iron and nickel and minor amounts of lighter elements, such as oxygen or sulfur.

P wave The first type of *seismic wave* to arrive at a *seismograph* from the *focus* of an *earthquake*; a type of *compressional wave*.

Pangaea A supercontinent that coalesced in the late Paleozoic *era* and comprised all present continents, then began to break up in the Mesozoic era.

partial melting Incomplete melting of a *rock* that occurs because the *minerals* that compose it melt at different temperatures.

pegmatite A *vein* of extremely coarse-grained *granite*, crystallized from a water-rich *magma* in the late stages of solidification, that cuts across much finer grained *country rock* and may contain rich concentrations of rare *minerals*.

peridotite A coarse-grained, dark greenish gray, *ultramafic intrusive igneous rock* composed primarily of olivine with smaller amounts of pyroxene and other minerals such as spinel or garnet; the dominant rock in Earth's *mantle* and the source rock of *basaltic magmas*.

period A division of geologic time representing one subdivision of an *era*.

permeability The ability of a solid to allow fluids to pass through it.

phase change A transformation of a *rock*'s *crystal* structure (but probably not its chemical composition) by changing conditions of temperature and pressure, signaled by a change in *seismic wave* velocity.

photosynthesis The process by which organisms such as plants and algae use energy from sunlight to convert water and carbon dioxide into carbohydrates and oxygen.

phyllite A *foliated rock* that is intermediate in metamorphic grade between *slate* and *schist*, containing small *crystals* of mica and chlorite that give it a more or less glossy sheen.

physical weathering *Weathering* in which solid *rock* is fragmented by mechanical processes that do not change its chemical composition. (Compare *chemical weathering*.)

plate tectonic system The global *geosystem* that includes the convecting *mantle* and its overlying mosaic of lithospheric *plates*.

plate tectonics The theory that describes and explains the creation and destruction of Earth's lithospheric *plates* and their movement over Earth's surface. (From the Greek *tekton*, meaning "builder.")

playa lake A permanent or temporary lake in an arid mountain *valley* or *basin*, where dissolved *minerals* may be concentrated and *precipitated* as the water evaporates.

pluton A large *igneous* intrusion, ranging in size from a cubic kilometer to hundreds of cubic kilometers, formed deep in the *crust*.

point bar A curved sandbar deposited along the inside bank of a *stream*, where the current is weakest.

polymorph One of two or more alternative possible *crystal* structures for a single chemical compound; for example, the *minerals* quartz and cristobalite are polymorphs of silica (SiO_2).

porosity The percentage of a *rock*'s volume consisting of open pores between particles.

porphyry An *igneous rock* of mixed *texture* in which large *crystals* (phenocrysts) "float" in a predominantly fine-grained matrix.

positive feedback A process in which one action produces an effect (the feedback) that tends to enhance the original action and amplify change in the system. (Compare *negative feedback*.)

potable Pertaining to water that tastes agreeable and is not dangerous to human health.

precipitation (1) A deposit on Earth's surface of condensed atmospheric water vapor in the form of rain, snow, sleet, hail, or mist. (2) The condensation of a solid from a solution during a chemical reaction.

principle of faunal succession A stratigraphic principle stating that the *sedimentary rock* strata in an *outcrop* contain distinct *fossils* in a definite sequence.

principle of original horizontality A stratigraphic principle stating that *sediments* are deposited as essentially horizontal beds.

principle of superposition A stratigraphic principle stating that each *sedimentary rock* stratum in a tectonically undisturbed sequence is younger than the one beneath it and older than the one above it.

principle of uniformitarianism A principle stating that the processes we see in action on Earth today have worked in much the same way throughout the geologic past.

pumice A volcanic *rock*, usually *rhyolitic* in composition, containing numerous cavities (vesicles) that remain after trapped gas has escaped from solidifying *lava*.

pyroclast A *rock* fragment ejected into the air by a volcanic eruption. (See also *bomb, volcanic ash*.)

pyroclastic flow A glowing cloud of hot ash, dust, and gases ejected by a *volcanic eruption* that rolls downhill at high speeds.

rain shadow An area of low rainfall on the leeward slope of a mountain range.

recharge The *infiltration* of water into any subsurface *rock formation*.

recurrence interval The average time between large *earthquakes* at a particular location; according to the *elastic rebound theory*, the time required to accumulate the strain that will be released by *fault slipping* in a future earthquake.

reef A moundlike or ridgelike organic structure constructed of the carbonate skeletons and shells of millions of marine organisms.

regional metamorphism *Metamorphism* caused by high pressures and temperatures that extend over large regions; typical of

convergent boundaries where two continents collide. (Compare *contact metamorphism*.)

relative age The age of one geologic event in relation to another. (Compare *absolute age*.)

relative humidity The amount of water vapor in the air, expressed as a percentage of the total amount of water the air could hold at the same temperature if it were saturated.

renewable resource A *natural resource* that is produced at a rate rapid enough to match the rate at which human civilization is using it up; for example, wood. (Compare *nonrenewable resource*.)

reserve The supply of a *natural resource* that has already been discovered and can be exploited economically and legally at the present time. (Compare *resource*.)

reservoir See *geochemical reservoir*.

residence time The average time an *atom* of a particular element spends in a *geochemical reservoir* before leaving it.

rhyolite A light brown to gray, fine-grained *felsic igneous rock*; the *extrusive* equivalent of *granite*.

rift basin A *sedimentary basin* that develops at a *divergent boundary* at an early stage of plate separation as the stretching and thinning of the continental *crust* results in subsidence. (Compare *thermal subsidence basin*.)

ripple A very small ridge of *sand* or *silt* whose long dimension is at right angles to the current that formed it.

risk The damage that is likely to be caused when a *natural hazard* strikes a particular location, usually measured in lost lives and dollars.

river A major branch of a *stream* system.

rock A naturally occurring solid aggregate of minerals or, in some cases, nonmineral solid matter.

rock cycle The set of geologic processes that convert *rocks* of each of the three major types—*igneous, sedimentary,* and *metamorphic*—into the other two types.

Rodinia A supercontinent older than *Pangaea* that formed about 1.1 billion years ago and began to break up about 750 million years ago.

runoff The sum of all *precipitation* that flows over the land surface, including not only *streams* but also the fraction that temporarily infiltrates near-surface *soil* and *rock* and then flows back to the surface.

S wave The second type of *seismic wave* to arrive at a *seismograph* from the *focus* of an *earthquake*; a type of *shear wave*. S waves cannot travel through liquids or gases.

salinity The total amount of dissolved substances in a given volume of water.

saltation The transportation of *sand* or smaller *sediment* particles by a current in such a manner that they move along in a series of short intermittent jumps.

sand A *siliciclastic sediment* consisting of medium-sized particles, ranging from 0.062 to 2 mm in diameter.

sandstone The lithified equivalent of *sand*.

saturated zone The level below the *groundwater table*, in which the pores of *soil* or *rock* are completely filled with water. (Also called the *phreatic zone*.) (Compare *unsaturated zone*.)

schist An intermediate-grade *metamorphic rock* characterized by pervasive coarse, wavy *foliation* known as schistosity.

scientific method A general procedure, based on systematic observations and experiments, by which scientists propose and test *hypotheses* that explain some aspect of how the physical universe works.

seafloor spreading The mechanism by which new oceanic *crust* is formed at a *spreading center* on the crest of a mid-ocean ridge. As two plates move apart, *magma* wells up into the *rift* between them to form new crust, which spreads laterally away from the rift and is replaced continually by newer crust.

sediment Material deposited on Earth's surface by physical agents (wind, water, and ice), chemical agents (*precipitation* from *oceans, lakes,* and *rivers*), or biological agents (organisms, living and dead).

sedimentary basin A region where the combination of sedimentation and *subsidence* has formed thick accumulations of *sediment* and *sedimentary rock.*

sedimentary environment A geographic location characterized by a particular combination of climate conditions and physical, chemical, and biological processes.

sedimentary rock A *rock* formed by the burial and *diagenesis* of layers of *sediment.*

sedimentary structure Any kind of *bedding* or other feature (such as *cross-bedding, graded bedding,* or *ripples*) formed at the time of *sediment* deposition.

seismic hazard The intensity of shaking and ground disruption by *earthquakes* that can be expected over the long term at some specified location.

seismic risk The *earthquake* damage that can be expected over the long term in a specified region, usually measured in average dollar losses per year.

seismic tomography A technique that uses differences in the travel times of *seismic waves* produced by *earthquakes* and recorded on *seismographs* to construct three-dimensional images of Earth's interior.

seismic wave A ground vibration produced by an *earthquake.* (See also *P wave, S wave, surface wave.*) (From the Greek *seismos,* meaning "earthquake.")

seismicity map A map that shows the *epicenters* of *earthquakes* recorded around the world or in a particular region over a given period of time.

seismograph An instrument that records the *seismic waves* generated by *earthquakes.*

shadow zone (1) A zone beyond 105° from the *focus* of an *earthquake* where *S waves* are not recorded because they are not transmitted through Earth's liquid *outer core.* (2) A zone at angular distances of 105° to 142° from the focus of an earthquake where *P waves* are not recorded because they are refracted downward into the core and emerge at greater distances after the delay caused by their detour through the core.

shale A fine-grained *sedimentary rock* composed of *silt* plus a significant component of *clay,* which causes it to break readily along *bedding* planes.

shearing force A force that pushes two sides of a body in opposite directions. (Compare *compressive force, tensional force.*)

shield volcano A broad, shield-shaped *volcano* many tens of kilometers in circumference and more than 2 km high, built by successive flows of *basaltic lava* from a central vent.

shoreline The line where the ocean surface meets the land surface.

siliciclastic sediment *Sediment* formed from clastic particles produced by the *weathering* of *rocks* and physically deposited by running water, wind, or ice. (From the Greek *klastos,* meaning "broken.")

siliciclastic sedimentary environment A *sedimentary environment* dominated by *siliciclastic sediments.*

sill A sheetlike *concordant igneous intrusion* formed by the injection of *magma* between parallel layers of bedded *country rock.* (Compare *dike.*)

silt A *siliciclastic sediment* in which most of the particles are between 0.0039 and 0.062 mm in diameter.

siltstone A *sedimentary rock* that contains mostly *silt* and looks similar to *mudstone* or very fine grained *sandstone;* the lithified equivalent of silt.

sinkhole A small, steep depression in the land surface formed when the thin roof of a *limestone* cave collapses suddenly.

slate A fine-grained *foliated rock* that is easily split into thin sheets, formed primarily by low-grade *metamorphism* of *shale.*

soil An intricate combination of *weathered rock* and organic material.

solar energy Energy derived from the Sun.

solar nebula According to the *nebular hypothesis,* a disk of gas and dust that surrounded the proto-Sun from which the planets of the solar system formed.

sorting The tendency for variations in current velocity to segregate *sediments* according to size.

specific gravity The weight of a substance divided by the weight of an equal volume of pure water at 4°C. (Compare *density.*)

spit A narrow extension of a *beach* formed by *longshore currents* that carry *sand* to its downcurrent end.

spreading center A *divergent boundary,* marked by a rift at the crest of a mid-ocean ridge, where new oceanic *crust* is formed by *seafloor spreading.*

stabilization wedge A strategy for reducing carbon emissions by 1 gigaton per year in the next 50 years relative to a business-as-usual scenario. About eight stabilization wedges will be necessary to stabilize carbon emissions at current levels.

stock A *pluton* less than 100 km^2 in area.

storm surge A dome of seawater formed by a *hurricane* that rises above the level of the surrounding ocean surface.

stratigraphic succession A chronologically ordered set of *rock* strata.

stratigraphy The description, correlation, and classification of strata in *sedimentary rocks.*

stratosphere The cold, dry layer of the atmosphere above the *troposphere* that extends from about 11 to 50 km in altitude. (Compare *troposphere.*)

stratospheric ozone depletion The thinning of the layer of ozone in the *stratosphere,* which shields Earth from cell-damaging ultraviolet radiation, caused by chemical reactions involving *anthropogenic* compounds such as chlorofluorocarbons (CFCs).

stratovolcano A concave-shaped *volcano* formed from alternating layers of *lava* flows and beds of *pyroclasts.*

streak The *color* of the fine deposit of *mineral* powder left on an abrasive surface when a mineral is scraped across it.

stream Any body of water, large or small, that flows over the land surface.

stream valley The entire area between the tops of the slopes on both sides of a *stream.*

strike The compass direction of a line formed by the intersection of a rock layer's surface with a horizontal surface.

strike-slip fault A *fault* on which the relative movement of the opposing blocks of *rock* has been horizontal, parallel to the *strike* of the fault plane.

subduction The sinking of oceanic *lithosphere* beneath overriding oceanic or continental lithosphere at a *convergent plate boundary.*

surface wave A type of *seismic wave* that travels around Earth's surface from the *focus* of an *earthquake* and arrives at a *seismograph* later than *S waves.*

suspended load All the material temporarily or permanently suspended in the flow of a current. (Compare *bed load.*)

sustainable development Development that meets the needs of the present without compromising the ability of future generations to meet their own needs.

syncline A trough-like fold of layered *rocks* that contains younger rock layers in the core of the fold. (Compare *anticline.*)

tensional force A force that stretches a body and tends to pull it apart. (Compare *compressive force, shearing force.*)

texture The sizes and shapes of a *rock's mineral crystals* and the way they are put together.

thermal subsidence basin A *sedimentary basin* that develops in the later stages of plate separation as *lithosphere* that was thinned and heated during the earlier rifting stage cools, becomes more dense, and subsides below sea level. (Compare *rift basin.*)

thermohaline circulation A global three-dimensional oceanic circulation pattern driven by differences in the temperature and salinity—and therefore in the density—of ocean waters.

thrust fault A low-angled reverse *fault*—one with a dip of less than 45°.

tide The twice-daily rise and fall of the ocean caused by the gravitational attraction between Earth and the Moon.

till Unstratified and poorly sorted *drift* deposited directly by a melting glacier, containing particles of all sizes from *clay* to boulders.

tillite The *lithified* equivalent of *till.*

topography The general configuration of varying heights that gives shape to Earth's surface, which is measured with respect to sea level.

transform fault A plate boundary at which the plates slide horizontally past each other and *lithosphere* is neither created nor destroyed.

troposphere The lowest layer of the atmosphere, which has an average thickness of about 11 km, contains about three-fourths of the atmosphere's mass, and convects vigorously due to the uneven heating of Earth's surface by the Sun. (From the Greek *tropos,* meaning "turn" or "mix.") (Compare *stratosphere.*)

tsunami A fast-moving sea wave, generated by an *earthquake* that lifts the seafloor, that propagates across the ocean and increases in size when it reaches the shore.

tuff A volcanic *rock* formed by the *lithification* of small *pyroclasts.* (Compare *breccia.*)

turbidity current A *turbulent flow* of water carrying a *suspended load* of mud that flows down the *continental slope* beneath the overlying clear water.

turbulent flow Fluid movement in which streamlines mix, cross, and form swirls and eddies. (Compare *laminar flow.*)

twentieth-century warming The rise in Earth's average surface temperature by about 0.6°C between the end of the nineteenth century and the beginning of the twenty-first.

ultramafic rock An *igneous rock* consisting primarily of mafic *minerals* and containing less than 10 percent feldspar. (Compare *felsic rock, mafic rock.*)

unconformity A surface between two *rock* layers in a *stratigraphic succession* that were laid down with a time gap between them.

unsaturated zone The level above the *groundwater table,* in which the pores of the *soil* or *rock* are not completely filled with water. (Also called the *vadose zone.*) (Compare *saturated zone.*)

U-shaped valley A deep *valley* with steep upper walls that grade into a flat floor; the typical shape of a valley eroded by a *glacier.*

valley glacier A river of ice that forms in the cold heights of a mountain range, where snow accumulates, then moves downslope, either flowing down an existing stream valley or carving out a new valley. (Compare *continental glacier.*)

vein A sheetlike deposit of *minerals precipitated* in fractures or *joints* in *country rock,* often by a *hydrothermal solution.*

viscosity A measure of a fluid's resistance to flow.

volcanic ash *Pyroclasts* less than 2 mm in diameter, usually glass, that form when escaping gases force a fine spray of *magma* from a *volcano.* (Compare with *bomb.*)

volcanic eruption A geologic event that deposits molten rock from a planet's interior onto its surface and spews gases and solid materials into its atmosphere.

volcano A hill or mountain constructed from the accumulation of erupted *lava* and *pyroclasts.*

wave-cut terrace A level surface formed by wave *erosion* of a rocky *shoreline* beneath the surf zone, which may be visible at low *tide.*

weathering The processes by which *rocks* are broken down at Earth's surface to produce *sediment* particles. (See also *chemical weathering, physical weathering.*)

Wilson cycle The sequence of tectonic events in the formation and breakup of supercontinents that comprises (1) rifting during the breakup of a supercontinent, (2) *passive margin* cooling and *sediment* accumulation during *seafloor spreading* and ocean opening, (3) *magmatic addition* and *accretion* during *subduction* and ocean closure, and (4) *orogeny* during the continent-continent collision that forms the next supercontinent.

INDEX

Boldface indicates a Glossary term   *Italics* indicate a figure   *t* indicates a table

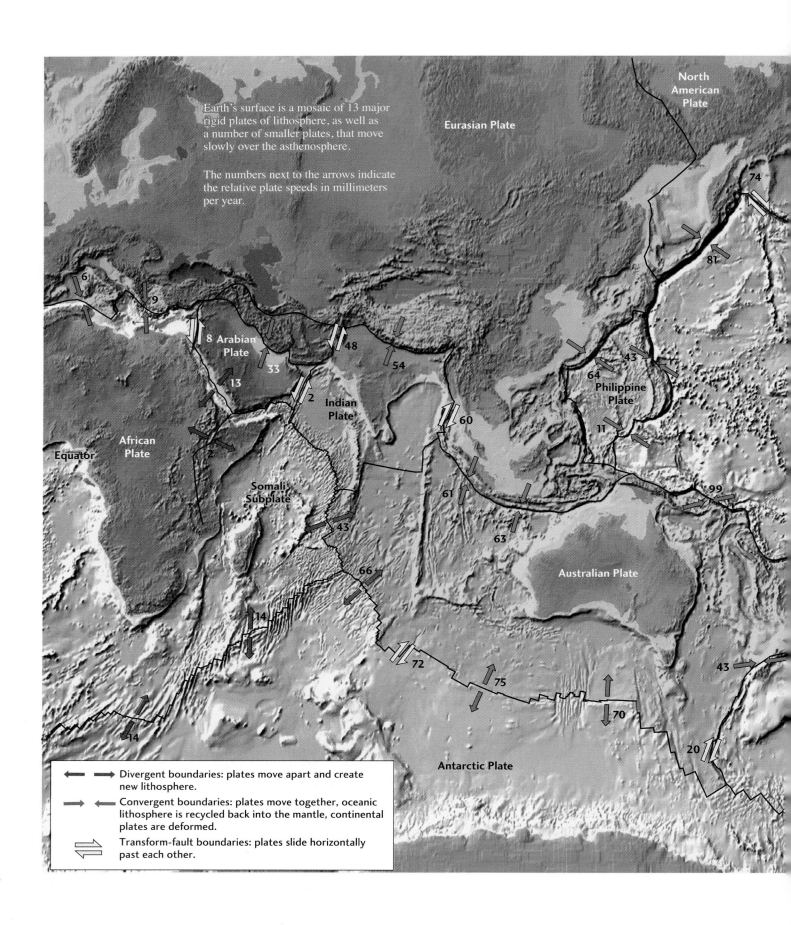

Juan de Fuca Plate

North American Plate

Eurasian Plate

African Plate

Caribbean Plate

Cocos Plate

Pacific Plate

Nazca Plate

South American Plate

Antarctic Plate

73

63

47

34

50

64

11

89

12

24

22

18

27

31

118

50

138

84

72

150

55

79

92

80

73

35

34

81

64

48

18

50